# THE LITTLE OXFORD Thesaurus

*Compiled by*

ALAN SPOONER

CLARENDON PRESS · OXFORD

1993

*Oxford University Press, Walton Street, Oxford* OX2 6DP

*Oxford New York Toronto*
*Delhi Bombay Calcutta Madras Karachi*
*Kuala Lumpur Singapore Hong Kong Tokyo*
*Nairobi Dar es Salaam Cape Town*
*Melbourne Auckland Madrid*

*and associated companies in*
*Berlin Ibadan*

*Oxford is a trade mark of Oxford University Press*

*Published in the United States*
*by Oxford University Press Inc., New York*

*First published 1992 as* The Oxford Minireference Thesaurus

*This edition first published 1993*



*British Library Cataloguing in Publication Data*
*Data available*

*Library of Congress Cataloging in Publication Data*
*Data available*

*ISBN 0-19-869221-8*

*Typeset by Wyvern Typesetting Ltd.*
*Printed in Great Britain by*
*Clays Ltd.*
*Bungay, Suffolk*

# Preface

A THESAURUS helps you find the words you need to express yourself more effectively and more interestingly. This thesaurus is designed to combine within its small format maximum ease of use with maximum helpfulness. It includes a larger range of synonyms and other information than might be expected in a book of this size. Headwords are arranged in a simple alphabetical sequence, and the organization of each entry is straightforward and largely self-explanatory. Normally, you will find what you want under the headword you look up, though cross-references may be necessary if you need opposites, or if you want a larger range of words to choose from.

A thesaurus should be used with caution. Firstly, no synonym list should be regarded as 'complete'. Many could be extended, some almost indefinitely. Consider, for example, the variety of words we might substitute for (say) *good* or *pleasant*. Secondly, seldom in English are two words completely interchangeable. So-called synonyms may convey distinct nuances of meaning, or belong to different contexts, or carry different signals about the writer or intended reader – and so on. I hope, therefore, that this volume will be a useful resource, but not just as a lifeless repository of 'words to use': my main hope is that a thesaurus – even a small one like this – will prompt us to *think* about language, and enable us to exploit more fully our own knowledge and understanding of its complex processes.

ALAN SPOONER

*The Nottingham Trent University*
*May 1993*

# *The Little Oxford Thesaurus*

COMPILER AND EDITOR-IN-CHIEF
Alan Spooner

MANAGING EDITOR
Sara Tulloch

ASSISTANT EDITORS
Anne Knight    Christine Cowley

# Using the thesaurus

In this thesaurus you will find

**Headwords**

The words you want to look up are printed in bold and arranged in a single alphabetical sequence. In addition, there may be sub-heads in bold at the end of main entries for derived forms and phrases.

**Synonyms**

Synonyms are listed alphabetically, except that distinct senses of a headword are numbered and treated separately.

Under some headwords, in addition to the lists of synonyms given there, a cross-reference printed in SMALL CAPITALS takes you to another entry to provide an extended range of synonyms. These cross-references are marked by the arrowhead symbol ▷.

**Related words**

Lists of words which are not synonyms but which have a common relationship to the headword (eg, kinds of vehicle listed under *vehicle*) are printed in italic, flagged by the symbol □.

**Antonyms**

Cross-references printed in SMALL CAPITALS introduce you to lists of opposites. These cross-references are preceded by the abbreviation *Opp.*

**Part-of-speech labels**

Part-of-speech labels are given throughout. (See list of abbreviations.) Under each headword, uses as *adjective, adverb, noun,* and *verb* are separated by the symbol •.

**Illustrative phrases**

Meanings of less obvious senses are indicated by illustrative phrases printed in *italic*.

**Usage warnings**

Usage markers in *italic* precede words which are normally informal, derogatory, etc. (See list of abbreviations.)

# Abbreviations used in this thesaurus

**Parts of speech**

| | |
|---|---|
| *adj* | adjective |
| *adv* | adverb |
| *int* | interjection |
| *n* | noun |
| *prep* | preposition |
| *vb* | verb |

**Other abbreviations**

| | |
|---|---|
| *derog* | normally used in a derogatory, negative, or uncomplimentary sense |
| *fem* | feminine |
| *inf* | normally used informally |
| *joc* | normally jocular or joking |
| *old use* | old-fashioned or obsolete |
| *opp* | opposites, antonyms |
| *plur* | plural |
| *poet* | poetic |
| *sl* | slang |
| *Amer* | word or phrase usually regarded as American usage. |
| *Fr* | word or phrase common in English contexts, but still identifiably French. |
| *Ger* | ditto German |
| *Gr* | ditto Greek |
| *It* | ditto Italian |
| *Lat* | ditto Latin |
| *Scot* | word or phrase usually regarded as Scottish usage. |
| ▷ | This symbol shows that you will find relevant information if you go to the word indicated. |

# A

**abandon** *vb* **1** evacuate, leave, quit, vacate, withdraw from. **2** break with, desert, *inf* dump, forsake, *inf* give someone the brush-off, jilt, leave behind, *inf* leave in the lurch, maroon, renounce, repudiate, strand, *inf* throw over, *inf* wash your hands of. **3** *abandon a claim.* abdicate, cancel, cede, *sl* chuck in, discard, discontinue, disown, *inf* ditch, drop, finish, forfeit, forgo, give up, postpone, relinquish, resign, scrap, surrender, waive, yield.

**abbey** *n* cathedral, church, convent, friary, monastery, nunnery, priory.

**abbreviate** *vb* abridge, compress, condense, curtail, cut, digest, edit, précis, prune, reduce, shorten, summarize, trim, truncate. *Opp* LENGTHEN.

**abdicate** *vb* renounce the throne, *inf* step down. ▷ ABANDON, RESIGN.

**abduct** *vb* carry off, kidnap, *inf* make away with, seize.

**abhor** *vb* detest, execrate, loathe, recoil from, shudder at. ▷ HATE.

**abhorrent** *adj* abominable, detestable, disgusting, execrable, horrible, loathsome, nauseating, obnoxious, odious, offensive, repellent, repugnant, repulsive, revolting. ▷ HATEFUL. *Opp* ATTRACTIVE.

**abide** *vb* **1** accept, bear, brook, endure, put up with, stand, *inf* stomach, submit to, suffer, tolerate. **2** ▷ STAY. **abide by** ▷ OBEY.

**ability** *n* aptitude, bent, brains, capability, capacity, cleverness, competence, expertise, facility, faculty, flair, genius, gift, intelligence, knack, *inf* know-how, knowledge, means, potential, potentiality, power, proficiency, prowess, qualification, resources, scope, skill, strength, talent, training, wit.

**ablaze** *adj* afire, aflame, aglow, alight, blazing, burning, fiery, flaming, incandescent, lit up, on fire, raging. ▷ BRILLIANT.

**able** *adj* **1** accomplished, adept, capable, clever, competent, effective, efficient, experienced, expert, gifted, *inf* handy, intelligent, masterly, practised, proficient, qualified, skilful, skilled, strong, talented, trained. *Opp* INCOMPETENT. **2** allowed, at liberty, authorized, available, eligible, equipped, fit, free, permitted, prepared, ready, willing. *Opp* UNABLE.

**abnormal** *adj* aberrant, anomalous, atypical, *inf* bent, bizarre, curious, deformed, deviant, distorted, eccentric, erratic, exceptional, extraordinary, freak, funny, heretical, idiosyncratic, irregular, *inf* kinky, malformed, odd, peculiar, perverse, perverted, queer, singular, strange, uncharacteristic, uncommon, unexpected, unnatural, unorthodox, unrepresentative, untypical, unusual, wayward, weird. *Opp* NORMAL.

**abolish** *vb* abrogate, annul, cancel, delete, destroy, dispense with, do away with, eliminate, end, eradicate, finish, *inf* get rid of, invalidate, liquidate, nullify, overturn, put an end to, quash, repeal, remove, rescind, revoke, suppress, terminate, withdraw. *Opp* CREATE.

**abominable** *adj* abhorrent, appalling, atrocious, awful, base, beastly, brutal, contemptible, cruel, despicable, detestable, disgusting, distasteful, dreadful, execrable, foul, hateful, heinous, horrible, immoral, inhuman, inhumane, loathsome, nasty, nauseating, obnoxious, odious, offensive, repellent, repugnant, repulsive, revolting, terrible, vile. ▷ UNPLEASANT. *Opp* PLEASANT.

**abort** *vb* **1** be born prematurely, die, miscarry. **2** *abort take-off.* call off, end, halt, nullify, stop, terminate.

**abortion** *n* **1** miscarriage, premature birth, termination of pregnancy. **2** ▷ MONSTER.

**abortive** *adj* fruitless, futile, ineffective, ineffectual, pointless, stillborn, unavailing, unfruitful, unproductive, unsuccessful, useless, vain. *Opp* SUCCESSFUL.

**abound** *vb* be plentiful, flourish, prevail, swarm, teem, thrive.

**abrasive** *adj* biting, caustic, galling, grating, harsh, hurtful, irritating, rough, sharp. ▷ UNKIND. *Opp* KIND.

**abridge** *vb* abbreviate, compress, condense, curtail, cut, digest, edit, précis, prune, reduce, shorten, summarize, telescope, truncate. *Opp* EXPAND.

**abridged** *adj* abbreviated, bowdlerized, censored, compact, concise, condensed, cut, edited, *inf* potted, shortened, telescoped, truncated.

**abrupt** *adj* **1** disconnected, hasty, headlong, hurried, precipitate, quick, rapid, sudden, swift, unexpected, unforeseen, unpredicted. **2** *an abrupt drop.* precipitous, sharp, sheer, steep. **3** *an abrupt manner.* blunt, brisk, brusque, curt, discourteous, gruff, impolite, rude, snappy, terse, unceremonious, uncivil, ungracious. *Opp* GENTLE, GRADUAL.

**absent** *adj* **1** away, *sl* bunking off, gone, missing, off, out, playing truant, *inf* skiving. *Opp* PRESENT. **2** ▷ ABSENT-MINDED.

**absent-minded** *adj* absent, absorbed, abstracted, careless, day-dreaming, distracted, dreamy, far-away, forgetful, heedless, impractical, inattentive, oblivious, preoccupied, scatterbrained, thoughtless, unaware, unheeding, unthinking, vague, withdrawn, wool-gathering. *Opp* ALERT.

**absolute** *adj* **1** categorical, certain, complete, conclusive, decided, definite, downright, entire, full, genuine, implicit, inalienable, indubitable, infallible, *inf* out-and-out, perfect, positive, pure, sheer, supreme, sure, thorough, total, unadulterated, unalloyed, unambiguous, unconditional, unequivocal, unmitigated, unmixed, unqualified, unquestionable, unreserved, unrestricted, utter. **2** *absolute ruler.* almighty, autocratic, despotic, dictatorial, omnipotent, sovereign, totalitarian, tyrannical, undemocratic. **3** *absolute opposites. inf* dead, diametrical, exact, precise.

**absorb** *vb* **1** assimilate, consume, devour, digest, drink in, fill up with, hold, imbibe, incorporate, ingest, mop up, receive, retain, soak up, suck up, take in, utilize. *Opp* EMIT. **2** *absorb a blow.* cushion, deaden, lessen, reduce, soften. **3** *absorb a person.* captivate, engage, engross, enthrall, fascinate, involve, occupy, preoccupy. ▷ INTEREST. **absorbed** ▷ INTERESTED.

**absorbent** *adj* absorptive, permeable, pervious, porous, spongy. *Opp* IMPERVIOUS.

**absorbing** *adj* engrossing, fascinating, gripping, riveting, spellbinding. ▷ INTERESTING.

**abstain** *vb*

**abstain from** avoid, cease, decline, deny yourself, desist from, eschew, forgo, give up, go without, refrain from, refuse, reject, renounce, resist, shun, withhold from.

**abstemious** *adj* ascetic, austere, frugal, moderate, restrained, self-denying, self-disciplined, sober, sparing, teetotal, temperate. *Opp* SELF-INDULGENT.

**abstract** *adj* **1** abstruse, academic, hypothetical, indefinite, intangible, intellectual, metaphysical, notional, philosophical, theoretical, unpractical, unreal, unrealistic. *Opp* CONCRETE. **2** *abstract art.* non-pictorial, non-representational, symbolic. ● *n* digest, outline, précis, résumé, summary, synopsis.

**abstruse** *adj* complex, cryptic, deep, devious, difficult, enigmatic, esoteric, hard, incomprehensible, mysterious, mystical, obscure, perplexing, problematical, profound, puzzling, recherché, recondite, unfathomable. *Opp* OBVIOUS.

**absurd** *adj* anomalous, crazy, daft, eccentric, farcical, foolish, grotesque, idiotic, illogical, incongruous, irrational, laughable, ludicrous, nonsensical, outlandish, paradoxical, preposterous, ridiculous, risible, senseless, silly, stupid, surreal, unreasonable, untenable, zany. ▷ FUNNY, MAD. *Opp* RATIONAL.

**abundant** *adj* ample, bounteous, bountiful, copious, excessive, flourishing, full, generous, lavish, liberal, luxuriant, overflowing, plenteous, plentiful, prodigal, profuse, rampant, rank, rich, well-supplied. *Opp* SCARCE.

**abuse** *n* **1** assault, cruel treatment, ill-treatment, maltreatment, misappropriation, misuse, perversion. **2** *verbal abuse.* curse, execration, imprecation, insult, invective, obloquy, obscenity, slander, vilification, vituperation. ● *vb* **1** batter, damage, exploit, harm, hurt, ill-treat, injure, maltreat, manhandle, misemploy, misuse, molest, rape, spoil, treat roughly. **2** *abuse verbally.* affront, berate, be rude to, *inf* call names, castigate, criticize, curse, defame, denigrate, disparage, insult, inveigh against, libel, malign, revile, slander, *inf* slate, *inf* smear, sneer at, swear at, traduce, upbraid, vilify, vituperate, wrong. ▷ COMPLIMENT.

**abusive** *adj* acrimonious, angry, censorious, contemptuous, critical, cruel, defamatory, denigrating, derisive, derogatory, disparaging, harsh, hurtful, impolite, injurious, insulting, libellous, obscene, offensive, opprobrious, pejorative, profane, rude, scathing, scornful, scurrilous, slanderous, vituperative. *Opp* KIND, POLITE.

**abysmal** *adj* **1** bottomless, boundless, deep, extreme, immeasurable, incalculable, infinite, profound, unfathomable, vast. **2** ▷ BAD.

**abyss** *n inf* bottomless pit, chasm, crater, fissure, gap, gulf, hole, opening, pit, rift, void.

**academic** *adj* **1** educational, pedagogical, scholastic. **2** bookish, brainy, clever, erudite, highbrow, intelligent, learned, scholarly, studious, well-read. **3** *academic study.* abstract, conjectural, hypothetical, impractical, intellectual, notional, pure, speculative, theoretical, unpractical. • *n inf* egghead, highbrow, intellectual, scholar, thinker.

**accelerate** *vb* 1 *inf* do a spurt, *inf* get a move on, go faster, hasten, increase speed, pick up speed, quicken, speed up. 2 bring on, expedite, promote, spur on, step up, stimulate.

**accent** *n* 1 brogue, cadence, dialect, enunciation, inflection, intonation, pronunciation, sound, speech pattern, tone. 2 accentuation, beat, emphasis, force, prominence, pulse, rhythm, stress.

**accept** *vb* 1 acquire, get, *inf* jump at, receive, take, welcome. 2 acknowledge, admit, assume, bear, put up with, reconcile yourself to, resign yourself to, submit to, suffer, tolerate, undertake. 3 *accept an argument.* abide by, accede to, acquiesce in, adopt, agree to, approve, believe in, be reconciled to, consent to, defer to, grant, recognize, *inf* stomach, *inf* swallow, take in, *inf* wear, yield to. *Opp* REJECT.

**acceptable** *adj* 1 agreeable, appreciated, gratifying, pleasant, pleasing, welcome, worthwhile. 2 adequate, admissible, appropriate, moderate, passable, satisfactory, suitable, tolerable, unexceptionable. *Opp* UNACCEPTABLE.

**acceptance** *n* acquiescence, agreement, approval, consent, willingness. *Opp* REFUSAL.

**accepted** *adj* acknowledged, agreed, axiomatic, canonical, common, indisputable, recognized, standard, undeniable, undisputed, universal, unquestioned. *Opp* CONTROVERSIAL.

**accessible** *adj* approachable, at hand, attainable, available, close, convenient, *inf* get-at-able, *inf* handy, reachable, ready, within reach. *Opp* INACCESSIBLE.

**accessory** *n* 1 addition, adjunct, appendage, attachment, component, extension, extra, fitting. 2 ▷ ACCOMPLICE.

**accident** *n* 1 blunder, chance, coincidence, contingency, fate, fluke, fortune, hazard, luck, misadventure, mischance, misfortune, mishap, mistake, *inf* pot luck, serendipity. 2 calamity, catastrophe, collision, *inf* contretemps, crash, derailment, disaster, *inf* pile-up, *sl* shunt, wreck.

**accidental** *adj* adventitious, arbitrary, casual, chance, coincidental, *inf* fluky, fortuitous, fortunate, haphazard, inadvertent, lucky, random, unconscious, unexpected, unforeseen, unfortunate, unintended, unintentional, unlooked-for, unlucky, unplanned, unpremeditated. *Opp* INTENTIONAL.

**acclaim** *vb* applaud, celebrate, cheer, clap, commend, exalt, extol, hail, honour, laud, praise, salute, welcome.

**accommodate** *vb* 1 aid, assist, equip, fit, furnish, help, oblige, provide, serve, suit, supply. 2 *accommodate guests.* billet, board, cater for, entertain, harbour, hold, house, lodge, provide for, *inf* put up, quarter, shelter, take in. 3 *accommodate yourself to new surroundings.* accustom, adapt, reconcile. **accommodating** ▷ CONSIDERATE.

**accommodation** *n* board, home, housing, lodgings, pied-à-terre, premises, shelter. □ *apartment, barracks,* inf *bedsit, bedsitter, billet, boarding house,* inf *digs, flat, guest house, hall of residence, hos-*

*tel, hotel, inn, lodge, married quarters, motel, pension, rooms, self-catering, timeshare, youth hostel.* ▷ HOUSE.

**accompany** *vb* 1 attend, chaperon, conduct, convoy, escort, follow, go with, guard, guide, look after, partner, squire, *inf* tag along with, travel with, usher. 2 be associated with, be linked with, belong with, coexist with, coincide with, complement, occur with, supplement.

**accompanying** *adj* associated, attached, attendant, complementary, concomitant, connected, related.

**accomplice** *n* abettor, accessory, ally, assistant, associate, collaborator, colleague, confederate, conspirator, helper, *inf* henchman, partner.

**accomplish** *vb* achieve, attain, *inf* bring off, carry off, carry out, carry through, complete, conclude, consummate, discharge, do successfully, effect, execute, finish, fulfil, perform, realize, succeed in.

**accomplished** *adj* adept, expert, gifted, polished, proficient, skilful, talented.

**accomplishment** *n* ability, attainment, expertise, gift, skill, talent.

**accord** *n* agreement, concord, congruence, harmony, rapport, unanimity, understanding.

**account** *n* 1 bill, calculation, check, computation, invoice, receipt, reckoning, *inf* score, statement, tally. 2 chronicle, commentary, description, diary, explanation, history, log, memoir, narration, narrative, portrayal, record, report, statement, story, tale, version, *inf* write-up. 3 *of no account*. advantage, benefit, concern, consequence, consideration, importance, interest, merit, profit, significance, standing, use, value, worth. **account for** ▷ EXPLAIN.

**accumulate** *vb* accrue, agglomerate, aggregate, amass, assemble, bring together, build up, collect, come together, gather, grow, heap up, hoard, increase, mass, multiply, pile up, stack up, *inf* stash away, stockpile, store up. *Opp* DISPERSE.

**accumulation** *n inf* build-up, collection, conglomeration, gathering, growth, heap, hoard, mass, pile, stock, stockpile, store, supply.

**accurate** *adj* authentic, careful, certain, correct, exact, factual, faithful, faultless, meticulous, minute, nice, perfect, precise, reliable, right, scrupulous, sound, *inf* spot-on, strict, sure, true, truthful, unerring, veracious. *Opp* INACCURATE.

**accusation** *n* allegation, alleged offence, charge, citation, complaint, denunciation, impeachment, indictment, summons.

**accuse** *vb* arraign, attack, blame, bring charges against, censure, charge, condemn, denounce, hold responsible, impeach, impugn, incriminate, indict, inform against, make allegations against, *inf* point the finger at, prosecute, summons, tax. *Opp* DEFEND.

**accustomed** *adj* common, conventional, customary, established, expected, familiar, habitual, normal, ordinary, prevailing, regular, routine, set, traditional, usual, wonted. **get accustomed** ▷ ADAPT.

**ache** *n* anguish, discomfort, distress, hurt, pain, pang, smart, soreness, suffering, throbbing, twinge. • *vb* 1 be painful, be sore, hurt, pound, smart, sting, suffer, throb. 2 ▷ DESIRE.

**achieve** *vb* 1 accomplish, attain, bring off, carry out, complete, conclude, consummate, discharge, do successfully, effect, engineer, execute, finish, fulfil, manage, perform, succeed in. 2 *achieve fame.* acquire, earn, gain, get, obtain, procure, reach, score, win.

**acid** *adj* sharp, sour, stinging, tangy, tart, vinegary. *Opp* BLAND, SWEET.

**acknowledge** *vb* 1 accede, accept, acquiesce, admit, affirm, agree, allow, concede, confess, confirm, declare, endorse, grant, own up to, profess, yield. *Opp* DENY. 2 *acknowledge a greeting.* answer, notice, react to, reply to, respond to, return. 3 *acknowledge a friend.* greet, hail, recognize, salute, *inf* say hello to. *Opp* IGNORE.

**acme** *n* apex, crown, culmination, height, highest point, maximum, peak, pinnacle, summit, top, zenith. *Opp* NADIR.

**acquaint** *vb* advise, announce, apprise, brief, disclose, divulge, enlighten, inform, make aware, make familiar, notify, reveal, tell.

**acquaintance** *n* 1 awareness, familiarity, knowledge, understanding. 2 ▷ FRIEND.

**acquire** *vb* buy, come by, earn, get, obtain, procure, purchase, receive, secure.

**acquisition** *n* accession, addition, *inf* buy, gain, possession, prize, property, purchase.

**acquit** *vb* absolve, clear, declare innocent, discharge, dismiss, exculpate, excuse, exonerate, find innocent, free, *inf* let off, liberate, release, reprieve, set free, vindicate. *Opp* CONDEMN. **acquit yourself** ▷ BEHAVE.

**acrid** *adj* bitter, caustic, harsh, pungent, sharp, unpleasant.

**acrimonious** *adj* abusive, acerbic, angry, bad-tempered, bitter, caustic, censorious, churlish, cutting, hostile, hot-tempered, ill-natured, ill-tempered, irascible, mordant, peevish, petulant, quarrelsome, rancorous, sarcastic, sharp, spiteful, tart, testy, venomous, virulent, waspish. *Opp* PEACEABLE.

**act** *n* 1 achievement, action, deed, effort, enterprise, exploit, feat, move, operation, proceeding, step, undertaking. 2 *act of parliament.* bill [= *draft act*], decree, edict, law, order, regulation, statute. 3 *a stage act.* item, performance, routine, sketch, turn. • *vb* 1 behave, carry on, conduct yourself, deport yourself. 2 function, have an effect, operate, serve, take effect, work. 3 *Act now!* do something, get involved, make a move, react, take steps. ▷ BEGIN. 4 *act a role.* appear (as), assume the character of, *derog* camp it up, characterize, dramatize, enact, *derog* ham it up, imitate, impersonate, mime, mimic, *derog* overact, perform, personify, play, portray, pose as, represent, seem to be, simulate. ▷ PRETEND.

**acting** *adj* deputy, interim, provisional, stand-by, stopgap, substitute, surrogate, temporary, vice-.

**action** *n* 1 act, deed, effort, endeavour, enterprise, exploit, feat, measure, performance, proceeding, process, step, undertaking, work. 2 activity, drama, energy, enterprise, excitement, exercise, exertion, initiative, liveliness, motion, movement, vigour, vitality. 3 *action of a play.* events, happenings, incidents, story. 4 *action of a watch.* functioning, mechanism, operation, working, works. **5** *military action*. ▷ BATTLE.

**activate** *vb* actuate, animate, arouse, energize, excite, fire, galvanize, *inf* get going, initiate, mobilize, motivate, prompt, rouse, set in motion, set off, start, stimulate, stir, trigger.

**active** *adj* **1** agile, animated, brisk, bustling, busy, dynamic, energetic, enterprising, enthusiastic, functioning, hyperactive, live, lively, militant, moving, nimble, *inf* on the go, restless, spirited, sprightly, strenuous, vigorous, vital, vivacious, working. **2** *active support*. assiduous, committed, dedicated, devoted, diligent, employed, engaged, enthusiastic, hard-working, industrious, involved, occupied, sedulous, staunch, zealous. *Opp* INACTIVE.

**activity** *n* **1** action, animation, bustle, commotion, energy, excitement, hurly-burly, hustle, industry, life, liveliness, motion, movement, stir. **2** hobby, interest, job, occupation, pastime, project, pursuit, scheme, task, undertaking, venture. ▷ WORK.

**actor, actress** *n* artist, artiste, lead, leading lady, performer, player, star, supporting actor, trouper, walk-on part. ▷ ENTERTAINER. **actors** cast, company, troupe.

**actual** *adj* authentic, bona fide, certain, confirmed, corporeal, current, definite, existing, factual, genuine, indisputable, in existence, legitimate, living, material, real, realistic, tangible, true, truthful, unquestionable, verifiable. *Opp* IMAGINARY.

**acute** *adj* **1** narrow, pointed, sharp. **2** *acute pain*. cutting, excruciating, exquisite, extreme, fierce, intense, keen, piercing, racking, severe, sharp, shooting, sudden, violent. **3** *an acute mind*. alert, analytical, astute, canny, *inf* cute, discerning, incisive, intelligent, keen, observant, penetrating, perceptive, percipient, perspicacious, quick, sharp, shrewd, *inf* smart, subtle. ▷ CLEVER. **4** *an acute problem*. compelling, crucial, decisive, immediate, important, overwhelming, pressing, serious, urgent, vital. **5** *an acute illness*. critical, sudden. *Opp* CHRONIC, DULL, STUPID.

**adapt** *vb* **1** acclimatize, accommodate, accustom, adjust, attune, become conditioned, become hardened, become inured, fit, get accustomed (to), get used (to), habituate, harmonize, orientate, reconcile, suit, tailor, turn. **2** *adapt to a new use*. alter, amend, change, convert, metamorphose, modify, process, rearrange, rebuild, reconstruct, refashion, remake, remodel, reorganize, reshape, transform, vary. ▷ EDIT.

**add** *vb* annex, append, attach, combine, integrate, join, put together, *inf* tack on, unite. *Opp* DEDUCT. **add to** ▷ INCREASE. **add up (to)** ▷ TOTAL.

**addict** *n* **1** alcoholic, *sl* dopefiend, *sl* junkie, *inf* user. **2** *a TV addict*. ▷ ENTHUSIAST.

**addiction** *n* compulsion, craving, dependence, fixation, habit, obsession.

**addition** *n* **1** adding up, calculation, computation, reckoning, totalling, *inf* totting up. **2** accession, accessory, accretion, addendum, additive, adjunct, admixture, afterthought, amplification, annexe, appendage, appendix, appurtenance, attachment, continuation, development, enlargement, expansion, extension, extra, increase, increment, postscript, supplement.

**additional** *adj* added, extra, further, increased, more, new, other, spare, supplementary.

**address** *n* **1** directions, location, whereabouts. **2** *deliver an address.* discourse, disquisition, harangue, homily, lecture, oration, sermon, speech, talk. • *vb* **1** accost, apostrophize, approach, *inf* buttonhole, engage in conversation, greet, hail, salute, speak to, talk to. **2** *address an audience.* give a speech to, harangue, lecture. **address yourself to** ▷ TACKLE.

**adept** *adj* accomplished, clever, competent, expert, gifted, practised, proficient. ▷ SKILFUL. *Opp* UNSKILFUL.

**adequate** *adj* acceptable, all right, average, competent, fair, fitting, good enough, middling, *inf* OK, passable, presentable, respectable, satisfactory, *inf* so-so, sufficient, suitable, tolerable. *Opp* INADEQUATE.

**adhere** *vb* bind, bond, cement, cling, fuse, glue, gum, paste, stick. ▷ FASTEN.

**adherent** *n* aficionado, devotee, disciple, fan, follower, *inf* hanger-on, supporter.

**adhesive** *adj* glued, gluey, gummed, self-adhesive, sticky.

**adjoining** *adj* abutting, adjacent, bordering, closest, contiguous, juxtaposed, nearest, neighbouring, next, touching. *Opp* DISTANT.

**adjourn** *vb* break off, defer, discontinue, dissolve, interrupt, postpone, prorogue, put off, suspend.

**adjournment** *n* break, delay, interruption, pause, postponement, prorogation, recess, stay, stoppage, suspension.

**adjust** *vb* **1** adapt, alter, amend, arrange, balance, change, convert, correct, modify, position, put right, rectify, refashion, regulate, remake, remodel, reorganize, reshape, set, set to rights, tailor, temper, tune, vary. **2** acclimatize, accommodate, accustom, conform, fit, habituate, harmonize, reconcile yourself.

**administer** *vb* **1** administrate, command, conduct affairs, control, direct, govern, head, lead, manage, organize, oversee, preside over, regulate, rule, run, superintend, supervise. **2** *administer justice.* apply, carry out, execute, implement, prosecute. **3** *administer medicine.* deal out, dispense, distribute, dole out, give, hand out, measure out, mete out, provide, supply.

**administrator** *n* boss, bureaucrat, civil servant, controller, director, executive, head, manager, managing director, *derog* mandarin, organizer, superintendent. ▷ CHIEF.

**admirable** *adj* awe-inspiring, commendable, creditable, deserving, enjoyable, estimable, excellent, exemplary, fine, great, honourable, laudable, likeable, lovable, marvellous, meritorious, pleasing, praiseworthy, valued, wonderful, worthy. *Opp* CONTEMPTIBLE.

**admiration** *n* appreciation, approval, awe, commendation, esteem, hero-worship, high regard, honour, praise, respect. *Opp* CONTEMPT.

**admire** *vb* applaud, appreciate, approve of, be delighted by, commend, enjoy, esteem, have a high opinion of, hero-worship, honour, idolize, laud, like, look up to, love, marvel at, praise, respect, revere,

think highly of, value, venerate, wonder at. ▷ LOVE. *Opp* HATE.
**admiring** ▷ COMPLIMENTARY, RESPECTFUL.

**admission** *n* **1** access, admittance, entrance, entrée, entry. **2** acceptance, acknowledgement, affirmation, agreement, avowal, concession, confession, declaration, disclosure, profession, revelation. *Opp* DENIAL.

**admit** *vb* **1** accept, allow in, grant access, let in, provide a place (in), receive, take in. **2** *admit guilt.* accept, acknowledge, agree, allow, concede, confess, declare, disclose, divulge, grant, own up, profess, recognize, reveal, say reluctantly. *Opp* DENY.

**adolescence** *n* boyhood, girlhood, growing up, puberty, *inf* your teens, youth.

**adolescent** *adj* boyish, girlish, immature, juvenile, pubescent, puerile, teenage, youthful. • *n* boy, girl, juvenile, minor, *inf* teenager, youngster, youth.

**adopt** *vb* **1** accept, appropriate, approve, back, champion, choose, embrace, endorse, espouse, follow, *inf* go for, patronize, support, take on, take up. **2** befriend, foster, stand by, take in, *inf* take under your wing.

**adore** *vb* adulate, dote on, glorify, honour, idolize, love, revere, reverence, venerate, worship. ▷ ADMIRE. *Opp* HATE.

**adorn** *vb* beautify, decorate, embellish, garnish, grace, ornament, trim.

**adrift** *adj* **1** afloat, anchorless, drifting, floating. **2** aimless, astray, directionless, lost, purposeless, rootless.

**adult** *adj* developed, full-grown, full-size, grown-up, marriageable, mature, nubile, of age. *Opp* IMMATURE.

**adulterate** *vb* alloy, contaminate, corrupt, debase, defile, dilute, *inf* doctor, pollute, taint, thin, water down, weaken.

**advance** *n* betterment, development, evolution, forward movement, growth, headway, improvement, progress. • *vb* **1** approach, bear down, come near, forge ahead, gain ground, go forward, make headway, make progress, *inf* make strides, move forward, press ahead, press on, proceed, progress, *inf* push on. *Opp* RETREAT. **2** *science advances.* develop, evolve, grow, improve, increase, prosper, thrive. **3** *advance your career.* accelerate, assist, benefit, boost, expedite, facilitate, further, help the progress of, promote. *Opp* HINDER. **4** *advance a theory.* adduce, cite, furnish, give, present, propose, submit, suggest. **5** *advance money.* lend, loan, offer, pay, proffer, provide, supply. *Opp* WITHHOLD.

**advanced** *adj* **1** latest, modern, sophisticated, ultra-modern, up-to-date. **2** *advanced ideas.* avant-garde, contemporary, experimental, forward-looking, futuristic, imaginative, innovative, inventive, new, novel, original, pioneering, progressive, revolutionary, trend-setting, unconventional, unheard-of, *inf* way-out. **3** *advanced maths.* complex, complicated, difficult, hard, higher. **4** *advanced for her age.* grown-up, mature, precocious, sophisticated, well-developed. *Opp* BACKWARD, BASIC, OLD.

**advantage** *n* **1** aid, asset, assistance, benefit, boon, convenience, gain, help, profit, service, use, usefulness. 2 *have an advantage.* dominance, edge, *inf* head start, superiority. **take advantage of** ▷ EXPLOIT.

**advantageous** *adj* beneficial, constructive, favourable, gainful, helpful, invaluable, positive, profitable, salutary, useful, valuable, worthwhile. ▷ GOOD. *Opp* USELESS.

**adventure** *n* 1 chance, enterprise, escapade, exploit, feat, gamble, incident, occurrence, operation, risk, undertaking, venture. 2 danger, excitement, hazard.

**adventurous** *adj* 1 audacious, bold, brave, courageous, daredevil, daring, enterprising, *derog* foolhardy, heroic, intrepid, *derog* rash, *derog* reckless, valiant, venturesome. 2 *an adventurous trip.* challenging, dangerous, difficult, eventful, exciting, hazardous, perilous, risky. *Opp* UNADVENTUROUS.

**adversary** *n* antagonist, attacker, enemy, foe, opponent, rival. *Opp* FRIEND.

**adverse** *adj* 1 antagonistic, attacking, censorious, critical, derogatory, disapproving, hostile, hurtful, inimical, negative, uncomplimentary, unfavourable, unfriendly, unkind, unsympathetic. 2 *adverse conditions.* contrary, deleterious, detrimental, disadvantageous, harmful, inappropriate, inauspicious, opposing, prejudicial, uncongenial, unfortunate, unpropitious. *Opp* FAVOURABLE.

**advertise** *vb* announce, broadcast, display, flaunt, make known, market, merchandise, notify, *inf* plug, proclaim, promote, promulgate, publicize, *inf* push, show off, *inf* spotlight, tout.

**advertisement** *n inf* advert, announcement, bill, *inf* blurb, *TV* break, circular, commercial, handout, leaflet, notice, placard, *inf* plug, poster, promotion, publicity, *old use* puff, sign, *inf* small ad.

**advice** *n* admonition, caution, counsel, guidance, help, opinion, recommendation, suggestion, tip, view, warning. ▷ NEWS.

**advisable** *adj* expedient, judicious, politic, prudent, recommended, sensible. ▷ WISE. *Opp* SILLY.

**advise** *vb* 1 admonish, advocate, caution, counsel, encourage, enjoin, exhort, guide, instruct, prescribe, recommend, suggest, urge, warn. 2 ▷ INFORM.

**adviser** *n* cicerone, confidant(e), consultant, counsellor, guide, mentor.

**advocate** *n* 1 apologist, backer, champion, proponent, supporter. 2 ▷ LAWYER. • *vb* argue for, back, champion, endorse, favour, recommend, speak for, uphold.

**aerodrome** *n* airfield, airport, airstrip, landing-strip.

**aesthetic** *adj* artistic, beautiful, cultivated, in good taste, sensitive, tasteful. *Opp* UGLY.

**affair** *n* 1 activity, business, concern, issue, interest, matter, operation, project, question, subject, topic, transaction, undertaking. 2 circumstance, episode, event, happening, incident, occasion, occurrence, proceeding, thing. 3 *love affair.* affaire, amour, attachment, intrigue, involvement, liaison, relationship, romance.

**affect** *vb* 1 act on, agitate, alter, attack, change, concern, disturb, grieve, have an effect on, have an impact on, *inf* hit, impinge on, impress, influence, modify, move, pertain to, perturb, relate to, stir, touch, transform, trouble, upset. **2** *affect an accent.* adopt, assume, feign, *inf* put on. ▷ PRETEND.

**affectation** *n* artificiality, insincerity, mannerism, posturing, pretension. ▷ PRETENCE.

**affected** *adj* **1** artificial, contrived, insincere, mannered, *inf* put on, studied, unnatural. ▷ PRETENTIOUS. **2** *affected by disease.* afflicted, attacked, damaged, distressed, hurt, infected, injured, poisoned, stricken, troubled.

**affection** *n* amity, attachment, feeling, fondness, friendliness, friendship, liking, partiality, regard, *inf* soft spot, tenderness, warmth. ▷ LOVE. *Opp* HATRED.

**affectionate** *adj* caring, doting, fond, kind, tender, warm. ▷ LOVING. *Opp* ALOOF.

**affinity** *n* closeness, compatibility, fondness, kinship, like-mindedness, likeness, liking, rapport, relationship, resemblance, similarity, sympathy.

**affirm** *vb* assert, attest, aver, avow, confirm, declare, maintain, state, swear, testify.

**affirmation** *n* assertion, avowal, confirmation, declaration, oath, promise, pronouncement, statement, testimony.

**affirmative** *adj* agreeing, assenting, concurring, confirming, consenting, positive. *Opp* NEGATIVE.

**afflict** *vb* affect, annoy, bedevil, beset, bother, burden, cause suffering to, distress, grieve, harass, harm, hurt, oppress, pain, pester, plague, rack, torment, torture, trouble, try, vex, worry, wound.

**affluent** *adj* **1** flourishing, *inf* flush, *sl* loaded, moneyed, *usu derog* plutocratic, prosperous, rich, wealthy, *inf* well-heeled, well-off, well-to-do. **2** *affluent life-style.* expensive, gracious, lavish, luxurious, opulent, pampered, self-indulgent, sumptuous. *Opp* POOR.

**afford** *vb* **1** be rich enough, find enough, have the means, manage to give, sacrifice, spare, *inf* stand. **2** ▷ PROVIDE.

**afloat** *adj* aboard, adrift, at sea, floating, on board ship, under sail.

**afraid** *adj* **1** aghast, agitated, alarmed, anxious, apprehensive, *inf* chicken, cowardly, cowed, craven, daunted, diffident, faint-hearted, fearful, frightened, hesitant, horrified, horror-struck, intimidated, jittery, nervous, panicky, panic-stricken, pusillanimous, reluctant, scared, terrified, terror-stricken, timid, timorous, trembling, uneasy, unheroic, *inf* windy, *inf* yellow. *Opp* CONFIDENT, FEARLESS. **2** [*inf*] *I'm afraid I'm late.* apologetic, regretful, sorry, unhappy. **be afraid** ▷ FEAR.

**afterthought** *n* addendum, addition, appendix, extra, postscript.

**age** *n* **1** advancing years, decrepitude, dotage, old age, senescence, senility. **2** *a bygone age.* days, epoch, era, generation, period, time. **3** [*inf*] *ages ago.* aeon, lifetime, long time. • *vb* decline, degenerate, develop, grow older, look older, mature, mellow, ripen. **aged** ▷ OLD.

**agenda** *n* list, plan, programme, schedule, timetable.

**agent** *n* broker, delegate, emissary, envoy, executor, *old use* functionary, go-between, intermediary, mediator, middleman, negotiator, proxy, representative, spokesman, spokeswoman, surrogate, trustee.

**aggravate** *vb* **1** add to, augment, compound, exacerbate, exaggerate, heighten, increase, inflame, intensify, magnify, make more serious, make worse, worsen. *Opp* ALLEVIATE. **2** [Some think this use wrong.] ▷ ANNOY.

**aggressive** *adj* antagonistic, assertive, attacking, bellicose, belligerent, bullying, *sl* butch, contentious, destructive, hostile, jingoistic, *sl* macho, militant, offensive, provocative, pugnacious, pushful, *inf* pushy, quarrelsome, violent, warlike, zealous. *Opp* DEFENSIVE, PEACEABLE.

**aggressor** *n* assailant, attacker, belligerent, instigator, invader.

**agile** *adj* acrobatic, active, adroit, deft, fleet, graceful, limber, lissom, lithe, lively, mobile, nimble, quick-moving, sprightly, spry, supple, swift. *Opp* CLUMSY, SLOW.

**agitate** *vb* **1** beat, churn, convulse, ferment, froth up, ruffle, shake, stimulate, stir, toss, work up. **2** alarm, arouse, confuse, discomfit, disconcert, disturb, excite, fluster, incite, perturb, rouse, shake up, stir up, trouble, unnerve, unsettle, upset, worry. *Opp* CALM. **agitated** ▷ EXCITED, NERVOUS.

**agitator** *n Fr* agent provocateur, demagogue, firebrand, rabble-rouser, revolutionary, trouble-maker.

**agonize** *vb* be in agony, hurt, labour, struggle, suffer, worry, wrestle.

**agony** *n* anguish, distress, suffering, torment, torture. ▷ PAIN.

**agree** *vb* **1** accede, accept, acknowledge, acquiesce, admit, allow, assent, be willing, concede, consent, covenant, grant, make a contract, pledge yourself, promise, undertake. **2** accord, be unanimous, be united, coincide, concur, conform, correspond, fit, get on, harmonize, match, *inf* see eye to eye, suit. *Opp* DISAGREE. **agree on** ▷ CHOOSE. **agree with** ▷ ENDORSE.

**agreeable** *adj* acceptable, delightful, enjoyable, nice. ▷ PLEASANT. *Opp* DISAGREEABLE.

**agreement** *n* **1** accord, affinity, compatibility, compliance, concord, conformity, congruence, consensus, consent, consistency, correspondence, harmony, similarity, sympathy, unanimity, unity. **2** acceptance, alliance, armistice, arrangement, bargain, bond, compact, concordat, contract, convention, covenant, deal, *Fr* entente, pact, pledge, protocol, settlement, treaty, truce, understanding. *Opp* DISAGREEMENT.

**agricultural** *adj* **1** agrarian, bucolic, pastoral, rural. **2** *agricultural land.* cultivated, farmed, planted, productive, tilled.

**agriculture** *n* agronomy, crofting, cultivation, farming, growing, husbandry, tilling.

**aground** *adj* beached, grounded, helpless, high-and-dry, marooned, shipwrecked, stranded, stuck.

**aid** *n* advice, assistance, avail, backing, benefit, collaboration, contribution, cooperation, donation, encouragement, funding, grant, guidance, help, loan, patronage, prop, relief, sponsorship, subsidy, succour, support. • *vb* abet, assist, back, befriend, benefit, collaborate with, contribute to, co-operate with, encourage, facilitate, forward, help, *inf* lend a hand, profit, promote, prop up, *inf* rally round, relieve, subsidize, succour, support, sustain.

**ailing** *adj* diseased, feeble, infirm, poorly, sick, suffering, unwell, weak. ▷ ILL.

**ailment** *n* affliction, disease, disorder, infirmity, malady, sickness. ▷ ILLNESS.

**aim** *n* ambition, aspiration, cause, design, desire, destination, direction, dream, end, focus, goal, hope, intent, intention, mark, object, objective, plan, purpose, target, wish. • *vb* **1** address, beam, direct, fire at, focus, level, line up, point, send, sight, take aim, train, turn, zero in on. **2** *aim to win.* aspire, attempt, design, endeavour, essay, intend, mean, plan, propose, resolve, seek, strive, try, want, wish.

**aimless** *adj* chance, directionless, pointless, purposeless, rambling, random, undisciplined, unfocused, wayward. *Opp* PURPOSEFUL.

**air** *n* **1** airspace, atmosphere, ether, heavens, sky, *poet* welkin. **2** *fresh air.* breath, breeze, draught, oxygen, waft, wind, *poet* zephyr. **3** *air of authority.* ambience, appearance, aspect, aura, bearing, character, demeanour, effect, feeling, impression, look, manner, mien, mood, quality, style. • *vb* **1** aerate, dry off, freshen, refresh, ventilate. **2** *air opinions.* articulate, disclose, display, exhibit, express, give vent to, make known, make public, put into words, show off, vent, voice.

**aircraft** *n old use* flying-machine. □ *aeroplane, airliner, airship, balloon, biplane, bomber, delta-wing, dirigible, fighter, flying boat, glider, gunship, hang-glider, helicopter, jet, jumbo, jump-jet, microlight, monoplane, plane, seaplane, turboprop, VTOL (vertical take-off and landing).*

**airman** *n* aviator, flier, pilot.

**airport** *n* aerodrome, airfield, air strip, heliport, landing-strip, runway.

**airy** *adj* blowy, breezy, draughty, fresh, open, spacious, ventilated. *Opp* STUFFY.

**aisle** *n* corridor, gangway, passage, passageway.

**akin** *adj* allied, related, similar.

**alarm** *n* **1** alert, signal, warning. □ *alarm-clock, bell, fire-alarm, gong, siren, tocsin, whistle.* **2** anxiety, apprehension, consternation, dismay, distress, fright, nervousness, panic, trepidation, uneasiness. ▷ FEAR. • *vb* agitate, daunt, dismay, distress, disturb, panic, *inf* put the wind up, scare, shock, startle, surprise, unnerve, upset, worry. ▷ FRIGHTEN. *Opp* REASSURE.

**alcohol** *n sl* bevvy, *inf* booze, drink, hard stuff, intoxicant, liquor, spirits, wine.

**alcoholic** *adj* brewed, distilled, fermented, *sl* hard, inebriating, intoxicating, spirituous, *inf* strong. • *n* addict, dipsomaniac, drunkard, inebriate, toper. *Opp* TEETOTALLER.

**alert** *adj* active, agile, alive (to), attentive, awake, careful, circumspect, eagle-eyed, heedful, lively, observant, on the alert, on the lookout, *inf* on the qui vive, on the watch, on your guard, on your toes, perceptive, quick, ready, responsive, sensitive, sharp-eyed, vigilant, wary, watchful, wide-awake. *Opp* ABSENT-MINDED, INATTENTIVE. • *vb* advise, alarm, caution, forewarn, give the alarm, inform, make aware, notify, signal, tip off, warn.

**alibi** *n* excuse, explanation.

**alien** *adj* exotic, extra-terrestrial, foreign, outlandish, remote, strange, unfamiliar. • *n* foreigner, newcomer, outsider, stranger.

**alight** *adj* ablaze, afire, aflame, blazing, bright, burning, fiery, ignited, illuminated, lit up, live, on fire, shining. • *vb* come down, come to rest, descend, disembark, dismount, get down, get off, land, perch, settle, touch down.

**align** *vb* **1** arrange in line, line up, place in line, straighten up. **2** *align with the opposition.* affiliate, agree, ally, associate, co-operate, join, side, sympathize.

**alike** *adj* akin, analogous, close, cognate, comparable, corresponding, equivalent, identical, indistinguishable, like, matching, parallel, related, resembling, similar, the same, twin, uniform. *Opp* DISSIMILAR.

**alive** *adj* **1** active, animate, breathing, existent, existing, extant, flourishing, in existence, live, living, *old use* quick, surviving. **2** *alive to new ideas.* ▷ ALERT. *Opp* DEAD.

**allay** *vb* alleviate, assuage, calm, check, compose, diminish, ease, lessen, lull, mitigate, moderate, mollify, pacify, quell, quench, quiet, quieten, reduce, relieve, slake (*thirst*), soften, soothe, subdue. *Opp* STIMULATE.

**allegation** *n* accusation, assertion, charge, claim, complaint, declaration, deposition, statement, testimony.

**allege** *vb* adduce, affirm, assert, asseverate, attest, aver, avow, claim, contend, declare, depose, insist, maintain, make a charge, plead, state.

**allegiance** *n* devotion, duty, faithfulness, *old use* fealty, fidelity, loyalty, obedience.

**allergic** *adj* antagonistic, antipathetic, averse, disinclined, hostile, incompatible (with), opposed.

**alleviate** *vb* abate, allay, ameliorate, assuage, check, diminish, ease, lessen, lighten, make lighter, mitigate, moderate, pacify, palliate, quell, quench, reduce, relieve, slake (*thirst*), soften, soothe, subdue, temper. *Opp* AGGRAVATE.

**alliance** *n* affiliation, agreement, association, bloc, bond, cartel, coalition, combination, compact, concordat, confederation, connection, consortium, covenant, entente, federation, guild, league, marriage, pact, partnership, relationship, syndicate, treaty, understanding, union.

**allot** *vb* allocate, allow, apportion, assign, award, deal out, *inf* dish out, *inf* dole out, dispense, distribute, divide out, give out, grant, mete out, provide, ration, set aside, share out.

**allow** *vb* **1** approve, authorize, bear, consent to, enable, endure, grant permission for, let, license, permit, *inf* put up with, sanction, *inf* stand, suffer, support, tolerate. *Opp* FORBID. **2** acknowledge, admit, concede, grant, own. **3** ▷ ALLOT. *Opp* DENY.

**allowance** *n* **1** allocation, allotment, amount, measure, portion, quota, ration, share. **2** alimony, annuity, grant, maintenance, payment, pension, pocket money, subsistence. **3** *allowance on the full price.* deduction, discount, rebate, reduction, remittance, subsidy. **make allowances for** ▷ TOLERATE.

**alloy** *n* admixture, aggregate, amalgam, blend, combination, composite, compound, fusion, mixture.

**allude** *vb* **allude to** hint at, make an allusion to, mention, refer to, speak of, suggest, touch on.

**allure** *vb* attract, beguile, bewitch, cajole, charm, coax, decoy, draw, entice, fascinate, inveigle, lead on, lure, magnetize, persuade, seduce, tempt.

**allusion** *n* hint, mention, reference, suggestion.

**ally** *n* abettor, accessory, accomplice, associate, backer, collaborator, colleague, companion, comrade, confederate, friend, helper, helpmate, *inf* mate, partner, supporter. *Opp* ENEMY. • *vb* affiliate, amalgamate, associate, band together, collaborate, combine, confederate, cooperate, form an alliance, fraternize, join, join forces, league, *inf* link up, marry, merge, side, *inf* team up, unite.

**almighty** *adj* **1** all-powerful, omnipotent, supreme. **2** ▷ BIG.

**almost** *adv* about, all but, approximately, around, as good as, just about, nearly, not quite, practically, virtually.

**alone** *adj* apart, by yourself, deserted, desolate, forlorn, friendless, isolated, lonely, lonesome, on your own, separate, single, solitary, solo, unaccompanied, unassisted.

**aloof** *adj* chilly, cold, cool, detached, disinterested, dispassionate, distant, formal, frigid, haughty, impressive, inaccessible, indifferent, remote, reserved, reticent, self-contained, self-possessed, *inf* standoffish, supercilious, unapproachable, unconcerned, undemonstrative, unemotional, unforthcoming, unfriendly, uninvolved, unresponsive, unsociable, unsympathetic. *Opp* FRIENDLY, SOCIABLE.

**aloud** *adv* audibly, clearly, distinctly, out loud.

**also** *adv* additionally, besides, furthermore, in addition, moreover, *joc* to boot, too.

**alter** *vb* adapt, adjust, amend, change, convert, edit, emend, enlarge, modify, reconstruct, reduce, reform, remake, remodel, reorganize, reshape, revise, transform, vary.

**alteration** *n* adaptation, adjustment, amendment, change, conversion, difference, modification, reorganization, revision, transformation.

**alternate** *vb* come alternately, follow each other, interchange, oscillate, replace each other, rotate, *inf* see-saw, substitute for each other, take turns.

**alternative** *n* **1** choice, option, selection. **2** back-up, replacement, substitute.

**altitude** *n* elevation, height.

**altogether** *adv* absolutely, completely, entirely, fully, perfectly, quite, thoroughly, totally, utterly, wholly.

**always** *adv* consistently, constantly, continually, continuously, endlessly, eternally, everlastingly, evermore, forever, invariably, perpetually, persistently, regularly, repeatedly, unceasingly, unfailingly, unremittingly.

**amalgamate** *vb* affiliate, ally, associate, band together, blend, coalesce, combine, come together, compound, confederate, form an alliance, fuse, integrate, join, join forces, league, *inf* link up, marry, merge, mix, put together, synthesize, *inf* team up, unite. *Opp* SPLIT.

**amateur** *adj* inexperienced, lay, unpaid, unqualified. ▷ AMATEURISH. • *n* dabbler, dilettante, enthusiast, layman, non-professional. *Opp* PROFESSIONAL.

**amateurish** *adj* clumsy, crude, *inf* do-it-yourself, incompetent, inept, inexpert, *inf* rough-and-ready, second-rate, shoddy, unpolished, unprofessional, unskilful, unskilled, untrained. *Opp* SKILLED.

**amaze** *vb* astonish, astound, awe, bewilder, confound, confuse, daze, disconcert, dumbfound, *inf* flabbergast, perplex, *inf* rock, shock, stagger, startle, stun, stupefy, surprise. **amazed** ▷ SURPRISED.

**amazing** *adj* astonishing, astounding, awe-inspiring, breathtaking, exceptional, exciting, extraordinary, *inf* fantastic, incredible, miraculous, notable, phenomenal, prodigious, remarkable, *inf* sensational, shocking, special, staggering, startling, stunning, stupendous, unusual, *inf* wonderful. *Opp* ORDINARY.

**ambassador** *n* agent, attaché, *Fr* chargé d'affaires, consul, diplomat, emissary, envoy, legate, nuncio, plenipotentiary, representative.

**ambiguous** *adj* ambivalent, confusing, enigmatic, equivocal, indefinite, indeterminate, puzzling, uncertain, unclear, vague, woolly. ▷ UNCERTAIN. *Opp* DEFINITE.

**ambition** *n* **1** commitment, drive, energy, enterprise, enthusiasm, *inf* go, initiative, *inf* push, pushfulness, self-assertion, thrust, zeal. **2** aim, aspiration, desire, dream, goal, hope, ideal, intention, object, objective, target, wish.

**ambitious** *adj* **1** assertive, committed, eager, energetic, enterprising, enthusiastic, go-ahead, *inf* go-getting, hard-working, industrious, keen, *inf* pushy, zealous. **2** *ambitious ideas*. *inf* big, far-reaching, grand, grandiose, large-scale, unrealistic. *Opp* APATHETIC.

**ambivalent** *adj* ambiguous, backhanded (*compliment*), confusing, doubtful, equivocal, inconclusive, inconsistent, indefinite, self-contradictory, *inf* two-faced, unclear, uncommitted, unresolved, unsettled. ▷ UNCERTAIN.

**ambush** *n* ambuscade, attack, *snare, surprise attack, trap*. • *vb* attack, ensnare, entrap, intercept, lie in wait for, pounce on, surprise, swoop on, trap, waylay.

**amenable** *adj* accommodating, acquiescent, adaptable, agreeable, biddable, complaisant, compliant, co-operative, deferential, docile, open-minded, persuadable, responsive, submissive, tractable, willing. *Opp* OBSTINATE.

**amend** *vb* adapt, adjust, alter, ameliorate, change, convert, correct, edit, emend, improve, make better, mend, modify, put right, rectify, reform, remedy, reorganize, reshape, revise, transform, vary.

**amiable** *adj* affable, agreeable, amicable, friendly, genial, good-natured, kind-hearted, kindly, likeable, well-disposed. *Opp* UNFRIENDLY.

**ammunition** *n* buckshot, bullet, cartridge, grenade, missile, projectile, round, shell, shrapnel.

**amoral** *adj* lax, loose, unethical, unprincipled, without standards. ▷ IMMORAL. *Opp* MORAL.

**amorous** *adj* affectionate, ardent, carnal, doting, enamoured, erotic, fond, impassioned, loving, lustful, passionate, *sl* randy, sexual, *inf* sexy. *Opp* COLD.

**amount** *n* aggregate, bulk, entirety, extent, lot, mass, measure, quantity, quantum, reckoning, size, sum, supply, total, value, volume, whole. • *vb* **amount to** add up to, aggregate, be equivalent to, come to, equal, make, mean, total.

**ample** *adj* abundant, bountiful, broad, capacious, commodious, considerable, copious, extensive, fruitful, generous, great, large, lavish, liberal, munificent, plentiful, profuse, roomy, spacious, substantial, unstinting, voluminous. ▷ BIG, PLENTY. *Opp* INSUFFICIENT.

**amplify** *vb* **1** add to, augment, broaden, develop, dilate upon, elaborate, enlarge, expand, expatiate on, extend, fill out, lengthen, make fuller, make longer, supplement. **2** *amplify sound.* boost, heighten, increase, intensify, magnify, make louder, raise the volume. *Opp* DECREASE.

**amputate** *vb* chop off, cut off, dock, lop off, poll, pollard, remove, sever, truncate. ▷ CUT.

**amuse** *vb* absorb, beguile, cheer (up), delight, divert, engross, enliven, entertain, gladden, interest, involve, make laugh, occupy, please, raise a smile, *inf* tickle. *Opp* BORE. **amusing** ▷ ENJOYABLE, FUNNY.

**amusement** *n* **1** delight, enjoyment, fun, hilarity, laughter, mirth. ▷ MERRIMENT. **2** distraction, diversion, entertainment, game, hobby, interest, joke, leisure activity, pastime, play, pleasure, recreation, sport.

**anaemic** *adj* bloodless, colourless, feeble, frail, pale, pallid, pasty, sallow, sickly, unhealthy, wan, weak.

**analogy** *n* comparison, likeness, metaphor, parallel, resemblance, similarity, simile.

**analyse** *vb* anatomize, assay, break down, criticize, dissect, evaluate, examine, interpret, investigate, scrutinize, separate out, take apart, test.

**analysis** *n* breakdown, critique, dissection, enquiry, evaluation, examination, interpretation, investigation, *inf* post-mortem, scrutiny, study, test. *Opp* SYNTHESIS.

**analytical** *adj* analytic, critical, *inf* in-depth, inquiring, investigative, logical, methodical, penetrating, questioning, rational, searching, systematic. *Opp* SUPERFICIAL.

**anarchy** *n* bedlam, chaos, confusion, disorder, disorganization, insurrection, lawlessness, misgovernment, misrule, mutiny, pandemonium, riot. *Opp* ORDER.

**ancestor** *n* antecedent, forebear, forefather, forerunner, precursor, predecessor, progenitor.

**ancestry** *n* blood, derivation, descent, extraction, family, genealogy, heredity, line, lineage, origin, parentage, pedigree, roots, stock, strain.

**anchor** *vb* berth, make fast, moor, secure, tie up. ▷ FASTEN.

**anchorage** *n* harbour, haven, marina, moorings, port, refuge, sanctuary, shelter.

**ancient** *adj* **1** aged, antediluvian, antiquated, antique, archaic, elderly, fossilized, obsolete, old, old-fashioned, outmoded, out-of-date, passé, superannuated, time-worn, venerable. **2** *ancient times.* bygone, earlier, early, former, *poet* immemorial, *inf* olden, past, prehistoric, primeval, primitive, primordial, remote, *old use* of yore. *Opp* MODERN.

**angel** *n* archangel, cherub, divine messenger, seraph.

**angelic** *adj* **1** beatific, blessed, celestial, cherubic, divine, ethereal, heavenly, holy, seraphic, spiritual. **2** *angelic behaviour*. exemplary, innocent, pious, pure, saintly, unworldly, virtuous. ▷ GOOD. *Opp* DEVILISH.

**anger** *n* angry feelings, annoyance, antagonism, bitterness, choler, displeasure, exasperation, fury, hostility, indignation, ire, irritability, outrage, passion, pique, rage, rancour, resentment, spleen, tantrum, temper, vexation, wrath. • *vb inf* aggravate, antagonize, *sl* bug, displease, *inf* drive mad, enrage, exasperate, incense, incite, inflame, infuriate, irritate, madden, make angry, *inf* make someone's blood boil, *inf* needle, outrage, pique, provoke, *inf* rile, vex. ▷ ANNOY. *Opp* PACIFY.

**angle** *n* **1** bend, corner, crook, nook, point. **2** *a new angle*. approach, outlook, perspective, point of view, position, slant, standpoint, viewpoint. • *vb* bend, bevel, chamfer, slant, turn, twist.

**angry** *adj inf* aerated, apoplectic, bad-tempered, bitter, *inf* bristling, *inf* choked, choleric, cross, disgruntled, enraged, exasperated, excited, fiery, fuming, furious, heated, hostile, *inf* hot under the collar, ill-tempered, incensed, indignant, infuriated, *inf* in high dudgeon, irascible, irate, livid, mad, outraged, provoked, raging, *inf* ratty, raving, resentful, riled, seething, smouldering, *inf* sore, splenetic, *sl* steamed up, stormy, tempestuous, vexed, *inf* ugly, *inf* up in arms, wild, wrathful. ▷ ANNOYED. *Opp* CALM. **be angry**, become angry *inf* be in a paddy, *inf* blow up, boil, bridle, bristle, flare up, *inf* fly off the handle, fulminate, fume, *inf* get steamed up, lose your temper, rage, rant, rave, *inf* see red, seethe, snap, storm. **make angry** ▷ ANGER.

**anguish** *n* agony, anxiety, distress, grief, heartache, misery, pain, sorrow, suffering, torment, torture, tribulation, woe.

**angular** *adj* bent, crooked, indented, jagged, sharp-cornered, zigzag. *Opp* STRAIGHT.

**animal** *adj* beastly, bestial, brutish, carnal, fleshly, inhuman, instinctive, physical, savage, sensual, subhuman, wild. • *n* beast, being, brute, creature, *plur* fauna, organism, *plur* wildlife.
□ *amphibian, arachnid, biped, carnivore, herbivore, insect, invertebrate, mammal, marsupial, mollusc, monster, omnivore, pet, quadruped, reptile, rodent, scavenger, vertebrate.* ▷ BIRD, FISH, INSECT.
□ *aardvark, antelope, ape, armadillo, baboon, badger, bear, beaver, bison, buffalo, camel, caribou, cat, chamois, cheetah, chimpanzee, chinchilla, chipmunk, coypu, deer, dog, dolphin, donkey, dormouse, dromedary, elephant, elk, ermine, ferret, fox, frog, gazelle, gerbil, gibbon, giraffe, gnu, goat, gorilla, grizzly bear, guinea-pig, hamster, hare, hedgehog, hippopotamus, horse, hyena, ibex, impala, jackal, jaguar, jerboa, kangaroo, koala, lemming, lemur, leopard, lion, llama, lynx, marmoset, marmot, marten, mink, mongoose, monkey, moose, mouse, musquash, ocelot, octopus, opossum, orang-utan, otter, panda, panther, pig, platypus, polar bear, pole-cat, porcupine, porpoise, rabbit, rat, reindeer, rhinoceros, roe, salamander, scorpion, seal,*

*sea-lion, sheep, shrew, skunk, snake, spider, squirrel, stoat, tapir, tiger, toad, vole, wallaby, walrus, weasel, whale, wildebeest, wolf, wolverine, wombat, yak, zebra.*

**animate** *adj* alive, breathing, conscious, feeling, live, living, sentient. ▷ ANIMATED. *Opp* INANIMATE. • *vb* activate, arouse, brighten up, *inf* buck up, cheer up, encourage, energize, enliven, excite, exhilarate, fire, galvanize, incite, inspire, invigorate, kindle, liven up, make lively, move, *inf* pep up, *inf* perk up, quicken, rejuvenate, revitalize, revive, rouse, spark, spur, stimulate, stir, urge, vitalize.

**animated** *adj* active, alive, bright, brisk, bubbling, busy, cheerful, eager, ebullient, energetic, enthusiastic, excited, exuberant, gay, impassioned, lively, passionate, quick, spirited, sprightly, vibrant, vigorous, vivacious, zestful. *Opp* LETHARGIC.

**animation** *n* activity, briskness, eagerness, ebullience, energy, enthusiasm, excitement, exhilaration, gaiety, high spirits, life, liveliness, *inf* pep, sparkle, spirit, sprightliness, verve, vigour, vitality, vivacity, zest. *Opp* LETHARGY.

**animosity** *n* acerbity, acrimony, animus, antagonism, antipathy, asperity, aversion, bad blood, bitterness, dislike, enmity, grudge, hate, hatred, hostility, ill will, loathing, malevolence, malice, malignancy, malignity, odium, rancour, resentment, sarcasm, sharpness, sourness, spite, unfriendliness, venom, vindictiveness, virulence. *Opp* FRIENDLINESS.

**annex** *vb* acquire, appropriate, conquer, occupy, purloin, seize, take over, usurp.

**annihilate** *vb* abolish, destroy, eliminate, eradicate, erase, exterminate, extinguish, extirpate, *inf* finish off, *inf* kill off, *inf* liquidate, nullify, obliterate, raze, slaughter, wipe out.

**annotation** *n* comment, commentary, elucidation, explanation, footnote, gloss, interpretation, note.

**announce** *vb* **1** advertise, broadcast, declare, disclose, divulge, give notice of, intimate, make public, notify, proclaim, promulgate, propound, publicize, publish, put out, report, reveal, state. **2** *announce a speaker.* introduce, lead into, preface, present.

**announcement** *n* advertisement, bulletin, communiqué, declaration, disclosure, intimation, notification, proclamation, promulgation, publication, report, revelation, statement.

**announcer** *n* anchorman, anchorwoman, broadcaster, commentator, compère, disc jockey, DJ, *poet* harbinger, herald, master of ceremonies, *inf* MC, messenger, newscaster, newsreader, reporter, town crier.

**annoy** *vb* *inf* aggravate, antagonize, *inf* badger, be an annoyance to, bother, *sl* bug, chagrin, displease, distress, drive mad, exasperate, fret, gall, *inf* get at, *inf* get on your nerves, grate, harass, harry, infuriate, irk, irritate, jar, madden, make cross, molest, *inf* needle, *inf* nettle, offend, peeve, pester, pique, *inf* plague, provoke, put out, rankle, rile, *inf* rub up the wrong way, ruffle, *inf* spite, tease, trouble, try (someone's patience), upset, vex, worry. ▷ ANGER. *Opp* PLEASE.

**annoyance** *n* **1** chagrin, crossness, displeasure, exasperation, irritation, pique, vexation.

▷ ANGER. **2** *Noise is an annoyance.* *inf* aggravation, bother, harassment, irritant, nuisance, offence, *inf* pain in the neck, pest, provocation, worry.

**annoyed** *adj* chagrined, cross, displeased, exasperated, *inf* huffy, irritated, jaundiced, *inf* miffed, *inf* needled, *inf* nettled, offended, *inf* peeved, piqued, *inf* put out, *inf* riled, *inf* shirty, *inf* sore, upset, vexed. ▷ ANGRY. *Opp* PLEASED. **be annoyed** *inf* go off in a huff, take offence, *inf* take umbrage.

**annoying** *adj* *inf* aggravating, bothersome, displeasing, exasperating, galling, grating, inconvenient, infuriating, irksome, irritating, jarring, maddening, offensive, provocative, provoking, tiresome, troublesome, trying, upsetting, vexatious, vexing, wearisome, worrying.

**anoint** *vb* **1** embrocate, grease, lubricate, oil, rub, smear. **2** bless, consecrate, dedicate, hallow, sanctify.

**anonymous** *adj* **1** incognito, nameless, unacknowledged, unidentified, unknown, unnamed, unspecified, unsung. **2** *anonymous letters.* unattributed, unsigned. **3** *anonymous style.* characterless, impersonal, nondescript, unidentifiable, unrecognizable, unremarkable.

**answer** *n* **1** acknowledgement, *inf* comeback, reaction, rejoinder, reply, response, retort, riposte. **2** explanation, outcome, solution. **3** *answer to a charge.* countercharge, defence, plea, rebuttal, refutation, vindication. • *vb* **1** acknowledge, give an answer, react, rejoin, reply, respond, retort, return. **2** explain, resolve, solve. **3** *answer a charge.* counter, defend yourself against, disprove, rebut, refute. **4** *answer a need.* correspond to, echo, fit, match up to, meet, satisfy, serve, suffice, suit. **answer back** ▷ ARGUE.

**antagonism** *n* antipathy, dissension, enmity, friction, opposition, rancour, rivalry, strife. ▷ HOSTILITY.

**antagonize** *vb* alienate, anger, annoy, embitter, estrange, irritate, make an enemy of, offend, provoke, *inf* put off, upset.

**anthem** *n* canticle, chant, chorale, hymn, introit, paean, psalm.

**anthology** *n* collection, compendium, compilation, digest, miscellany, selection, treasury.

**anticipate** *vb* **1** forestall, obviate, preclude, pre-empt, prevent. **2** [Many think this use incorrect.] ▷ FORESEE.

**anticlimax** *n* bathos, *inf* comedown, *inf* damp squib, disappointment, *inf* let-down.

**antics** *n* buffoonery, capers, clowning, escapades, foolery, fooling, *inf* larking-about, pranks, *inf* skylarking, tomfoolery, tricks.

**antidote** *n* antitoxin, corrective, countermeasure, cure, drug, neutralizing agent, remedy.

**antiquarian** *n* antiquary, antiques expert, collector, dealer.

**antiquated** *adj* aged, anachronistic, ancient, antediluvian, archaic, dated, medieval, obsolete, old, old-fashioned, *inf* out, outdated, outmoded, out-of-date, passé, *inf* past it, *inf* prehistoric, *inf* primeval, primitive, quaint, superannuated, unfashionable. ▷ ANTIQUE. *Opp* NEW.

**antique** *adj* antiquarian, collectible, historic, old-fashioned, traditional, veteran, vintage. ▷ ANTIQUATED. • *n* bygone, collectible, collector's item, curio, curiosity, *Fr* objet d'art, rarity.

**antiquity** *n* classical times, days gone by, former times, *inf* olden days, the past.

**antiseptic** *adj* aseptic, clean, disinfectant, disinfected, germ free, germicidal, hygienic, medicated, sanitized, sterile, sterilized, sterilizing, unpolluted.

**antisocial** *adj* alienated, anarchic, disagreeable, disorderly, disruptive, misanthropic, nasty, obnoxious, offensive, rebellious, rude, troublesome, uncooperative, undisciplined, unruly, unsociable. ▷ UNFRIENDLY. *Opp* SOCIABLE.

**anxiety** *n* **1** angst, apprehension, concern, disquiet, distress, doubt, dread, fear, foreboding, fretfulness, misgiving, nervousness, qualm, scruple, strain, stress, tension, uncertainty, unease, worry. **2** *anxiety to succeed.* desire, eagerness, enthusiasm, impatience, keenness, longing, solicitude, willingness.

**anxious** **1** afraid, agitated, alarmed, apprehensive, concerned, distracted, distraught, distressed, disturbed, edgy, fearful, *inf* fraught, fretful, *inf* jittery, nervous, *inf* nervy, *inf* on edge, overwrought, perturbed, restless, tense, troubled, uneasy, upset, watchful, worried. **2** *anxious to succeed.* avid, careful, desirous, *inf* desperate, *inf* dying, eager, impatient, intent, *inf* itching, keen, longing, solicitous, willing, yearning. **be anxious** ▷ WORRY.

**apathetic** *adj* casual, cool, dispassionate, dull, emotionless, half-hearted, impassive, inactive, indifferent, indolent, languid, lethargic, listless, passive, phlegmatic, slow, sluggish, tepid, torpid, unambitious, uncommitted, unconcerned, unenterprising, unenthusiastic, unfeeling, uninterested, uninvolved, unmotivated, unresponsive. *Opp* ENTHUSIASTIC.

**apathy** *n* coolness, inactivity, indifference, lassitude, lethargy, listlessness, passivity, torpor. *Opp* ENTHUSIASM.

**apex** *n* **1** crest, crown, head, peak, pinnacle, point, summit, tip, top, vertex. **2** *apex of your career.* acme, apogee, climax, consummation, crowning moment, culmination, height, high point, zenith. *Opp* NADIR.

**aphrodisiac** *adj* arousing, erotic, *inf* sexy, stimulating.

**apologetic** *adj* ashamed, blushing, conscience-stricken, contrite, penitent, red-faced, regretful, remorseful, repentant, rueful, sorry. *Opp* UNREPENTANT.

**apologize** *vb* ask pardon, be apologetic, express regret, make an apology, repent, say sorry.

**apology** *n* acknowledgement, confession, defence, excuse, explanation, justification, plea.

**apostle** *n* crusader, disciple, evangelist, follower, messenger, missionary, preacher, propagandist, proselytizer, teacher.

**appal** *vb* alarm, disgust, dismay, distress, harrow, horrify, nauseate, outrage, revolt, shock, sicken, terrify, unnerve. ▷ FRIGHTEN. **appalling** ▷ ATROCIOUS, BAD, FRIGHTENING.

**apparatus** *n* appliance, *inf* contraption, device, equipment, gadget, *inf* gear, implement, instrument, machine, machinery, mechanism, *inf* set-up, system, *inf* tackle, tool, utensil.

**apparent** *adj* blatant, clear, conspicuous, detectable, discernible, evident, manifest, noticeable, observable, obvious, ostensible, overt, patent, perceptible, recognizable, self-explanatory, unconcealed, unmistakable, visible. *Opp* HIDDEN.

**apparition** *n* chimera, ghost, hallucination, illusion, manifestation, phantasm, phantom, presence, shade, spectre, spirit, *inf* spook, vision, wraith.

**appeal** *n* **1** application, call, *Fr* cri de coeur, cry, entreaty, petition, plea, prayer, request, solicitation, supplication. **2** allure, attractiveness, charisma, charm, *inf* pull, seductiveness. • *vb* ask earnestly, beg, beseech, call, canvass, cry out, entreat, implore, invoke, petition, plead, pray, request, solicit, supplicate. **appeal to** ▷ ATTRACT.

**appear** *vb* **1** arise, arrive, attend, begin, be published, be revealed, be seen, *inf* bob up, come, come into view, come out, *inf* crop up, enter, develop, emerge, *inf* heave into sight, loom, materialize, occur, originate, show, *inf* show up, spring up, surface, turn up. **2** *I appear to be wrong.* look, seem, transpire, turn out. **3** *appear in a play.* ▷ PERFORM.

**appearance** *n* **1** arrival, advent, emergence, presence, rise. **2** *a smart appearance.* air, aspect, bearing, demeanour, exterior, impression, likeness, look, mien, semblance.

**appease** *vb* assuage, calm, conciliate, humour, mollify, pacify, placate, propitiate, quiet, reconcile, satisfy, soothe, *inf* sweeten, tranquillize, win over. *Opp* ANGER.

**appendix** *n* addendum, addition, annexe, codicil, epilogue, postscript, rider, supplement.

**appetite** *n* craving, demand, desire, eagerness, fondness, greed, hankering, hunger, keenness, longing, lust, passion, predilection, proclivity, relish, *inf* stomach, taste, thirst, urge, willingness, wish, yearning, *inf* yen, zeal, zest.

**appetizing** *adj* delicious, *inf* moreish, mouthwatering, tasty, tempting.

**applaud** *vb* acclaim, approve, *inf* bring the house down, cheer, clap, commend, compliment, congratulate, eulogize, extol, *inf* give someone a hand, give someone an ovation, hail, laud, praise, salute. *Opp* CRITICIZE.

**applause** *n* acclaim, acclamation, approval, cheering, clapping, éclat, ovation, plaudits. ▷ PRAISE.

**appliance** *n* apparatus, contraption, device, gadget, implement, instrument, machine, mechanism, tool, utensil.

**applicant** *n* aspirant, candidate, competitor, entrant, interviewee, participant, postulant.

**apply** *vb* **1** administer, affix, bring into contact, lay on, put on, rub on, spread, stick. ▷ FASTEN. **2** *rules apply to all.* appertain, be relevant, have a bearing (on), pertain, refer, relate. **3** *apply common sense.* bring into use, employ, exercise, implement, practise, use, utilize, wield. **apply for** ▷ REQUEST. **apply yourself** ▷ CONCENTRATE.

**appoint** *vb* **1** arrange, authorize, decide on, determine, establish, fix, ordain, prescribe, settle. **2** *appoint you to do a job.* assign, choose, co-opt, delegate, depute, designate, detail, elect, make an appointment, name, nominate, *inf* plump for, select, settle on, vote for.

**appointment** *n* **1** arrangement, assignation, consultation, date, engagement, fixture, interview, meeting, rendezvous, session, *old* *use* tryst. **2** choice, choosing, commissioning, election, naming, nomination, selection. **3** job, office, place, position, post, situation.

**appreciate** *vb* **1** admire, applaud, approve of, be grateful for, be sensitive to, cherish, commend, enjoy, esteem, favour, find worthwhile, like, praise, prize, rate highly, regard highly, respect, sympathize with, treasure, value, welcome. **2** *appreciate the facts.* acknowledge, apprehend, comprehend, know, realize, recognize, see, understand. **3** *value appreciates.* build up, escalate, gain, go up, grow, improve, increase, inflate, mount, rise, soar, strengthen. *Opp* DEPRECIATE, DESPISE, DISREGARD.

**apprehensive** *adj* afraid, concerned, disturbed, edgy, fearful, *inf* jittery, nervous, *inf* nervy, *inf* on edge, troubled, uneasy, worried. ▷ ANXIOUS. *Opp* FEARLESS.

**apprentice** *n* beginner, learner, novice, probationer, pupil, starter, tiro, trainee.

**approach** *n* **1** advance, advent, arrival, coming, movement, nearing. **2** access, doorway, entrance, entry, passage, road, way in. **3** *your approach to work.* attitude, course, manner, means, method, mode, *Lat* modus operandi, procedure, style, system, technique, way. **4** *an approach for help.* appeal, application, invitation, offer, overture, proposal, proposition. • *vb* **1** advance, bear down, catch up, come near, draw near, gain (on), loom, move towards, near, progress. *Opp* RETREAT. **2** *approach a task.* ▷ BEGIN. **3** *approach someone for help.* ▷ CONTACT.

**approachable** *adj* accessible, affable, informal, kind, open, relaxed, sympathetic, *inf* unstuffy, well-disposed. ▷ FRIENDLY. *Opp* ALOOF.

**appropriate** *adj* applicable, apposite, apropos, apt, becoming, befitting, compatible, correct, decorous, deserved, due, felicitous, fit, fitting, germane, happy, just, *old use* meet, opportune, pertinent, proper, relevant, right, seasonable, seemly, suitable, tactful, tasteful, timely, well-judged, well-suited, well-timed. *Opp* INAPPROPRIATE. • *vb* annex, arrogate, commandeer, confiscate, expropriate, gain control of, *inf* hijack, requisition, seize, take, take over, usurp. ▷ STEAL.

**approval** *n* **1** acclaim, acclamation, admiration, applause, appreciation, approbation, commendation, esteem, favour, liking, plaudits, praise, regard, respect, support. *Opp* DISAPPROVAL. **2** acceptance, acquiescence, agreement, assent, authorization, *inf* blessing, confirmation, consent, endorsement, *inf* go-ahead, *inf* green light, licence, mandate, *inf* OK, permission, ratification, sanction, seal, stamp, support, *inf* thumbs up, validation. *Opp* REFUSAL.

**approve** *vb* accede to, accept, affirm, agree to, allow, assent to, authorize, *inf* back, *inf* bless, confirm, consent to, countenance, endorse, *inf* give your blessing to, *inf* go along with, pass, permit, ratify, *inf* rubber-stamp, sanction, sign, subscribe to, support, tolerate, uphold, validate. *Opp* REFUSE, VETO. approve of ▷ ADMIRE.

**approximate** *adj* close, estimated, imprecise, inexact, loose, near, rough. *Opp* EXACT. • *vb* **approximate to** approach, be close to, be similar to, border on, come near to, equal roughly, look like, resemble, simulate, verge on.

**approximately** *adv* about, approaching, around, *Lat* circa, close to, just about, loosely, more or less, nearly, *inf* nigh on, *inf* pushing, roughly, round about.

**aptitude** *n* ability, bent, capability, facility, fitness, flair, gift, suitability, talent. ▷ SKILL.

**arbitrary** *adj* **1** capricious, casual, chance, erratic, fanciful, illogical, indiscriminate, irrational, random, subjective, unplanned, unpredictable, unreasonable, whimsical, wilful. *Opp* METHODICAL. **2** *arbitrary rule.* absolute, autocratic, despotic, dictatorial, high-handed, imperious, summary, tyrannical, tyrannous, uncompromising.

**arbitrate** *vb* adjudicate, decide the outcome, intercede, judge, make peace, mediate, negotiate, pass judgement, referee, settle, umpire.

**arbitration** *n* adjudication, *inf* good offices, intercession, judgement, mediation, negotiation, settlement.

**arbitrator** *n* adjudicator, arbiter, go-between, intermediary, judge, mediator, middleman, negotiator, ombudsman, peacemaker, referee, *inf* trouble-shooter, umpire.

**arch** *n* arc, archway, bridge, vault. • *vb* arc, bend, bow. ▷ CURVE.

**archetype** *n* classic, example, ideal, model, original, paradigm, pattern, precursor, prototype, standard.

**archives** *n* annals, chronicles, documents, history, libraries, memorials, museums, papers, records, registers.

**ardent** *adj* eager, enthusiastic, fervent, hot, impassioned, intense, keen, passionate, warm, zealous. *Opp* APATHETIC.

**arduous** *adj* backbreaking, demanding, exhausting, gruelling, heavy, herculean, laborious, onerous, punishing, rigorous, severe, strenuous, taxing, tiring, tough, uphill. ▷ DIFFICULT. *Opp* EASY.

**area** *n* **1** acreage, breadth, expanse, extent, patch, sheet, size, space, square-footage, stretch, surface, tract, width. **2** district, environment, environs, locality, neighbourhood, part, precinct, province, quarter, region, sector, terrain, territory, vicinity, zone. **3** *an area of study.* field, sphere, subject.

**argue** *vb* **1** answer back, *inf* bandy words, bargain, bicker, debate, deliberate, demur, differ, disagree, discuss, dispute, dissent, expostulate, fall out, feud, fight, haggle, have an argument, *inf* have words, object, protest, quarrel, remonstrate, *inf* row, spar, squabble, take exception, wrangle. **2** *argue a case.* assert, claim, contend, demonstrate, hold, maintain, make a case, plead, prove, reason, show, suggest.

**argument** *n* **1** altercation, bickering, clash, conflict, controversy, difference (of opinion), disagreement, dispute, expostulation, feud, fight, protest, quarrel, remonstration, row, *inf* set-to, squabble, *inf* tiff, wrangle. **2** consultation, debate, defence, deliberation, dialectic, discussion, exposition, polemic. **3** *argument of a lecture.* abstract, case, contention, gist, hypothesis, idea, outline, plot, reasoning, summary, synopsis, theme, thesis, view.

**arid** *adj* **1** barren, desert, dry, fruitless, infertile, lifeless, parched, sterile, torrid, unproductive, waste, waterless. *Opp* FRUITFUL. **2** *arid work.* boring, dreary, dull, pointless, tedious, uninspired, uninteresting, vapid.

**arise** *vb* come up, crop up, get up, rise. ▷ APPEAR.

**aristocrat** *n* grandee, lady, lord, noble, nobleman, noblewoman, patrician. ▷ PEER.

**aristocratic** *adj inf* blueblooded, courtly, élite, gentle, highborn, lordly, noble, patrician, princely, royal, thoroughbred, titled, upper class.

**arm** *n* appendage, bough, branch, extension, limb, offshoot, projection. • *vb* equip, fortify, furnish, provide, supply. **arms** ▷ WEAPON(S).

**armed services** *plur n* force, forces, troops. □ *air force, army, militia, navy.* □ *cavalry, infantry.* □ *battalion, brigade, cohort, company, corps, foreign legion, garrison, legion, patrol, platoon, rearguard, regiment, reinforcements, squad, squadron, task-force, vanguard.* ▷ FIGHTER, RANK, SOLDIER.

**armistice** *n* agreement, cease-fire, peace, treaty, truce.

**armoury** *n* ammunition-dump, arsenal, depot, magazine, ordnance depot, stockpile.

**aroma** *n* bouquet, fragrance, odour, perfume, redolence, savour, scent, smell, whiff.

**arouse** *vb* awaken, call forth, encourage, foment, foster, kindle, provoke, quicken, stimulate, stir up, *inf* whip up. ▷ CAUSE. *Opp* ALLAY.

**arrange** *vb* **1** adjust, align, array, categorize, classify, collate, display, dispose, distribute, grade, group, lay out, line up, marshal, order, organize, *inf* pigeon-hole, position, put in order, range, rank, set out, sift, sort (out), space out, systematize, tabulate, tidy up. **2** *arrange a party.* bring about, contrive, coordinate, devise, manage, organize, plan, prepare, see to, settle, set up. **3** *arrange music.* adapt, harmonize, orchestrate, score, set.

**arrangement** *n* **1** adjustment, alignment, design, disposition, distribution, grouping, layout, marshalling, organization, planning, positioning, setting out, spacing, tabulation. ▷ ARRAY. **2** agreement, bargain, compact, contract, deal, pact, scheme, settlement, terms, understanding. **3** *musical arrangement.* adaptation, harmonization, orchestration, setting, version.

**array** *n* arrangement, assemblage, collection, demonstration, display, exhibition, formation, *inf* line-up, muster, panoply, parade, presentation, show, spectacle. • *vb* **1** adorn, apparel, attire, clothe, deck, decorate, dress, equip, fit out, garb, rig out, robe, wrap. **2** ▷ ARRANGE.

**arrest** *n* apprehension, capture, detention, seizure. • *vb* **1** bar, block, check, delay, end, halt, hinder, impede, inhibit, interrupt, obstruct, prevent, restrain, retard, slow, stem, stop. **2** *arrest a suspect.* apprehend, *inf* book, capture, catch, *inf* collar, detain, have up, hold, *inf* nab, *inf* nick, *inf* pinch, *inf* run in, seize, take into custody, take prisoner.

**arrival** *n* **1** advent, appearance, approach, coming, entrance, homecoming, landing, return, touchdown. **2** *new arrivals.* caller, newcomer, visitor.

**arrive** *vb* **1** appear, come, disembark, drive up, drop in, enter, get in, land, make an entrance, *inf* roll in, *inf* roll up, show up, touch down, turn up. **2** ▷ SUCCEED. **arrive at** ▷ REACH.

**arrogant** *adj* boastful, brash, brazen, bumptious, cavalier, *inf* cocky, conceited, condescending, disdainful, egotistical, haughty, *inf* high and mighty, high-handed, imperious, impudent, insolent, lofty, lordly, overbearing, patronizing, pompous, presumptuous, proud, scornful, self-admiring, self-important, smug, snobbish, *inf* snooty, *inf* stuck-up, supercilious, superior, vain. *Opp* MODEST.

**arsonist** *n* fire-raiser, incendiary, pyromaniac.

**art** *n* **1** aptitude, artistry, cleverness, craft, craftsmanship, dexterity, expertise, facility, knack, proficiency, skilfulness, skill, talent, technique, touch, trick. **2** artwork, craft, fine art. □ *architecture, batik, carpentry, cloisonné, collage, crochet, drawing, embroidery, enamelling, engraving, etching, fashion design, graphics, handicraft, illustration, jewellery, knitting, linocut, lithography, marquetry, metalwork, modelling, monoprint, needlework, origami, painting, patchwork, photography, pottery, printmaking, sculpture, sewing, sketching, spinning, weaving, woodcut, woodwork.*

**artful** *adj* astute, canny, clever, crafty, cunning, deceitful, designing, devious, *inf* fly, *inf* foxy, ingenious, knowing, scheming, shrewd, skilful, sly, smart, sophisticated, subtle, tricky, wily. *Opp* NAÏVE.

**article** **1** item, object, thing. **2** *magazine article.* ▷ WRITING.

**articulate** *adj* clear, coherent, comprehensible, distinct, eloquent, expressive, fluent, *derog* glib, intelligible, lucid, understandable, vocal. *Opp* INARTICULATE. • *vb* ▷ SPEAK.

**articulated** *adj* bending, flexible, hinged, jointed.

**artificial** *adj* **1** fabricated, made-up, man-made, manufactured, synthetic, unnatural. **2** *artificial style.* affected, assumed, bogus, concocted, contrived, counterfeit, factitious, fake, false, feigned, forced, imitation, insincere, laboured, mock, *inf* phoney, pretended, pseudo, *inf* put on, sham, simulated, spurious, unreal. *Opp* NATURAL.

**artist** *n* craftsman, craftswoman. □ *architect, carpenter, cartoonist, commercial artist, designer, draughtsman, draughtswoman, engraver, goldsmith, graphic designer, illustrator, mason, painter, photographer, potter, printer, sculptor, silversmith, smith, weaver.* ▷ ENTERTAINER, MUSICIAN, PERFORMER.

**artistic** *adj* aesthetic, attractive, beautiful, creative, cultured, decorative, *inf* designer, imaginative, ornamental, tasteful. *Opp* UGLY.

**ascend** *vb* climb, come up, defy gravity, fly, go up, levitate, lift off, make an ascent, mount, move up, rise, scale, slope up, soar, take off. *Opp* DESCEND.

**ascent** *n* ascension, climb, gradient, hill, incline, ramp, rise, slope. *Opp* DESCENT.

**ascertain** *vb* confirm, determine, discover, establish, find out, identify, learn, make certain, make sure, settle, verify.

**ascetic** *adj* abstemious, austere, celibate, chaste, frugal, harsh, hermit-like, plain, puritanical, restrained, rigorous, selfcontrolled, self-denying, selfdisciplined, severe, spartan, strict, temperate. *Opp* SELF-INDULGENT.

**ash** *n* burnt remains, cinders, clinker, embers.

**ashamed** *adj* **1** abashed, apologetic, chagrined, chastened, conscience-stricken, contrite, discomfited, distressed, guilty, humbled, humiliated, mortified, penitent, red-faced, remorseful, repentant, rueful, shamefaced, sorry, upset. **2** *ashamed of your nakedness.* bashful, blushing, demure, diffident, embarrassed, modest, prudish, self-conscious, sheepish, shy. *Opp* SHAMELESS.

**ask** *vb* appeal, apply, badger, beg, beseech, catechize, crave, demand, enquire, entreat, implore, importune, inquire, interrogate, invite, petition, plead, pose a question, pray, press, query, question, quiz, request, require, seek, solicit, sue, supplicate. **ask for** ▷ ATTRACT.

**asleep** *adj* comatose, *inf* dead to the world, dormant, dozing, *inf* fast off, hibernating, inactive, inattentive, *inf* in the land of nod, *sl* kipping, napping, *inf* off, *inf* out like a light, resting, sedated, sleeping, slumbering, snoozing, *inf* sound off, unconscious, under sedation. ▷ NUMB. *Opp* AWAKE.

**aspect** *n* **1** angle, attribute, characteristic, circumstance, detail, element, facet, feature, quality, side, standpoint, viewpoint. **2** air, appearance, attitude, bearing, countenance, demeanour, expression, face, look, manner, mien, visage. **3** *a southern aspect.* direction, orientation, outlook, position, prospect, situation, view.

**asperity** *n* abrasiveness, acerbity, acidity, acrimony, astringency, bitterness, churlishness, crossness, harshness, hostility, irascibility, irritability, peevishness, rancour, roughness, severity, sharpness, sourness, venom, virulence. *Opp* MILDNESS.

**aspiration** *n* aim, ambition, craving, desire, dream, goal, hope, longing, objective, purpose, wish, yearning.

**aspire** *vb* **aspire to** aim for, crave, desire, dream of, hope for, long for, pursue, seek, set your sights on, strive after, want, wish for, yearn for. **aspiring** ▷ POTENTIAL.

**assail** *vb* assault, bombard, pelt, set on. ▷ ATTACK.

**assault** *n* battery, *inf* GBH, mugging, rape. ▷ ATTACK. • *vb* abuse, assail, *inf* beat up, *inf* do over, fall on, fight, fly at, jump on, lash out at, *inf* lay into, mob, molest, mug, *inf* pitch into, pounce on, rape, rush at, set about, set on, strike at, violate, *inf* wade into. ▷ ATTACK.

**assemble** *vb* **1** come together, congregate, convene, converge, crowd, flock, gather, group, herd, join up, meet, rally round, swarm, throng round. **2** accumulate, amass, bring together, collect, gather, get together, marshal, mobilize, muster, pile up, rally, round up. **3** build, construct, erect, fabricate, fit together, make, manufacture, piece together, produce, put together. *Opp* DISMANTLE, DISPERSE.

**assembly** *n* assemblage, conclave, conference, congregation, congress, convention, convocation, council, gathering, meeting, parliament, rally, synod. ▷ CROWD.

**assent** *n* acceptance, accord, acquiescence, agreement, approbation, approval, compliance, consent,

*inf* go-ahead, permission, sanction, willingness. *Opp* REFUSAL.
• *vb* accede, accept, acquiesce, agree, approve, be willing, comply, concede, concur, consent, express agreement, give assent, say 'yes', submit, yield.
*Opp* REFUSE.

**assert** *vb* affirm, allege, argue, asseverate, attest, claim, contend, declare, emphasize, insist, maintain, proclaim, profess, protest, state, stress, swear, testify. **assert yourself** ▷ INSIST.

**assertive** *adj* aggressive, assured, authoritative, bold, *inf* bossy, certain, confident, decided, decisive, definite, dogmatic, domineering, emphatic, firm, forceful, insistent, peremptory, *derog* opinionated, positive, *derog* pushy, self-assured, strong, strong-willed, *derog* stubborn, uncompromising.
*Opp* SUBMISSIVE.

**assess** *vb* appraise, assay (*metal*), calculate, compute, consider, determine, estimate, evaluate, fix, gauge, judge, price, reckon, review, *inf* size up, value, weigh up, work out.

**asset** *n* advantage, aid, benefit, blessing, boon, *inf* godsend, good, help, profit, resource, strength, support. **assets** capital, effects, estate, funds, goods, holdings, means, money, possessions, property, resources, savings, securities, valuables, wealth, *inf* worldly goods.

**assign** *vb* **1** allocate, allot, apportion, consign, dispense, distribute, give, hand over, share out.
**2** *assign to a job.* appoint, authorize, delegate, designate, nominate, ordain, prescribe, put down, select, specify, stipulate. **3** *assign my success to luck.* accredit, ascribe, attribute, credit.

**assignment** *n* chore, duty, errand, job, mission, obligation, post, project, responsibility, task.
▷ WORK.

**assist** *vb* abet, advance, aid, back, benefit, boost, collaborate, cooperate, facilitate, further, help, *inf* lend a hand, promote, *inf* rally round, reinforce, relieve, second, serve, succour, support, sustain, work with. *Opp* HINDER.

**assistance** *n* aid, backing, benefit, collaboration, contribution, cooperation, encouragement, help, patronage, reinforcement, relief, sponsorship, subsidy, succour, support. *Opp* HINDRANCE.

**assistant** *n* abettor, accessory, accomplice, acolyte, aide, ally, associate, auxiliary, backer, collaborator, colleague, companion, comrade, confederate, deputy, helper, helpmate, *inf* henchman, mainstay, *derog* minion, partner, *inf* right-hand man, *inf* right-hand woman, second, second-in-command, stand-by, subordinate, supporter.

**associate** *n* ▷ ASSISTANT, FRIEND.
• *vb* **1** ally yourself, be friends, combine, consort, fraternize, *inf* gang up, *inf* go around (with), *sl* hang out (with), *inf* hob nob (with), keep company, join up, link up, make friends, mingle, mix, side, socialize.
*Opp* DISSOCIATE. **2** *associate snow with winter.* bracket together, connect, put together, relate, *inf* tie up.

**association** *n* affiliation, alliance, amalgamation, body, brotherhood, cartel, clique, club, coalition, combination, company, confederation, consortium, cooperative, corporation, federation, fellowship, group, league, marriage, merger, organization, partnership, party, society, syndicate, trust, union. ▷ FRIENDSHIP.

**assorted** *adj* different, differing, *old use* divers, diverse, heterogeneous, manifold, miscellaneous, mixed, motley, multifarious, sundry, varied, various.

**assortment** *n* agglomeration, array, choice, collection, diversity, farrago, jumble, medley, mélange, miscellany, *inf* mishmash, *inf* mixed bag, mixture, pot-pourri, range, selection, variety.

**assume** *vb* **1** believe, deduce, expect, guess, *inf* have a hunch, have no doubt, imagine, infer, presume, presuppose, suppose, surmise, suspect, take for granted, think, understand. **2** *assume duties.* accept, embrace, take on, undertake. **3** *assume an air of.* acquire, adopt, affect, don, dress up in, fake, feign, pretend, put on, simulate, try on, wear.

**assumption** *n* belief, conjecture, expectation, guess, hypothesis, premise, premiss, supposition, surmise, theory.

**assurance** *n* commitment, guarantee, oath, pledge, promise, vow, undertaking, word (of honour).

**assure** *vb* convince, give a promise, guarantee, make sure, persuade, pledge, promise, reassure, swear, vow. **assured** ▷ CONFIDENT.

**astonish** *vb* amaze, astound, baffle, bewilder, confound, daze, *inf* dazzle, dumbfound, electrify, *inf* flabbergast, leave speechless, nonplus, shock, stagger, startle, stun, stupefy, surprise, take aback, take by surprise, *inf* take your breath away, *sl* wow. **astonishing** ▷ AMAZING.

**astound** *vb* ▷ ASTONISH.

**astray** *adv* adrift, amiss, awry, lost, off course, *inf* off the rails, wide of the mark, wrong.

**astute** *adj* acute, adroit, artful, canny, clever, crafty, cunning, discerning, *inf* fly, *inf* foxy, guileful, ingenious, intelligent, knowing, observant, perceptive, perspicacious, sagacious, sharp, shrewd, sly, subtle, wily. *Opp* STUPID.

**asylum** *n* cover, haven, refuge, retreat, safety, sanctuary, shelter.

**asymmetrical** *adj* awry, crooked, distorted, irregular, lop-sided, unbalanced, uneven, *inf* wonky. *Opp* SYMMETRICAL.

**atheist** *n* heathen, pagan, sceptic, unbeliever.

**athletic** *adj* acrobatic, active, energetic, fit, muscular, powerful, robust, sinewy, *inf* sporty, *inf* strapping, strong, sturdy, vigorous, well-built, wiry. *Opp* WEAK.

**athletics** *n* field events, track events. □ *cross-country, decathlon, discus, high jump, hurdles, javelin, long jump, marathon, pentathlon, pole-vault, relay, running, shot, sprint, triple jump.*

**atmosphere** *n* **1** aerospace, air, ether, heavens, ionosphere, sky, stratosphere, troposphere. **2** ambience, aura, character, climate, environment, feeling, mood, spirit, tone, *inf* vibes, vibrations.

**atom** *n inf* bit, crumb, grain, iota, jot, molecule, morsel, particle, scrap, speck, spot, trace.

**atone** *vb* answer, be punished, compensate, do penance, expiate, make amends, make reparation, make up (for), pay the penalty, pay the price, recompense, redeem yourself, redress.

**atrocious** *adj* abominable, appalling, barbaric, bloodthirsty, brutal, brutish, callous, cruel, diabolical, dreadful, evil, execrable, fiendish, frightful, grim, gruesome, hateful, heartless, heinous, hideous, horrendous, horrible, horrific, horrifying, inhuman, merciless, monstrous, nauseating, revolting, sadistic, savage, shocking, sickening, terrible, vicious, vile, villainous, wicked.

**atrocity** *n* crime, cruelty, enormity, offence, outrage. ▷ EVIL.

**attach** *vb* **1** add, affix, anchor, append, bind, combine, connect, couple, fix, join, link, secure, stick, tie, unite, weld. ▷ FASTEN. *Opp* DETACH. **2** ascribe, assign, associate, attribute, impute, place, relate to. **attached** ▷ LOVING.

**attack** *n* **1** aggression, ambush, assault, battery, blitz, bombardment, broadside, cannonade, charge, counter-attack, foray, incursion, invasion, offensive, onset, onslaught, pre-emptive strike, raid, rush, sortie, strike. **2** *verbal attack*. abuse, censure, criticism, diatribe, impugnment, invective, outburst, tirade. **3** *attack of coughing*. bout, convulsion, fit, outbreak, paroxysm, seizure, spasm, stroke, *inf* turn. ● *vb* **1** ambush, assail, assault, *inf* beat up, *inf* blast, bombard, charge, counterattack, descend on, *inf* do over, engage, fall on, fight, fly at, invade, jump on, lash out at, *inf* lay into, mob, mug, *inf* pitch into, pounce on, raid, rush, set about, set on, storm, strike at, *inf* wade into. **2** *attack verbally*. abuse, censure, criticize, denounce, impugn, inveigh against, libel, malign, round on, slander, snipe at, traduce, vilify. *Opp* DEFEND. **3** *attack a task*. ▷ BEGIN.

**attacker** *n* aggressor, assailant, critic, detractor, enemy, intruder, invader, mugger, opponent, persecutor, raider, slanderer. ▷ FIGHTER.

**attain** *vb* accomplish, achieve, acquire, arrive at, complete, earn, fulfil, gain, get, grasp, *inf* make, obtain, *inf* pull (something) off, procure, reach, realize, secure, touch, win.

**attempt** *n* assault, bid, effort, endeavour, *inf* go, start, try, undertaking. ● *vb* aim, aspire, do your best, endeavour, essay, exert yourself, *inf* have a crack, *inf* have a go, make a bid, make an assault, make an effort, put yourself out, seek, *inf* spare no effort, strive, *inf* sweat blood, tackle, try, undertake, venture.

**attend** *vb* **1** appear, be present, go (to), frequent, present yourself, *inf* put in an appearance, visit. **2** accompany, chaperon, conduct, escort, follow, guard, usher. **3** *attend carefully*. concentrate, follow, hear, heed, listen, mark, mind, note, notice, observe, pay attention, think, watch. **attend to** assist, care for, help, look after, mind, minister to, nurse, see to, take care of, tend, wait on.

**attendant** *n* assistant, escort, helper, usher. ▷ SERVANT.

**attention** *n* **1** alertness, awareness, care, concentration, concern, diligence, heed, notice, recognition, thought, vigilance. **2** *kind attention*. attentiveness, civility, consideration, courtesy, gallantry, good manners, kindness, politeness, regard, respect, thoughtfulness.

**attentive** *adj* **1** alert, awake, concentrating, heedful, intent, observant, watchful. *Opp* INATTENTIVE. **2** ▷ POLITE. *Opp* RUDE.

**attire** *n* accoutrements, apparel, array, clothes, clothing, costume, dress, finery, garb, garments, *inf* gear, *old use* habit, outfit, raiment, wear, *old use* weeds. • *vb* ▷ DRESS.

**attitude** *n* **1** air, approach, aspect, bearing, behaviour, carriage, demeanour, disposition, frame of mind, manner, mien, mood, posture, stance. **2** *political attitudes.* approach, belief, feeling, opinion, orientation, outlook, position, standpoint, thought, view, viewpoint.

**attract** *vb* **1** allure, appeal to, beguile, bewitch, bring in, captivate, charm, decoy, enchant, entice, fascinate, *sl* get someone going, interest, inveigle, lure, magnetize, seduce, tempt, *sl* turn someone on. **2** *a magnet attracts iron.* drag, draw, pull, tug at. **3** *attract attention.* ask for, cause, court, encourage, generate, incite, induce, invite, provoke, seek out, *inf* stir up. *Opp* REPEL.

**attractive** *adj* adorable, alluring, appealing, appetizing, becoming, bewitching, captivating, *inf* catchy (*tune*), charming, *inf* cute, delightful, desirable, disarming, enchanting, endearing, engaging, enticing, enviable, fascinating, fetching, flattering, glamorous, good-looking, gorgeous, handsome, hypnotic, interesting, inviting, irresistible, lovable, lovely, magnetic, personable, pleasing, prepossessing, pretty, quaint, seductive, sought-after, stunning, *inf* taking, tasteful, tempting, winning, winsome. ▷ BEAUTIFUL. *Opp* REPULSIVE.

**attribute** *n* characteristic, feature, property, quality, trait. • *vb* accredit, ascribe, assign, blame, charge, credit, impute, put down, refer, trace back.

**audacious** *adj* adventurous, courageous, daring, fearless, *derog* foolhardy, intrepid, *derog* rash, *derog* reckless, venturesome. ▷ BOLD. *Opp* TIMID.

**audacity** *n* boldness, *inf* cheek, effrontery, forwardness, impertinence, impudence, presumptuousness, rashness, *inf* sauce, temerity. ▷ COURAGE.

**audible** *adj* clear, detectable, distinct, high, loud, noisy, recognizable. *Opp* INAUDIBLE.

**audience** *n* assembly, congregation, crowd, gathering, house, listeners, meeting, onlookers, *TV* ratings, spectators, *inf* turn-out, viewers.

**auditorium** *n* assembly room, concert-hall, hall, theatre.

**augment** *vb* add to, amplify, boost, eke out, enlarge, expand, extend, fill out, grow, increase, intensify, magnify, make larger, multiply, raise, reinforce, strengthen, supplement, swell. *Opp* DECREASE.

**augur** *vb* bode, forebode, foreshadow, forewarn, give an omen, herald, portend, predict, promise, prophesy, signal.

**augury** *n* forecast, forewarning, omen, portent, prophecy, sign, warning.

**auspicious** *adj* favourable, *inf* hopeful, lucky, positive, promising, propitious. *Opp* OMINOUS.

**austere** *adj* **1** abstemious, ascetic, chaste, cold, economical, exacting, forbidding, formal, frugal, grave, hard, harsh, hermit-like, parsimonious, puritanical, restrained, rigorous, self-denying, self-disciplined, serious, severe, sober, spartan, stern, *inf* strait-laced, strict, thrifty, unpampered. **2** *austere dress*. modest, plain, simple, unadorned, unfussy. *Opp* LUXURIOUS, ORNATE.

**authentic** *adj* accurate, actual, bona fide, certain, dependable, factual, genuine, honest, legitimate, original, real, reliable, true, trustworthy, truthful, undisputed, valid, veracious. ▷ AUTHORITATIVE. *Opp* FALSE.

**authenticate** *vb* certify, confirm, corroborate, endorse, substantiate, validate, verify.

**author** *n* **1** composer, dramatist, novelist, playwright, poet, scriptwriter. ▷ WRITER. **2** architect, begetter, creator, designer, father, founder, initiator, inventor, maker, mover, organizer, originator, parent, planner, prime mover, producer.

**authoritarian** *adj* autocratic, *inf* bossy, despotic, dictatorial, dogmatic, domineering, strict, tyrannical.

**authoritative** *adj* approved, certified, definitive, dependable, official, recognized, sanctioned, scholarly. ▷ AUTHENTIC.

**authority** *n* **1** approval, authorization, consent, licence, mandate, permission, permit, sanction, warrant. **2** charge, command, control, domination, force, influence, jurisdiction, might, power, prerogative, right, sovereignty, supremacy, sway, weight. **3** *authority on wine*. *inf* boffin, *inf* buff, connoisseur, expert, scholar, specialist. **the authorities** administration, government, management, officialdom, *inf* powers that be.

**authorize** *vb* accede to, agree to, allow, approve, *inf* back, commission, consent to, empower, endorse, entitle, legalize, license, make official, mandate, *inf* OK, pass, permit, ratify, *inf* rubber-stamp, sanction, sign the order, sign the warrant, validate. **authorized** ▷ OFFICIAL.

**automatic** *adj* **1** conditioned, habitual, impulsive, instinctive, involuntary, natural, reflex, spontaneous, unconscious, unintentional, unthinking. **2** automated, computerized, electronic, mechanical, programmable, programmed, robotic, self-regulating, unmanned.

**autonomous** *adj* free, independent, self-determining, self-governing, sovereign.

**auxiliary** *adj* additional, ancillary, assisting, *inf* back-up, emergency, extra, helping, reserve, secondary, spare, subordinate, subsidiary, substitute, supplementary, supporting, supportive.

**available** *adj* accessible, at hand, convenient, disposable, free, handy, obtainable, procurable, ready, to hand, uncommitted, unengaged, unused, usable. *Opp* INACCESSIBLE.

**avaricious** *adj* acquisitive, covetous, grasping, greedy, mercenary, miserly.

**avenge** *vb* exact punishment, *inf* get your own back, repay, requite, take revenge.

**average** *adj* common, commonplace, everyday, mediocre, medium, middling, moderate, normal, regular, *inf* run of the mill, typical, unexceptional, usual. ▷ ORDINARY. *Opp* EXCEPTIONAL. • *n* mean, mid-point, norm, standard. • *vb* equalize, even out, normalize, standardize.

**averse** *adj* antipathetic, disinclined, hostile, opposed, reluctant, resistant, unwilling.

**aversion** *n* antagonism, antipathy, dislike, distaste, hostility, reluctance, repugnance, unwillingness. ▷ HATRED.

**avert** *vb* change the course of, deflect, draw off, fend off, parry, prevent, stave off, turn aside, turn away, ward off.

**avoid** *vb* abstain from, be absent from, *inf* beg the question, *inf* bypass, circumvent, dodge, *inf* duck, elude, escape, eschew, evade, fend off, find a way round, get out of the way of, *inf* get round, *inf* give a wide berth to, help (*can't help it*), ignore, keep away from, keep clear of, refrain from, run away from, shirk, shun, side-step, skirt round, *inf* skive off, steer clear of. *Opp* SEEK.

**await** *vb* be ready for, expect, hope for, lie in wait for, look out for, wait for.

**awake** *adj* **1** aware, conscious, insomniac, open-eyed, restless, sleepless, *inf* tossing and turning, wakeful, wide awake. **2** ▷ ALERT. *Opp* ASLEEP.

**awaken** *vb* alert, animate, arouse, awake, call, excite, kindle, revive, rouse, stimulate, stir up, wake, waken.

**award** *n* badge, cap, cup, decoration, endowment, grant, medal, prize, reward, scholarship, trophy. • *vb* accord, allot, assign, bestow, confer, decorate with, endow, give, grant, hand over, present.

**aware** *adj* acquainted, alive (to), appreciative, attentive, cognizant, conscious, conversant, familiar, heedful, informed, knowledgeable, mindful, observant, responsive, sensible, sensitive, versed. *Opp* IGNORANT, INSENSITIVE.

**awe** *n* admiration, amazement, apprehension, dread, fear, respect, reverence, terror, veneration, wonder.

**awe-inspiring** *adj* awesome, *old use* awful, breathtaking, dramatic, grand, imposing, impressive, magnificent, marvellous, overwhelming, solemn, *inf* stunning, stupendous, sublime, wondrous. ▷ FRIGHTENING, WONDERFUL. *Opp* INSIGNIFICANT.

**awful** *adj* **1** ▷ AWE-INSPIRING. **2** *awful weather*. ▷ BAD.

**awkward** *adj* **1** blundering, bungling, clumsy, gauche, gawky, *inf* ham-fisted, inelegant, inept, inexpert, maladroit, uncoordinated, ungainly, ungraceful, unskilful, wooden. **2** *an awkward load*. bulky, cumbersome, inconvenient, unmanageable, unwieldy. **3** *an awkward problem*. annoying, difficult, perplexing, thorny, *inf* ticklish, troublesome, trying, vexatious, vexing. **4** *an awkward silence*. embarrassing, touchy, tricky, uncomfortable, uneasy. **5** *awkward children*. *inf* bloody-minded, *sl* bolshie, defiant, disobedient, disobliging, exasperating, intractable, misbehaving, naughty, obstinate, perverse, *inf* prickly, rebellious, refractory, rude, stubborn, touchy, uncooperative, undisciplined, unruly, wayward. *Opp* COOPERATIVE, EASY, NEAT.

**awning** *n* canopy, flysheet, screen, shade, shelter, tarpaulin.

**axe** *n* battleaxe, chopper, cleaver, hatchet, tomahawk. • *vb* cancel, cut, discharge, discontinue, dismiss, eliminate, get rid of, *inf* give the chop to, make redundant, rationalize, remove, sack, terminate, withdraw.

**axle** *n* rod, shaft, spindle.

# B

**baby** *n* babe, child, infant, newborn, toddler.

**babyish** *adj* childish, immature, infantile, juvenile, puerile, simple. *Opp* MATURE.

**back** *adj* dorsal, end, hind, hinder, hindmost, last, rear, rearmost. • *n* 1 end, hindquarters, posterior, rear, stern, tail, tail-end. 2 reverse, verso. *Opp* FRONT. • *vb* 1 back away, back off, back-pedal, backtrack, *inf* beat a retreat, give way, go backwards, move back, recede, recoil, retire, retreat, reverse. *Opp* ADVANCE. 2 ▷ SUPPORT. **back down** ▷ RETREAT. **back out** ▷ WITHDRAW.

**backer** *n* advocate, *inf* angel, benefactor, patron, promoter, sponsor, supporter.

**background** *n* 1 circumstances, context, history, *inf* lead-up, setting, surroundings. 2 breeding, culture, education, experience, grounding, milieu, tradition, training, upbringing.

**backing** *n* 1 aid, approval, assistance, encouragement, endorsement, funding, grant, help, investment, loan, patronage, sponsorship, subsidy, support. 2 *musical backing*. accompaniment, orchestration, scoring.

**backward** *adj* 1 regressive, retreating, retrograde, retrogressive, reverse. 2 afraid, bashful, coy, diffident, hesitant, inhibited, modest, reluctant, reserved, reticent, self-effacing, shy, timid, unassertive, unforthcoming. 3 *a backward pupil*. disadvantaged, handicapped, immature, late-starting, retarded, slow, subnormal, underdeveloped, undeveloped. *Opp* FORWARD.

**bad** *adj* [*Bad* describes anything we don't like. Possible synonyms are almost limitless.] 1 *bad men, deeds*. abhorrent, base, beastly, blameworthy, corrupt, criminal, cruel, dangerous, delinquent, deplorable, depraved, detestable, evil, guilty, immoral, infamous, malevolent, malicious, malignant, mean, mischievous, nasty, naughty, offensive, regrettable, reprehensible, rotten, shameful, sinful, unworthy, vicious, vile, villainous, wicked, wrong. 2 *a bad accident*. appalling, awful, calamitous, dire, disastrous, distressing, dreadful, frightful, ghastly, grave, hair-raising, hideous, horrible, painful, serious, severe, shocking, terrible, unfortunate, unpleasant, violent. 3 *bad driving, work*. abominable, abysmal, appalling, atrocious, awful, cheap, *inf* chronic, defective, deficient, diabolical, disgraceful, dreadful, egregious, execrable, faulty, feeble, *inf* grotty, hopeless, imperfect, inadequate, incompetent, incorrect, ineffective, inefficient, inferior, *inf* lousy, pitiful, poor, *inf* ropy, shoddy, *inf* sorry, substandard, unsound, unsatisfactory, useless, weak, worthless. 4 *bad conditions*. adverse, deleterious, detrimental, discouraging, *inf* frightful, harmful, harsh, hostile, inappropriate, inauspicious, prejudicial, uncongenial, unfortunate, unhelpful, unpropitious. 5 *bad smell*. decayed, decomposing, diseased, foul, loathsome, mildewed, mouldy, nauseating, noxious, objectionable, obnoxious, odious, offensive, polluted, putrid, rancid, repellent, repulsive, revolting, sickening, rotten, smelly, sour, spoiled, tainted, vile. **6** *I feel bad.* ▷ ILL. *Opp* GOOD.

**badge** *n* chevron, crest, device, emblem, insignia, logo, mark, medal, sign, symbol, token.

**bad-tempered** *adj* acrimonious, angry, bilious, cantankerous, churlish, crabbed, cross, *inf* crotchety, disgruntled, disobliging, dyspeptic, fretful, gruff, grumbling, grumpy, hostile, hot-tempered, ill-humoured, ill-tempered, irascible, irritable, malevolent, malign, moody, morose, peevish, petulant, quarrelsome, querulous, rude, scowling, short-tempered, shrewish, snappy, *inf* stroppy, sulky, sullen, testy, truculent, unfriendly, unsympathetic. *Opp* GOOD-TEMPERED.

**baffle** *vb* **1** *inf* bamboozle, bemuse, bewilder, confound, confuse, defeat, *inf* floor, *inf* flummox, foil, frustrate, mystify, perplex, puzzle, *inf* stump, thwart. **baffling** ▷ INEXPLICABLE.

**bag** *n* basket, carrier, carrier-bag, case, handbag, haversack, holdall, reticule, rucksack, sack, satchel, shopping-bag, shoulder-bag. ▷ BAGGAGE. • *vb* capture, catch, ensnare, snare.

**baggage** *n* accoutrements, bags, belongings, *inf* gear, impedimenta, paraphernalia. ▷ LUGGAGE.

**bait** *n* allurement, attraction, bribe, carrot, decoy, enticement, inducement, lure, temptation. • *vb* annoy, goad, harass, hound, jeer at, *inf* needle, persecute, pester, provoke, tease, torment.

**balance** *n* **1** scales, weighing-machine. **2** equilibrium, equipoise, poise, stability, steadiness. **3** correspondence, equality, equivalence, evenness, parity, symmetry. **4** *spend a bit & save the balance.* difference, excess, remainder, residue, rest, surplus. • *vb* **1** cancel out, compensate for, counteract, counterbalance, counterpoise, equalize, even up, level, make steady, match, neutralize, offset, parallel, stabilize, steady. **2** keep balanced, keep in equilibrium, poise, steady, support. **balanced** ▷ EVEN, IMPARTIAL, STABLE.

**bald** *adj* **1** baldheaded, bare, hairless, smooth, thin on top. **2** *bald truth.* direct, forthright, plain, simple, stark, straightforward, unadorned, uncompromising.

**bale** *n* bunch, bundle, pack, package, truss. • *vb* **bale out** eject, escape, jump out, parachute down.

**ball** *n* **1** drop, globe, globule, orb, shot, sphere, spheroid. **2** dance, disco, party, social.

**balloon** *n* airship, dirigible, hot-air balloon. • *vb* ▷ BILLOW.

**ballot** *n* election, plebiscite, poll, referendum, vote.

**ban** *n* boycott, embargo, interdiction, moratorium, prohibition, proscription, taboo, veto. • *vb* banish, bar, debar, disallow, exclude, forbid, interdict, make illegal, ostracize, outlaw, prevent, prohibit, proscribe, put a ban on, restrict, stop, suppress, veto. *Opp* PERMIT.

**banal** *adj* boring, clichéd, cliché-ridden, commonplace, *inf* corny, dull, hackneyed, humdrum, obvious, *inf* old hat, ordinary, overused, pedestrian, platitudinous, predictable, stereotyped, trite, unimaginative, uninteresting, unoriginal, vapid. *Opp* INTERESTING.

**band** *n* **1** belt, border, fillet, hoop, line, loop, ribbon, ring, strip, stripe, swathe. **2** association, body, clique, club, company, crew, flock, gang, herd, horde, party, society, troop. ▷ GROUP. **3** [*music*] ensemble, group, orchestra.

**bandage** *n* dressing, gauze, lint, plaster.

**bandit** *n* brigand, buccaneer, desperado, footpad, gangster, gunman, highwayman, hijacker, marauder, outlaw, pirate, robber, thief.

**bandy** *adj* bandy-legged, bowed, bow-legged. • *vb bandy words.* exchange, interchange, pass, swap, throw, toss. ▷ ARGUE.

**bang** *n* **1** blow, *inf* box, bump, collision, cuff, knock, punch, slam, smack, stroke, thump, wallop, whack. ▷ HIT. **2** blast, boom, clap, crash, explosion, pop, report, thud, thump. ▷ SOUND.

**banish** *vb* deport, drive out, eject, evict, excommunicate, exile, expatriate, expel, ostracize, oust, outlaw, rusticate, send away, ship away, transport. **2** ban, bar, *inf* black, debar, eliminate, exclude, forbid, get rid of, make illegal, prohibit, proscribe, put an embargo on, remove, restrict, stop, suppress, veto.

**bank** *n* **1** camber, declivity, dike, earthwork, embankment, gradient, incline, mound, ramp, rampart, ridge, rise, slope, tilt. **2** *river bank.* brink, edge, margin, shore, side. **3** *bank of controls.* array, collection, display, file, group, line, panel, rank, row, series. • *vb* **1** cant, heel, incline, lean, list, pitch, slant, slope, tilt, tip. **2** *bank money.* deposit, save.

**bankrupt** *adj inf* broke, failed, *sl* gone bust, gone into liquidation, insolvent, ruined, spent up, wound up. ▷ POOR. *Opp* SOLVENT.

**banner** *n* banderole, colours, ensign, flag, pennant, pennon, standard, streamer.

**banquet** *n inf* binge, *sl* blow-out, dinner, feast, repast, *inf* spread. ▷ MEAL.

**banter** *n* badinage, chaffing, joking, persiflage, pleasantry, raillery, repartee, ribbing, ridicule, teasing, word-play.

**bar** *n* **1** beam, girder, pole, rail, railing, rod, shaft, stake, stick, strut. **2** barricade, barrier, check, deterrent, hindrance, impediment, obstacle, obstruction. **3** band, belt, line, streak, strip, stripe. **4** *bar of soap.* block, cake, chunk, hunk, ingot, lump, nugget, piece, slab, wedge. **5** *drink in a bar.* café, canteen, counter, inn, lounge, pub, public house, saloon, taproom, tavern, wine bar. • *vb* **1** ban, banish, debar, exclude, forbid to enter, keep out, ostracize, outlaw, prevent from entering, prohibit, proscribe. **2** *bar the way.* arrest, block, check, deter, halt, hinder, impede, obstruct, prevent, stop, thwart.

**barbarian** *adj* ▷ BARBARIC. • *n* boor, churl, heathen, hun, ignoramus, lout, pagan, philistine, savage, vandal, *sl* yob.

**barbaric** *adj* barbarous, brutal, brutish, crude, inhuman, primitive, rough, savage, uncivil, uncivilized, uncultivated, wild. ▷ CRUEL. *Opp* CIVILIZED.

**bare** *adj* **1** bald, denuded, exposed, naked, nude, stark-naked, stripped, unclad, unclothed, uncovered, undressed. **2** *bare moor.* barren, bleak, desolate, featureless, open, treeless, unwooded, windswept. **3** *bare trees.* defoliated, leafless, shorn. **4** *a bare room.* austere, empty, plain, simple, unadorned, undecorated, unfurnished, vacant. **5** *a bare wall.* blank, clean, unmarked. **6** *bare facts.* direct, explicit, hard, honest, literal, open, plain, straightforward, unconcealed, undisguised, unembellished. 7 *the bare minimum.* basic, essential, just adequate, just sufficient, minimal, minimum. • *vb* betray, bring to light, disclose, expose, lay bare, make known, publish, reveal, show, uncover, undress, unmask, unveil.

**bargain** *n* 1 agreement, arrangement, compact, contract, covenant, deal, negotiation, pact, pledge, promise, settlement, transaction, treaty, understanding. 2 *bargain in the sales.* *inf* giveaway, good buy, good deal, loss-leader, reduced item, *inf* snip, special offer. • *vb* argue, barter, discuss terms, do a deal, haggle, negotiate. **bargain for** ▷ EXPECT.

**bark** *vb* 1 growl, yap. 2 *bark your shin.* abrade, chafe, graze, rub, score, scrape, scratch.

**barmaid, barman** *ns* attendant, server, steward, stewardess, waiter, waitress.

**barracks** *n* accommodation, billet, camp, garrison, lodging, quarters.

**barrage** *n* 1 ▷ BARRIER. 2 *barrage of gunfire.* assault, attack, battery, bombardment, cannonade, fusillade, gunfire, onslaught, salvo, storm, volley.

**barrel** *n* butt, cask, churn, cistern, drum, hogshead, keg, tank, tub, tun, water-butt.

**barren** *adj* 1 arid, bare, desert, desolate, dried-up, dry, empty, infertile, lifeless, non-productive, treeless, uncultivated, unproductive, unprofitable, untilled, useless, waste. 2 childless, fruitless, infertile, sterile, sterilized, unfruitful. *Opp* FERTILE.

**barricade** *n* ▷ BARRIER. • *vb* bar, block off, defend, obstruct.

**barrier** *n* 1 bar, barrage, barricade, blockade, boom, bulwark, dam, earthwork, fortification, embankment, fence, frontier, hurdle, obstacle, obstruction, palisade, railing, rampart, stockade, wall. 2 *barrier to progress.* check, drawback, handicap, hindrance, impediment, limitation, restriction, stumbling-block.

**barter** *vb* bargain, deal, exchange, negotiate, swap, trade, traffic.

**base** *adj* contemptible, cowardly, degrading, depraved, despicable, detestable, dishonourable, evil, ignoble, immoral, inferior, low, mean, scandalous, selfish, shabby, shameful, sordid, undignified, unworthy, vulgar, vile. ▷ WICKED. • *n* 1 basis, bed, bedrock, bottom, core, essentials, foot, footing, foundation, fundamentals, groundwork, infrastructure, pedestal, plinth, rest, root, stand, substructure, support, underpinning. 2 camp, centre, depot, headquarters, post, starting-point, station. • *vb* build, construct, establish, found, ground, locate, position, post, secure, set up, station.

**basement** *n* cellar, crypt, vault.

**bashful** *adj* abashed, backward, blushing, coy, demure, diffident, embarrassed, faint-hearted, inhibited, meek, modest, nervous, reserved, reticent, retiring, self-conscious, self-effacing, shamefaced, sheepish, shy, timid, timorous, uneasy, unforthcoming. *Opp* ASSERTIVE.

**basic** *adj* central, chief, crucial, elementary, essential, foremost, fundamental, important, intrinsic, key, main, necessary, primary, principal, radical, underlying, vital. *Opp* UNIMPORTANT.

**basin** *n* bath, bowl, container, dish, pool, sink, stoup.

**basis** *n* base, core, footing, foundation, ground, infrastructure, premise, principle, starting-point, support, underpinning.

**bask** *vb* enjoy, feel pleasure, glory, lie, lounge, luxuriate, relax, sunbathe, wallow.

**basket** *n* bag, hamper, pannier, punnet, skip, trug.

**bastard** *n* illegitimate child, *old use* love-child, natural child.

**bat** *n* club, racket, racquet.

**bath** *n* douche, jacuzzi, pool, sauna, shower, *inf* soak, *inf* tub, wash.

**bathe** *vb* **1** clean, cleanse, immerse, moisten, rinse, soak, steep, swill, wash. **2** *bathe in the sea.* go swimming, paddle, plunge, splash about, swim, *inf* take a dip.

**bathos** *n* anticlimax, *inf* come-down, disappointment, *inf* let-down. *Opp* CLIMAX.

**baton** *n* cane, club, cudgel, rod, staff, stick, truncheon.

**batter** *vb* beat, bludgeon, cudgel, keep hitting, pound. ▷ HIT.

**battery** *n* **1** artillery-unit, emplacement. **2** *electric battery.* accumulator, cell. **3** *assault and battery.* assault, attack, *inf* beating-up, blows, mugging, onslaught, thrashing, violence.

**battle** *n* action, air-raid, Armageddon, attack, blitz, brush, campaign, clash, combat, conflict, contest, confrontation, crusade, *inf* dogfight, encounter, engagement, fight, fray, hostilities, offensive, pitched battle, pre-emptive strike, quarrel, *inf* shoot-out, siege, skirmish, strife, struggle, war, warfare. • *vb* ▷ FIGHT.

**battlefield** *n* arena, battleground, theatre of war.

**bawdy** *adj* broad, earthy, erotic, lusty, *inf* naughty, racy, *inf* raunchy, ribald, *inf* sexy, *inf* spicy. [*derog synonyms*] *inf* blue, coarse, dirty, immoral, improper, indecent, indecorous, indelicate, lascivious, lecherous, lewd, licentious, obscene, pornographic, prurient, risqué, rude, salacious, smutty, suggestive, titillating, vulgar. *Opp* PROPER.

**bawl** *vb* cry, roar, shout, thunder, wail, yell, yelp.

**bay** *n* **1** bight, cove, creek, estuary, fjord, gulf, harbour, indentation, inlet, ria, sound. **2** alcove, booth, compartment, niche, nook, opening, recess.

**bazaar** *n* auction, boot-sale, bring-and-buy, fair, fête, jumble sale, market, sale.

**be** *vb* **1** be alive, breathe, endure, exist, live. **2** *be here all day.* continue, dwell, inhabit, keep going, last, occupy a position, persist, remain, stay, survive. **3** *the next event will be tomorrow.* arise, befall, come about, happen, occur, take place. **4** *want to be a writer.* become, develop into.

**beach** *n* bank, coast, coastline, foreshore, littoral, sand, sands, seashore, seaside, shore, *poet* strand.

**beacon** *n* bonfire, fire, flare, light, lighthouse, pharos, signal.

**bead** *n* blob, drip, drop, droplet, globule, jewel, pearl.

**beaker** *n* cup, glass, goblet, jar, mug, tankard, tumbler.

**beam** *n* **1** bar, board, boom, brace, girder, joist, plank, post, rafter, spar, stanchion, stud, support, timber. **2** *beam of light.* gleam, pencil, ray, shaft, stream. • *vb* **1** aim, broadcast, direct, emit, radiate, send out, shine, transmit. **2** *beam* *happily*. grin, laugh, look radiant, radiate happiness, smile.

**bear** *vb* **1** carry, hold, prop up, shoulder, support, sustain, take. **2** *bear an inscription*. display, exhibit, have, possess, show. **3** *bear gifts*. bring, carry, convey, deliver, fetch, move, take, transfer, transport. **4** *bear pain*. abide, accept, brook, cope with, endure, live with, permit, *inf* put up with, reconcile yourself to, *inf* stand, *inf* stomach, suffer, sustain, tolerate, undergo. **5** *bear children, fruit*. breed, *old use* bring forth, develop, engender, generate, give birth to, produce, spawn, yield. **bear out** ▷ CONFIRM. **bear up** ▷ SURVIVE. **bear witness** ▷ TESTIFY.

**bearable** *adj* acceptable, endurable, supportable, survivable, sustainable, tolerable.

**bearing** *n* **1** air, appearance, aspect, attitude, behaviour, carriage, demeanour, deportment, look, manner, mien, poise, posture, presence, stance, style. **2** *evidence had no bearing*. applicability, application, connection, import, pertinence, reference, relation, relationship, relevance, significance. **bearings** aim, course, direction, line, location, orientation, path, position, road, sense of direction, tack, track, way, whereabouts.

**beast** *n* brute, creature, monster, savage. ▷ ANIMAL.

**beastly** *adj* abominable, barbaric, bestial, brutal, cruel, savage. ▷ VILE.

**beat** *n* **1** accent, pulse, rhythm, stress, tempo, throb. **2** *policeman's beat*. course, itinerary, journey, path, rounds, route, way.
• *vb* **1** batter, bludgeon, buffet, cane, clout, cudgel, flail, flog, hammer, knock about, lash, *inf* lay into, manhandle, pound, punch, scourge, strike, *inf* tan, thrash, thump, trounce, *inf* wallop, whack, whip. ▷ HIT. **2** *beat eggs*. agitate, blend, froth up, knead, mix, pound, stir, whip, whisk. **3** *heart was beating*. flutter, palpitate, pound, pulsate, race, throb, thump. **5** *beat an opponent*. best, conquer, crush, defeat, excel, get the better of, *inf* lick, master, outclass, outdistance, outdo, outpace, outrun, outwit, overcome, overpower, overthrow, overwhelm, rout, subdue, surpass, *inf* thrash, trounce, vanquish, win against, worst. **beat up** ▷ ATTACK.

**beautiful** *adj* admirable, aesthetic, alluring, appealing, artistic, attractive, becoming, bewitching, brilliant, captivating, charming, *old use* comely, dainty, decorative, delightful, elegant, enjoyable, exquisite, *old use* fair, fascinating, fetching, fine, good-looking, glamorous, glorious, gorgeous, graceful, handsome, irresistible, lovely, magnificent, neat, picturesque, pleasing, pretty, pulchritudinous, quaint, radiant, ravishing, scenic, seductive, sensuous, sexy, spectacular, splendid, stunning, superb, tasteful, tempting. *Opp* UGLY.

**beautify** *vb* adorn, bedeck, deck, decorate, embellish, garnish, make beautiful, ornament, prettify, *derog* tart up, *inf* titivate. *Opp* DISFIGURE.

**beauty** *n* allure, appeal, attractiveness, charm, elegance, fascination, glamour, glory, grace, handsomeness, loveliness, magnificence, picturesqueness, prettiness, pulchritude, radiance, splendour.

**becalmed** *adj* helpless, idle, motionless, still, unmoving.

**beckon** *vb* gesture, motion, signal, summon, wave.

**become** *vb* **1** be transformed into, change into, develop into, grow into, mature into, metamorphose into, turn into. **2** *Red becomes you.* be appropriate to, be becoming to, befit, enhance, fit, flatter, harmonize with, set off, suit. **becoming** ▷ ATTRACTIVE, SUITABLE.

**bed** *n* **1** resting-place. □ *air-bed, berth, bunk, cot, couch, couchette, cradle, crib, divan, four-poster, hammock, pallet, palliasse, truckle bed, waterbed.* **2** *bed of concrete.* base, foundation, groundwork, layer, substratum. **3** *river bed.* bottom, channel, course, watercourse. **4** *flower bed.* border, garden, patch, plot.

**bedclothes** *n* bedding, bed linen. □ *bedspread, blanket, bolster, continental quilt, counterpane, coverlet, duvet, eiderdown, electric-blanket, mattress, pillow, pillowcase, pillowslip, quilt, sheet, sleeping-bag.*

**bedraggled** *adj* dirty, dishevelled, drenched, messy, muddy, scruffy, sodden, soiled, stained, unkempt, untidy, wet, wringing. *Opp* SMART.

**beer** *n* ale, bitter, lager, mild, porter, stout.

**befall** *vb* be the outcome, *old use* betide, chance, come about, *inf* crop up, eventuate, happen, occur, take place, *inf* transpire.

**before** *adv* already, earlier, in advance, previously, sooner.

**befriend** *vb inf* chat up, *inf* gang up with, get to know, make friends with, make the acquaintance of, *inf* pal up with.

**beg** *vb* **1** *inf* cadge, scrounge, solicit, sponge. **2** *beg a favour.* ask, beseech, cajole, crave, entreat, implore, importune, petition, plead, pray, request, supplicate, wheedle.

**beget** *vb* breed, bring about, cause, create, engender, father, generate, give rise to, procreate, produce, propagate, result in, sire, spawn.

**beggar** *n* cadger, destitute person, down-and-out, homeless person, mendicant, pauper, ragamuffin, scrounger, sponger, tramp, vagrant. ▷ POOR.

**begin** *vb* **1** activate, approach, attack, be first, broach, commence, conceive, create, embark on, enter into, found, *inf* get cracking, *inf* get going, inaugurate, initiate, inspire, instigate, introduce, kindle, launch, lay the foundations, lead off, move into, move off, open, originate, pioneer, precipitate, provoke, set about, set in motion, set off, set out, set up, *inf* spark off, start, *inf* take steps, take the initiative, take up, touch off, trigger off, undertake. **2** *Spring begins gradually.* appear, arise, break out, come into existence, crop up, emerge, get going, happen, materialize, originate, spring up. *Opp* END.

**beginner** *n* **1** creator, founder, initiator, inspiration, instigator, originator, pioneer. **2** *only a beginner.* apprentice, fresher, greenhorn, inexperienced person, initiate, learner, novice, recruit, starter, tiro, trainee.

**beginning** *n* **1** birth, commencement, conception, creation, dawn, embryo, emergence, establishment, foundation, genesis, germ, inauguration, inception, initiation, instigation, introduction, launch, onset, opening, origin, outset, point of departure, rise, source, start, starting-point, threshold. **2** *beginning of a book.*

preface, prelude, prologue. *Opp* END.

**begrudge** *vb* be bitter about, covet, envy, grudge, mind, object to, resent.

**behave** *vb* 1 acquit yourself, act, *inf* carry on, comport yourself, conduct yourself, function, operate, perform, react, respond, run, work. 2 *told to behave.* act properly, be good, be on best behaviour.

**behaviour** *n* actions, attitude, bearing, comportment, conduct, courtesy, dealings, demeanour, deportment, manners, performance, reaction, response, ways.

**behead** *vb* decapitate, guillotine.

**behold** *vb* descry, discern, espy, look at, note, notice, see, set eyes on, view.

**being** *n* 1 actuality, essence, existence, life, living, reality, solidity, substance. 2 animal, creature, individual, person, spirit, soul.

**belated** *adj* behindhand, delayed, last-minute, late, overdue, posthumous, tardy, unpunctual.

**belch** *vb* 1 break wind, *inf* burp, emit wind. 2 *belch smoke.* discharge, emit, erupt, fume, gush, send out, smoke, spew out, vomit.

**belief** *n* 1 acceptance, assent, assurance, certainty, confidence, credence, reliance, security, sureness, trust. 2 *religious belief.* attitude, conviction, creed, doctrine, dogma, ethos, faith, feeling, ideology, morality, notion, opinion, persuasion, principles, religion, standards, tenets, theories, views. *Opp* SCEPTICISM.

**believe** *vb* 1 accept, be certain of, count on, credit, depend on, endorse, have faith in, reckon on, rely on, subscribe to, *inf* swallow, swear by, trust. *Opp* DISBELIEVE. 2 assume, consider, *inf* dare say, feel, gather, guess, hold, imagine, judge, know, maintain, postulate, presume, speculate, suppose, take it for granted, think. **make believe** ▷ IMAGINE.

**believer** *n* adherent, devotee, disciple, fanatic, follower, proselyte, supporter, upholder, zealot. *Opp* ATHEIST.

**belittle** *vb* be unimpressed by, criticize, decry, denigrate, deprecate, depreciate, detract from, disparage, minimize, *inf* play down, slight, speak slightingly of, underrate, undervalue. *Opp* EXAGGERATE, FLATTER, PRAISE.

**bell** *n* alarm, carillon, chime, knell, peal, signal. ▷ RING.

**belligerent** *adj* aggressive, antagonistic, argumentative, bellicose, bullying, combative, contentious, defiant, disputatious, fierce, hawkish, hostile, jingoistic, martial, militant, militaristic, provocative, pugnacious, quarrelsome, violent, warlike, warmongering, warring. ▷ UNFRIENDLY. *Opp* PEACEABLE.

**belong** *vb* 1 be owned (by), go (with), pertain (to), relate (to). 2 be at home, feel welcome, have a place. 3 *belong to a club.* be affiliated with, be a member of, be connected with, *inf* be in with.

**belongings** *n* chattels, effects, *inf* gear, goods, impedimenta, possessions, property, things.

**belt** *n* 1 band, circle, loop. 2 *belt round the waist.* cincture, cummerbund, girdle, girth, sash, strap, waistband, *old use* zone. 3 *green belt.* area, district, line, stretch, strip, swathe, tract, zone.

**bemuse** *vb* befuddle, bewilder, confuse, mix up, muddle, perplex, puzzle, stupefy.

**bench** *n* 1 form, pew, seat, settle. 2 counter, table, work-bench, work-table. 3 *magistrate's bench.* court, courtroom, judge, magistrate, tribunal.

**bend** *n* angle, arc, bow, corner, crank, crook, curvature, curve, flexure, loop, turn, turning, twist, zigzag. • *vb* 1 arch, be flexible, bow, buckle, coil, contort, crook, curl, curve, deflect, distort, divert, flex, fold, *inf* give, loop, mould, refract, shape, turn, twist, warp, wind, yield. 2 bow, crouch, curtsy, duck, genuflect, kneel, lean, stoop.

**benefactor** *n inf* angel, backer, *derog* do-gooder, donor, *inf* fairy godmother, patron, philanthropist, promoter, sponsor, supporter, well-wisher.

**beneficial** *adj* advantageous, benign, constructive, favourable, fruitful, good, health-giving, healthy, helpful, improving, nourishing, nutritious, positive, productive, profitable, rewarding, salubrious, salutary, supportive, useful, valuable, wholesome. *Opp* HARMFUL.

**beneficiary** *n* heir, heiress, inheritor, legatee, recipient, successor (*to title*).

**benefit** *n* 1 advantage, asset, blessing, boon, convenience, gain, good thing, help, privilege, prize, profit, service, use. *Opp* DISADVANTAGE. 2 *unemployment benefit.* aid, allowance, assistance, *inf* dole, *inf* hand-out, grant, payment, social security, welfare. • *vb* advance, advantage, aid, assist, better, boost, do good to, enhance, further, help, improve, profit, promote, serve.

**benevolent** *adj* altruistic, beneficent, benign, caring, charitable, compassionate, considerate, friendly, generous, helpful, humane, humanitarian, kind-hearted, kindly, liberal, magnanimous, merciful, philanthropic, supportive, sympathetic, unselfish, warm-hearted. ▷ KIND. *Opp* UNKIND.

**benign** *adj* gentle, harmless, kind. ▷ BENEFICIAL, BENEVOLENT.

**bent** *adj* 1 angled, arched, bowed, buckled, coiled, contorted, crooked, curved, distorted, folded, hunched, looped, twisted, warped. 2 [*inf*] *a bent dealer.* corrupt, criminal, dishonest, illegal, immoral, untrustworthy, wicked. *Opp* HONEST, STRAIGHT. • *n* ▷ APTITUDE, BIAS.

**bequeath** *vb* endow, hand down, leave, make over, pass on, settle, will.

**bequest** *n* endowment, gift, inheritance, legacy, settlement.

**bereavement** *n* death, loss.

**bereft** *adj* deprived, destitute, devoid, lacking, robbed, wanting.

**berserk** *adj inf* beside yourself, crazed, crazy, demented, deranged, frantic, frenetic, frenzied, furious, infuriated, insane, mad, maniacal, rabid, violent, wild. *Opp* CALM. **go berserk** ▷ RAGE, RAMPAGE.

**berth** *n* 1 bed, bunk, hammock. 2 *berth for ships.* anchorage, dock, harbour, haven, landing-stage, moorings, pier, port, quay, slipway, wharf. • *vb* anchor, dock, drop anchor, land, moor, tie up. **give a wide berth to** ▷ AVOID.

**beseech** *vb* ask, beg, entreat, implore, importune, plead, supplicate.

**besiege** *vb* beleaguer, beset, blockade, cut off, encircle, encompass, hem in, isolate, pester, plague, siege, surround.

**best** *adj* choicest, excellent, finest, first-class, foremost, incomparable, leading, matchless, optimum, outstanding, pre-eminient, superlative, supreme, top, unequalled, unrivalled, unsurpassed.

**bestial** *adj* animal, beast-like, beastly, brutal, brutish, inhuman, subhuman. ▷ SAVAGE.

**bestow** *vb* award, confer, donate, give, grant, present.

**bet** *n inf* flutter, gamble, *inf* punt, speculation, stake, wager. • *vb* bid, chance, do the pools, enter a lottery, gamble, *inf* have a flutter, hazard, lay bets, *inf* punt, risk, speculate, stake, venture, wager.

**betray** *vb* **1** be a Judas to, be a traitor to, be false to, cheat, conspire against, deceive, denounce, desert, double-cross, give away, *inf* grass on, incriminate, inform against, inform on, jilt, let down, *inf* rat on, report, *inf* sell down the river, sell out, *inf* shop, *inf* tell tales about, *inf* turn Queen's evidence on. **2** *betray secrets.* disclose, divulge, expose, give away, indicate, let out, let slip, manifest, reveal, show, tell.

**better** *adj* **1** preferable, recommended, superior. **2** convalescent, cured, fitter, healed, healthier, improved, *inf* on the mend, progressing, recovered, recovering, restored. • *vb* ▷ IMPROVE, SURPASS.

**beware** *vb* avoid, be alert, be careful, be cautious, be on your guard, guard (against), heed, keep clear (of), look out, mind, shun, steer away (from), take care, take heed, take precautions, watch out, *inf* watch your step.

**bewilder** *vb* baffle, *inf* bam boozle, bemuse, confound, confuse, daze, disconcert, disorientate, distract, floor, *inf* flummox, mislead, muddle, mystify, perplex, puzzle, stump.

**bewitch** *vb* captivate, cast a spell on, charm, enchant, enrapture, fascinate, spellbind.

**bias** *n* **1** aptitude, bent, inclination, leaning, liking, partiality, penchant, predilection, predisposition, preference, proclivity, proneness, propensity, tendency. **2** [*derog*] bigotry, chauvinism, favouritism, imbalance, injustice, nepotism, one-sidedness, partiality, partisanship, prejudice, racism, sexism, unfairness. • *vb* ▷ INFLUENCE.

**biased** *adj* bigoted, blinkered, chauvinistic, distorted, emotive, influenced, interested, jaundiced, loaded, one-sided, partial, partisan, prejudiced, racist, sexist, slanted, tendentious, unfair, unjust, warped. *Opp* UNBIASED.

**bicycle** *n inf* bike, cycle, penny-farthing, *inf* push-bike, racer, tandem, *inf* two-wheeler.

**bid** *n* **1** offer, price, proposal, proposition, tender. **2** *a bid to win.* attempt, *inf* crack, effort, endeavour, *inf* go, try, venture. • *vb* **1** make an offer, offer, proffer, propose, tender. **2** ▷ COMMAND.

**big** *adj* **1** above average, *inf* almighty, ample, astronomical, bold, broad, Brobdingnagian, bulky, burly, capacious, colossal, commodious, considerable, elephantine, enormous, extensive, fat, formidable, gargantuan, generous, giant, gigantic, grand, great, gross, heavy, hefty, high, huge, *inf* hulking, husky, immeasurable, immense, impressive, *inf* jumbo, *inf* king-sized, large, largish, lofty, long, mammoth, massive, mighty, monstrous, monumental, mountainous, overgrown, oversized, prodigious, roomy, sizeable, spacious, stupendous, substantial, swingeing (*increase*), tall, *inf* terrific, thick, *inf* thumping, tidy (*sum*), titanic, towering, *inf* tremendous, vast, voluminous, weighty, *inf* whacking, *inf* whopping, wide. **2** *a big decision*. grave, important, influential, leading, main, major, momentous, notable, powerful, prime, principal, prominent, serious, significant. **3** *a big number*. ▷ INFINITE. **4** *a big name*. ▷ FAMOUS. **5** *a big noise*. ▷ LOUD. *Opp* SMALL.

**bigot** *n* chauvinist, fanatic, prejudiced person, racist, sexist, zealot.

**bigoted** *adj* intolerant, one-sided, partial, prejudiced. ▷ BIASED.

**bill** *n* **1** account, invoice, receipt, statement, tally. **2** advertisement, broadsheet, bulletin, circular, handbill, handout, leaflet, notice, placard, poster, sheet. **3** *a Parliamentary bill*. draft law, proposed law. **4** *a bird's bill*. beak, mandible.

**billow** *vb* balloon, belly, bulge, fill out, heave, puff out, rise, roll, surge, swell, undulate.

**bind** *vb* **1** attach, clamp, combine, connect, fuse, hitch, hold together, join, lash, link, rope, secure, strap, tie, truss, unify, unite, weld. ▷ FASTEN. **2** *bind a wound*. bandage, cover, dress, encase, swathe, wrap. **3** *bound to obey*. compel, constrain, force, necessitate, oblige, require. **binding** ▷ COMPULSORY, FORMAL.

**biography** *n* autobiography, life, life-story, memoirs, recollections. ▷ WRITING.

**bird** *n inf* birdie, chick, cock, *joc* feathered friend, fledgling, fowl, hen, nestling. □ *gamebird, plur poultry, seabird, wader, waterfowl, wildfowl.* □ *albatross, auk, bittern, blackbird, budgerigar, bullfinch, bunting, bustard, buzzard, canary, cassowary, chaffinch, chiff-chaff, chough, cockatoo, coot, cormorant, corncrake, crane, crow, cuckoo, curlew, dabchick, dipper, dove, duck, dunnock, eagle, egret, emu, falcon, finch, flamingo, flycatcher, fulmar, goldcrest, goldfinch, goose, grebe, greenfinch, grouse, gull, hawk, heron, hoopoe, hornbill, humming bird, ibis, jackdaw, jay, kingfisher, kestrel, kite, kiwi, kookaburra, lapwing, lark, linnet, macaw, magpie, martin, mina bird, moorhen, nightingale, nightjar, nuthatch, oriole, osprey, ostrich, ousel, owl, parakeet, parrot, partridge, peacock, peewit, pelican, penguin, peregrine, petrel, pheasant, pigeon, pipit, plover, ptarmigan, puffin, quail, raven, redbreast, redstart, robin, rook, sandpiper, seagull, shearwater, shelduck, shrike, skua, skylark, snipe, sparrow, sparrowhawk, spoonbill, starling, stonechat, stork, swallow, swan, swift, teal, tern, thrush, tit, toucan, turkey, turtle-dove, vulture, wagtail, warbler, waxwing, wheatear, woodcock, woodpecker, wren, yellowhammer.*

**birth** *n* **1** childbirth, confinement, delivery, labour, nativity, parturition. **2** ancestry, background, blood, breeding, derivation, descent, extraction, family, genealogy, line, lineage, parentage, pedigree, race, stock, strain. **3** ▷ BEGINNING. **give birth** bear, calve, farrow, foal. ▷ BEGIN.

**bisect** *vb* cross, cut in half, divide, halve, intersect.

**bit** *n* **1** atom, bite, block, chip, chunk, crumb, division, dollop, fraction, fragment, gobbet, grain, helping, hunk, iota, lump, modicum, morsel, mouthful, part, particle, piece, portion, sample, scrap, section, segment, share, slab, slice, snippet, soupçon, speck, spot, taste, titbit, trace. 2 *Wait a bit.* flash, instant, *inf* jiffy, minute, moment, second, *inf* tick, time, while.

**bite** *n* 1 nip, pinch, sting. 2 *a bite to eat.* morsel, mouthful, nibble, snack, taste. ▷ BIT. • *vb* 1 champ, chew, crunch, cut into, gnaw, masticate, munch, nibble, nip, rend, snap, tear at, wound. 2 *An insect bit me.* pierce, sting. 3 *The screw won't bite.* grip, hold.

**bitter** *adj* 1 acid, acrid, harsh, sharp, sour, unpleasant. 2 *a bitter experience.* calamitous, dire, distasteful, distressing, galling, hateful, heartbreaking, painful, poignant, sorrowful, unhappy, unwelcome, upsetting. 3 *bitter remarks.* acrimonious, acerbic, angry, cruel, cynical, embittered, envious, hostile, jaundiced, jealous, malicious, rancorous, resentful, savage, sharp, spiteful, stinging, vicious, violent, waspish. 4 *a bitter wind.* biting, cold, fierce, freezing, perishing, piercing, raw. *Opp* KIND, MILD, PLEASANT.

**bizarre** *adj* curious, eccentric, fantastic, freakish, grotesque, odd, outlandish, outré, surreal, weird. ▷ STRANGE. *Opp* ORDINARY.

**black** *adj* blackish, coal-black, dark, dusky, ebony, funereal, gloomy, inky, jet, jet-black, moonless, murky, pitch-black, pitch-dark, raven, sable, sooty, starless, unlit. • *vb* 1 blacken, polish. 2 ▷ BLACKLIST.

**blackleg** *n sl* scab, strikebreaker, traitor.

**blacklist** *vb* ban, bar, blackball, boycott, debar, disallow, exclude, ostracize, preclude, proscribe, put an embargo on, refuse to handle, repudiate, snub, veto.

**blade** *n* dagger, edge, knife, razor, scalpel, vane. ▷ SWORD.

**blame** *n* accountability, accusation, castigation, censure, charge, complaint, condemnation, criticism, culpability, fault, guilt, imputation, incrimination, liability, onus, *inf* rap, recrimination, reprimand, reproach, reproof, responsibility, *inf* stick, stricture. • *vb* accuse, admonish, censure, charge, chide, condemn, criticize, denounce, *inf* get at, hold responsible, incriminate, rebuke, reprehend, reprimand, reproach, reprove, round on, scold, tax, upbraid. *Opp* EXCUSE.

**blameless** *adj* faultless, guiltless, innocent, irreproachable, moral, unimpeachable, upright. *Opp* GUILTY.

**bland** *adj* affable, amiable, banal, boring, calm, characterless, dull, flat, gentle, insipid, mild, nondescript, smooth, soft, soothing, suave, tasteless, trite, unappetizing, unexciting, uninspiring, uninteresting, vapid, watery, weak, *inf* wishy-washy. *Opp* INTERESTING.

**blank** *adj* 1 bare, clean, clear, empty, plain, spotless, unadorned, unmarked, unused, void. 2 *a blank look.* apathetic, baffled, baffling, dead, *inf* deadpan, emotionless, expressionless, featureless, glazed, immobile, impassive, inane, inscrutable, lifeless, poker-faced, uncomprehending, unresponsive, vacant, vacuous. • *n* 1 emptiness, nothingness, vacuity, vacuum, void. 2 *blanks on a form.* box, break, gap, line, space.

**blaspheme** *vb* curse, execrate, imprecate, profane, swear.

**blasphemous** *adj* disrespectful, godless, impious, irreligious, irreverent, profane, sacrilegious, sinful, ungodly, wicked. *Opp* REVERENT.

**blast** *n* **1** gale, gust, wind. **2** blare, din, noise, racket, roar, sound. **3** ▷ EXPLOSION. • *vb* ▷ ATTACK, EXPLODE. **blast off** ▷ LAUNCH.

**blatant** *adj* apparent, bare-faced, bold, brazen, conspicuous, evident, flagrant, glaring, obtrusive, obvious, open, overt, shameless, stark, unconcealed, undisguised, unmistakable, visible. *Opp* HIDDEN.

**blaze** *n* conflagration, fire, flame, flare-up, holocaust, inferno, outburst. • *vb* burn, erupt, flame, flare.

**bleach** *vb* blanch, discolour, etiolate, fade, lighten, pale, peroxide (*hair*), whiten.

**bleak** *adj* bare, barren, blasted, cheerless, chilly, cold, comfortless, depressing, desolate, dismal, dreary, exposed, grim, hopeless, joyless, sombre, uncomfortable, unpromising, windswept, wintry. *Opp* COMFORTABLE, WARM.

**bleary** *adj* blurred, *inf* blurry, cloudy, dim, filmy, fogged, foggy, fuzzy, hazy, indistinct, misty, murky, obscured, smeary, unclear, watery. *Opp* CLEAR.

**blemish** *n* blotch, blot, chip, crack, defect, deformity, disfigurement, eyesore, fault, flaw, imperfection, mark, mess, smudge, speck, stain, ugliness. □ *birthmark, blackhead, blister, callus, corn, freckle, mole, naevus, pimple, pustule, scar, spot, verruca, wart, whitlow,* sl *zit.* • *vb* deface, disfigure, flaw, mar, mark, scar, spoil, stain, tarnish.

**blend** *n* alloy, amalgam, amalgamation, combination, composite, compound, concoction, fusion, mélange, mix, mixture, synthesis, union. • *vb* **1** amalgamate, coalesce, combine, commingle, compound, fuse, harmonize, integrate, intermingle, intermix, meld, merge, mingle, synthesize, unite. **2** *blend in a bowl.* beat, mix, stir together, whip, whisk.

**bless** *vb* **1** anoint, consecrate, dedicate, grace, hallow, make sacred, ordain, sanctify. **2** *bless God's name.* adore, exalt, extol, glorify, magnify, praise. *Opp* CURSE.

**blessed** *adj* **1** adored, divine, hallowed, holy, revered, sacred, sanctified. **2** ▷ HAPPY.

**blessing** **1** benediction, consecration, grace, prayer. **2** approbation, approval, backing, concurrence, consent, leave, permission, sanction, support. **3** *Warmth is a blessing.* advantage, asset, benefit, boon, comfort, convenience, *inf* godsend, help. *Opp* CURSE, MISFORTUNE.

**blight** *n* affliction, ailment, *old use* bane, cancer, canker, curse, decay, disease, evil, illness, infestation, misfortune, *old use* pestilence, plague, pollution, rot, scourge, sickness, trouble. • *vb* ▷ SPOIL.

**blind** *adj* **1** blinded, eyeless, sightless, unseeing. □ *astigmatic, colour-blind, long-sighted, myopic, near-sighted, short-sighted, suffering from cataract, suffering from glaucoma, visually handicapped.* **2** *blind devotion.* blinkered, heedless, ignorant, inattentive, indifferent, indiscriminate, insensible, insensitive, irrational, mindless, oblivious, prejudiced, unaware, unobservant, unreasoning. • *n* awning, cover, curtain, screen, shade, shutters. • *vb* **1** dazzle, make blind. **2** ▷ DECEIVE.

**blink** *vb* coruscate, flash, flicker, flutter, gleam, glimmer, shimmer, sparkle, twinkle, wink.

**bliss** *n* blessedness, delight, ecstasy, euphoria, felicity, gladness, glee, happiness, heaven, joy, paradise, rapture. ▷ PLEASURE. *Opp* MISERY.

**bloated** *adj* dilated, distended, enlarged, inflated, puffy, swollen.

**block** *n* **1** bar, brick, cake, chock, chunk, hunk, ingot, lump, mass, piece, slab. **2** ▷ BLOCKAGE. • *vb* **1** bar, barricade, *inf* bung up, choke, clog, close, congest, constrict, dam, fill, impede, jam, obstruct, plug, stop up. **2** *block a plan.* deter, halt, hamper, hinder, hold back, prevent, prohibit, resist, *inf* scotch, *inf* stonewall, stop, thwart.

**blockage** *n* barrier, block, bottleneck, congestion, constriction, delay, hang-up, hindrance, impediment, jam, obstacle, obstruction, resistance, stoppage.

**blond, blonde** *adj* bleached, fair, flaxen, golden, light, platinum, silvery, yellow.

**bloodshed** *n* bloodletting, butchery, carnage, killing, massacre, murder, slaughter, slaying, violence.

**bloodthirsty** *adj* barbaric, brutal, feral, ferocious, fierce, homicidal, inhuman, murderous, pitiless, ruthless, sadistic, sanguinary, savage, vicious, violent, warlike. ▷ CRUEL. *Opp* HUMANE.

**bloody** *adj* **1** bleeding, bloodstained, raw. **2** *a bloody battle.* cruel, gory, sanguinary. ▷ BLOODTHIRSTY.

**bloom** *n* **1** blossom, bud, floret, flower. **2** *bloom of youth.* beauty, blush, flush, glow, prime. • *vb* be healthy, blossom, *poet* blow, bud, burgeon, *inf* come out, develop, flourish, flower, grow, open, prosper, sprout, thrive. *Opp* FADE.

**blot** *n* **1** blob, blotch, mark, smear, smirch, smudge, *inf* splodge, spot, stain. **2** *blot on the landscape.* blemish, defect, eyesore, fault, flaw, ugliness. • *vb* bespatter, blemish, blotch, blur, disfigure, mar, mark, smudge, spoil, spot, stain. **blot out** ▷ OBLITERATE. **blot your copybook** ▷ MISBEHAVE.

**blotchy** *adj* blemished, brindled, discoloured, inflamed, marked, patchy, smudged, spotty, streaked, uneven.

**blow** *n* **1** bang, bash, *inf* belt, *inf* biff, box (*ears*), buffet, bump, clip, clout, clump, concussion, hit, jolt, knock, punch, rap, slap, *inf* slosh, smack, *inf* sock, stroke, swat, swipe, thump, thwack, wallop, welt, whack, whop. **2** *a sad blow.* affliction, *inf* bombshell, calamity, disappointment, disaster, misfortune, shock, surprise, upset. • *vb* blast, breathe, exhale, fan, puff, waft, whine, whirl, whistle. **blow up 1** dilate, enlarge, expand, fill, inflate, pump up. **2** exaggerate, magnify, make worse, overstate. **3** blast, bomb, burst, detonate, dynamite, erupt, explode, go off, set off, shatter. **4** [*inf*] erupt, get angry, lose your temper, rage.

**blue** *adj* **1** aquamarine, azure, cerulean, cobalt, indigo, navy, sapphire, sky-blue, turquoise, ultramarine. **2** ▷ BAWDY. **3** ▷ SAD. • *vb* ▷ SQUANDER.

**blueprint** *n* basis, design, draft, model, outline, pattern, pilot, plan, project, proposal, prototype, scheme.

**bluff** *vb* cozen, deceive, delude, dupe, fool, hoodwink, mislead.

**blunder** *n inf* boob, *inf* botch, *inf* clanger, *sl* cock-up, error, fault, *Fr* faux pas, gaffe, howler, indiscretion, miscalculation, misjudgement, mistake, slip, slip-up, solecism. • *vb* be clumsy, *inf* botch up, bumble, bungle, *inf* drop a clanger, err, flounder, *inf* foul up, *sl* goof, go wrong, *inf* make a hash of something, make a mistake, mess up, miscalculate, misjudge, *inf* put your foot in it, slip up, stumble.

**blunt** *adj* **1** dull, rounded, thick, unpointed, unsharpened, worn. *Opp* SHARP. **2** *blunt criticism.* abrupt, bluff, brusque, candid, curt, direct, downright, forthright, frank, honest, insensitive, outspoken, plain-spoken, rude, straightforward, tactless, unceremonious, undiplomatic. *Opp* TACTFUL. • *vb* abate, allay, anaesthetize, dampen, deaden, desensitize, dull, lessen, numb, soften, take the edge off, weaken. *Opp* SHARPEN.

**blur** *vb* bedim, befog, blear, cloud, conceal, confuse, darken, dim, fog, mask, muddle, obscure, smear, unfocus.

**blurred** *adj* bleary, blurry, clouded, cloudy, confused, dim, faint, foggy, fuzzy, hazy, ill-defined, indefinite, indistinct, misty, nebulous, out of focus, smoky, unclear, unfocused, vague. *Opp* CLEAR.

**blurt** *vb* **blurt out** be indiscreet, *inf* blab, burst out with, come out with, cry out, disclose, divulge, exclaim, *inf* give the game away, let out, let slip, reveal, *inf* spill the beans, tell, utter.

**blush** *vb* be ashamed, colour, flush, glow, go red, redden.

**blustering** *adj* angry, boasting, boisterous, bragging, bullying, crowing, defiant, domineering, hectoring, noisy, ranting, self-assertive, showing-off, storming, swaggering, threatening, vaunting, violent. *Opp* MODEST.

**blustery** *adj* gusty, squally, unsettled, windy.

**board** *n* **1** blockboard, chipboard, clapboard, panel, plank, plywood, scantling, sheet, slab, slat, timber, weatherboard. **2** *board of directors.* cabinet, committee, council, directorate, jury, panel. • *vb* **1** accommodate, billet, feed, house, lodge, put up, quarter, stay. **2** *board a bus.* catch, embark (on), enter, get on, go on board.

**boast** *vb sl* be all mouth, *inf* blow your own trumpet, bluster, brag, crow, exaggerate, gloat, praise yourself, *sl* shoot a line, show off, *inf* sing your own praises, swagger, *inf* swank, *inf* talk big, vaunt.

**boaster** *n inf* big-head, *inf* big mouth, braggadocio, braggart, *inf* loudmouth, *inf* poser, show-off, swaggerer, swank.

**boastful** *adj inf* big-headed, bragging, *inf* cocky, conceited, egotistical, ostentatious, proud, puffed up, swaggering, swanky, swollen-headed, vain, vainglorious. *Opp* MODEST.

**boat** *n* craft, ship. ▷ VESSEL.

**boatman** *n* bargee, coxswain, ferryman, gondolier, lighterman, oarsman, rower, waterman, yachtsman. ▷ SAILOR.

**bob** *vb* be agitated, bounce, dance, hop, jerk, jig about, jolt, jump, leap, move about, nod, oscillate, shake, toss about, twitch. **bob up** ▷ APPEAR.

**body** *n* **1** anatomy, being, build, figure, form, frame, individual, physique, shape, substance, torso, trunk. **2** cadaver, carcass, corpse, mortal remains, mummy, relics, remains, *sl* stiff. **3** association, band, committee, company, corporation, society. ▷ GROUP. 4 *body of material*. accumulation, agglomeration, collection, corpus, mass.

**bodyguard** *n* defender, guard, minder, protector.

**bog** *n* fen, marsh, marshland, mire, morass, mudflats, peat bog, quagmire, quicksands, salt-marsh, *old use* slough, swamp, wetlands. **get bogged down** be hindered, get into difficulties, get stuck, grind to a halt, sink.

**bogus** *adj* counterfeit, fake, false, fictitious, fraudulent, imitation, *inf* phoney, sham, spurious. *Opp* GENUINE.

**Bohemian** *adj inf* arty, *old use* beatnik, bizarre, eccentric, hippie, informal, nonconformist, off-beat, unconventional, unorthodox, *inf* way-out, weird.

**boil** *n* abscess, blister, carbuncle, chilblain, eruption, gathering, gumboil, inflammation, pimple, pock, pustule, sore, spot, tumour, ulcer, *sl* zit. • *vb* **1** cook, heat, simmer, stew. **2** bubble, effervesce, foam, seethe, steam. **3** ▷ RAGE.

**boisterous** *adj* animated, disorderly, exuberant, irrepressible, lively, loud, noisy, obstreperous, riotous, rollicking, rough, rowdy, stormy, tempestuous, tumultuous, undisciplined, unrestrained, unruly, uproarious, wild. *Opp* CALM.

**bold** *adj* **1** adventurous, audacious, brave, confident, courageous, daredevil, daring, dauntless, enterprising, fearless, *derog* foolhardy, forceful, gallant, hardy, heroic, intrepid, *inf* plucky, *derog* rash, *derog* reckless, resolute, self-confident, unafraid, valiant, valorous, venturesome. **2** [*derog*] *a bold request*. brash, brazen, *inf* cheeky, forward, fresh, impertinent, impudent, insolent, pert, presumptuous, rude, saucy, shameless, unashamed. **3** *bold colours, writing*. big, bright, clear, conspicuous, eye-catching, large, obvious, prominent, pronounced, showy, striking, strong, vivid. *Opp* FAINT, TIMID.

**bolster** *n* cushion, pillow. • *vb* ▷ SUPPORT.

**bolt** *n* **1** arrow, dart, missile, projectile. **2** peg, pin, rivet, rod, screw. **3** *bolt on a door*. bar, catch, fastening, latch, lock. • *vb* **1** bar, close, fasten, latch, lock, secure. **2** *The animals bolted*. abscond, dart away, dash away, escape, flee, fly, run off, rush off. **3** *bolt food*. ▷ EAT. **bolt from the blue** ▷ SURPRISE.

**bomb** *n* bombshell, explosive. ▷ WEAPON. • *vb* ▷ BOMBARD.

**bombard** *vb* **1** assail, assault, attack, batter, blast, blitz, bomb, fire at, pelt, pound, shell, shoot at, strafe. **2** badger, beset, harass, importune, pester, plague.

**bombardment** *n* attack, barrage, blast, blitz, broadside, burst, cannonade, discharge, fusillade, hail, salvo, volley.

**bombastic** *adj* extravagant, grandiloquent, grandiose, high-flown, inflated, magniloquent, pompous, turgid.

**bond** *n* **1** chain, cord, fastening, fetters, handcuffs, manacles, restraints, rope, shackles. **2** *bond of friendship*. affiliation, affinity, attachment, connection, link, relationship, tie, unity. **3** *a legal bond*. agreement, compact, contract, covenant, guarantee, legal document, pledge, promise, word. • *vb* ▷ STICK.

**bondage** *n* enslavement, serfdom, servitude, slavery, subjection, thraldom, vassalage.

**bonus** *n* **1** bounty, commission, dividend, gift, gratuity, hand-out, honorarium, largesse, payment, *inf* perk, reward, supplement, tip. **2** addition, advantage, benefit, extra, *inf* plus.

**bony** *adj* angular, emaciated, gangling, gawky, lanky, lean, scraggy, scrawny, skinny, thin, ungainly. *Opp* GRACEFUL, PLUMP.

**book** *n* booklet, copy, edition, hardback, paperback, publication, tome, volume, work. □ *album, annual, anthology, atlas, bestiary,* old use *chap-book, compendium, concordance, diary, dictionary, digest, directory, encyclopaedia, fiction, gazetteer, guidebook, handbook, hymnal, hymn-book, jotter, ledger, lexicon, libretto, manual, manuscript, missal, notebook, omnibus, picture-book, prayer-book, primer, psalter, reading book, reference book,* music *score, scrap-book, scroll, sketch-book, text-book, thesaurus, vade mecum.* ▷ WRITING. • *vb* **1** *book for speeding.* arrest, take your name, write down details. **2** *book in advance.* arrange, buy, engage, order, organize, reserve, sign up.

**booklet** *n* brochure, leaflet, pamphlet, paperback.

**boom** *n* **1** bang, blast, crash, explosion, reverberation, roar, rumble. ▷ SOUND. **2** *boom in trade.* bonanza, boost, expansion, growth, improvement, increase, spurt, upsurge, upturn. ▷ PROSPERITY. **3** *boom across a river.* ▷ BARRIER. • *vb* **1** ▷ SOUND. **2** ▷ PROSPER.

**boorish** *adj* barbarian, ignorant, ill-bred, ill-mannered, loutish, oafish, philistine, uncultured, vulgar. *Opp* CULTURED.

**boost** *n* aid, encouragement, fillip, impetus, help, lift, push, stimulus. • *vb* advance, aid, assist, augment, bolster, build up, buoy up, encourage, enhance, enlarge, expand, foster, further, give an impetus to, heighten, help, improve, increase, inspire, lift, promote, push up, raise, support, sustain. *Opp* DEPRESS.

**booth** *n* box, carrel, compartment, cubicle, hut, kiosk, stall, stand.

**booty** *n* contraband, gains, haul, loot, pickings, pillage, plunder, spoils, *inf* swag, takings, trophies, winnings.

**border** *n* **1** brim, brink, edge, edging, frame, frieze, frill, fringe, hem, margin, perimeter, periphery, rim, surround, verge. **2** borderline, boundary, frontier, limit. **3** *flower border.* bed, herbaceous border. • *vb* abut on, adjoin, be adjacent to, be alongside, join, share a border with, touch.

**bore** *vb* **1** burrow, drill, mine, penetrate, sink, tunnel. ▷ PIERCE. **2** *bore listeners.* alienate, depress, jade, *inf* leave cold, tire, *inf* turn off, weary. *Opp* INTEREST.

**boring** *adj* arid, commonplace, dead, dreary, dry, dull, flat, humdrum, long-winded, monotonous, prolix, repetitious, repetitive, soporific, stale, tedious, tiresome, trite, uneventful, unexciting, uninspiring, uninteresting, vapid, wearisome, wordy. *Opp* INTERESTING.

**born** *adj* congenital, genuine, instinctive, natural, untaught.

**borrow** *vb* adopt, appropriate, be lent, *inf* cadge, copy, crib, make use of, obtain, pirate, plagiarize, *inf* scrounge, *inf* sponge, take, use, usurp. *Opp* LEND.

**boss** *n* employer, head. ▷ CHIEF.

**bossy** *adj* aggressive, assertive, authoritarian, autocratic, bullying, despotic, dictatorial, domineering, exacting, hectoring, high-handed, imperious, lordly, magisterial, masterful, officious, oppressive, overbearing, peremptory, *inf* pushy, self-assertive, tyrannical. *Opp* SERVILE.

**bother** *n* **1** ado, difficulty, disorder, disturbance, fuss, *inf* hassle, problem, *inf* to-do. **2** annoyance, inconvenience, irritation, nuisance, pest, trouble, worry. • *vb* **1** annoy, bewilder, concern, confuse, disconcert, dismay, disturb, exasperate, harass, *inf* hassle, inconvenience, irk, irritate, molest, nag, perturb, pester, plague, trouble, upset, vex, worry. **2** be concerned, be worried, care, mind, take trouble. ▷ TROUBLE.

**bottle** *n* flask. □ *carafe, carboy, decanter, flagon, jar, jeroboam, magnum, phial, pitcher, vial, wine-bottle.* **bottle up** ▷ SUPPRESS.

**bottom** *adj* deepest, least, lowest, minimum. • *n* **1** base, bed, depth, floor, foot, foundation, lowest point, nadir, pedestal, substructure, underneath, underside. *Opp* TOP. **2** basis, essence, grounds, heart, origin, root, source. **3** *your bottom. vulg* arse, backside, behind, *inf* bum, buttocks, *joc* posterior, rear, rump, seat, *inf* sit-upon.

**bottomless** *adj* deep, immeasurable, unfathomable, unplumbable.

**bounce** *vb* bob, bound, bump, jump, leap, move about, rebound, recoil, ricochet, spring.

**bound** *adj* **1** certain, committed, compelled, constrained, destined, doomed, duty-bound, fated, forced, obligated, obliged, pledged, required, sure. **2** *bound with rope.* ▷ BIND. • *vb* bob, bounce, caper, frisk, frolic, gambol, hop, hurdle, jump, leap, pounce, romp, skip, spring, vault. **bound for** aimed at, directed towards, going to, heading for, making for, off to, travelling towards.

**boundary** *n* border, borderline, bounds, brink, circumference, confines, demarcation, edge, end, extremity, fringe, frontier, interface, limit, margin, perimeter, threshold, verge.

**boundless** *adj* endless, everlasting, immeasurable, incalculable, inexhaustible, infinite, limitless, unbounded, unconfined, unflagging, unlimited, unrestricted, untold. ▷ VAST. *Opp* FINITE.

**bounty** *n* alms, altruism, beneficence, benevolence, charity, generosity, giving, goodness, kindness, largesse, liberality, munificence, philanthropy, unselfishness.

**bouquet** *n* **1** arrangement, bunch, buttonhole, corsage, garland, nosegay, posy, spray, wreath. **2** *bouquet of wine.* ▷ SMELL.

**bout** *n* **1** attack, fit, period, run, spell, stint, stretch, time, turn. **2** battle, combat, competition, contest, encounter, engagement, fight, match, round, *inf* set-to, struggle.

**bow** *vb* **1** bend, bob, curtsy, genuflect, incline, kowtow, nod, prostrate yourself, salaam, stoop. **2** ▷ SUBMIT.

**bowels** *n* **1** entrails, guts, *inf* innards, insides, intestines, viscera, vitals. **2** core, depths, heart, inside.

**bower** *n* alcove, arbour, bay, gazebo, grotto, hideaway, pavilion, pergola, recess, retreat, sanctuary, shelter, summer-house.

**bowl** *n* basin, bath, casserole, container, dish, pan, pie-dish, tureen. • *vb* fling, hurl, lob, pitch, throw, toss.

**box** *n* carton, case, chest, container, crate. □ *bin, caddy, canister, cartridge, casket, coffer, coffin, pack, package, punnet, tea-chest, tin, trunk.* • *vb inf* engage in fisticuffs, fight, punch, scrap, spar. ▷ HIT.

**boxer** *n* prize-fighter, sparring partner. □ *bantamweight, cruiserweight, featherweight, flyweight, heavyweight, lightweight, middleweight, welterweight.*

**boy** *n derog* brat, *inf* kid, lad, schoolboy, son, *derog* stripling, *derog* urchin, youngster, youth.

**boycott** *n* ban, blacklist, embargo, prohibition. • *vb* avoid, black, blackball, blacklist, exclude, *inf* give the cold-shoulder to, ignore, make unwelcome, ostracize, outlaw, prohibit, spurn, stay away from.

**bracing** *adj* crisp, exhilarating, health-giving, invigorating, refreshing, restorative, stimulating, tonic.

**brag** *vb* crow, gloat, show off. ▷ BOAST.

**brain** *n* cerebrum, *inf* grey matter, intellect, intelligence, mind, *inf* nous, reason, sense, understanding, wisdom, wit.

**brainwash** *vb* condition, indoctrinate, re-educate.

**branch** *n* **1** arm, bough, limb, prong, shoot, sprig, stem, twig. **2** department, division, office, offshoot, part, ramification, section, subdivision, wing. • *vb* diverge, divide, fork, ramify, split, subdivide. **branch out** ▷ DIVERSIFY.

**brand** *n* kind, label, line, make, sort, trademark, type, variety. • *vb* **1** burn, identify, label, mark, scar, stamp, tag. **2** censure, characterize, denounce, discredit, stigmatize, vilify.

**brash** *adj* brazen, bumptious, insolent, rash, reckless, rude, self-assertive. ▷ ARROGANT.

**bravado** *n* arrogance, bluster, braggadocio, machismo, swagger.

**brave** *adj* adventurous, audacious, bold, chivalrous, cool, courageous, daring, dauntless, determined, fearless, gallant, game, *sl* gutsy, heroic, indomitable, intrepid, lion-hearted, *derog* macho, noble, *inf* plucky, resolute, spirited, stalwart, stoical, stout-hearted, tough, unafraid, uncomplaining, undaunted, unshrinking, valiant, valorous, venturesome. *Opp* COWARDLY.

**bravery** *n* audacity, boldness, *sl* bottle, courage, daring, dauntlessness, determination, fearlessness, fibre, firmness, fortitude, gallantry, *inf* grit, *inf* guts, heroism, intrepidity, mettle, *inf* nerve, *inf* pluck, prowess, resolution, spirit, *sl* spunk, stoicism, tenacity, valour, will-power. *Opp* COWARDICE.

**brawl** *n* affray, altercation, *inf* bust-up, clash, *inf* dust-up, fracas, fray, *inf* free-for-all, mêlée, *inf* punch-up, quarrel, row, scrap, scuffle, *inf* set-to, tussle. • *vb* ▷ FIGHT.

**brazen** *adj* barefaced, blatant, cheeky, defiant, flagrant, impertinent, impudent, insolent, rude, shameless, unabashed, unashamed. *Opp* SHAMEFACED.

**breach** *n* **1** aperture, break, chasm, crack, fissure, gap, hole, opening, rent, space, split. **2** alienation, difference, disagreement, divorce, drifting apart, estrangement, quarrel, rift, rupture, schism, separation, split.

3 *breach of law*. contravention, failure, infringement, offence, transgression, violation.

**bread** *n* □ *brioche, cob, croissant, French bread, loaf, roll, stick of bread, toast*. ▷ FOOD.

**break** *n* 1 breach, breakage, burst, chink, cleft, crack, crevice, cut, fissure, fracture, gap, gash, hole, leak, opening, rent, rift, rupture, slit, split, tear. 2 *break from work*. *inf* breather, breathing-space, hiatus, interlude, intermission, interval, *inf* let-up, lull, pause, respite, rest, tea-break. 3 *break in service*. disruption, halt, interruption, lapse, suspension. • *vb* 1 breach, burst, *inf* bust, chip, crack, crumple, crush, damage, demolish, fracture, fragment, knock down, ruin, shatter, shiver, smash, *inf* smash to smithereens, snap, splinter, split, squash, wreck. ▷ DESTROY. 2 *break the law*. contravene, defy, disobey, disregard, fail to observe, flout, go back on, infringe, transgress, violate. 3 *break a record*. beat, better, do more than, exceed, excel, go beyond, outdo, outstrip, pass, surpass. **break down** ▷ ANALYSE, DEMOLISH. **break in** ▷ INTERRUPT, INTRUDE. **break off** ▷ FINISH. **break out** ▷ ESCAPE. **break through** ▷ PENETRATE. **break up** ▷ DISINTEGRATE.

**breakdown** *n* 1 collapse, destruction, disintegration,downfall, failure, fault, hitch, malfunction, ruin, stoppage. 2 analysis, classification, detailing, dissection, itemization, *inf* rundown.

**breakthrough** *n* advance, development, discovery, find, improvement, innovation, invention, leap forward, progress, revolution, success.

**breakwater** *n* groyne, jetty, mole, pier, sea-defence.

**breath** *n* breeze, gust, murmur, pant, puff, sigh, stir, waft, whiff, whisper.

**breathe** *vb* 1 exhale, inhale, pant, puff, respire, suspire. 2 hint, let out, tell, whisper.

**breathless** *adj* exhausted, gasping, out of breath, panting, *inf* puffed, *inf* puffing and blowing, tired out, wheezy, winded.

**breed** *n* ancestry, clan, family, kind, line, lineage, nation, pedigree, progeny, race, sort, species, stock, strain, type, variety. • *vb* 1 bear young ones, beget young ones, increase, multiply, procreate, produce young, propagate (*plants*), raise young ones, reproduce. 2 *breed contempt*. arouse, cause, create, cultivate, develop, engender, foster, generate, induce, nourish, nurture, occasion.

**breeze** *n* air-current, breath, draught, waft, wind, *poet* zephyr.

**breezy** *adj* airy, *inf* blowy, draughty, fresh, gusty, windy.

**brevity** *n* briefness, compactness, compression, conciseness, concision, curtness, economy, incisiveness, pithiness, shortness, succinctness, terseness.

**brew** *n* blend, compound, concoction, drink, hash, infusion, liquor, mixture, potion, preparation, punch, stew. • *vb* 1 boil, cook, ferment, infuse, make, simmer, steep, stew. 2 *brew mischief*. concoct, contrive, *inf* cook up, develop, devise, foment, hatch, plan, plot, prepare, scheme, stir up.

**bribe** *n sl* backhander, bribery, *inf* carrot, enticement, *sl* graft, gratuity, incentive, inducement, *inf* payola, protection money, *inf* sweetener, tip. • *vb* buy off, corrupt, entice, *inf* grease your palm, influence, offer a bribe, pervert, reward, suborn, tempt, tip.

**brick** *n* block, breeze-block, cube, set, sett, stone.

**bridge** *n* arch, connection, crossing, link, span, way over. □ *aqueduct, Bailey bridge, drawbridge, flyover, footbridge, overpass, pontoon bridge, suspension bridge, swing bridge, viaduct.* • *vb* connect, cross, fill, join, link, pass over, span, straddle, tie together, traverse, unite.

**bridle** *vb* check, control, curb, restrain.

**brief** *adj* **1** cursory, ephemeral, evanescent, fast, fleeting, hasty, limited, little, momentary, passing, quick, sharp, short, short-lived, temporary, transient, transitory. **2** *brief comment.* abbreviated, abridged, compact, compendious, compressed, concise, condensed, crisp, curt, curtailed, incisive, laconic, pithy, shortened, succinct, terse, thumbnail, to the point. *Opp* LONG. • *n* **1** advice, briefing, data, description, directions, information, instructions, orders, outline, plan. **2** *a barrister's brief.* argument, case, defence, dossier, summary. • *vb* advise, coach, direct, enlighten, *inf* fill someone in, give someone the facts, guide, inform, instruct, prepare, prime, *inf* put someone in the picture.

**briefs** *n* camiknickers, knickers, panties, pants, shorts, trunks, underpants.

**brigand** *n* bandit, buccaneer, desperado, footpad, gangster, highwayman, marauder, outlaw, pirate, robber, ruffian, thief.

**bright** *adj* **1** ablaze, aglow, alight, beaming, blazing, burnished, colourful, dazzling, flashing, *derog* flashy, fresh, *derog* gaudy, glaring, gleaming, glistening, glittering, glossy, glowing, incandescent, lambent, light, luminous, lustrous, pellucid, polished, radiant, refulgent, resplendent, scintillating, shimmering, shining, shiny, showy, sparkling, twinkling, vivid. **2** *bright sky.* clear, cloudless, fair, sunny. **3** *bright prospects.* auspicious, favourable, good, hopeful, optimistic, rosy. **4** *a bright smile.* ▷ CHEERFUL. **5** *bright ideas.* ▷ CLEVER. *Opp* DULL.

**brighten** *vb* **1** cheer (up), enliven, gladden, illuminate, light up, liven up, *inf* perk up, revitalize, smarten up. **2** *The sky brightened.* become sunny, clear up, lighten.

**brilliant** *adj* **1** coruscating, dazzling, glaring, glittering, glorious, intense, resplendent, scintillating, shining, showy, sparkling, splendid, vivid. ▷ BRIGHT. *Opp* DULL. **2** [*inf*] *a brilliant game.* ▷ EXCELLENT.

**brim** *n* brink, circumference, edge, limit, lip, margin, perimeter, periphery, rim, top, verge.

**bring** *vb* **1** bear, carry, convey, deliver, fetch, take, transfer, transport. **2** *bring a friend.* accompany, conduct, escort, guide, lead, usher. **3** *The play brought great applause.* attract, cause, create, draw, earn, engender, generate, get, give rise to, induce, lead to, occasion, produce, prompt, provoke, result in. **bring about** ▷ CREATE. **bring in** ▷ EARN, INTRODUCE. **bring off** ▷ ACHIEVE. **bring on** ▷ ACCELERATE, CAUSE. **bring out** ▷ EMPHASIZE, PRODUCE. **bring up** ▷ EDUCATE, RAISE.

**brink** *n* bank, border, boundary, brim, circumference, edge, fringe, limit, lip, margin, perimeter, periphery, rim, skirt, threshold, verge.

**brisk** *adj* 1 active, alert, animated, bright, businesslike, bustling, busy, crisp, decisive, energetic, fast, keen, lively, nimble, quick, rapid, *inf* snappy, *inf* spanking (*pace*), speedy, spirited, sprightly, spry, vigorous. *Opp* LEISURELY. 2 *a brisk wind.* bracing, enlivening, fresh, invigorating, refreshing, stimulating.

**bristle** *n* barb, hair, prickle, quill, spine, stubble, thorn, whisker, wire. • *vb* become angry, become defensive, become indignant, bridle, flare up.

**brittle** *adj* breakable, crackly, crisp, crumbling, delicate, easily broken, fragile, frail, frangible, weak. *Opp* FLEXIBLE, RESILIENT.

**broad** *adj* 1 ample, capacious, expansive, extensive, great, large, open, roomy, spacious, sweeping, vast, wide. 2 *broad daylight.* clear, full, open, plain, undisguised. 3 *broad outline.* general, imprecise, indefinite, inexact, non-specific, sweeping, undetailed, vague. 4 *broad tastes.* all-embracing, catholic, comprehensive, eclectic, encyclopaedic, universal, wide-ranging. 5 *broad humour.* bawdy, *sl* blue, coarse, earthy, improper, impure, indecent, indelicate, racy, ribald, suggestive, vulgar. ▷ BROAD-MINDED. *Opp* FINITE, NARROW.

**broadcast** *n* programme, relay, show, telecast, transmission. • *vb* 1 advertise, announce, circulate, disseminate, make known, make public, proclaim, promulgate, publish, relay, report, send out, spread about, televise, transmit. 2 *broadcast seed.* scatter, sow at random.

**broadcaster** *n* anchor-man, announcer, commentator, compère, disc jockey, DJ, linkman, newsreader, presenter. ▷ ENTERTAINER.

**broaden** *vb* branch out, build up, develop, diversify, enlarge, expand, extend, increase, open up, spread, widen. *Opp* LIMIT.

**broad-minded** *adj* all-embracing, balanced, broad, catholic, comprehensive, cosmopolitan, eclectic, enlightened, liberal, open-minded, permissive, tolerant, unbiased, unbigoted, unprejudiced, unshockable. *Opp* NARROW-MINDED.

**brochure** *n* booklet, broadsheet, catalogue, circular, folder, handbill, leaflet, pamphlet, prospectus, tract.

**brooch** *n* badge, clasp, clip, fastening.

**brood** *n* children, clutch (*of eggs*), family, issue, litter, offspring, progeny, young. • *vb* 1 hatch, incubate, sit on. 2 *brood over mistakes.* agonize, dwell (on), *inf* eat your heart out, fret, mope, sulk, worry. ▷ THINK.

**brook** *n* beck, burn, channel, *poet* rill, rivulet, runnel, stream, watercourse. • *vb* ▷ TOLERATE.

**browbeat** *vb* badger, bully, coerce, cow, hector, intimidate, tyrannize. ▷ FRIGHTEN.

**brown** *adj* beige, bronze, buff, chestnut, chocolate, dun, fawn, khaki, ochre, russet, sepia, tan, tawny, terracotta, umber. • *vb* bronze, burn, colour, grill, tan, toast.

**browse** *vb* 1 crop grass, eat, feed, graze, pasture. 2 *browse in a book.* dip in, flick through, leaf through, look through, peruse, read here and there, scan, skim, thumb through.

**bruise** *n* black eye, bump, contusion, discoloration, *inf* shiner, welt. • *vb* blacken, crush, damage, discolour, injure, knock, mark. ▷ WOUND.

**brush** *n* 1 besom, broom. 2 *brush with police.* ▷ CONFLICT. • *vb* 1 comb, groom, scrub, sweep, tidy, whisk. 2 *just brushed the gatepost.* graze, touch. **brush aside** ▷ DISMISS, DISREGARD. **brush-off** ▷ REBUFF. **brush up** ▷ REVISE.

**brutal** *adj* atrocious, barbaric, barbarous, beastly, bestial, bloodthirsty, bloody, brutish, callous, cold-blooded, cruel, dehumanized, ferocious, hard-hearted, heartless, inhuman, inhumane, merciless, murderous, pitiless, remorseless, ruthless, sadistic, savage, uncivilized, unfeeling, vicious, violent, wild. ▷ UNKIND. *Opp* HUMANE.

**brutalize** *vb* dehumanize, harden, inure, make brutal.

**brute** *adj* crude, irrational, mindless, physical, rough, stupid, unfeeling, unthinking. ▷ BRUTISH. • *n* 1 beast, creature, dumb animal. ▷ ANIMAL. 2 [*inf*] *a cruel brute.* barbarian, bully, devil, lout, monster, ruffian, sadist, savage, swine.

**brutish** *adj* animal, barbaric, barbarous, beastly, bestial, boorish, brutal, coarse, cold-blooded, crude, cruel, *inf* gross, inhuman, insensitive, loutish, mindless, savage, senseless, stupid, subhuman, uncouth, unintelligent, unthinking. *Opp* HUMANE.

**bubble** *n* air-pocket, blister, hollow, vesicle. • *vb* boil, effervesce, fizz, fizzle, foam, froth, gurgle, seethe, sparkle. **bubbles** effervescence, fizz, foam, froth, head, lather, suds.

**bubbly** *adj* carbonated, effervescent, fizzy, foaming, seething, sparkling. ▷ LIVELY.

**buccaneer** *n* adventurer, bandit, brigand, corsair, marauder, pirate, privateer, robber.

**bucket** *n* can, pail, scuttle, tub.

**buckle** *n* catch, clasp, clip, fastener, fastening, hasp. • *vb* 1 clasp, clip, do up, fasten, hitch up, hook up, secure. 2 bend, bulge, cave in, collapse, contort, crumple, curve, dent, distort, fold, twist, warp.

**bud** *n* shoot, sprout. • *vb* begin to grow, burgeon, develop, shoot, sprout. **budding** ▷ POTENTIAL, PROMISING.

**budge** *vb* 1 change position, give way, move, shift, stir, yield. 2 *can't budge him.* alter, change, dislodge, influence, move, persuade, propel, push, remove, shift, sway.

**budget** *n* accounts, allocation of funds, allowance, estimate, financial planning, funds, means, resources. • *vb* allocate money, allot resources, allow (for), estimate expenditure, plan your spending, provide (for), ration your spending.

**buff** *n* ▷ ENTHUSIAST. • *vb* burnish, clean, polish, rub, shine, smooth.

**buffer** *n* bulwark, bumper, cushion, fender, pad, safeguard, screen, shield, shock-absorber.

**buffet** *n* 1 bar, café, cafeteria, counter, snack-bar. 2 *a stand-up buffet.* ▷ MEAL. • *vb* ▷ HIT.

**bug** *n* 1 ▷ INSECT, MICROBE. 2 *bug in a computer program.* breakdown, defect, error, failing, fault, flaw, *inf* gremlin, imperfection, malfunction, mistake, *inf* snarl-up, virus. • *vb* 1 intercept, interfere with, listen in to, spy on, tap. 2 [*sl*] *Untidiness bugs me.* ▷ ANNOY.

**build** *vb* assemble, construct, develop, erect, fabricate, form, found, *inf* knock together, make, put together, put up, raise, rear, set up. **build up** ▷ INTENSIFY.

**builder** *n* bricklayer, construction worker, labourer.

**building** *n* construction, edifice, erection, piece of architecture, *inf* pile, premises, structure. □ *arcade, barn, barracks, basilica, boat-house, bungalow, cabin, castle, cathedral, chapel, chateau, church, cinema, college, complex, cottage, dovecote, factory, farm-house, flats, fort, fortress, garage, gazebo, gymnasium, hall, hangar, hotel, house, inn, library, light-house, mansion, mausoleum, mill, monastery, monument, mosque, museum, observatory, outbuilding, outhouse, pagoda, palace, pavilion, pier, power-station, prison, pub, public house, restaurant, school, shed, shop, silo, skyscraper, stable, storehouse, studio, summer-house, synagogue, temple, theatre, tower, villa, warehouse, windmill.*

**bulb** *n* **1** corm, tuber. □ *amaryllis, bluebell, crocus, daffodil, freesia, hyacinth, lily, narcissus, snowdrop, tulip.* **2** *electric bulb.* lamp, light.

**bulbous** *adj* bloated, bulging, convex, distended, ovoid, pear-shaped, pot-bellied, rotund, rounded, spherical, swollen, tuberous.

**bulge** *n* bump, distension, hump, knob, lump, projection, protrusion, protuberance, rise, swelling. • *vb* belly, billow, dilate, distend, enlarge, expand, project, protrude, stick out, swell.

**bulk** *n* **1** amplitude, bigness, body, dimensions, extent, immensity, largeness, magnitude, mass, size, substance, volume, weight. **2** *the bulk of the work. inf* best part, greater part, majority, preponderance.

**bulky** *adj* awkward, chunky, cumbersome, large, unwieldy. ▷ BIG.

**bulletin** *n* account, announcement, communication, communiqué, dispatch, message, news-flash, notice, proclamation, report, statement.

**bullion** *n* bar, ingot, nugget, solid gold, solid silver.

**bull's-eye** *n* bull, centre, mark, middle, target.

**bully** *vb* bludgeon, browbeat, coerce, cow, domineer, frighten, harass, hector, intimidate, oppress, persecute, *inf* pick on, *inf* push around, terrorize, threaten, torment, tyrannize.

**bulwark** *n* defence, earthwork, fortification, parapet, protection, rampart, redoubt, wall. ▷ BARRIER.

**bump** *n* **1** bang, blow, buffet, collision, crash, knock, smash, thud, thump. **2** bulge, distension, hump, knob, lump, projection, protrusion, protuberance, rise, swelling, tumescence, welt. • *vb* **1** bang, collide with, crash into, jar, knock, ram, slam, smash into, strike, thump, wallop. ▷ HIT. **2** bounce, jerk, jolt, shake. **bump into** ▷ MEET. **bump off** ▷ KILL.

**bumptious** *adj* arrogant, *inf* big-headed, boastful, brash, *inf* cocky, conceited, egotistic, forward, immodest, officious, over-bearing, over-confident, pompous, presumptuous, pretentious, *inf* pushy, self-assertive, self-important, smug, *inf* stuck-up, *inf* snooty, swaggering, vain, vain-glorious, vaunting. *Opp* MODEST.

**bumpy** *adj* **1** bouncy, jarring, jerky, jolting. **2** *a bumpy road.* broken, irregular, jagged, knobbly, lumpy, pitted, rocky, rough, rutted, stony, uneven. *Opp* SMOOTH.

**bunch** *n* **1** batch, bundle, clump, cluster, collection, heap, lot, number, pack, quantity, set, sheaf, tuft. 2 *bunch of flowers.* bouquet, posy, spray. 3 [*inf*] *bunch of friends.* band, crowd, gang, gathering, mob, party, team, troop. ▷ GROUP. • *vb* assemble, cluster, collect, congregate, crowd, flock, gather, group, herd, huddle, mass, pack. *Opp* DISPERSE.

**bundle** *n* bag, bale, bunch, carton, collection, pack, package, packet, parcel, sheaf, truss. • *vb* bale, bind, enclose, fasten, pack, package, roll, tie, truss, wrap. **bundle out** ▷ EJECT.

**bung** *n* cork, plug, stopper. • *vb* ▷ THROW.

**bungle** *vb* blunder, botch, *sl* cock up, *inf* foul up, fluff, *inf* make a hash of, *inf* make a mess of, *inf* mess up, mismanage, *inf* muck up, *inf* muff, ruin, *inf* screw up, spoil.

**buoy** *n* beacon, float, marker, mooring buoy, signal. • *vb* **buoy up** ▷ RAISE.

**buoyant** *adj* 1 floating, light. 2 *a buoyant mood.* ▷ CHEERFUL.

**burden** *n* 1 cargo, encumbrance, load, weight. 2 *burden of guilt.* affliction, albatross, anxiety, care, cross, duty, handicap, millstone, obligation, onus, problem, responsibility, sorrow, trial, trouble, worry. • *vb* afflict, bother, encumber, hamper, handicap, impose on, load (with), *inf* lumber (with), oppress, overload (with), *inf* saddle (with), strain, tax, trouble, weigh down, worry.

**burdensome** *adj* bothersome, difficult, exacting, hard, heavy, onerous, oppressive, taxing, tiring, troublesome, trying, wearisome, wearying, weighty, worrying. *Opp* EASY.

**bureau** 1 desk, writing-desk. 2 *travel bureau.* agency, counter, department, office, service.

**bureaucracy** *n* administration, government, officialdom, paperwork, *inf* red tape, regulations.

**burglar** *n* cat-burglar, housebreaker, intruder, robber. ▷ THIEF.

**burglary** *n* break-in, forcible entry, house-breaking, larceny, pilfering, robbery, stealing, theft, thieving.

**burgle** *vb* break in, pilfer, rob. ▷ STEAL.

**burial** *n* entombment, funeral, interment, obsequies.

**burlesque** *n* caricature, imitation, mockery, parody, pastiche, satire, *inf* send-up, spoof, *inf* take-off, travesty.

**burly** *adj* athletic, beefy, brawny, heavy, hefty, hulking, husky, muscular, powerful, stocky, stout, *inf* strapping, strong, sturdy, thickset, tough, well-built. *Opp* THIN.

**burn** *n* blister, charring. • *vb* 1 be alight, blaze, flame, flare, flash, flicker, glow, smoke, smoulder, spark, sparkle. 2 carbonize, consume, cremate, destroy by fire, ignite, incinerate, kindle, light, reduce to ashes, set fire to, set on fire. 3 *burn your skin.* blister, brand, char, scald, scorch, sear, shrivel, singe, sting, toast. ▷ FIRE, HEAT.

**burning** *adj* 1 ablaze, afire, aflame, alight, blazing, flaming, glowing, incandescent, lit up, on fire, raging, smouldering. 2 *burning pain.* biting, blistering, boiling, fiery, hot, inflamed, scalding, scorching, searing, smarting, stinging. 3 *burning chemicals.* acid, caustic, corrosive. 4 *a burning smell.* acrid, pungent, reeking, scorching, smoky. **5** *a burning desire.* acute, ardent, consuming, eager, fervent, flaming, frenzied, heated, impassioned, intense, passionate, red-hot, vehement. **6** *a burning issue.* crucial, important, pertinent, pressing, relevant, urgent, vital.

**burrow** *n* earth, excavation, hole, retreat, set, shelter, tunnel, warren. • *vb* delve, dig, excavate, mine, tunnel.

**burst** *vb* **1** break, crack, disintegrate, erupt, explode, force open, give way, open suddenly, part suddenly, puncture, rupture, shatter, split, tear. **2** ▷ RUSH.

**bury** *vb* cover, embed, enclose, engulf, entomb, immerse, implant, insert, inter, lay to rest, plant, put away, secrete, sink, submerge. ▷ HIDE.

**bus** *n old use* charabanc, coach, double-decker, minibus, *old use* omnibus.

**bushy** *adj* bristling, bristly, dense, fluffy, fuzzy, hairy, luxuriant, rough, shaggy, spreading, sticking out, tangled, thick, thick-growing, unruly, untidy.

**business** *n* **1** affair, concern, duty, function, issue, matter, obligation, problem, question, responsibility, subject, task, topic. **2** calling, career, craft, employment, industry, job, line of work, occupation, profession, pursuit, trade, vocation, work. **3** buying and selling, commerce, dealings, industry, marketing, merchandising, selling, trade, trading, transactions. **4** company, concern, corporation, enterprise, establishment, firm, organization, *inf* outfit, partnership, practice, *inf* set-up, venture.

**businesslike** *adj* careful, efficient, hard-headed, logical, methodical, neat, orderly, practical, professional, prompt, systematic, well-organized. *Opp* DISORGANIZED.

**businessman, businesswoman** *ns* dealer, entrepreneur, executive, financier, industrialist, magnate, manager, merchant, trader, tycoon.

**bustle** *n* activity, agitation, commotion, excitement, flurry, fuss, haste, hurly-burly, hurry, hustle, movement, restlessness, scurry, stir, *inf* to-do, *inf* toing and froing. • *vb* dart, dash, fuss, hasten, hurry, hustle, make haste, move busily, rush, scamper, scramble, scurry, scuttle, *inf* tear, whirl.

**busy** *adj* **1** active, assiduous, bustling about, committed, dedicated, diligent, employed, energetic, engaged, engrossed, *inf* hard at it, immersed, industrious, involved, keen, occupied, *inf* on the go, pottering, preoccupied, slaving, *inf* tied up, tireless, *inf* up to your eyes, working. *Opp* IDLE. **2** *busy shops.* bustling, frantic, full, hectic, lively.

**busybody** *n* gossip, meddler, *inf* Nosey Parker, scandalmonger, snooper, spy. **be a busybody** ▷ INTERFERE.

**butt** *n* **1** haft, handle, shaft, stock. **2** *water butt.* barrel, cask, water-butt. **3** *cigar butt.* end, remains, remnant, stub. **4** *butt of ridicule.* end, mark, object, subject, target, victim. • *vb* buffet, bump, jab, knock, poke, prod, punch, push, ram, shove, strike, thump. ▷ HIT. **butt in** ▷ INTERRUPT.

**buttocks** *n vulg* arse, backside, behind, bottom, *vulg* bum, *Amer* butt, *Fr* derrière, fundament, haunches, hindquarters, *joc* posterior, rear, rump, seat.

**buttress** *n* pier, prop, support. • *vb* brace, prop up, reinforce, shore up, strengthen, support.

**buxom** *adj* ample, bosomy, *vulg* chesty, full-figured, healthy-looking, plump, robust, rounded, voluptuous. *Opp* THIN.

**buy** *vb* acquire, come by, gain, get, get on hire purchase, *inf* invest in, obtain, pay for, procure, purchase. *Opp* SELL.

**buyer** *n* client, consumer, customer, puchaser, shopper.

**bypass** *vb* avoid, circumvent, dodge, evade, find a way round, get out of, go round, ignore, neglect, omit, sidestep, skirt.

**by-product** *n* adjunct, complement, consequence, corollary, repercussion, result, side-effect.

**bystander** *n* eyewitness, looker-on, observer, onlooker, passer-by, spectator, watcher, witness.

# C

**cabin** *n* **1** bothy, chalet, cottage, hut, lodge, shack, shanty, shed, shelter. **2** *cabin on a ship.* berth, compartment, deck-house, quarters.

**cable** *n* **1** chain, cord, flex, guy, hawser, lead, line, mooring, rope, wire. **2** *news by cable.* message, telegram, wire.

**cacophonous** *adj* atonal, discordant, dissonant, harsh, noisy, unmusical. *Opp* HARMONIOUS.

**cacophony** *n* atonality, caterwauling, din, discord, disharmony, dissonance, harshness, jangle, noise, racket, row, rumpus, tumult. *Opp* HARMONY.

**cadence** *n* accent, beat, inflection, intonation, lilt, metre, pattern, rhythm, rise and fall, sound, stress, tune.

**cadet** *n* beginner, learner, recruit, tiro, trainee.

**cadge** *vb* ask, beg, scrounge, sponge.

**café** *n* bar, bistro, brasserie, buffet, cafeteria, canteen, coffee bar, coffee house, coffee shop, diner, restaurant, snack-bar, take-away, tea-room, tea-shop.

**cage** *n* aviary, coop, enclosure, hutch, pen, pound. • *vb* ▷ CONFINE.

**cajole** *vb inf* butter up, coax, flatter, inveigle, persuade, seduce.

**cake** *n* **1** bun, gateau. **2** *cake of soap.* bar, block, chunk, cube, loaf, lump, mass, piece, slab. • *vb* **1** coat, clog, cover, encrust, make dirty, make muddy. **2** coagulate, congeal, consolidate, dry, harden, solidify, thicken.

**calamitous** *adj* awful, cataclysmic, catastrophic, deadly, devastating, dire, disastrous, distressful, dreadful, fatal, ghastly, ruinous, serious, terrible, tragic, unfortunate, unlucky, woeful.

**calamity** *n* accident, affliction, cataclysm, catastrophe, disaster, misadventure, mischance, misfortune, mishap, tragedy, tribulation.

**calculate** *vb* add up, ascertain, assess, compute, count, determine, do sums, enumerate, estimate, evaluate, figure out, find out, gauge, judge, reckon, total, value, weigh, work out. **calculated** ▷ DELIBERATE. **calculating** ▷ CRAFTY.

**calibre** *n* **1** bore, diameter, gauge, measure, size. **2** ability, capability, capacity, character, competence, distinction, excellence, genius, gifts, importance, merit, proficiency, quality, skill, stature, talent, worth.

**call** *n* **1** bellow, cry, exclamation, roar, scream, shout, yell. **2** bidding, invitation, signal, summons. **3** *social call.* stay, stop, visit. **4** *no call for it.* cause, demand, excuse, justification, need, occasion, request, requirement. • *vb* **1** bellow, clamour, cry out, exclaim, hail, roar, shout, yell. **2** *call on friends.* drop in, socialize, visit. **3** *called her 'Jane'.* baptize, christen, dub, name. **4** *play called 'Lear'.* entitle, title. **5** *call me at 7.* arouse, awaken, get someone up, rouse, wake, waken. **6** *call a meeting.* convene, gather, invite, order, summon. **7** *call by phone.* contact, dial, phone, ring, telephone. **call for** ▷ FETCH, REQUEST. **call off** ▷ CANCEL. **call someone names** ▷ INSULT.

**calligraphy** *n* copperplate, handwriting, illumination, lettering, penmanship, script.

**calling** *n* business, career, employment, job, line of work, métier, occupation, profession, pursuit, trade, vocation, work.

**callous** *adj* apathetic, cold, cold-hearted, cool, dispassionate, hard-bitten, *inf* hard-boiled, hardened, hard-hearted, *inf* hard-nosed, heartless, inhuman, insensitive, merciless, pitiless, ruthless, *inf* thick-skinned, uncaring, unconcerned, unemotional, unfeeling, unsympathetic. ▷ CRUEL. *Opp* SENSITIVE.

**callow** *adj* adolescent, *inf* born yesterday, *inf* green, immature, inexperienced, innocent, juvenile, naïve, raw, unsophisticated, *inf* wet behind the ears, young. *Opp* MATURE.

**calm** *adj* **1** airless, even, flat, glassy, *poet* halcyon (*days*), like a millpond, motionless, placid, quiet, slow-moving, smooth, still, unclouded, unwrinkled, windless. **2** collected, *derog* complacent, composed, controlled, cool, dispassionate, equable, impassive, imperturbable, *inf* laid-back, level-headed, moderate, pacific, passionless, patient, peaceful, poised, quiet, relaxed, restful, restrained, sedate, self-possessed, sensible, serene, tranquil, undemonstrative, unemotional, unexcitable, *inf* unflappable, unhurried, unperturbed, unruffled, untroubled. *Opp* EXCITABLE, STORMY. • *n* flat sea, peace, quietness, stillness, tranquillity. ▷ CALMNESS. • *vb* appease, compose, control, cool, lull, mollify, pacify, placate, quieten, sedate, settle down, smooth, sober down, soothe, tranquillize. *Opp* DISTURB.

**calmness** *n derog* complacency, composure, equability, equanimity, imperturbability, level-headedness, peace of mind, sangfroid, self-possession, serenity, *inf* unflappability. *Opp* ANXIETY, EXCITEMENT.

**camouflage** *n* blind, cloak, concealment, cover, disguise, façade, front, guise, mask, pretence, protective colouring, screen, veil. • *vb* cloak, conceal, cover up, disguise, hide, mask, obscure, screen, veil.

**camp** *n* bivouac, camping-ground, campsite, encampment, settlement.

**campaign** *n* action, battle, crusade, drive, effort, fight, manoeuvre, movement, offensive, operation, push, struggle, war.

**campus** *n* grounds, setting, site.

**canal** *n* channel, waterway.

**cancel** *vb* abandon, abolish, abort, abrogate, annul, call off, countermand, cross out, delete, drop, eliminate, erase, expunge, frank (*stamps*), give up, invalidate, override, overrule, postpone, quash, repeal, repudiate, rescind, revoke, scrap, *inf* scrub, wipe out, write off. **cancel out** ▷ NEUTRALIZE.

**cancer** *n* canker, carcinoma, growth, malignancy, melanoma, tumour.

**candid** *adj* blunt, direct, fair, forthright, frank, honest, ingenuous, just, *inf* no-nonsense, objective, open, out-spoken, plain, sincere, straight, straightforward, transparent, true, truthful, unbiased, undisguised, unequivocal, unflattering, unprejudiced. *Opp* INSINCERE.

**candidate** *n* applicant, aspirant, competitor, contender, contestant, entrant, nominee, *inf* possibility, pretender (*to throne*), runner, suitor.

**cane** *n* bamboo, rod, stick. • *vb* ▷ THRASH.

**canoe** *n* dug-out, kayak.

**canopy** *n* awning, cover, covering, shade, shelter, umbrella.

**canvass** *n* campaign, census, enquiry, examination, investigation, market research, opinion poll, poll, probe, scrutiny, survey. • *vb* ask for, campaign, *inf* drum up support, electioneer, seek, solicit.

**canyon** *n* defile, gap, gorge, gulch, pass, ravine, valley.

**cap** *n* covering, lid, top. ▷ HAT. • *vb* ▷ COVER.

**capable** *adj* able, accomplished, adept, clever, competent, effective, effectual, efficient, experienced, expert, gifted, *inf* handy, intelligent, masterly, practised, proficient, qualified, skilful, skilled, talented, trained. *Opp* INCAPABLE. **capable of** apt to, disposed to, equal to, liable to.

**capacity** *n* **1** content, dimensions, magnitude, room, size, volume. **2** ability, acumen, capability, cleverness, competence, intelligence, potential, power, skill, talent, wit. **3** *in an official capacity.* appointment, duty, function, job, office, place, position, post, province, responsibility, role.

**cape** *n* **1** cloak, coat, cope, mantle, robe, shawl, wrap. **2** head, headland, peninsula, point, promontory.

**caper** *vb* bound, cavort, dance, frisk, frolic, gambol, hop, jig about, jump, leap, play, prance, romp, skip, spring.

**capital** *adj* **1** chief, controlling, first, foremost, important, leading, main, paramount, pre-eminent, primary, principal. **2** *capital letters.* big, block, initial, large, upper-case. **3** ▷ EXCELLENT. • *n* **1** chief city, centre of government. **2** assets, cash, finance, funds, investments, money, principal, property, *sl* (the) ready, resources, riches, savings, stock, wealth, *inf* the wherewithal.

**capitulate** *vb* acquiesce, be defeated, concede, desist, fall, give in, relent, submit, succumb, surrender, *inf* throw in the towel, yield.

**capricious** *adj* changeable, erratic, fanciful, fickle, fitful, flighty, impulsive, inconstant, mercurial, moody, quirky, uncertain, unpredictable, unreliable, unstable, variable, wayward, whimsical. *Opp* STEADY.

**capsize** *vb* flip over, invert, keel over, overturn, tip over, turn over, *inf* turn turtle, turn upside down.

**capsule** *n* lozenge, medicine, pill, tablet.

**captain** *n* **1** boss, chief, head, leader. **2** commander, master, officer in charge, pilot, skipper.

**caption** *n* description, explanation, heading, headline, superscription, title.

**captivate** *vb* attract, beguile, bewitch, charm, delight, enamour, enchant, enrapture, enslave, ensnare, enthral, entrance, fascinate, hypnotize, infatuate, mesmerize, seduce, *inf* steal your heart, *inf* turn your head, win. *Opp* DISGUST.

**captive** *adj* caged, captured, chained, confined, detained, enslaved, ensnared, fettered, gaoled, imprisoned, incarcerated, jailed, restricted, secure, taken prisoner, *inf* under lock and key. *Opp* FREE. • *n* convict, detainee, hostage, internee, prisoner, slave.

**captivity** *n* bondage, confinement, custody, detention, duress, imprisonment, incarceration, internment, protective custody, remand, restraint, servitude, slavery. ▷ PRISON. *Opp* FREEDOM.

**capture** *n* apprehension, arrest, seizure. • *vb* apprehend, arrest, *inf* bag, bind, catch, *inf* collar, corner, ensnare, entrap, *inf* get, *inf* nab, net, *inf* nick, overpower, secure, seize, snare, take prisoner, trap. ▷ CONQUER. *Opp* LIBERATE.

**car** *n* automobile, *inf* banger, *joc* bus, *joc* jalopy, motor, motor car, *sl* wheels. □ *cab, convertible, coupé, Dormobile, estate, fastback, hatchback, jeep, Land Rover, limousine, Mini, panda car, patrol car, police car, saloon, shooting brake, sports car, taxi, tourer.* ▷ VEHICLE.

**carcass** *n* **1** body, cadaver, corpse, meat, remains. **2** *carcass of a car.* framework, hulk, remains, shell, skeleton, structure.

**card** *n* cardboard, pasteboard. □ *bank card, birthday card, business card, calling card, credit card, get-well card, greetings card, identity card, invitation, membership card, notelet, picture postcard, playing-card, postcard, union card, Valentine, visiting card.*

**care** *n* **1** attention, carefulness, caution, circumspection, concentration, concern, diligence, exactness, forethought, heed, interest, meticulousness, pains, prudence, solicitude, thoroughness, thought, vigilance, watchfulness. **2** anxiety, burden, concern, difficulty, hardship, problem, responsibility, sorrow, stress, tribulation, vexation, woe, worry. ▷ TROUBLE. **3** *left in my care.* charge, control, custody, guardianship, keeping, management, protection, safekeeping, ward. • *vb* be troubled, bother, concern yourself, mind, worry. **care for** ▷ LOVE, TEND.

**career** *n* business, calling, craft, employment, job, livelihood, living, métier, occupation, profession, trade, vocation, work. • *vb* ▷ RUSH.

**carefree** *adj* **1** blasé, casual, cheery, contented, debonair, easy, easy-going, happy-go-lucky, indifferent, insouciant, *inf* laid-back, light-hearted, nonchalant, relaxed, unconcerned, unworried. ▷ HAPPY. **2** *carefree holiday.* leisured, peaceful, quiet, relaxing, restful, trouble-free, untroubled. *Opp* ANXIOUS.

**careful** *adj* **1** alert, attentive, cautious, chary, circumspect, heedful, mindful, observant, prudent, solicitous, thoughtful, vigilant, wary, watchful. **2** *careful work.* accurate, conscientious, deliberate, diligent, exhaustive, fastidious, *derog* fussy, judicious, methodical, meticulous, neat, orderly, organized, painstaking, particular, precise, punctilious, responsible, rigorous, scrupulous, systematic, thorough, well-organized. *Opp* CARELESS. **be careful** ▷ BEWARE.

**careless** *adj* **1** absent-minded, heedless, ill-considered, imprudent, inattentive, incautious, inconsiderate, irresponsible, negligent, rash, reckless, thoughtless, uncaring, unguarded, unthinking, unwary. **2** *careless work.* casual, confused, cursory, disorganized, hasty, imprecise, inaccurate, jumbled, messy, perfunctory, scatter-brained, shoddy, slapdash, slipshod, *inf* sloppy, slovenly, thoughtless, untidy. *Opp* CAREFUL.

**carelessness** *n* haste, inattention, irresponsibility, negligence, recklessness, *inf* sloppiness, slovenliness, thoughtlessness, untidiness. *Opp* CARE.

**caress** *vb* cuddle, embrace, fondle, hug, kiss, make love to, *sl* neck with, nuzzle, pat, pet, rub against, smooth, stroke, touch.

**caretaker** *n* custodian, janitor, keeper, porter, superintendent, warden, watchman.

**careworn** *adj* gaunt, grim, haggard. ▷ WEARY.

**cargo** *n* consignment, freight, goods, lading (*bill of lading*), load, merchandise, payload, shipment.

**caricature** *n* burlesque, cartoon, parody, satire, *inf* send-up, spoof, *inf* take-off, travesty. • *vb* burlesque, distort, exaggerate, imitate, lampoon, make fun of, mimic, mock, overact, overdo, parody, ridicule, satirize, *inf* send up, *inf* take off.

**caring** *n* concern, kindness, nursing, solicitude.

**carnage** *n* blood-bath, bloodshed, butchery, havoc, holocaust, killing, massacre, pogrom, shambles, slaughter.

**carnal** *adj* animal, bodily, erotic, fleshly, natural, physical, sensual, sexual. ▷ LUSTFUL. *Opp* SPIRITUAL.

**carnival** *n* celebration, fair, festival, festivity, fête, fiesta, fun and games, gala, jamboree, merrymaking, pageant, parade, procession, revelry, show, spectacle.

**carp** *vb* cavil, find fault, *inf* go on, *inf* gripe, grumble, object, pick holes, quibble, *inf* split hairs, whinge. ▷ COMPLAIN.

**carpentry** *n* joinery, woodwork.

**carriage** *n* **1** coach. ▷ VEHICLE. **2** bearing, comportment, demeanour, gait, manner, mien, posture, presence, stance.

**carrier** *n* **1** bearer, conveyor, courier, delivery-man, delivery-woman, dispatch rider, errand-boy, errand-girl, haulier, messenger, porter, postman, runner. **2** *carrier of a disease.* contact, host, transmitter.

**carry** *vb* **1** bring, *inf* cart, communicate, ferry, fetch, haul, lead, lift, *inf* lug, manhandle, move, relay, remove, ship, shoulder, take, transfer, transmit, transport. ▷ CONVEY. **2** *carry weight.* bear, hold up, maintain, support. **3** *carry a penalty.* demand, entail, involve, lead to, occasion, require, result in. **carry on** ▷ CONTINUE. **carry out** ▷ DO.

**cart** *n* barrow, dray, truck, wagon, wheelbarrow. • *vb* ▷ CARRY.

**carton** *n* box, cartridge, case, container, pack, package, packet.

**cartoon** *n* animation, caricature, comic strip, drawing, sketch.

**cartridge** *n* 1 canister, capsule, case, cassette, container, cylinder, tube. 2 *cartridge for a gun.* magazine, round, shell.

**carve** *vb* 1 slice. ▷ CUT. 2 *carve stone. inf* chip away at, chisel, engrave, fashion, hew, incise, sculpture, shape.

**cascade** *n* cataract, deluge, flood, gush, torrent, waterfall.
• *vb* ▷ POUR.

**case** *n* 1 box, cabinet, carton, casket, chest, container, crate, pack, packaging, suitcase, trunk. ▷ LUGGAGE. 2 *case of mistaken identity.* example, illustration, instance, occurrence, specimen, state of affairs. 3 *rules don't apply in his case.* circumstances, condition, context, plight, predicament, situation, state. 4 *a legal case.* action, argument, cause, dispute, inquiry, investigation, lawsuit, suit.

**cash** *n* banknotes, bills, change, coins, currency, *inf* dough, funds, hard money, legal tender, money, notes, *inf* (the) ready, *inf* the wherewithal. • *vb* exchange for cash, realize, sell. **cash in on** ▷ PROFIT.

**cashier** *n* accountant, banker, check-out person, clerk, teller, treasurer. • *vb* ▷ DISMISS.

**cask** *n* barrel, butt, hogshead, tub, tun, vat.

**cast** *n* 1 ▷ SCULPTURE. 2 *cast of a play.* characters, company, dramatis personae, performers, players, troupe. • *vb* 1 bowl, chuck, drop, fling, hurl, impel, launch, lob, pelt, pitch, project, scatter, shy, sling, throw, toss. 2 *cast a sculpture.* form, found, mould, shape. ▷ SCULPTURE. **cast off** ▷ SHED, UNTIE.

**castaway** *adj* abandoned, deserted, exiled, marooned, rejected, shipwrecked, stranded.

**caste** *n* class, degree, estate, grade, level, position, rank, standing, station, status, stratum.

**castigate** *vb* censure, chasten, chastise, *old use* chide, correct, discipline, lash, punish, rebuke, reprimand, scold, *inf* tell off. ▷ CRITICIZE.

**castle** *n* château, citadel, fort, fortress, mansion, palace, stately home, stronghold, tower.

**castrate** *vb* emasculate, geld, neuter, spay, sterilize, unsex.

**casual** *adj* 1 accidental, chance, erratic, fortuitous, incidental, irregular, promiscuous, random, serendipitous, sporadic, unexpected, unforeseen, unintentional, unplanned, unpremeditated, unstructured, unsystematic. *Opp* DELIBERATE. 2 *casual attitude.* apathetic, blasé, careless, *inf* couldn't-care-less, easy-going, *inf* free-and-easy, lackadaisical, *inf* laid-back, lax, negligent, nonchalant, offhand, relaxed, *inf* slap-happy, *inf* throwaway, unconcerned, unenthusiastic, unimportant, unprofessional. *Opp* ENTHUSIASTIC. 3 *casual clothes.* comfortable, informal. *Opp* FORMAL.

**casualty** *n* dead person, death, fatality, injured person, injury, loss, victim, wounded person.

**cat** *n* kitten, *inf* moggy, *inf* pussy, tabby, tom, tomcat.

**catacombs** *n* crypt, sepulchre, tomb, underground passage, vault.

**catalogue** *n* brochure, directory, index, inventory, list, record, register, roll, schedule, table.
• *vb* classify, codify, file, index,

list, make an inventory of, record, register, tabulate.

**catapult** *vb* fire, fling, hurl, launch. ▷ THROW.

**cataract** *n* cascade, falls, rapids, torrent, waterfall.

**catastrophe** *n* blow, calamity, cataclysm, crushing blow, *débâcle, devastation, disaster, fiasco, holocaust, mischance, mishap, ruin, ruination, tragedy, upheaval.* ▷ MISFORTUNE.

**catch** *n* **1** bag, booty, capture, haul, net, prey, prize, take. **2** *suspected a catch.* difficulty, disadvantage, drawback, obstacle, problem, snag, trap, trick. **3** *catch on a door.* bolt, clasp, clip, fastener, fastening, hasp, hook, latch, lock. • *vb* **1** clutch, ensnare, entrap, grab, grasp, grip, hang on to, hold, hook, net, seize, snare, snatch, take, tangle, trap. **2** *catch a thief.* apprehend, arrest, capture, *inf* cop, corner, detect, discover, expose, intercept, *inf* nab, *inf* nobble, stop, surprise, take by surprise, unmask. **3** *caught me unawares.* come upon, discover, find, surprise. **4** *catch a bus.* be in time for, get on. **5** *catch a cold.* become infected by, contract, get. **catch on** ▷ SUCCEED, UNDERSTAND. **catch-phrase** ▷ SAYING. **catch-22** ▷ DILEMMA. **catch up** ▷ OVERTAKE.

**catching** *adj* communicable, contagious, infectious, spreading, transmissible, transmittable.

**catchy** *adj* attractive, haunting, memorable, popular, singable, tuneful.

**categorical** *adj* absolute, authoritative, certain, complete, decided, decisive, definite, direct, dogmatic, downright, emphatic, explicit, express, firm, forceful, *inf* out-and-out, positive, strong, total, unambiguous, unconditional, unequivocal, unmitigated, unqualified, unreserved, utter, vigorous. *Opp* TENTATIVE.

**category** *n* class, classification, division, grade, group, head, heading, kind, order, rank, ranking, section, sector, set, sort, type, variety.

**cater** *vb* cook, make arrangements, minister, provide, provision, serve, supply.

**catholic** *adj* all-embracing, all-inclusive, broad, broad-minded, comprehensive, cosmopolitan, eclectic, general, liberal, universal, varied, wide, wide-ranging.

**cattle** *plur n* beef, bullocks, bulls, calves, cows, heifers, livestock, oxen, steers, stock.

**catty** *adj inf* bitchy, ill-natured, malevolent, malicious, mean, nasty, rancorous, sly, spiteful, venomous, vicious. ▷ UNKIND. *Opp* KIND.

**cause** *n* **1** basis, beginning, genesis, grounds, motivation, motive, occasion, origin, reason, root, source, spring, stimulus. **2** agent, author, *old use* begetter, creator, initiator, inspiration, inventor, originator, producer. **3** *cause of his lateness.* excuse, explanation, pretext, reason. **4** *a good cause.* aim, belief, concern, end, ideal, object, purpose, undertaking. • *vb* **1** arouse, awaken, begin, bring about, bring on, create, effect, effectuate, engender, foment, generate, give rise to, incite, kindle, lead to, occasion, precipitate, produce, provoke, result in, set off, spark off, stimulate, trigger off, *inf* whip up. **2** compel, force, induce, motivate.

**caustic** *adj* **1** acid, astringent, burning, corrosive, destructive. **2** *caustic criticism.* acidulous, acrimonious, biting, bitter, critical, cutting, mordant, pungent, sarcastic, scathing, severe, sharp, stinging, trenchant, virulent, waspish. *Opp* MILD.

**caution** *n* 1 alertness, attentiveness, care, carefulness, circumspection, discretion, forethought, heed, heedfulness, prudence, vigilance, wariness, watchfulness. 2 *let off with a caution.* admonition, caveat, *inf* dressing-down, injunction, reprimand, *inf* talking-to, *inf* ticking-off, warning. • *vb* 1 advise, alert, counsel, forewarn, inform, *inf* tip off, warn. 2 *cautioned by the police.* admonish, censure, give a warning, reprehend, reprimand, *inf* tell off, *inf* tick off.

**cautious** *adj* 1 alert, attentive, careful, heedful, prudent, scrupulous, vigilant, watchful. 2 *cautious comments. inf* cagey, calculating, chary, circumspect, deliberate, discreet, gingerly, grudging, guarded, hesitant, judicious, non-committal, restrained, suspicious, tactful, tentative, unadventurous, wary, watchful. *Opp* RECKLESS.

**cavalcade** *n* march-past, parade, procession, spectacle, troop.

**cave** *n* cavern, cavity, den, grotto, hole, pothole, underground chamber. • *vb* **cave in** ▷ COLLAPSE, SURRENDER.

**cavity** *n* cave, crater, dent, hole, hollow, pit.

**cease** *vb* break off, call a halt, conclude, cut off, desist, discontinue, end, finish, halt, *inf* kick (*a habit*), *inf* knock off, *inf* lay off, leave off, *inf* pack in, *inf* pack up, refrain, stop, terminate. *Opp* BEGIN.

**ceaseless** *adj* chronic, constant, continual, continuous, endless, everlasting, incessant, interminable, never-ending, non-stop, permanent, perpetual, persistent, relentless, unending, unremitting, untiring. *Opp* INTERMITTENT, TEMPORARY.

**celebrate** *vb* 1 be happy, have a celebration, let yourself go, *inf* live it up, make merry, *inf* paint the town red, rejoice, revel, *old use* wassail. 2 *celebrate an anniversary.* commemorate, hold, honour, keep, observe, remember. 3 *celebrate a wedding.* officiate at, solemnize. **celebrated** ▷ FAMOUS.

**celebration** *n* banquet, binge, carnival, commemoration, feast, festivity, *inf* jamboree, *joc* jollification, merry-making, observance, *inf* orgy, party, *inf* rave-up, revelry, *church* service, *inf* shindig, solemnization. □ *anniversary, birthday, festival, fête, gala, jubilee, remembrance, reunion, wedding.*

**celebrity** *n* 1 ▷ FAME. 2 big name, *inf* bigwig, dignitary, famous person, idol, notability, personality, public figure, star, superstar, VIP, worthy.

**celestial** *adj* 1 astronomical, cosmic, galactic, interplanetary, interstellar, starry, stellar, universal. 2 *celestial beings.* angelic, blissful, divine, ethereal, godlike, heavenly, seraphic, spiritual, sublime, supernatural, transcendental, visionary.

**celibacy** *n* bachelorhood, chastity, continence, purity, self-restraint, spinsterhood, virginity.

**celibate** *adj* abstinent, chaste, continent, immaculate, single, unmarried, unwedded, virgin. • *n* bachelor, spinster, virgin.

**cell** *n* cavity, chamber, compartment, cubicle, den, enclosure, living space, prison, room, space, unit.

**cellar** *n* basement, crypt, vault, wine-cellar.

**cemetery** *n* burial-ground, churchyard, graveyard, necropolis.

**censor** *vb* amend, ban, bowdlerize, *inf* clean up, cut, edit, exclude, expurgate, forbid, prohibit, remove.

**censorious** *adj* fault-finding, *inf* holier-than-thou, judgemental, moralistic, Pharisaical, self-righteous. ▷ CRITICAL.

**censure** *n* accusation, admonition, blame, castigation, condemnation, criticism, denunciation, diatribe, disapproval, *inf* dressing-down, harangue, rebuke, reprimand, reproach, reprobation, reproof, *inf* slating, stricture, *inf* talking-to, *inf* telling-off, tirade, verbal attack, vituperation. • *vb* admonish, berate, blame, *inf* carpet, castigate, caution, chide, condemn, criticize, denounce, lecture, rebuke, reproach, reprove, scold, take to task, *sl* tear (someone) off a strip, *inf* tell off, *inf* tick off, upbraid.

**census** *n* count, survey, tally.

**central** *adj* **1** focal, inner, innermost, interior, medial, middle. **2** *central facts.* chief, crucial, essential, fundamental, important, key, main, major, overriding, pivotal, primary, principal, vital. *Opp* PERIPHERAL.

**centralize** *vb* amalgamate, bring together, concentrate, rationalize, streamline, unify. *Opp* DISPERSE.

**centre** *n* bull's-eye, core, focal point, focus, heart, hub, inside, interior, kernel, middle, mid-point, nucleus, pivot. *Opp* PERIMETER. • *vb* concentrate, converge, focus.

**centrifugal** *adj* dispersing, diverging, moving outwards, scattering, spreading. *Opp* CENTRIPETAL.

**centripetal** *adj* converging. *Opp* CENTRIFUGAL.

**cereal** *n* corn, grain. □ *barley, corn on the cob, maize, millet, oats, rice, rye, sweet corn, wheat.*

**ceremonial** *adj* celebratory, dignified, liturgical, majestic, official, ritual, ritualistic, solemn, stately. ▷ FORMAL. *Opp* INFORMAL.

**ceremonious** *adj* civil, courteous, courtly, dignified, formal, grand, *derog* pompous, proper, punctilious, *derog* starchy. ▷ POLITE. *Opp* CASUAL.

**ceremony** *n* **1** celebration, commemoration, *inf* do, event, formal occasion, function, occasion, parade, reception, rite, ritual, service, solemnity. **2** ceremonial, decorum, etiquette, formality, grandeur, pageantry, pomp, pomp and circumstance, protocol, ritual, spectacle.

**certain** *adj* **1** adamant, assured, confident, constant, convinced, decided, determined, firm, invariable, positive, resolved, satisfied, settled, stable, steady, sure, undoubting, unshakable, unwavering. **2** *certain proof.* absolute, authenticated, categorical, certified, clear, clear-cut, conclusive, convincing, definite, dependable, established, genuine, guaranteed, incontestable, incontrovertible, indubitable, infallible, irrefutable, known, official, plain, reliable, settled, sure, true, trustworthy, unarguable, undeniable, undisputed, undoubted, unmistakable, unquestionable, valid, verifiable. **3** *certain disaster.* destined, fated, guaranteed, imminent, inescapable, inevitable, inexorable, predestined, predictable, unavoidable.

4 *certain to pay up*. bound, compelled, obliged, required, sure. 5 *certain people*. individual, particular, some, specific, unnamed, unspecified. *Opp* UNCERTAIN. **be certain** ▷ KNOW. **for certain** ▷ DEFINITELY. **make certain** ▷ ENSURE.

**certainty** *n* 1 actuality, certain fact, *inf* foregone conclusion, foreseeable outcome, inevitability, necessity, *inf* sure thing. 2 assertiveness, assurance, authority, certitude, confidence, conviction, knowledge, positiveness, proof, sureness, truth, validity. *Opp* DOUBT.

**certificate** *n* authorization, award, credentials, degree, diploma, document, guarantee, licence, pass, permit, qualification, warrant.

**certify** *vb* 1 affirm, asseverate, attest, authenticate, aver, avow, bear witness, confirm, declare, endorse, guarantee, notify, sign, swear, testify, verify, vouch, vouchsafe, warrant, witness. 2 *certify as competent*. authorize, charter, commission, franchise, license, recognize, validate.

**chain** *n* 1 bonds, coupling, fetters, handcuffs, irons, links, manacles, shackles. 2 *chain of events*. column, combination, concatenation, cordon, line, progression, row, sequence, series, set, string, succession, train. • *vb* bind, clap in irons, fetter, handcuff, link, manacle, shackle, tether, tie. ▷ FASTEN.

**chair** *n* armchair, carver, deckchair, dining-chair, easy chair, recliner, rocking-chair, throne. ▷ SEAT. • *vb* ▷ PRESIDE.

**chairperson** *n* chair, chairman, chairwoman, convenor, director, leader, moderator, organizer, president, speaker.

**challenge** *vb* 1 accost, confront, *inf* have a go at, take on, tax. 2 *challenge to duel*. dare, defy, *old use* demand satisfaction, provoke, summon. 3 *challenge a decision*. argue against, call in doubt, contest, dispute, dissent from, impugn, object to, oppose, protest against, query, question, take exception to.

**challenging** *adj* inspiring, stimulating, testing, thought-provoking, worthwhile. ▷ DIFFICULT. *Opp* EASY.

**chamber** *n* cavity, cell, compartment, niche, nook, space. ▷ ROOM.

**champion** *adj* great, leading, record-breaking, supreme, top, unrivalled, victorious, winning, world-beating. • *n* 1 conqueror, hero, medallist, prize-winner, record-breaker, superman, superwoman, titleholder, victor, winner. 2 *champion of the poor*. backer, defender, guardian, patron, protector, supporter, upholder, vindicator. 3 [*old use*] *champion in lists*. challenger, contender, contestant, fighter, knight, warrior. • *vb* ▷ SUPPORT.

**championship** *n* competition, contest, series, tournament.

**chance** *adj* accidental, adventitious, casual, coincidental, *inf* fluky, fortuitous, fortunate, haphazard, inadvertent, incidental, lucky, random, unexpected, unforeseen, unfortunate, unlooked-for, unplanned, unpremeditated. *Opp* DELIBERATE. • *n* 1 accident, coincidence, destiny, fate, fluke, fortune, gamble, hazard, luck, misfortune, serendipity. 2 *chance of rain*. danger, liability, likelihood, possibility, probability, prospect, risk. 3 occasion, opportunity, time, turn. • *vb* 1 ▷ RISK. 2 ▷ HAPPEN.

**chancy** *adj* dangerous, *inf* dicey, *inf* dodgy, hazardous, *inf* iffy, insecure, precarious, risky, speculative, ticklish, tricky, uncertain, unpredictable, unsafe. *Opp* SAFE.

**change** *n* 1 adaptation, adjustment, alteration, break, conversion, deterioration, development, difference, diversion, improvement, innovation, metamorphosis, modification, modulation, mutation, new look, rearrangement, refinement, reformation, reorganization, revolution, shift, substitution, swing, transfiguration, transformation, transition, translation, transmogrification, transmutation, transposition, *inf* turn-about, U-turn, variation, variety, vicissitude. 2 *small change*. ▷ CASH. • *vb* 1 acclimatize, accommodate, accustom, adapt, adjust, affect, alter, amend, convert, diversify, influence, modify, process, rearrange, reconstruct, refashion, reform, remodel, reorganize, reshape, restyle, tailor, transfigure, transform, translate, transmogrify, transmute, vary. 2 *opinions change*. alter, be transformed, *inf* chop and change, develop, fluctuate, metamorphose, move on, mutate, shift, vary. 3 *change one thing for another*. alternate, displace, exchange, replace, substitute, switch, swop, transpose. 4 *change money*. barter, convert, trade in. **change into** ▷ BECOME. **change someone's mind** ▷ CONVERT. **change your mind** ▷ RECONSIDER.

**changeable** *adj* capricious, chequered (*career*), erratic, fickle, fitful, fluctuating, fluid, inconsistent, inconstant, irregular, mercurial, mutable, protean, shifting, temperamental, uncertain, unpredictable, unreliable, unsettled, unstable, unsteady, *inf* up and down, vacillating, variable, varying, volatile, wavering. *Opp* CONSTANT.

**channel** *n* 1 aqueduct, canal, conduit, course, dike, ditch, duct, groove, gully, gutter, moat, overflow, pipe, sluice, sound, strait, trench, trough, watercourse, waterway. ▷ STREAM. 2 avenue, means, medium, path, route, way. 3 *TV channel*. *inf* side, station, waveband, wavelength. • *vb* conduct, convey, direct, guide, lead, pass on, route, send, transmit.

**chant** *n* hymn, plainsong, psalm. ▷ SONG. • *vb* intone. ▷ SING.

**chaos** *n* anarchy, bedlam, confusion, disorder, disorganization, lawlessness, mayhem, muddle, pandemonium, shambles, tumult, turmoil. *Opp* ORDER.

**chaotic** *adj* anarchic, confused, deranged, disordered, disorderly, disorganized, haphazard, *inf* haywire, *inf* higgledy-piggledy, jumbled, lawless, muddled, rebellious, riotous, *inf* shambolic, *inf* topsy-turvy, tumultuous, uncontrolled, ungovernable, unruly, untidy, *inf* upside-down. *Opp* ORDERLY.

**char** *vb* blacken, brown, burn, carbonize, scorch, sear, singe.

**character** *n* 1 distinctiveness, flavour, idiosyncracy, individuality, integrity, peculiarity, quality, stamp, taste, uniqueness. ▷ CHARACTERISTIC. 2 *a forceful character*. attitude, constitution, disposition, individuality, make-up, manner, nature, personality, reputation, temper, temperament. 3 *a famous character*. figure human being, individual, person, personality, *inf* type. 4 *She's a* *character! inf* case, comedian, comic, eccentric, *inf* nut-case, oddity, *derog* weirdo. 5 *character in a play*. part, persona, portrayal, role. 6 *written characters*. cipher, figure, hieroglyphic, ideogram, letter, mark, rune, sign, symbol, type.

**characteristic** *adj* 1 [*of an individual*] distinctive, distinguishing, essential, idiosyncratic, individual, particular, peculiar, recognizable, singular, special, specific, symptomatic, unique. 2 [*of a kind*] representative, typical. • *n* attribute, distinguishing feature, feature, hallmark, idiosyncracy, mark, peculiarity, property, quality, symptom, trait.

**characterize** *vb* brand, delineate, depict, describe, differentiate, distinguish, draw, identify, individualize, mark, portray, present, recognize, typify.

**charade** *n* absurdity, deceit, deception, fabrication, farce, make-believe, masquerade, mockery, *inf* play-acting, pose, pretence, *inf* put-up job, sham.

**charge** *n* 1 cost, expenditure, expense, fare, fee, payment, postage, price, rate, terms, toll, value. 2 *in my charge*. care, command, control, custody, guardianship, jurisdiction, keeping, protection, responsibility, safe-keeping, supervision, trust. 3 *criminal charges*. accusation, allegation, imputation, indictment. 4 *cavalry charge*. action, assault, attack, drive, incursion, invasion, offensive, onslaught, raid, rush, sally, sortie, strike. • *vb* 1 ask for, debit, exact, levy, make you pay, require. 2 accuse, blame, impeach, indict, prosecute, tax. 3 *charge with a duty*. burden, commit, empower, entrust, give, impose on. 4 *charged us to do our best*. ask, command, direct, enjoin, exhort, instruct. 5 *charge an enemy*. assail, assault, attack, *inf* fall on, rush, set on, storm, *inf* wade into.

**charitable** *adj* bountiful, generous, humanitarian, liberal, munificent, open-handed, philanthropic, unsparing. ▷ KIND. *Opp* MEAN.

**charity** *n* 1 affection, altruism, benevolence, bounty, caring, compassion, consideration, generosity, goodness, helpfulness, humanity, kindness, love, mercy, philanthropy, self-sacrifice, sympathy, tender-heartedness, unselfishness, warm-heartedness. 2 *old use* alms, alms-giving, bounty, donation, financial support, gift, *inf* handout, largesse, offering, patronage, poor relief. 3 good cause, the needy, the poor.

**charm** *n* 1 allure, appeal, attractiveness, charisma, fascination, hypnotic power, lovable nature, lure, magic, magnetism, power, pull, seductiveness. ▷ BEAUTY. 2 *magic charm*. curse, enchantment, incantation, magic, mumbo-jumbo, sorcery, spell, witchcraft, wizardry. 3 *charm on a bracelet*. amulet, lucky charm, mascot, ornament, talisman, trinket. • *vb* allure, attract, beguile, bewitch, cajole, captivate, cast a spell on, decoy, delight, disarm, enchant, enrapture, enthral, entrance, fascinate, hold spellbound, hypnotize, intrigue, lure, mesmerize, please, seduce, soothe, win over. **charming** ▷ ATTRACTIVE.

**chart** *n* diagram, graph, map, plan, sketch-map, table.

**charter** *vb* 1 employ, engage, hire, lease, rent. 2 ▷ CERTIFY.

**chase** *vb* drive, follow, go after, hound, hunt, pursue, run after, track, trail.

**chasm** *n* abyss, canyon, cleft, crater, crevasse, drop, fissure, gap, gulf, hole, hollow, opening, pit, ravine, rift, split, void.

**chaste** *adj* **1** abstinent, celibate, *inf* clean, continent, good, immaculate, inexperienced, innocent, moral, pure, sinless, uncorrupted, undefiled, unmarried, virgin, virginal, virtuous. *Opp* IMMORAL **2** *chaste dress*. austere, becoming, decent, decorous, maidenly, modest, plain, restrained, severe, simple, tasteful, unadorned. *Opp* INDECENT.

**chasten** *vb* **1** restrain, subdue. ▷ HUMILIATE. **2** ▷ CHASTISE.

**chastise** *vb* castigate, chasten, correct, discipline, penalize, rebuke, scold. ▷ PUNISH, REPRIMAND.

**chastity** *n* abstinence, celibacy, continence, innocence, integrity, maidenhood, morality, purity, restraint, sinlessness, virginity, virtue. *Opp* LUST.

**chat** *n* chatter, *inf* chin-wag, *inf* chit-chat, conversation, gossip, *inf* heart-to-heart. • *vb* chatter, converse, gossip, *inf* natter, prattle. ▷ TALK. **chat up** ▷ WOO.

**chauvinist** *n* bigot, *inf* MCP (= *male chauvinist pig*), patriot, sexist, xenophobe.

**cheap** *adj* **1** bargain, budget, cut-price, *inf* dirt-cheap, discount, economical, economy, fair, inexpensive, *inf* knock-down, low-priced, reasonable, reduced, *inf* rock-bottom, sale, under-priced. **2** *cheap quality*. base, inferior, poor, second-rate, shoddy, *inf* tatty, tawdry, *inf* tinny, *inf* trashy, worthless. **3** *a cheap insult*. contemptible, crude, despicable, facile, glib, ill-bred, ill-mannered, mean, silly, tasteless, unworthy, vulgar. *Opp* EXPENSIVE, WORTHY.

**cheapen** *vb* belittle, debase, degrade, demean, devalue, discredit, downgrade, lower the tone (of), popularize, prostitute, vulgarize.

**cheat** *n* **1** charlatan, cheater, *inf* con-man, counterfeiter, deceiver, double-crosser, extortioner, forger, fraud, hoaxer, impersonator, impostor, mountebank, *inf* phoney, *inf* quack, racketeer, rogue, *inf* shark, swindler, trickster, *inf* twister. **2** artifice, bluff, chicanery, *inf* con, confidence trick, deceit, deception, *sl* fiddle, fraud, hoax, imposture, lie, misrepresentation, pretence, *inf* put-up job, *inf* racket, *inf* rip-off, ruse, sham, swindle, *inf* swizz, treachery, trick. • *vb* **1** bamboozle, beguile, bilk, *inf* con, deceive, defraud, *sl* diddle, *inf* do, double-cross, dupe, *sl* fiddle, *inf* fleece, fool, hoax, hoodwink, outwit, *inf* rip off, rob, *inf* short-change, swindle, take in, trick. **2** *cheat in an exam*. copy, crib, plagiarize.

**check** *adj* ▷ CHEQUERED. • *n* **1** break, delay, halt, hesitation, hiatus, interruption, pause, stop, stoppage, suspension. **2** *medical check*. check-up, examination, *inf* going-over, inspection, investigation, *inf* once-over, scrutiny, test. • *vb* **1** arrest, bar, block, bridle, control, curb, delay, foil, govern, halt, hamper, hinder, hold back, impede, inhibit, keep in check, obstruct, regulate, rein, repress, restrain, retard, slow down, stem, stop, stunt (*growth*), thwart. **2** *check answers*. *Amer* check out, compare, cross-check, examine, inspect, investigate, monitor, research, scrutinize, test, verify.

**cheek** *n* audacity, boldness, brazenness, effrontery, impertinence, impudence, insolence, presumptuousness, rudeness, shamelessness, temerity.

**cheeky** *adj* arrogant, audacious, bold, brazen, cool, discourteous, disrespectful, flippant, forward, impertinent, impolite, impudent, insolent, insulting, irreverent, mocking, pert, presumptuous, rude, *inf* saucy, shameless, *inf* tongue-in-cheek. *Opp* RESPECTFUL.

**cheer** *n* **1** acclamation, applause, cry of approval, encouragement, hurrah, ovation, shout of approval. **2** ▷ HAPPINESS. • *vb* **1** acclaim, applaud, clap, encourage, shout, yell. *Opp* JEER. **2** comfort, console, delight, encourage, exhilarate, gladden, make cheerful, please, solace, uplift. *Opp* SADDEN. **cheer someone up** ▷ COMFORT, ENTERTAIN. **cheer up** ▷ BRIGHTEN. **Cheer up!** *inf* buck up, look happy, *inf* perk up, smile, *sl* snap out of it, take heart.

**cheerful** *adj* animated, bouncy, bright, buoyant, cheery, *inf* chirpy, contented, convivial, delighted, elated, festive, gay, genial, glad, gleeful, good-humoured, hearty, hopeful, jaunty, jocund, jolly, jovial, joyful, joyous, jubilant, laughing, light, light-hearted, lively, merry, optimistic, *inf* perky, pleased, positive, rapturous, sparkling, spirited, sprightly, sunny, warm-hearted. ▷ HAPPY. *Opp* BAD-TEMPERED, CHEERLESS.

**cheerless** *adj* bleak, comfortless, dark, depressing, desolate, dingy, disconsolate, dismal, drab, dreary, dull, forbidding, forlorn, frowning, funereal, gloomy, grim, joyless, lacklustre, melancholy, miserable, mournful, sober, sombre, sullen, sunless, uncongenial, unhappy, uninviting, unpleasant, unpromising, woeful, wretched. ▷ SAD. *Opp* CHEERFUL.

**chemical** *n* compound, element, substance.

**chemist** *n old use* apothecary, *Amer* drug-store, pharmacist, pharmacy.

**chequered** *adj* **1** check, criss-cross, in squares, like a chessboard, patchwork, tartan, tessellated. **2** *chequered career.* ▷ CHANGEABLE.

**cherish** *vb* be fond of, care for, cosset, foster, hold dear, keep safe, look after, love, nourish, nurse, nurture, prize, protect, treasure, value.

**chest** *n* **1** box, caddy, case, casket, coffer, crate, strongbox, trunk. **2** breast, rib-cage, thorax.

**chew** *vb* bite, champ, crunch, gnaw, grind, masticate, munch, nibble. ▷ EAT. **chew over** ▷ CONSIDER.

**chick** *n* fledgling, nestling.

**chicken** *n* bantam, broiler, cockerel, fowl, hen, pullet, rooster.

**chief** *adj* **1** arch, best, first, greatest, head, highest, in charge, leading, major, most experienced, most honoured, most important, oldest, outstanding, premier, principal, senior, supreme, top, unequalled, unrivalled. **2** *chief facts.* basic, cardinal, central, dominant, especial, essential, foremost, fundamental, high-priority, indispensable, key, main, necessary, overriding, paramount, predominant, primary, prime, salient, significant, substantial, uppermost, vital, weighty. *Opp* UNIMPORTANT.

• *n* administrator, authority-figure, *inf* bigwig, *inf* boss, captain, chairperson, chieftain, commander, commanding officer, commissioner, controller, director, employer, executive, foreman, forewoman, *inf* gaffer, *Amer inf* godfather, governor, head, king, leader, manager, managing director, master, mistress, *inf* number one, officer, organizer, overseer, owner, president, principal, proprietor, ring-leader, ruler, superintendent, supervisor, *inf* supremo.

**chiefly** *adv* especially, essentially, generally, in particular, mainly, mostly, particularly, predominantly, primarily, principally, usually.

**child** *n* **1** adolescent, *inf* babe, baby, *Scot* bairn, *inf* bambino, boy, *derog* brat, girl, *derog* guttersnipe, infant, juvenile, *inf* kid, lad, lass, minor, newborn, *inf* nipper, offspring, *inf* stripling, toddler, *inf* tot, *derog* urchin, youngster, youth. **2** daughter, descendant, heir, issue, offspring, progeny, son.

**childhood** *n* adolescence, babyhood, boyhood, girlhood, infancy, minority, schooldays, *inf* teens, youth.

**childish** *adj* babyish, credulous, foolish, immature, infantile, juvenile, puerile. ▷ SILLY. *Opp* MATURE.

**childlike** *adj* artless, frank, *inf* green, guileless, ingenuous, innocent, naïve, natural, simple, trustful, unaffected, unsophisticated. *Opp* ARTFUL.

**chill** *n* ▷ COLD. • *vb* cool, freeze, keep cold, make cold, refrigerate. *Opp* WARM.

**chilly** *adj* **1** cold, cool, crisp, fresh, frosty, icy, *inf* nippy, *inf* parky, raw, sharp, wintry. **2** *a chilly greeting*. aloof, cool, dispassionate, frigid, hostile, ill-disposed, remote, reserved, *inf* standoffish, unforthcoming, unfriendly, unresponsive, unsympathetic, unwelcoming. *Opp* WARM.

**chime** *n* carillon, peal, striking, tintinnabulation, tolling.
• *vb* ▷ RING.

**chimney** *n* flue, funnel, smoke-stack.

**china** *n* porcelain. ▷ CROCKERY.

**chink** *n* **1** cleft, crack, cranny, crevice, cut, fissure, gap, opening, rift, slit, slot, space, split.
**2** ▷ SOUND.

**chip** *n* **1** bit, flake, fleck, fragment, piece, scrap, shard, shaving, shiver, slice, sliver, splinter, wedge. **2** *a chip in a cup*. crack, damage, flaw, gash, nick, notch, scratch, snick. • *vb* break, crack, damage, gash, nick, notch, scratch, splinter. **chip away** ▷ CHISEL. **chip in** ▷ CONTRIBUTE, INTERRUPT.

**chisel** *vb* carve, *inf* chip away, cut, engrave, fashion, model, sculpture, shape.

**chivalrous** *adj* bold, brave, chivalric, courageous, courteous, courtly, gallant, generous, gentlemanly, heroic, honourable, knightly, noble, polite, respectable, true, trustworthy, valiant, valorous, worthy. *Opp* COWARDLY, RUDE.

**choice** *adj* ▷ EXCELLENT. • *n* **1** alternative, dilemma, need to choose, option. **2** *make your choice*. choosing, decision, election, liking, nomination, pick, preference, say, vote. **3** *a choice of food*. array, assortment, diversity, miscellany, mixture, range, selection, variety.

**choke** *vb* 1 asphyxiate, garrotte, smother, stifle, strangle, suffocate, throttle. 2 *choke in smoke.* cough, gag, gasp, retch. 3 *choked with traffic.* block, *inf* bung up, clog, close, congest, constrict, dam, fill, jam, obstruct, smother, stop up. **choke back** ▷ SUPPRESS.

**choose** *vb* adopt, agree on, appoint, decide on, determine on, distinguish, draw lots for, elect, establish, fix on, identify, isolate, name, nominate, opt for, pick out, *inf* plump for, prefer, select, settle on, show a preference for, single out, vote for.

**choosy** *adj* dainty, discerning, discriminating, exacting, fastidious, finical, finicky, fussy, *inf* hard to please, nice, particular, pernickety, *inf* picky, selective. *Opp* INDIFFERENT.

**chop** *vb* cleave, cut, hack, hew, lop, slash, split. ▷ CUT. **chop and change** ▷ CHANGE.

**chopper** *n* axe, cleaver.

**choppy** *adj* roughish, ruffled, turbulent, uneven, wavy. *Opp* SMOOTH.

**chore** *n* burden, drudgery, duty, errand, job, task, work.

**chorus** *n* 1 choir, choral society, vocal ensemble. 2 *join in the chorus.* refrain, response.

**christen** *vb* anoint, baptize, call, dub, name.

**chronic** *adj* 1 ceaseless, constant, continuing, deep-rooted, habitual, incessant, incurable, ineradicable, ingrained, lasting, lifelong, lingering, long-lasting, long-lived, long-standing, never-ending, non-stop, permanent, persistent, unending. *Opp* ACUTE, TEMPORARY. 2 [*inf*] *chronic driving.* ▷ BAD.

**chronicle** *n* account, annals, archive, chronology, description, diary, history, journal, narrative, record, register, saga, story.

**chronological** *adj* consecutive, in order, sequential.

**chronology** *n* 1 almanac, calendar, diary, journal, log, schedule, timetable. 2 *establish the chronology.* dating, order, sequence, timing.

**chubby** *adj* buxom, dumpy, plump, podgy, portly, rotund, round, stout, tubby. ▷ FAT. *Opp* THIN.

**chunk** *n* bar, block, brick, chuck, *inf* dollop, hunk, lump, mass, piece, portion, slab, wad, wedge, *inf* wodge.

**church** *n* abbey, basilica, cathedral, chapel, convent, monastery, nunnery, parish church, priory.

**churchyard** *n* burial-ground, cemetery, graveyard.

**chute** *n* channel, incline, ramp, rapid, slide, slope.

**cinema** *n* films, *old use* flicks, *Amer* motion pictures, *inf* movies, *inf* pictures.

**circle** *n* 1 annulus, band, circlet, disc, hoop, ring. □ *belt, circuit, circulation, circumference, circumnavigation, coil, cordon, curl, curve, cycle, ellipse, girdle, globe, gyration, lap, loop, orb, orbit, oval, revolution, rotation, round, sphere, spiral, tour, turn, wheel, whirl, whorl.* 2 circle of friends. association, band, body, clique, club, company, fellowship, fraternity, gang, party, set, society. ▷ GROUP. • *vb* 1 circulate, circumnavigate, circumscribe, coil, compass, corkscrew, curl, curve, go round, gyrate, loop, orbit, pirouette, pivot, reel, revolve, rotate, spin, spiral, swirl, swivel, tour, turn, wheel, whirl, wind. 2 *trees circle the lawn.* encircle, enclose, encompass, girdle, hem in, ring, skirt, surround.

**circuit** *n* journey round, lap, orbit, revolution, tour.

**circuitous** *adj* curving, devious, indirect, labyrinthine, meandering, oblique, rambling, roundabout, serpentine, tortuous, twisting, winding, zigzag. *Opp* DIRECT.

**circular** *adj* **1** annular, discoid, ringlike, round. **2** *circular conversation.* circumlocutory, cyclic, periphrastic, repeating, repetitive, roundabout, tautologous.
• *n* advertisement, leaflet, letter, notice, pamphlet.

**circulate** *vb* **1** go round, move about, move round, orbit. ▷ CIRCLE. **2** *circulate gossip.* advertise, disseminate, distribute, issue, make known, noise abroad, promulgate, publicize, publish, *inf* put about, send round, spread about.

**circulation** *n* **1** flow, movement, pumping, recycling. **2** broadcasting, diffusion, dissemination, distribution, promulgation, publication, spreading, transmission. **3** *newspaper circulation.* distribution, sales-figures.

**circumference** *n* border, boundary, circuit, edge, exterior, fringe, limit, margin, outline, outside, perimeter, periphery, rim, verge.

**circumstance** *n* affair, event, happening, incident, occasion, occurrence. **circumstances** **1** background, causes, conditions, considerations, context, contingencies, details, factors, facts, influences, particulars, position, situation, state of affairs, surroundings. **2** finances, income, resources.

**circumstantial** *adj* conjectural, deduced, inferred, unprovable. *Opp* PROVABLE.

**cistern** *n* bath, container, reservoir, tank.

**citadel** *n* acropolis, bastion, castle, fort, fortification, fortress, garrison, stronghold, tower.

**cite** *vb* adduce, advance, *inf* bring up, enumerate, mention, name, quote, *inf* reel off, refer to, specify.

**citizen** *n* burgess, commoner, denizen, dweller, freeman, householder, inhabitant, national, native, passport-holder, ratepayer, resident, subject, taxpayer, voter.

**city** *n* capital, conurbation, metropolis, town, urban district.

**civil** *adj* **1** affable, civilized, considerate, courteous, obliging, respectful, urbane, well-bred, well-mannered. ▷ POLITE. *Opp* IMPOLITE. **2** *civil administration.* civilian, domestic, internal, national. ▷ MILITARY. **3** *civil liberties.* communal, public, social, state. **civil rights** freedom, human rights, legal rights, liberty, political rights. **civil servant** administrator, bureaucrat, *derog* mandarin.

**civilization** *n* achievements, attainments, culture, customs, mores, organization, refinement, sophistication, urbanity, urbanization.

**civilize** *vb* cultivate, domesticate, educate, enlighten, humanize, improve, make better, organize, refine, socialize, urbanize.

**civilized** *adj* advanced, cultivated, cultured, democratic, developed, domesticated, educated, enlightened, humane, orderly, polite, refined, sociable, social, sophisticated, urbane, urbanized, well-behaved, well-run. *Opp* UNCIVILIZED.

**claim** *vb* **1** ask for, collect, command, demand, exact, insist on, request, require, take. **2** affirm, allege, argue, assert, attest, contend, declare, insist, maintain, pretend, profess, state.

**clairvoyant** *adj* extra-sensory, oracular, prophetic, psychic, telepathic. • *n* fortune-teller, oracle, prophet, seer, sibyl, soothsayer.

**clamber** *vb* climb, crawl, move awkwardly, scramble.

**clammy** *adj* close, damp, dank, humid, moist, muggy, slimy, sticky, sweaty, wet.

**clamour** *n* babel, commotion, din, hubbub, hullabaloo, noise, outcry, racket, row, screeching, shouting, storm, uproar. • *vb* call out, cry out, exclaim, shout, yell.

**clan** *n* family, house, tribe.

**clannish** *adj derog* cliquish, close, close-knit, insular, isolated, narrow, united.

**clap** *n* bang, crack, crash, report, smack. ▷ SOUND. • *vb* **1** applaud, *sl* put your hands together, show approval. **2** *clap on the back*. ▷ HIT.

**clarify** *vb* **1** clear up, define, elucidate, explain, explicate, gloss, illuminate, make clear, simplify, *inf* spell out, throw light on. *Opp* CONFUSE. **2** *clarify wine*. cleanse, clear, filter, purify, refine. *Opp* CLOUD.

**clash** *vb* **1** bang, clang, clank, crash, resonate, ring. ▷ SOUND. **2** ▷ CONFLICT. **3** *The events clashed*. ▷ COINCIDE.

**clasp** *n* **1** brooch, buckle, catch, clip, fastener, fastening, hasp, hook, pin. **2** cuddle, embrace, grasp, grip, hold, hug. • *vb* **1** ▷ FASTEN. **2** cling to, clutch, embrace, enfold, grasp, grip, hold, hug, squeeze. **3** *clasp your hands*. hold together, wring.

**class** *n* **1** category, classification, division, domain, genre, genus, grade, group, kind, league, order, quality, rank, set, sort, species, sphere, type. **2** *social class*. caste, degree, descent, extraction, grouping, lineage, pedigree, standing, station, status. □ *aristocracy, bourgeoisie, commoners, (the) commons, gentry, lower class, middle class, nobility, proletariat, ruling class, serfs, upper class, upper-middle class, (the) workers, working class*. **3** class in school. band, form, *Amer* grade, group, set, stream, year. • *vb* ▷ CLASSIFY.

**classic** *adj* **1** abiding, ageless, deathless, enduring, established, exemplary, flawless, ideal, immortal, lasting, legendary, masterly, memorable, notable, outstanding, perfect, time-honoured, undying, unforgettable, *inf* vintage. ▷ EXCELLENT. *Opp* COMMONPLACE, EPHEMERAL. **2** *a classic case*. archetypal, characteristic, copybook, definitive, model, paradigmatic, regular, standard, typical, usual. *Opp* UNUSUAL. • *n* masterpiece, masterwork, model.

**classical** *adj* **1** ancient, Attic, Greek, Hellenic, Latin, Roman. **2** *classical style*. austere, dignified, elegant, pure, restrained, simple, symmetrical, well-proportioned. **3** *classical music*. established, harmonious, highbrow, serious.

**classification** *n* categorization, codification, ordering, organization, systematization, tabulation, taxonomy. ▷ CLASS.

**classify** *vb* arrange, bracket together, catalogue, categorize, class, grade, group, order, organize, *inf* pigeon-hole, put into sets, sort, systematize, tabulate. **classified** ▷ SECRET.

**clause** *n* article, condition, item, paragraph, part, passage, provision, proviso, section, subsection.

**claw** *n* nail, talon. • *vb* graze, injure, lacerate, maul, rip, scrape, scratch, slash, tear.

**clean** *adj* **1** decontaminated, dirt-free, disinfected, hygienic, immaculate, laundered, perfect, polished, sanitary, scrubbed, spotless, sterile, sterilized, tidy, unadulterated, unsoiled, unstained, unsullied, washed, wholesome. **2** *clean water*. clarified, clear, distilled, fresh, pure, purified, unpolluted. **3** *clean paper*. blank, new, plain, uncreased, unmarked, untouched, unused. **4** *a clean edge*. neat, regular, smooth, straight, tidy. **5** *a clean fight*. chivalrous, fair, honest, honourable, sporting, sportsmanlike. **6** *clean fun*. chaste, decent, good, innocent, moral, respectable, upright, virtuous. *Opp* DIRTY. • *vb* cleanse, clear up, tidy up, wash. □ *bath, bathe, brush, buff, decontaminate, deodorize, disinfect, dry-clean, dust, filter, flush, groom, hoover, launder, mop, polish, purge, purify, rinse, sand-blast, sanitize, scour, scrape, scrub, shampoo, shower, soap, sponge, spring-clean, spruce up, sterilize, swab, sweep, swill, vacuum, wipe, wring out*. *Opp* CONTAMINATE. **make a clean breast of** ▷ CONFESS.

**clean-shaven** *adj* beardless, shaved, shaven, shorn, smooth.

**clear** *adj* **1** clean, colourless, crystalline, glassy, limpid, pellucid, pure, transparent. **2** *clear weather*. cloudless, fair, fine, sunny, starlit, unclouded. *Opp* CLOUDY. **3** *clear colours*. bright, lustrous, shining, sparkling, strong, vivid. **4** *clear conscience*. blameless, easy, guiltless, innocent, quiet, satisfied, sinless, undisturbed, untarnished, untroubled, unworried. **5** *clear handwriting*. bold, clean, definite, distinct, explicit, focused, legible, positive, recognizable, sharp, simple, visible, well-defined. **6** *clear sound*. audible, clarion (*call*), distinct, penetrating, sharp. **7** *clear instructions*. clear-cut, coherent, comprehensible, explicit, intelligible, lucid, perspicuous, precise, specific, straightforward, unambiguous, understandable, unequivocal, well-presented. **8** *clear case of cheating*. apparent, blatant, clear-cut, conspicuous, evident, glaring, indisputable, manifest, noticeable, obvious, palpable, perceptible, plain, pronounced, straightforward, unconcealed, undisguised, unmistakable. *Opp* UNCERTAIN. **9** *clear space*. empty, free, open, passable, uncluttered, uncrowded, unhampered, unhindered, unimpeded, unobstructed. • *vb* **1** disappear, evaporate, fade, melt away, vanish. **2** become clear, brighten, clarify, lighten, uncloud. **3** clean, make clean, make transparent, polish, wipe. **4** *clear weeds*. disentangle, eliminate, get rid of, remove, strip. **5** *clear a drain*. clean out, free, loosen, open up, unblock, unclog. **6** *clear of blame*. absolve, acquit, exculpate, excuse, exonerate, free, *inf* let off, liberate, release, vindicate. **7** *clear a building*. empty, evacuate. **8** *clear a fence*. bound over, jump, leap over, pass over, spring over, vault. **clear away** ▷ REMOVE. **clear off** ▷ DEPART. **clear up** ▷ CLEAN, EXPLAIN.

**clearing** *n* gap, glade, opening, space.

**cleave** *vb* divide, halve, rive, slit, split. ▷ CUT.

**clench** *vb* **1** clamp up, close tightly, double up, grit (*your teeth*), squeeze tightly. **2** clasp, grasp, grip, hold.

**clergyman** *n* archbishop, ayatollah, bishop, canon, cardinal, chaplain, churchman, cleric, curate, deacon, *fem* deaconess, dean, divine, ecclesiastic, evangelist, friar, guru, imam, *inf* man of the cloth, minister, missionary, monk, padre, parson, pastor, preacher, prebend, prelate, priest, rabbi, rector, vicar. *Opp* LAYMAN.

**clerical** *adj* **1** *clerical and administrative work.* office, secretarial, *inf* white-collar. **2** *a clerical collar* canonical, ecclesiastical, episcopal, ministerial, monastic, pastoral, priestly, rabbinical, sacerdotal, spiritual.

**clerk** *n* assistant, bookkeeper, computer operator, copyist, filing clerk, office boy, office girl, office worker, *inf* pen-pusher, receptionist, recorder, scribe, secretary, shorthand-typist, stenographer, typist, word-processor operator.

**clever** *adj* able, academic, accomplished, acute, adept, adroit, apt, artful, artistic, astute, *inf* brainy, bright, brilliant, canny, capable, *derog* crafty, creative, *derog* cunning, *inf* cute, *inf* deep, deft, dextrous, discerning, expert, *derog* foxy, gifted, guileful, *inf* handy, imaginative, ingenious, intellectual, intelligent, inventive, judicious, keen, knowing, knowledgeable, observant, penetrating, perceptive, percipient, perspicacious, precocious, quick, quick-witted, rational, resourceful, sagacious, sensible, sharp, shrewd, skilful, skilled, slick, *derog* sly, smart, subtle, talented, *derog* wily, wise, witty. *Opp* STUPID, UNSKILFUL. **clever person** *inf* egghead, expert, genius, *derog* know-all, mastermind, prodigy, sage, *derog* smart alec, *derog* smart-arse, virtuoso, wizard.

**cleverness** *n* ability, acuteness, astuteness, brilliance, *derog* cunning, expertise, ingenuity, intellect, intelligence, mastery, quickness, sagacity, sharpness, shrewdness, skill, subtlety, talent, wisdom, wit. *Opp* STUPIDITY.

**cliché** *n* banality, *inf* chestnut, commonplace, hackneyed phrase, platitude, stereotype, truism, well-worn phrase.

**client** *n* plur clientele, consumer, customer, patient, patron, shopper, user.

**cliff** *n* bluff, crag, escarpment, precipice, rock-face, scar, sheer drop.

**climate** *n* **1** ▷ WEATHER. **2** *climate of opinion.* ambience, atmosphere, aura, disposition, environment, feeling, mood, spirit, temper, trend.

**climax** *n* **1** acme, apex, apogee, crisis, culmination, head, highlight, high point, peak, summit, zenith. *Opp* BATHOS. **2** *sexual climax.* orgasm.

**climb** *n* ascent, grade, gradient, hill, incline, pitch, rise, slope. • *vb* **1** ascend, clamber up, defy gravity, go up, levitate, lift off, mount, move up, scale, shin up, soar, swarm up, take off. **2** incline, rise, slope up. **3** *climb a mountain.* conquer, reach the top of. **climb down** ▷ DESCEND.

**clinch** *vb* agree, close, complete, conclude, confirm, decide, determine, finalize, make certain of, ratify, secure, settle, shake hands on, sign, verify.

**cling** *vb* adhere, attach, fasten, fix, hold fast, stick. **cling to** ▷ EMBRACE.

**clinic** *n* health centre, infirmary, medical centre, sick-bay, surgery.

**clip** *n* **1** ▷ FASTENER. **2** *clip from a film.* bit, cutting, excerpt, extract, fragment, part, passage, portion, quotation, section, snippet, trailer. • *vb* **1** pin, staple. ▷ FASTEN. **2** crop, dock, prune, shear, snip, trim. ▷ CUT.

**cloak** *n* **1** cape, cope, mantle, poncho, robe, wrap. ▷ COAT. **2** ▷ COVER. • *vb* cover, disguise, mantle, mask, screen, shroud, veil, wrap. ▷ HIDE.

**clock** *n* time-piece. □ *alarm-clock, chronometer, dial, digital clock, grandfather clock, hourglass, pendulum clock, sundial, watch.*

**clog** *vb* block, *inf* bung up, choke, close, congest, dam, fill, impede, jam, obstruct, plug, stop up.

**close** *adj* **1** accessible, adjacent, adjoining, at hand, convenient, handy, near, neighbouring, point-blank. **2** *close friends.* affectionate, attached, dear, devoted, familiar, fond, friendly, intimate, loving, *inf* thick. **3** *close comparison.* alike, analogous, comparable, compatible, corresponding, related, resembling, similar. **4** *a close crowd.* compact, compressed, congested, cramped, crowded, dense, *inf* jam-packed, packed, thick. **5** *close scrutiny.* attentive, careful, concentrated, detailed, minute, painstaking, precise, rigorous, searching, thorough. **6** *close with information.* confidential, private, reserved, reticent, secretive, taciturn. **7** *close with money.* illiberal, mean, *inf* mingy, miserly, niggardly, parsimonious, penurious, stingy, tight, tight-fisted, ungenerous. **8** *close atmosphere.* airless, confined, fuggy, humid, muggy, oppressive, stale, stifling, stuffy, suffocating, sweltering, unventilated, warm. *Opp* DISTANT, OPEN. • *n* **1** cessation, completion, conclusion, culmination, end, finish, stop, termination. **2** cadence, coda, finale. **3** *close of a play.* denouement, last act. • *vb* **1** bolt, fasten, lock, make inaccessible, padlock, put out of bounds, seal, secure, shut. **2** *close a road.* bar, barricade, block, make impassable, obstruct, seal off, stop up. **3** *close proceedings.* complete, conclude, culminate, discontinue, end, finish, stop, terminate, *inf* wind up. **4** *close a gap.* fill, join up, make smaller, reduce, shorten. *Opp* OPEN.

**closed** *adj* **1** fastened, locked, sealed, shut. **2** completed, concluded, done with, ended, finished, over, resolved, settled, tied up.

**clot** *n* embolism, lump, mass, thrombosis. • *vb* coagulate, coalesce, congeal, curdle, make lumps, set, solidify, stiffen, thicken.

**cloth** *n* fabric, material, stuff, textile. □ *astrakhan, bouclé, brocade, broderie anglaise, buckram, calico, cambric, candlewick, canvas, cashmere, cheesecloth, chenille, chiffon, chintz, corduroy, cotton, crepe, cretonne, damask, denim, dimity, drill, drugget, elastic, felt, flannel, flannelette, gabardine, gauze, georgette, gingham, hessian, holland, lace, lamé, lawn, linen, lint, mohair, moiré, moquette, muslin, nankeen, nylon, oilcloth, oilskin, organdie, organza, patchwork, piqué, plaid, plissé, plush, polycotton, polyester, poplin, rayon, sackcloth, sacking, sailcloth, sarsenet, sateen, satin, satinette, seersucker, serge, silk, stockinet, taffeta, tapes-*

*try, tartan, terry, ticking, tulle, tussore, tweed, velour, velvet, velveteen, viscose, voile, winceyette, wool, worsted.*

**clothe** *vb* accoutre, apparel, array, attire, cover, deck, drape, dress, fit out, garb, *inf* kit out, outfit, robe, swathe, wrap up. *Opp* STRIP. **clothe yourself in** ▷ WEAR.

**clothes** *plur n* apparel, attire, *inf* clobber, clothing, costume, dress, ensemble, finery, garb, garments, *inf* gear, *inf* get-up, outfit, *old use* raiment, *inf* rig-out, *sl* togs, trousseau, underclothes, uniform, vestments, wardrobe, wear, weeds. □ *anorak, apron, blazer, blouse, bodice, breeches, caftan, cagoule, cape, cardigan, cassock, chemise, chuddar, cloak, coat, crinoline, culottes, décolletage, doublet, dress, dressing-gown, duffel coat, dungarees, frock, gaiters, gauntlet, glove, gown, greatcoat, gym-slip, habit, housecoat, jacket, jeans, jerkin, jersey, jodhpurs, jumper, kilt, knickers, leg-warmers, leotard, livery, loincloth, lounge suit, mackintosh, mantle, miniskirt, muffler, neck-tie, négligé, nightclothes, nightdress, oilskins, overalls, overcoat, pants, parka, pinafore, poncho, pullover, pyjamas, raincoat, robe, rompers, sari, sarong, scarf, shawl, shirt, shorts, singlet, skirt, slacks, smock, sock, sou'wester, spats, stocking, stole, suit, surplice, sweater, sweatshirt, tail-coat, tie, tights, trousers, trunks, t-shirt, tunic, tutu, uniform, waistcoat, wet-suit, wind-cheater, wrap, yashmak.* ▷ HAT, SHOE, UNDERCLOTHES.

**cloud** *n* billow, haze, mist, rain cloud, storm cloud. • *vb* blur, conceal, cover, darken, dull, eclipse, enshroud, hide, mantle, mist up, obfuscate, obscure, screen, shroud, veil.

**cloudless** *adj* bright, clear, starlit, sunny, unclouded. *Opp* CLOUDY.

**cloudy** *adj* **1** dark, dismal, dull, gloomy, grey, leaden, lowering, overcast, sullen, sunless. *Opp* CLOUDLESS. **2** *cloudy windows.* blurred, blurry, dim, misty, opaque, steamy, unclear. **3** *cloudy liquid.* hazy, milky, muddy, murky. *Opp* CLEAR.

**clown** *n* buffoon, comedian, comic, fool, funnyman, jester, joker. ▷ IDIOT.

**club** *n* **1** bat, baton, bludgeon, cosh, cudgel, mace, staff, stick, truncheon. **2** association, brotherhood, circle, company, federation, fellowship, fraternity, group, guild, league, order, organization, party, set, sisterhood, society, sorority, union. • *vb* ▷ HIT. **club together** ▷ COMBINE.

**clue** *n* hint, idea, indication, indicator, inkling, key, lead, pointer, sign, suggestion, suspicion, tip, tip-off, trace.

**clump** *n* bunch, bundle, cluster, collection, mass, shock (*of hair*), thicket, tuft. ▷ GROUP.

**clumsy** *adj* **1** awkward, blundering, bumbling, bungling, fumbling, gangling, gawky, graceless, *inf* ham-fisted, heavy-handed, hulking, inelegant, lumbering, maladroit, shambling, uncoordinated, ungainly, ungraceful, unskilful. *Opp* SKILFUL. **2** amateurish, badly-made, bulky, cumbersome, heavy, inconvenient, inelegant, large, ponderous, rough, shapeless, unmanageable, unwieldy. *Opp* NEAT. **3** *a clumsy remark.* boorish, gauche, ill-judged, inappropriate, indelicate, indiscreet, inept, insensitive, tactless, uncouth, undiplomatic, unsubtle, unsuitable.

**cluster** *n* assembly, batch, bunch, clump, collection, crowd, gathering, knot. ▷ GROUP. • *vb* ▷ GATHER.

**clutch** *n* clasp, control, evil embrace, grasp, grip, hold, possession, power. • *vb* catch, clasp, cling to, grab, grasp, grip, hang on to, hold on to, seize, snatch, take hold of.

**clutter** *n* chaos, confusion, disorder, jumble, junk, litter, lumber, mess, mix-up, muddle, odds and ends, rubbish, tangle, untidiness. • *vb* be scattered about, fill, lie about, litter, make untidy, *inf* mess up, muddle, strew.

**coach** *n* **1** bus, carriage, *old use* charabanc. **2** *games coach.* instructor, teacher, trainer, tutor. • *vb* direct, drill, exercise, guide, instruct, prepare, teach, train, tutor.

**coagulate** *vb* clot, congeal, curdle, *inf* jell, set, solidify, stiffen, thicken.

**coarse** *adj* **1** bristly, gritty, hairy, harsh, lumpy, prickly, rough, scratchy, sharp, stony, uneven, unfinished. *Opp* FINE, SOFT. **2** *coarse language.* bawdy, blasphemous, boorish, common, crude, earthy, foul, immodest, impolite, improper, impure, indecent, indelicate, offensive, ribald, rude, smutty, uncouth, unrefined, vulgar. *Opp* REFINED.

**coast** *n* beach, coastline, littoral, seaboard, seashore, seaside, shore. • *vb* cruise, drift, free-wheel, glide, sail, skim, slide, slip.

**coastal** *adj* maritime, nautical, naval, seaside.

**coat** *n* **1** □ *anorak, blazer, cagoule, cardigan, dinner-jacket, doublet, duffel coat, greatcoat, jacket, jerkin, mackintosh, overcoat, raincoat, tail-coat, tunic, tuxedo, waistcoat, wind-cheater.* **2** *an animal's coat.* fleece, fur, hair, hide, pelt, skin. **3** *coat of paint.* coating, cover, film, finish, glaze, layer, membrane, overlay, patina, sheet, veneer, wash. ▷ COVERING. • *vb* ▷ COVER. **coat of arms** ▷ CREST.

**coax** *vb* allure, beguile, cajole, charm, decoy, entice, induce, inveigle, manipulate, persuade, tempt, urge, wheedle.

**cobble** *vb* **cobble together** botch, knock up, make, mend, patch up, put together.

**code** *n* **1** etiquette, laws, manners, regulations, rule-book, rules, system. **2** *message in code.* cipher, secret language, signals, sign-system.

**coerce** *vb* bludgeon, browbeat, bully, compel, constrain, dragoon, force, frighten, intimidate, press-gang, pressurize, terrorize.

**coercion** *n* browbeating, brute force, bullying, compulsion, conscription, constraint, duress, force, intimidation, physical force pressure, *inf* strong-arm tactics, threats.

**coffer** *n* box, cabinet, case, casket, chest, crate, trunk.

**cog** *n* ratchet, sprocket, tooth.

**cogent** *adj* compelling, conclusive convincing, effective, forceful, forcible, indisputable, irresistible, logical, persuasive, potent, powerful, rational, sound, strong, unanswerable, weighty, well-argued. ▷ COHERENT. *Opp* IRRATIONAL.

**cohere** *vb* bind, cake, cling together, coalesce, combine, consolidate, fuse, hang together, hold together, join, stick together, unite.

**coherent** *adj* articulate, cohering, cohesive, connected, consistent, integrated, logical, lucid, orderly, organized, rational, reasonable, reasoned, sound, structured, systematic, unified, united, well-ordered, well-structured. ▷ COGENT. *Opp* INCOHERENT.

**coil** *n* circle, convolution, corkscrew, curl, helix, kink, loop, ring, roll, screw, spiral, twirl, twist, vortex, whirl, whorl. • *vb* bend, curl, entwine, loop, roll, snake, spiral, turn, twine, twirl, twist, wind, writhe.

**coin** *n* **1** bit, piece. **2** [*plur*] cash, change, coppers, loose change, silver, small change. ▷ MONEY. • *vb* **1** forge, make, mint, mould, stamp. **2** *coin a name.* conceive, concoct, create, devise, dream up, fabricate, hatch, introduce, invent, make up, originate, produce, think up.

**coincide** *vb* accord, agree, be congruent, be identical, be in unison, be the same, clash, coexist, come together, concur, correspond, fall together, happen together, harmonize, line up, match, square, synchronize, tally.

**coincidence** *n* **1** accord, agreement, coexistence, concurrence, conformity, congruence, congruity, correspondence, harmony, similarity. **2** *meet by coincidence.* accident, chance, fluke, luck.

**cold** *adj* **1** arctic, biting, bitter, bleak, chill, chilly, cool, crisp, cutting, draughty, freezing, fresh, frosty, glacial, heatless, ice-cold, icy, inclement, keen, *inf* nippy, numbing, *inf* parky, penetrating, perishing, piercing, polar, raw, shivery, Siberian, snowy, unheated, wintry. **2** *cold hands.* blue with cold, chilled, dead, frostbitten, frozen, numbed, shivering, shivery. **3** *a cold heart.* aloof, apathetic, callous, cold-blooded, cool, cruel, distant, frigid, hard, hard-hearted, heartless, indifferent, inhospitable, inhuman, insensitive, passionless, phlegmatic, reserved, standoffish, stony, uncaring, unconcerned, undemonstrative, unemotional, unenthusiastic, unfeeling, unkind, unresponsive, unsympathetic. ▷ UNFRIENDLY. *Opp* HOT, KIND. • *n* **1** chill, coldness, coolness, freshness, iciness, low temperature, wintriness. *Opp* HEAT. **2** *cold in the head.* catarrh, *inf* flu, influenza, *inf* the sniffles. **feel the cold** freeze, quiver, shake, shiver, shudder, suffer from hypothermia, tremble.

**cold-blooded** *adj* barbaric, brutal, callous, hard-hearted, inhuman, inhumane, merciless, pitiless, ruthless, savage. ▷ CRUEL. *Opp* HUMANE.

**cold-hearted** *adj* apathetic, cool, dispassionate, frigid, heartless, impassive, impersonal, indifferent, insensitive, thick-skinned, uncaring, unemotional, unfeeling, unkind, unresponsive, unsympathetic. ▷ UNFRIENDLY. *Opp* FRIENDLY.

**collaborate** *vb* **1** band together, cooperate, join forces, *inf* pull together, team up, work together. **2** [*derog*] collude, connive, conspire, join the opposition, *inf* rat, turn traitor.

**collaboration** *n* **1** association, concerted effort, cooperation, partnership, tandem, teamwork. **2** [*derog*] collusion, connivance, conspiracy, treachery.

**collaborator** *n* **1** accomplice, ally, assistant, associate, co-author, colleague, confederate, fellow-worker, helper, helpmate, partner, *joc* partner-in-crime, teammate. 2 *collaborator with an enemy*. blackleg, *inf* Judas, quisling, *inf* scab, traitor, turncoat.

**collapse** *n* break-down, break-up, cave-in, destruction, downfall, end, fall, ruin, ruination, subsidence, wreck. • *vb* 1 break down, break up, buckle, cave in, crumble, crumple, deflate, disintegrate, double up, fall apart, fall down, fall in, fold up, give in, *inf* go west, sink, subside, tumble down. 2 *collapse in the heat*. become ill, *inf* bite the dust, black out, *inf* crack up, faint, founder, *inf* go under, *inf* keel over, pass out, *old use* swoon. 3 *sales collapsed*. become less, crash, deteriorate, diminish, drop, fail, slump, worsen.

**collapsible** *adj* adjustable, folding, retractable, telescopic.

**colleague** *n* associate, business partner, fellow-worker. ▷ COLLABORATOR.

**collect** *vb* 1 accumulate, agglomerate, aggregate, amass, assemble, bring together, cluster, come together, concentrate, congregate, convene, converge, crowd, forgather, garner, gather, group, harvest, heap, hoard, lay up, muster, pile up, put by, rally, reserve, save, scrape together, stack up, stockpile, store. *Opp* DISPERSE. 2 *collect money for charity*. be given, raise, secure, take. 3 *collect goods from a shop*. acquire, bring, fetch, get, load up, obtain, pick up. **collected** ▷ CALM.

**collection** *n* 1 accumulation, array, assemblage, assortment, cluster, conglomeration, heap, hoard, mass, pile, set, stack, store. ▷ GROUP. 2 *old use* alms-giving, flag-day, free-will offering, offertory, voluntary contributions, *inf* whip-round.

**collective** *adj* combined, common, composite, co-operative, corporate, democratic, group, joint, shared, unified, united. *Opp* INDIVIDUAL.

**college** *n* academy, conservatory, institute, polytechnic, school, university.

**collide** *vb* **collide with** bump into, cannon into, crash into, knock, meet, run into, slam into, smash into, strike, touch. ▷ HIT.

**collision** *n* accident, bump, clash, crash, head-on collision, impact, knock, pile-up, scrape, smash, wreck.

**colloquial** *adj* chatty, conversational, everyday, informal, slangy, vernacular. *Opp* FORMAL.

**colonist** *n* colonizer, explorer, pioneer, settler. *Opp* NATIVE.

**colonize** *vb* occupy, people, populate, settle in, subjugate.

**colony** *n* 1 dependency, dominion, possession, protectorate, province, settlement, territory. 2 ▷ GROUP.

**colossal** *adj* Brobdingnagian, elephantine, enormous, gargantuan, giant, gigantic, herculean, huge, immense, *inf* jumbo, mammoth, massive, mighty, monstrous, monumental, prodigious, titanic, towering, vast. ▷ BIG. *Opp* SMALL.

**colour** *n* 1 coloration, colouring, hue, pigment, pigmentation, shade, tincture, tinge, tint, tone. □ *amber, azure, beige, black, blue, bronze, brown, buff, carroty, cherry, chestnut, chocolate, cobalt, cream, crimson, dun, fawn, gilt, gold, golden, green, grey, indigo, ivory, jet-black, khaki, lavender, maroon, mauve, navy blue, ochre, olive, orange, pink, puce, purple, red, rosy, russet, sandy, scarlet, silver, tan, tawny, turquoise, vermil-*

*ion, violet, white, yellow.* **2** *colour in your cheeks.* bloom, blush, flush, glow, rosiness, ruddiness. • *vb* **1** colour-wash, crayon, dye, paint, pigment, shade, stain, tinge, tint. **2** blush, bronze, brown, burn, flush, redden, tan. *Opp* FADE. **3** *coloured by prejudice.* affect, bias, distort, impinge on, influence, pervert, prejudice, slant, sway. **colours** ▷ FLAG.

**colourful** *adj* **1** bright, brilliant, chromatic, gaudy, irridescent, multicoloured, psychedelic, showy, vibrant. **2** *colourful personality.* dashing, distinctive, dynamic, eccentric, energetic, exciting, flamboyant, flashy, florid, glamorous, unusual, vigorous. **3** *colourful description.* graphic, interesting, lively, picturesque, rich, stimulating, striking, telling, vivid. *Opp* COLOURLESS.

**colouring** *n* colourant, dye, pigment, pigmentation, stain, tincture. ▷ COLOUR.

**colourless** *adj* **1** albino, ashen, blanched, faded, grey, monochrome, neutral, pale, pallid, sickly, wan, *inf* washed out, waxen. ▷ WHITE. **2** bland, boring, characterless, dingy, dismal, dowdy, drab, dreary, dull, insipid, lacklustre, lifeless, ordinary, tame, uninspiring, uninteresting, vacuous, vapid. *Opp* COLOURFUL.

**column** *n* **1** pilaster, pile, pillar, pole, post, prop, shaft, support, upright. **2** *newspaper column.* article, feature, leader, leading article, piece. **3** *column of soldiers.* cavalcade, file, line, procession, queue, rank, row, string, train.

**comb** *vb* **1** arrange, groom, neaten, smarten up, spruce up, tidy, untangle. **2** *comb the house.* hunt through, ransack, rummage through, scour, search thoroughly.

**combat** *n* action, battle, bout, clash, conflict, contest, duel, encounter, engagement, fight, skirmish, struggle, war, warfare. • *vb* battle against, contend against, contest, counter, defy, face up to, grapple with, oppose, resist, stand up to, strive against, struggle against, tackle, withstand. ▷ FIGHT.

**combination** *n* aggregate, alloy, amalgam, blend, compound, concoction, concurrence, conjunction, fusion, marriage, mix, mixture, synthesis, unification. **2** alliance, amalgamation, association, coalition, confederacy, confederation, consortium, conspiracy, federation, grouping, link-up, merger, partnership, syndicate, union.

**combine** *vb* **1** add together, amalgamate, bind, blend, bring together, compound, fuse, incorporate, integrate, intertwine, interweave, join, link, *inf* lump together, marry, merge, mingle, mix, pool, put together, synthesize, unify, unite. *Opp* DIVIDE. **2** *combine as a team.* ally, associate, band together, club together, coalesce, connect, cooperate, form an alliance, gang together, gang up, join forces, team up. *Opp* DISPERSE.

**combustible** *adj* flammable, inflammable. *Opp* INCOMBUSTIBLE.

**come** *vb* **1** advance, appear, approach, arrive, draw near, enter, get to, move (towards), near, reach, visit. **2** *take what comes.* happen, materialize, occur, put in an appearance, show up. **come about** ▷ HAPPEN. **come across** ▷ FIND. **come apart** ▷ DISINTEGRATE. **come clean** ▷ CONFESS. **come out with** ▷ SAY. **come round** ▷ RECOVER. **come up** ▷ ARISE. **come upon** ▷ FIND.

**comedian** *n* buffoon, clown, comic, fool, humorist, jester, joker, wag. ▷ ENTERTAINER.

**comedy** *n* buffoonery, clowning, facetiousness, farce, hilarity, humour, jesting, joking, satire, slapstick, wit.

**comfort** *n* **1** aid, cheer, consolation, encouragement, help, moral support, reassurance, relief, solace, succour, sympathy. **2** *living in comfort.* abundance, affluence, contentment, cosiness, ease, luxury, opulence, plenty, relaxation, well-being. *Opp* DISCOMFORT, POVERTY. • *vb* assuage, calm, cheer up, console, ease, encourage, gladden, hearten, help, reassure, relieve, solace, soothe, succour, sympathize with.

**comfortable** *adj* **1** *inf* comfy, convenient, cosy, easy, padded, reassuring, relaxing, roomy, snug, soft, upholstered, warm. **2** *comfortable clothes.* informal, loose-fitting, well-fitting, well-made. **3** *a comfortable life.* affluent, agreeable, contented, happy, homely, luxurious, pleasant, prosperous, relaxed, restful, serene, tranquil, untroubled, well-off. *Opp* UNCOMFORTABLE.

**comic** *adj* absurd, amusing, comical, diverting, droll, facetious, farcical, funny, hilarious, humorous, hysterical, jocular, joking, laughable, ludicrous, *inf* priceless, *inf* rich, ridiculous, sarcastic, sardonic, satirical, side-splitting, silly, uproarious, waggish, witty. *Opp* SERIOUS. • *n* **1** ▷ COMEDIAN. **2** ▷ MAGAZINE.

**command** *n* **1** behest, bidding, commandment, decree, directive, edict, injunction, instruction, mandate, order, requirement, ultimatum, writ. **2** authority, charge, control, direction, government, jurisdiction, management, oversight, power, rule, sovereignty, supervision, sway. **3** *command of a language.* grasp, knowledge, mastery. • *vb* **1** adjure, *old use* bid, charge, compel, decree, demand, direct, enjoin, instruct, ordain, order, prescribe, request, require. **2** *command a ship.* administer, be in charge of, control, direct, govern, have authority over, head, lead, manage, reign over, rule, supervise.

**commandeer** *vb* appropriate, confiscate, hijack, impound, requisition, seize, sequester, take over.

**commander** *n* captain, commandant, commanding-officer, general, head, leader, officer-in-charge. ▷ CHIEF.

**commemorate** *vb* be a memorial to, be a reminder of, celebrate, honour, immortalize, keep alive the memory of, memorialize, pay your respects to, pay homage to, pay tribute to, remember, salute, solemnize.

**commence** *vb* embark on, enter on, inaugurate, initiate, launch, open, set off, set out, set up, start. ▷ BEGIN. *Opp* FINISH.

**commend** *vb* acclaim, applaud, approve of, compliment, congratulate, eulogize, extol, praise, recommend. *Opp* CRITICIZE.

**commendable** *adj* admirable, creditable, deserving, laudable, meritorious, praiseworthy, worthwhile. ▷ GOOD. *Opp* DEPLORABLE.

**comment** *vb* animadversion, annotation, clarification, commentary, criticism, elucidation, explanation, footnote, gloss, interjection, interpolation, mention, note, observation, opinion, reaction, reference, remark, statement. • *vb* animadvert, criticize, elucidate, explain, interject, interpolate, interpose, mention, note, observe, opine, remark, say, state.

**commentary** *n* **1** account, broadcast, description, report. **2** *commentary on a poem.* analysis, criticism, critique, discourse, elucidation, explanation, interpretation, notes, review.

**commentator** *n* announcer, broadcaster, journalist, reporter.

**commerce** *n* business, buying and selling, dealings, financial transactions, marketing, merchandising, trade, trading, traffic, trafficking.

**commercial** *adj* business, economic, financial, mercantile, monetary, money-making, pecuniary, profitable, profit-making, trading. • *n inf* advert, advertisement, *inf* break, *inf* plug.

**commiserate** *vb* be sorry (for), be sympathetic, comfort, condole, console, feel (for), grieve, mourn, show sympathy (for), sympathize. *Opp* CONGRATULATE.

**commission** *n* **1** appointment, promotion, warrant. **2** *commission to do a job.* booking, order, request. **3** *commission on a sale.* allowance, *inf* cut, fee, percentage, *inf* rake-off, reward. **4** ▷ COMMITTEE.

**commit** *vb* **1** be guilty of, carry out, do, enact, execute, perform, perpetrate. **2** *commit to safe-keeping.* consign, deliver, deposit, entrust, give, hand over, put away, transfer. **commit yourself** ▷ PROMISE.

**commitment** *n* **1** assurance, duty, guarantee, liability, pledge, promise, undertaking, vow, word. **2** *commitment to a cause.* adherence, dedication, determination, devotion, involvement, loyalty, zeal. **3** *social commitments.* appointment, arrangement, engagement.

**committed** *adj* active, ardent, *inf* card-carrying, dedicated, devoted, earnest, enthusiastic, fervent, firm, keen, passionate, resolute, single-minded, staunch, unwavering, wholehearted, zealous. *Opp* APATHETIC.

**committee** *n* body, council, panel. □ *assembly, board, cabinet, caucus, commission, convention, junta, jury, parliament, quango, synod, think-tank, working party.* ▷ GROUP, MEETING.

**common** *adj* **1** average, *inf* common or garden, conventional, customary, daily, everyday, familiar, frequent, habitual, normal, ordinary, plain, popular, prevalent, regular, routine, *inf* run-of-the-mill, standard, stock, traditional, typical, undistinguished, unexceptional, unsurprising, usual, well-known, widespread, workaday. ▷ COMMONPLACE. **2** *common knowledge.* accepted, collective, communal, general, joint, mutual, open, popular, public, shared, universal. **3** *the common people.* lower class, lowly, plebeian, proletarian. **4** [*inf*] *Don't be common!* boorish, churlish, coarse, crude, disreputable, ill-bred, inferior, loutish, low, rude, uncouth, unrefined, vulgar, *inf* yobbish. *Opp* ARISTOCRATIC, DISTINCTIVE, UNUSUAL. • *n* heath, park, parkland.

**commonplace** *adj* banal, boring, forgettable, hackneyed, humdrum, mediocre, obvious, ordinary, pedestrian, plain, platitudinous, predictable, prosaic, routine, standard, trite, unexciting, unremarkable. ▷ COMMON. *Opp* MEMORABLE. • *n* ▷ PLATITUDE.

**commotion** *n inf* ado, agitation, *inf* bedlam, bother, brawl, *inf* brouhaha, *inf* bust-up, chaos, clamour, confusion, contretemps, din, disorder, disturbance, excitement, ferment, flurry, fracas, fray, furore, fuss, hubbub, hullabaloo, incident, *inf* kerfuffle, noise, *inf* palaver, pandemonium, *inf* punch-up, quarrel, racket, riot, row, rumpus, sensation, *inf* shemozzle, *inf* stir, *inf* to-do, tumult, turbulence, turmoil, unrest, upheaval, uproar, upset.

**communal** *adj* collective, common, general, joint, mutual, open, public, shared. *Opp* PRIVATE.

**communicate** *vb* **1** commune, confer, converse, correspond, discuss, get in touch, interrelate, make contact, speak, talk, write (to). **2** *communicate information.* advise, announce, broadcast, convey, declare, disclose, disseminate, divulge, express, get across, impart, indicate, inform, intimate, make known, mention, network, notify, pass on, proclaim, promulgate, publish, put across, put over, relay, report, reveal, say, show, speak, spread, state, transfer, transmit, write. **3** *communicate a disease.* give, infect someone with, pass on, spread, transfer, transmit. **4** *The passage communicates with the kitchen.* be connected, lead (to).

**communication** *n* **1** communicating, communion, contact, interaction, *old use* intercourse. □ *announcement, bulletin, cable, card, communiqué, conversation, correspondence, dialogue, directive, dispatch, document, fax, gossip,* inf *grapevine, information, intelligence, intimation, letter,* inf *memo, memorandum, message, news, note, notice, proclamation, report, rumour, signal, statement, talk, telegram, transmission, wire, word of mouth, writing.* □ *CB, computer, intercom, radar, telegraph, telephone, teleprinter, walkie-talkie.* **2** *mass communication.* mass media, the media. □ *advertising, broadcasting, cable television, magazines, newspapers, the press, radio, satellite, telecommunication, television.*

**communicative** *adj* articulate, *inf* chatty, frank, informative, open, out-going, responsive, sociable. ▷ TALKATIVE. *Opp* SECRETIVE.

**community** *n* colony, commonwealth, commune, country, kibbutz, nation, society, state. ▷ GROUP.

**commute** *vb* **1** adjust, alter, curtail, decrease, lessen, lighten, mitigate, reduce, shorten. **2** ▷ TRAVEL.

**compact** *adj* **1** close-packed, compacted, compressed, consolidated, dense, firm, heavy, packed, solid, tight-packed. *Opp* LOOSE. **2** handy, neat, portable, small. **3** abbreviated, abridged, brief, compendious, compressed, concentrated, condensed, short, small, succinct, terse. ▷ CONCISE. *Opp* LARGE. • *n* ▷ AGREEMENT.

**companion** *n* accomplice, assistant, associate, chaperone, colleague, comrade, confederate, confidant(e), consort, *inf* crony, escort, fellow, follower, *inf* henchman, mate, partner, stalwart. ▷ FRIEND, HELPER.

**company** *n* **1** companionship, fellowship, friendship, society. **2** [*inf*] *company for tea.* callers, guests, visitors. **3** *mixed company.* assemblage, association, band, body, circle, club, community, coterie, crew, crowd, ensemble, entourage, gang, gathering, society, throng, troop, troupe (*of actors*). **4** *trading company*. business, cartel, concern, conglomerate, consortium, corporation, establishment, firm, house, line, organization, partnership, *inf* set-up, syndicate, union. ▷ GROUP.

**comparable** *adj* analogous, cognate, commensurate, compatible, corresponding, equal, equivalent, matching, parallel, proportionate, related, similar, twin. *Opp* DISSIMILAR.

**compare** *vb* check, contrast, correlate, draw parallels (between), equate, juxtapose, liken, make comparisons, make connections (between), measure (against), relate (to), set side by side, weigh (against). **compare with** ▷ EQUAL.

**comparison** *n* analogy, comparability, contrast, correlation, difference, distinction, juxtaposition, likeness, parallel, relationship, resemblance, similarity.

**compartment** *n* alcove, area, bay, berth, booth, cell, chamber, *inf* cubbyhole, cubicle, division, hole, kiosk, locker, niche, nook, partition, pigeonhole, section, slot, space, subdivision.

**compatible** *adj* **1** harmonious, like-minded, similar, well-matched. ▷ FRIENDLY. **2** *compatible claims*. accordant, congruent, consistent, consonant, matching, reconcilable. *Opp* INCOMPATIBLE.

**compel** *vb* bind, bully, coerce, constrain, dragoon, drive, exact, force, impel, make, necessitate, oblige, order, press, press-gang, pressurize, require, *inf* shanghai, urge.

**compendium** *n* abridgement, abstract, anthology, collection, condensation, digest, handbook, summary.

**compensate** *vb* **1** atone, *inf* cough up, expiate, indemnify, make amends, make good, make reparation, make restitution, make up for, pay back, pay compensation, recompense, redress, reimburse, remunerate, repay, requite. **2** counterbalance, counterpoise, even up, neutralize, offset.

**compensation** *n* amends, damages, indemnity, recompense, refund, reimbursement, reparation, repayment, restitution.

**compère** *n* anchor-man, announcer, disc jockey, host, hostess, linkman, Master of Ceremonies, MC, presenter.

**compete** *vb* **1** be a contestant, enter, participate, perform, take part, take up the challenge. **2** be in competition, conflict, contend, emulate, oppose, rival, strive, struggle, undercut, vie. ▷ FIGHT. *Opp* CO-OPERATE. **compete with** ▷ RIVAL.

**competent** *adj* able, acceptable, accomplished, adept, adequate, capable, clever, effective, effectual, efficient, experienced, expert, fit, *inf* handy, practical, proficient, qualified, satisfactory, skilful, skilled, trained, workmanlike, worthwhile. *Opp* INCOMPETENT.

**competition** *n* **1** competitiveness, conflict, contention, emulation, rivalry, struggle. **2** challenge, championship, contest, event, game, heat, match, quiz, race, rally, series, tournament, trial.

**competitive** *adj* **1** aggressive, antagonistic, combative, contentious, cut-throat, hard-fought, keen, lively, sporting, well-fought. **2** *competitive prices*. average, comparable with others, fair, moderate, reasonable, similar to others.

**competitor** *n* adversary, antagonist, candidate, challenger, contender, contestant, entrant, finalist, opponent, participant, rival.

**compile** *vb* accumulate, amass, arrange, assemble, collate, collect, compose, edit, gather, marshal, organize, put together.

**complain** *vb inf* beef, *inf* bellyache, *sl* bind, carp, cavil, find fault, fuss, *inf* gripe, groan, *inf* grouch, grouse, grumble, lament, moan, object, protest, wail, whine, *inf* whinge. *Opp* PRAISE. **complain about** ▷ CRITICIZE.

**complaint** *n* **1** accusation, *inf* beef, charge, condemnation, criticism, grievance, *inf* gripe, grouse, grumble, moan, objection, protest, stricture, whine, whinge. **2** *a medical complaint.* affliction, ailment, disease, disorder, infection, malady, malaise, sickness, upset. ▷ ILLNESS.

**complaisant** *adj* accommodating, acquiescent, amenable, biddable, compliant, cooperative, deferential, docile, obedient, obliging, pliant, polite, submissive, tractable, willing. *Opp* OBSTINATE.

**complement** *n* **1** completion, *inf* finishing touch, perfection. **2** *a full complement.* aggregate, capacity, quota, sum, total. • *vb* add to, complete, make whole, perfect, round off, top up.

**complementary** *adj* interdependent, matching, reciprocal, toning, twin.

**complete** *adj* **1** comprehensive, entire, exhaustive, full, intact, total, unabbreviated, unabridged, uncut, unedited, unexpurgated, whole. **2** accomplished, achieved, completed, concluded, done, ended, finished, over. ▷ PERFECT. **3** *a complete disaster.* absolute, arrant, downright, extreme, *inf* out-and-out, outright, pure, rank, sheer, thorough, thoroughgoing, total, unmitigated, unmixed, unqualified, utter, *inf* wholesale. *Opp* INCOMPLETE. • *vb* **1** accomplish, achieve, carry out, clinch, close, conclude, crown, do, end, finalize, finish, fulfil, perfect, perform, round off, terminate, *inf* top off, *inf* wind up. **2** *complete forms.* answer, fill in.

**complex** *adj* complicated, composite, compound, convoluted, elaborate, *inf* fiddly, heterogeneous, intricate, involved, *inf* knotty (*problem*), labyrinthine, manifold, mixed, multifarious, multiple, multiplex, ornate, perplexing, problematical, sophisticated, tortuous, *inf* tricky. *Opp* SIMPLE.

**complexion** *n* appearance, colour, colouring, look, pigmentation, skin, texture.

**complicate** *vb* compound, confound, confuse, elaborate, entangle, make complicated, mix up, muddle, *inf* screw up, *inf* snarl up, tangle, twist. *Opp* SIMPLIFY. **complicated** ▷ COMPLEX.

**complication** *n* complexity, confusion, convolution, difficulty, dilemma, intricacy, *inf* mix-up, obstacle, problem, ramification, set-back, snag, tangle.

**compliment** *n* accolade, admiration, appreciation, approval, commendation, congratulations, encomium, eulogy, felicitations, flattery, honour, panegyric, plaudits, praise, testimonial, tribute. • *vb* applaud, commend, congratulate, *inf* crack up, eulogize, extol, felicitate, flatter, give credit, laud, pay homage to, praise, salute, speak highly of. *Opp* INSULT.

**complimentary** *adj* admiring, appreciative, approving, commendatory, congratulatory, encomiastic, eulogistic, favourable, flattering, *derog* fulsome, generous, laudatory, panegyrical, rapturous, supportive. *Opp* ABUSIVE, CONTEMPTUOUS, CRITICAL.

**comply** *vb* abide (by), accede, accord, acquiesce, adhere (to), agree, assent, be in accordance, coincide, concur, consent, correspond (to), defer, fall in (with), fit in, follow, fulfil, harmonize, keep (to), match, meet, obey, observe, perform, respect, satisfy, square (with), submit, suit, yield. ▷ CONFORM. *Opp* DEFY.

**component** *n* bit, constituent, element, essential part, ingredient, item, part, piece, *inf* spare, spare part, unit.

**compose** *vb* **1** build, compile, constitute, construct, fashion, form, frame, make, put together. **2** *compose music.* arrange, create, devise, imagine, make up, produce, write. **3** *compose yourself.* calm, control, pacify, quieten, soothe, tranquillize. be composed of ▷ COMPRISE. **composed** ▷ CALM.

**composition** *n* **1** assembly, constitution, creation, establishment, formation, formulation, *inf* make-up, setting up. **2** balance, configuration, layout, organization, structure. **3** *a literary composition.* article, essay, story. ▷ WRITING. **4** *a musical composition.* opus, piece, work. ▷ MUSIC.

**compound** *adj* complex, complicated, composite, intricate, involved, multiple. *Opp* SIMPLE. • *n* **1** alloy, amalgam, blend, combination, composite, composition, fusion, mixture, synthesis. **2** *compound for cattle. Amer* corral, enclosure, pen, run. • *vb* ▷ COMBINE, COMPLICATE.

**comprehend** *vb* appreciate, apprehend, conceive, discern, fathom, follow, grasp, know, perceive, realize, see, take in, *inf* twig, understand.

**comprehensible** *adj* clear, easy, intelligible, lucid, meaningful, plain, self-explanatory, simple, straightforward, understandable. *Opp* INCOMPREHENSIBLE.

**comprehensive** *adj* all-embracing, broad, catholic, compendious, complete, detailed, encyclopaedic, exhaustive, extensive, far-reaching, full, inclusive, indiscriminate, sweeping, thorough, total, universal, wholesale, wide-ranging. *Opp* SELECTIVE.

**compress** *vb* abbreviate, abridge, compact, concentrate, condense, constrict, contract, cram, crush, flatten, *inf* jam, précis, press, shorten, squash, squeeze, stuff, summarize, telescope, truncate. *Opp* EXPAND. **compressed** ▷ COMPACT, CONCISE.

**comprise** *vb* be composed of, comprehend, consist of, contain, cover, embody, embrace, include, incorporate, involve.

**compromise** *n* bargain, concession, *inf* give-and-take, *inf* halfway house, middle course, middle way, settlement. • *vb* **1** concede a point, go to arbitration, make concessions, meet halfway, negotiate a settlement, reach a formula, settle, *inf* split the difference, strike a balance. **2** *compromise your reputation.* damage, discredit, disgrace, dishonour, imperil, jeopardize, prejudice, risk, undermine, weaken. **compromising** ▷ SHAMEFUL.

**compulsion** *n* **1** coercion, duress, force, necessity, restriction, restraint. **2** *compulsion to smoke*. addiction, drive, habit, impulse, pressure, urge.

**compulsive** *adj* **1** besetting, compelling, driving, instinctive, involuntary, irresistible, overpowering, overwhelming, powerful, uncontrollable, urgent. **2** *compulsive drinker*. addicted, habitual, incorrigible, incurable, obsessive, persistent.

**compulsory** *adj* binding, contractual, *Fr* de rigueur, enforceable, essential, imperative, imposed, incumbent, indispensable, inescapable, mandatory, necessary, obligatory, official, prescribed, required, requisite, set, statutory, stipulated, unavoidable. *Opp* OPTIONAL.

**compunction** *n* contrition, hesitation, pang of conscience, qualm, regret, remorse, scruple, self-reproach.

**compute** *vb* add up, ascertain, assess, calculate, count, determine, estimate, evaluate, measure, reckon, total, work out.

**computer** *n* mainframe, micro, microcomputer, mini-computer, PC, personal computer, robot, word-processor.

**comrade** *n* associate, colleague, companion. ▷ FRIEND.

**conceal** *vb* blot out, bury, camouflage, cloak, cover up, disguise, envelop, gloss over, hide, hush up, keep dark, keep quiet, keep secret, mask, obscure, screen, secrete, suppress, veil. *Opp* REVEAL. **concealed** ▷ HIDDEN.

**concede** *vb* accept, acknowledge, admit, agree, allow, confess, grant, make a concession, own, profess, recognize. **concede defeat** capitulate, *inf* cave in, cede, give in, resign, submit, surrender, yield.

**conceit** *n* arrogance, boastfulness, egotism, self-admiration, self-esteem, self-love, vanity. ▷ PRIDE.

**conceited** *adj* arrogant, *inf* bigheaded, boastful, bumptious, *inf* cocksure, *inf* cocky, egocentric, egotistic(al), grand, haughty, *inf* high and mighty, immodest, narcissistic, overweening, pleased with yourself, proud, self-centred, self-important, self-satisfied, smug, snobbish, *inf* snooty, *inf* stuck-up, supercilious, *inf* swollen-headed, *inf* toffee-nosed, vain, vainglorious. *Opp* MODEST.

**conceive** *vb* **1** become pregnant. **2** *conceive a plan*. conjure up, contrive, create, design, devise, *inf* dream up, envisage, evolve, form, formulate, frame, germinate, hatch, imagine, initiate, invent, make up, originate, plan, plot, produce, realize, suggest, think up, visualize, work out. ▷ THINK.

**concentrate** *n* distillation, essence, extract. • *vb* **1** apply yourself, attend, be absorbed, be attentive, engross yourself, think, work hard. **2** accumulate, centralize, centre, cluster, collect, congregate, converge, crowd, focus, gather, mass. *Opp* DISPERSE. **3** *concentrate a liquid*. condense, reduce, thicken. *Opp* DILUTE. **concentrated** **1** ▷ INTENSIVE. **2** condensed, evaporated, reduced, strong, thick, undiluted.

**conception** *n* **1** begetting, conceiving, fathering, fertilization, genesis, impregnation, initiation, origin. ▷ BEGINNING. **2** ▷ IDEA.

**concern** *n* **1** attention, care, charge, consideration, heed, interest, regard. **2** *no concern of yours*. affair, business, involvement, matter, problem, responsibility, task.

3 *matter for concern*. anxiety, burden, disquiet, distress, fear, malaise, solicitude, worry. 4 *business concern*. business, company, corporation, establishment, enterprise, firm, organization.
• *vb* affect, be important to, be relevant to, interest, involve, matter to, pertain to, refer to, relate to.

**concerned** *adj* 1 *concerned parents*. bothered, caring, distressed, disturbed, fearful, perturbed, solicitous, touched, troubled, uneasy, unhappy, upset, worried. ▷ ANXIOUS. 2 *the people concerned*. connected, implicated, interested, involved, referred to, relevant. ▷ RESPONSIBLE.

**concerning** *prep* about, apropos of, germane to, involving, re, regarding, relating to, relevant to, with reference to, with regard to.

**concert** *n* performance, programme, show. ▷ ENTERTAINMENT, MUSIC.

**concerted** *adj* collaborative, collective, combined, cooperative, joint, mutual, shared, united.

**concession** *n* adjustment, allowance, reduction.

**concise** *adj* brief, compact, compendious, compressed, concentrated, condensed, epigrammatic, laconic, pithy, short, small, succinct, terse. ▷ ABRIDGED. *Opp* DIFFUSE.

**conclude** *vb* 1 cease, close, complete, culminate, end, finish, round off, stop, terminate. 2 assume, decide, deduce, gather, infer, judge, reckon, suppose, surmise. ▷ THINK.

**conclusion** *n* 1 close, completion, culmination, end, epilogue, finale, finish, peroration, rounding-off, termination. 2 answer, belief, decision, deduction, inference, interpretation, judgement, opinion, outcome, resolution, result, solution, upshot, verdict.

**conclusive** *adj* certain, convincing, decisive, definite, persuasive, unambiguous, unanswerable, unequivocal, unquestionable. *Opp* INCONCLUSIVE.

**concoct** *vb* contrive, cook up, counterfeit, devise, fabricate, feign, formulate, hatch, invent, make up, plan, prepare, put together, think up.

**concord** *n* agreement, euphony, harmony, peace.

**concrete** *adj* actual, definite, existing, factual, firm, material, objective, palpable, physical, real, solid, substantial, tactile, tangible, touchable, visible. *Opp* ABSTRACT.

**concur** *vb* accede, accord, agree, assent. ▷ COMPLY.

**concurrent** *adj* coexisting, coinciding, concomitant, contemporaneous, contemporary, overlapping, parallel, simultaneous, synchronous.

**condemn** *vb* 1 blame, castigate, censure, criticize, damn, decry, denounce, deplore, deprecate, disapprove of, disparage, execrate, rebuke, reprehend, reprove, revile, *inf* slam, *inf* slate, upbraid. *Opp* COMMEND. 2 convict, find guilty, judge, pass judgement, prove guilty, punish, sentence. *Opp* ACQUIT.

**condense** *vb* 1 abbreviate, abridge, compress, contract, curtail, précis, reduce, shorten, summarize, synopsize. *Opp* EXPAND. 2 *condense a liquid*. concentrate, distil, reduce, solidify, thicken. *Opp* DILUTE.

**condensation** *n* haze, mist, precipitation, *inf* steam, water-drops.

**condescend** *vb* deign, demean yourself, humble yourself, lower yourself, stoop. **condescending** ▷ HAUGHTY.

**condition** *n* **1** case, circumstance, *inf* fettle, fitness, form, health, *inf* nick, order, shape, situation, state, *inf* trim, working order. **2** limitation, obligation, prerequisite, proviso, qualification, requirement, requisite, restriction, stipulation, terms. **3** *medical condition.* ▷ ILLNESS. • *vb* acclimatize, accustom, brainwash, educate, mould, prepare, re-educate, *inf* soften up, teach, train.

**conditional** *adj* dependent, limited, provisional, qualified, restricted, safeguarded, *inf* with strings attached. *Opp* UNCONDITIONAL.

**condone** *vb* allow, connive at, disregard, endorse, excuse, forgive, ignore, let someone off, overlook, pardon, tolerate.

**conducive** *adj* advantageous, beneficial, encouraging, favourable, helpful, supportive. **be conducive to** ▷ ENCOURAGE.

**conduct** *n* **1** actions, attitude, bearing, behaviour, comportment, demeanour, deportment, manners, ways. **2** *conduct of affairs.* administration, control, direction, discharge, government, guidance, handling, leading, management, operation, organization, regulation, running, supervision. • *vb* **1** administer, be in charge of, chair, command, control, direct, escort, govern, handle, head, look after, manage, organize, oversee, preside over, regulate, rule, run, steer, superintend, supervise, usher. **2** *conduct me home.* accompany, escort, guide, lead, pilot, take, usher. **3** *conduct electricity.* carry, channel, convey, transmit. **conduct yourself** ▷ BEHAVE.

**confer** *vb* **1** accord, award, bestow, give, grant, honour with, impart, invest, present. **2** compare notes, consult, converse, debate, deliberate, discourse, discuss, exchange ideas, *inf* put your heads together, seek advice. ▷ TALK.

**conference** *n* colloquium, congress, consultation, convention, council, deliberation, discussion, forum, seminar, symposium. ▷ MEETING.

**confess** *vb* acknowledge, admit, be truthful, *inf* come clean, concede, disclose, divulge, *inf* make a clean breast (of), own up, unbosom yourself, unburden yourself.

**confession** *n* acknowledgement, admission, declaration, disclosure, expression, profession, revelation.

**confide** *vb* consult, speak confidentially, *inf* spill the beans, open your heart, *inf* tell all, tell secrets, unbosom yourself, trust.

**confidence** *n* **1** belief, certainty, credence, faith, hope, optimism, positiveness, reliance, trust. **2** aplomb, assurance, boldness, composure, conviction, firmness, nerve, panache, self-assurance, self-confidence, self-possession, self-reliance, spirit, verve. *Opp* DOUBT, HESITATION. **have confidence in** ▷ TRUST.

**confident** *adj* **1** certain, convinced, hopeful, optimistic, positive, sanguine, sure, trusting. **2** *a confident person.* assertive, assured, bold, *derog* cocksure, composed, cool, definite, fearless, secure, self-assured, self-confident, self-possessed, self-reliant, unafraid. *Opp* DOUBTFUL.

**confidential** *adj* **1** classified, *inf* hush-hush, intimate, *inf* off the record, personal, private, restricted, secret, suppressed, top secret. **2** *confidential secretary*. personal, private, trusted.

**confine** *vb* bind, box in, cage, circumscribe, constrain, *inf* coop up, cordon off, cramp, curb, detain, enclose, gaol, hedge in, hem in, *inf* hold down, immure, incarcerate, isolate, keep in, limit, localize, restrain, restrict, rope off, shut in, shut up, surround, wall up. ▷ IMPRISON. *Opp* FREE.

**confirm** *vb* **1** authenticate, back up, bear out, corroborate, demonstrate, endorse, establish, fortify, give credence to, justify, lend force to, prove, reinforce, settle, show, strengthen, substantiate, support, underline, witness to, vindicate. **2** *confirm a deal*. authorize, *inf* clinch, formalize, guarantee, make legal, make official, ratify, sanction, validate, verify.

**confiscate** *vb* appropriate, commandeer, expropriate, impound, remove, seize, sequester, sequestrate, take away, take possession of.

**conflict** *n* **1** antagonism, antipathy, contention, contradiction, difference, disagreement, discord, dissension, friction, hostility, incompatibility, inconsistency, opposition, strife. **2** altercation, battle, *inf* brush, clash, combat, confrontation, contest, dispute, encounter, engagement, feud, fight, quarrel, row, *inf* set-to, skirmish, struggle, war, warfare, wrangle. • *vb* **1** *inf* be at odds, be at variance, be incompatible, clash, compete, contend, contradict, contrast, *inf* cross swords, differ, disagree, oppose each other. ▷ FIGHT, QUARREL.

**conform** *vb* acquiesce, agree, be good, behave conventionally, blend in, *inf* do what you are told, fit in, *inf* keep in step, obey, *inf* see eye to eye, *inf* toe the line. ▷ COMPLY.

**conformist** *n* conventional person, traditionalist, yes-man. *Opp* REBEL.

**conformity** *n* complaisance, compliance, conventionality, obedience, orthodoxy, submission, uniformity.

**confront** *vb* accost, argue with, attack, brave, challenge, defy, encounter, face up to, meet, oppose, resist, stand up to, take on, withstand. *Opp* AVOID.

**confuse** *vb* **1** disarrange, disorder, distort, entangle, garble, jumble, *inf* mess up, mingle, mix up, muddle, tangle, *inf* throw into disarray, upset. **2** *rules confuse me*. agitate, baffle, befuddle, bemuse, bewilder, confound, disconcert, disorientate, distract, *inf* flummox, fluster, mislead, mystify, perplex, puzzle, *inf* rattle, *inf* throw. **3** *confuse twins*. fail to distinguish. **confusing** ▷ PUZZLING.

**confused** *adj* **1** chaotic, disordered, disorderly, disorganized, *inf* higgledy-piggledy, jumbled, messy, mixed up, muddled, *inf* screwed-up, *sl* shambolic, *inf* topsy-turvy, twisted. **2** *confused ideas*. aimless, contradictory, disconnected, disjointed, garbled, incoherent, inconsistent, irrational, misleading, obscure, rambling, unclear, unsound, unstructured, woolly. **3** *confused mind*. addled, addle-headed, baffled, bewildered, dazed, disorientated, distracted, flustered, fuddled, *inf* in a tizzy, inebriated, muddle-headed, *inf* muzzy, mystified, nonplussed, perplexed, puzzled. ▷ MAD. *Opp* ORDERLY.

**confusion** *n* 1 *inf* ado, anarchy, bedlam, bother, chaos, clutter, commotion, confusion, din, disorder, disorganization, disturbance, fuss, hubbub, hullabaloo, jumble, maelstrom, *inf* mayhem, mêlée, mess, *inf* mix-up, muddle, pandemonium, racket, riot, rumpus, shambles, tumult, turbulence, turmoil, upheaval, uproar, welter, whirl. 2 *mental confusion.* bemusement, bewilderment, disorientation, distraction, mystification, perplexity, puzzlement. *Opp* ORDER.

**congeal** *vb* clot, coagulate, coalesce, condense, curdle, freeze, harden, *inf* jell, set, solidify, stiffen, thicken.

**congenial** *adj* acceptable, agreeable, amicable, companionable, compatible, genial, kindly, suitable, sympathetic, understanding, well-suited. ▷ FRIENDLY. *Opp* UNCONGENIAL.

**congenital** *adj* hereditary, inborn, inbred, inherent, inherited, innate, natural.

**congested** *adj* blocked, choked, clogged, crammed, crowded, full, jammed, obstructed, overcrowded, stuffed. *Opp* CLEAR.

**congratulate** *vb* applaud, compliment, felicitate, praise.

**congregate** *vb* assemble, cluster, collect, come together, convene, converge, crowd, forgather, gather, get together, mass, meet, muster, rally, rendezvous, swarm, throng. ▷ GROUP.

**conjure** *vb* bewitch, charm, compel, enchant, invoke, raise, rouse, summon. **conjure up** ▷ PRODUCE.

**conjuring** *n* illusions, legerdemain, magic, sleight of hand, tricks, wizardry.

**connect** *vb* 1 attach, combine, couple, engage, fix, interlock, join, link, put on, switch on, tie, turn on, unite. ▷ FASTEN. 2 associate, bracket together, compare, make a connection between, put together, relate, tie up. *Opp* SEPARATE.

**connection** *n* affinity, association, bond, coherence, contact, correlation, correspondence, interrelationship, link, relationship, relevance, tie, *inf* tie-up, unity. *Opp* SEPARATION.

**conquer** *vb* 1 annex, beat, best, capture, checkmate, crush, defeat, get the better of, humble, *inf* lick, master, occupy, outdo, overcome, overpower, overrun, overthrow, overwhelm, possess, prevail over, quell, rout, seize, silence, subdue, subject, subjugate, succeed against, surmount, take, *inf* thrash, triumph over, vanquish, worst. ▷ WIN. 2 *conquer a mountain.* climb, reach the top of.

**conquest** *n* annexation, appropriation, capture, defeat, domination, invasion, occupation, overthrow, subjection, subjugation, *inf* takeover. ▷ VICTORY.

**conscience** *n* compunction, ethics, honour, fairness, misgivings, morality, morals, principles, qualms, reservations, scruples, standards.

**conscientious** *adj* accurate, attentive, careful, diligent, dutiful, exact, hard-working, high-minded, honest, meticulous, painstaking, particular, punctilious, responsible, rigorous, scrupulous, serious, thorough. *Opp* CARELESS.

**conscious** *adj* 1 alert, awake, aware, compos mentis, sensible. 2 *a conscious act.* calculated, deliberate, intended, intentional, knowing, planned, premeditated, self-conscious, studied, voluntary, waking, wilful. *Opp* UNCONSCIOUS.

**consecrate** *vb* bless, dedicate, devote, hallow, make sacred, sanctify. *Opp* DESECRATE.

**consecutive** *adj* continuous, following, one after the other, running (*3 days running*), sequential, succeeding, successive.

**consent** *n* acquiescence, agreement, approval, assent, concurrence, imprimatur, permission, seal of approval. • *vb* accede, acquiesce, agree, approve, comply, concede, concur, conform, submit, undertake, yield. *Opp* REFUSE. **consent to** ▷ ALLOW.

**consequence** *n* **1** aftermath, by-product, corollary, effect, end, *inf* follow-up, issue, outcome, repercussion, result, sequel, side-effect, upshot. **2** *of no consequence*. account, concern, importance, moment, note, significance, value, weight.

**consequent** *adj* consequential, ensuing, following, resultant, resulting, subsequent.

**conservation** *n* careful management, economy, good husbandry, maintenance, preservation, protection, safeguarding, saving, upkeep. *Opp* DESTRUCTION.

**conservationist** *n* ecologist, environmentalist, *inf* green, preservationist.

**conservative** *adj* **1** conventional, die-hard, hidebound, moderate, narrow-minded, old-fashioned, reactionary, sober, traditional, unadventurous. **2** *conservative estimate*. cautious, moderate, reasonable, understated, unexaggerated. **3** *conservative politics*. right-of-centre, right-wing, Tory. *Opp* PROGRESSIVE. • *n* conformist, die-hard, reactionary, right-winger, Tory, traditionalist.

**conserve** *vb* be economical with, hold in reserve, keep, look after, maintain, preserve, protect, safeguard, save, store up, use sparingly. *Opp* DESTROY, WASTE.

**consider** *vb* **1** *inf* chew over, cogitate, contemplate, deliberate, discuss, examine, meditate, mull over, muse, ponder, puzzle over, reflect, ruminate, study, *inf* turn over, weigh up. ▷ THINK. **2** believe, deem, judge, reckon.

**considerable** *adj* appreciable, big, biggish, comfortable, fairly important, fairly large, noteworthy, noticeable, perceptible, reasonable, respectable, significant, sizeable, substantial, *inf* tidy (amount), tolerable, worthwhile. *Opp* NEGLIGIBLE.

**considerate** *adj* accommodating, altruistic, attentive, caring, charitable, cooperative, friendly, generous, gracious, helpful, kind, kind-hearted, kindly, neighbourly, obliging, polite, sensitive, solicitous, sympathetic, tactful, thoughtful, unselfish. *Opp* SELFISH.

**consign** *vb* commit, convey, deliver, devote, entrust, give, hand over, pass on, relegate, send, ship, transfer.

**consignment** *n* batch, cargo, delivery, load, lorry-load, shipment, van-load.

**consist** *vb* **consist of** add up to, amount to, be composed of, be made of, comprise, contain, embody, include, incorporate, involve.

**consistent** *adj* **1** constant, dependable, faithful, predictable, regular, reliable, stable, steadfast, steady, unchanging, undeviating, unfailing, uniform, unvarying.

2 *The stories are consistent.* accordant, compatible, congruous, consonant, in accordance, in agreement, in harmony, of a piece. *Opp* INCONSISTENT.

**console** *vb* calm, cheer, comfort, ease, encourage, hearten, relieve, solace, soothe, succour, sympathize with.

**consolidate** *vb* make secure, make strong, reinforce, stabilize, strengthen. *Opp* WEAKEN.

**consort** *vb* **consort with** accompany, associate with, befriend, be friends with, be seen with, fraternize with, *inf* gang up with, keep company with, mix with.

**conspicuous** *adj* apparent, blatant, clear, discernible, distinguished, dominant, eminent, evident, flagrant, glaring, impressive, manifest, marked, notable, noticeable, obtrusive, obvious, ostentatious, outstanding, patent, perceptible, plain, prominent, pronounced, self-evident, shining (*example*), showy, striking, unconcealed, unmistakable, visible. *Opp* INCONSPICUOUS.

**conspiracy** *n* cabal, collusion, connivance, *inf* frame-up, insider dealing, intrigue, machinations, plot, *inf* racket, scheme, stratagem, treason.

**conspirator** *n* plotter, schemer, traitor, *inf* wheeler-dealer.

**conspire** *vb* be in league, collude, combine, connive, cooperate, hatch a plot, have designs, intrigue, plot, scheme.

**constant** *adj* **1** ceaseless, chronic, consistent, continual, continuous, endless, eternal, everlasting, fixed, immutable, incessant, invariable, neverending, non-stop, permanent, perpetual, persistent, predictable, regular, relentless, repeated, stable, steady, sustained, unbroken, unchanging, unending, unflagging, uniform, uninterrupted, unremitting, unvarying. **2** *a constant friend.* dedicated, dependable, determined, devoted, faithful, firm, indefatigable, loyal, reliable, resolute, staunch, steadfast, tireless, true, trustworthy, trusty, unswerving, unwavering. *Opp* CHANGEABLE.

**constitute** *vb* appoint, bring together, compose, comprise, create, establish, form, found, inaugurate, make (up), set up.

**construct** *vb* assemble, build, create, engineer, erect, fabricate, fashion, fit together, form, *inf* knock together, make, manufacture, pitch (*tent*), produce, put together, put up, set up. *Opp* DEMOLISH.

**construction** *n* **1** assembly, building, creation, erecting, erection, manufacture, production, putting-up, setting-up. **2** building, edifice, erection, structure.

**constructive** *adj* advantageous, beneficial, cooperative, creative, helpful, positive, practical, productive, useful, valuable, worthwhile. *Opp* DESTRUCTIVE.

**consult** *vb* confer, debate, discuss, exchange views, *inf* put your heads together, refer (to), seek advice, speak (to), *inf* talk things over. ▷ QUESTION.

**consume** *vb* **1** devour, digest, drink, eat, *inf* gobble up, *inf* guzzle, *inf* put away, swallow. **2** *consume energy.* absorb, deplete, drain, eat into, employ, exhaust, expend, swallow up, use up, utilize.

**contact** *n* connection, join, junction, touch, union. ▷ COMMUNICATION. • *vb* apply to, approach, call on, communicate with, correspond with, *inf* drop a line to, get hold of, get in touch with, make overtures to, notify, phone, ring, sound out, speak to, talk to, telephone.

**contagious** *adj* catching, communicable, infectious, spreading, transmittable.

**contain** *vb* **1** accommodate, enclose, hold. **2** be composed of, comprise, consist of, embody, embrace, include, incorporate, involve. **3** *contain your anger*. check, control, curb, hold back, keep back, limit, repress, restrain, stifle.

**container** *n* holder, receptacle, repository, vessel. ▷ BAG, BARREL, BOTTLE, BOWL, BOX, CUP, DISH, GLASS, LUGGAGE, POT.

**contaminate** *vb* adulterate, befoul, corrupt, debase, defile, dirty, foul, infect, poison, pollute, soil, spoil, stain, sully, taint. *Opp* PURIFY.

**contemplate** *vb* **1** eye, gaze at, look at, observe, regard, stare at, survey, view, watch. ▷ SEE. **2** cogitate, consider, day-dream, deliberate, examine, meditate, mull over, muse, plan, ponder, reflect, ruminate, study, work out. ▷ THINK. **3** envisage, expect, intend, propose.

**contemporary** *adj* **1** *contemporary with me at school*. coeval, coexistent, coinciding, concurrent, contemporaneous, simultaneous, synchronous. **2** *contemporary music*. current, fashionable, the latest, modern, newest, novel, present-day, *inf* trendy, topical, up-to-date, *inf* with-it.

**contempt** *n* abhorrence, contumely, derision, detestation, disdain, disgust, dislike, disrespect, loathing, ridicule, scorn. ▷ HATRED. *Opp* ADMIRATION. **feel contempt for** ▷ DESPISE.

**contemptible** *adj* base, beneath contempt, despicable, detestable, discreditable, disgraceful, dishonourable, disreputable, ignominious, inferior, loathsome, low-down, mean, odious, pitiful, *inf* shabby, shameful, worthless, wretched. ▷ HATEFUL. *Opp* ADMIRABLE.

**contemptuous** *adj* arrogant, belittling, condescending, derisive, disdainful, dismissive, disrespectful, haughty, *inf* holier-than-thou, imperious, insolent, insulting, jeering, lofty, patronizing, sarcastic, scathing, scornful, sneering, *sl* snide, snobbish, *inf* snooty, *sl* snotty, supercilious, superior, withering. *Opp* RESPECTFUL. **be contemptuous of** ▷ DESPISE.

**contend** *vb* **1** compete, contest, cope, dispute, grapple, oppose, rival, strive, struggle, vie. ▷ FIGHT, QUARREL. **2** *contend that you're innocent*. affirm, allege, argue, assert, claim, declare, maintain, plead.

**content** *adj* ▷ CONTENTED. • *n* **1** constituent, element, ingredient, part. **2** ▷ CONTENTMENT. • *vb* ▷ SATISFY.

**contented** *adj* cheerful, comfortable, *derog* complacent, content, fulfilled, gratified, peaceful, pleased, relaxed, satisfied, serene, smiling, smug, uncomplaining, untroubled, well-fed. ▷ HAPPY. *Opp* DISSATISFIED.

**contentment** *n* comfort, content, contentedness, ease, fulfilment, relaxation, satisfaction, serenity, smugness, tranquillity, well-being. ▷ HAPPINESS. *Opp* DISSATISFACTION.

**contest** *n* ▷ COMPETITION, FIGHT. • *vb* **1** compete for, contend for, fight for, *inf* make a bid for, strive for, struggle for, take up the challenge of, vie for. **2** *contest a decision.* argue against, challenge, debate, dispute, doubt, oppose, query, question, refute, resist.

**contestant** *n* candidate, competitor, contender, entrant, opponent, participant, player, rival.

**context** *n* background, environment, frame of reference, framework, milieu, position, setting, situation, surroundings.

**continual** *adj* eternal, everlasting, frequent, limitless, ongoing, perennial, perpetual, recurrent, regular, repeated. ▷ CONTINUOUS. *Opp* OCCASIONAL.

**continuation** *n* **1** continuance, extension, maintenance, prolongation, protraction, resumption. **2** addition, appendix, postscript, sequel, supplement.

**continue** *vb* **1** carry on, endure, go on, keep on, last, linger, persevere, persist, proceed, pursue, remain, stay, *inf* stick at, survive, sustain. **2** *continue after lunch. inf* pick up the threads, recommence, restart, resume. **3** *continue a series.* extend, keep going, keep up, lengthen, maintain, prolong.

**continuous** *adj* ceaseless, constant, continuing, endless, incessant, interminable, lasting, never-ending, non-stop, permanent, persistent, relentless, *inf* round-the-clock, *inf* solid, sustained, unbroken, unceasing, unending, uninterrupted, unremitting. ▷ CHRONIC, CONTINUAL. *Opp* INTERMITTENT.

**contour** *n* curve, form, outline, relief, shape.

**contract** *n* agreement, bargain, bond, commitment, compact, concordat, covenant, deal, indenture, lease, pact, settlement, treaty, understanding, undertaking. • *vb* **1** become denser, become smaller, close up, condense, decrease, diminish, draw together, dwindle, fall away, lessen, narrow, reduce, shrink, shrivel, slim down, thin out, wither. *Opp* EXPAND. **2** agree, arrange, close a deal, covenant, negotiate a deal, promise, sign an agreement, undertake. **3** *contract a disease.* become infected by, catch, develop, get.

**contraction** *n* **1** diminution, narrowing, shortening, shrinkage, shrivelling. **2** abbreviation, diminutive, shortened form.

**contradict** *vb* argue with, challenge, confute, controvert, deny, disagree with, dispute, gainsay, impugn, oppose, speak against.

**contradictory** *adj* antithetical, conflicting, contrary, different, discrepant, incompatible, inconsistent, irreconcilable, opposed, opposite. *Opp* COMPATIBLE.

**contraption** *n* apparatus, contrivance, device, gadget, invention, machine, mechanism.

**contrary** *adj* **1** conflicting, contradictory, converse, different, opposed, opposite, other, reverse. **2** *contrary winds.* adverse, hostile, inimical, opposing, unfavourable. **3** *a contrary child.* awkward, cantankerous, defiant, difficult, disobedient, disobliging, disruptive, intractable, obstinate, perverse, rebellious, *inf* stroppy, stubborn, subversive, uncooperative, unhelpful, wayward, wilful. *Opp* HELPFUL.

**contrast** *n* antithesis, comparison, difference, differentiation, disparity, dissimilarity, distinction, divergence, foil, opposition. *Opp* SIMILARITY. • *vb* **1** compare, differentiate, discriminate, distinguish, emphasize differences, make a distinction, set one against the other. **2** be set off (by), clash, conflict, deviate (from), differ (from). **contrasting** ▷ DISSIMILAR.

**contribute** *vb* add, bestow, *inf* chip in, donate, *inf* fork out, furnish, give, present, provide, put up, subscribe, supply. **contribute to** ▷ SUPPORT.

**contribution** *n* **1** donation, fee, gift, grant, *inf* hand-out, offering, payment, sponsorship, subscription. **2** addition, encouragement, input, support. ▷ HELP.

**contributor** *n* **1** backer, benefactor, donor, giver, helper, patron, sponsor, subscriber, supporter. **2** ▷ WRITER.

**control** *n* **1** administration, authority, charge, command, curb, direction, discipline, government, grip, guidance, influence, jurisdiction, leadership, management, mastery, orderliness, organization, oversight, power, regulation, restraint, rule, strictness, supervision, supremacy, sway. **2** button, dial, handle, key, lever, switch. • *vb* **1** administer, *inf* be at the helm, be in charge, *inf* boss, command, conduct, cope with, deal with, direct, dominate, engineer, govern, guide, handle, have control of, lead, look after, manage, manipulate, order about, oversee, regiment, regulate, rule, run, superintend, supervise. **2** *control animals.* check, confine, contain, curb, hold back, keep in check, master, repress, restrain, subdue, supress.

**controversial** *adj* **1** arguable, controvertible, debatable, disputable, doubtful, problematical, questionable. *Opp* ACCEPTED. **2** argumentative, contentious, dialectic, litigious, polemical, provocative.

**controversy** *n* altercation, argument, confrontation, contention, debate, disagreement, dispute, dissension, issue, polemic, quarrel, war of words, wrangle.

**convalesce** *vb* get better, improve, make progress, mend, recover, recuperate, regain strength.

**convalescent** *adj* getting better, healing, improving, making progress, *inf* on the mend, recovering, recuperating.

**convene** *vb* bring together, call, convoke, summon. ▷ GATHER.

**convenient** *adj* accessible, appropriate, at hand, available, commodious, expedient, handy, helpful, labour-saving, nearby, neat, opportune, serviceable, suitable, timely, usable, useful. *Opp* INCONVENIENT.

**convention** *n* **1** custom, etiquette, formality, matter of form, practice, rule, tradition. **2** ▷ ASSEMBLY.

**conventional** *adj* **1** accepted, accustomed, commonplace, correct, customary, decorous, expected, formal, habitual, mainstream, orthodox, prevalent, received, *inf* run-of-the-mill, standard, straight, traditional, unadventurous, unimaginative, unoriginal, unsurprising. ▷ ORDINARY. **2** [*derog*] bourgeois, conservative, hidebound, pedestrian, reactionary, rigid, stereotyped, *inf* stuffy. *Opp* UNCONVENTIONAL.

**converge** *vb* coincide, combine, come together, join, link up, meet, merge, unite. *Opp* DIVERGE.

**conversation** *n inf* chat, *inf* chin-wag, colloquy, communication, conference, dialogue, discourse, discussion, exchange of views, gossip, *inf* heart-to-heart, intercourse, *inf* natter, palaver, phone-call, *inf* powwow, tête-à-tête. ▷ TALK.

**convert** *vb* change someone's mind, convince, persuade, re-educate, reform, regenerate, rehabilitate, save, win over. ▷ CHANGE.

**convey** *vb* **1** bear, bring, carry, conduct, deliver, export, ferry, fetch, forward, import, move, send, shift, ship, shuttle, take, taxi, transfer, transport. **2** *convey a message.* communicate, disclose, impart, imply, indicate, mean, relay, reveal, signify, tell, transmit.

**convict** *n* condemned person, criminal, culprit, felon, malefactor, prisoner, wrongdoer. • *vb* condemn, declare guilty, prove guilty, sentence. *Opp* ACQUIT.

**conviction** *n* **1** assurance, certainty, confidence, firmness. **2** *religious conviction.* belief, creed, faith, opinion, persuasion, position, principle, tenet, view.

**convince** *vb* assure, *inf* bring round, convert, persuade, prove to, reassure, satisfy, sway, win over.

**convincing** ▷ PERSUASIVE.

**convulsion** *n* **1** disturbance, eruption, outburst, tremor, turbulence, upheaval. **2** [*medical*] attack, fit, paroxysm, seizure, spasm.

**convulsive** *adj* jerky, shaking, spasmodic, *inf* twitchy, uncontrolled, uncoordinated, violent, wrenching.

**cook** *vb* concoct, heat up, make, prepare, warm up. □ *bake, barbecue, boil, braise, brew, broil, casserole, coddle, fry, grill, pickle, poach, roast, sauté, scramble, simmer, steam, stew, toast.* **cook up** ▷ PLOT.

**cooking** *n* baking, catering, cookery, cuisine.

**cool** *adj* **1** chilled, chilly, coldish, iced, refreshing, unheated. ▷ COLD. *Opp* HOT. **2** calm, collected, composed, dignified, elegant, *inf* laid-back, level-headed, phlegmatic, quiet, relaxed, self-possessed, sensible, serene, unexcited, unflustered, unruffled, urbane. ▷ BRAVE. **3** [*derog*] aloof, apathetic, cold-blooded, dispassionate, distant, frigid, half-hearted, indifferent, lukewarm, negative, offhand, reserved, *inf* stand-offish, unconcerned, unemotional, unenthusiastic, unfriendly, uninvolved, unresponsive, unsociable, unwelcoming. *Opp* PASSIONATE. **4** [*inf*] *cool customer.* ▷ INSOLENT. • *vb* **1** chill, freeze, ice, refrigerate. *Opp* HEAT. **2** *cool your enthusiasm.* abate, allay, assuage, calm, dampen, diminish, lessen, moderate, *inf* pour cold water on, quiet, temper. *Opp* INFLAME.

**cooperate** *vb* act in concert, collaborate, combine, conspire, help each other, *inf* join forces, *inf* pitch in, *inf* play along, *inf* play ball, *inf* pull together, support each other, unite, work as a team, work together. *Opp* COMPETE.

**cooperation** *n* assistance, collaboration, cooperative effort, coordination, help, joint action, mutual support, teamwork. *Opp* COMPETITION.

**cooperative** *adj* **1** accommodating, comradely, constructive, hard-working, helpful to each other, keen, obliging, supportive, united, willing, working as a team. **2** *cooperative effort.* collective, combined, communal, concerted, coordinated, corporate, joint, shared.

**cope** *vb* get by, make do, manage, survive, win through. **cope with** ▷ ENDURE, MANAGE.

**copious** *adj* abundant, ample, bountiful, extravagant, generous, great, huge, inexhaustible, large, lavish, liberal, luxuriant, overflowing, plentiful, profuse, unsparing, unstinting. *Opp* SCARCE.

**copy** *n* **1** carbon copy, clone, counterfeit, double, duplicate, facsimile, fake, forgery, imitation, likeness, model, pattern, photocopy, print, replica, representation, reproduction, tracing, transcript, twin, Xerox. **2** *copy of a book*. edition, volume. • *vb* **1** borrow, counterfeit, crib, duplicate, emulate, follow, forge, imitate, photocopy, plagiarize, print, repeat, reproduce, simulate, transcribe. **2** ape, imitate, impersonate, mimic, parrot.

**cord** *n* cable, catgut, lace, line, rope, strand, string, twine, wire.

**cordon** *n* barrier, chain, fence, line, ring, row. **cordon off** ▷ ISOLATE.

**core** *n* **1** centre, heart, inside, middle, nucleus. **2** *core of a problem*. central issue, crux, essence, gist, heart, kernel, *sl* nitty-gritty, nub.

**cork** *n* bung, plug, stopper.

**corner** *n* **1** angle, crook, joint. **2** bend, crossroads, intersection, junction, turn, turning. **3** *a quiet corner*. hideaway, hiding-place, hole, niche, nook, recess, retreat. • *vb* capture, catch, trap.

**corporation** *n* **1** company, concern, enterprise, firm, organization. **2** council, local government.

**corpse** *n* body, cadaver, carcass, mortal remains, remains, skeleton, *sl* stiff.

**correct** *adj* **1** accurate, authentic, confirmed, exact, factual, faithful, faultless, flawless, genuine, literal, precise, reliable, right, strict, true, truthful, verified. **2** *correct manners*. acceptable, appropriate, fitting, just, normal, proper, regular, standard, suitable, tactful, unexceptionable, well-mannered. *Opp* WRONG. • *vb* **1** adjust, alter, cure, *inf* debug, put right, rectify, redress, remedy, repair. **2** *correct pupils' work*. assess, mark. **3** ▷ REPRIMAND.

**correspond** *vb* accord, agree, be congruous, be consistent, coincide, concur, conform, correlate, fit, harmonize, match, parallel, square, tally. **corresponding** ▷ EQUIVALENT. **correspond with** communicate with, send letters to, write to.

**correspondence** *n* letters, memoranda, *inf* memos, messages, notes, writings.

**correspondent** *n* contributor, journalist, reporter. ▷ WRITER.

**corridor** *n* hall, hallway, passage, passageway.

**corrode** *vb* **1** consume, eat into, erode, oxidize, rot, rust, tarnish. **2** crumble, deteriorate, disintegrate, tarnish.

**corrugated** *adj* creased, *inf* crinkly, fluted, furrowed, lined, puckered, ribbed, ridged, wrinkled.

**corrupt** *adj inf* bent, bribable, criminal, crooked, debauched, decadent, degenerate, depraved, *inf* dirty, dishonest, dishonourable, dissolute, evil, false, fraudulent, illegal, immoral, iniquitous, low, perverted, profligate, rotten, sinful, unethical, unprincipled, unscrupulous, unsound, untrustworthy, venal, vicious, wicked. *Opp* HONEST. • *vb* **1** bribe, divert, *inf* fix, influence, pervert, suborn, subvert. **2** *corrupt the innocent.* debauch, deprave, lead astray, make corrupt, tempt, seduce.

**cosmetics** *n* make-up, toiletries.

**cosmic** *adj* boundless, endless, infinite, limitless, universal.

**cosmopolitan** *adj* international, multicultural, sophisticated, urbane. *Opp* PROVINCIAL.

**cost** *n* amount, charge, expenditure, expense, fare, figure, outlay, payment, price, rate, tariff, value. • *vb* be valued at, be worth, fetch, go for, realize, sell for, *inf* set you back.

**costume** *n* apparel, attire, clothing, dress, fancy-dress, garb, garments, *inf* get-up, livery, outfit, period dress, raiment, robes, uniform, vestments. ▷ CLOTHES.

**cosy** *adj* comfortable, *inf* comfy, easy, homely, intimate, reassuring, relaxing, restful, secure, snug, soft, warm. *Opp* UNCOMFORTABLE.

**council** *n* committee, conclave, convention, convocation, corporation, gathering, meeting. ▷ ASSEMBLY.

**counsel** *n* ▷ LAWYER. • *vb* advise, discuss (with), give help, guide, listen to your views.

**count** *vb* **1** add up, calculate, check, compute, enumerate, estimate, figure out, keep account of, *inf* notch up, number, reckon, score, take stock of, tell, total, *inf* tot up, work out. **2** be important, have significance, matter, signify. **count on** ▷ EXPECT.

**countenance** *n* air, appearance, aspect, demeanour, expression, face, features, look, visage. • *vb* ▷ APPROVE.

**counter** *n* **1** bar, sales-point, service-point, table. **2** chip, disc, marker, piece, token. • *vb* answer, *inf* come back at, contradict, defend yourself against, hit back at, parry, react to, rebut, refute, reply to, resist, ward off.

**counteract** *vb* act against, annul, be an antidote to, cancel out, counterbalance, fight against, foil, invalidate, militate against, negate, neutralize, offset, oppose, resist, thwart, withstand, work against.

**counterbalance** *vb* balance, compensate for, counteract, counterpoise, counterweight, equalize.

**counterfeit** *adj* artificial, bogus, copied, ersatz, fake, false, feigned, forged, fraudulent, imitation, make-believe, meretricious, pastiche, *inf* phoney, *inf* pretend, *inf* pseudo, sham, simulated, spurious, synthetic. *Opp* GENUINE. • *vb* copy, fake, falsify, feign, forge, imitate, pretend, *inf* put on, sham, simulate.

**countless** *adj* endless, immeasurable, incalculable, infinite, innumerable, limitless, many, measureless, myriad, numberless, numerous, unnumbered, untold. *Opp* ▷ FINITE.

**country** *n* **1** canton, commonwealth, domain, empire, kingdom, land, nation, people, power, principality, realm, state, territory. **2** *open country.* countryside, green belt, landscape, scenery.

**couple** *n* brace, duo, pair, twosome. • *vb* **1** combine, connect, fasten, hitch, join, link, match, pair, unite, yoke. ▷ MATE.

**coupon** *n* tear-off slip, ticket, token, voucher.

**courage** *n* audacity, boldness, *sl* bottle, bravery, daring, dauntlessness, determination, fearlessness, fibre, firmness, fortitude, gallantry, *inf* grit, *inf* guts, heroism, indomitability, intrepidity, mettle, *inf* nerve, patience, *inf* pluck, prowess, resolution, spirit, *sl* spunk, stoicism, tenacity, valour, will-power. *Opp* COWARDICE.

**courageous** *adj* audacious, bold, brave, cool, daring, dauntless, determined, fearless, gallant, game, *sl* gutsy, heroic, indomitable, intrepid, lion-hearted, noble, *inf* plucky, resolute, spirited, stalwart, stoical, stout-hearted, tough, unafraid, uncomplaining, undaunted, unshrinking, valiant, valorous. *Opp* COWARDLY.

**course** *n* **1** bearings, circuit, direction, line, orbit, path, route, track, way. **2** *course of events.* advance, continuation, development, movement, passage, passing, progress, progression, succession. **3** *course of lectures.* curriculum, programme, schedule, sequence, series, syllabus.

**court** *n* **1** assizes, bench, court martial, high court, lawcourt, magistrates' court. **2** entourage, followers, palace, retinue. **3** ▷ COURTYARD. • *vb* **1** *inf* ask for, attract, invite, provoke, seek, solicit. **2** date, *inf* go out with, make advances to, make love to, try to win, woo.

**courteous** *adj* civil, considerate, gentlemanly, ladylike, urbane, well-bred, well-mannered. ▷ POLITE.

**courtier** *n* attendant, follower, lady, lord, noble, page, steward.

**courtyard** *n* court, enclosure, forecourt, patio, *inf* quad, quadrangle, yard.

**cover** *n* **1** ▷ COVERING. **2** binding, case, dust-jacket, envelope, file, folder, portfolio, wrapper. **3** camouflage, cloak, concealment, cover-up, deception, disguise, façade, front, hiding-place, mask, pretence, refuge, sanctuary, shelter, smokescreen. **3** *air cover.* defence, guard, protection, support. • *vb* **1** blot out, bury, camouflage, cap, carpet, cloak, clothe, cloud, coat, conceal, curtain, disguise, drape, dress, encase, enclose, enshroud, envelop, face, hide, hood, mantle, mask, obscure, overlay, overspread, plaster, protect, screen, shade, sheathe, shield, shroud, spread over, surface, tile, veil, veneer, wrap up. **2** *cover expenses.* be enough for, match, meet, pay for, suffice for. **3** *talk covered many subjects.* comprise, contain, deal with, embrace, encompass, include, incorporate, involve, treat.

**covering** *n* blanket, canopy, cap, carpet, casing, cladding, cloak, coat, coating, cocoon, cover, crust, facing, film, incrustation, layer, lid, mantle, outside, pall, rind, roof, screen, sheath, sheet, shell, shield, shroud, skin, surface, tarpaulin, top, veil, veneer, wrapping. ▷ BEDCLOTHES.

**coward** *n inf* chicken, craven, deserter, runaway, *inf* wimp.

**cowardice** *n* cowardliness, desertion, evasion, faint-heartedness, *inf* funk, shirking, spinelessness, timidity. ▷ FEAR. *Opp* COURAGE.

**cowardly** *adj* abject, afraid, base, chicken-hearted, cowering, craven, dastardly, faint-hearted, fearful, *inf* gutless, *inf* lily-livered, pusillanimous, spineless, submissive, timid, timorous, unchivalrous, ungallant, unheroic, *inf* wimpish, *sl* yellow. ▷ FRIGHTENED. *Opp* COURAGEOUS.

**cower** *vb* cringe, crouch, flinch, grovel, hide, quail, shiver, shrink, skulk, tremble.

**coy** *adj* arch, bashful, coquettish, demure, diffident, embarrassed, evasive, hesitant, modest, reserved, reticent, retiring, self-conscious, sheepish, shy, timid, unforthcoming. *Opp* BOLD.

**crack** *n* **1** breach, break, chink, chip, cleavage, cleft, cranny, craze, crevice, fissure, flaw, fracture, gap, opening, rift, rupture, slit, split. **2** ▷ JOKE. • *vb* **1** break, chip, fracture, snap, splinter, split. **2** ▷ HIT, SOUND. **crack up** ▷ DISINTEGRATE.

**craft** *n* **1** handicraft, job, skilled work, technique, trade. ▷ CRAFTSMANSHIP, CUNNING. **2** *a sea-going craft.* ▷ VESSEL. • *vb* ▷ MAKE.

**craftsmanship** *n* art, artistry, cleverness, craft, dexterity, expertise, handiwork, knack, *inf* know-how, workmanship. ▷ SKILL.

**crafty** *adj* artful, astute, calculating, canny, cheating, clever, conniving, cunning, deceitful, designing, devious, *inf* dodgy, *inf* foxy, furtive, guileful, ingenious, knowing, machiavellian, manipulative, scheming, shifty, shrewd, sly, *inf* sneaky, tricky, wily. *Opp* HONEST, NAÏVE.

**craggy** *adj* jagged, rocky, rough, rugged, steep, uneven.

**cram** *vb* **1** compress, crowd, crush, fill, force, jam, overcrowd, overfill, pack, press, squeeze, stuff. **2** ▷ STUDY.

**cramped** *adj* close, crowded, narrow, restricted, tight, uncomfortable. *Opp* ROOMY.

**crane** *n* davit, derrick, hoist.

**crash** *n* **1** bang, boom, clash, explosion. ▷ SOUND. **2** accident, bump, collision, derailment, disaster, impact, knock, pile-up, smash, wreck. **3** *crash on the stock market* collapse, depression, failure, fall. • *vb* **1** bump, collide, knock, lurch pitch, smash. ▷ HIT. **2** collapse, crash-dive, dive, fall, plummet, plunge, topple.

**crate** *n* box, carton, case, packing-case, tea-chest.

**crater** *n* abyss, cavity, chasm, hole, hollow, opening, pit.

**crawl** *vb* **1** clamber, creep, edge, inch, slither, squirm, worm, wriggle. **2** [*inf*] be obsequious, cringe, fawn, flatter, grovel, *sl* suck up, toady.

**craze** *n* diversion, enthusiasm, fad, fashion, infatuation, mania, novelty, obsession, passion, pastime, rage, *inf* thing, trend, vogue.

**crazy** *adj* **1** berserk, crazed, delirious, demented, deranged, frantic, frenzied, hysterical, insane, lunatic, *inf* potty, *inf* scatty, unbalanced, unhinged, wild. ▷ MAD. **2** *crazy comedy.* daft, eccentric, farcical, idiotic, *inf* knockabout, ludicrous, ridiculous, *sl* wacky, zany. ▷ ABSURD. **3** *crazy ideas.* confused, foolish, ill-considered, illogical, impractical, irrational, senseless, silly, unrealistic, unreasonable, unwise. ▷ STUPID. **4** ▷ ENTHUSIASTIC. *Opp* SENSIBLE.

**creamy** *adj* milky, oily, rich, smooth, thick, velvety.

**crease** *n* corrugation, crinkle, fold, furrow, groove, line, pleat, pucker, ridge, ruck, tuck, wrinkle. • *vb* crimp, crinkle, crumple, crush, fold, furrow, pleat, pucker, ridge, ruck, rumple, wrinkle.

**create** *vb old use* beget, begin, be the creator of, breed, bring about, bring into existence, build, cause, compose, conceive, concoct, constitute, construct, design, devise, *inf* dream up, engender, engineer, establish, father, forge, form, found, generate, give rise to, hatch, imagine, institute, invent, make up, manufacture, occasion, originate, produce, set up, shape, sire, think up. ▷ MAKE. *Opp* DESTROY.

**creation** *n* 1 beginning, birth, building, conception, constitution, construction, establishing, formation, foundation, generation, genesis, inception, institution, making, origin, procreation, production, shaping. 2 achievement, brainchild, concept, effort, handiwork, invention, product, work of art. *Opp* DESTRUCTION.

**creative** *adj* artistic, clever, fecund, fertile, imaginative, ingenious, inspired, inventive, original, positive, productive, resourceful, talented. *Opp* DESTRUCTIVE.

**creator** *n* architect, artist, author, begetter, builder, composer, craftsman, designer, deviser, discoverer, initiator, inventor, maker, manufacturer, originator, painter, parent, photographer, potter, producer, sculptor, smith, weaver, writer.

**creature** *n* beast, being, brute, mortal being, organism. ▷ ANIMAL.

**credentials** *n* authorization, documents, identity card, licence, passport, permit, proof of identity, warrant.

**credible** *adj* believable, conceivable, convincing, imaginable, likely, persuasive, plausible, possible, reasonable, tenable, thinkable, trustworthy. *Opp* INCREDIBLE.

**credit** *n* approval, commendation, distinction, esteem, fame, glory, honour, *inf* kudos, merit, praise, prestige, recognition, reputation, status, tribute. • *vb* 1 accept, believe, *inf* buy, count on, depend on, endorse, have faith in, reckon on, rely on, subscribe to, *inf* swallow, swear by, trust. *Opp* DOUBT. 2 *credit you with sense.* ascribe to, assign to, attach to, attribute to. 3 *credit £10 to my account.* add, enter. *Opp* DEBIT.

**creditable** *adj* admirable, commendable, estimable, good, honourable, laudable, meritorious, praiseworthy, respectable, well thought of, worthy. *Opp* UNWORTHY.

**credulous** *adj* easily taken in, *inf* green, gullible, *inf* soft, trusting, unsuspecting. ▷ NAÏVE. *Opp* SCEPTICAL.

**creed** *n* belief, conviction, doctrine, dogma, faith, principle, teaching, tenet.

**creek** *n* bay, cove, estuary, harbour, inlet.

**creep** *vb* crawl, edge, inch, move quietly, move slowly, pussyfoot, slink, slip, slither, sneak, steal, tiptoe, worm, wriggle, writhe.

**creepy** *adj* disturbing, eerie, frightening, ghostly, hair-raising, macabre, ominous, scary, sinister, spine-chilling, *inf* spooky, supernatural, threatening, uncanny, unearthly, weird.

**crest** *n* 1 comb, plume, tuft. 2 *crest of a hill.* apex, brow, crown, head, peak, pinnacle, ridge, summit, top. 3 badge, coat of arms, design, device, emblem, heraldic device, insignia, seal, shield, sign, symbol.

**crevice** *n* break, chink, cleft, crack, cranny, fissure, furrow, groove, rift, slit, split.

**crew** *n* band, company, gang, party, team. ▷ GROUP.

**crime** *n* delinquency, dishonesty, *old use* felony, illegality, lawbreaking, lawlessness, misconduct, misdeed, misdemeanour, offence, *inf* racket, sin, transgression of the law, violation, wrongdoing. □ *abduction, arson, assassination, blackmail, burglary, extortion, hijacking, hooliganism, kidnapping, manslaughter, misappropriation, mugging, murder, pilfering, piracy, poaching, rape, robbery, shop-lifting, smuggling, stealing, terrorism, theft, treason, vandalism.*

**criminal** *adj inf* bent, corrupt, *inf* crooked, culpable, dishonest, felonious, illegal, illicit, indictable, lawless, nefarious, *inf* shady, unlawful. ▷ WICKED, WRONG. *Opp* LAWFUL. • *n inf* baddy, convict, *inf* crook, culprit, delinquent, desperado, felon, gangster, knave, lawbreaker, malefactor, miscreant, offender, outlaw, recidivist, ruffian, scoundrel, *old use* transgressor, villain, wrongdoer. □ *bandit, brigand, buccaneer, defaulter, gunman, highwayman,* sl *hoodlum, hooligan, pickpocket, racketeer, receiver, swindler, thug,* ▷ CRIME.

**cringe** *vb* blench, cower, crouch, dodge, duck, flinch, grovel, quail, quiver, recoil, shrink back, shy away, tremble, wince.

**cripple** *vb* **1** disable, dislocate, fracture, hamper, hamstring, incapacitate, lame, maim, mutilate, paralyse, weaken. **2** damage, make useless, put out of action, sabotage, spoil. **crippled** ▷ HANDICAPPED.

**crisis** *n* calamity, catastrophe, climax, critical moment, danger, difficulty, disaster, emergency, predicament, problem, turning point.

**crisp** *adj* **1** breakable, brittle, crackly, crispy, crunchy, fragile, friable, hard and dry. **2** ▷ BRACING, BRISK.

**criterion** *n* measure, principle, standard, touchstone, yardstick.

**critic** *n* **1** analyst, authority, commentator, judge, pundit, reviewer. **2** attacker, detractor.

**critical** *adj* **1** captious, carping, censorious, criticizing, deprecatory, depreciatory, derogatory, disapproving, disparaging, fault-finding, hypercritical, judgemental, *inf* nit-picking, *derog* Pharisaical, scathing, slighting, uncomplimentary, unfavourable. *Opp* COMPLIMENTARY. **2** analytical, discerning, discriminating, intelligent, judicious, perceptive, probing, sharp. **3** *critical moment.* basic, crucial, dangerous, decisive, important, key, momentous, pivotal, vital. *Opp* UNIMPORTANT.

**criticism** *n* **1** censure, condemnation, diatribe, disapproval, disparagement, reprimand, reproach, stricture, tirade, verbal attack. **2** *literary criticism.* analysis, appraisal, appreciation, assessment, commentary, critique, elucidation, evaluation, judgement, valuation.

**criticize** *vb* **1** belittle, berate, blame, carp, *inf* cast aspersions on, castigate, censure, *old use* chide, condemn, complain about, decry, disapprove of, disparage, fault, find fault with, *inf* flay, *inf* get at, impugn, *inf* knock, *inf* lash, *inf* pan, *inf* pick holes in, *inf* pitch into, *inf* rap, rate, rebuke, reprimand, satirize, scold, *inf* slam, *inf* slate, snipe at. *Opp* PRAISE. **2** analyse, appraise, assess, evaluate, discuss, judge, review.

**crockery** *n* ceramics, china, crocks, dishes, earthenware, porcelain, pottery, tableware. □ *basin, bowl, coffee-cup, coffee-pot, cup, din-*

*ner plate, dish, jug, milk-jug, mug, plate,* Amer *platter, pot, sauceboat, saucer, serving dish, side plate, soup bowl, sugar-bowl, teacup, teapot,* old use *trencher, tureen.*

**crook** *n* 1 angle, bend, corner, hook. 2 ▷ CRIMINAL.

**crooked** *adj* 1 angled, askew, awry, bendy, bent, bowed, contorted, curved, curving, deformed, gnarled, lopsided, misshapen, off-centre, tortuous, twisted, twisty, warped, winding, zigzag. ▷ INDIRECT. 2 ▷ CRIMINAL.

**crop** *n* gathering, harvest, produce, sowing, vintage, yield. • *vb* bite off, browse, clip, graze, nibble, shear, snip, trim. ▷ CUT. **crop up** ▷ ARISE.

**cross** *adj* bad-tempered, cantankerous, crotchety, *inf* grumpy, ill-tempered, irascible, irate, irritable, peevish, short-tempered, testy, tetchy, upset, vexed. ▷ ANGRY, ANNOYED. *Opp* GOOD-TEMPERED. • *n* 1 intersection, X. 2 *a cross to bear.* affliction, burden, difficulty, grief, misfortune, problem, sorrow, trial, tribulation, trouble, worry. 3 *cross of breeds.* amalgam, blend, combination, cross-breed, half-way house, hybrid, mixture, mongrel. • *vb* 1 criss-cross, intersect, meet, zigzag. 2 *cross a river.* bridge, ford, go across, pass over, span, traverse. 3 *cross someone.* annoy, block, frustrate, hinder, impede, interfere with, oppose, stand in the way of, thwart. **cross out** ▷ CANCEL. **cross swords** ▷ CONFLICT.

**crossing** *vb* 1 bridge, causeway, flyover, ford, level-crossing, overpass, pedestrian crossing, pelican crossing, subway, stepping-stones, underpass, zebra crossing. 2 *sea crossing.* ▷ JOURNEY.

**crossroads** *n* interchange, intersection, junction.

**crouch** *vb* bend, bow, cower, cringe, duck, kneel, squat, stoop.

**crowd** *n* 1 army, assemblage, assembly, bunch, circle, cluster, collection, company, crush, flock, gathering, horde, host, mass, mob, multitude, pack, press, rabble, swarm, throng. ▷ GROUP. 2 *a football crowd.* audience, gate, spectators. • *vb* assemble, bundle, cluster, collect, compress, congregate, cram, crush, flock, gather, get together, herd, huddle, jam, jostle, mass, muster, overcrowd, pack, *inf* pile, press, push, squeeze, swarm, throng.

**crowded** *adj* congested, cramped, full, jammed, *inf* jam-packed, jostling, overcrowded, overflowing, packed, swarming, teeming, thronging. *Opp* EMPTY.

**crown** *n* 1 circlet, coronet, diadem, tiara. 2 *crown of a hill.* apex, brow, head, peak, ridge, summit, top. • *vb* 1 anoint, appoint, enthrone, install. 2 cap, complete, conclude, consummate, culminate, finish off, perfect, round off, top.

**crucial** *adj* central, critical, decisive, essential, important, major, momentous, pivotal, serious. *Opp* UNIMPORTANT.

**crude** *adj* 1 natural, raw, unprocessed, unrefined. 2 *crude work.* amateurish, awkward, bungling, clumsy, inartistic, incompetent, inelegant, inept, makeshift, primitive, rough, rudimentary, unpolished, unskilful, unworkmanlike. *Opp* REFINED. 3 ▷ VULGAR.

**cruel** *adj* atrocious, barbaric, barbarous, beastly, bestial, bloodthirsty, bloody, brutal, callous, cold-blooded, cold-hearted, diabolical, ferocious, fiendish, fierce, flinty, grim, hard, hard-hearted, harsh, heartless, hellish, implacable, inexorable, inhuman, inhumane, malevolent, merciless, murderous, pitiless, relentless, remorseless, ruthless, sadistic, savage, severe, sharp, spiteful, stern, stony-hearted, tyrannical, unfeeling, unjust, unkind, unmerciful, unrelenting, vengeful, venomous, vicious, violent. *Opp* KIND.

**cruelty** *n* barbarity, bestiality, bloodthirstiness, brutality, callousness, cold-bloodedness, ferocity, hard-heartedness, heartlessness, inhumanity, malevolence, ruthlessness, sadism, savagery, unkindness, viciousness, violence.

**cruise** *n* sail, voyage. ▷ TRAVEL.

**crumb** *n* bit, bite, fragment, grain, morsel, particle, scrap, shred, sliver, speck.

**crumble** *vb* break into pieces, break up, crush, decay, decompose, deteriorate, disintegrate, fall apart, fragment, grind, perish, pound, powder, pulverize.

**crumbly** *adj* friable, granular, powdery. *Opp* SOLID.

**crumple** *vb* crease, crinkle, crush, dent, fold, mangle, pucker, rumple, wrinkle.

**crunch** *vb* break, champ, chew, crush, grind, masticate, munch, scrunch, smash, squash.

**crusade** *n* campaign, drive, holy war, jehad, movement, struggle, war.

**crush** *n* congestion, jam. ▷ CROWD. • *vb* **1** break, bruise, compress, crumple, crunch, grind, mangle, mash, pound, press, pulp, pulverize, shiver, smash, splinter, squash, squeeze. **2** *crush opponents.* humiliate, mortify, overwhelm, quash, rout, thrash, vanquish. ▷ CONQUER.

**crust** *n* incrustation, outer layer, outside, rind, scab, shell, skin, surface. ▷ COVERING.

**crux** *n* centre, core, crucial issue, essence, heart, nub.

**cry** *n* battle-cry, bellow, call, caterwaul, ejaculation, exclamation, hoot, howl, outcry, roar, scream, screech, shout, shriek, whoop, yell, yelp, yowl. • *vb* bawl, blubber, grizzle, howl, keen, shed tears, snivel, sob, wail, weep, whimper, whinge. **cry off** ▷ WITHDRAW. **cry out** ▷ SHOUT.

**crypt** *n* basement, catacomb, cellar, grave, sepulchre, tomb, undercroft, vault.

**cryptic** *adj* arcane, cabalistic, coded, concealed, enigmatic, esoteric, hidden, mysterious, mystical, obscure, occult, perplexing, puzzling, recondite, secret, unclear, unintelligible, veiled. *Opp* INTELLIGIBLE.

**cuddle** *vb* caress, clasp lovingly, dandle, embrace, fondle, hold closely, huddle against, hug, kiss, make love, nestle against, nurse, pet, snuggle up to.

**cudgel** *n* baton, bludgeon, cane, club, cosh, stick, truncheon. • *vb* batter, beat, bludgeon, cane, *inf* clobber, cosh, pound, pummel, thrash, thump, *inf* thwack. ▷ HIT.

**cue** *n* hint, prompt, reminder, sign, signal.

**culminate** *vb* build up to, climax, conclude, reach a finale, rise to a peak. ▷ END.

**culpable** *adj* blameworthy, criminal, guilty, knowing, liable, punishable, reprehensible, wrong. ▷ DELIBERATE. *Opp* INNOCENT.

**culprit** *n* delinquent, malefactor, miscreant, offender, troublemaker, wrongdoer. ▷ CRIMINAL.

**cult** *n* 1 craze, fan-club, fashion, following, devotees, party, school, trend, vogue. 2 *religious cult.* ▷ DENOMINATION.

**cultivate** *vb* 1 dig, farm, fertilize, hoe, manure, mulch, plough, prepare, rake, till, turn, work. 2 grow, plant, produce, raise, sow, take cuttings, tend. 3 *cultivate a friendship.* court, develop, encourage, foster, further, improve, promote, pursue, try to achieve.

**cultivated** *adj* 1 agricultural, farmed, planted, prepared, tilled. 2 ▷ CULTURED.

**cultivation** *n* agriculture, agronomy, breeding, culture, growing, farming, gardening, horticulture, husbandry, nurturing.

**cultural** *adj* aesthetic, artistic, civilized, civilizing, educational, elevating, enlightening, highbrow, improving, intellectual.

**culture** *n* 1 art, background, civilization, customs, education, learning, mores, traditions, way of life. 2 ▷ CULTIVATION.

**cultured** *adj* artistic, civilized, cultivated, discriminating, educated, elegant, erudite, highbrow, knowledgeable, polished, refined, scholarly, sophisticated, well-bred, well-educated, well-read. *Opp* IGNORANT.

**cunning** *adj* 1 artful, devious, dodgy, guileful, insidious, knowing, machiavellian, sly, subtle, tricky, wily. ▷ CRAFTY. 2 adroit, astute, ingenious, skilful. ▷ CLEVER. • *n* 1 artfulness, chicanery, craft, craftiness, deceit, deception, deviousness, duplicity, guile, slyness, trickery. 2 cleverness, expertise, ingenuity, skill.

**cup** *n* 1 beaker, bowl, chalice, glass, goblet, mug, tankard, teacup, tumbler, wine-glass. 2 award, prize, trophy.

**cupboard** *n* cabinet, chiffonier, closet, dresser, filing-cabinet, food-cupboard, larder, locker, sideboard, wardrobe.

**curable** *adj* operable, remediable, treatable. *Opp* INCURABLE.

**curb** *vb* bridle, check, contain, control, deter, hamper, hinder, hold back, impede, inhibit, limit, moderate, repress, restrain, restrict, subdue, suppress. *Opp* ENCOURAGE.

**curdle** *vb* clot, coagulate, congeal, go lumpy, go sour, thicken.

**cure** *n* 1 antidote, corrective, medication, nostrum, palliative, panacea, prescription, remedy, restorative, solution, therapy, treatment. ▷ MEDICINE. 2 deliverance, healing, recovery, recuperation, restoration, revival. • *vb* alleviate, correct, counteract, ease, *inf* fix, heal, help, mend, palliate, put right, rectify, relieve, remedy, repair, restore, solve, treat. *Opp* AGGRAVATE.

**curiosity** *n* inquisitiveness, interest, interference, meddling, nosiness, prying, snooping.

**curious** *adj* 1 inquiring, inquisitive, interested, probing, puzzled, questioning, searching. 2 interfering, intrusive, meddlesome, *inf* nosy, prying. 3 ▷ STRANGE. **be curious** ▷ PRY.

**curl** *n* bend, coil, curve, kink, loop, ringlet, scroll, spiral, swirl, turn, twist, wave, whorl. • *vb* 1 bend, coil, corkscrew, curve, entwine, loop, spiral, turn, twine, twist, wind, wreathe, writhe. 2 *curl your hair.* crimp, frizz, perm.

**curly** *adj* crimped, curled, curling, frizzy, fuzzy, kinky, permed, wavy. *Opp* STRAIGHT.

**current** *adj* **1** alive, contemporary, continuing, existing, extant, fashionable, living, modern, ongoing, present, present-day, prevailing, prevalent, reigning, remaining, surviving, *inf* trendy, up-to-date. **2** *current passport.* usable, valid. *Opp* OLD. • *n* course, draught, drift, flow, jet, river, stream, tide, trend, undercurrent, undertow.

**curriculum** *n* course, programme of study, syllabus.

**curse** *n* blasphemy, exclamation, expletive, imprecation, malediction, oath, obscenity, profanity, swearword. ▷ EVIL. *Opp* BLESSING. • *vb* blaspheme, damn, fulminate, swear, utter curses. *Opp* BLESS. **cursed** ▷ HATEFUL.

**cursory** *adj* brief, careless, casual, desultory, fleeting, hasty, hurried, perfunctory, quick, slapdash, superficial. *Opp* THOROUGH.

**curt** *adj* abrupt, blunt, brief, brusque, concise, crusty, gruff, laconic, monosyllabic, offhand, rude, sharp, short, snappy, succinct, tart, terse, unceremonious, uncommunicative, ungracious. ▷ RUDE. *Opp* EXPANSIVE.

**curtail** *vb* abbreviate, abridge, break off, contract, cut short, decrease, diminish, *inf* dock, guillotine, halt, lessen, lop, prune, reduce, restrict, shorten, stop, terminate, trim, truncate. *Opp* EXTEND.

**curtain** *n* blind, drape, drapery, hanging, screen. • *vb* drape, mask, screen, shroud, veil. ▷ HIDE.

**curtsy** *vb* bend the knee, bow, genuflect, salaam.

**curve** *n* arc, arch, bend, bow, bulge, camber, circle, convolution, corkscrew, crescent, curl, curvature, cycloid, loop, meander, spiral, swirl, trajectory, turn, twist, undulation, whorl. • *vb* arc, arch, bend, bow, bulge, camber, coil, corkscrew, curl, loop, meander, snake, spiral, swerve, swirl, turn, twist, wind. ▷ CIRCLE.

**curved** *adj* concave, convex, convoluted, crescent, crooked, curvilinear, curving, curvy, rounded, serpentine, shaped, sinuous, sweeping, swelling, tortuous, turned, undulating, whorled.

**cushion** *n* bean-bag, bolster, hassock, headrest, pad, pillow. • *vb* absorb, bolster, deaden, lessen, mitigate, muffle, protect from, reduce the effect of, soften, support.

**custodian** *n* caretaker, curator, guardian, keeper, overseer, superintendent, warden, warder, *inf* watch-dog, watchman.

**custody** *n* **1** care, charge, guardianship, keeping, observation, possession, preservation, protection, safe-keeping. **2** *in police custody.* captivity, confinement, detention, imprisonment, incarceration, remand.

**custom** *n* **1** convention, etiquette, fashion, form, formality, habit, institution, manner, observance, policy, practice, procedure, routine, tradition, usage, way, wont. **2** *A shop needs custom.* business, buyers, customers, patronage, support, trade.

**customary** *adj* accepted, accustomed, common, commonplace, conventional, established, everyday, expected, fashionable, general, habitual, normal, ordinary, popular, prevailing, regular, routine, traditional, typical, usual, wonted. *Opp* UNUSUAL.

**customer** *n* buyer, client, consumer, patron, purchaser, shopper. *Opp* SELLER.

**cut** *n* **1** gash, graze, groove, incision, laceration, nick, notch, opening, rent, rip, slash, slice, slit, snick, snip, split, stab, tear. ▷ INJURY. **2** *cut in prices.* cut-back, decrease, fall, lowering, reduction, saving. • *vb* **1** amputate, axe, carve, chip, chisel, chop, cleave, clip, crop, dice, dissect, divide, dock, engrave, fell, gash, gouge, grate, graze, guillotine, hack, halve, hew, incise, knife, lacerate, lance, lop, mince, mow, nick, notch, open, pare, pierce, poll, pollard, prune, reap, rive, saw, scalp, score, sever, share, shave, shear, shred, slash, slice, slit, snick, snip, split, stab, subdivide, trim, whittle, wound. **2** abbreviate, abridge, bowdlerize, censor, condense, curtail, digest, edit, précis, shorten, summarize, truncate. ▷ REDUCE. **cut and dried** ▷ DEFINITE. **cut in** ▷ INTERRUPT. **cut off** ▷ REMOVE, STOP. **cut short** ▷ CURTAIL.

**cutlery** *n inf* eating irons. □ *breadknife, butter-knife, carving knife, cheese knife, dessert-spoon, fish knife, fish fork, fork, knife, ladle, salad servers, spoon, steak knife, tablespoon, teaspoon.*

**cutter** *n* □ *axe, billhook, chisel, chopper, clippers, harvester, guillotine, lawnmower, mower, saw, scalpel, scissors, scythe, secateurs, shears, sickle.* ▷ KNIFE.

**cutting** *adj* acute, biting, caustic, incisive, keen, mordant, sarcastic, satirical, sharp, trenchant. ▷ HURTFUL.

**cycle** *n* **1** circle, repetition, revolution, rotation, round, sequence, series. **2** bicycle, *inf* bike, moped, *inf* motor bike, motor cycle, penny-farthing, scooter, tandem, tricycle. • *vb* ▷ TRAVEL.

**cyclic** *adj* circular, recurring, repeating, repetitive, rotating.

**cynical** *adj* doubting, *inf* hard, incredulous, misanthropic, mocking, negative, pessimistic, questioning, sceptical, sneering. *Opp* OPTIMISTIC.

# D

**dabble** *vb* **1** dip, paddle, splash, wet. **2** *dabble in a hobby.* potter about, tinker, work casually.

**dabbler** *n* amateur, dilettante, potterer.

**dagger** *n* bayonet, blade, *old use* dirk, knife, kris, poniard, stiletto.

**daily** *adj* diurnal, everyday, quotidian, regular.

**dainty** *adj* **1** charming, delicate, exquisite, fine, graceful, meticulous, neat, nice, pretty, skilful. **2** choosy, discriminating, fastidious, finicky, fussy, genteel, mincing, sensitive, squeamish, well-mannered. **3** *a dainty morsel.* appealing, appetizing, choice, delectable, delicious. *Opp* CLUMSY, GROSS.

**dally** *vb* dawdle, delay, *inf* dilly-dally, hang about, idle, linger, loaf, loiter, play about, procrastinate, saunter, *old use* tarry, waste time.

**dam** *n* bank, barrage, barrier, dike, embankment, wall, weir. • *vb* block, check, hold back, obstruct, restrict, stanch, stem, stop.

**damage** *n* destruction, devastation, harm, havoc, hurt, injury, loss, mutilation, sabotage. • *vb* 1 blemish, break, buckle, burst, *inf* bust, chip, crack, cripple, deface, destroy, disable, disfigure, *inf* do mischief to, flaw, fracture, harm, hurt, immobilize, impair, incapacitate, injure, make inoperative, make useless, mar, mark, mutilate, *inf* play havoc with, ruin, rupture, sabotage, scar, scratch, spoil, strain, vandalize, warp, weaken, wound, wreck. **damaged** ▷ FAULTY. **damages** ▷ COMPENSATION. **damaging** ▷ HARMFUL.

**damn** *vb* attack, berate, castigate, censure, condemn, criticize, curse, denounce, doom, execrate, sentence, swear at.

**damnation** *n* doom, everlasting fire, hell, perdition, ruin. *Opp* SALVATION.

**damp** *adj* clammy, dank, dewy, dripping, drizzly, foggy, humid, misty, moist, muggy, perspiring, rainy, soggy, steamy, sticky, sweaty, unaired, unventilated, wet, wettish. *Opp* DRY. • *vb* 1 dampen, humidify, moisten, sprinkle. 2 ▷ DISCOURAGE.

**dance** *n* choreography, dancing. □ *ball, barn-dance, ceilidh, disco, discothηque,* inf *hop,* inf *knees-up, party,* inf *shindy, social, square dance.* □ *ballet, ballroom dancing, break-dancing, country dancing, disco dancing, flamenco dancing, folk dancing, Latin-American dancing, limbo dancing, morris dancing, old-time dancing, tap-dancing.* □ *bolero, cancan, conga, fandango, fling, foxtrot, gavotte, hornpipe, jig, mazurka, minuet, polka, polonaise, quadrille, quickstep, reel, rumba, square dance, tango, waltz.* • *vb* caper, cavort, frisk, frolic, gambol, hop about, jig, jive, jump, leap, prance, rock, skip, *joc* trip the light fantastic, whirl.

**danger** *n* 1 crisis, distress, hazard, insecurity, jeopardy, menace, peril, pitfall, trouble, uncertainty. 2 *danger of frost.* chance, liability, possibility, risk, threat.

**dangerous** *adj* 1 alarming, breakneck, *inf* chancy, critical, destructive, explosive, grave, *sl* hairy, harmful, hazardous, insecure, menacing, *inf* nasty, noxious, perilous, precarious, reckless, risky, threatening, toxic, uncertain, unsafe. 2 *dangerous men.* desperate, ruthless, treacherous, unmanageable, unpredictable, violent, volatile, wild. *Opp* HARMLESS.

**dangle** *vb* be suspended, depend, droop, flap, hang, sway, swing, trail, wave about.

**dank** *adj* chilly, clammy, damp, moist, unaired.

**dappled** *adj* blotchy, brindled, dotted, flecked, freckled, marbled, motley, mottled, particoloured, patchy, pied, speckled, spotted, stippled, streaked, varicoloured, variegated.

**dare** *vb* 1 gamble, have the courage, risk, take a chance, venture. 2 challenge, defy, provoke, taunt. **daring** ▷ BOLD.

**dark** *adj* 1 black, blackish, cheerless, clouded, cloudy, coal-black, dim, dingy, dismal, drab, dreary, dull, dusky, funereal, gloomy, glowering, glum, grim, inky, moonless, murky, overcast, pitch-black, pitch-dark, *poet* sable, shadowy, shady, sombre, starless, stygian, sullen, sunless, tenebrous, unilluminated, unlighted, unlit. 2 *dark colours.* dense, heavy, strong. 3 *dark complexion.* black, brown, dark-skinned, dusky, swarthy, tanned. **4** ▷ HIDDEN, MYSTERIOUS. *Opp* LIGHT, PALE.

**darken** *vb* **1** become overcast, cloud over. **2** blacken, dim, eclipse, obscure, overshadow, shade. *Opp* LIGHTEN.

**darling** *n inf* apple of your eye, beloved, *inf* blue-eyed boy, dear, dearest, favourite, honey, love, loved one, pet, sweet, sweetheart, true love.

**dart** *n* arrow, bolt, missile, shaft. • *vb* bound, fling, flit, fly, hurtle, leap, move suddenly, shoot, spring, *inf* whiz, *inf* zip. ▷ DASH.

**dash** *n* **1** chase, race, run, rush, sprint, spurt. • *vb* **1** bolt, chase, dart, fly, hasten, hurry, move quickly, race, run, rush, speed, sprint, tear, *inf* zoom. **2** ▷ HIT.

**dashing** *adj* animated, dapper, dynamic, elegant, lively, smart, spirited, stylish, vigorous.

**data** *plur n* details, evidence, facts, figures, information, statistics.

**date** *n* **1** day. ▷ TIME. **2** *date with a friend*. appointment, assignation, engagement, fixture, meeting, rendezvous. **out-of-date** ▷ OBSOLETE. **up-to-date** ▷ MODERN.

**daunt** *vb* alarm, depress, deter, discourage, dishearten, dismay, intimidate, overawe, put off, unnerve. ▷ FRIGHTEN. *Opp* ENCOURAGE.

**dawdle** *vb* be slow, dally, delay, *inf* dilly-dally, hang about, idle, lag behind, linger, loaf about, loiter, move slowly, straggle, *inf* take it easy, *inf* take your time, trail behind. *Opp* HURRY.

**dawn** *n* day-break, first light, *inf* peep of day, sunrise. ▷ BEGINNING.

**day** *n* **1** daylight, daytime, light. **2** age, epoch, era, period, time.

**day-dream** *n* dream, fantasy, hope, illusion, meditation, pipe-dream, reverie, vision, wool-gathering. • *vb* dream, fantasize, imagine, meditate.

**daze** *vb* benumb, paralyse, shock, stun, stupefy. ▷ AMAZE.

**dazzle** *vb* blind, confuse, disorientate. **dazzling** ▷ BRILLIANT.

**dead** *adj* **1** cold, dead and buried, deceased, departed, *inf* done for, inanimate, inert, killed, late, lifeless, perished, rigid, stiff. *Opp* ALIVE. **2** *dead language*. died out, extinct, obsolete. **3** *dead with cold*. deadened, insensitive, numb, paralysed, without feeling. **4** *dead battery, engine*. burnt out, defunct, flat, inoperative, not going, not working, no use, out of order, unresponsive, used up, useless, worn out. **5** *a dead party*. boring, dull, moribund, slow, uninteresting. *Opp* LIVELY. **6** *dead centre*. ▷ EXACT. **dead person** ▷ CORPSE. **dead to the world** ▷ ASLEEP.

**deaden** *vb* **1** anaesthetize, desensitize, dull, numb, paralyse. **2** blunt, check, cushion, damp, diminish, hush, lessen, mitigate, muffle, mute, quieten, reduce, smother, soften, stifle, suppress, weaken.

**deadlock** *n* halt, impasse, stalemate, standstill, stop, stoppage, tie.

**deadly** *adj* dangerous, destructive, fatal, lethal, mortal, noxious, terminal. ▷ HARMFUL. *Opp* HARMLESS.

**deafen** *vb* make deaf, overwhelm.

**deafening** ▷ LOUD.

**deal** *n* **1** agreement, arrangement, bargain, contract, pact, settlement, transaction, understanding. **2** amount, quantity, volume. • *vb* **1** allot, apportion, assign, dispense, distribute, divide, *inf* dole out, give out, share out. **2** *deal someone a blow.* administer, apply, deliver, give, inflict, mete out. **3** *deal in stocks and shares.* buy and sell, do business, trade, traffic. deal with ▷ MANAGE, TREAT.

**dealer** *n* agent, broker, distributor, merchant, retailer, shopkeeper, stockist, supplier, trader, tradesman, vendor, wholesaler.

**dear** *adj* **1** adored, beloved, close, darling, intimate, loved, precious, treasured, valued, venerated. ▷ LOVABLE. *Opp* HATEFUL. **2** costly, exorbitant, expensive, high-priced, over-priced, *inf* pricey. *Opp* CHEAP. • *n* ▷ DARLING.

**death** *n* **1** decease, demise, dying, loss, passing. ▷ END. **2** casualty, fatality. **put to death** ▷ EXECUTE.

**debase** *vb* belittle, commercialize, degrade, demean, depreciate, devalue, diminish, lower the tone of, pollute, reduce the value of, ruin, soil, spoil, sully, vulgarize.

**debatable** *adj* arguable, contentious, controversial, controvertible, disputable, doubtful, dubious, moot (*point*), open to doubt, open to question, problematical, questionable, uncertain, unsettled, unsure. *Opp* CERTAIN.

**debate** *n* argument, conference, consultation, controversy, deliberation, dialectic, discussion, disputation, dispute, polemic. • *vb* argue, *inf* chew over, consider, deliberate, discuss, dispute, *inf* mull over, question, reflect on, weigh up, wrangle.

**debit** *vb* cancel, remove, subtract, take away. *Opp* CREDIT.

**debris** *n* bits, detritus, flotsam, fragments, litter, pieces, remains, rubbish, rubble, ruins, waste, wreckage.

**debt** *n* account, arrears, bill, debit, dues, indebtedness, liability, obligation, score, what you owe. **in debt** bankrupt, defaulting, insolvent. ▷ POOR.

**decadent** *adj* corrupt, debased, debauched, declining, degenerate, dissolute, immoral, self-indulgent. *Opp* MORAL.

**decay** *vb* atrophy, break down, corrode, crumble, decompose, degenerate, deteriorate, disintegrate, dissolve, fall apart, fester, go bad, go off, mortify, moulder, oxidize, perish, putrefy, rot, shrivel, spoil, waste away, weaken, wither.

**deceit** *n* artifice, cheating, chicanery, craftiness, cunning, deceitfulness, dishonesty, dissimulation, double-dealing, duplicity, guile, hypocrisy, insincerity, lying, misrepresentation, pretence, sham, slyness, treachery, trickery, underhandedness, untruthfulness. ▷ DECEPTION. *Opp* HONESTY.

**deceitful** *adj* cheating, crafty, cunning, deceiving, deceptive, designing, dishonest, double-dealing, duplicitous, false, fraudulent, furtive, hypocritical, insincere, lying, secretive, shifty, sneaky, treacherous, *inf* tricky, *inf* two-faced, underhand, unfaithful, untrustworthy, wily. *Opp* HONEST.

**deceive** *vb inf* bamboozle, be an impostor, beguile, betray, blind, bluff, cheat, *inf* con, defraud, delude, *inf* diddle, double-cross, dupe, fool, *inf* fox, *inf* have on, hoax, hoodwink, *inf* kid, *inf* lead on, lie, mislead, mystify, *inf* outsmart, outwit, pretend, swindle, *inf* take for a ride, *inf* take in, trick.

**decelerate** *vb* brake, decrease speed, go slower, lose speed, slow down. *Opp* ACCELERATE.

**decent** *adj* **1** acceptable, appropriate, becoming, befitting, chaste, courteous, decorous, delicate, fitting, honourable, modest, polite, presentable, proper, pure, respectable, seemly, sensitive, suitable, tasteful. *Opp* INDECENT. **2** [*inf*] *a decent meal.* agreeable, nice, pleasant, satisfactory. ▷ GOOD. *Opp* BAD.

**deception** *n* bluff, cheat, *inf* con, confidence trick, cover-up, deceit, fake, feint, *inf* fiddle, fraud, hoax, imposture, lie, pretence, ruse, sham, stratagem, subterfuge, swindle, trick, wile. ▷ DECEIT.

**deceptive** *adj* ambiguous, deceiving, delusive, dishonest, distorted, equivocal, evasive, fallacious, false, fraudulent, illusory, insincere, lying, mendacious, misleading, specious, spurious, treacherous, unreliable, wrong. *Opp* GENUINE.

**decide** *vb* adjudicate, arbitrate, choose, conclude, determine, elect, fix on, judge, make up your mind, opt for, pick, reach a decision, resolve, select, settle. **decided** ▷ DEFINITE.

**decipher** *vb* disentangle, *inf* figure out, read, work out. ▷ DECODE.

**decision** *n* conclusion, decree, finding, judgement, outcome, result, ruling, verdict.

**decisive** *adj* **1** conclusive, convincing, crucial, final, influential, positive, significant. **2** *decisive action.* certain, confident, decided, definite, determined, firm, forceful, forthright, incisive, resolute, strong-minded, sure, unhesitating. *Opp* TENTATIVE.

**declaration** *n* affirmation, announcement, assertion, avowal, confirmation, deposition, disclosure, edict, manifesto, notice, proclamation, profession, promulgation, pronouncement, protestation, revelation, statement, testimony.

**declare** *vb* affirm, announce, assert, attest, avow, broadcast, certify, claim, confirm, contend, disclose, emphasize, insist, maintain, make known, proclaim, profess, pronounce, protest, report, reveal, show, state, swear, testify, *inf* trumpet forth, witness. ▷ SAY.

**decline** *n* decrease, degeneration, deterioration, diminuendo, downturn, drop, fall, falling off, loss, recession, reduction, slump, worsening. • *vb* **1** decrease, degenerate, deteriorate, die away, diminish, drop away, dwindle, ebb, fail, fall off, flag, lessen, peter out, reduce, shrink, sink, slacken, subside, tail off, taper off, wane, weaken, wilt, worsen. *Opp* IMPROVE. **2** *decline an invitation.* abstain from, forgo, refuse, reject, *inf* turn down, veto. *Opp* ACCEPT.

**decode** *vb inf* crack, decipher, explain, figure out, interpret, make out, read, solve, understand, unravel, unscramble.

**decompose** *vb* break down, decay, disintegrate, go off, moulder, putrefy, rot.

**decorate** *vb* **1** adorn, array, beautify, *old use* bedeck, colour, deck, *inf* do up, embellish, embroider, festoon, garnish, make beautiful, ornament, paint, paper, *derog* prettify, refurbish, renovate, smarten up, spruce up, *derog* tart up, trim, wallpaper. **2** give a medal to, honour, reward.

**decoration** **1** accessories, adornment, arabesque, elaboration, embellishment, finery, flourishes, ornament, ornamentation, trappings, trimmings. **2** award, badge, colours, medal, order, ribbon, star.

**decorative** *adj* elaborate, fancy, non-functional, ornamental, ornate. *Opp* FUNCTIONAL.

**decorous** *adj* appropriate, becoming, befitting, correct, dignified, fitting, genteel, polite, presentable, proper, refined, respectable, sedate, seemly, staid, suitable, well-behaved. ▷ DECENT. *Opp* INDECOROUS.

**decorum** *n* correctness, decency, dignity, etiquette, good form, good manners, gravity, modesty, politeness, propriety, protocol, respectability, seemliness.

**decoy** *n* bait, distraction, diversion, enticement, inducement, lure, red herring, stool-pigeon, trap. • *vb* allure, attract, bait, draw, entice, inveigle, lead, lure, seduce, tempt, trick.

**decrease** *n* abatement, contraction, curtailment, cut, cut-back, decline, de-escalation, diminuendo, diminution, downturn, drop, dwindling, easing-off, ebb, fall, falling off, lessening, lowering, reduction, shrinkage, wane. *Opp* INCREASE. • *vb* **1** abate, curtail, cut, ease off, lower, reduce, slim down, turn down. **2** condense, contract, decline, die away, diminish, dwindle, fall off, lessen, peter out, shrink, slacken, subside, *inf* tail off, taper off, wane. *Opp* INCREASE.

**decree** *n* act, command, declaration, dictate, dictum, directive, edict, enactment, fiat, injunction, judgement, law, mandate, order, ordinance, proclamation, promulgation, regulation, ruling, statute. • *vb* command, decide, declare, determine, dictate, direct, ordain, order, prescribe, proclaim, promulgate, pronounce, rule.

**decrepit** *adj* battered, broken down, derelict, dilapidated, feeble, frail, infirm, ramshackle, tumble-down, weak, worn out. ▷ OLD.

**dedicate** *vb* **1** commit, consecrate, devote, give, hallow, pledge, sanctify, set apart. **2** *dedicate a book*. address, inscribe. **dedicate**
▷ KEEN, LOYAL.

**dedication** *n* **1** adherence, allegi ance, commitment, devotion, enthusiasm, faithfulness, fidelity, loyalty, single-mindedness, zeal. **2** inscription.

**deduce** *vb* conclude, divine, draw the conclusion, extrapolate, gather, glean, infer, *inf* put two and two together, reason, surmis *sl* suss out, understand, work out

**deduct** *vb inf* knock off, subtract, take away. *Opp* ADD.

**deduction** *n* **1** allowance, decrease, diminution, discount, reduction, removal, subtraction, withdrawal. **2** conclusion, finding inference, reasoning, result.

**deed** *n* **1** accomplishment, achiev ment, act, action, adventure, effort, endeavour, enterprise, exploit, feat, performance, stunt, undertaking. **2** ▷ DOCUMENT.

**deep** *adj* **1** abyssal, bottomless, chasmic, fathomless, profound, unfathomable, unplumbed, yawn ing. **2** *deep feelings*. earnest, extreme, genuine, heartfelt, intense, serious, sincere. **3** *deep i thought*. absorbed, concentrating engrossed, immersed, lost, preoccupied, rapt, thoughtful. **4** *deep matters*. abstruse, arcane, esoteri intellectual, learned, obscure, recondite. ▷ DIFFICULT. **5** *deep sleep*. heavy, sound. **6** *deep colour* dark, rich, strong, vivid. **7** *deep sound*. bass, booming, growling, low, low-pitched, resonant, rever berating, sonorous. *Opp* SHALLOW SUPERFICIAL, THIN.

**deface** *vb* blemish, damage, disfigure, harm, impair, injure, mar, mutilate, ruin, spoil, vandalize.

**defeat** *n* beating, conquest, downfall, *inf* drubbing, failure, humiliation, *inf* licking, overthrow, *inf* put-down, rebuff, repulse, reverse, rout, setback, subjugation, thrashing, trouncing. *Opp* VICTORY. • *vb* baulk, beat, best, be victorious over, check, checkmate, *inf* clobber, confound, conquer, crush, destroy, *inf* flatten, foil, frustrate, get the better of, *sl* hammer, *inf* lay low, *inf* lick, master, outdo, outvote, outwit, overcome, overpower, overthrow, overwhelm, prevail over, put down, quell, repulse, rout, ruin, *inf* smash, stop, subdue, subjugate, suppress, *inf* thrash, thwart, triumph over, trounce, vanquish, whip, win a victory over. *Opp* LOSE. **be defeated** ▷ LOSE. **defeated** ▷ UNSUCCESSFUL.

**defect** *n* blemish, *computing* bug, deficiency, error, failing, fault, flaw, imperfection, inadequacy, irregularity, lack, mark, mistake, shortcoming, shortfall, spot, stain, want, weakness, weak point. • *vb* change sides, desert, go over.

**defective** *adj* broken, deficient, faulty, flawed, *inf* gone wrong, imperfect, incomplete, *inf* on the blink, unsatisfactory, wanting, weak. *Opp* PERFECT.

**defence** *n* **1** cover, deterrence, guard, protection, safeguard, security, shelter, shield. ▷ BARRIER. **2** alibi, apologia, apology, case, excuse, explanation, justification, plea, testimony, vindication.

**defenceless** *adj* exposed, helpless, impotent, insecure, powerless, unguarded, unprotected, vulnerable, weak.

**defend** *vb* **1** cover, fight for, fortify, guard, keep safe, preserve, protect, safeguard, screen, secure, shelter, shield, *inf* stick up for, watch over. **2** argue for, champion, justify, plead for, speak up for, stand by, stand up for, support, uphold, vindicate. *Opp* ATTACK.

**defendant** *n* accused, appellant, offender, prisoner.

**defensive** *adj* **1** cautious, defending, protective, wary, watchful. **2** apologetic, faint-hearted, self-justifying. *Opp* AGGRESSIVE.

**defer** *vb* **1** adjourn, delay, hold over, lay aside, postpone, prorogue (*parliament*), put off, *inf* shelve, suspend. **2** ▷ YIELD.

**deference** *n* acquiescence, compliance, obedience, submission. ▷ RESPECT.

**defiant** *adj* aggressive, antagonistic, belligerent, bold, brazen, challenging, daring, disobedient, headstrong, insolent, insubordinate, mutinous, obstinate, rebellious, recalcitrant, refractory, self-willed, stubborn, truculent, uncooperative, unruly, unyielding. *Opp* COOPERATIVE.

**deficient** *adj* defective, inadequate, insufficient, lacking, meagre, scanty, scarce, short, sketchy, unsatisfactory, wanting, weak. *Opp* ADEQUATE, EXCESSIVE.

**defile** *vb* contaminate, corrupt, degrade, desecrate, dirty, dishonour, foul, infect, make dirty, poison, pollute, soil, stain, sully, taint, tarnish.

**define** *vb* **1** be the boundary of, bound, circumscribe, delineate, demarcate, describe, determine, fix, limit, mark off, mark out, outline, specify. **2** *define a word.* clarify, explain, formulate, give the meaning of, interpret, spell out.

**definite** *adj* apparent, assured, categorical, certain, clear, clear-cut, confident, confirmed, cut-and-dried, decided, determined, discernible, distinct, emphatic, exact, explicit, express, fixed, incisive, marked, noticeable, obvious, particular, perceptible, plain, positive, precise, pronounced, settled, specific, sure, unambiguous, unequivocal, unmistakable, well-defined. *Opp* VAGUE.

**definitely** *adv* beyond doubt, certainly, doubtless, for certain, indubitably, positively, surely, unquestionably, without doubt, without fail.

**definition** *n* **1** clarification, elucidation, explanation, interpretation. **2** clarity, clearness, focus, precision, sharpness.

**definitive** *adj* agreed, authoritative, complete, conclusive, correct, decisive, final, last (*word*), official, permanent, reliable, settled, standard, ultimate, unconditional. *Opp* PROVISIONAL.

**deflect** *vb* avert, deviate, divert, fend off, head off, intercept, parry, prevent, sidetrack, swerve, switch, turn aside, veer, ward off.

**deformed** *adj* bent, buckled, contorted, crippled, crooked, defaced, disfigured, distorted, gnarled, grotesque, malformed, mangled, misshapen, mutilated, twisted, ugly, warped.

**defraud** *vb inf* con, *inf* diddle, embezzle, *inf* fleece, rob, swindle. ▷ CHEAT.

**deft** *adj* adept, adroit, agile, clever, dextrous, expert, handy, neat, *inf* nifty, nimble, proficient, quick, skilful. *Opp* CLUMSY.

**defy** *vb* **1** challenge, confront, dare, disobey, face up to, flout, *inf* kick against, rebel against, refuse to obey, resist, stand up to, withstand. **2** baffle, beat, defeat, elude, foil, frustrate, repel, repulse, resist, thwart, withstand.

**degenerate** *adj* ▷ CORRUPT. • *vb* become worse, decline, deteriorate, *inf* go to the dogs, regress, retrogress, sink, slip, weaken, worsen. *Opp* IMPROVE.

**degrade** *vb* **1** cashier, demote, depose, downgrade. **2** abase, brutalize, cheapen, corrupt, debase, dehumanize, deprave, desensitize, dishonour, harden, humiliate, mortify. **degrading** ▷ SHAMEFUL.

**degree** *n* **1** calibre, class, grade, order, position, rank, standard, standing, station, status. **2** extent, intensity, level, measure.

**deify** *vb* idolize, treat as a god, venerate, worship.

**deign** *vb* concede, condescend, demean yourself, lower yourself, stoop, vouchsafe.

**deity** *n* creator, divinity, god, goddess, godhead, idol, immortal, power, spirit, supreme being.

**dejected** *adj* depressed, disconsolate, dispirited, down, downcast, downhearted, heavy-hearted, in low spirits. ▷ SAD.

**delay** *n* check, deferment, deferral, filibuster, hiatus, hitch, hold-up, interruption, moratorium, pause, postponement, setback, stay (*of execution*), stoppage, wait. • *vb* **1** check, defer, detain, halt, hinder, hold over, hold up, impede, keep back, keep waiting, make late, obstruct, postpone, put back, put off, retard, set back, slow down, stay, stop, suspend.

**2** be late, be slow, *inf* bide your time, dally, dawdle, *inf* dilly-dally, *inf* drag your feet, *inf* get bogged down, hang about, hang back, hang fire, hesitate, lag, linger, loiter, mark time, pause, *inf* play for time, procrastinate, stall, *old use* tarry, temporize, vacillate, wait. *Opp* HURRY.

**delegate** *n* agent, ambassador, emissary, envoy, go-between, legate, messenger, nuncio, plenipotentiary, representative, spokesperson. • *vb* appoint, assign, authorize, charge, commission, depute, designate, empower, entrust, mandate, nominate.

**delegation** *n* commission, deputation, mission.

**delete** *vb* blot out, cancel, cross out, cut out, edit out, efface, eliminate, eradicate, erase, expunge, obliterate, remove, rub out, strike out, wipe out.

**deliberate** *adj* **1** arranged, calculated, cold-blooded, conscious, contrived, culpable, designed, intended, intentional, knowing, malicious, organized, planned, prearranged, preconceived, premeditated, prepared, purposeful, studied, thought out, wilful, worked out. **2** careful, cautious, circumspect, considered, diligent, measured, methodical, orderly, painstaking, regular, slow, thoughtful, unhurried, watchful. *Opp* HASTY, INSTINCTIVE. • *vb* ▷ THINK.

**delicacy** *n* **1** accuracy, care, cleverness, daintiness, discrimination, exquisiteness, fineness, finesse, fragility, intricacy, precision, sensitivity, subtlety, tact. **2** *delicacies to eat.* rarity, speciality, treat.

**delicate** *adj* **1** dainty, diaphanous, easily broken, easily damaged, elegant, exquisite, fine, flimsy, fragile, frail, gauzy, gentle, feathery, intricate, light, sensitive, slender, soft, tender. *Opp* TOUGH. **2** *delicate work.* accurate, careful, clever, deft, precise, skilled. *Opp* CLUMSY. **3** *delicate flavour, colour.* faint, mild, muted, pale, slight, subtle. **4** *delicate health.* feeble, puny, sickly, squeamish, unhealthy, weak. **5** *delicate problem.* awkward, confidential, embarrassing, private, problematical, prudish, *inf* sticky, ticklish, touchy. **6** *delicate handling.* considerate, diplomatic, discreet, judicious, prudent, sensitive, tactful. *Opp* CRUDE.

**delicious** *adj* appetizing, choice, delectable, enjoyable, luscious, *inf* mouth-watering, *inf* nice, palatable, savoury, *inf* scrumptious, succulent, tasty, tempting, toothsome, *sl* yummy.

**delight** *n* bliss, delectation, ecstasy, enchantment, enjoyment, felicity, gratification, happiness, joy, paradise, pleasure, rapture, satisfaction. • *vb* amuse, bewitch, captivate, charm, cheer, divert, enchant, enrapture, entertain, enthral, entrance, fascinate, gladden, gratify, please, ravish, thrill, transport. *Opp* DISMAY. **delighted** ▷ HAPPY, PLEASED.

**delightful** *adj* agreeable, attractive, captivating, charming, congenial, delectable, diverting, enjoyable, *inf* nice, pleasant, pleasing, pleasurable, rewarding, satisfying, spell-binding. ▷ BEAUTIFUL.

**delinquent** *n* culprit, defaulter, hooligan, lawbreaker, malefactor, miscreant, offender, roughneck, ruffian, *inf* tear-away, vandal, wrongdoer, young offender. ▷ CRIMINAL.

**delirious** *adj inf* beside yourself, crazy, demented, deranged, distracted, ecstatic, excited, feverish, frantic, frenzied, hysterical, incoherent, irrational, light-headed, rambling, wild. ▷ DRUNK, MAD. *Opp* SANE, SOBER.

**deliver** *vb* **1** bear, bring, carry, cart, convey, distribute, give out, hand over, make over, present, purvey, supply, surrender, take round, transfer, transport, turn over. **2** *deliver a lecture*. announce, broadcast, express, give, make, read. ▷ SPEAK. **3** *deliver a blow*. administer, aim, deal, direct, fire, inflict, launch, strike, throw. ▷ HIT. **4** ▷ RESCUE.

**delivery** *n* **1** conveyance, dispatch, distribution, shipment, transmission, transportation. **2** *a delivery of goods*. batch, consignment. **3** *delivery of a speech*. enunciation, execution, implementation, performance, presentation. **4** childbirth, confinement, parturition.

**deluge** *n* downpour, flood, inundation, rainfall, rainstorm, rush, spate. • *vb* drown, engulf, flood, inundate, overwhelm, submerge, swamp.

**delusion** *n* dream, fantasy, hallucination, illusion, mirage, misconception, mistake, self-deception.

**delve** *vb* burrow, dig, explore, investigate, probe, research, search.

**demand** *n old use* behest, claim, command, desire, expectation, importunity, insistence, need, order, request, requirement, requisition, want. • *vb* call for, claim, cry out for, exact, expect, insist on, necessitate, order, request, require, requisition, want. ▷ ASK. **demanding** ▷ DIFFICULT, IMPORTUNATE. **in demand** ▷ POPULAR.

**demean** *vb* abase, cheapen, debase, degrade, disgrace, humble, humiliate, lower, make (yourself) cheap, *inf* put (yourself) down, sacrifice (your) pride, undervalue. **demeaning** ▷ SHAMEFUL.

**democratic** *adj* **1** classless, egalitarian. **2** chosen, elected, elective, popular, representative. *Opp* TOTALITARIAN.

**demolish** *vb* break down, bulldoze, dismantle, flatten, knock down, level, pull down, raze, tear down, topple, undo, wreck. ▷ DESTROY. *Opp* BUILD.

**demon** *n* devil, evil spirit, fiend, goblin, imp, spirit.

**demonstrable** *adj* conclusive, confirmable, evident, incontrovertible, indisputable, irrefutable, palpable, positive, provable, undeniable, unquestionable, verifiable.

**demonstrate** *vb* **1** confirm, describe, display, embody, establish, evince, exemplify, exhibit, explain, expound, express, illustrate, indicate, manifest, prove, represent, show, substantiate, teach, typify, verify. **2** lobby, march, parade, picket, protest, rally.

**demonstration** *n* **1** confirmation, description, display, evidence, exhibition, experiment, expression, illustration, indication, manifestation, presentation, proof, representation, show, substantiation, test, trial, verification. **2** *inf* demo, march, parade, picket, protest, rally, sit-in, vigil.

**demonstrative** *adj* affectionate, effusive, emotional, fulsome, loving, open, uninhibited, unre-

›rved, unrestrained.
*›pp* RETICENT.

**›mote** *vb* downgrade, put down, ›duce, relegate. *Opp* PROMOTE.

**›mure** *adj* bashful, coy, diffident, ›odest, prim, quiet, reserved, reti›nt, retiring, sedate, shy, sober, ›aid. *Opp* CONCEITED.

**›n** *n* hideaway, hide-out, hiding›lace, hole, lair, private place, ›etreat, sanctuary, secret place, ›helter.

**›nial** *n* abnegation, contradic›on, disavowal, disclaimer, nega›on, refusal, refutation, rejection, ›nunciation, repudiation, veto. *›pp* ADMISSION.

**›nigrate** *vb* belittle, blacken the ›eputation of, criticize, decry, dis›arage, impugn, malign, *inf* put ›own, *inf* run down, sneer at, ›peak slightingly of, traduce, *›f* turn your nose up, vilify. › DESPISE. *Opp* PRAISE.

**›nomination** *n* 1 category, ›lass, classification, designation, ›ind, size, sort, species, type, ›alue. 2 church, communion, ›reed, cult, order, persuasion, ›chism, school, sect.

**›note** *vb* be the sign for, desig›ate, express, indicate, mean, rep›esent, signal, signify, stand for, ›ymbolize.

**›nouement** *n* climax, *inf* pay-off, ›esolution, solution, *inf* sorting ›ut, *inf* tidying up, unravelling. › END.

**›nounce** *vb* accuse, attack verb›lly, betray, blame, brand, cen›ure, complain about, condemn, ›riticize, declaim against, decry, ›lminate against, *inf* hold forth ›gainst, impugn, incriminate, ›nform against, inveigh against, ›illory, report, reveal, stigmatize, *inf* tell off, vilify, vituperate. *Opp* PRAISE.

**dense** *adj* 1 close, compact, concentrated, heavy, impassable, impenetrable, *inf* jam-packed, lush, massed, packed, solid, thick, tight, viscous. *Opp* THIN. 2 ▷ STUPID.

**dent** *n* concavity, depression, dimple, dint, dip, hollow, indentation, pit. • *vb* bend, buckle, crumple, knock in.

**denude** *vb* bare, defoliate, deforest, expose, remove, strip, unclothe, uncover. *Opp* CLOTHE.

**deny** *vb* 1 contradict, controvert, disagree with, disclaim, disown, dispute, gainsay, negate, oppose, rebuff, refute, reject, repudiate. *Opp* AGREE. 2 begrudge, deprive of, disallow, refuse, withhold. *Opp* GRANT. **deny yourself** ▷ ABSTAIN.

**depart** *vb* 1 abscond, begin a journey, *inf* check out, *inf* clear off, decamp, disappear, embark, emigrate, escape, exit, go away, *sl* hit the road, leave, make off, *inf* make tracks, *inf* make yourself scarce, migrate, move away, move off, *inf* push off, quit, retire, retreat, run away, run off, *sl* scarper, *sl* scram, set forth, set off, set out, start, take your leave, vanish, withdraw. 2 ▷ DEVIATE. **departed** ▷ DEAD.

**department** *n* 1 branch, division, office, part, section, sector, subdivision, unit. 2 [*inf*] *not my department.* area, concern, domain, field, function, job, line, province, responsibility, specialism, sphere.

**departure** *n* disappearance, embarkation, escape, exit, exodus, going, retirement, retreat, withdrawal. *Opp* ARRIVAL.

**depend** *vb* **depend on** *inf* bank on, be dependent on, count on, hinge on, need, pivot on, put your faith in, *inf* reckon on, rely on, rest on, trust.

**dependable** *adj* conscientious, consistent, faithful, honest, regular, reliable, safe, sound, steady, true, trustworthy, unfailing. *Opp* UNRELIABLE.

**dependence** *n* **1** confidence, need, reliance, trust. **2** ▷ ADDICTION.

**dependent** *adj* **dependent on** **1** conditional on, connected with, controlled by, determined by, liable to, relative to, subject to, vulnerable to. *Opp* INDEPENDENT. **2** *dependent on drugs.* addicted to, enslaved by, *inf* hooked on, reliant on.

**depict** *vb* delineate, describe, draw, illustrate, narrate, outline, paint, picture, portray, represent, reproduce, show, sketch.

**deplete** *vb* consume, cut, decrease, drain, lessen, reduce, use up. *Opp* INCREASE.

**deplorable** *adj* awful, blameworthy, discreditable, disgraceful, disreputable, dreadful, execrable, lamentable, regrettable, reprehensible, scandalous, shameful, shocking, unfortunate, unworthy. ▷ BAD. *Opp* COMMENDABLE.

**deplore** *vb* **1** grieve for, lament, mourn, regret. **2** ▷ CONDEMN.

**deploy** *vb* arrange, bring into action, distribute, manage, position, use systematically, utilize.

**deport** *vb* banish, exile, expatriate, expel, remove, send abroad, transport.

**depose** *vb* demote, dethrone, dismiss, displace, get rid of, oust, remove, *inf* topple.

**deposit** *n* **1** advance payment, down-payment, initial payment, part-payment, retainer, security stake. **2** accumulation, alluvium dregs, layer, lees, precipitate, se ment, silt, sludge. • *vb* **1** drop, *inf* dump, lay down, leave, *inf* park, place, precipitate, put down, set down. **2** *deposit money* bank, pay in, save.

**depot** *n* **1** arsenal, base, cache, depository, dump, hoard, store, storehouse. **2** *bus depot.* garage, headquarters, station, terminus.

**deprave** *vb* brutalize, corrupt, debase, degrade, influence, pervert. **depraved** ▷ CORRUPT.

**depreciate** *vb* **1** become less, decrease, deflate, drop, fall, go down, lessen, lower, reduce, slump, weaken. *Opp* APPRECIATE **2** ▷ DISPARAGE.

**depress** *vb* **1** burden, cast down discourage, dishearten, dismay, dispirit, enervate, grieve, lower the spirits of, make sad, oppress sadden, tire, upset, weary. *Opp* CHEER. **2** *depress the market* bring down, deflate, make less a ive, push down, undermine, weaken. *Opp* BOOST. **depressed, depressing** ▷ SAD.

**depression** *n* **1** *inf* blues, dejection, desolation, despair, despon ency, gloom, glumness, heavine hopelessness, low spirits, melan choly, misery, pessimism, sadne weariness. *Opp* HAPPINESS. **2** cavity, concavity, dent, dimple dip, excavation, hole, hollow, impression, indentation, pit, recess, rut, sunken area. *Opp* BUMP. **3** *economic depression* decline, hard times, recession, slump. *Opp* BOOM, HIGH.

**deprive** *vb* **deprive of** deny, di possess of, prevent from using, refuse, rob of, starve of, strip of,

take away, withdraw, withhold.
**deprived** ▷ POOR.

**deputize** *vb* **deputize for** act as deputy, act as stand-in for, cover for, do the job of, replace, represent, stand in for, substitute for, take over from, understudy.

**deputy** *n* agent, ambassador, assistant, delegate, emissary, *inf* fill-in, locum, proxy, relief, replacement, representative, reserve, second-in-command, spokesperson, *inf* stand-in, substitute, supply, surrogate, understudy, vice-captain, vice-president.

**derelict** *adj* abandoned, broken down, decrepit, deserted, desolate, dilapidated, forgotten, forlorn, forsaken, neglected, overgrown, ruined, run-down, tumbledown, uncared-for, untended.

**derivation** *n* ancestry, descent, etymology, extraction, origin, root. ▷ BEGINNING.

**derive** *vb* acquire, borrow, collect, crib, draw, extract, gain, gather, get, glean, *inf* lift, obtain, pick up, procure, receive, secure, take. **be derived** ▷ ORIGINATE.

**descend** *vb* **1** climb down, come down, drop, fall, go down, move down, plummet, plunge, sink, swoop down. **2** decline, dip, incline, slant, slope. **3** alight, disembark, dismount, get down, get off. *Opp* ASCEND. **be descended** ▷ ORIGINATE. **descend on** ▷ ATTACK.

**descendant** *n* child, heir, scion, successor. *Opp* ANCESTOR. **descendants** family, issue, line, lineage, offspring, posterity, progeny, *old use* seed.

**descent** *n* **1** declivity, dip, drop, fall, incline, slant, slope, way down. *Opp* ASCENT. **2** *aristocratic descent*. ancestry, background, blood, derivation, extraction, family, genealogy, heredity, lineage, origin, parentage, pedigree, stock, strain.

**describe** *vb* **1** characterize, define, delineate, depict, detail, explain, express, give an account of, narrate, outline, portray, present, recount, relate, report, represent, sketch, speak of, tell about. **2** *describe a circle*. draw, mark out, trace.

**description** *n* account, characterization, commentary, definition, delineation, depiction, explanation, narration, outline, portrait, portrayal, report, representation, sketch, story, word-picture.

**descriptive** *adj* colourful, detailed, explanatory, expressive, graphic, illustrative, pictorial, vivid.

**desecrate** *vb* abuse, contaminate, corrupt, debase, defile, degrade, dishonour, pervert, pollute, profane, treat blasphemously, treat disrespectfully, treat irreverently, vandalize, violate, vitiate. *Opp* REVERE.

**desert** *adj* arid, barren, desolate, dry, infertile, isolated, lonely, sterile, uncultivated, unfrequented, uninhabited, waterless, wild. *Opp* FERTILE. • *n* dust bowl, wasteland, wilderness. • *vb* **1** abandon, betray, forsake, give up, jilt, leave, *inf* leave in the lurch, maroon, quit, *inf* rat on, renounce, strand, vacate, *inf* walk out on, *inf* wash your hands of. **2** abscond, decamp, defect, go absent, run away. **deserted** ▷ EMPTY, LONELY.

**deserter** *n* absconder, absentee, apostate, backslider, betrayer, defector, escapee, fugitive, outlaw, renegade, runaway, traitor, truant, turncoat.

**deserve** *vb* be good enough for, be worthy of, earn, justify, merit, rate, warrant. **deserving** ▷ WORTHY.

**design** *n* **1** blueprint, conception, draft, drawing, model, pattern, plan, proposal, prototype, sketch. **2** mark, style, type, version. **3** arrangement, composition, configuration, form, pattern, shape. **4** *wander without design.* aim, end, goal, intention, object, objective, purpose, scheme. • *vb* conceive, construct, contrive, create, delineate, devise, draft, draw, draw up, fashion, form, intend, invent, lay out, make, map out, originate, outline, plan, plot, project, propose, scheme, shape, sketch, think up. **designing** ▷ CRAFTY. **have designs** ▷ PLOT.

**designer** *n* architect, artist, author, contriver, creator, deviser, inventor, originator.

**desire** *n* **1** ache, ambition, appetite, craving, fancy, hankering, hunger, *inf* itch, longing, requirement, thirst, urge, want, wish, yearning, *inf* yen. **2** avarice, covetousness, cupidity, greed, miserliness, rapacity. **3** *sexual desire.* ardour, libido, love, lust, passion. • *vb* ache for, ask for, aspire to, covet, crave, dream of, fancy, hanker after, *inf* have a yen for, hope for, hunger for, *inf* itch for, like, long for, lust after, need, pine for, prefer, pursue, *inf* set your heart on, set your sights on, strive after, thirst for, want, wish for, yearn for.

**desolate** *adj* **1** abandoned, bare, barren, benighted, bleak, cheerless, depressing, deserted, dismal, dreary, empty, forsaken, gloomy, *inf* god-forsaken, inhospitable, isolated, lonely, remote, unfrequented, uninhabited, wild, windswept. **2** bereft, companionless, dejected, depressed, despairing, disconsolate, distressed, forlorn, forsaken, inconsolable, lonely, melancholy, miserable, neglected, solitary, suicidal, wretched. ▷ SAD. *Opp* CHEERFUL.

**despair** *n* anguish, dejection, depression, desperation, despondency, hopelessness, pessimism, resignation, wretchedness. ▷ MISERY. • *vb* give in, give up, lose heart, lose hope, quit, surrender. *Opp* HOPE.

**desperate** *adj* **1** *inf* at your wits' end, beyond hope, despairing, inconsolable, wretched. **2** *desperate situation.* acute, bad, critical, dangerous, drastic, grave, hopeless, irretrievable, pressing, serious, severe, urgent. **3** *desperate criminals.* dangerous, foolhardy, impetuous, rash, reckless, violent, wild. **4** ▷ ANXIOUS.

**despise** *vb* be contemptuous of, condemn, deride, disapprove of, disdain, feel contempt for, hate, have a low opinion of, look down on, *inf* put down, scorn, sneer at, spurn, undervalue. ▷ DENIGRATE. *Opp* ADMIRE.

**despondent** *adj* dejected, depressed, discouraged, disheartened, down, downcast, *inf* down in the mouth, melancholy, morose, pessimistic, sad, sorrowful. ▷ MISERABLE.

**despotic** *adj* absolute, arbitrary, authoritarian, autocratic, dictatorial, domineering, oppressive, totalitarian, tyrannical. *Opp* DEMOCRATIC.

**destination** *n* goal, objective, purpose, stopping-place, target, terminus.

**destined** *adj* **1** foreordained, ineluctable, inescapable, inevitable, intended, ordained, predestined, predetermined, preordained, unavoidable. **2** *destined to fail.* bound, certain, doomed, fated, meant.

**destiny** *n* chance, doom, fate, fortune, karma, kismet, lot, luck, providence.

**destitute** *adj* bankrupt, deprived, down-and-out, homeless, impecunious, impoverished, indigent, insolvent, needy, penniless, poverty-stricken, *inf* skint. ▷ POOR. *Opp* WEALTHY.

**destroy** *vb* abolish, annihilate, blast, break down, burst, *inf* bust, crush, *inf* decimate, demolish, devastate, devour, dismantle, dispose of, do away with, eliminate, eradicate, erase, exterminate, extinguish, extirpate, finish off, flatten, fragment, get rid of, knock down, lay waste, level, liquidate, make useless, nullify, pull down, pulverize, put out of existence, raze, root out, ruin, sabotage, sack, scuttle, shatter, smash, stamp out, undo, uproot, vaporize, wipe out, wreck, write off. ▷ DEFEAT, END, KILL. *Opp* CONSERVE, CREATE.

**destruction** *n* annihilation, damage, *inf* decimation, demolition, depredation, devastation, elimination, end, eradication, erasure, extermination, extinction, extirpation, havoc, holocaust, liquidation, overthrow, pulling down, ruin, ruination, shattering, smashing, undoing, uprooting, wiping out, wrecking. ▷ KILLING. *Opp* CONSERVATION, CREATION.

**destructive** *adj* adverse, antagonistic, baleful, baneful, calamitous, catastrophic, damaging, dangerous, deadly, deleterious, detrimental, devastating, disastrous, fatal, harmful, injurious, internecine, lethal, malignant, negative, pernicious, pestilential, ruinous, violent. *Opp* CONSTRUCTIVE.

**detach** *vb* cut loose, cut off, disconnect, disengage, disentangle, divide, free, isolate, part, pull off, release, remove, segregate, separate, sever, take off, tear off, uncouple, undo, unfasten, unfix, unhitch. *Opp* ATTACH. **detached** ▷ ALOOF, IMPARTIAL, SEPARATE.

**detail** *n* aspect, circumstance, complexity, complication, component, element, fact, factor, feature, ingredient, intricacy, item, *plur* minutiae, nicety, particular, point, refinement, respect, specific, technicality.

**detailed** *adj inf* blow-by-blow, complete, complex, comprehensive, descriptive, exact, exhaustive, full, *derog* fussy, giving all details, *derog* hair-splitting, intricate, itemized, minute, particularized, specific. *Opp* GENERAL.

**detain** *vb* **1** arrest, capture, confine, gaol, hold, imprison, intern. **2** buttonhole, delay, hinder, hold up, impede, keep, keep waiting, restrain, retard, slow, stop, waylay.

**detect** *vb* ascertain, become aware of, diagnose, discern, discover, expose, feel, *inf* ferret out, find, hear, identify, locate, note, notice, observe, perceive, *inf* put your finger on, recognize, reveal, scent, see, sense, sight, smell, sniff out, spot, spy, taste, track down, uncover, unearth, unmask.

**detective** *n* investigator, policeman, policewoman, *inf* private eye, sleuth, *inf* snooper.

**detention** *n* captivity, confinement, custody, imprisonment, incarceration, internment.

**deter** *vb* check, daunt, discourage, dismay, dissuade, frighten off, hinder, impede, intimidate, obstruct, prevent, put off, repel, send away, stop, *inf* turn off, warn off. *Opp* ENCOURAGE.

**deteriorate** *vb* crumble, decay, decline, degenerate, depreciate, disintegrate, fall off, get worse, *inf* go downhill, lapse, relapse, slip, weaken, worsen. *Opp* IMPROVE.

**determination** *n inf* backbone, commitment, courage, dedication, doggedness, drive, firmness, fortitude, *inf* grit, *inf* guts, perseverance, persistence, pertinacity, resoluteness, resolution, resolve, single-mindedness, spirit, steadfastness, *derog* stubbornness, tenacity, will-power.

**determine** *vb* **1** arbitrate, clinch, conclude, decide, establish, find out, identify, judge, settle. **2** choose, decide on, fix on, resolve, select. **3** *What determined your choice?* affect, condition, dictate, govern, influence, regulate.

**determined** *adj* adamant, assertive, bent (*on success*), certain, convinced, decided, decisive, definite, dogged, firm, insistent, intent, *derog* obstinate, persistent, pertinacious, purposeful, resolute, resolved, single-minded, steadfast, strong-minded, strong-willed, *derog* stubborn, sure, tenacious, tough, unwavering. *Opp* IRRESOLUTE.

**deterrent** *n* barrier, caution, check, curb, difficulty, discouragement, disincentive, dissuasion, hindrance, impediment, obstacle, restraint, threat, *inf* turn-off, warning. *Opp* ENCOURAGEMENT.

**detest** *vb* abhor, abominate, despise, execrate, loathe. ▷ HATE.

**detour** *n* deviation, diversion, indirect route, roundabout route. **make a detour** ▷ DEVIATE.

**detract** *vb* **detract from** diminish, lessen, lower, reduce, take away from.

**detrimental** *adj* damaging, deleterious, disadvantageous, harmful, hurtful, inimical, injurious, prejudicial, unfavourable. *Opp* ADVANTAGEOUS.

**devastate** *vb* **1** damage severely, demolish, destroy, flatten, lay waste, level, obliterate, overwhelm, ravage, raze, ruin, sack, waste, wreck. **2** ▷ DISMAY.

**develop** *vb* **1** advance, age, arise, *inf* blow up, come into existence, evolve, get better, grow, flourish, improve, mature, move on, progress, ripen. *Opp* REGRESS. **2** *develop habits.* acquire, contract, cultivate, evolve, foster, get, pick up. **3** *develop ideas.* amplify, augment, elaborate, enlarge on, expatiate on, unfold, work up. **4** *business developed.* branch out, build up, diversify, enlarge, expand, extend, increase, swell.

**development** *n* **1** advance, betterment, change, enlargement, evolution, expansion, extension, *inf* forward march, furtherance, gain, growth, improvement, increase, progress, promotion, regeneration, reinforcement, spread. **2** happening, incident, occurrence, outcome, result, upshot. **3** *industrial development.* building, conversion, exploitation, use.

**deviate** *vb* branch off, depart, digress, diverge, divert, drift, err, go astray, go round, make a detour, stray, swerve, turn aside, turn off, vary, veer, wander.

**device** *n* **1** apparatus, appliance, contraption, contrivance, gadget, implement, instrument, invention, machine, tool, utensil. **2** dodge, expedient, gambit, gimmick, manoeuvre, plan, ploy, ruse, scheme, stratagem, stunt, tactic, trick, wile. **3** *heraldic device*. badge, crest, design, figure, logo, motif, shield, sign, symbol, token.

**devil** *n* demon, fiend, imp, spirit. **The Devil** the Adversary, Beelzebub, the Evil One, Lucifer, Mephistopheles, *inf* Old Nick, the Prince of Darkness, Satan.

**devilish** *adj* demoniac(al), demonic, diabolic(al), fiendish, hellish, impish, infernal, inhuman, Mephisophelian, satanic. ▷ EVIL. *Opp* ANGELIC.

**devious** *adj* **1** circuitous, crooked, deviating, indirect, periphrastic, rambling, round-about, sinuous, tortuous, wandering, winding. **2** [*derog*] calculating, cunning, deceitful, evasive, insincere, misleading, scheming, *inf* slippery, sly, sneaky, treacherous, underhand, wily. ▷ DISHONEST. *Opp* DIRECT.

**devise** *vb* arrange, conceive, concoct, contrive, *inf* cook up, create, design, engineer, form, formulate, frame, imagine, invent, make up, plan, plot, prepare, project, scheme, think out, think up, work out.

**devoted** *adj* committed, dedicated, enthusiastic, faithful, loving, staunch, true, unswerving, wholehearted, zealous. ▷ LOYAL. *Opp* DISLOYAL, HALF-HEARTED.

**devotee** *n* *inf* addict, aficionado, *inf* buff, enthusiast, fan, follower, *sl* freak, supporter.

**devotion** *n* allegiance, attachment, commitment, dedication, devotedness, enthusiasm, fanaticism, fervour, loyalty, zeal. ▷ LOVE, PIETY.

**devour** *vb* consume, demolish, eat up, engulf, swallow up, take in. ▷ DESTROY, EAT.

**devout** *adj* god-fearing, godly, holy, religious, sincere, spiritual. ▷ PIOUS. *Opp* IRRELIGIOUS.

**dexterous** *adj* adroit, agile, deft, nimble, quick, sharp, skilful. ▷ CLEVER. *Opp* CLUMSY.

**diabolical** *adj* evil, fiendish, inhuman, satanic, wicked. ▷ DEVILISH. *Opp* SAINTLY.

**diagnose** *vb* detect, determine, distinguish, find, identify, isolate, name, pinpoint, recognize.

**diagnosis** *n* analysis, conclusion, explanation, identification, interpretation, opinion, pronouncement, verdict.

**diagram** *n* chart, drawing, figure, flow-chart, graph, illustration, outline, picture, plan, representation, sketch, table.

**dial** *n* clock, digital display, face, instrument, pointer, speedometer.

**dialect** *n* accent, argot, brogue, cant, creole, idiom, jargon, language, patois, phraseology, pronunciation, register, slang, speech, tongue, vernacular.

**dialogue** *n* *inf* chat, *inf* chin-wag, colloquy, communication, conference, conversation, debate, discourse, discussion, duologue, exchange, interchange, *old use* intercourse, meeting, oral communication, talk, *inf* tête-à-tête.

**diary** *n* annals, appointment book, calendar, chronicle, engagement book, journal, log, record.

**dictate** *vb* **1** read aloud, speak slowly. **2** command, decree, direct, enforce, give orders, impose, *inf* lay down the law, make the rules, ordain, order, prescribe, state categorically.

**dictator** *n* autocrat, *inf* Big Brother, despot, tyrant. ▷ RULER.

**dictatorial** *adj* absolute, arbitrary, authoritarian, autocratic, *inf* bossy, despotic, dogmatic, dominant, domineering, illiberal, imperious, intolerant, omnipotent, oppressive, overbearing, repressive, totalitarian, tyrannical, undemocratic. *Opp* DEMOCRATIC.

**dictionary** *n* concordance, glossary, lexicon, thesaurus, vocabulary, wordbook.

**didactic** *adj* instructive, lecturing, pedagogic, pedantic.

**die** *vb* **1** *inf* bite the dust, *inf* breathe your last, cease to exist, come to the end, decease, depart, expire, fall, *inf* give up the ghost, *sl* kick the bucket, lay down your life, lose your life, pass away, *sl* peg out, perish, *sl* pop off, *sl* snuff it, starve. **2** decline, decrease, die away, disappear, droop, dwindle, ebb, end, fade, fail, fizzle out, go out, languish, lessen, peter out, stop, subside, vanish, wane, weaken, wilt, wither.

**diet** *n* fare, food, intake, nourishment, nutriment, nutrition, sustenance. • *vb* abstain, *inf* cut down, deny yourself, fast, lose weight, ration yourself, reduce, slim.

**differ** *vb* **1** be different, be distinct, contrast, deviate, diverge, show differences, vary. **2** argue, be at odds, be at variance, clash, conflict, contradict, disagree, dispute, dissent, fall out, *inf* have a difference, oppose each other, quarrel, take issue with each other. *Opp* AGREE.

**difference** *n* **1** alteration, change, comparison, contrast, development, deviation, differential, differentiation, discrepancy, disparity, dissimilarity, distinction, diversity, incompatibility, incongruity, inconsistency, modification, nuance, unlikeness, variation, variety. *Opp* SIMILARITY. **2** argument, clash, conflict, controversy, debate, disagreement, disharmony, dispute, dissent, quarrel, strife, tiff, wrangle. *Opp* AGREEMENT.

**different** *adj* **1** assorted, clashing, conflicting, contradictory, contrasting, deviating, discordant, discrepant, disparate, dissimilar, distinguishable, divergent, diverse, heterogeneous, ill-matched, incompatible, inconsistent, miscellaneous, mixed, multifarious, opposed, opposite, *inf* poles apart, several, sundry, unlike, varied, various. *Opp* SIMILAR. **2** abnormal, altered, anomalous, atypical, bizarre, changed, distinct, distinctive, eccentric, extraordinary, fresh, individual, irregular, new, original, particular, peculiar, personal, revolutionary, separate, singular, special, specific, strange, uncommon, unconventional, unique, unorthodox, unusual. *Opp* CONVENTIONAL.

**differentiate** *vb* contrast, discriminate, distinguish, tell apart.

**difficult** *adj* **1** abstruse, advanced, baffling, complex, complicated, deep, *inf* dodgy, enigmatic, hard, intractable, intricate, involved, *inf* knotty, *inf* nasty, obscure, perplexing, problematical, *inf* thorny, ticklish, tricky. **2** arduous, awkward, backbreaking, burdensome, challenging, daunting, demanding, exacting, exhausting, formidable, gruelling, heavy, herculean, *inf* killing, laborious, onerous, punishing, rigorous, severe, strenuous, taxing, tough, uphill. **3** *difficult children.*

annoying, disruptive, fussy, headstrong, intractable, obstinate, obstreperous, refractory, stubborn, tiresome, troublesome, trying, uncooperative, unfriendly, unhelpful, unresponsive, unruly. *Opp* COOPERATIVE, EASY.

**difficulty** *n* adversity, challenge, complication, dilemma, embarrassment, enigma, *inf* fix, *inf* hang-up, hardship, *inf* hiccup, hindrance, hurdle, impediment, *inf* jam, *inf* mess, obstacle, perplexity, *inf* pickle, pitfall, plight, predicament, problem, puzzle, quandary, snag, *inf* spot, straits, *inf* stumbling-block, tribulation, trouble, *inf* vexed question.

**diffident** *adj* backward, bashful, coy, distrustful, doubtful, fearful, hesitant, hesitating, inhibited, insecure, introvert, meek, modest, nervous, private, reluctant, reserved, retiring, self-effacing, sheepish, shrinking, shy, tentative, timid, timorous, unadventurous, unassuming, underconfident, unsure, withdrawn. *Opp* CONFIDENT.

**diffuse** *adj* digressive, discursive, long-winded, loose, meandering, rambling, spread out, unstructured, vague, *inf* waffly, wandering. ▷ WORDY. *Opp* CONCISE. • *vb* ▷ SPREAD.

**dig** *vb* **1** burrow, delve, excavate, gouge, hollow, mine, quarry, scoop, tunnel. **2** cultivate, fork over, *inf* grub up, till, trench, turn over. **3** jab, nudge, poke, prod, punch, shove, thrust. **dig out** ▷ FIND. **dig up** disinter, exhume.

**digest** *n* ▷ SUMMARY. • *vb* **1** absorb, assimilate, dissolve, ingest, process, utilize. ▷ EAT. **2** consider, ponder, study, take in, understand.

**digit** *n* **1** figure, integer, number, numeral. **2** finger, toe.

**dignified** *adj* august, becoming, calm, courtly, decorous, distinguished, elegant, exalted, formal, grand, grave, imposing, impressive, lofty, lordly, majestic, noble, proper, refined, regal, sedate, serious, sober, solemn, stately, tasteful, upright. ▷ PROUD. *Opp* UNBECOMING.

**dignitary** *n* *inf* high-up, important person, luminary, notable, official, *inf* VIP, worthy.

**dignity** *n* calmness, courtliness, decorum, elegance, eminence, formality, glory, grandeur, *Lat* gravitas, gravity, greatness, honour, importance, majesty, nobility, propriety, regality, respectability, seriousness, sobriety, solemnity, stateliness. ▷ PRIDE.

**digress** *vb* depart, deviate, diverge, drift, get off the subject, *inf* go off at a tangent, *inf* lose the thread, ramble, stray, veer, wander.

**dilapidated** *adj* badly maintained, broken down, crumbling, decayed, decrepit, derelict, falling apart, falling down, in disrepair, in ruins, neglected, ramshackle, rickety, ruined, *inf* run-down, shaky, tottering, tumbledown, uncared-for.

**dilemma** *n* *inf* catch-22, deadlock, difficulty, doubt, embarrassment, *inf* fix, impasse, *inf* jam, *inf* mess, *inf* pickle, plight, predicament, problem, quandary, *inf* spot, stalemate.

**diligent** *adj* assiduous, busy, careful, conscientious, constant, devoted, earnest, energetic, hard-working, indefatigable, industrious, meticulous, painstaking, persevering, persistent, pertinacious, punctilious, scrupulous, sedulous, studious, thorough, tireless. *Opp* LAZY.

**dilute** *vb* adulterate, reduce the strength of, thin, water down, weaken. *Opp* CONCENTRATE.

**dim** *adj* 1 bleary, blurred, clouded, cloudy, dark, dingy, dull, faint, fogged, foggy, fuzzy, gloomy, grey, hazy, ill-defined, imperceptible, indistinct, indistinguishable, misty, murky, nebulous, obscure, obscured, pale, shadowy, sombre, unclear, vague, weak. 2 ▷ STUPID. *Opp* BRIGHT. • *vb* 1 blacken, cloud, darken, dull, make dim, mask, obscure, shade, shroud. 2 become dim, fade, go out, lose brightness, lower. *Opp* BRIGHTEN. **take a dim view** ▷ DISAPPROVE.

**dimensions** *plur n* capacity, extent, magnitude, measurements, proportions, scale, scope, size. ▷ MEASUREMENT.

**diminish** *vb* 1 abate, become less, contract, curtail, decline, decrease, depreciate, die down, dwindle, ease off, ebb, fade, lessen, *inf* let up, lower, peter out, recede, reduce, shorten, shrink, shrivel, slow down, subside, wane, *inf* wind down. ▷ CUT. *Opp* INCREASE. 2 belittle, cheapen, demean, deprecate, devalue, disparage, minimize, undervalue. *Opp* EXAGGERATE.

**diminutive** *adj* microscopic, midget, miniature, minuscule, minute, tiny, undersized. ▷ SMALL.

**din** *n* blaring, clamour, clangour, clatter, commotion, crash, hubbub, hullabaloo, noise, outcry, pandemonium, racket, roar, row, rumpus, shouting, tumult, uproar. ▷ SOUND.

**dingy** *adj* colourless, dark, depressing, dim, dirty, discoloured, dismal, drab, dreary, dull, faded, gloomy, grimy, murky, old, seedy, shabby, smoky, soiled, sooty, worn. *Opp* BRIGHT.

**dining-room** *n* cafeteria, carvery, refectory, restaurant.

**dinner** *n* banquet, feast. ▷ MEAL.

**dip** *n* 1 concavity, declivity, dent, depression, fall, hole, hollow, incline, slope. 2 *dip in the sea.* bathe, dive, immersion, plunge, soaking, swim. • *vb* 1 decline, descend, dive, fall, go down, sag, sink, slope down, slump, subside. 2 douse, drop, duck, dunk, immerse, lower, plunge, submerge. **take a dip** ▷ BATHE.

**diplomacy** *n* adroitness, delicacy, discretion, finesse, negotiation, skill, tact, tactfulness.

**diplomat** *n* ambassador, consul, government representative, negotiator, official, peacemaker, politician, representative, tactician.

**diplomatic** *adj* careful, considerate, delicate, discreet, judicious, polite, politic, prudent, sensitive, subtle, tactful, thoughtful, understanding. *Opp* TACTLESS.

**direct** *adj* 1 non-stop, shortest, straight, unbroken, undeviating, uninterrupted, unswerving. 2 blunt, candid, categorical, clear, decided, explicit, express, forthright, frank, honest, open, outspoken, plain, point-blank, sincere, straightforward, *derog* tactless, to the point, unambiguous, uncomplicated, *derog* undiplomatic, unequivocal, uninhibited, unqualified, unreserved. 3 *direct experience.* empirical, firsthand, *inf* from the horse's mouth, personal. 4 *direct opposites.* absolute, complete, diametrical, exact, head-on, *inf* out-and-out, utter. *Opp* INDIRECT. • *vb* 1 address, escort, guide, indicate the way, point, route, send, show the way, tell the way, usher. **2** aim, focus, level, target, train, turn. **3** administer, be in charge of, command, conduct, control, govern, handle, lead, manage, mastermind, oversee, regulate, rule, run, stage-manage, superintend, supervise, take charge of. **4** *direct someone to do something.* advise, bid, charge, command, counsel, enjoin, instruct, order, require, tell.

**direction** *n* aim, approach, (compass) bearing, course, orientation, path, point of the compass, road, route, tack, track, way. **directions** guidance, guidelines, instructions, orders, plans.

**director** *n* administrator, *inf* boss, executive, governor, manager, managing director, organizer, president, principal. ▷ CHIEF.

**directory** *n* catalogue, index, list, register.

**dirt** *n* **1** dust, excrement, filth, garbage, grime, impurity, mess, mire, muck, ooze, ordure, pollution, slime, sludge, smut, soot, stain. ▷ OBSCENITY, RUBBISH. **2** clay, earth, loam, mud, soil.

**dirty** *adj* **1** befouled, begrimed, besmirched, bespattered, black, dingy, dusty, filthy, foul, grimy, grubby, marked, messy, mucky, muddy, nasty, scruffy, shabby, slatternly, smeary, smudged, soiled, sooty, sordid, spotted, squalid, stained, sullied, tarnished, travel-stained, uncared for, unclean, untidy, unwashed. **2** *dirty water.* cloudy, contaminated, impure, muddy, murky, poisoned, polluted, tainted, untreated. **3** *dirty tactics.* dishonest, dishonourable, illegal, *inf* low-down, mean, rough, treacherous, unfair, ungentlemanly, unscrupulous, unsporting, unsportsmanlike. ▷ CORRUPT. **4** *dirty talk.* coarse, crude, improper, indecent, offensive, rude, smutty, vulgar. ▷ OBSCENE. *Opp* CLEAN. • *vb* befoul, foul, make dirty, mark, *inf* mess up, smear, smudge, soil, spatter, spot, stain, streak, tarnish. ▷ DEFILE. *Opp* CLEAN.

**disability** *n* affliction, complaint, defect, disablement, handicap, impairment, incapacity, infirmity, weakness.

**disable** *vb* cripple, damage, debilitate, enfeeble, *inf* hamstring, handicap, immobilize, impair, incapacitate, injure, lame, maim, make useless, mutilate, paralyse, put out of action, ruin, weaken. **disabled** ▷ HANDICAPPED.

**disadvantage** *n* drawback, handicap, hardship, hindrance, impediment, inconvenience, liability, *inf* minus, nuisance, privation, snag, trouble, weakness.

**disagree** *vb* argue, bicker, clash, conflict, contend, differ, dispute, dissent, diverge, fall out, fight, quarrel, squabble, wrangle. **disagree with** ▷ OPPOSE.

**disagreeable** *adj* disgusting, distasteful, nasty, objectionable, obnoxious, offensive, *inf* off-putting, repellent, sickening, unsavoury. ▷ UNPLEASANT. *Opp* PLEASANT.

**disagreement** *n* altercation, argument, clash, conflict, contention, controversy, debate, difference, discrepancy, disharmony, disparity, dispute, dissension, dissent, divergence, incompatibility, inconsistency, misunderstanding, opposition, quarrel, squabble, strife, *inf* tiff, variance, wrangle. *Opp* AGREEMENT.

**disappear** *vb* **1** become invisible, cease to exist, clear, die out, disperse, dissolve, dwindle, ebb, evanesce, evaporate, fade, melt away, recede, vanish, vaporize, wane. ▷ DIE. **2** depart, escape, flee, fly, go, pass out of sight, run away, walk away, withdraw. *Opp* APPEAR.

**disappoint** *vb* be worse than expected, chagrin, *inf* dash your hopes, disenchant, disillusion, dismay, displease, dissatisfy, fail to satisfy, *inf* let down, upset, vex. ▷ FRUSTRATE. *Opp* SATISFY. **disappointed** disillusioned, frustrated, *inf* let down, unsatisfied. ▷ SAD.

**disapproval** *n* anger, censure, condemnation, criticism, disapprobation, disfavour, dislike, displeasure, dissatisfaction, hostility, reprimand, reproach. *Opp* APPROVAL.

**disapprove** *vb* **disapprove of** be displeased by, belittle, blame, censure, condemn, criticize, denounce, deplore, deprecate, dislike, disparage, frown on, jeer at, look askance at, make unwelcome, object to, regret, reject, *inf* take a dim view of, take exception to. *Opp* APPROVE. **disapproving** ▷ CRITICAL.

**disarm** *vb* **1** demilitarize, demobilize, disband troops, make powerless, take weapons from. **2** charm, mollify, pacify, placate.

**disaster** *n* accident, act of God, blow, calamity, cataclysm, catastrophe, crash, débacle, failure, fiasco, *inf* flop, *inf* mess-up, misadventure, mischance, misfortune, mishap, reverse, tragedy, *inf* wash-out. *Opp* SUCCESS.

**disastrous** *adj* appalling, awful, calamitous, cataclysmic, catastrophic, crippling, destructive, devastating, dire, dreadful, fatal, ruinous, terrible, tragic. *Opp* SUCCESSFUL.

**disbelieve** *vb* be sceptical of, discount, discredit, doubt, have no faith in, mistrust, reject, suspect. *Opp* BELIEVE. **disbelieving** ▷ INCREDULOUS.

**disc** *n* **1** circle, counter, plate, token. **2** album, CD, LP, record, single. **3** [*computing*] CD-ROM, disk, diskette, floppy disk, hard disk.

**discard** *vb* abandon, cast off, *inf* chuck away, dispense with, dispose of, *inf* ditch, dump, eliminate, get rid of, jettison, junk, reject, scrap, shed, throw away, toss out.

**discern** *vb* be aware of, be sensitive to, detect, discover, discriminate, distinguish, make out, mark, notice, observe, perceive, recognize, spy. ▷ SEE. **discerning** ▷ PERCEPTIVE.

**discernible** *adj* detectable, distinguishable, measurable, perceptible. ▷ NOTICEABLE.

**discharge** *n* **1** release, dismissal. **2** emission, excretion, ooze, pus, secretion, suppuration. • *vb* **1** belch, eject, emit, expel, exude, give off, give out, pour out, produce, release, secrete, send out, spew, spit out. **2** *discharge guns.* detonate, explode, fire, let off, shoot. **3** *discharge employees.* dismiss, fire, make redundant, remove, sack, throw out. **4** *discharge a prisoner.* absolve, acquit, allow to leave, clear, dismiss, excuse, exonerate, free, let off, liberate, pardon, release. **5** *discharge duties.* accomplish, carry out, execute, fulfil, perform.

**disciple** *n* acolyte, adherent, admirer, apostle, apprentice, devotee, follower, learner, proselyte, pupil, scholar, student, supporter.

**disciplinarian** *n* authoritarian, autocrat, despot, dictator, *inf* hard-liner, *inf* hard taskmaster, martinet, *inf* slave-driver, *inf* stickler, tyrant.

**discipline** *n* 1 control, drilling, indoctrination, instruction, management, strictness, system, training. 2 good behaviour, obedience, order, orderliness, routine, self-control, self-restraint. • *vb* 1 break in, coach, control, drill, educate, govern, indoctrinate, instruct, keep in check, manage, restrain, school, train. 2 castigate, chasten, chastise, correct, penalize, punish, rebuke, reprimand, reprove, scold. **disciplined** ▷ OBEDIENT.

**disclaim** *vb* deny, disown, forswear, reject, renounce, repudiate. *Opp* ACKNOWLEDGE.

**disclose** *vb* divulge, expose, let out, make known. ▷ REVEAL.

**discolour** *vb* bleach, dirty, fade, mark, spoil the colour of, stain, tarnish, tinge.

**discomfort** *n* ache, care, difficulty, distress, hardship, inconvenience, irritation, soreness, uncomfortableness, uneasiness. ▷ PAIN. *Opp* COMFORT.

**disconcert** *vb* agitate, bewilder, confuse, discomfit, distract, disturb, fluster, nonplus, perplex, *inf* put off, puzzle, *inf* rattle, ruffle, throw off balance, trouble, unsettle, upset, worry. *Opp* REASSURE.

**disconnect** *vb* break off, cut off, detach, disengage, divide, part, sever, switch off, take away, turn off, uncouple, undo, unhitch, unhook, unplug. **disconnected** ▷ INCOHERENT.

**discontented** *adj* annoyed, disgruntled, displeased, dissatisfied, *inf* fed up, restless, sulky, unhappy, unsettled.

**discord** *n* 1 argument, conflict, contention, difference of opinion, disagreement, disharmony, dispute, friction, incompatibility, strife. ▷ QUARREL. 2 [*music*] cacophony, clash, jangle. ▷ NOISE. *Opp* HARMONY.

**discordant** *adj* 1 conflicting, contrary, differing, disagreeing, dissimilar, divergent, incompatible, incongruous, inconsistent, opposed, opposite. ▷ QUARRELSOME. 2 atonal, cacophanous, clashing, dissonant, grating, grinding, harsh, jangling, jarring, shrill, strident, tuneless, unmusical. *Opp* HARMONIOUS.

**discount** *n* abatement, allowance, concession, cut, deduction, *inf* mark-down, rebate, reduction. • *vb* disbelieve, dismiss, disregard, gloss over, ignore, overlook, reject.

**discourage** *vb* 1 cow, damp, dampen, daunt, demoralize, depress, disenchant, dishearten, dismay, dispirit, frighten, inhibit, intimidate, overawe, *inf* put down, *inf* put off, scare, *inf* throw cold water on, unman, unnerve. 2 *discourage vandalism.* check, deflect, deter, dissuade, hinder, prevent, put an end to, repress, restrain, slow down, stop, suppress. *Opp* ENCOURAGE.

**discouragement** *n* constraint, *inf* damper, deterrent, disincentive, hindrance, impediment, obstacle, restraint, setback. *Opp* ENCOURAGEMENT.

**discourse** *n* 1 ▷ CONVERSATION. 2 dissertation, essay, monograph, paper, speech, thesis, treatise. ▷ WRITING. • *vb* ▷ SPEAK.

**discover** *vb* ascertain, bring to light, come across, detect, *inf* dig up, disclose, *inf* dredge up, explore, expose, *inf* ferret out, find, hit on, identify, learn, light upon, locate, notice, observe, perceive, recognize, reveal, search out, spot, *sl* sus out, track down, turn up, uncover, unearth. ▷ INVENT. *Opp* HIDE.

**discoverer** *n* creator, explorer, finder, initiator, inventor, originator, pioneer, traveller.

**discovery** *n* breakthrough, conception, detection, disclosure, exploration, *inf* find, innovation, invention, recognition, revelation.

**discredit** *vb* attack, calumniate, challenge, defame, disbelieve, disgrace, dishonour, disprove, *inf* explode, prove false, raise doubts about, refuse to believe, ruin the reputation of, show up, slander, slur, smear, vilify.

**discreet** *adj* careful, cautious, chary, circumspect, considerate, delicate, diplomatic, guarded, judicious, low-key, mild, muted, polite, politic, prudent, restrained, sensitive, soft, subdued, tactful, thoughtful, understated, wary. *Opp* INDISCREET.

**discrepancy** *n* conflict, difference, disparity, dissimilarity, divergence, incompatibility, incongruity, inconsistency, variance. *Opp* SIMILARITY.

**discretion** *n* circumspection, diplomacy, good sense, judgement, maturity, prudence, responsibility, sensitivity, tact, wisdom. *Opp* TACTLESSNESS.

**discriminate** *vb* **1** differentiate, distinguish, draw a distinction, separate, tell apart. **2** be biased, be intolerant, be prejudiced, show discrimination. **discriminating** ▷ PERCEPTIVE.

**discrimination** *n* **1** discernment good taste, insight, judgement, perceptiveness, refinement, selectivity, subtlety, taste. **2** [*derog*] bias, bigotry, chauvinism, favouritism, intolerance, male chauvinism, prejudice, racialism, racism, sexism, unfairness. *Opp* IMPARTIALITY.

**discuss** *vb* argue about, confer about, consider, consult about, debate, deliberate, examine, *inf* put heads together about, talk about, *inf* weigh up the pros and cons of, write about. ▷ TALK.

**discussion** *n* argument, colloquy, confabulation, conference, consideration, consultation, conversation, debate, deliberation, dialogue, discourse, examination, exchange of views, *inf* powwow, symposium. ▷ TALK.

**disdainful** *adj* contemptuous, jeering, mocking, scornful, sneering, supercilious, superior. ▷ PROUD.

**disease** *n* affliction, ailment, blight, *inf* bug, complaint, *inf* condition, contagion, disorder, infection, infirmity, malady, plague, sickness. ▷ ILLNESS.

**diseased** *adj* ailing, infirm, sick, unwell. ▷ ILL.

**disembark** *vb* alight, debark, detrain, get off, go ashore, land. *Opp* EMBARK.

**disfigure** *vb* blemish, damage, deface, deform, distort, impair, injure, make ugly, mar, mutilate, ruin, scar, spoil. *Opp* BEAUTIFY.

**disgrace** *n* **1** blot, contumely, degradation, discredit, dishonour, disrepute, embarrassment, humiliation, ignominy, obloquy, odium, opprobrium, scandal, shame, slur, stain, stigma. **2** ▷ OUTRAGE.

**disgraceful** *adj* contemptible, degrading, dishonourable, embarrassing, humiliating, ignominious, shameful, shaming, wicked. ▷ BAD.

**disgruntled** *adj* annoyed, cross, disaffected, disappointed, discontented, dissatisfied, *inf* fed up, grumpy, moody, sulky, sullen. ▷ BAD-TEMPERED.

**disguise** *n* camouflage, cloak, costume, cover, fancy dress, front, *inf* get-up, impersonation, make-up, mask, pretence, smokescreen. • *vb* blend into the background, camouflage, conceal, cover up, dress up, falsify, gloss over, hide, make inconspicuous, mask, misrepresent, screen, shroud, veil. **disguise yourself as** ▷ IMPERSONATE.

**disgust** *n* abhorrence, antipathy, aversion, contempt, detestation, dislike, distaste, hatred, loathing, nausea, outrage, repugnance, repulsion, revulsion, sickness. • *vb* appal, be distasteful to, displease, horrify, nauseate, offend, outrage, put off, repel, revolt, sicken, shock, *inf* turn your stomach. *Opp* PLEASE. **disgusting** ▷ HATEFUL.

**dish** *n* **1** basin, bowl, casserole, container, plate, *old use* platter, tureen. **2** concoction, food, item on the menu, recipe. **dish out** ▷ DISTRIBUTE. **dish up** ▷ SERVE.

**dishearten** *vb* depress, deter, discourage, dismay, put off, sadden. *Opp* ENCOURAGE. **disheartened** ▷ SAD.

**dishevelled** *adj* bedraggled, disarranged, disordered, knotted, matted, messy, ruffled, rumpled, *inf* scruffy, slovenly, tangled, tousled, uncombed, unkempt, untidy. *Opp* NEAT.

**dishonest** *adj inf* bent, cheating, corrupt, criminal, crooked, deceitful, deceiving, deceptive, devious, dishonourable, disreputable, false, fraudulent, hypocritical, immoral, insincere, lying, mendacious, misleading, perfidious, *inf* shady, *inf* slippery, specious, swindling, thieving, treacherous, *inf* two-faced, *inf* underhand, unethical, unprincipled, unscrupulous, untrustworthy, untruthful. *Opp* HONEST.

**dishonour** *n inf* black mark, blot, degradation, discredit, disgrace, humiliation, ignominy, indignity, loss of face, obloquy, opprobrium, reproach, scandal, shame, slander, slur, stain, stigma. *Opp* HONOUR. • *vb* **1** abuse, affront, debase, defile, degrade, disgrace, offend, profane, shame, slight. **2** ▷ RAPE.

**dishonourable** *adj* base, blameworthy, compromising, despicable, discreditable, disgraceful, disgusting, dishonest, disloyal, disreputable, ignoble, ignominious, improper, infamous, mean, outrageous, perfidious, reprehensible, scandalous, shabby, shameful, shameless, treacherous, unchivalrous, unethical, unprincipled, unscrupulous, untrustworthy, unworthy, wicked. ▷ CORRUPT. *Opp* HONOURABLE.

**disillusion** *vb* disabuse, disappoint, disenchant, enlighten, reveal the truth to, undeceive.

**disinfect** *vb* cauterize, chlorinate, clean, cleanse, decontaminate, fumigate, purge, purify, sanitize, sterilize.

**disinfectant** *n* antiseptic, decontaminant, fumigant, germicide.

**disinherit** *vb* cut off, cut out of a will, deprive someone of his/her birthright, deprive someone of his/her inheritance.

**disintegrate** *vb* break into pieces, break up, come apart, crack up, crumble, decay, decompose, degenerate, deteriorate, fall apart, lose coherence, moulder, rot, shatter, smash, splinter.

**disinterested** *adj* detached, dispassionate, impartial, impersonal, neutral, objective, unbiased, uninvolved, unprejudiced. *Opp* BIASED.

**disjointed** *adj* aimless, broken up, confused, desultory, disconnected, dislocated, disordered, disunited, divided, incoherent, jumbled, loose, mixed up, muddled, rambling, separate, split up, unconnected, uncoordinated, wandering. *Opp* COHERENT.

**dislike** *n* animus, antagonism, antipathy, aversion, contempt, detestation, disapproval, disfavour, disgust, distaste, hatred, hostility, ill will, loathing, repugnance, revulsion. • *vb* avoid, despise, detest, disapprove of, feel dislike for, scorn, *inf* take against. ▷ HATE. *Opp* LOVE.

**dislocate** *vb* disengage, disjoint, displace, misplace, *inf* put out, put out of joint.

**disloyal** *adj* apostate, faithless, false, insincere, perfidious, recreant, renegade, seditious, subversive, treacherous, treasonable, *inf* two-faced, unfaithful, unreliable, untrue, untrustworthy. *Opp* LOYAL.

**disloyalty** *n* betrayal, double-dealing, duplicity, faithlessness, falseness, inconstancy, infidelity, perfidy, treachery, treason, unfaithfulness. *Opp* LOYALTY.

**dismal** *adj* bleak, cheerless, depressing, dreary, dull, funereal, gloomy, grey, grim, joyless, miserable, sombre, wretched. ▷ SAD.

**dismantle** *vb* demolish, knock down, strike, strip down, take apart, take down. *Opp* ASSEMBLE.

**dismay** *n* agitation, alarm, anxiety, apprehension, astonishment, consternation, depression, disappointment, discouragement, distress, dread, gloom, horror, pessimism, surprise. ▷ FEAR. • *vb* alarm, appal, daunt, depress, devastate, disappoint, discompose, discourage, disgust, dishearten, dispirit, distress, horrify, scare, shock, take aback, unnerve. ▷ FRIGHTEN. *Opp* PLEASE.

**dismiss** *vb* **1** disband, discard, free, let go, *inf* pack off, release, send away, *inf* send packing. **2** belittle, brush aside, discount, disregard, drop, give up, *inf* pooh-pooh, reject, repudiate, set aside, shelve, shrug off, wave aside. **3** *dismiss a worker.* *inf* axe, banish, cashier, disband, discharge, *inf* fire, get rid of, give notice to, *inf* give someone his/her cards, give the push to, lay off, make redundant, sack.

**disobedient** *adj* anarchic, contrary, defiant, delinquent, disorderly, disruptive, fractious, headstrong, insubordinate, intractable, mutinous, obdurate, obstinate, obstreperous, perverse, rebellious, recalcitrant, refractory, riotous, selfwilled, stubborn, uncontrollable, undisciplined, ungovernable, unmanageable, unruly, wayward, wild, wilful. ▷ NAUGHTY. *Opp* OBEDIENT.

**disobey** *vb* **1** be disobedient, mutiny, protest, rebel, revolt, rise up, strike. **2** break, contravene, defy, disregard, flout, ignore, infringe, oppose, rebel against, resist, transgress, violate. *Opp* OBEY.

**disorder** *n* **1** anarchy, chaos, clamour, confusion, disarray, disorderliness, disorganization, disturbance, fighting, fracas, fuss, jumble, lawlessness, mess, muddle, rumpus, *inf* shambles, tangle, tumult, untidiness, uproar. ▷ COMMOTION. *Opp* ORDER. **2** ▷ ILLNESS.

**disorderly** *adj* ▷ DISOBEDIENT, DISORGANIZED.

**disorganized** *adj* aimless, careless, chaotic, confused, disorderly, haphazard, illogical, jumbled, messy, muddled, rambling, scatter-brained, *inf* slapdash, *inf* slipshod, *inf* sloppy, slovenly, straggling, unmethodical, unplanned, unstructured, unsystematic, untidy. *Opp* SYSTEMATIC.

**disown** *vb* cast off, disclaim knowledge of, renounce, repudiate.

**disparage** *vb* belittle, demean, depreciate, discredit, insult, *inf* put down, slight, undervalue. ▷ CRITICIZE. **disparaging** ▷ UNCOMPLIMENTARY.

**dispassionate** *adj* calm, composed, cool, equable, even-tempered, level-headed, sober. ▷ IMPARTIAL, UNEMOTIONAL. *Opp* EMOTIONAL.

**dispatch** *n* bulletin, communiqué, document, letter, message, report. • *vb* **1** consign, convey, forward, mail, post, send, ship, transmit. **2** ▷ KILL.

**dispense** *vb* **1** allocate, allot, apportion, assign, deal out, disburse, distribute, dole out, give out, issue, measure out, mete out, parcel out, provide, ration out, share. **2** *dispense medicine.* make up, prepare, supply. **dispense with** ▷ OMIT, REMOVE.

**disperse** *vb* **1** break up, decentralize, devolve, disband, dismiss, dispel, dissipate, distribute, divide up, drive away, send away, send in different directions, separate, spread, stray. *Opp* GATHER. **2** disappear, dissolve, melt away, scatter, spread out, vanish.

**displace** *vb* **1** disarrange, dislocate, dislodge, disturb, misplace, move, put out of place, shift. **2** crowd out, depose, dispossess, evict, expel, oust, replace, succeed, supersede, supplant, take the place of, unseat, usurp.

**display** *n* **1** array, demonstration, exhibition, manifestation, pageant, parade, presentation, show, spectacle. **2** ceremony, ostentation, pageantry, pomp, showing off. • *vb* advertise, air, betray, demonstrate, disclose, exhibit, expose, flaunt, flourish, give evidence of, parade, present, produce, put on show, reveal, set out, show, show off, unfold, unfurl, unveil, vaunt. *Opp* HIDE.

**displease** *vb* anger, offend, *inf* put out, upset. ▷ ANNOY.

**disposable** *adj* **1** at your disposal, available, spendable, usable. **2** biodegradable, expendable, non-returnable, replaceable, *inf* throw-away.

**dispose** *vb* adjust, arrange, array, distribute, group, order, organize, place, position, put, set out, situate. **disposed** ▷ LIABLE. **dispose of** ▷ DESTROY, DISCARD.

**disproportionate** *adj* excessive, incommensurate, incongruous, inequitable, inordinate, out of proportion, unbalanced, uneven, unreasonable. *Opp* PROPORTIONAL.

**disprove** *vb* confute, contradict, controvert, demolish, discredit, *inf* explode, invalidate, negate, rebut, refute, show to be wrong. *Opp* PROVE.

**dispute** *n* ▷ QUARREL. • *vb* argue against, challenge, contest, contradict, controvert, deny, disagree with, doubt, fault, gainsay, impugn, object to, oppose, *inf* pick holes in, quarrel with, query, question, raise doubts about, take exception to. ▷ DEBATE. *Opp* ACCEPT.

**disqualify** *vb* bar, debar, declare ineligible, exclude, preclude, prohibit, reject, turn down.

**disregard** *vb* brush aside, despise, discount, dismiss, disobey, exclude, *inf* fly in the face of, forget, ignore, leave out, *inf* make light of, miss out, neglect, omit, overlook, pass over, pay no attention to, *inf* pooh-pooh, reject, shrug off, skip, slight, snub, turn a blind eye to. *Opp* HEED.

**disreputable** *adj* dishonest, dishonourable, *inf* dodgy, dubious, infamous, questionable, raffish, *inf* shady, suspect, suspicious, unconventional, unreliable, unsound, untrustworthy. *Opp* REPUTABLE.

**disrespectful** *adj* bad-mannered, blasphemous, derisive, discourteous, disparaging, impolite, impudent, inconsiderate, insolent, insulting, irreverent, mocking, scornful, uncivil, uncomplimentary, unmannerly. ▷ RUDE. *Opp* RESPECTFUL.

**disrupt** *vb* agitate, break up, confuse, disconcert, dislocate, disorder, disturb, interfere with, interrupt, intrude on, spoil, throw into disorder, unsettle, upset.

**dissatisfaction** *n* annoyance, chagrin, disappointment, discontentment, dismay, displeasure, disquiet, exasperation, frustration, irritation, malaise, mortification, regret, unhappiness. *Opp* SATISFACTION.

**dissatisfied** *adj* disaffected, disappointed, discontented, disgruntled, displeased, fed up, frustrated, unfulfilled, unsatisfied. ▷ UNHAPPY. *Opp* CONTENTED.

**dissident** *n derog* agitator, apostate, dissenter, independent thinker, non-conformer, protester, rebel, recusant, *inf* refusenik, revolutionary. *Opp* CONFORMIST.

**dissimilar** *adj* antithetical, clashing, conflicting, contrasting, different, disparate, distinct, distinguishable, divergent, diverse, heterogeneous, incompatible, irreconcilable, opposite, unlike, unrelated, various. *Opp* SIMILAR.

**dissipate** *vb* **1** break up, diffuse, disappear, disperse, scatter. **2** distribute, fritter away, spread about, squander, throw away, use up, waste. **dissipated** ▷ IMMORAL.

**dissociate** *vb* back away, cut off, detach, disengage, distance, divorce, isolate, segregate. ▷ SEPARATE. *Opp* ASSOCIATE.

**dissolve** *vb* **1** become liquid, decompose, deliquesce, dematerialize, diffuse, disappear, disintegrate, disperse, liquefy, melt away, vanish. **2** *dissolve a meeting.* adjourn, break up, cancel, disband, dismiss, divorce, end, sever, split up, suspend, terminate, *inf* wind up.

**dissuade** *vb* dissuade from advise against, argue out of, deter from, discourage from, persuade not to, put off, remonstrate against, warn against. *Opp* PERSUADE.

**distance** *n* **1** breadth, extent, gap, *inf* haul, interval, journey, length, measurement, mileage, range, reach, separation, space, span, stretch, width. **2** aloofness, coolness, haughtiness, isolation, remoteness, separation, *inf* standoffishness, unfriendliness.
• *vb* **distance yourself** be unfriendly, detach yourself, dissociate yourself, keep away, keep

your distance, remove yourself, separate yourself, set yourself apart, stay away. *Opp* INVOLVE.

**distant** *adj* **1** far, far-away, far-flung, *inf* god-forsaken, inaccessible, outlying, out-of-the-way, remote, removed. *Opp* CLOSE. **2** aloof, cool, formal, frigid, haughty, reserved, reticent, stiff, unapproachable, unenthusiastic, unfriendly, withdrawn. *Opp* FRIENDLY.

**distasteful** *adj* disgusting, displeasing, nasty, nauseating, objectionable, offensive, *inf* off-putting, repugnant, revolting, unpalatable. ▷ UNPLEASANT. *Opp* PLEASANT.

**distinct** *adj* **1** apparent, clear, clear-cut, definite, evident, noticeable, obvious, palpable, patent, perceptible, plain, precise, recognizable, sharp, unambiguous, unequivocal, unmistakable, visible, well-defined. *Opp* INDISTINCT. **2** contrasting, detached, different, discrete, dissimilar, distinguishable, individual, separate, special, *Lat* sui generis, unconnected, unique.

**distinction** *n* **1** contrast, difference, differentiation, discrimination, dissimilarity, distinctiveness, dividing line, division, individuality, particularity, peculiarity, separation. *Opp* SIMILARITY. **2** *distinction of being first.* celebrity, credit, eminence, excellence, fame, glory, greatness, honour, importance, merit, prestige, renown, reputation, superiority.

**distinctive** *adj* characteristic, different, distinguishing, idiosyncratic, individual, inimitable, original, peculiar, personal, singular, special, striking, typical, uncommon, unique. *Opp* COMMON.

**distinguish** *vb* **1** choose, decide, differentiate, discriminate, judge, make a distinction, separate, tell apart. **2** ascertain, determine, discern, know, make out, perceive, pick out, recognize, see, single out, tell. **distinguished** ▷ FAMOUS.

**distort** *vb* **1** bend, buckle, contort, deform, misshape, twist, warp, wrench. **2** alter, exaggerate, falsify, garble, misrepresent, pervert, slant, tamper with, twist, violate. **distorted** ▷ GNARLED, FALSE.

**distract** *vb* bewilder, bother, confound, confuse, deflect, disconcert, distress, divert, harass, mystify, perplex, puzzle, rattle, sidetrack, trouble, worry. **distracted** ▷ DISTRAUGHT, MAD.

**distraction** *n* **1** disturbance, diversion, interference, interruption, temptation, *inf* upset. **2** agitation, befuddlement, bewilderment, confusion, delirium, frenzy, insanity, madness. **3** ▷ DIVERSION.

**distraught** *adj* agitated, *inf* beside yourself, distracted, distressed, disturbed, emotional, excited, frantic, hysterical, overcome, overwrought, troubled, upset, worked up. ▷ ANXIOUS. *Opp* CALM.

**distress** *n* adversity, affliction, angst, anguish, anxiety, danger, desolation, difficulty, discomfort, dismay, fright, grief, heartache, misery, pain, poverty, privation, sadness, sorrow, stress, suffering, torment, tribulation, trouble, unhappiness, woe, worry, wretchedness. ▷ PAIN. • *vb* afflict, alarm, bother, *inf* cut up, dismay, disturb, frighten, grieve, harass, harrow, hurt, make miserable, oppress, pain, perplex, perturb, plague, sadden, scare, shake, shock, terrify, torment, torture, trouble, upset, vex, worry, wound. *Opp* COMFORT.

**distribute** *vb* allocate, allot, apportion, arrange, assign, circulate, deal out, deliver, *inf* dish out, dispense, disperse, dispose of, disseminate, divide out, *inf* dole out, give out, hand round, issue, mete out, partition, pass round, scatter, share out, spread, strew, take round. *Opp* COLLECT.

**district** *n* area, community, department, division, locality, neighbourhood, parish, part, partition, precinct, province, quarter, region, sector, territory, vicinity, ward, zone.

**distrust** *vb* be distrustful of, disbelieve, doubt, have misgivings about, have qualms about, mistrust, question, suspect. *Opp* TRUST.

**distrustful** *adj* cautious, chary, cynical, disbelieving, distrusting, doubtful, dubious, sceptical, suspicious, uncertain, uneasy, unsure, wary. *Opp* TRUSTFUL.

**disturb** *vb* **1** agitate, alarm, annoy, bother, discompose, disrupt, distract, distress, excite, fluster, frighten, hassle, interrupt, intrude on, perturb, pester, ruffle, scare, shake, startle, stir up, trouble, unsettle, upset, worry. **2** confuse, disorder, interfere with, jumble up, *inf* mess about with, move, muddle, rearrange, reorganize. **disturbed** ▷ DISTRAUGHT.

**disturbance** *n* disruption, interference, upheaval, upset. ▷ COMMOTION.

**disunited** *adj* divided, opposed, polarized, split. *Opp* UNITED.

**disunity** *n* difference, disagreement, discord, disharmony, disintegration, division, fragmentation, incoherence, opposition, polarization. *Opp* UNITY.

**disused** *adj* abandoned, archaic, closed, dead, discarded, discontinued, idle, neglected, obsolete, superannuated, unused, withdrawn. ▷ OLD. *Opp* CURRENT.

**ditch** *n* aqueduct, channel, dike, drain, gully, gutter, moat, trench, watercourse. • *vb* ▷ ABANDON.

**dive** *vb* crash-dive, descend, dip, drop, duck, fall, go snorkelling, go under, jump, leap, nosedive, pitch, plummet, plunge, sink, submerge, subside, swoop.

**diverge** *vb* branch, deviate, divide, fork, go off at a tangent, part, radiate, ramify, separate, split, spread, subdivide. ▷ DIFFER. *Opp* CONVERGE.

**diverse** *adj* assorted, different, dissimilar, distinct, divergent, diversified, heterogeneous, miscellaneous, mixed, multifarious, varied, various.

**diversify** *vb* branch out, broaden out, develop, divide, enlarge, expand, extend, spread out, vary.

**diversion** *n* **1** detour, deviation. **2** amusement, distraction, entertainment, fun, game, hobby, interest, pastime, play, recreation, relaxation, sport.

**divert** *vb* **1** alter, avert, change direction, deflect, deviate, rechannel, redirect, reroute, shunt, sidetrack, switch, turn aside. **2** amuse, beguile, cheer up, delight, distract, engage, entertain, keep happy, occupy, recreate, regale. **diverting** ▷ FUNNY.

**divide** *vb* **1** branch, detach, diverge, fork, move apart, part, separate, sunder. **2** allocate, allot, apportion, break up, cut up, deal out, dispense, distribute, dole out, give out, halve, measure out, mete out, parcel out, pass round, share out. **3** *divide a party*. cause disagreement in, disunite, polarize, split. 4 *divide into sets*. arrange, categorize, classify, grade, group, sort out, subdivide. *Opp* GATHER, UNITE.

**divine** *adj* angelic, celestial, godlike, hallowed, heavenly, holy, immortal, mystical, religious, sacred, saintly, seraphic, spiritual, superhuman, supernatural, transcendental. *Opp* MORTAL. • *n* ▷ CLERGYMAN. • *vb* ▷ PROPHESY.

**divinity** *n* 1 ▷ GOD. 2 religion, religious studies, theology.

**division** *n* 1 allocation, allotment, apportionment, cutting up, dividing, partition, segmentation, separation, splitting. 2 disagreement, discord, disunity, feud, quarrel, rupture, schism, split. 3 alcove, compartment, part, recess, section, segment. 4 *division between rooms, lands*. border, borderline, boundary line, demarcation, divider, dividing wall, fence, frontier, margin, partition, screen. 5 *division of a business*. branch, department, section, subdivision, unit.

**divorce** *n* annulment, *inf* break-up, decree nisi, dissolution, separation, *inf* split-up. • *vb* annul marriage, dissolve marriage, part, separate, *inf* split up.

**dizziness** *n* faintness, giddiness, light-headedness, vertigo.

**dizzy** *adj* bewildered, confused, dazed, faint, giddy, light-headed, muddled, reeling, shaky, swimming, unsteady, *inf* woozy.

**do** *vb* 1 accomplish, achieve, bring about, carry out, cause, commit, complete, effect, execute, finish, fulfil, implement, initiate, instigate, organize, perform, produce, undertake. 2 *do the garden*. arrange, attend to, cope with, deal with, handle, look after, manage, work at. 3 *do sums*. answer, give your mind to, puzzle out, solve, think out, work out. 4 *Will this do?* be acceptable, be enough, be satisfactory, be sufficient, be suitable, satisfy, serve, suffice. 5 *Do as you like*. act, behave, conduct yourself, perform. **do away with** ▷ ABOLISH. **do up** ▷ DECORATE, FASTEN.

**docile** *adj* cooperative, domesticated, obedient, submissive, tractable. ▷ TAME.

**dock** *n* berth, boatyard, dockyard, dry dock, harbour, haven, jetty, landing-stage, marina, pier, port, quay, slipway, wharf. • *vb* 1 anchor, berth, drop anchor, land, moor, put in, tie up. 2 ▷ CUT.

**doctor** *n* general practitioner, *inf* GP, *inf* medic, medical officer, medical practitioner, *inf* MO, physician, *derog* quack, surgeon.

**doctrine** *n* axiom, belief, conviction, *Lat* credo, creed, dogma, maxim, orthodoxy, postulate, precept, principle, teaching, tenet, theory, thesis.

**document** *n* certificate, charter, chronicle, deed, diploma, form, instrument, legal document, licence, manuscript, *inf* MS, paper, parchment, passport, policy, printout, record, typescript, visa, warrant, will. • *vb* ▷ RECORD.

**documentary** *adj* 1 authenticated, chronicled, recorded, substantiated, written. 2 factual, historical, non-fiction, real life.

**dodge** *n* contrivance, device, knack, manoeuvre, ploy, *sl* racket, ruse, scheme, stratagem, subterfuge, trick, *inf* wheeze. • *vb* 1 avoid, duck, elude, escape, evade, fend off, move out of the way, sidestep, swerve, turn away, veer, weave. 2 *dodge work*. shirk, *inf* skive, *inf* wriggle out of. 3 *dodge a question*. equivocate, fudge, hedge, quibble, *inf* waffle.

**dog** *n* bitch, *inf* bow-wow, *derog* cur, dingo, hound, mongrel, pedigree, pup, puppy, whelp. • *vb* ▷ FOLLOW.

**dogma** *n* article of faith, belief, conviction, creed, doctrine, orthodoxy, precept, principle, teaching, tenet, truth.

**dogmatic** *adj* assertive, arbitrary, authoritarian, authoritative, categorical, certain, dictatorial, doctrinaire, *inf* hard-line, hidebound, imperious, inflexible, intolerant, legalistic, narrow-minded, obdurate, opinionated, pontifical, positive. ▷ STUBBORN. *Opp* AMENABLE.

**dole** *n* [*inf*] benefit, income support, social security, unemployment benefit. **dole out** ▷ DISTRIBUTE. **on the dole** ▷ UNEMPLOYED.

**doll** *n inf* dolly, figure, marionette, puppet, rag doll.

**domestic** *adj* 1 family, household, in the home, private. 2 *domestic air service*. indigenous, inland, internal, national.

**domesticated** *adj* house-broken, house-trained, tame, tamed, trained. *Opp* WILD.

**dominant** *adj* 1 biggest, chief, commanding, conspicuous, eye-catching, highest, imposing, largest, main, major, obvious, outstanding, pre-eminent, prevailing, primary, principal, tallest, uppermost, widespread. 2 ascendant, controlling, dominating, domineering, governing, influential, leading, powerful, predominant, presiding, reigning, ruling, supreme.

**dominate** *vb* 1 be dominant, be in the majority, control, direct, govern, influence, lead, manage, master, monopolize, outnumber, preponderate, prevail, rule, subjugate, take control, tyrannize. 2 dwarf, look down on, overshadow, tower over.

**domineering** *adj* authoritarian, autocratic, *inf* bossy, despotic, dictatorial, high-handed, oppressive, overbearing, *inf* pushy, strict, tyrannical. *Opp* SUBMISSIVE.

**donate** *vb* contribute, give, grant, hand over, make a donation, present, subscribe, supply.

**donation** *n* alms, contribution, freewill offering, gift, offering, present, subscription.

**donor** *n* backer, benefactor, contributor, giver, philanthropist, provider, sponsor, supplier, supporter.

**doom** *n* destiny, end, fate, fortune, karma, kismet, lot.

**doomed** *adj* 1 condemned, destined, fated, intended, ordained, predestined. 2 *a doomed enterprise*. accursed, bedevilled, cursed, damned, hopeless, ill-fated, ill-starred, luckless, star-crossed, unlucky.

**door** *n* barrier, doorway, entrance, exit, French window, gate, gateway, opening, portal, postern, revolving door, swing door, way out.

**dormant** *adj* 1 asleep, comatose, hibernating, inactive, inert, passive, quiescent, quiet, resting, sleeping. 2 *dormant talent*. hidden, latent, potential, unrevealed, untapped, unused. *Opp* ACTIVE.

**dose** *n* amount, dosage, measure, portion, prescribed amount, quantity. • *vb* administer, dispense, prescribe.

**dossier** *n* file, folder, records, set of documents.

**dot** *n* decimal point, fleck, full stop, iota, jot, mark, point, speck, spot. • *vb* fleck, mark with dots, punctuate, speckle, spot, stipple.

**dote** *vb* **dote on** adore, idolize, worship. ▷ LOVE.

**double** *adj* coupled, doubled, dual, duple, duplicated, paired, twin, twofold, two-ply. • *n* clone, copy, counterpart, *Ger* doppelgänger, duplicate, *inf* look-alike, opposite, *inf* spitting image, twin. • *vb* duplicate, increase, multiply by two, reduplicate, repeat. **double back** ▷ RETURN. **double up** ▷ COLLAPSE.

**double-cross** *vb* cheat, deceive, let down, trick. ▷ BETRAY.

**doubt** *n* **1** agnosticism, anxiety, apprehension, confusion, cynicism, diffidence, disbelief, disquiet, distrust, fear, hesitation, incredulity, indecision, misgiving, mistrust, perplexity, qualm, reservation, scepticism, suspicion, worry. **2** *doubt about meaning.* ambiguity, difficulty, dilemma, problem, query, question, uncertainty. *Opp* CERTAINTY. • *vb* be dubious, be sceptical about, disbelieve, distrust, fear, feel uncertain about, have doubts about, have misgivings about, have reservations about, hesitate, lack confidence, mistrust, query, question, suspect. *Opp* TRUST.

**doubtful** *adj* **1** agnostic, cynical, diffident, disbelieving, distrustful, dubious, hesitant, incredulous, sceptical, suspicious, tentative, uncertain, unclear, unconvinced, undecided, unsure. **2** *a doubtful decision.* ambiguous, debatable, dubious, equivocal, *inf* iffy, inconclusive, problematical, questionable, suspect, vague, worrying. **3** *a doubtful ally.* irresolute, uncommitted, unreliable, untrustworthy, vacillating, wavering. *Opp* CERTAIN, DEPENDABLE.

**dowdy** *adj* colourless, dingy, drab, dull, *inf* frumpish, old-fashioned, shabby, *inf* sloppy, slovenly, *inf* tatty, unattractive, unstylish. *Opp* SMART.

**downfall** *n* collapse, defeat, overthrow, ruin, undoing.

**downhearted** *adj* dejected, depressed, discouraged, *inf* down, downcast, miserable, unhappy. ▷ SAD.

**downward** *adj* declining, descending, downhill, easy, falling, going down. *Opp* UPWARD.

**downy** *adj* feathery, fleecy, fluffy, furry, fuzzy, soft, velvety, woolly.

**drab** *adj* cheerless, colourless, dingy, dismal, dowdy, dreary, dull, flat, gloomy, grey, grimy, lacklustre, shabby, sombre, unattractive, uninteresting. *Opp* BRIGHT.

**draft** *n* **1** first version, notes, outline, plan, rough version, sketch. **2** *bank draft.* cheque, order, postal order. • *vb* block out, compose, delineate, draw up, outline, plan, prepare, put together, sketch out, work out, write a draft of.

**drag** *vb* **1** draw, haul, lug, pull, tow, trail, tug. **2** *time drags.* be boring, crawl, creep, go slowly, linger, loiter, lose momentum, move slowly, pass slowly.

**drain** *n* channel, conduit, culvert, dike, ditch, drainage, drainpipe, duct, gutter, outlet, pipe, sewer, trench, water-course. • *vb* **1** bleed, clear, draw off, dry out, empty, evacuate, extract, pump out, remove, tap, take off. **2** drip, ebb, leak out, ooze, seep, strain, trickle. **3** *drain resources.* consume, deplete, exhaust, sap, spend, use up.

**drama** *n* **1** acting, dramatics, dramaturgy, histrionics, improvisation, stagecraft, theatre, theatricals, thespian arts. **2** comedy, dramatization, farce, melodrama, musical, opera, operetta, pantomime, performance, play, production, screenplay, script, show, stage version, TV version, tragedy. **3** *real-life drama.* action, crisis, excitement, suspense, turmoil.

**dramatic** *adj* **1** histrionic, stage, theatrical, thespian. **2** *dramatic gestures.* exaggerated, flamboyant, large, overdone, showy. **3** ▷ EXCITING.

**dramatist** *n* dramaturge, playwright, scriptwriter.

**dramatize** *vb* **1** adapt, make into a play. **2** exaggerate, make too much of, overdo, overplay, overstate.

**drape** *n old use* arras, curtain, drapery, hanging, screen, tapestry, valance. • *vb* cover, decorate, festoon, hang, swathe.

**drastic** *adj* desperate, dire, draconian, extreme, far-reaching, forceful, harsh, radical, rigorous, severe, strong, vigorous.

**draught** *n* **1** breeze, current, movement, puff, wind. **2** *draught of ale.* dose, drink, gulp, measure, pull, swallow, *inf* swig.

**draw** *n* **1** attraction, enticement, lure, *inf* pull. **2** dead-heat, deadlock, stalemate, tie. **3** competition, lottery, raffle. • *vb* **1** drag, haul, lug, pull, tow, tug. **2** *draw a crowd.* allure, attract, bring in, coax, entice, invite, lure, persuade, pull in, win over. **3** *draw a sword.* extract, remove, take out, unsheathe, withdraw. **4** *draw lots.* choose, pick, select. **5** *draw a conclusion.* arrive at, come to, deduce, formulate, infer, work out. **6** *draw water.* drain, let (*blood*), pour, pump, syphon, tap. **7** *draw 1-1.* be equal, finish equal, tie. **8** *draw pictures.* depict, map out, mark out, outline, paint, pen, pencil, portray, represent, sketch, trace. **draw out** ▷ EXTEND. **draw up** ▷ DRAFT, HALT.

**drawback** *n* defect, difficulty, disadvantage, hindrance, hurdle, impediment, obstacle, obstruction, problem, snag, stumbling block.

**drawing** *n* cartoon, design, graphics, illustration, outline, sketch. ▷ PICTURE.

**dread** *n* anxiety, apprehension, awe, *inf* cold feet, dismay, fear, *inf* the jitters, nervousness, perturbation, qualm, trepidation, uneasiness, worry. • *vb* be afraid of, shrink from, view with horror. ▷ FEAR.

**dreadful** *adj* alarming, appalling, awful, dire, distressing, evil, fearful, frightful, ghastly, grisly, gruesome, harrowing, hideous, horrible, horrifying, indescribable, monstrous, shocking, terrible, tragic, unspeakable, upsetting, wicked. ▷ BAD, FRIGHTENING.

**dream** *n* **1** daydream, delusion, fantasy, hallucination, illusion, mirage, nightmare, reverie, trance, vision. **2** ambition, aspiration, ideal, pipe-dream, wish. • *vb* conjure up, daydream, fancy, fantasize, hallucinate, have a vision, imagine, think. **dream up** ▷ INVENT.

**dreary** *adj* bleak, boring, cheerless, depressing, dismal, dull, gloomy, joyless, sombre, uninteresting. ▷ MISERABLE.

**dregs** *n* deposit, grounds (*of coffee*), lees, precipitate, remains, residue, sediment.

**drench** *vb* douse, drown, flood, inundate, saturate, soak, souse, steep, wet thoroughly.

**dress** *n* **1** apparel, attire, clothing, costume, garb, garments, *inf* gear, *inf* get-up, outfit, *old use* raiment. ▷ CLOTHES. **2** frock, gown, robe, shift. • *vb* **1** array, attire, clothe, cover, fit out, provide clothes for, put clothes on, robe. **2** *dress a wound.* attend to, bandage, bind up, care for, put a dressing on, tend, treat. *Opp* UNCOVER.

**dressing** *n* bandage, compress, plaster, poultice.

**dribble** *vb* **1** drool, slaver, slobber. **2** drip, flow, leak, ooze, run, seep, trickle.

**drift** *n* **1** accumulation, bank, dune, heap, mound, pile, ridge. **2** *drift of a speech.* ▷ GIST. • *vb* **1** be carried, coast, float, meander, move casually, move slowly, ramble, roam, rove, stray, waft, walk aimlessly, wander. **2** *snow drifts.* accumulate, gather, make drifts, pile up.

**drill** *n* **1** discipline, exercises, instruction, practice, *sl* square-bashing, training. • *vb* **1** coach, discipline, exercise, indoctrinate, instruct, practise, rehearse, school, teach, train. **2** bore, penetrate, perforate, pierce.

**drink** *n* **1** beverage, *inf* bevvy, *inf* cuppa, *inf* dram, draught, glass, *inf* gulp, *inf* night-cap, *inf* nip, pint, *joc* potation, sip, swallow, swig, *inf* tipple, tot. **2** alcohol, *inf* booze, *joc* grog, *joc* liquid refreshment, liquor. □ *ale, beer, bourbon, brandy, champagne, chartreuse, cider, cocktail, Cognac, crème de menthe, gin, Kirsch, lager, mead, perry,* inf *plonk, port, punch, rum, schnapps, shandy, sherry, vermouth, vodka, whisky, wine.* • *vb* **1** gulp, guzzle, imbibe, *inf* knock back, lap, partake of, *old use* quaff, sip, suck, swallow, swig, *inf* swill. **2** *inf* booze, carouse, get drunk, *inf* indulge, tipple, tope.

**drip** *n* bead, dribble, drop, leak, splash, spot, tear, trickle. • *vb* dribble, drizzle, drop, fall in drips, leak, plop, splash, sprinkle, trickle, weep.

**drive** *n* **1** excursion, jaunt, journey, outing, ride, run, *inf* spin, trip. **2** aggressiveness, ambition, determination, energy, enterprise, enthusiasm, *inf* get-up-and-go, impetus, industry, initiative, keenness, motivation, persistence, *inf* push, vigour, vim, zeal. **3** campaign, crusade, effort. • *vb* **1** bang, dig, hammer, hit, impel, knock, plunge, prod, push, ram, sink, stab, strike, thrust. **2** coerce, compel, constrain, force, oblige, press, urge. **3** *drive a car.* control, direct, guide, handle, herd, manage, pilot, propel, send, steer. ▷ TRAVEL. **drive out** ▷ EXPEL.

**droop** *vb* be limp, bend, dangle, fall, flop, hang, sag, slump, wilt, wither.

**drop** *n* **1** bead, blob, bubble, dab, drip, droplet, globule, pearl, spot, tear. **2** dash, *inf* nip, small quantity, *inf* tot. **3** *a steep drop.* declivity, descent, dive, escarpment, fall, incline, plunge, precipice, scarp. **4** *a drop in price.* cut, decrease, reduction, slump. *Opp* RISE. • *vb* **1** collapse, descend, dip, dive, fall, go down, jump down, lower, nosedive, plummet, *inf* plump, plunge, sink, slump, subside, swoop, tumble. **2** *drop from a team.* eliminate, exclude, leave out, omit. **3** *drop a friend.* abandon, desert, discard, *inf* dump, forsake, give up, jilt, leave, reject, scrap, shed.

**drop behind** ▷ LAG. **drop in on** ▷ VISIT. **drop off** ▷ SLEEP.

**drown** *vb* **1** engulf, flood, immerse, submerge, swamp. ▷ KILL. **2** *noise drowned my voice.* be louder than, overpower, overwhelm, silence.

**drowsy** *adj* dozing, dozy, heavy-eyed, listless, *inf* nodding off, sleepy, sluggish, somnolent, soporific, tired, weary. *Opp* LIVELY.

**drudgery** *n* chore, donkey-work, *inf* grind, labour, slavery, *inf* slog, toil, travail. ▷ WORK.

**drug** *n* **1** cure, medicament, medication, medicine, *old use* physic, remedy, treatment. **2** *inf* dope, narcotic, opiate. □ *analgesic, antidepressant, barbiturate, hallucinogen, pain-killer, sedative, stimulant, tonic, tranquillizer.* □ *caffeine, cannabis, cocaine, digitalis, hashish, heroin, insulin, laudanum, marijuana, morphia, nicotine, opium, phenobarbitone, quinine.* • *vb* anaesthetize, *inf* dope, dose, give a drug to, *inf* knock out, medicate, poison, sedate, stupefy, tranquillize, treat.

**drum** *n* **1** ▷ BARREL. **2** □ *bass-drum, bongo-drum, kettle-drum, side-drum, snare-drum, tambour, tenor-drum,* plur *timpani, tom-tom.*

**drunk** *adj* delirious, fuddled, incapable, inebriate, inebriated, intoxicated, maudlin. [*slang*] blotto, bombed, boozed-up, canned, high, legless, merry, paralytic, pickled, pie-eyed, pissed, plastered, sloshed, soused, sozzled, stoned, tanked, tiddly, tight, tipsy. *Opp* SOBER.

**drunkard** *n* alcoholic, *inf* boozer, *sl* dipso, dipsomaniac, drunk, *inf* sot, tippler, toper, *sl* wino. *Opp* TEETOTALLER.

**dry** *adj* **1** arid, baked, barren, dead, dehydrated, desiccated, moistureless, parched, scorched, shrivelled, sterile, thirsty, waterless. *Opp* WET. **2** *a dry book.* boring, dreary, dull, flat, prosaic, stale, tedious, tiresome, uninspired, uninteresting. *Opp* LIVELY. **3** *dry humour. inf* dead-pan, droll, expressionless, laconic, lugubrious, unsmiling. • *vb* become dry, dehumidify, dehydrate, desiccate, go hard, make dry, parch, shrivel, wilt, wither.

**dual** *adj* binary, coupled, double, duplicate, linked, paired, twin.

**dubious** *adj* **1** ▷ DOUBTFUL. **2** *a dubious character. inf* fishy, *inf* shady, suspect, suspicious, unreliable, untrustworthy.

**duck** *vb* **1** avoid, bend, bob down, crouch, dip down, dodge, evade, sidestep, stoop, swerve, take evasive action. **2** immerse, plunge, push under, submerge.

**due** *adj* **1** in arrears, outstanding, owed, owing, payable, unpaid. **2** *due consideration.* adequate, appropriate, decent, deserved, expected, fitting, just, mature, merited, proper, requisite, right, rightful, scheduled, sufficient, suitable, wellearned. • *n* deserts, entitlement, merits, reward, rights. **dues** ▷ DUTY.

**dull** *adj* **1** dim, dingy, dowdy, drab, dreary, faded, flat, gloomy, lacklustre, lifeless, matt, plain, shabby, sombre, subdued. **2** *a dull sky.* cloudy, dismal, grey, heavy, leaden, murky, overcast, sullen, sunless. **3** *a dull sound.* deadened, indistinct, muffled, muted. **4** *a dull student.* dense, dim, dim-witted, obtuse, slow, *inf* thick, unimaginative, unintelligent, unresponsive. ▷ STUPID. **5** *a dull edge.* blunt, blunted, unsharpened. **6** *dull talk.*

boring, commonplace, dry, monotonous, prosaic, stodgy, tame, tedious, unexciting, uninteresting. *Opp* BRIGHT, SHARP.

**dumb** *adj* inarticulate, *inf* mum, mute, silent, speechless, tongue-tied, unable to speak.

**dummy** *n* **1** copy, counterfeit, duplicate, imitation, mock-up, model, reproduction, sample, sham, simulation, substitute, toy. **2** doll, figure, manikin, puppet.

**dump** *n* **1** junkyard, rubbish-heap, tip. **2** *arms dump.* arsenal, cache, depot, hoard, store. • *vb* deposit, discard, dispose of, *inf* ditch, drop, empty out, get rid of, jettison, offload, *inf* park, place, put down, reject, scrap, throw away, throw down, tip, unload.

**dune** *n* drift, hillock, hummock, mound, sand-dune.

**dungeon** *n* old use donjon, gaol, keep, lock-up, oubliette, pit, prison, vault.

**duplicate** *adj* alternative, copied, corresponding, identical, matching, second, twin. • *n* carbon copy, clone, copy, double, facsimile, imitation, likeness, *inf* look-alike, match, photocopy, photostat, replica, reproduction, twin, Xerox. • *vb* copy, do again, double up on, photocopy, print, repeat, reproduce, Xerox.

**durable** *adj* enduring, hard-wearing, heavy-duty, indestructible, long-lasting, permanent, resilient, stout, strong, substantial, thick, tough. *Opp* IMPERMANENT, WEAK.

**dusk** *n* evening, gloaming, gloom, sundown, sunset, twilight.

**dust** *n* dirt, grime, grit, particles, powder.

**dusty** *adj* **1** chalky, crumbly, dry, fine, friable, gritty, powdery, sandy, sooty. **2** *a dusty room.* dirty, filthy, grimy, grubby, mucky, uncleaned, unswept.

**dutiful** *adj* attentive, careful, compliant, conscientious, devoted, diligent, faithful, hard-working, loyal, obedient, obliging, punctilious, reliable, responsible, scrupulous, thorough, trustworthy, willing. *Opp* IRRESPONSIBLE.

**duty** *n* **1** allegiance, faithfulness, loyalty, obedience, obligation, onus, responsibility, service. **2** assignment, business, charge, chore, function, job, office, role, stint, task, work. **3** charge, customs, dues, fee, impost, levy, tariff, tax, toll.

**dwarf** *adj* ▷ SMALL. • *n* midget, pigmy. • *vb* dominate, look bigger than, overshadow, tower over.

**dwell** *vb* abide, be accommodated, live, lodge, reside, stay. **dwell in** ▷ INHABIT.

**dwelling** *n* abode, domicile, habitation, home, lodging, quarters, residence. ▷ HOUSE.

**dying** *adj* declining, expiring, fading, failing, moribund, obsolescent. *Opp* ALIVE.

**dynamic** *adj* active, committed, driving, eager, energetic, enterprising, enthusiastic, forceful, *inf* go-ahead, *derog* go-getting, high-powered, lively, motivated, powerful, pushful, *derog* pushy, spirited, vigorous, zealous. *Opp* APATHETIC.

# E

**eager** *adj* agog, animated, anxious (*to please*), ardent, avid, bursting, committed, craving, desirous, earnest, enthusiastic, excited, fervent, fervid, hungry, impatient, intent, interested, *inf* itching, keen, *inf* keyed up, longing, motivated, passionate, *inf* raring (*to* go), voracious, yearning, zealous. *Opp* APATHETIC.

**eagerness** *n* alacrity, anxiety, appetite, ardour, avidity, commitment, desire, earnestness, enthusiasm, excitement, fervour, hunger, impatience, intentness, interest, keenness, longing, motivation, passion, thirst, zeal. *Opp* APATHY.

**early** *adj* **1** advance, ahead of time, before time, first, forward, premature. *Opp* LATE. **2** ancient, antiquated, initial, original, primeval, primitive. ▷ OLD. *Opp* RECENT.

**earn** *vb* **1** be paid, *inf* bring in, *inf* clear, draw, fetch in, gain, get, *inf* gross, make, make a profit of, net, obtain, pocket, realize, receive, *inf* take home, work for, yield. **2** attain, be worthy of, deserve, merit, qualify for, warrant, win.

**earnest** *adj* **1** assiduous, committed, conscientious, dedicated, determined, devoted, diligent, eager, hard-working, industrious, involved, purposeful, resolved, zealous. *Opp* CASUAL. **2** grave, heartfelt, impassioned, serious, sincere, sober, solemn, thoughtful, well-meant.

**earnings** *n* income, salary, stipend, wages. ▷ PAY.

**earth** *n* clay, dirt, ground, humus, land, loam, soil, topsoil.

**earthenware** *n* ceramics, china, crockery, *inf* crocks, porcelain, pots, pottery.

**earthly** *adj* corporeal, human, material, materialistic, mortal, mundane, physical, secular, temporal, terrestrial, worldly. *Opp* SPIRITUAL.

**earthquake** *n* quake, shock, tremor, upheaval.

**earthy** *adj* bawdy, coarse, crude, down to earth, frank, lusty, ribald, uninhibited. ▷ OBSCENE.

**ease** *n* **1** aplomb, calmness, comfort, composure, contentment, enjoyment, happiness, leisure, luxury, peace, quiet, relaxation, repose, rest, serenity, tranquillity. **2** dexterity, easiness, effortlessness, facility, nonchalance, simplicity, skill, speed, straightforwardness. *Opp* DIFFICULTY. • *vb* **1** allay, alleviate, assuage, calm, comfort, decrease, lessen, lighten, mitigate, moderate, pacify, quell, quieten, reduce, relax, relieve, slacken, soothe, tranquillize. **2** edge, guide, inch, manoeuvre, move gradually, slide, slip, steer.

**easy** *adj* **1** carefree, comfortable, contented, cosy, *inf* cushy, effortless, leisurely, light, painless, peaceful, pleasant, relaxed, relaxing, restful, serene, soft, tranquil, undemanding, unexacting, unhurried, untroubled. **2** clear, elementary, facile, foolproof, *inf* idiot-proof, manageable, plain, simple, straightforward, uncomplicated, understandable, user-friendly. **3** ▷ EASYGOING. *Opp* DIFFICULT.

**easygoing** *adj* accommodating, affable, amenable, calm, carefree, casual, cheerful, docile, even-tempered, flexible, forbearing, *inf* free and easy, friendly, genial, *inf* happy-go-lucky, indulgent, informal, *inf* laid-back, *derog* lax, lenient, liberal, mellow, natural, nonchalant, open, patient, permissive, placid, relaxed, tolerant, unexcitable, unruffled, *derog* weak. *Opp* STRICT.

**eat** *vb* consume, devour, digest, feed on, ingest, live on, *old use* partake of, swallow. □ *bite, bolt, champ, chew, crunch, gnaw, gobble, gorge, gormandize, graze, grind, gulp, guzzle,* inf *make a pig of yourself, masticate, munch, nibble, overeat, peck,* inf *scoff,* inf *slurp,* inf *stuff (yourself), taste,* inf *tuck in,* inf *wolf.* □ *banquet, breakfast, dine, feast, lunch, snack,* old use *sup.* **eat away, eat into** ▷ ERODE.

**eatable** *adj* digestible, edible, fit to eat, good, palatable, safe to eat, wholesome. ▷ TASTY. *Opp* INEDIBLE.

**ebb** *vb* fall, flow back, go down, recede, retreat, subside. ▷ DECLINE.

**eccentric** *adj* **1** aberrant, abnormal, anomalous, atypical, bizarre, cranky, curious, freakish, grotesque, idiosyncratic, *sl* kinky, odd, outlandish, out of the ordinary, peculiar, preposterous, quaint, queer, quirky, singular, strange, unconventional, unusual, *sl* wacky, *inf* way-out, *inf* weird, *inf* zany. ▷ ABSURD, MAD. **2** *eccentric circles.* irregular, off-centre. • *n inf* character, *inf* crackpot, crank, *inf* freak, individualist, nonconformist, *inf* oddball, oddity, *inf* weirdie, *inf* weirdo.

**echo** *vb* **1** resound, reverberate, ring, sound again. **2** ape, copy, duplicate, emulate, imitate, mimic, mirror, reiterate, repeat, reproduce, say again.

**eclipse** *vb* **1** block out, blot out, cloud, darken, dim, extinguish, obscure, veil. ▷ COVER. **2** excel, outdo, outshine, overshadow, *inf* put in the shade, surpass, top.

**economic** *adj* budgetary, business, financial, fiscal, monetary, money-making, trading.

**economical** *adj* **1** careful, cheeseparing, frugal, parsimonious, provident, prudent, sparing, thrifty. ▷ MISERLY. *Opp* WASTEFUL. **2** *an economical meal.* cheap, cost-effective, inexpensive, low-priced, money-saving, reasonable, *inf* value-for-money. *Opp* EXPENSIVE.

**economize** *vb* be economical, cut back, retrench, save, *inf* scrimp, skimp, spend less, *inf* tighten your belt. *Opp* SQUANDER.

**economy** *n* **1** frugality, *derog* meanness, *derog* miserliness, parsimony, providence, prudence, saving, thrift. *Opp* WASTE. **2** *the national economy.* budget, economic affairs, wealth. **3** ▷ BREVITY.

**ecstasy** *n* bliss, delight, delirium, elation, enthusiasm, euphoria, exaltation, fervour, frenzy, gratification, happiness, joy, rapture, thrill, trance, *old use* transport.

**ecstatic** *adj* blissful, delighted, delirious, elated, enraptured, enthusiastic, euphoric, exhilarated, exultant, fervent, frenzied, gleeful, joyful, orgasmic, overjoyed, *inf* over the moon, rapturous, transported. ▷ HAPPY.

**eddy** *n* circular movement, maelstrom, swirl, vortex, whirl, whirlpool, whirlwind. • *vb* move in circles, spin, swirl, turn, whirl.

**edge** *n* **1** border, boundary, brim, brink, circumference, frame, kerb, limit, lip, margin, outline, perimeter, periphery, rim, side, verge. **2** *edge of town.* outlying parts, outskirts, suburbs. **3** *edge on a knife.* acuteness, keenness, sharpness. **4** *edge of a curtain.* edging, fringe, hem, selvage. • *vb* **1** bind, border, fringe, hem, make an edge for, trim. **2** *edge away.* crawl, creep, inch, move stealthily, sidle, slink, steal, work your way, worm.

**edible** *adj* digestible, eatable, fit to eat, palatable, safe to eat, wholesome. ▷ TASTY. *Opp* INEDIBLE.

**edit** *vb* adapt, alter, amend, arrange, assemble, compile, get ready, modify, organize, prepare, put together, select, supervise the production of. □ *abridge, annotate, bowdlerize, censor, clean up, condense, copy-edit, correct, cut, dub, emend, expurgate, format, polish, proof-read, rearrange, rephrase, revise, rewrite, select, shorten, splice* (film).

**edition** *n* **1** copy, issue, number. **2** impression, printing, print-run, publication, version.

**educate** *vb* bring up, civilize, coach, counsel, cultivate, discipline, drill, edify, enlighten, guide, improve, inculcate, indoctrinate, inform, instruct, lecture, nurture, rear, school, teach, train, tutor.

**educated** *adj* cultured, enlightened, erudite, knowledgeable, learned, literate, numerate, sophisticated, trained, well-bred, well-read.

**education** *n* coaching, curriculum, enlightenment, guidance, indoctrination, instruction, schooling, syllabus, teaching, training, tuition. □ *academy, college, conservatory, polytechnic, sixth-form college, tertiary college, university.* ▷ SCHOOL, TEACHING.

**eerie** *adj inf* creepy, frightening, ghostly, mysterious, *inf* scary, spectral, *inf* spooky, strange, uncanny, unearthly, unnatural, weird.

**effect** *n* **1** aftermath, conclusion, consequence, impact, influence, issue, outcome, repercussion, result, sequel, upshot. **2** feeling, illusion, impression, sensation, sense. • *vb* accomplish, achieve, bring about, bring in, carry out, cause, create, effectuate, enforce, execute, implement, initiate, make, produce, put into effect, secure.

**effective** *adj* **1** able, capable, competent, effectual, efficacious, functional, impressive, potent, powerful, productive, proficient, real, serviceable, strong, successful, useful, worthwhile. ▷ EFFICIENT. **2** *an effective argument.* cogent, compelling, convincing, meaningful, persuasive, striking, telling. *Opp* INEFFECTIVE.

**effeminate** *adj* camp, effete, girlish, *inf* pansy, *inf* sissy, unmanly, weak, womanish. *Opp* MANLY.

**effervesce** *vb* bubble, ferment, fizz, foam, froth, sparkle.

**effervescent** *adj* bubbling, bubbly, carbonated, fizzy, foaming, frothy, gassy, sparkling.

**efficient** *adj* businesslike, cost-effective, economic, productive, streamlined, thrifty. ▷ EFFECTIVE. *Opp* INEFFICIENT.

**effort** *n* **1** application, diligence, *inf* elbow grease, endeavour, exertion, industry, labour, pains, strain, stress, striving, struggle, toil, *old use* travail, trouble, work. **2** *a brave effort.* attempt, endeavour, go, try, venture. **3** *a successful effort.* accomplishment, achievement, exploit, feat, job, outcome, product, production, result.

**effusive** *adj* demonstrative, ebullient, enthusiastic, exuberant, fulsome, gushing, lavish, *inf* over the top, profuse, voluble. *Opp* RETICENT.

**egoism** *n* egocentricity, egotism, narcissism, pride, self-centredness, self-importance, self-interest, selfishness, self-love, self-regard, vanity.

**egotistical** *adj* egocentric, self-admiring, self-centred, selfish. ▷ CONCEITED.

**eject** *vb* **1** banish, *inf* boot out, *inf* bundle out, deport, discharge, dismiss, drive out, evict, exile, expel, get rid of, *inf* kick out, oust, push out, put out, remove, sack, send out, shoot out, *inf* shove out, throw out, turn out. **2** ▷ EMIT.

**elaborate** *adj* **1** complex, complicated, detailed, exhaustive, intricate, involved, meticulous, minute, painstaking, thorough, well worked out. **2** *elaborate décor.* baroque, busy, Byzantine, decorative, fancy, fantastic, fussy, grotesque, intricate, ornamental, ornamented, ornate, rococo, showy. *Opp* SIMPLE. • *vb* add to, adorn, amplify, complicate, decorate, develop, embellish, enlarge on, enrich, expand, expatiate on, fill out, flesh out, give details of, improve on, ornament. *Opp* SIMPLIFY.

**elapse** *vb* go by, lapse, pass, slip by.

**elastic** *adj* bendy, bouncy, ductile, expandable, flexible, plastic, pliable, pliant, resilient, rubbery, *inf* springy, stretchable, *inf* stretchy, yielding. *Opp* RIGID.

**elderly** *adj* ageing, *inf* getting on, oldish. ▷ OLD.

**elect** *adj* [*goes after noun*] *president elect.* [*synonyms used after noun*] designate, to be; [*synonyms before noun*] chosen, elected, prospective, selected. • *vb* adopt, appoint, choose, name, nominate, opt for, pick, select, vote for.

**election** *n* ballot, choice, plebiscite, poll, referendum, selection, vote, voting.

**electioneer** *vb* campaign, canvass.

**electorate** *n* constituents, electors, voters.

**electric** *adj* **1** battery-operated, electrical, mains-operated. **2** *electric atmosphere.* electrifying. ▷ EXCITING.

**electricity** *n* current, energy, power, power supply.

**elegant** *adj* artistic, beautiful, chic, courtly, cultivated, dapper, debonair, dignified, exquisite, fashionable, fine, genteel, graceful, gracious, handsome, luxurious, modish, noble, pleasing, *inf* plush, *inf* posh, refined, smart, soigné(e), sophisticated, splendid, stately, stylish, suave, tasteful, urbane, well-bred. *Opp* INELEGANT.

**elegy** *n* dirge, lament, requiem.

**element** *n* **1** component, constituent, detail, essential, factor, feature, fragment, hint, ingredient, part, piece, small amount, trace, unit. **2** *in your element.* domain, environment, habitat, medium, sphere, territory. **elements** ▷ RUDIMENTS, WEATHER.

**elementary** *adj* basic, early, first, fundamental, initial, introductory, primary, principal, rudimentary, simple, straightforward, uncomplicated, understandable. ▷ EASY. *Opp* ADVANCED.

**elevate** *vb* exalt, hold up, lift, make higher, promote, rear. ▷ RAISE. **elevated** ▷ HIGH, NOBLE.

**elicit** *vb* bring out, call forth, derive, draw out, evoke, extort, extract, get, obtain, wrest, wring.

**eligible** *adj* acceptable, allowed, appropriate, authorized, available, competent, equipped, fit, fitting, proper, qualified, suitable, worthy. *Opp* INELIGIBLE.

**eliminate** *vb* 1 abolish, annihilate, delete, destroy, dispense with, do away with, eject, end, eradicate, exterminate, extinguish, finish off, get rid of, put an end to, remove, stamp out. ▷ KILL. **2** cut out, drop, exclude, knock out, leave out, omit, reject.

**élite** *n* aristocracy, best, *inf* cream, first-class people, flower, meritocracy, nobility, top people, *inf* upper crust.

**eloquent** *adj* articulate, expressive, fluent, forceful, *derog* glib, moving, persuasive, plausible, powerful, unfaltering. *Opp* INARTICULATE.

**elude** *vb* avoid, circumvent, dodge, *inf* duck, escape, evade, foil, get away from, *inf* give (someone) the slip, shake off, slip away from.

**elusive** *adj* **1** *inf* always on the move, evasive, fugitive, hard to find, slippery. **2** *elusive meaning.* ambiguous, baffling, deceptive, hard to pin down, indefinable, intangible, puzzling, shifting.

**emaciated** *adj* anorectic, atrophied, bony, cadaverous, gaunt, haggard, shrivelled, skeletal, skinny, starved, underfed, undernourished, wasted away, wizened. ▷ THIN.

**emancipate** *vb* deliver, discharge, enfranchise, free, give rights to, let go, liberate, loose, manumit, release, set free, unchain. *Opp* ENSLAVE.

**embankment** *n* bank, causeway, dam, earthwork, mound, rampart.

**embark** *vb* board, depart, go aboard, leave, set out. *Opp* DISEMBARK. **embark on** ▷ BEGIN.

**embarrass** *vb* abash, chagrin, confuse, discomfit, discompose, disconcert, discountenance, disgrace, distress, fluster, humiliate, *inf* make you blush, mortify, *inf* put you on the spot, shame, *inf* show up, upset. **embarrassed** ▷ ASHAMED. **embarrassing** ▷ AWKWARD, SHAMEFUL.

**embellish** *vb* adorn, beautify, deck, decorate, embroider, garnish, ornament, *sl* tart up, *inf* titivate. ▷ ELABORATE.

**embezzle** *vb* appropriate, misapply, misappropriate, peculate, *inf* put your hand in the till, take fraudulently. ▷ STEAL.

**embezzlement** *n* fraud, misappropriation, misuse of funds, peculation, stealing, theft.

**embittered** *adj* acid, bitter, disillusioned, envious, rancorous, resentful, sour. ▷ ANGRY.

**emblem** *n* badge, crest, device, image, insignia, mark, regalia, seal, sign, symbol, token.

**embody** *vb* **1** exemplify, express, incarnate, manifest, personify, reify, represent, stand for, symbolize. **2** bring together, combine, comprise, embrace, enclose, gather together, include, incorporate, integrate, involve, take in, unite.

**embrace** *vb* **1** clasp, cling to, cuddle, enfold, fondle, grasp, hold, hug, kiss, snuggle up to. **2** *embrace new ideas.* accept, espouse, receive, take on, welcome. **3** ▷ EMBODY.

**embryonic** *adj* early, immature, just beginning, rudimentary, underdeveloped, undeveloped, unformed. *Opp* MATURE.

**emerge** *vb* appear, arise, be revealed, come out, come to light, come to notice, emanate, *old* *use* issue forth, leak out, *inf* pop up, proceed, surface, transpire, *inf* turn out.

**emergency** *n* crisis, danger, difficulty, exigency, predicament, serious situation.

**emigrate** *vb* depart, go abroad, leave, quit, relocate, resettle, set out.

**eminent** *adj* august, celebrated, conspicuous, distinguished, elevated, esteemed, exalted, familiar, famous, great, high-ranking, honoured, illustrious, important, notable, noted, noteworthy, outstanding, pre-eminent, prominent, renowned, well-known. *Opp* LOWLY.

**emit** *vb* belch, discharge, disgorge, ejaculate, eject, exhale, expel, exude, give off, give out, issue, radiate, send out, spew out, spout, transmit, vent, vomit.

**emotion** *n* agitation, excitement, feeling, fervour, passion, sentiment, warmth. ▷ ANGER, LOVE, etc.

**emotional** *adj* **1** ardent, demonstrative, enthusiastic, excited, fervent, fiery, heated, hot-headed, impassioned, intense, irrational, moved, passionate, romantic, stirred, touched, warm-hearted, *inf* worked up. ▷ ANGRY, LOVING, etc. **2** *emotional language.* affecting, biased, emotive, heartfelt, heart-rending, inflammatory, loaded, moving, pathetic, poignant, prejudiced, provocative, sentimental, stirring, subjective, tear-jerking, tender, touching. *Opp* UNEMOTIONAL.

**emphasis** *n* accent, attention, force, gravity, importance, intensity, priority, prominence, strength, stress, urgency, weight.

**emphasize** *vb* accent, accentuate, bring out, dwell on, focus on, foreground, give emphasis to, highlight, impress, insist on, make obvious, *inf* play up, point up, *inf* press home, *inf* rub it in, show clearly, spotlight, stress, underline, underscore.

**emphatic** *adj* affirmative, assertive, categorical, confident, dogmatic, definite, firm, forceful, insistent, positive, pronounced, resolute, strong, uncompromising, unequivocal. *Opp* TENTATIVE.

**empirical** *adj* experiential, experimental, observed, practical, pragmatic. *Opp* THEORETICAL.

**employ** *vb* **1** commission, engage, enlist, have on the payroll, hire, pay, sign up, take on, use the services of. **2** apply, use, utilize.

**employed** *adj* active, busy, earning, engaged, hired, involved, in work, occupied, practising, working. ▷ BUSY. *Opp* UNEMPLOYED.

**employee** *n old use* hand, *inf* underling, worker. **employees** staff, workforce.

**employer** *n* boss, chief, *inf* gaffer, *inf* governor, head, manager, owner, proprietor, taskmaster.

**employment** *n* business, calling, craft, job, line, livelihood, living, métier, occupation, profession, pursuit, trade, vocation, work.

**empty** *adj* **1** bare, blank, clean, clear, deserted, desolate, forsaken, vacant, hollow, unfilled, unfurnished, uninhabited, unladen, unoccupied, unused, vacant, void. *Opp* FULL. **2** *empty threats.* futile, idle, impotent, ineffective, insincere, meaningless, pointless, purposeless, senseless, silly, unreal, worthless. • *vb* clear, discharge, drain, eject, evacuate, exhaust, pour out, remove, take out, unload, vacate, void. *Opp* FILL.

**enable** *vb* aid, allow, approve, assist, authorize, charter, empower, entitle, equip, facilitate, franchise, help, license, make it possible, permit, provide the means, qualify, sanction. *Opp* PREVENT.

**enchant** *vb* allure, beguile, bewitch, captivate, cast a spell on, charm, delight, enrapture, enthral, entrance, fascinate, hypnotize, mesmerize, spellbind. **enchanting** ▷ ATTRACTIVE.

**enchantment** *n* charm, conjuration, magic, sorcery, spell, witchcraft, wizardry. ▷ DELIGHT.

**enclose** *vb* bound, box, cage, case, cocoon, conceal, confine, contain, cover, encase, encircle, encompass, enfold, envelop, fence in, hedge in, hem in, immure, insert, limit, package, parcel up, pen, restrict, ring, secure, sheathe, shut in, shut up, surround, wall in, wall up, wrap. ▷ IMPRISON.

**enclosure** *n* **1** arena, cage, compound, coop, corral, court, courtyard, farmyard, field, fold, paddock, pen, pound, ring, run, sheepfold, stockade, sty, yard. **2** *enclosure in an envelope.* contents, inclusion, insertion.

**encounter** *n* **1** confrontation, meeting. **2** [*military*] battle, brush, clash, dispute, skirmish, struggle. ▷ FIGHT. • *vb* chance upon, clash with, come upon, confront, contend with, *inf* cross swords with, face, grapple with, happen upon, have an encounter with, meet, *inf* run into.

**encourage** *vb* **1** abet, advocate, animate, applaud, cheer, *inf* egg on, embolden, give hope to, hearten, incite, inspire, invite, persuade, prompt, rally, reassure, rouse, spur on, support, urge. **2** *encourage sales.* aid, be an incentive to, be conducive to, boost, engender, foster, further, generate, help, increase, induce, promote, stimulate. *Opp* DISCOURAGE.

**encouragement** *n* applause, approval, boost, cheer, exhortation, incentive, incitement, inspiration, reassurance, *inf* shot in the arm, stimulation, stimulus, support. *Opp* DISCOURAGEMENT.

**encouraging** *adj* comforting, heartening, hopeful, inspiring, optimistic, positive, promising, reassuring. ▷ FAVOURABLE.

**encroach** *vb* enter, impinge, infringe, intrude, invade, make inroads, trespass, violate.

**end** *n* **1** boundary, edge, extreme, extremity, limit, pole, tip. **2** cessation, close, coda, completion, conclusion, culmination, curtain (*of play*), denouement (*of plot*), ending, expiration, expiry, finale, finish, *inf* pay-off, resolution. **3** *journey's end.* destination, home, termination, terminus. **4** *end of a queue.* back, rear, tail. **5** *end of your life.* destruction, destiny, doom, extinction, fate, passing, ruin. ▷ DEATH. **6** *an end in view.* aim, aspiration, consequence, design, effect, intention, objective, outcome, plan, purpose, result, upshot. *Opp* BEGINNING. • *vb* **1** abolish, break off, bring to an end, complete, conclude, cut off, destroy, discontinue, *inf* drop, eliminate, exterminate, finalize, *inf* get rid of, halt, phase out, *inf* put an end to, *inf* round off, ruin, scotch, terminate, *inf* wind up. **2** break up, cease, close, come to an end, culminate, die, disappear, expire, fade away, finish, *inf* pack up, reach a climax, stop. *Opp* BEGIN.

**endanger** *vb* expose to risk, imperil, jeopardize, put at risk, threaten. *Opp* PROTECT.

**endearing** *adj* appealing, attractive, captivating, charming, disarming, enchanting, engaging, likable, lovable, sweet, winning, winsome. *Opp* REPULSIVE.

**endeavour** *vb* aim, aspire, attempt, do your best, exert yourself, strive, try.

**endless** *adj* **1** boundless, immeasurable, inexhaustible, infinite, limitless, measureless, unbounded, unfailing, unlimited. **2** abiding, ceaseless, constant, continual, continuous, enduring, eternal, everlasting, immortal, incessant, interminable, never-ending, nonstop, perpetual, persistent, unbroken, undying, unending, uninterrupted.

**endorse** *vb* **1** advocate, agree with, approve, authorize, *inf* back, condone, confirm, *inf* OK, sanction, set your seal of approval to, subscribe to, support. **2** *endorse a cheque.* countersign, sign.

**endurance** *n* determination, fortitude, patience, perseverance, persistence, pertinacity, resolution, stamina, staying-power, strength, tenacity.

**endure** *vb* **1** carry on, continue, exist, last, live on, persevere, persist, prevail, remain, stay, survive. **2** bear, cope with, experience, go through, *inf* put up with, stand, *inf* stick, *inf* stomach, submit to, suffer, *sl* sweat it out, tolerate, undergo, weather, withstand.
**enduring** ▷ ENDLESS.

**enemy** *n* adversary, antagonist, assailant, attacker, competitor, foe, opponent, opposition, the other side, rival, *inf* them. *Opp* FRIEND.

**energetic** *adj* active, animated, brisk, dynamic, enthusiastic, fast, forceful, hard-working, high-powered, indefatigable, lively, powerful, quick-moving, spirited, strenuous, tireless, unflagging, vigorous, zestful. *Opp* LETHARGIC.

**energy** *n* **1** animation, ardour, *inf* dash, drive, dynamism, élan, enthusiasm, exertion, fire, force, forcefulness, *inf* get-up-and-go, *inf* go, life, liveliness, might, *inf* pep, spirit, stamina, strength, verve, vigour, *inf* vim, vitality, vivacity, zeal, zest. *Opp* LETHARGY. **2** fuel, power.

**enforce** *vb* administer, apply, carry out, compel, execute, implement, impose, inflict, insist on, prosecute, put into effect, require, stress. *Opp* WAIVE.

**engage** *vb* **1** contract with, employ, enlist, hire, recruit, sign up, take on. **2** *cogs engage.* bite, fit together, interlock. **3** *engage to do something.* ▷ PROMISE. **4** *engaged me in gossip.* ▷ OCCUPY. **5** *engage in sport.* ▷ PARTICIPATE.

**engaged** *adj* **1** affianced, betrothed, *old use* plighted, *old use* promised, *old use* spoken for. **2** ▷ BUSY.

**engagement** *n* **1** betrothal, promise to marry, *old use* troth. **2** *social engagements.* appointment, arrangement, commitment, date, fixture, meeting, obligation, rendezvous. **3** ▷ BATTLE.

**engine** *n* **1** machine, motor. □ *diesel, electric, internal-combustion, jet, outboard, petrol, steam, turbine, turbo-jet, turbo-prop.* **2** locomotive.

**engineer** *n* mechanic, technician. • *vb* ▷ CONSTRUCT, DEVISE.

**engrave** *vb* carve, chisel, etch, inscribe. ▷ CUT.

**enigma** *n* conundrum, mystery, *inf* poser, problem, puzzle, riddle.

**enjoy** *vb* **1** admire, appreciate, bask in, be happy in, delight in, *inf* go in for, indulge in, *inf* lap up, luxuriate in, rejoice in, relish, revel in, savour, take pleasure from, take pleasure in. ▷ LIKE. **2** benefit from, experience, have, take advantage of, use. **enjoy yourself** celebrate, *inf* gad about, *inf* have a fling, have a good time, make merry.

**enjoyable** *adj* agreeable, amusing, delicious, delightful, diverting, entertaining, gratifying, likeable, *inf* nice, pleasurable, rewarding, satisfying. ▷ PLEASANT. *Opp* UNPLEASANT.

**enlarge** *vb* amplify, augment, blow up, broaden, build up, develop, dilate, distend, diversify, elongate, expand, extend, fill out, grow, increase, inflate, lengthen, magnify, multiply, spread, stretch, swell, supplement, wax, widen. *Opp* DECREASE. **enlarge on** ▷ ELABORATE.

**enlighten** *vb* edify, illuminate, inform, make aware. ▷ TEACH.

**enlist** *vb* **1** conscript, engage, enrol, impress, muster, recruit, sign up. **2** *enlist in the army*. enrol, enter, join up, register, sign on, volunteer. **3** *enlist help*. ▷ OBTAIN.

**enliven** *vb* animate, arouse, brighten, cheer up, energize, inspire, *inf* pep up, quicken, rouse, stimulate, vitalize, wake up.

**enormous** *adj* Brobdingnagian, colossal, elephantine, gargantuan, giant, gigantic, gross, huge, hulking, immense, *inf* jumbo, mammoth, massive, mighty, monstrous, mountainous, prodigious, stupendous, titanic, towering, tremendous, vast. ▷ BIG. *Opp* SMALL.

**enough** *adj* adequate, ample, as much as necessary, sufficient.

**enquire** *vb* ask, beg, demand, entreat, implore, inquire, query, question, quiz, request. **enquire about** ▷ INVESTIGATE.

**enrage** *vb* incense, inflame, infuriate, madden, provoke. ▷ ANGER.

**enslave** *vb* disenfranchise, dominate, make slaves of, subject, subjugate, take away the rights of. *Opp* EMANCIPATE.

**ensure** *vb* confirm, guarantee, make certain, make sure, secure.

**entail** *vb* call for, demand, give rise to, involve, lead to, necessitate, require.

**enter** *vb* **1** arrive, come in, get in, go in, infiltrate, invade, move in, step in. *Opp* DEPART. **2** dig into, penetrate, pierce, puncture, push into. **3** *enter a contest*. engage in, enlist in, enrol in, *inf* go in for, join, participate in, sign up for, take part in, take up, volunteer for. **4** *enter names on a list*. add, inscribe, insert, note down, put down, record, register, set down, sign, write. *Opp* REMOVE. **enter into** ▷ BEGIN.

**enterprise** *n* **1** adventure, effort, endeavour, operation, programme, project, undertaking, venture. **2** adventurousness, ambition, boldness, courage, daring, determination, drive, energy, *inf* get-up-and-go, initiative, *inf* push. **3** business, company, concern, firm, organization.

**enterprising** *adj* adventurous, ambitious, bold, courageous, daring, determined, eager, energetic, enthusiastic, *inf* go-ahead, *derog* go-getting, hard-working, imaginative, indefatigable, industrious, intrepid, keen, purposeful, *inf* pushful, *derog* pushy, resource-

ful, spirited, venturesome, vigorous, zealous.
*Opp* UNADVENTUROUS.

**entertain** *vb* **1** amuse, cheer up, delight, divert, keep amused, make laugh, occupy, please, regale, *inf* tickle. *Opp* BORE. **2** *entertain friends.* accommodate, be host to, be hostess to, cater for, give hospitality to, *inf* put up, receive, treat, welcome. **3** *entertain an idea.* accept, agree to, approve, consent to, consider, contemplate, harbour, support, take seriously. *Opp* IGNORE. **entertaining** ▷ INTERESTING.

**entertainer** *n* artist, artiste, performer. □ *acrobat, actor, actress, ballerina, broadcaster, busker, clown, comedian, comic, compère, conjurer, dancer, disc jockey, DJ, impersonator, jester, juggler, lion-tamer, magician, matador, mime artist, minstrel, singer, stunt man, toreador, trapeze artist, trouper, ventriloquist.* ▷ MUSICIAN.

**entertainment** *n* **1** amusement, distraction, diversion, enjoyment, fun, night-life, pastime, play, pleasure, recreation, sport. **2** divertissement, exhibition, extravaganza, performance, presentation, production, show, spectacle. □ *ballet, bullfight, cabaret, casino, ceilidh, cinema, circus, concert, dance, disco, discothèque, fair, firework display, flower show, gymkhana, motor show, nightclub, pageant, pantomime, play, radio, recital, recitation, revue, rodeo, son et lumière, tattoo, television, variety show, waxworks, zoo.* ▷ DANCE, DRAMA, MUSIC, SPORT.

**enthusiasm** *n* **1** ambition, ardour, avidity, commitment, drive, eagerness, excitement, exuberance, *derog* fanaticism, fervour, gusto, keenness, panache, passion, relish, spirit, verve, zeal, zest. *Opp* APATHY. **2** craze, diversion, *inf* fad, hobby, interest, passion, pastime.

**enthusiast** *n* addict, adherent, admirer, aficionado, *inf* buff, champion, devotee, fan, fanatic, *sl* fiend, *sl* freak, lover, supporter, zealot.

**enthusiastic** *adj* ambitious, ardent, avid, committed, *inf* crazy, delighted, devoted, eager, earnest, ebullient, energetic, excited, exuberant, fervent, fervid, hearty, impassioned, interested, involved, irrepressible, keen, lively, *inf* mad (about), *inf* mad keen, motivated, optimistic, passionate, positive, rapturous, raring (*to go*), spirited, unqualified, unstinting, vigorous, wholehearted, zealous. *Opp* APATHETIC. **be enthusiastic** enthuse, get excited, *inf* go into raptures, *inf* go overboard, rave.

**entice** *vb* allure, attract, cajole, coax, decoy, inveigle, lead on, lure, persuade, seduce, tempt, trap, wheedle.

**entire** *adj* complete, full, intact, sound, total, unbroken, undivided, uninterrupted, whole.

**entitle** *vb* **1** call, christen, designate, dub, name, style, term, title. **2** *A licence entitles you to drive.* allow, authorize, empower, enable, justify, license, permit, qualify, warrant.

**entitlement** *n* claim, ownership, prerogative, right, title.

**entity** *n* article, being, object, organism, thing, whole.

**entrails** *n* bowels, guts, *inf* innards, inner organs, *inf* insides, intestines, viscera.

**entrance** *n* **1** access, admission, admittance. **2** appearance, arrival, coming, entry. **3** door, doorway, gate, gateway, ingress, opening,

portal, turnstile, way in.
**4** ante-room, entrance hall, foyer, lobby, passage, passageway, porch, vestibule. *Opp* EXIT.

**entrant** *n* applicant, candidate, competitor, contender, contestant, entry, participant, player, rival.

**entreat** *vb* ask, beg, beseech, implore, importune, petition, sue, supplicate. ▷ REQUEST.

**entry** *n* **1** insertion, item, jotting, listing, note, record.
**2** ▷ ENTRANCE. **3** ▷ ENTRANT.

**envelop** *vb* cloak, cover, enclose, enfold, enshroud, enwrap, shroud, swathe, veil, wrap. ▷ HIDE.

**envelope** *n* cover, sheath, wrapper, wrapping.

**enviable** *adj* attractive, covetable, desirable, favourable, sought-after.

**envious** *adj* begrudging, bitter, covetous, dissatisfied, *inf* green-eyed, *inf* green with envy, grudging, jaundiced, jealous, resentful.

**environment** *n* circumstances, conditions, context, ecosystem, environs, habitat, location, milieu, setting, situation, surroundings, territory.

**envisage** *vb* anticipate, contemplate, dream of, envision, fancy, forecast, foresee, imagine, picture, predict, visualize.

**envy** *n* bitterness, covetousness, cupidity, desire, discontent, dissatisfaction, ill-will, jealousy, longing, resentment. • *vb* begrudge, grudge, resent.

**ephemeral** *adj* brief, evanescent, fleeting, fugitive, impermanent, momentary, passing, short-lived, temporary, transient, transitory. *Opp* PERMANENT.

**epidemic** *adj* general, pandemic, prevalent, spreading, universal, widespread. • *n* outbreak, pestilence, plague, rash, upsurge.

**episode** *n* **1** affair, event, happening, incident, matter, occurrence.
**2** chapter, instalment, part, passage, scene, section.

**epitome** *n* **1** archetype, embodiment, essence, exemplar, incarnation, personification, quintessence, representation, type.
**2** ▷ SUMMARY.

**equal** *adj* balanced, coextensive, commensurate, congruent, correspondent, egalitarian, even, fair, identical, indistinguishable, interchangeable, level, like, matched, matching, proportionate, regular, the same, symmetrical, uniform.
▷ EQUIVALENT. *Opp* UNEQUAL.
• *n* clone, compeer, counterpart, equivalent, fellow, peer, twin. • *vb*
**1** balance, correspond to, draw with, tie with. **2** *No one equals Caruso.* be in the same class as, compare with, match, parallel, resemble, rival, vie with.

**equality** *n* **1** balance, congruence, correspondence, equivalence, identity, similarity, uniformity.
*Opp* BIAS. **2** *social equality.* egalitarianism, even-handedness, fairness, justice, parity. *Opp* INEQUALITY.

**equalize** *vb* balance, catch up, compensate, even up, level, make equal, match, regularize, *inf* square, standardize.

**equate** *vb* assume to be equal, compare, juxtapose, liken, match, parallel, set side by side.

**equilibrium** *n* balance, equanimity, equipoise, evenness, poise, stability, steadiness, symmetry.

**equip** *vb* accoutre, arm, array, attire, caparison, clothe, dress, fit out, fit up, furnish, *inf* kit out, outfit, provide, stock, supply.

**equipment** *n* accoutrements, apparatus, appurtenances, *sl* clobber, furnishings, *inf* gear, *inf* hardware, implements, instruments, kit, machinery, materials, outfit, paraphernalia, plant, *inf* rig, *inf* stuff, supplies, tackle, *inf* things, tools, trappings. ▷ CLOTHES.

**equivalent** *adj* alike, analogous, comparable, corresponding, fair, interchangeable, parallel, proportionate, *Lat* pro rata, similar, synonymous. ▷ EQUAL.

**equivocal** *adj* ambiguous, circumlocutory, equivocating, evasive, noncommittal, oblique, periphrastic, questionable, roundabout, suspect.

**equivocate** *vb inf* beat about the bush, be equivocal, dodge the issue, fence, *inf* have it both ways, hedge, prevaricate, quibble, waffle.

**era** *n* age, date, day, epoch, period, time.

**eradicate** *vb* eliminate, erase, get rid of, root out, uproot. ▷ DESTROY.

**erase** *vb* cancel, cross out, delete, eradicate, expunge, efface, obliterate, rub out, wipe away, wipe off. ▷ REMOVE.

**erect** *adj* perpendicular, rigid, standing, straight, upright, vertical. • *vb* build, construct, elevate, establish, lift up, make upright, pitch (*a tent*), put up, raise, set up.

**erode** *vb* abrade, corrode, eat away, eat into, gnaw away, grind down, wash away, wear away.

**erotic** *adj* amatory, amorous, aphrodisiac, arousing, lubricious, lustful, *sl* randy, *sl* raunchy, seductive, sensual, venereal, voluptuous. ▷ SEXY.

**err** *vb* be mistaken, be naughty, *sl* boob, *inf* get it wrong, go astray, go wrong, misbehave, miscalculate, sin, *inf* slip up, transgress.

**errand** *n* assignment, commission, duty, job, journey, mission, task, trip.

**erratic** *adj* aberrant, capricious, changeable, fickle, fitful, fluctuating, inconsistent, irregular, shifting, spasmodic, sporadic, uneven, unpredictable, unreliable, unstable, unsteady, variable, wayward. *Opp* REGULAR. **2** aimless, directionless, haphazard, meandering, wandering.

**error** *n inf* bloomer, blunder, *sl* boob, *Lat* corrigendum, *Lat* erratum, fallacy, falsehood, fault, flaw, gaffe, *inf* howler, inaccuracy, inconsistency, inexactitude, lapse, misapprehension, miscalculation, misconception, misprint, mistake, misunderstanding, omission, oversight, sin, *inf* slip-up, solecism, transgression, *old use* trespass, wrongdoing. ▷ WRONG.

**erupt** *vb* be discharged, be emitted, belch, break out, burst out, explode, gush, issue, pour out, shoot out, spew, spout, spurt, vomit.

**eruption** *n* burst, discharge, emission, explosion, outbreak, outburst, rash.

**escapade** *n* adventure, exploit, *inf* lark, mischief, practical joke, prank, scrape, stunt.

**escape** *n* **1** bolt, breakout, departure, flight, flit, getaway, jailbreak, retreat, running away. **2** discharge, emission, leak, leakage, seepage. **3** *escape from reality.* avoidance, distraction, diversion, escapism, evasion, relaxation, relief. • *vb* **1** abscond, *inf* beat it, bolt, break free, break out, *inf* cut and run, decamp, disappear, *sl* do a bunk, elope, flee, fly, get away, *inf* give someone the slip, *sl* scarper, run away, slip away, *inf* slip the net, *inf* take to your heels, *inf* turn tail. **2** discharge, drain, leak, ooze, pour out, run out, seep. **3** *escape the nasty jobs.* avoid, dodge, duck, elude, evade, get away from, shirk, *sl* skive off.

**escapism** *n* day-dreaming, fantasy, pretence, unreality, wishful thinking.

**escort** *n* **1** bodyguard, convoy, guard, guide, pilot, protection, protector, safe-conduct. **2** *royal escort.* attendant, entourage, retinue, train. **3** *escort at a dance.* chaperon, companion, *inf* date, partner. • *vb* accompany, attend, chaperon, conduct, guard, *inf* keep an eye on, *inf* keep tabs on, look after, protect, shepherd, stay with, usher, watch.

**essence** *n* **1** centre, character, core, cornerstone, crux, essential quality, heart, kernel, life, meaning, nature, pith, quiddity, quintessence, soul, spirit, substance. **2** concentrate, decoction, elixir, extract, flavouring, fragrance, perfume, scent, tincture.

**essential** *adj* basic, characteristic, chief, crucial, elementary, fundamental, important, indispensable, inherent, innate, intrinsic, irreplaceable, key, leading, main, necessary, primary, principal, quintessential, requisite, vital. *Opp* INESSENTIAL.

**establish** *vb* **1** base, begin, constitute, construct, create, decree, found, form, inaugurate, initiate, institute, introduce, organize, originate, set up, start. **2** *establish yourself in a job.* confirm, ensconce, entrench, install, lodge, secure, settle, station. **3** *establish facts.* accept, agree, authenticate, certify, confirm, corroborate, decide, demonstrate, fix, prove, ratify, recognize, show to be true, substantiate, verify.

**established** *adj* deep-rooted, deep-seated, indelible, ineradicable, ingrained, long-lasting, long-standing, permanent, proven, reliable, respected, rooted, secure, traditional, well-known, well-tried. *Opp* NEW.

**establishment** *n* **1** composition, constitution, creation, formation, foundation, inauguration, inception, institution, introduction, setting up. **2** *a well-run establishment.* business, company, concern, enterprise, factory, household, institution, office, organization, shop.

**estate** *n* **1** area, development, domain, land. **2** assets, belongings, capital, chattels, effects, fortune, goods, inheritance, lands, possessions, property, wealth.

**esteem** *n* admiration, credit, estimation, favour, honour, regard, respect, reverence, veneration. • *vb* ▷ RESPECT.

**estimate** *n* appraisal, approximation, assessment, calculation, conjecture, estimation, evaluation, guess, *inf* guesstimate, judgement, opinion, price, quotation, reckoning, specification, valuation. • *vb* appraise, assess, calculate, compute, conjecture, consider, count up, evaluate, gauge, guess, judge, project, reckon, surmise, think out, weigh up, work out.

**estimation** *n* appraisal, appreciation, assessment, calculation, computation, consideration, estimate, evaluation, judgement, opinion, rating, view.

**estuary** *n* creek, *Scot* firth, fjord, inlet, *Scot* loch, river mouth.

**eternal** *adj* ceaseless, deathless, endless, everlasting, heavenly, immeasurable, immortal, infinite, lasting, limitless, measureless, never-ending, permanent, perpetual, timeless, unchanging, undying, unending, unlimited. ▷ CONTINUAL. *Opp* OCCASIONAL, TRANSIENT.

**eternity** *n* afterlife, eternal life, immortality, infinity, perpetuity.

**ethical** *adj* decent, fair, good, honest, just, moral, noble, principled, righteous, upright, virtuous. *Opp* IMMORAL.

**ethnic** *adj* cultural, folk, national, racial, traditional, tribal.

**etiquette** *n* ceremony, civility, code of behaviour, conventions, courtesy, decency, decorum, form, formalities, manners, politeness, propriety, protocol, rules of behaviour, standards of behaviour.

**evacuate** *vb* **1** clear, deplete, drain, move out, remove, send away, void. **2** abandon, decamp from, desert, empty, forsake, leave, pull out of, quit, relinquish, vacate, withdraw from.

**evade** *vb* **1** avoid, *inf* chicken out of, circumvent, dodge, duck, elude, escape from, fend off, flinch from, get away from, shirk, shrink from, shun, sidestep, *inf* skive, steer clear of, turn your back on. **2** *evade a question.* fudge, hedge, parry. ▷ EQUIVOCATE. *Opp* CONFRONT.

**evaluate** *vb* apprise, assess, calculate value of, estimate, judge, value, weigh up.

**evaporate** *vb* dehydrate, desiccate, disappear, disperse, dissipate, dissolve, dry up, evanesce, melt away, vanish, vaporize.

**evasive** *adj* ambiguous, *inf* cagey, circumlocutory, deceptive, devious, disingenuous, equivocal, equivocating, inconclusive, indecisive, indirect, *inf* jesuitical, misleading, noncommittal, oblique, prevaricating, roundabout, *inf* shifty, sophistical, uninformative. *Opp* DIRECT.

**even** *adj* **1** flat, flush, horizontal, level, plane, smooth, straight, true. **2** *even pulse.* consistent, constant, equalized, measured, metrical, monotonous, proportional, regular, rhythmical, symmetrical, unbroken, uniform, unvarying. **3** *even scores.* balanced, equal, identical, level, matching, the same. **4** ▷ EVEN-TEMPERED. *Opp* IRREGULAR. **even out** ▷ FLATTEN. **even up** ▷ EQUALIZE. **get even** ▷ RETALIATE.

**evening** *n* dusk, *poet* eventide, *poet* gloaming, nightfall, sundown, sunset, twilight.

**event** *n* **1** affair, business, chance, circumstance, contingency, episode, eventuality, experience, happening, incident, occurrence. **2** conclusion, consequence, effect, issue, outcome, result, upshot. **3** activity, ceremony, entertainment, function, occasion. **4** *sporting event.* bout, championship, competition, contest, engagement, fixture, game, match, meeting, tournament.

**even-tempered** *adj* balanced, calm, composed, cool, equable, even, impassive, imperturbable, pacific, peaceable, peaceful, placid, poised, reliable, self-possessed, serene, stable, steady, tranquil, unemotional, unexcitable, unruffled. *Opp* EXCITABLE.

**eventual** *adj* concluding, consequent, destined, due, ensuing, expected, final, last, overall, probable, resultant, resulting, ultimate.

**everlasting** *adj* ceaseless, deathless, endless, eternal, immortal, incorruptible, infinite, lasting, limitless, measureless, never-ending, permanent, perpetual, persistent, timeless, unchanging, undying, unending. *Opp* TRANSIENT.

**evermore** *adv* always, eternally, for ever, unceasingly.

**evict** *vb* dislodge, dispossess, eject, expel, *sl* give (someone) the boot, *inf* kick out, oust, put out, remove, throw out, *inf* turf out, turn out.

**evidence** *n* attestation, certification, confirmation, corroboration, data, demonstration, deposition, documentation, facts, grounds, information, proof, sign, statement, statistics, substantiation, testimony. **give evidence** ▷ TESTIFY.

**evident** *adj* apparent, certain, clear, discernible, manifest, noticeable, obvious, palpable, patent, perceptible, plain, self-explanatory, unambiguous, undeniable, unmistakable, visible. *Opp* UNCERTAIN.

**evil** *adj* **1** amoral, atrocious, base, black-hearted, blasphemous, corrupt, criminal, cruel, depraved, devilish, diabolical, dishonest, fiendish, foul, harmful, hateful, heinous, hellish, immoral, impious, infamous, iniquitous, irreligious, machiavellian, malevolent, malicious, malignant, nefarious, pernicious, perverted, reprobate, satanic, sinful, sinister, treacherous, ungodly, unprincipled, unrighteous, vicious, vile, villainous, wicked, wrong. ▷ BAD. *Opp* GOOD. **2** *evil smell.* foul, nasty, pestilential, poisonous, troublesome, unspeakable, vile. ▷ UNPLEASANT. *Opp* PLEASANT. • *n* **1** amorality, blasphemy, corruption, criminality, cruelty, depravity, dishonesty, fiendishness, heinousness, immorality, impiety, iniquity, *old use* knavery, malevolence, malice, mischief, pain, sin, sinfulness, suffering, treachery, turpitude, ungodliness, unrighteousness, vice, viciousness, villainy, wickedness, wrongdoing. ▷ CRIME. **2** *Poverty is an evil.* affliction, bane, calamity, catastrophe, curse, disaster, enormity, hardship, harm, ill, misfortune, wrong.

**evocative** *adj* atmospheric, convincing, descriptive, emotive, graphic, imaginative, provoking, realistic, stimulating, suggestive, vivid.

**evoke** *vb* arouse, awaken, call up, conjure up, elicit, excite, inspire, invoke, kindle, produce, provoke, raise, rouse, stimulate, stir up, suggest, summon up.

**evolution** *n* advance, development, emergence, formation, growth, improvement, maturation, maturing, progress, unfolding.

**evolve** *vb* derive, descend, develop, emerge, grow, improve, mature, modify gradually, progress, unfold.

**exact** *adj* **1** accurate, correct, dead (*centre*), detailed, faithful, faultless, flawless, meticulous, painstaking, precise, punctilious, right, rigorous, scrupulous, specific, *inf* spot-on, strict, true, truthful, veracious. *Opp* IMPRECISE. **2** *exact copy.* identical, indistinguishable, literal, perfect. • *vb* claim, compel, demand, enforce, extort, extract, get, impose, insist on, obtain, require. **exacting** ▷ DIFFICULT.

**exaggerate** *vb* **1** amplify, embellish, embroider, enlarge, inflate, *inf* lay it on thick, magnify, make too much of, maximize, overdo, overemphasize, overestimate, overstate, *inf* pile it on, *inf* play up. *Opp* MINIMIZE. 2 ▷ CARICATURE. **exaggerated** ▷ EXCESSIVE.

**exalt** *vb* boost, elevate, lift, promote, raise, uplift. ▷ PRAISE. **exalted** ▷ HIGH.

**examination** *n* **1** analysis, appraisal, assessment, audit, catechism, *inf* exam, inspection, investigation, *inf* oral, paper, post-mortem, review, scrutiny, study, survey, test, *inf* viva, *Lat* viva voce. **2** [*medical*] *inf* check-up, scan. **3** *police examination*. cross-examination, enquiry, inquiry, inquisition, interrogation, probe, questioning, trial.

**examine** *vb* **1** analyse, appraise, audit (*accounts*), check, *inf* check out, explore, inquire into, inspect, investigate, peruse, probe, research, scan, scrutinize, sift, sort out, study, *sl* sus out, test, vet, weigh up. **2** *examine a witness*. catechize, cross-examine, cross-question, *inf* grill, interrogate, *inf* pump, question, sound out, try.

**example** *n* **1** case, illustration, instance, occurrence, sample, specimen. **2** *example to follow*. ideal, lesson, model, paragon, pattern, prototype. **make an example of** ▷ PUNISH.

**exasperate** *vb inf* aggravate, drive mad, gall, infuriate, irk, irritate, *inf* needle, pique, provoke, rile, vex. ▷ ANNOY.

**excavate** *vb* burrow, dig, gouge out, hollow out, mine, scoop out, unearth.

**exceed** *vb* beat, be more than, do more than, go beyond, go over, outnumber, outshine, outstrip, overstep, overtake, pass, transcend. ▷ EXCEL.

**exceedingly** *adv* amazingly, especially, exceptionally, excessively, extraordinarily, extremely, outstandingly, specially, unusually, very.

**excel** *vb* beat, be excellent, better, do best, eclipse, outclass, outdo, outshine, shine, stand out, surpass, top. ▷ EXCEED.

**excellent** *adj inf* ace, admirable, *inf* brilliant, *old use* capital, champion, choice, consummate, *sl* cracking, distinguished, esteemed, estimable, exceptional, exemplary, extraordinary, *inf* fabulous, *inf* fantastic, fine, first-class, first-rate, flawless, gorgeous, great, high-class, ideal, impressive, magnificent, marvellous, model, notable, outstanding, perfect, *inf* phenomenal, remarkable, *inf* smashing, splendid, sterling, *inf* stunning, *inf* super, superb, superlative, supreme, surpassing, *inf* terrific, *inf* tip-top, *old use* top-hole, *inf* top-notch, top-ranking, *inf* tremendous, unequalled, wonderful. *Opp* BAD.

**except** *vb* exclude, leave out, omit.

**exception** *n* **1** exclusion, omission, rejection. **2** abnormality, anomaly, departure, deviation, eccentricity, freak, irregularity, oddity, peculiarity, quirk, rarity. **take exception** ▷ OBJECT.

**exceptional** *adj* **1** aberrant, abnormal, anomalous, atypical, curious, deviant, eccentric, extraordinary, extreme, isolated, memorable, notable, odd, out-of-the-ordinary, peculiar, phenomenal, quirky, rare, remarkable, singular, solitary, special, strange, surprising, uncommon, unconventional, unexpected, unheard-of, unique, unparalleled, unprecedented, unpredictable, untypical, unusual. 2 ▷ EXCELLENT. *Opp* ORDINARY.

**excerpt** *n* citation, clip, extract, fragment, highlight, part, passage, quotation, section, selection.

**excess** *n* 1 abundance, glut, overabundance, overflow, *inf* overkill, profit, redundancy, superabundance, superfluity, surfeit, surplus. *Opp* SCARCITY. 2 debauchery, dissipation, extravagance, intemperance, over-indulgence, profligacy, wastefulness. *Opp* MODERATION.

**excessive** *adj* 1 disproportionate, exaggerated, extravagant, extreme, fanatical, immoderate, inordinate, intemperate, needless, overdone, prodigal, profligate, profuse, superfluous, undue, unnecessary, unneeded, wasteful. ▷ HUGE. *Opp* INADEQUATE. 2 *excessive prices.* exorbitant, extortionate, unjustifiable, unrealistic, unreasonable. *Opp* MODERATE.

**exchange** *n* deal, interchange, reciprocity, replacement, substitution, *inf* swap, switch.
• *vb* bargain, barter, change, convert (*currency*), interchange, reciprocate, replace, substitute, *inf* swap, switch, *inf* swop, trade, trade in, traffic. **exchange words** ▷ TALK.

**excitable** *adj inf* bubbly, chattery, edgy, emotional, explosive, fidgety, fiery, highly-strung, hot-tempered, irrepressible, jumpy, lively, mercurial, nervous, passionate, quick-tempered, restive, temperamental, unstable, volatile. *Opp* CALM.

**excite** *vb* 1 agitate, amaze, animate, arouse, awaken, discompose, disturb, elate, electrify, enthral, exhilarate, fluster, *inf* get going, incite, inflame, interest, intoxicate, make excited, move, perturb, provoke, rouse, stimulate, stir up, thrill, titillate, *inf* turn on, upset, urge, *inf* wind up, *inf* work up. 2 *excite interest.* activate, cause, elicit, encourage, engender, evoke, fire, generate, kindle, motivate, produce, set off, whet. *Opp* CALM.

**excited** *adj* agitated, boisterous, delirious, eager, enthusiastic, excitable, exuberant, feverish, frantic, frenzied, heated, *inf* het up, hysterical, impassioned, intoxicated, lively, moved, nervous, overwrought, restless, spirited, vivacious, wild. *Opp* APATHETIC.

**exciting** *adj* cliff-hanging, dramatic, electric, electrifying, eventful, fast-moving, galvanizing, gripping, heady, hair-raising, inspiring, intoxicating, *inf* nail-biting, provocative, riveting, rousing, sensational, spectacular, spine-tingling, stimulating, stirring, suspenseful, tense, thrilling. ▷ AMAZING. *Opp* BORING.

**excitement** *n* action, activity, adventure, agitation, animation, commotion, delirium, drama, eagerness, enthusiasm, furore, fuss, heat, intensity, *inf* kicks, passion, stimulation, suspense, tension, thrill, unrest.

**exclaim** *vb* bawl, bellow, blurt out, call, cry out, *old use* ejaculate, shout, utter, vociferate, yell. ▷ SAY.

**exclamation** *n* bellow, call, cry, *old use* ejaculation, expletive, interjection, oath, shout, swear-word, utterance, vociferation, yell.

**exclude** *vb* ban, banish, bar, blacklist, debar, disallow, disown, eject, except, excommunicate, expel, forbid, interdict, keep out, leave out, lock out, omit, ostracize, oust, outlaw, prohibit, proscribe, put an embargo on, refuse, reject, repudiate, rule out, shut out, veto. ▷ REMOVE. *Opp* INCLUDE.

**exclusive** *adj* 1 limiting, restricted, sole, unique, unshared. 2 *an exclusive club.* clannish, classy, closed, fashionable, *sl* posh, private, restrictive, select, selective, snobbish, *inf* up-market.

**excreta** *plur n* droppings, dung, excrement, faeces, manure, sewage, waste matter.

**excrete** *vb* defecate, evacuate the bowels, go to the lavatory, relieve yourself.

**excursion** *n* cruise, expedition, jaunt, journey, outing, ramble, tour, trip, voyage. ▷ TRAVEL.

**excuse** *n* alibi, apology, defence, explanation, extenuation, justification, mitigation, palliation, plea, pretext, rationalization, reason, vindication. • *vb* 1 apologize for, condone, disregard, explain away, forgive, ignore, justify, mitigate, overlook, pardon, pass over, sanction, tolerate, vindicate, warrant. 2 absolve, acquit, clear, discharge, exculpate, exempt, exonerate, free, let off, *inf* let off the hook, liberate, release. *Opp* BLAME.

**execute** *vb* 1 accomplish, achieve, bring off, carry out, complete, discharge, do, effect, enact, finish, implement, perform, *inf* pull off. 2 kill, put to death. □ *behead, burn, crucify, decapitate, electrocute, garrotte, gas, guillotine, hang, lynch, shoot, stone.*

**executive** *n* administrator, *inf* boss, director, manager, officer. ▷ CHIEF.

**exemplary** *adj* admirable, commendable, faultless, flawless, ideal, model, perfect, praiseworthy, unexceptionable.

**exemplify** *vb* demonstrate, depict, embody, illustrate, personify, represent, show, symbolize, typify.

**exempt** *vb* except, exclude, excuse, free, let off, *inf* let off the hook, liberate, release, spare.

**exercise** *n* 1 action, activity, aerobics, callisthenics, effort, exertion, games, gymnastics, PE, sport, *inf* warm-up, *inf* work-out. 2 *military exercises.* discipline, drill, manoeuvres, operation, practice, training. • *vb* 1 apply, bring to bear, display, effect, employ, execute, exert, expend, implement, put to use, show, use, utilize, wield. 2 *exercise your body.* discipline, drill, exert, jog, keep fit, practise, train, *inf* work out. 3 ▷ WORRY.

**exertion** *n* action, effort, endeavour, strain, striving, struggle. ▷ WORK.

**exhaust** *n* discharge, effluent, emission, fumes, gases, smoke. • *vb* 1 consume, deplete, dissipate, drain, dry up, empty, expend, finish off, *inf* run through, sap, spend, use up, void. 2 debilitate, enervate, *inf* fag, fatigue, prostrate, tax, tire, wear out, weary.

**exhausted** ▷ BREATHLESS, WEARY.

**exhausting** *adj* arduous, backbreaking, crippling, debilitating, demanding, difficult, enervating, fatiguing, gruelling, hard, laborious, punishing, severe, strenuous, taxing, tiring, wearying.

**exhaustion** *n* debility, fatigue, lassitude, tiredness, weakness, weariness.

**exhaustive** *adj inf* all-out, careful, comprehensive, full-scale, intensive, meticulous, thorough. *Opp* INCOMPLETE.

**exhibit** *vb* 1 arrange, display, offer, present, put up, set up, show. 2 *exhibit knowledge.* air, betray, brandish, demonstrate, disclose, evidence, express, *derog* flaunt, indicate, manifest, *derog* parade, reveal, *derog* show off. *Opp* HIDE.

**exhibition** *n* demonstration, display, *inf* expo, exposition, presentation, show.

**exhilarating** *adj* bracing, cheering, enlivening, exciting, invigorating, refreshing, rejuvenating, stimulating, tonic, uplifting. ▷ HAPPY.

**exhort** *vb* advise, encourage, harangue, *inf* give a pep talk to, lecture, sermonize, urge.

**exile** *n* **1** banishment, deportation, expatriation, expulsion, transportation. **2** deportee, displaced person, émigré, expatriate, outcast, refugee, wanderer. • *vb* ban, banish, bar, deport, drive out, eject, evict, expatriate, expel, oust, send away, transport.

**exist** *vb* **1** be, be found, be in existence, be real, happen, occur. **2** abide, continue, endure, hold out, keep going, last, live, remain alive, subsist, survive. **existing** ▷ ACTUAL, CURRENT, LIVING.

**existence** *n* actuality, being, continuance, life, living, persistence, reality, survival.

**exit** *n* **1** barrier, door, doorway, egress, gate, gateway, opening, portal, way out. **2** *a hurried exit.* departure, escape, evacuation, exodus, flight, leave-taking, retreat, withdrawal. • *vb* ▷ DEPART.

**exorbitant** *adj* disproportionate, excessive, extortionate, extravagant, high, inordinate, outrageous, profiteering, prohibitive, *inf* sky-high, *inf* steep, *inf* stiff, *inf* swingeing, top, unjustifiable, unrealistic, unreasonable, unwarranted. ▷ EXPENSIVE. *Opp* REASONABLE.

**exotic** *adj* **1** alien, faraway, foreign, remote, romantic, unfamiliar, wonderful. **2** bizarre, colourful, different, exciting, extraordinary, foreign-looking, novel, odd, outlandish, peculiar, rare, singular, strange, striking, unfamiliar, unusual, weird. *Opp* ORDINARY.

**expand** *vb* **1** amplify, augment, broaden, build up, develop, diversify, elaborate, enlarge, extend, fill out, heighten, increase, make bigger, make longer, prolong. **2** become bigger, dilate, distend, grow, increase, lengthen, open out, stretch, swell, thicken, widen. *Opp* CONTRACT.

**expanse** *n* area, breadth, extent, range, sheet, space, spread, stretch, sweep, surface, tract.

**expansive** *adj* **1** affable, amiable, communicative, effusive, extrovert, friendly, genial, open, outgoing, sociable, well-disposed. ▷ TALKATIVE. *Opp* TACITURN. **2** ▷ BROAD. *Opp* NARROW.

**expect** *vb* **1** anticipate, await, bank on, bargain for, be prepared for, contemplate, count on, envisage, forecast, foresee, have faith in, hope for, imagine, look forward to, plan for, predict, prophesy, reckon on, wait for. **2** *expect obedience.* consider necessary, demand, insist on, look for, rely on, require, want. **3** *I expect he'll come.* assume, believe, conjecture, guess, imagine, judge, presume, presuppose, suppose, surmise, think. **expected** ▷ PREDICTABLE.

**expectant** *adj* **1** eager, hopeful, *inf* keyed up, *inf* on tenterhooks, optimistic, ready. **2** *inf* expecting, pregnant.

**expedient** *adj* advantageous, advisable, appropriate, apropos, beneficial, convenient, desirable, helpful, judicious, opportune, politic, practical, pragmatic, profitable, propitious, prudent, right, sensible, suitable, to your advantage, useful, worthwhile. • *n* contrivance, device, *inf* dodge, manoeuvre, means, measure, method, *inf* ploy, recourse, resort, ruse, scheme, stratagem, tactics.

**expedition** *n* crusade, excursion, exploration, journey, mission, pilgrimage, quest, raid, safari, tour, trek, trip, undertaking, voyage.

**expel** *vb* **1** ban, banish, cast out, *inf* chuck out, dismiss, drive out, eject, evict, exile, exorcise, *inf* fire, *inf* kick out, oust, remove, *inf* sack, send away, throw out, turn out, *inf* turf out. **2** *expel fumes*. belch, discharge, emit, exhale, give out, push out, send out, spew out.

**expend** *vb* consume, disburse, *sl* dish out, employ, pay out, spend, use.

**expendable** *adj* disposable, inessential, insignificant, replaceable, *inf* throw-away, unimportant.

**expense** *n* charge, cost, disbursement, expenditure, fee, outgoings, outlay, overheads, payment, price, rate, spending.

**expensive** *adj* costly, dear, generous, high-priced, over-priced, precious, *inf* pricey, *inf* steep, *inf* up-market, valuable. ▷ EXORBITANT. *Opp* CHEAP.

**experience** *n* **1** familiarity, involvement, observation, participation, practice, taking part. **2** background, expertise, *inf* know-how, knowledge, *Fr* savoir faire, skill, understanding, wisdom. **3** *a nasty experience*. adventure, circumstance, episode, event, happening, incident, occurrence, ordeal, trial. • *vb* encounter, endure, face, go through, have a taste of, know, meet, practise, sample, suffer, test out, try, undergo. **experienced** ▷ EXPERT.

**experiment** *n* demonstration, investigation, *inf* practical, proof, research, test, trial, try-out. • *vb* do experiments, examine, investigate, make tests, probe, research, test, try out.

**experimental** *adj* **1** exploratory, on trial, pilot, provisional, tentative, trial. **2** *experimental evidence*. empirical, experiential, proved, tested.

**expert** *adj* able, *inf* ace, *inf* brilliant, capable, competent, *inf* crack, experienced, knowing, knowledgeable, master, masterly, practised, professional, proficient, qualified, skilful, skilled, sophisticated, specialized, trained, well-versed, worldly-wise. ▷ CLEVER. *Opp* UNSKILFUL. • *n inf* ace, authority, connoisseur, *inf* dab hand, genius, *derog* know-all, master, *inf* old hand, professional, pundit, specialist, veteran, virtuoso, *derog* wiseacre, *inf* wizard. *Opp* AMATEUR.

**expertise** *n* adroitness, dexterity, expertness, judgement, *inf* know-how, knowledge, *Fr* savoir faire, skill.

**expire** *vb* become invalid, cease, come to an end, discontinue, finish, *inf* run out, terminate. ▷ DIE.

**explain** *vb* **1** clarify, clear up, decipher, decode, define, demonstrate, describe, disentangle, elucidate, expound, *inf* get across, *inf* get over, gloss, illustrate, interpret, make clear, make plain, provide an explanation, resolve, shed light on, simplify, solve, *inf* sort out, spell out, teach, translate, unravel. **2** *explain a mistake*. account for, excuse, give reasons for, justify, legitimatize, legitimize, make excuses for, rationalize, vindicate.

**explanation** *n* **1** account, analysis, clarification, definition, demonstration, description, elucidation, exegesis, explication, exposition, gloss, illustration, interpretation, key, meaning, rubric, significance, solution, translation. **2** cause, excuse, justification, motivation, motive, rationalization, reason, vindication.

**explanatory** *adj* descriptive, expository, helpful, illuminating, illustrative, interpretive, revelatory.

**explicit** *adj* categorical, clear, definite, detailed, direct, exact, express, frank, graphic, manifest, open, outspoken, patent, plain, positive, precise, put into words, said, specific, *inf* spelt out, spoken, stated, straightforward, unambiguous, unconcealed, unequivocal, unhidden, unreserved, well-defined. *Opp* IMPLICIT.

**explode** *vb* **1** backfire, blast, blow up, burst, detonate, erupt, go off, make an explosion, set off, shatter. **2** *explode a theory*. debunk, destroy, discredit, disprove, put an end to, rebut, refute, reject.

**exploit** *n* achievement, adventure, attainment, deed, enterprise, feat. • *vb* **1** build on, capitalize on, *inf* cash in on, develop, make capital out of, make use of, profit by, profit from, trade on, work on, use, utilize. **2** *exploit people*. *inf* bleed, enslave, ill-treat, impose on, keep down, manipulate, *inf* milk, misuse, oppress, *inf* rip off, *inf* squeeze dry, take advantage of, treat unfairly, withhold rights from.

**explore** *vb* **1** break new ground, probe, prospect, reconnoitre, scout, search, survey, tour, travel through. **2** *explore a problem*. analyse, examine, inspect, investigate, look into, probe, research, scrutinize, study.

**explosion** *n* **1** bang, blast, boom, burst, clap, crack, detonation, discharge, eruption, firing, report. **2** *explosion of anger*. fit, outbreak, outburst, *inf* paddy, paroxysm, spasm.

**explosive** *adj* dangerous, highly-charged, liable to explode, sensitive, unstable, volatile. *Opp* STABLE. • *n* cordite, dynamite, gelignite, gunpowder, TNT.

**exponent** *n* **1** executant, interpreter, performer, player. **2** advocate, champion, defender, expounder, presenter, propagandist, proponent, supporter, upholder.

**expose** *vb* bare, betray, dig up, disclose, display, exhibit, lay bare, reveal, show (up), uncover, unearth, unmask. ▷ REVEAL. *Opp* HIDE.

**express** *vb* air, articulate, disclose, give vent to, make known, phrase, put into words, release, vent, ventilate, voice, word. ▷ COMMUNICATE.

**expression** *n* **1** cliché, formula, phrase, phraseology, remark, statement, term, turn of phrase, usage, utterance, wording. ▷ SAYING. **2** articulation, confession, declaration, disclosure, revelation, statement. **3** *expression in your voice*. accent, depth, emotion, expressiveness, feeling, intensity, intonation, nuance, pathos, sensibility, sensitivity, sympathy, tone, understanding. **4** *expression on your face*. air, appearance, aspect, countenance, face, look, mien. □ *beam, frown, glare, glower, grimace, grin, laugh, leer, long face, lour,* *lower, poker-face, pout, scowl, smile, smirk, sneer, wince, yawn.*

**xpressionless** *adj* blank, *inf* dead-pan, emotionless, empty, glassy, impassive, inscrutable, poker-faced, straight-faced, uncommunicative, wooden. **2** boring, dull, flat, monotonous, uninspiring, unmodulated, unvarying. *Opp* EXPRESSIVE.

**xpressive** *adj* **1** indicative, meaningful, mobile, revealing, sensitive, significant, striking, suggestive, telling. **2** articulate, eloquent, lively, modulated, varied. *Opp* EXPRESSIONLESS.

**xquisite** *adj* delicate, elegant, fine, intricate, refined, skilful, well-crafted. ▷ BEAUTIFUL. *Opp* CRUDE.

**xtend** *vb* **1** add to, broaden, build up, develop, draw out, enlarge, expand, increase, keep going, lengthen, make longer, open up, pad out, perpetuate, prolong, protract, *inf* spin out, spread, stretch, widen. **2** *extend a deadline.* defer, delay, postpone, put back, put off. **3** *extend your hand.* give, hold out, offer, outstretch, present, proffer, put out, raise, reach out, stick out, stretch out. **4** *The garden extends to the fence.* continue, range, reach.

**xtensive** *adj* broad, comprehensive, expansive, far-ranging, far-reaching, sweeping, vast, wide, widespread. ▷ LARGE.

**xtent** *n* amount, area, bounds, breadth, compass, degree, dimensions, distance, expanse, length, limit, magnitude, measure, measurement, proportions, quantity, range, reach, scale, scope, size, space, spread, sweep, width.

**xterior** *adj* external, outer, outside, outward, superficial. • *n* coating, covering, façade, front, outside, shell, skin, surface. *Opp* INTERIOR.

**exterminate** *vb* annihilate, destroy, eliminate, eradicate, extirpate, get rid of, obliterate, put an end to, root out, terminate. ▷ KILL.

**external** *adj* exterior, outer, outside, outward, superficial. *Opp* INTERNAL.

**extinct** *adj* burnt out, dead, defunct, died out, exterminated, extinguished, gone, inactive, vanished. ▷ OLD. *Opp* LIVING.

**extinguish** *vb* blow out, damp down, douse, put out, quench, slake, smother, snuff out, switch off. ▷ DESTROY. *Opp* KINDLE.

**extort** *vb* blackmail, bully, coerce, exact, extract, force, obtain by force.

**extra** *adj* accessory, added, additional, ancillary, auxiliary, excess, further, left-over, more, other, reserve, spare, superfluous, supernumerary, supplementary, surplus, temporary, unneeded, unused, unwanted.

**extract** *n* **1** concentrate, concentration, decoction, distillation, essence, quintessence. **2** abstract, citation, *inf* clip, clipping, cutting, excerpt, passage, quotation, selection. • *vb* **1** draw out, extricate, pull out, remove, take out, withdraw. **2** *extract a confession.* extort, force out, *inf* winkle out, *inf* worm out, wrench, wrest, wring. **3** *extract what you need.* choose, cull, derive, distil, gather, glean, quote, select. ▷ OBTAIN.

**extraordinary** *adj* abnormal, amazing, astonishing, astounding, awe-inspiring, bizarre, breathtaking, curious, exceptional, extreme, fantastic, *inf* funny, incredible, marvellous, miraculous, mysterious, mystical, notable, noteworthy, odd, outstanding, peculiar, *inf* phenomenal, prodigious, queer, rare, remarkable, *inf* sensational, signal, singular, special, staggering, strange, striking, stunning, stupendous, surprising, *inf* unbelievable, uncommon, unheard-of, unimaginable, unique, unprecedented, unusual, *inf* weird, wonderful. *Opp* ORDINARY.

**extravagance** *n* excess, immoderation, improvidence, lavishness, overindulgence, overspending, prodigality, profligacy, self-indulgence, wastefulness. *Opp* ECONOMY.

**extravagant** *adj* exaggerated, excessive, flamboyant, grandiose, immoderate, improvident, lavish, outrageous, overblown, overdone, pretentious, prodigal, profligate, profuse, reckless, self-indulgent, *inf* showy, spendthrift, uneconomical, unreasonable, unthrifty, wasteful. ▷ EXPENSIVE. *Opp* ECONOMICAL.

**extreme** *adj* **1** acute, drastic, excessive, greatest, intensest, maximum, severest, *inf* terrific, utmost. ▷ EXTRAORDINARY. **2** distant, endmost, farthest, furthest, furthermost, last, outermost, remotest, ultimate, uttermost. **3** *extreme opinions.* absolute, avant-garde, exaggerated, extravagant, extremist, fanatical, *inf* hard-line, immoderate, intemperate, intransigent, left-wing, militant, obsessive, outrageous, radical, right-wing, uncompromising, *inf* way-out, zealous. • *n* bottom, bounds, edge, end, extremity, left wing, limit, maximum, minimum, opposite, pole, right wing, top, ultimate.

**extroverted** *adj* active, confident, exhibitionist, outgoing, positive. ▷ SOCIABLE. *Opp* INTROVERTED.

**exuberant** *adj* **1** animated, boisterous, *inf* bubbly, buoyant, eager ebullient, effervescent, energetic, enthusiastic, excited, exhilarated, exultant, high-spirited, irrepressible, lively, spirited, sprightly, vivacious. ▷ CHEERFUL. **2** *exuberant decoration.* baroque, exaggerated, highly-decorated, ornate, overdone, rich, rococo. **3** *exuberant growth.* abundant, copious, lush, luxuriant, overflowing, profuse, rank, teeming. *Opp* AUSTERE.

**exultant** *adj* delighted, ecstatic, elated, joyful, jubilant, *inf* on top of the world, overjoyed, rejoicing. ▷ EXUBERANT.

**eye** *n* **1** eyeball, *inf* peeper. **2** discernment, perception, sight, vision. • *vb* contemplate, examine inspect, look at, observe, regard, scrutinize, study, watch. ▷ SEE.

**eye-witness** *n* bystander, looker-on, observer, onlooker, passer-by, spectator, watcher, witness.

# F

**fabric** *n* **1** material, stuff, textile. ▷ CLOTH. **2** *fabric of a building.* constitution, construction, framework, make-up, structure, substance.

**fabulous** *adj* **1** fabled, fairy-tale, fanciful, fictitious, imaginary, legendary, mythical, story-book. **2** ▷ EXCELLENT.

**face** *n* **1** appearance, countenance features, lineaments, look, *sl* mug *old use* physiognomy, visage.

▷ EXPRESSION. **2** *face of building*. aspect, covering, exterior, façade, facet, front, outside, side, surface. • *vb* **1** be opposite, front, look towards, overlook. **2** *face danger*. appear before, brave, come to terms with, confront, cope with, defy, encounter, experience, face up to, meet, oppose, square up to, stand up to, tackle. **3** *face a wall with plaster*. clad, coat, cover, dress, finish, overlay, sheathe, veneer.

**facetious** *adj* cheeky, flippant, impudent, irreverent. ▷ FUNNY.

**facile** *adj* **1** cheap, easy, effortless, hasty, obvious, quick, simple, superficial, unconsidered. **2** *facile talker*. fluent, glib, insincere, plausible, ready, shallow, slick, *inf* smooth.

**facility** *n* **1** adroitness, alacrity, ease, expertise, fluency, *derog* glibness, skill, smoothness. **2** *a useful facility*. amenity, convenience, help, provision, resource, service.

**fact** *n* actuality, certainty, *Fr* fait accompli, reality, truth. *Opp* FICTION. **the facts** circumstances, data, details, evidence, information, *sl* the lowdown, particulars, statistics.

**factor** *n* aspect, cause, circumstance, component, consideration, constituent, contingency, detail, determinant, element, fact, influence, ingredient, item, parameter, part, particular.

**factory** *n* assembly line, forge, foundry, manufacturing plant, mill, plant, refinery, shop-floor, works, workshop.

**factual** *adj* **1** accurate, *Lat* bona fide, circumstantial, correct, demonstrable, empirical, faithful, genuine, matter-of-fact, objective, plain, prosaic, provable, realistic, straightforward, true, unadorned, unbiased, undistorted, unemotional, unimaginative, unvarnished, valid, verifiable, well-documented. *Opp* FALSE. **2** *a factual film*. biographical, documentary, historical, real-life. *Opp* FICTIONAL.

**faculty** *n* ability, aptitude, capability, capacity, flair, genius, gift, knack, power, talent.

**fade** *vb* **1** blanch, bleach, darken, dim, discolour, dull, etiolate, grow pale, whiten. *Opp* BRIGHTEN. **2** become less, decline, decrease, diminish, disappear, dwindle, evanesce, fail, melt away, vanish, wane, weaken. **3** *flowers fade*. droop, flag, perish, shrivel, wilt, wither.

**fail** *vb* **1** abort, be a failure, be unsuccessful, break down, close down, come to an end, *inf* come to grief, come to nothing, *sl* conk out, *inf* crash, cut out, fall through, *inf* fizzle out, *inf* flop, *inf* fold, fold up, founder, give up, go bankrupt, *inf* go bust, go out of business, meet with disaster, miscarry, misfire, *inf* miss out, peter out, stop working. **2** *fail in health*. decay, decline, deteriorate, diminish, disappear, dwindle, ebb, fade, get worse, give out, melt away, vanish, wane, weaken. **3** *fail to do something*. forget, neglect, omit. **4** *fail someone*. abandon, disappoint, *inf* let down. *Opp* IMPROVE, SUCCEED.

**failing** *n* blemish, defect, fault, flaw, foible, imperfection, shortcoming, weakness, weak spot.

**failure** *n* **1** abandonment, defeat, disappointment, disaster, downfall, fiasco, *inf* flop, loss, miscarriage, *inf* wash-out, wreck. **2** breakdown, collapse, crash, stoppage. **3** *failure to do your duty*.

dereliction, neglect, omission, remissness. *Opp* SUCCESS.

**faint** *adj* **1** blurred, blurry, dim, faded, feeble, hazy, ill-defined, indistinct, misty, muzzy, pale, pastel (*colours*), shadowy, unclear, vague. **2** *faint smell.* delicate, slight. **3** *faint sounds.* distant, hushed, low, muffled, muted, soft, stifled, subdued, thin, weak. **4** *faint in the head.* dizzy, exhausted, feeble, giddy, light-headed, unsteady, vertigious, weak, *inf* woozy. *Opp* CLEAR, STRONG. • *vb* become unconscious, black out, collapse, *inf* flake out, *inf* keel over, pass out, swoon.

**fair** *adj* **1** blond, blonde, flaxen, golden, light, yellow. **2** *fair weather.* bright, clear, clement, cloudless, dry, favourable, fine, pleasant, sunny. *Opp* DARK. **3** *a fair decision.* disinterested, even-handed, fair-minded, honest, honourable, impartial, just, lawful, legitimate, nonpartisan, open-minded, proper, right, unbiased, unprejudiced, upright. *Opp* UNJUST. **4** *a fair standard.* acceptable, adequate, average, indifferent, mediocre, middling, moderate, ordinary, passable, reasonable, respectable, satisfactory, *inf* so-so, tolerable. *Opp* UNACCEPTABLE. **5** ▷ BEAUTIFUL. • *n* **1** amusement-park, fair-ground, fun-fair. **2** bazaar, carnival, exhibition, festival, fête, gala, market, sale, show.

**fairly** *adv* moderately, pretty, quite, rather, reasonably, somewhat, tolerably, up to a point.

**faith** *n* **1** assurance, belief, certitude, confidence, credence, reliance, sureness, trust. *Opp* DOUBT. **2** conviction, creed, devotion, doctrine, dogma, persuasion, religion.

**faithful** *adj* **1** constant, dependable, devoted, dutiful, honest, loyal, reliable, staunch, steadfast, trusted, trusty, trustworthy, unswerving. **2** *a faithful account.* accurate, close, consistent, exact, factual, literal, precise. ▷ TRUE. *Opp* FALSE.

**fake** *adj* artificial, bogus, concocted, counterfeit, ersatz, factitious, false, fictitious, forged, fraudulent, imitation, invented, made-up, mock, *sl* phoney, pretended, sham, simulated, spurious, synthetic, trumped-up, unfounded, unreal. *Opp* GENUINE. • *n* **1** copy, counterfeit, duplicate, forgery, hoax, imitation, replica, reproduction, sham, simulation. **2** charlatan, cheat, fraud, hoaxer, humbug, impostor, mountebank, *sl* phoney, quack. • *vb* affect, copy, counterfeit, dissemble, falsify, feign, forge, fudge, imitate, make believe, mock up, pretend, put on, reproduce, sham, simulate.

**fall** *n* **1** collapse, crash, decline, decrease, depreciation, descent, dip, dive, downswing, downturn, drop, lowering, nosedive, plunge, reduction, slant, slump, tumble. **2** *fall of a fortress.* capitulation, capture, defeat, overthrow, seizure, submission, surrender. • *vb* **1** collapse, *inf* come a cropper, crash down, dive, drop down, founder, go down, keel over, overbalance, pitch, plummet, plunge, sink, slump, spiral, stumble, topple, trip over, tumble. **2** become less, become lower, decline, decrease, diminish, dwindle, ebb, lessen, subside. **3** descend, drop, fall away, slope down. **4** *curtains fell in folds.* be suspended, cascade, dangle, dip down, hang. **5** *silence fell.* come,come about, happen, occur, settle. **6** ▷ DIE.

7 ▷ SURRENDER. **fall apart** ▷ DISINTEGRATE. **fall back** ▷ RETREAT. **fall behind** ▷ LAG. **fall down, fall in** ▷ COLLAPSE. **fall off** ▷ DECLINE. **fall out** ▷ QUARREL. **fall through** ▷ FAIL.

**fallacy** *n* delusion, error, flaw, miscalculation, misconception, mistake, solecism.

**fallible** *adj* erring, frail, human, imperfect, liable to make mistakes, uncertain, unpredictable, unreliable, weak. *Opp* INFALLIBLE.

**fallow** *adj* dormant, resting, uncultivated, unplanted, unsown, unused.

**false** *adj* **1** deceptive, distorted, erroneous, fabricated, fallacious, faulty, fictitious, flawed, imprecise, inaccurate, incorrect, inexact, invalid, misleading, mistaken, spurious, unfactual, unsound, untrue, wrong. ▷ FAKE. **2** *false friends*. deceitful, dishonest, disloyal, double-dealing, double-faced, faithless, lying, treacherous, unfaithful, unreliable, untrustworthy. *Opp* TRUE. **false name** ▷ PSEUDONYM.

**falsehood** *n* fabrication, *inf* fib, fiction, lie, prevarication, *inf* story, untruth, *sl* whopper.

**falsify** *vb* alter, *inf* cook (*the books*), counterfeit, distort, exaggerate, fake, forge, *inf* fudge, imitate, misrepresent, mock up, oversimplify, pervert, simulate, slant, tamper with, tell lies about, twist.

**falter** *vb* **1** become weaker, flag, flinch, hesitate, hold back, lose confidence, pause, quail, stagger, stumble, totter, vacillate, waver. *Opp* PERSIST. **2** stammer, stutter. **faltering** ▷ HESITANT.

**fame** *n* acclaim, celebrity, distinction, eminence, glory, honour, illustriousness, importance, *inf* kudos, name, *derog* notoriety, pre-eminence, prestige, prominence, public esteem, renown, reputation, repute, *inf* stardom.

**familiar** *adj* **1** accustomed, common, conventional, current, customary, everyday, frequent, habitual, mundane, normal, ordinary, predictable, regular, routine, stock, traditional, usual, well-known. *Opp* STRANGE. **2** *familiar language*. *inf* chatty, close, confidential, *derog* forward, *inf* free-and-easy, *derog* impudent, informal, intimate, near, *derog* presumptuous, relaxed, sociable, unceremonious. ▷ FRIENDLY. *Opp* FORMAL. **familiar with** acquainted with, *inf* at home with, aware of, conscious of, expert in, informed about, knowledgeable about, trained in, versed in.

**family** *n* **1** brood, children, *inf* flesh and blood, generation, issue, kindred, *old use* kith and kin, litter, *inf* nearest and dearest, offspring, progeny, relations, relatives, *inf* tribe. **2** ancestry, blood, clan, dynasty, extraction, forebears, genealogy, house, line, lineage, pedigree, race, strain, tribe. □ *ancestor, descendant.* □ *aunt, brother, child, cousin, daughter, father, fiancé(e), forefather, foster-child, foster-parent, godchild, godparent, grandchild, grandparent, guardian, husband,* Amer *junior, kinsman, kinswoman, mother, nephew, next-of-kin, niece, parent, sibling, sister, son, step-child, step-parent, uncle, ward, widow, widower, wife.*

**famine** *n* dearth, hunger, lack, malnutrition, scarcity, shortage, starvation, want. *Opp* PLENTY.

**famished** *adj* craving, famishing, hungry, *inf* peckish, ravenous, starved, starving.

**famous** *adj* acclaimed, big, celebrated, distinguished, eminent, exalted, famed, glorious, great, historic, honoured, illustrious, important, legendary, lionized, notable, noted, *derog* notorious, outstanding, popular, prominent, proverbial, renowned, revered, time-honoured, venerable, well-known, world-famous. *Opp* UNKNOWN.

**fan** *n* **1** blower, extractor, propeller, ventilator. **2** *a soccer fan.* addict, admirer, aficionado, *inf* buff, devotee, enthusiast, fanatic, *inf* fiend, follower, *inf* freak, lover, supporter. ▷ FANATIC.

**fanatic** *n* activist, adherent, bigot, extremist, fiend, freak, maniac, militant, zealot.

**fanatical** *adj* bigoted, excessive, extreme, fervid, fervent, immoderate, irrational, maniacal, militant, obsessive, overenthusiastic, passionate, rabid, single-minded, zealous. *Opp* MODERATE.

**fanciful** *adj* capricious, chimerical, fancy, fantastic, illusory, imaginary, imagined, make-believe, unrealistic, whimsical.

**fancy** *adj* decorative, elaborate, embellished, embroidered, intricate, ornamented, ornate. ▷ FANCIFUL. • *n* ▷ IMAGINATION, WHIM. • *vb* **1** conjure up, dream of, envisage, imagine, picture, visualize. ▷ THINK. **2** be attracted to, crave, *inf* have a yen for, like, long for, prefer, want, wish for. ▷ DESIRE.

**fantastic** *adj* **1** absurd, amazing, elaborate, exaggerated, extraordinary, extravagant, fabulous, fanciful, far-fetched, grotesque, imaginative, implausible, incredible, odd, quaint, remarkable, rococo, strange, surreal, unbelievable, unlikely, unrealistic, weird. **2** ▷ EXCELLENT. *Opp* ORDINARY.

**fantasy** *n* chimera, day-dream, delusion, dream, fancy, hallucination, illusion, imagination, invention, make-believe, mirage, pipe-dream, reverie, vision. *Opp* REALITY.

**far** *adj* distant, far-away, far-off, outlying, remote. *Opp* NEAR.

**farcical** *adj* absurd, foolish, ludicrous, preposterous, ridiculous, silly. ▷ FUNNY.

**fare** *n* **1** charge, cost, fee, payment, price, ticket. **2** *festive fare.* ▷ FOOD.

**farewell** *adj* goodbye, last, leaving, parting, valedictory. • *n* departure, leave-taking, *inf* send-off, valediction. ▷ GOODBYE.

**farm** *n* farmhouse, farmstead, *old use* grange. □ *arable farm, croft, dairy farm, fish farm, fruit farm, livestock farm, organic farm, plantation, poultry farm, ranch, smallholding.*

**farming** *n* agriculture, agronomy, crofting, cultivation, food-production, husbandry.

**fascinate** *vb* allure, attract, beguile, bewitch, captivate, charm, delight, enchant, engross, enthral, entice, entrance, hypnotize, interest, mesmerize, rivet, spellbind. **fascinating** ▷ ATTRACTIVE.

**fashion** *n* **1** convention, manner, method, mode, way. **2** craze, cut, *inf* fad, line, look, pattern, rage, style, taste, trend, vogue.

**fashionable** *adj Fr* [à] la mode, chic, contemporary, current, elegant, *inf* in, in vogue, the latest, modern, modish, popular, smart, *inf* snazzy, sophisticated, stylish, tasteful, *inf* trendy, up-to-date, *inf* with it. *Opp* UNFASHIONABLE.

**fast** *adv* at full tilt, briskly, in no time, post-haste, quickly, rapidly, swiftly. • *adj* 1 breakneck, brisk, expeditious, express, hasty, headlong, high-speed, hurried, lively, *inf* nippy, precipitate, quick, rapid, smart, *inf* spanking, speedy, supersonic, swift, unhestitating. *Opp* SLOW. **2** *fast on the rocks.* attached, bound, fastened, firm, fixed, immobile, immovable, secure, tight. **3** *fast colours.* indelible, lasting, permanent, stable. **4** *fast living.* ▷ IMMORAL. • *vb* abstain, deny yourself, diet, go hungry, go without food, starve. *Opp* INDULGE.

**fasten** *vb* affix, anchor, attach, batten, bind, bolt, buckle, button, chain, clamp, clasp, cling, close, connect, couple, do up, fix, grip, hitch, hook, knot, join, lace, lash, latch on, link, lock, make fast, moor, nail, padlock, paste, peg, pin, rivet, rope, screw down, seal, secure, solder, staple, strap, tack, tape, tether, tie, unite, weld. ▷ STICK. *Opp* UNDO.

**fastener** *n* bond, connection, connector, coupling, fastening, link, linkage. □ *anchor, bolt, buckle, button, catch, chain, clamp, clasp, clip, dowel, dowel-pin, drawing-pin, glue, gum, hasp, hook, knot, lace, latch, lock, mooring, nail, padlock, painter, paste, peg, pin, rivet, rope, safety-pin, screw, seal, Sellotape, solder, staple, strap, string, tack, tape, tether, tie, toggle, Velcro, wedge, zip.*

**fastidious** *adj* choosy, dainty, delicate, discriminating, finical, finicky, fussy, hard to please, nice, particular, *inf* pernickety, *inf* picky, selective, squeamish.

**fat** *adj* 1 bloated, *inf* broad in the beam, bulky, chubby, corpulent, dumpy, flabby, fleshy, gross, heavy, massive, obese, overweight, paunchy, plump, podgy, portly, pot-bellied, pudgy, rotund, round, solid, squat, stocky, stout, thick, tubby, weighty, well-fed. ▷ BIG. **2** *fat meat.* fatty, greasy, oily. *Opp* LEAN. • *n* □ *adipose tissue, blubber, butter, dripping, grease, lard, margarine, oil, suet.*

**fatal** *adj* 1 deadly, final, incurable, lethal, malignant, mortal, terminal. **2** ▷ DISASTROUS.

**fatality** *n* casualty, death, loss.

**fate** *n* 1 chance, destiny, doom, fortune, karma, kismet, lot, luck, nemesis, *inf* powers above, predestination, providence, the stars. **2** death, demise, destruction, disaster, downfall, end, ruin.

**fated** *adj* certain, cursed, damned, decreed, destined, doomed, foreordained, inescapable, inevitable, intended, predestined, predetermined, preordained, sure.

**father** *n* begetter, *inf* dad, *inf* daddy, *inf* pa, *inf* papa, parent, *old use* pater, *inf* pop, sire.

**fatigue** *n* debility, exhaustion, feebleness, languor, lassitude, lethargy, tiredness, weakness, weariness. • *vb* debilitate, drain, enervate, exhaust, tire, weaken, weary.

**fatigued** ▷ WEARY.

**fault** *n* 1 blemish, defect, deficiency, demerit, failing, failure, fallacy, flaw, foible, frailty, imperfection, inaccuracy, malfunction, snag, weakness. **2** blunder, *inf* boob, error, failing, *Fr* faux pas, gaffe, *inf* howler, indiscretion, lapse, miscalculation, misconduct, misdeed, mistake, negligence, offence, omission, oversight, peccadillo, shortcoming, sin, slip, transgression, *old use* trespass, vice, wrongdoing. **3** *It was my fault.* accountability, blame, culpability, guilt, liability, responsibility. • *vb* ▷ CRITICIZE.

**faultless** *adj* accurate, correct, exemplary, flawless, ideal, in mint condition, irreproachable, sinless, unimpeachable. ▷ PERFECT. *Opp* FAULTY.

**faulty** *adj* broken, damaged, defective, deficient, flawed, illogical, imperfect, inaccurate, incomplete, incorrect, inoperative, invalid, not working, out of order, shop-soiled, unusable, useless. *Opp* FAULTLESS.

**favour** *n* **1** acceptance, approbation, approval, bias, favouritism, friendliness, goodwill, grace, liking, partiality, preference, support. **2** *Do me a favour.* benefit, courtesy, gift, good deed, good turn, indulgence, kindness, service. • *vb* **1** approve of, be in sympathy with, champion, choose, commend, esteem, *inf* fancy, *inf* go for, like, opt for, prefer, show favour to, think well of, value. *Opp* DISLIKE. **2** abet, advance, back, be advantageous to, befriend, forward, promote, support. ▷ HELP. *Opp* HINDER.

**favourable** *adj* **1** advantageous, appropriate, auspicious, beneficial, benign, convenient, following (*wind*), friendly, generous, helpful, kind, opportune, positive, promising, propitious, reassuring, suitable, supportive, sympathetic, understanding, well-disposed. **2** *a favourable review.* approving, commendatory, complimentary, congratulatory, encouraging, enthusiastic, laudatory. **3** *a favourable reputation.* agreeable, desirable, enviable, good, pleasing, satisfactory. *Opp* UNFAVOURABLE.

**favourite** *adj* beloved, best, choice, chosen, dearest, esteemed, ideal, liked, loved, popular, preferred, selected, well-liked. • *n* **1** choice, pick, preference. **2** *inf* apple of your eye, darling, idol, pet.

**fear** *n* alarm, anxiety, apprehension, apprehensiveness, awe, concern, consternation, cowardice, cravenness, diffidence, dismay, doubt, dread, faint-heartedness, foreboding, fright, *inf* funk, horror, misgiving, nervousness, panic, phobia, qualm, suspicion, terror, timidity, trepidation, uneasiness, worry. ▷ PHOBIA. *Opp* COURAGE. • *vb* be afraid of, dread, quail at, shrink from, suspect, tremble at, worry about.

**fearful** *adj* **1** alarmed, apprehensive, frightened, nervous, panic-stricken, scared, terrified, timid. ▷ AFRAID. *Opp* FEARLESS. **2** ▷ FEARSOME.

**fearless** *adj* bold, brave, dauntless, intrepid, resolute, stoical, unafraid, unconcerned, undaunted, valiant, valorous. ▷ COURAGEOUS. *Opp* FEARFUL.

**fearsome** *adj* appalling, awe-inspiring, awesome, daunting, dreadful, fearful, frightful, intimidating, terrible, terrifying. ▷ FRIGHTENING.

**feasible** *adj* **1** achievable, attainable, easy, possible, practicable, practical, realizable, viable, workable. *Opp* IMPRACTICAL. **2** *a feasible scenario.* credible, likely, plausible, reasonable. *Opp* IMPLAUSIBLE.

**feast** *n* banquet, *sl* blow-out, dinner, *inf* spread. ▷ MEAL. • *vb* dine, gorge, gormandize, *inf* wine and dine. ▷ EAT.

**feat** *n* accomplishment, achievement, act, action, attainment, deed, exploit, performance.

**feather** *n* plume, quill. **feathers** down, plumage.

**feathery** *adj* downy, fluffy, light, wispy.

**feature** *n* **1** aspect, attribute, characteristic, circumstance, detail, facet, hall mark, idiosyncrasy, mark, peculiarity, point, property, quality, trait. **2** *newspaper feature.* article, column, item, piece, report, story. • *vb* **1** emphasize, focus on, give prominence to, highlight, *inf* play up, present, promote, show up, *inf* spotlight, *inf* star, stress. **2** *feature in a film.* act, appear, figure, participate, perform, play a role, star, take a part. **features** ▷ FACE.

**fee** *n* bill, charge, cost, dues, emolument, fare, payment, price, remuneration, subscription, sum, tariff, terms, toll, wage.

**feeble** *adj* **1** ailing, debilitated, decrepit, delicate, enfeebled, exhausted, faint, fragile, frail, helpless, ill, impotent, inadequate, ineffective, infirm, languid, listless, poorly, powerless, puny, sickly, slight, useless, weak. *Opp* STRONG. **2** effete, feckless, hesitant, incompetent, indecisive, ineffectual, irresolute, *inf* namby-pamby, spineless, vacillating, weedy, wimpish, *inf* wishy-washy. **3** *feeble excuses.* flimsy, insubstantial, lame, paltry, poor, tame, thin, unconvincing.

**feed** *vb* **1** cater for, give food to, nourish, nurture, provender, provide for, provision, strengthen, suckle, support, sustain, *inf* wine and dine. **2** dine, eat, fare, graze, pasture. **feed on** ▷ EAT.

**feel** *vb* **1** caress, finger, fondle, handle, hold, manipulate, maul, *inf* paw, pet, stroke, touch. **2** *feel your way.* explore, fumble, grope. **3** *feel the cold.* be aware of, be conscious of, detect, discern, experience, know, notice, perceive, sense, suffer, undergo. **4** *It feels cold.* appear, give a feeling of, seem. **5** *feel something's true.* believe, consider, deem, guess, *inf* have a feeling, *inf* have a hunch, intuit, judge, think.

**feeling** *n* **1** sensation, sense of touch, sensitivity. **2** ardour, emotion, fervour, passion, sentiment, warmth. **3** *religious feelings.* attitude, belief, consciousness, guess, hunch, idea, impression, instinct, intuition, notion, opinion, perception, thought, view. **4** *a feeling for music.* fondness, responsiveness, sensibility, sympathy, understanding. **5** [*inf*] *a party feeling.* aura, atmosphere, mood, tone, *inf* vibrations.

**fell** *vb* bring down, chop down, cut down, flatten, *inf* floor, knock down, mow down, prostrate. ▷ KILL.

**female** *adj* ▷ FEMININE. *Opp* MALE. • *n* □ *aunt,* old use *damsel, daughter,* old use *débutante, fiancé, girl, girlfriend, grandmother, lady,* inf *lass, lesbian,* old use *maid,* old use *maiden,* old use *mistress, mother, niece, sister, spinster,* old use or sexist *wench, wife, woman.* □ *bitch, cow, doe, ewe, hen, lioness, mare, nanny-goat, sow, tigress, vixen.*

**feminine** *adj derog of men* effeminate, female, *derog* girlish, ladylike, womanly. *Opp* MASCULINE.

**fen** *n* bog, lowland, marsh, morass, quagmire, slough, swamp.

**fence** *n* barricade, barrier, fencing, hedge, hurdle, obstacle, paling, palisade, railing, rampart, stockade, wall, wire. • *vb* **1** bound, circumscribe, confine, coop up, encircle, enclose, hedge in, immure, pen, restrict, surround, wall in. **2** ▷ FIGHT.

**fend** *vb* **fend for yourself** care for yourself, do for yourself, *inf* get along, *inf* get by, look after yourself, *inf* scrape along, support yourself, survive. **fend off** ▷ REPEL.

**ferment** *n* ▷ COMMOTION. • *vb* **1** boil, bubble, effervesce, *inf* fizz, foam, froth, rise, seethe, work. **2** agitate, excite, foment, incite, instigate, provoke, rouse, stir up.

**ferocious** *adj* bestial, bloodthirsty, brutal, cruel, feral, fiendish, fierce, harsh, inhuman, merciless, murderous, pitiless, sadistic, savage, vicious, wild. *Opp* GENTLE.

**ferry** *n* ▷ VESSEL. • *vb* carry, export, fetch, import, shift, ship, shuttle, take across, taxi, transport. ▷ CONVEY.

**fertile** *adj* abundant, fecund, fertilized, flourishing, fruitful, lush, luxuriant, productive, prolific, rich, teeming, well-manured. *Opp* STERILE.

**fertilize** *vb* **1** impregnate, inseminate, pollinate. **2** cultivate, dress, enrich, feed, make fertile, manure, mulch, nourish, top-dress.

**fertilizer** *n* compost, dressing, dung, manure, mulch, nutrient.

**fervent** *adj* animated, ardent, avid, burning, committed, devout, eager, earnest, emotional, enthusiastic, excited, fanatical, fervid, fiery, frenzied, heated, impassioned, intense, keen, passionate, rapturous, spirited, vehement, vigorous, warm, wholehearted, zealous. *Opp* COOL.

**fervour** *n* ardour, eagerness, energy, enthusiasm, excitement, fervency, fire, heat, intensity, keenness, passion, sparkle, spirit, vehemence, vigour, warmth, zeal.

**fester** *vb* become infected, become inflamed, become poisoned, decay, discharge, gather, go bad, go septic, mortify, ooze, putrefy, rot, run, suppurate, ulcerate.

**festival** *n* anniversary, carnival, celebration, commemoration, fair, feast, fête, fiesta, gala, holiday, jamboree, jubilee. ▷ FESTIVITY.

**festive** *adj* celebratory, cheerful, cheery, convivial, gay, gleeful, jolly, jovial, joyful, joyous, light-hearted, merry, uproarious. ▷ HAPPY.

**festivity** *n* celebration, conviviality, entertainment, feasting, festive occasion, *inf* jollification, jollity, jubilation, merrymaking, merriment, mirth, rejoicing, revelry, revels. ▷ PARTY.

**fetch** *vb* **1** bear, bring, call for, carry, collect, convey, get, import, obtain, pick up, retrieve, transfer, transport. **2** *fetch a good price.* be bought for, bring in, earn, go for, make, produce, raise, realize, sell for. **fetching** ▷ ATTRACTIVE.

**feud** *n* animosity, antagonism, *inf* bad blood, conflict, dispute, enmity, grudge, hostility, rivalry, strife, vendetta. ▷ QUARREL.

**fever** *n* delirium, feverishness, high temperature.

**feverish** *adj* **1** burning, febrile, fevered, flushed, hot, inflamed, trembling. *Opp* COOL. **2** *feverish activity.* agitated, excited, frantic, frenetic, frenzied, hectic, hurried, impatient, passionate, restless.

**few** *adj inf* few and far between, hardly any, inadequate, infrequent, rare, scarce, sparse, sporadic, *inf* thin on the ground, uncommon. *Opp* MANY.

**fibre** *n* **1** filament, hair, strand, thread. **2** *moral fibre.* backbone, character, determination, spirit, tenacity, toughness. ▷ COURAGE.

**fickle** *adj* capricious, changeable, changing, disloyal, erratic, faithless, flighty, inconsistent, inconstant, mercurial, mutable, treacherous, undependable, unfaithful, unpredictable, unreliable, unstable, unsteady, *inf* up and down, vacillating, variable, volatile. *Opp* CONSTANT.

**fiction** *n* concoction, deception, fabrication, fantasy, figment of the imagination, flight of fancy, invention, lies, story-telling, *inf* tall story. ▷ WRITING. *Opp* FACT.

**fictional** *adj* fabulous, fanciful, imaginary, invented, legendary, made-up, make-believe, mythical, story-book. *Opp* FACTUAL.

**fictitious** *adj* apocryphal, assumed, fabricated, deceitful, fraudulent, imagined, invented, made-up, spurious, unreal, untrue. ▷ FALSE. *Opp* GENUINE.

**fiddle** *vb* interfere, meddle, play about, tamper. ▷ FIDGET. **fiddling** ▷ TRIVIAL.

**fidget** *vb* be restless, *inf* fiddle about, fret, frisk about, fuss, jerk about, *inf* jiggle, *inf* mess about, move restlessly, *inf* play about, shuffle, squirm, twitch, worry, wriggle about.

**fidgety** *adj* agitated, frisky, impatient, jittery, jumpy, nervous, on edge, restive, restless, *inf* twitchy, uneasy. *Opp* CALM.

**field** *n* **1** arable land, clearing, enclosure, grassland, *poet* glebe, green, *old use* mead, meadow, paddock, pasture. **2** *a games field.* arena, ground, pitch, playing-field, recreation ground, stadium. **3** *field of activity.* area, *inf* department, domain, province, sphere, subject, territory.

**fiend** *n* **1** demon, devil, evil spirit, goblin, hobgoblin, imp, Satan, spirit. **2** ▷ FANATIC.

**fierce** *adj* **1** angry, barbaric, barbarous, bloodthirsty, bloody, brutal, cold-blooded, cruel, dangerous, fearsome, ferocious, fiendish, fiery, homicidal, inhuman, merciless, murderous, pitiless, ruthless, sadistic, savage, untamed, vicious, violent, wild. **2** *fierce opposition.* active, aggressive, competitive, eager, heated, furious, intense, keen, passionate, relentless, strong, unrelenting. *Opp* GENTLE.

**fiery** *adj* **1** ablaze, afire, aflame, aglow, blazing, burning, fierce, flaming, glowing, heated, hot, incandescent, raging, red, red-hot. **2** *a fiery temper.* angry, ardent, choleric, excitable, fervent, furious, hot-headed, intense, irascible, irritable, livid, mad, passionate, touchy, violent. *Opp* COOL.

**fight** *n* action, affray, attack, battle, bout, brawl, *inf* brush, *inf* bust-up, clash, combat, competition, conflict, confrontation, contest, counter-attack, dispute, dogfight, duel, *inf* dustup, encounter, engagement, feud, *old use* fisticuffs, fracas, fray, *inf* free-for-all, hostilities, joust, match, mêlée, *inf* punch-up, raid, riot, rivalry, row, scramble, scrap, scrimmage, scuffle, *inf* set-to, skirmish, squabble, strife, struggle, tussle, war, wrangle. ▷ QUARREL. • *vb* **1** attack, battle, box, brawl, *inf* brush, clash, compete, conflict, contend, do battle, duel, engage, exchange blows, fence, feud, grapple, have a fight, joust, quarrel, row, scrap, scuffle, skirmish, spar, squabble, stand up (to), strive, struggle, *old use* tilt, tussle, wage war, wrestle. **2** *fight a decision.* campaign against, contest, defy, oppose, protest against, resist, take a stand against.

**fighter** *n* aggressor, antagonist, attacker, belligerent, campaigner, combatant, contender, contestant, defender. □ *archer, boxer,* inf *brawler, champion, duellist, freedom fighter, gladiator, guerrilla, gunman, knight, marine, marksman, mercenary, partisan, prize-fighter, pugilist, sniper, swordsman, terrorist, warrior, wrestler.* ▷ SOLDIER.

**figure** *n* **1** amount, cipher, digit, integer, number, numeral, sum, symbol, value. **2** diagram, drawing, graph, illustration, outline, picture, plate, representation. **3** *plump figure.* body, build, form, outline, physique, shape, silhouette. **4** *bronze figure.* ▷ SCULPTURE. **5** *well-known figure.* ▷ PERSON. • *vb* ▷ FEATURE. **figure out** ▷ CALCULATE, UNDERSTAND. **figures** ▷ STATISTICS.

**file** *n* **1** binder, box-file, case, cover, documentation, document-case, dossier, folder, portfolio, ring-binder. **2** *single file.* column, line, procession, queue, rank, row, stream, string, train. • *vb* **1** arrange, categorize, classify, enter, pigeon-hole, organize, put away, record, register, store, systematize. **2** *file through a door.* march, parade, proceed in a line, stream, troop.

**fill** *vb* **1** be full of, block, *inf* bung up, caulk, clog, close up, cram, crowd, flood, inflate, jam, load, obstruct, pack, plug, refill, replenish, seal, stock up, stop up, *inf* stuff, *inf* top up. *Opp* EMPTY. **2** *fill a need.* answer, fulfil, furnish, meet, provide, satisfy, supply. **3** *fill a post.* execute, hold, occupy, take over, take up. **fill out** ▷ SWELL.

**filling** *n* contents, *inf* innards, insides, padding, stuffing, wadding.

**film** *n* **1** coat, coating, cover, covering, haze, layer, membrane, mist, overlay, screen, sheet, skin, slick, tissue, veil. **2** cartoon, *old use* flick, motion picture, movie, picture, video, videotape.

**filter** *n* colander, gauze, membrane, mesh, riddle, screen, sieve, strainer. • *vb* clarify, filtrate, percolate, purify, refine, screen, sieve, sift, strain.

**filth** *n* decay, dirt, effluent, garbage, grime, *inf* gunge, impurity, muck, mud, ordure, pollution, putrescence, refuse, rubbish, scum, sewage, slime, sludge, trash. ▷ EXCRETA.

**filthy** *adj* **1** begrimed, caked, defiled, dirty, disgusting, dusty, foul, grimy, grubby, impure, messy, mucky, muddy, nasty, polluted, scummy, slimy, smelly, soiled, sooty, sordid, squalid, stinking, tainted, uncleaned, unkempt, unwashed, vile. **2** ▷ OBSCENE. *Opp* CLEAN.

**final** *adj* clinching, closing, concluding, conclusive, decisive, dying, end, eventual, finishing, last, settled, terminal, terminating, ultimate. *Opp* INITIAL.

**finalize** *vb* clinch, complete, conclude, settle, *inf* sew up, *inf* wrap up.

**finance** *n* accounting, banking, business, commerce, economics, investment, stocks and shares. • *vb* back, fund, guarantee, invest in, pay for, provide money for, subsidize, support, underwrite. **finances** assets, bank account, budget, capital, cash, funds, holdings, income, money, resources, wealth, *inf* the wherewithal.

**financial** *adj* economic, fiscal, monetary, pecuniary.

**find** *vb* **1** acquire, arrive at, become aware of, *inf* bump into, chance upon, come across, come upon, detect, diagnose, discover, dig out, dig up, encounter, espy, expose, *inf* ferret out, happen on, hit on, identify, learn, light on, locate, meet, note, notice, observe, *inf* put your finger on, reach, recognize, reveal, spot, stumble on, uncover, unearth. **2** *find lost property.* get back, recover, rediscover, regain, repossess, retrieve, trace, track down. **3** *found me a job.* give, pass on, procure, provide, supply. *Opp* LOSE.

**finding** *n* conclusion, decision, decree, judgement, pronouncement, verdict.

**fine** *adj* **1** admirable, beautiful, choice, classic, commendable, excellent, first-class, handsome, noble, select, superior, worthy. ▷ GOOD. **2** *fine workmanship.* consummate, craftsmanlike, meticulous, skilful, skilled. **3** *fine sand.* minute, powdery, soft. **4** *fine fabric.* dainty, delicate, exquisite, flimsy, fragile, silky. **5** *a fine point.* acute, keen, narrow, sharp, slender, slim, thin. **6** *a fine distinction.* fine-drawn, discriminating, hair-splitting, nice, precise, subtle. **7** *fine weather.* bright, clear, cloudless, dry, fair, nice, pleasant, sunny. • *n* charge, forfeit, penalty.

**finish** *n* **1** cessation, close, completion, conclusion, culmination, end, ending, finale, resolution, result, termination. **2** *finish on furniture.* appearance, completeness, gloss, lustre, patina, perfection, polish, shine, smoothness, surface, texture. • *vb* **1** accomplish, achieve, break off, bring to an end, cease, clinch, complete, conclude, discontinue, end, finalize, fulfil, halt, pack up, perfect, phase out, reach the end, round off, say goodbye, sign off, stop, take your leave, terminate, *inf* wind up, *inf* wrap up. **2** consume, drink up, eat up, empty, exhaust, expend, get through, *inf* polish off, *inf* say goodbye to, use up. **finish off** ▷ KILL.

**finite** *adj* bounded, calculable, controlled, countable, definable, defined, determinate, fixed, known, limited, measurable, numbered, rationed, restricted. *Opp* INFINITE.

**fire** *n* **1** blaze, burning, combustion, conflagration, flames, holocaust, inferno, pyre. **2** fireplace, grate, hearth. □ *boiler, bonfire, brazier, convector, electric fire, forge, furnace, gas fire, immersion-heater, incinerator, kiln, oven, radiator, stove.* **3** *fire in your veins.* ▷ PASSION. • *vb* **1** bake, burn, heat, ignite, kindle, light, put a light to, set alight, set fire to, spark off. **2** animate, awaken, enkindle, enliven, excite, incite, inflame, inspire, motivate, rouse, stimulate, stir. **3** *fire a gun or missile.* catapult, detonate, discharge, explode, launch, let off, propel, set off, shoot, trigger off. **4** *fire a worker.* dismiss, make redundant, sack, throw out. **fire at** ▷ BOMBARD. **hang fire** ▷ DELAY.

**fireproof** *adj* flameproof, incombustible, non-flammable. *Opp* INFLAMMABLE.

**fire-raiser** *n* arsonist, pyromaniac.

**firm** *adj* **1** compact, compressed, congealed, dense, hard, inelastic, inflexible, rigid, set, solid, stable, stiff, unyielding. **2** *firm on the* *rocks*. anchored, embedded, fast, fastened, fixed, immovable, secure, steady, tight. **3** *firm convictions*. adamant, decided, determined, dogged, obstinate, persistent, resolute, unshakeable, unwavering. **4** *a firm price*. agreed, settled, unchangeable. **5** *firm friends*. constant, dependable, devoted, faithful, loyal, reliable. • *n* business, company, concern, corporation, establishment, organization, partnership.

**first** *adj* **1** cardinal, chief, dominant, foremost, head, highest, key, leading, main, outstanding, paramount, predominant, primary, prime, prinicpal, top, uppermost. **2** *first steps*. basic, elementary, fundamental, initial, introductory, preliminary, rudimentary. **3** *first inhabitants*. aboriginal, archetypal, earliest, eldest, embryonic, oldest, original, primeval. **first-class, first-rate** ▷ EXCELLENT.

**fish** *n* □ *brill, brisling, carp, catfish, chub, cod, coelacanth, conger, cuttlefish, dab, dace, eel, flounder, goldfish, grayling, gudgeon, haddock, hake, halibut, herring, jellyfish, lamprey, ling, mackerel, minnow, mullet, perch, pike, pilchard, piranha, plaice, roach, salmon, sardine, sawfish, shark, skate, sole, sprat, squid, starfish, stickleback, sturgeon, swordfish,* inf *tiddler, trout, tuna, turbot, whitebait, whiting*. • *vb* angle, go fishing, trawl.

**fisher** *n* angler, fisherman, trawlerman.

**fit** *adj* **1** adapted, adequate, applicable, apposite, appropriate, apropos, apt, becoming, befitting, correct, decent, equipped, fitting, good enough, proper, right, satisfactory, seemly, sound, suitable, suited, timely. **2** able, capable, competent, in good form, on form, prepared, ready, strong, well enough. ▷ HEALTHY. *Opp* UNFIT. • *n* attack, bout, convulsion, eruption, explosion, outbreak, outburst, paroxysm, seizure, spasm, spell. • *vb* **1** accord with, become, be fitting for, conform with, correspond to, correspond with, go with, harmonize with, suit. **2** *fit things into place*. arrange, assemble, build, construct, dovetail, install, interlock, join, match, position, put in place, put together. **fit out, fit up** ▷ EQUIP.

**fix** *n inf* catch-22, corner, difficulty, dilemma, *inf* hole, *inf* jam, mess, *inf* pickle, plight, predicament, problem, quandary. • *vb* **1** attach, bind, connect, embed, implant, install, join, link, make firm, plant, position, secure, stabilize, stick. ▷ FASTEN. **2** *fix a price*. agree, appoint, arrange, arrive at, conclude, confirm, decide, define, establish, finalize, name, ordain, set, settle, sort out, specify. **3** *fix a broken window*. correct, make good, mend, put right, rectify, remedy, repair.

**fixture** *n* date, engagement, event, game, match, meeting.

**fizz** *vb* bubble, effervesce, fizzle, foam, froth, hiss, sizzle, sparkle, sputter.

**fizzy** *adj* bubbly, effervescent, foaming, sparkling.

**flag** *n* banner, bunting, colours, ensign, jack, pennant, pennon, standard, streamer. • *vb* **1** ▷ SIGNAL. **2** *enthusiasm flagged*. ▷ DECLINE.

**flake** *n* bit, chip, leaf, scale, scurf, shaving, slice, sliver, splinter, wafer.

**flame** *n* blaze, light, tongue. ▷ FIRE. • *vb* ▷ FLARE.

**flap** *vb* beat, flutter, oscillate, slap, sway, swing, thrash about, thresh about, wag, waggle, wave about.

**flare** *vb* 1 blaze, brighten, burst out, erupt, flame, shine. ▷ BURN. 2 ▷ WIDEN.

**flash** *vb* coruscate, dazzle, flicker, glare, glint, glitter, light up, reflect, scintillate, shine, spark, sparkle, twinkle. ▷ BURN.

**flat** *adj* 1 calm, even, horizontal, level, smooth, unbroken, unruffled. 2 outstretched, prone, prostrate, recumbent, spread-eagled, spread out, supine. 3 *a flat voice.* bland, boring, dead, dry, dull, featureless, insipid, lacklustre, lifeless, monotonous, spiritless, stale, tedious, tired, unexciting, uninteresting, unmodulated, unvarying. 4 *a flat tyre.* blown out, burst, deflated, punctured. • *n* apartment, bedsitter, flatlet, maisonette, penthouse, rooms, suite.

**flatten** *vb* 1 compress, even out, iron out, level out, press, roll, smooth, straighten. 2 crush, demolish, devastate, level, raze, run over, squash, trample. ▷ DESTROY. 3 *flatten an opponent.* fell, floor, knock down, prostrate. ▷ DEFEAT.

**flatter** *vb* be flattering to, *inf* butter up, compliment, court, curry favour with, fawn on, humour, *inf* play up to, praise, *sl* suck up to, *inf* toady to. *Opp* INSULT. **flattering** ▷ COMPLIMENTARY, OBSEQUIOUS.

**flatterer** *n inf* crawler, *inf* creep, groveller, lackey, sycophant, time-server, *inf* toady, *inf* yes-man.

**flattery** *n* adulation, blandishments, *inf* blarney, *inf* boot-licking, cajolery, fawning, *inf* flannel, insincerity, obsequiousness, servility, *inf* soft soap, sycophancy, unctuousness.

**flavour** *n* 1 savour, taste. ▷ FLAVOURING. 2 air, ambience, atmosphere, aura, character, characteristic, feel, feeling, property, quality, spirit, stamp, style. • *vb* add flavour to, add taste to, season, spice.

**flavouring** *n* additive, essence, extract, seasoning.

**flaw** *n* break, defect, error, fallacy, fault, imperfection, inaccuracy, loophole, mistake, shortcoming, slip, split, weakness. ▷ BLEMISH. **flawed** ▷ IMPERFECT.

**flawless** *adj* accurate, clean, faultless, immaculate, mint, pristine, sound, spotless, undamaged, unmarked. ▷ PERFECT. *Opp* IMPERFECT.

**flee** *vb* abscond, *inf* beat a retreat, *sl* beat it, bolt, clear off, cut and run, decamp, disappear, escape, fly, get away, hurry off, *inf* make a run for it, make off, retreat, run away, *sl* scarper, take flight, *inf* take to your heels, vanish, withdraw.

**fleet** *n* armada, convoy, flotilla, navy, squadron, task force.

**fleeting** *adj* brief, ephemeral, evanescent, fugitive, impermanent, momentary, mutable, passing, short, short-lived, temporary, transient, transitory. *Opp* PERMANENT.

**flesh** *n* carrion, fat, meat, muscle, tissue.

**flex** *n* cable, cord, extension, lead, wire. • *vb* ▷ BEND.

**flexible** *adj* 1 bendable, *inf* bendy, elastic, flexile, floppy, giving, limp, lithe, plastic, pliable, pliant, rubbery, soft, springy, stretchy, supple, whippy, willowy, yielding. 2 adjustable, alterable, fluid, mutable, open, provisional, variable. 3 *a flexible person.* accommodating, adaptable, amenable, compliant, conformable, co-operative, docile, easygoing, malleable, open-minded, responsive, tractable, willing. *Opp* RIGID.

**flicker** *vb* blink, flap, flutter, glimmer, gutter, quiver, shake, shimmer, sparkle, tremble, twinkle, vibrate, waver.

**flight** *n* **1** journey, trajectory. **2** ▷ ESCAPE.

**flimsy** *adj* **1** breakable, brittle, delicate, fine, fragile, frail, insubstantial, light, loose, slight, thin, weak. **2** *a flimsy building.* decrepit, dilapidated, gimcrack, jerry-built, makeshift, rickety, shaky, tottering, wobbly. **3** *a flimsy argument.* feeble, implausible, inadequate, superficial, trivial, unbelievable, unconvincing, unsatisfactory. *Opp* STRONG.

**flinch** *vb* blench, cower, cringe, dodge, draw back, duck, falter, jerk away, jump, quail, quake, recoil, shrink back, shy away, start, swerve, wince. **flinch from** ▷ EVADE.

**fling** *vb* bowl, *inf* bung, cast, *inf* chuck, heave, hurl, launch, lob, pelt, pitch, propel, send, *inf* shy, *inf* sling, throw, toss.

**flippant** *adj* cheeky, facetious, facile, *inf* flip, frivolous, light-hearted, shallow, superficial, thoughtless, unserious. *Opp* SERIOUS.

**flirt** *n female* coquette, *male* philanderer, *inf* tease. • *vb sl* chat someone up, lead someone on, make love, philander, toy with someone's affections.

**flirtatious** *adj* amorous, coquettish, flirty, *derog* philandering, playful, *derog* promiscuous, teasing.

**float** *vb* **1** be poised, be suspended, bob, drift, glide, hang, hover, sail, swim, waft. **2** *float a ship.* launch. *Opp* SINK.

**flock** *n* assembly, congregation, crowd, drove, gathering, herd, horde, multitude, swarm. ▷ GROUP. • *vb* ▷ GATHER.

**flog** *vb* beat, birch, cane, chastise, flagellate, flay, lash, scourge, thrash, whip. ▷ HIT.

**flood** *n* **1** cataract, deluge, downpour, flash-flood, inundation, overflow, rush, spate, stream, tidal wave, tide, torrent. **2** abundance, excess, glut, plethora, quantity, superfluity, surfeit, surge. • *vb* cover, deluge, drown, engulf, fill up, immerse, inundate, overflow, overwhelm, saturate, sink, submerge, swamp.

**floor** *n* **1** floorboards, flooring. **2** deck, level, storey, tier.

**flop** *vb* **1** collapse, dangle, droop, drop, fall, flag, flap about, hang down, sag, slump, topple, tumble, wilt. **2** ▷ FAIL.

**floppy** *adj* dangling, droopy, flabby, hanging, loose, limp, pliable, soft. ▷ FLEXIBLE. *Opp* RIGID.

**flounder** *vb* **1** blunder, flail, fumble, grope, move clumsily, plunge about, stagger, struggle, stumble, tumble, wallow. **2** falter, get confused, make mistakes, talk aimlessly.

**flourish** *n* ▷ GESTURE. • *vb* **1** be fruitful, be successful, bloom, blossom, boom, burgeon, develop, do well, flower, grow, increase, *inf* perk up, progress, prosper, strengthen, succeed, thrive. **2** *flourish an umbrella.* brandish, flaunt, gesture with, shake, swing, twirl, wag, wave, wield.

**flow** *n* cascade, course, current, drift, ebb, effusion, flood, gush, outpouring, spate, spurt, stream, tide, trickle. • *vb* bleed, cascade, course, dribble, drift, drip, ebb, flood, flush, glide, gush, issue, leak, move in a stream, ooze, overflow, pour, purl, ripple, roll, run, seep, spill, spring, spurt, squirt, stream, swirl, trickle, well, well up.

**flower** *n* **1** bloom, blossom, bud, floret, petal. □ *begonia, bluebell, buttercup, campanula, campion, candytuft, carnation, catkin, celandine, chrysanthemum, coltsfoot, columbine, cornflower, cowslip, crocus, crowfoot, cyclamen, daffodil, dahlia, daisy, dandelion, forget-me-not, foxglove, freesia, geranium, gladiolus, gypsophila, harebell, hollyhock, hyacinth, iris, jonquil, kingcup, lilac, lily, lupin, marguerite, marigold, montbretia, nasturtium, orchid, pansy, pelargonium, peony, periwinkle, petunia, phlox, pink, polyanthus, poppy, primrose, rhododendron, rose, saxifrage, scabious, scarlet pimpernel, snowdrop, speedwell, sunflower, tulip, violet, wallflower, water-lily.* • *vb* bloom, blossom, *poet* blow, bud, burgeon, come out, have flowers, open out, unfold. ▷ FLOURISH. **bunch of flowers** arrangement, bouquet, corsage, garland, posy, spray, wreath.

**fluctuate** *vb* alternate, be unsteady, change, go up and down, oscillate, seesaw, shift, swing, vacillate, vary, waver.

**fluent** *adj* articulate, effortless, eloquent, expressive, *derog* facile, felicitous, flowing, *derog* glib, natural, polished, ready, smooth, voluble, unhesitating. *Opp* HESITANT.

**fluff** *n* down, dust, feathers, floss, fuzz, thistledown.

**fluffy** *adj* downy, feathery, fibrous, fleecy, furry, fuzzy, hairy, light, silky, soft, velvety, wispy, woolly.

**fluid** *adj* **1** aqueous, flowing, gaseous, liquefied, liquid, melted, molten, running, *inf* runny, sloppy, watery. *Opp* SOLID. **2** *a fluid situation.* adjustable, alterable, changing, flexible, mutable, open, variable, undefined. • *n* gas, liquid, liquor, plasma, vapour.

**fluke** *n* accident, chance, serendipity, stroke of good luck, twist of fate.

**flush** *vb* **1** blush, colour, glow, go red, redden. **2** *flush a lavatory.* clean out, cleanse, flood, *inf* pull the plug, rinse out, wash out. **3** *flush from a hiding-place.* chase out, drive out, expel, send up.

**fluster** *vb* agitate, bewilder, bother, distract, flurry, perplex, put off, put out, *inf* rattle, *inf* throw, upset. ▷ CONFUSE.

**flutter** *vb* bat (*eyelid*), flap, flicker, flit, fluctuate, move agitatedly, oscillate, palpitate, quiver, shake, tremble, twitch, vacillate, vibrate, wave.

**fly** *vb* **1** ascend, flit, glide, hover, rise, sail, soar, swoop, take flight, take wing. **2** *fly a plane.* aviate, pilot, take off in. **3** *fly a flag.* display, flap, flutter, hang up, hoist, raise, show, wave. **4** *fly from danger.* flee, hurry, move quickly, run. ▷ ESCAPE. **fly at** ▷ ATTACK. **fly in the face of** ▷ DISREGARD.

**flying** *n* aeronautics, air-travel, aviation, flight, *inf* jetting.

**foam** *n* **1** bubbles, effervescence, froth, head (*on beer*), lather, scum, spume, suds. **2** sponge. • *vb* boil, bubble, effervesce, fizz, froth, lather, make foam.

**focus** *n* **1** clarity, correct adjustment, sharpness. **2** centre, core, focal point, heart, hub, pivot, target. • *vb* aim, centre, concentrate, direct attention, fix attention, home in, spotlight.

**fog** *n* bad visibility, cloud, haze, miasma, mist, smog, vapour.

**foggy** *adj* blurred, blurry, clouded, cloudy, dim, hazy, indistinct, misty, murky, obscure. *Opp* CLEAR.

**foil** *vb* baffle, block, check, circumvent, frustrate, halt, hamper, hinder, obstruct, outwit, prevent, stop, thwart. ▷ DEFEAT.

**foist** *vb inf* fob off, get rid of, impose, offload, palm off.

**fold** *n* **1** bend, corrugation, crease, crinkle, furrow, gather, hollow, knife-edge, line, pleat, pucker, wrinkle. **2** *fold for sheep.* ▷ ENCLOSURE. • *vb* **1** bend, crease, crimp, crinkle, double over, jack-knife, overlap, pleat, ply, pucker, tuck in, turn over. **2** close, collapse, let down, put down, shut. **3** *fold in your arms.* clasp, clip, embrace, enclose, enfold, entwine, envelop, hold, hug, wrap. **4** *business folded.* ▷ FAIL.

**folk** *n* clan, nation, people, the population, the public, race, society, tribe.

**follow** *vb* **1** accompany, chase, come after, dog, escort, go after, hound, hunt, keep pace with, pursue, replace, shadow, stalk, succeed, supersede, supplant, *inf* tag along with, tail, take the place of, track, trail. **2** *follow a path.* keep to, trace. **3** *follow rules.* abide by, adhere to, attend to, comply with, conform to, heed, honour, obey, observe, pay attention to, stick to, submit to, take notice of. **4** *follow my example.* adopt, be guided by, conform to, copy, imitate, mimic, mirror. **5** *follow an argument.* appreciate, comprehend, grasp, keep up with, take in, understand. **6** *follow football.* admire, be a fan of, keep abreast of, know about, take an interest in, support. **7** *It doesn't follow.* be inevitable, be logical, come about, ensue, happen, have the consequence, mean, result. **following** ▷ SUBSEQUENT.

**folly** *n* foolishness, insanity, lunacy, madness. ▷ STUPIDITY.

**foment** *vb* arouse, incite, instigate, kindle, provoke, rouse, stir up. ▷ STIMULATE.

**fond** *adj* **1** adoring, affectionate, caring, loving, tender, warm. **2** *a fond hope.* ▷ FOOLISH. **be fond of** ▷ LOVE.

**fondle** *vb* caress, cuddle, handle, pat, pet, snuggle, squeeze, touch.

**food** *n* aliment, *old use* bread, *old use* comestibles, cooking, cuisine, delicacies, diet, *inf* eatables, *inf* eats, fare, feed, fodder, foodstuff, forage, *inf* grub, *inf* junk food, *old use* meat, *sl* nosh, nourishment, nutriments, provender, provisions, rations, recipe, refreshments, sustenance, swill, *old use* tuck, *old use* viands, *old use* victuals. ▷ MEAL.

**fool** *n* **1** [*most synonyms inf*] ass, blockhead, booby, buffoon, dimwit, dope, dunce, dunderhead, dupe, fat-head, half-wit, ignoramus, mug, muggins, mutt, ninny, nit, nitwit, simpleton, sucker, twerp, wally. ▷ IDIOT. **2** clown, comedian, comic, coxcomb, entertainer, jester. • *vb inf* bamboozle, bluff, cheat, *inf* con, cozen, deceive, defraud, delude, dupe, fleece, gull, *inf* have on, hoax, hoodwink, *inf* kid, mislead, *inf* string along, swindle, take in, tease, trick. **fool about** ▷ MISBEHAVE.

**foolish** *adj* absurd, asinine, brainless, childish, crazy, daft, *inf* dopey, *inf* dotty, fatuous, feather-brained, feeble-minded, *old* *use* fond, frivolous, *inf* half-baked, hare-brained, idiotic, illogical, immature, inane, infantile, irrational, *inf* jokey, laughable, light-hearted, ludicrous, mad, meaningless, mindless, misguided, naïve, nonsensical, playful, pointless, preposterous, ridiculous, scatter-brained, *inf* scatty, senseless, shallow, silly, simple, simple-minded, simplistic, *inf* soppy, stupid, thoughtless, unintelligent, unreasonable, unsound, unwise, witless. *Opp* WISE.

**foot** *n* **1** claw, hoof, paw, trotter. **2** ▷ BASE.

**footprint** *n* footmark, spoor, track.

**forbid** *vb* ban, bar, debar, deny, deter, disallow, exclude, interdict, make illegal, outlaw, preclude, prevent, prohibit, proscribe, refuse, rule out, say no to, stop, veto. *Opp* ALLOW.

**forbidden** *adj* **1** against the law, taboo, unlawful, wrong. **2** *a forbidden area.* closed, out of bounds, restricted, secret.

**forbidding** *adj* gloomy, grim, menacing, ominous, stern, threatening, uninviting, unwelcoming. ▷ UNFRIENDLY. *Opp* FRIENDLY.

**force** *n* **1** aggression, *inf* arm-twisting, coercion, compulsion, constraint, drive, duress, effort, might, power, pressure, strength, vehemence, vigour, violence. **2** effect, energy, impact, intensity, momentum, shock. **3** *a military force.* army, body, group, troops. **4** *force of an argument.* cogency, effectiveness, persuasiveness, rightness, thrust, validity, weight. • *vb* **1** *inf* bulldoze, coerce, compel, constrain, drive, impel, impose on, make, oblige, order, press-gang, pressurize. **2** *force a door.* break open, burst open, prise open, smash, use force on, wrench. **3** *force something on someone.* impose, inflict.

**foreboding** *n* anxiety, apprehension, dread, fear, feeling, foreshadowing, forewarning, intimation, intuition, misgiving, omen, portent, premonition, presentiment, suspicion, warning, worry.

**forecast** *n* augury, expectation, outlook, prediction, prognosis, prognostication, projection, prophecy. • *vb* ▷ FORESEE.

**forefront** *n* avant-garde, front, lead, vanguard.

**foreign** *adj* **1** distant, exotic, far-away, outlandish, remote, strange, unfamiliar, unknown. **2** alien, external, immigrant, imported, incoming, international, outside, overseas, visiting. **3** *foreign ideas.* extraneous, odd, uncharacteristic, unnatural, untypical, unusual, unwanted. *Opp* NATIVE.

**foreigner** *n* alien, immigrant, newcomer, outsider, overseas visitor, stranger. *Opp* NATIVE.

**foremost** *adj* first, leading, main, primary, supreme. ▷ CHIEF.

**forerunner** *n* advance messenger, harbinger, herald, precursor, predecessor. ▷ ANCESTOR.

**foresee** *vb* anticipate, envisage, expect, forecast, picture. ▷ FORETELL.

**foresight** *n* anticipation, caution, far-sightedness, forethought, looking ahead, perspicacity, planning, preparation, prudence, readiness, vision.

**forest** *n* coppice, copse, jungle, plantation, trees, woodland, woods.

**foretaste** *n* advance warning, augury, example, foreknowledge, forewarning, indication, omen, premonition, preview, sample, specimen, *inf* tip-off, trailer, *inf* try-out.

**foretell** *vb* augur, *old use* bode, forebode, foreshadow, forewarn, give a foretaste of, herald, portend, predict, presage, prognosticate, prophesy, signify. ▷ FORESEE.

**forethought** *n* anticipation, caution, far-sightedness, foresight, looking ahead, perspicacity, planning, preparation, prudence, readiness, vision.

**forewarning** *n* advance warning, augury, omen, premonition, *inf* tip-off. ▷ FORETASTE.

**forfeit** *n* confiscation, damages, fee, fine, penalty, sequestration. • *vb* abandon, give up, let go, lose, pay up, relinquish, renounce, surrender.

**forge** *n* furnace, smithy, workshop. • *vb* **1** beat into shape, cast, construct, hammer out, manufacture, mould, shape, work. **2** coin, copy, counterfeit, fake, falsify, imitate, make illegally, reproduce. **forge ahead** ▷ ADVANCE.

**forgery** *n* copy, counterfeit, *inf* dud, fake, fraud, imitation, *inf* phoney, replica, reproduction.

**forget** *vb* **1** be forgetful, dismiss from your mind, disregard, fail to remember, ignore, leave out, lose track (of), miss out, neglect, omit, overlook, skip, suffer from amnesia, unlearn. **2** be without, leave behind, lose. *Opp* REMEMBER.

**forgetful** *adj* absent-minded, amnesiac, careless, distracted, inattentive, neglectful, negligent, oblivious, preoccupied, unconscious, unmindful, unreliable, vague, *inf* woolly-minded.

**forgivable** *adj* allowable, excusable, justifiable, negligible, pardonable, petty, understandable, venial. *Opp* UNFORGIVABLE.

**forgive** *vb* **1** absolve, acquit, clear, exculpate, excuse, exonerate, indulge, *inf* let off, pardon, spare. **2** *forgive a crime.* condone, ignore, make allowances for, overlook, pass over.

**forgiveness** *n* absolution, amnesty, clemency, compassion, exculpation, exoneration, grace, indulgence, leniency, mercy, pardon, reprieve, tolerance. *Opp* RETRIBUTION.

**forgiving** *adj* clement, compassionate, forbearing, generous, magnanimous, merciful, tolerant, understanding. ▷ KIND. *Opp* VENGEFUL.

**forgo** *vb* abandon, abstain from, do without, forswear, give up, go without, omit, pass up, relinquish, renounce, sacrifice, turn down, waive.

**forked** *adj* branched, cleft, divergent, divided, fork-like, pronged, split, V-shaped.

**forlorn** *adj* abandoned, alone, bereft, deserted, forsaken, friendless, lonely, outcast, solitary, unloved. ▷ SAD.

**form** *n* **1** appearance, arrangement, cast, character, configuration, design, format, framework, genre, guise, kind, manifestation, manner, model, mould, nature, pattern, plan, semblance, sort, species, structure, style, system, type, variety. **2** *human form.* anatomy, body, build, figure, frame, outline, physique, shape, silhouette. **3** *your form in school.* class, grade, group, level, set, stream, tutor-group. **4** *good form.* behaviour, convention, custom, etiquette, fashion, manners, practice. **5** *an application form.* document, paper. **6** *in good form.* condition, *inf* fettle, fitness, health, performance, spirits. 7 ▷ SEAT. • *vb* 1 bring into existence, cast, constitute, construct, create, design, establish, forge, found, give form to, make, model, mould, organize, produce, shape. 2 appear, arise, come into existence, develop, grow, materialize, take shape. 3 *form a team.* act as, compose, comprise, make up, serve as. 4 *form a habit.* acquire, cultivate, develop, get.

**formal** *adj* 1 aloof, ceremonial, ceremonious, conventional, cool, correct, customary, dignified, *inf* dressed-up, orthodox, *inf* posh, *derog* pretentious, proper, punctilious, ritualistic, solemn, sophisticated, stately, *inf* starchy, stiff, stiff-necked, unbending, unfriendly. 2 *formal language.* academic, impersonal, official, precise, reserved, specialist, stilted, technical, unemotional. 3 *a formal agreement.* binding, contractual, enforceable, legal, *inf* signed and sealed. 4 *a formal design.* calculated, geometrical, orderly, organized, regular, rigid, symmetrical. *Opp* INFORMAL.

**format** *n* appearance, design, layout, plan, shape, size, style.

**former** *adj* bygone, departed, ex-, last, late, old, one-time, past, previous, prior, recent. **the former** earlier, first, first-mentioned. *Opp* LATTER.

**formidable** *adj* awe-inspiring, awesome, challenging, daunting, difficult, dreadful, fearful, frightening, intimidating, large-scale, *inf* mind-boggling, onerous, overwhelming, prodigious, taxing. *Opp* EASY.

**formula** *n* 1 form of words, ritual, rubric, spell, wording. 2 *formula for success.* blueprint, method, prescription, procedure, recipe, rule, technique, way.

**formulate** *vb* 1 articulate, codify, define, express clearly, set out in detail, specify, systematize. 2 concoct, create, devise, evolve, form, invent, map out, originate, plan, work out.

**forsake** *vb* abandon, break off from, desert, forgo, forswear, give up, jettison, jilt, leave, quit, renounce, repudiate, surrender, throw over, *inf* turn your back on, vacate.

**fort** *n* camp, castle, citadel, fortification, fortress, garrison, stronghold, tower.

**forthright** *adj* blunt, candid, decisive, direct, outspoken, plain-speaking, straightforward, unequivocal, unhesitating, uninhibited. ▷ FRANK. *Opp* EVASIVE.

**fortify** *vb* 1 buttress, defend, garrison, protect, reinforce, secure against attack, shore up. 2 bolster, boost, brace, buoy up, cheer, embolden, encourage, hearten, invigorate, lift the morale of, reassure, stiffen the resolve of, strengthen, support, sustain. *Opp* WEAKEN.

**fortitude** *n* backbone, bravery, courage, determination, endurance, firmness, heroism, patience, resolution, stoicism, tenacity, valour, will-power. ▷ COURAGE. *Opp* COWARDICE.

**fortunate** *adj* auspicious, blessed, favourable, lucky, opportune, propitious, prosperous, providential, timely. ▷ HAPPY.

**fortune** *n* 1 accident, chance, destiny, fate, fortuity, karma, kismet, luck, providence. 2 affluence, assets, estate, holdings, inheritance, means, *inf* millions, money, opulence, *inf* pile, possessions, property, prosperity, riches, treasure, wealth.

**fortune-teller** *n* clairvoyant, crystal-gazer, futurologist, oracle, palmist, prophet, seer, soothsayer, star-gazer, sybil.

**forward** *adj* **1** advancing, front, frontal, head-first, leading, onward, progressive. **2** *forward planning*. advance, early, forward-looking, future, well-advanced. **3** *a forward child*. advanced, assertive, bold, brazen, cheeky, confident, familiar, *inf* fresh, impertinent, impudent, insolent, over-confident, precocious, presumptuous, pushful, *inf* pushy, shameless, uninhibited. *Opp* BACKWARD. • *vb* **1** dispatch, expedite, freight, post on, re-address, send, send on, ship, transmit, transport. **2** *forward your career*. accelerate, advance, encourage, facilitate, foster, further, hasten, help along, *inf* lend a helping hand to, promote, speed up, support. ▷ HELP. *Opp* HINDER.

**foster** *vb* **1** advance, cultivate, encourage, further, nurture, promote, stimulate. ▷ HELP. **2** *foster a child*. adopt, bring up, care for, look after, maintain, nourish, nurse, raise, rear, take care of.

**foul** *adj* **1** bad, contaminated, disagreeable, disgusting, fetid, filthy, hateful, impure, infected, loathsome, nasty, nauseating, nauseous, noisome, obnoxious, offensive, polluted, putrid, repellent, repugnant, repulsive, revolting, rotten, sickening, smelly, squalid, stinking, vile. ▷ DIRTY, SMELLING. **2** *foul crimes*. abhorrent, abominable, atrocious, beastly, cruel, evil, ingnominious, monstrous, scandalous, shameful, vicious, villainous, violent, wicked. **3** *foul language*. abusive, bawdy, blasphemous, coarse, common, crude, impolite, improper, indecent, insulting, licentious, offensive, rude, uncouth, vulgar. ▷ OBSCENE. **4** *foul weather*. foggy, rainy, rough, stormy, violent, windy. ▷ UNPLEASANT. **5** *foul play*. against the rules, dishonest, forbidden, illegal, invalid, prohibited, unfair, unsportsmanlike. *Opp* CLEAN, FAIR. • *n* infringement, violation. • *vb* ▷ DIRTY. **foul up** ▷ MUDDLE.

**found** *vb* **1** begin, bring about, create, endow, establish, fund, *inf* get going, inaugurate, initiate, institute, organize, originate, provide money for, raise, set up, start. **2** base, build, construct, erect, ground, rest, set.

**foundation** *n* **1** beginning, endowment, establishment, founding, inauguration, initiation, institution, organizing, setting up, starting. **2** base, basement, basis, bottom, cornerstone, foot, footing, substructure, underpinning. **3** *foundations of science*. basic principle, element, essential, fundamental, origin, *plur* rudiments.

**founder** *vb* abort, be wrecked, *inf* come to grief, fail, fall through, go down, miscarry, sink.

**fountain** *n* font, fount, fountainhead, jet, source, spout, spray, spring, well, well-spring.

**foyer** *n* ante-room, entrance, entrance hall, hall, lobby, reception.

**fraction** *n* division, part, portion, section, subdivision.

**fracture** *n* break, breakage, chip, cleavage, cleft, crack, fissure, gap, opening, rent, rift, rupture, split. • *vb* breach, break, cause a fracture in, chip, cleave, crack, rupture, separate, split, suffer a fracture in.

**fragile** *adj* ▷ FRAIL.

**fragment** *n* atom, bit, chip, crumb, *plur* debris, morsel, part, particle, piece, portion, remnant, scrap, shard, shiver, shred, sliver, *plur* smithereens, snippet, speck. • *vb* ▷ BREAK.

**fragmentary** *adj inf* bitty, broken, disconnected, disintegrated, disjointed, fragmented, imperfect, in bits, incoherent, incomplete, in fragments, partial, scattered, scrappy, sketchy, uncoordinated. *Opp* COMPLETE.

**fragrance** *n* aroma, bouquet, nose (*of wine*), odour, perfume, redolence, scent, smell.

**fragrant** *adj* aromatic, odorous, perfumed, redolent, scented, sweet-smelling.

**frail** *adj* breakable, brittle, dainty, delicate, easily damaged, feeble, flimsy, fragile, insubstantial, light, *derog* puny, rickety, slight, thin, unsound, unsteady, vulnerable, weak, *derog* weedy. ▷ ILL. *Opp* STRONG.

**frame** *n* **1** bodywork, chassis, construction, scaffolding, structure. ▷ FRAMEWORK. **2** *photo frame.* border, case, casing, edge, edging, mount, mounting. • *vb* **1** box in, enclose, mount, set off, surround. **2** ▷ COMPOSE. **frame of mind** ▷ ATTITUDE.

**framework** *n* bare bones, frame, outline, plan, shell, skeleton, support, trellis.

**frank** *adj* blunt, candid, direct, downright, explicit, forthright, genuine, *inf* heart-to-heart, honest, ingenuous, *inf* no-nonsense, open, outright, outspoken, plain, plain-spoken, revealing, serious, sincere, straightforward, straight from the heart, to the point, trustworthy, truthful, unconcealed, undisguised, unreserved. *Opp* INSINCERE.

**frantic** *adj* agitated, anxious, berserk, *inf* beside yourself, crazy, delirious, demented, deranged, desperate, distraught, excitable, feverish, *inf* fraught, frenetic, frenzied, furious, hectic, hurried, hysterical, mad, overwrought, panicky, rabid, uncontrollable, violent, wild, worked up. *Opp* CALM.

**fraud** *n* **1** cheating, chicanery, *inf* con-trick, counterfeit, deceit, deception, dishonesty, double-dealing, duplicity, fake, forgery, hoax, imposture, pretence, *inf* put-up job, ruse, sham, *inf* sharp practice, swindle, trick, trickery. **2** charlatan, cheat, *inf* con-man, hoaxer, humbug, impostor, mountebank, *sl* phoney, *inf* quack, rogue, scoundrel, swindler.

**fraudulent** *adj inf* bent, bogus, cheating, corrupt, counterfeit, criminal, *inf* crooked, deceitful, devious, *inf* dirty, dishonest, double-dealing, duplicitous, fake, false, forged, illegal, lying, *sl* phoney, sham, specious, swindling, underhand, unscrupulous. *Opp* HONEST.

**fray** *n* brawl, commotion, conflict, disturbance, fracas, mêlée, quarrel, rumpus. ▷ FIGHT.

**frayed** *adj* chafed, rough-edged, tattered, threadbare, unravelled, worn. ▷ RAGGED.

**freak** *adj* aberrant, abnormal, anomalous, atypical, bizarre, exceptional, extraordinary, freakish, odd, peculiar, queer, rare, unaccountable, unforeseeable, unpredictable, unusual, weird. *Opp* NORMAL. • *n* **1** aberration, abnormality, abortion, anomaly, curiosity, deformity, irregularity, monster, monstrosity, mutant, oddity, *inf* one-off, quirk, rarity, sport, variant. **2** ▷ FANATIC.

**free** *adj* **1** able, allowed, at leisure, at liberty, idle, independent, loose, not working, uncommitted, unconfined, unconstrained, unencumbered, unfixed, unrestrained, untrammelled. **2** *free from slavery.* emancipated, freeborn, let go, liberated, released, unchained, unfettered, unshackled. **3** *a free country.* autonomous, democratic, independent, self-governing, sovereign. **4** *free access.* accessible, clear, open, permitted, unhindered, unimpeded, unrestricted. **5** *free gifts.* complimentary, gratis, *sl* on the house, unasked-for, unsolicited, without charge. **6** *free space.* available, empty, uninhabited, unoccupied, vacant. **7** *free with money.* bounteous, casual, charitable, generous, lavish, liberal, munificent, ready, unstinting, willing. • *vb* **1** absolve, acquit, clear, deliver, discharge, disenthral, emancipate, enfranchise, exculpate, exonerate, let go, let off, let out, liberate, loose, make free, manumit, pardon, parole, ransom, release, reprieve, rescue, save, set free, spare, turn loose, unchain, unfetter, unleash, unlock, unloose. *Opp* CONFINE. **2** *free tangled ropes.* clear, disengage, disentangle, extricate, loose, unbind, undo, unknot, untie. *Opp* TANGLE. **free and easy** ▷ INFORMAL.

**freedom** *n* **1** autonomy, independence, liberty, self-determination, self-government, sovereignty. *Opp* CAPTIVITY. **2** deliverance, emancipation, exemption, immunity, liberation, release. **3** *freedom to choose.* ability, *Fr* carte blanche, discretion, free hand, latitude, leeway, leisure, licence, opportunity, permission, power, privilege, right, scope.

**freeze** *vb* **1** become ice, become solid, congeal, harden, ice over, ice up, solidify, stiffen. **2** chill, cool, make cold, numb. **3** *freeze food.* chill, deep-freeze, dry-freeze, ice, refrigerate. **4** *freeze the frame.* fix, hold, immobilize, keep still, paralyse, peg, petrify, stand still, stick, stop. **freezing** ▷ COLD.

**freight** *n* cargo, consignment, goods, haul, load, merchandise, payload, shipment.

**frenzy** *n* agitation, delirium, derangement, excitement, fever, fit, fury, hysteria, insanity, lunacy, madness, mania, outburst, paroxysm, passion, turmoil.

**frequent** *adj* common, constant, continual, countless, customary, everyday, familiar, habitual, incessant, innumerable, many, normal, numerous, ordinary, persistent, recurrent, recurring, regular, reiterative, repeated, usual. *Opp* INFREQUENT. • *vb* ▷ HAUNT.

**fresh** *adj* **1** additional, alternative, different, extra, just arrived, new, recent, supplementary, unfamiliar, up-to-date. **2** alert, energetic, healthy, invigorated, lively, *inf* perky, rested, revived, sprightly, spry, tingling, vigorous, vital. **3** *a fresh recruit.* callow, *inf* green, inexperienced, naïve, raw, unsophisticated, untried, *inf* wet behind the ears. **4** *fresh water.* clear, drinkable, potable, pure, refreshing, sweet, uncontaminated. **5** *fresh air.* airy, circulating, cool, unpolluted, ventilated. **6** *a fresh wind.* bracing, breezy, invigorating, moderate, sharp, stiff, strongish. **7** *fresh food.* healthy, natural, newly gathered, unprocessed, untreated, wholesome. **8** *fresh sheets.* clean, crisp, laundered, untouched, unused, washed-and-ironed. **9** *fresh colours.* bright, clean, glowing, just painted, renewed, restored, sparkling, unfaded, vivid. *Opp* OLD, STALE.

**fret** *vb* **1** agonize, be anxious, brood, lose sleep, worry. **2** ▷ ANNOY.

**fretful** *adj* anxious, distressed, disturbed, edgy, irritable, irritated, jittery, peevish, petulant, restless, testy, touchy, worried. ▷ BAD-TEMPERED. *Opp* CALM.

**friction** *n* **1** abrading, abrasion, attrition, chafing, fretting, grating, resistance, rubbing, scraping. **2** ▷ CONFLICT.

**friend** *n* acquaintance, associate, *inf* buddy, *inf* chum, companion, comrade, confidant(e), *inf* crony, intimate, *inf* mate, *inf* pal, partner, pen-friend, playfellow, playmate, supporter, well-wisher. ▷ ALLY, LOVER. *Opp* ENEMY. **be friends** ▷ ASSOCIATE. **make friends with** ▷ BEFRIEND.

**friendless** *adj* abandoned, alienated, alone, deserted, estranged, forlorn, forsaken, isolated, lonely, ostracized, shunned, shut out, solitary, unattached, unloved.

**friendliness** *n* benevolence, camaraderie, conviviality, devotion, esteem, familiarity, goodwill, helpfulness, hospitality, kindness, neighbourliness, regard, sociability, warmth. *Opp* HOSTILITY.

**friendly** *adj* accessible, affable, affectionate, agreeable, amiable, amicable, approachable, attached, benevolent, benign, *inf* chummy, civil, close, clubbable, companionable, compatible, comradely, conciliatory, congenial, convivial, cordial, demonstrative, expansive, favourable, genial, good-natured, gracious, helpful, hospitable, intimate, kind, kind-hearted, kindly, likeable, *inf* matey, neighbourly, outgoing, *inf* pally, sympathetic, tender, *inf* thick, warm, welcoming, well-disposed. ▷ FAMILIAR, LOVING, SOCIABLE. *Opp* UNFRIENDLY.

**friendship** *n* affection, alliance, amity, association, attachment, closeness, comradeship, fellowship, fondness, harmony, intimacy, rapport, relationship. ▷ FRIENDLINESS, LOVE. *Opp* HOSTILITY.

**fright** *n* **1** jolt, scare, shock, surprise. **2** alarm, apprehension, consternation, dismay, dread, fear, horror, panic, terror, trepidation.

**frighten** *vb* agitate, alarm, appal, browbeat, bully, cow, *inf* curdle your blood, daunt, dismay, distress, harrow, horrify, intimidate, make afraid, *inf* make your blood run cold, make your hair stand on end, menace, panic, persecute, *inf* petrify, *inf* put the wind up, scare, *inf* scare stiff, shake, shock, startle, terrify, terrorize, threaten, traumatize, tyrannize, unnerve, upset. ▷ DISCOURAGE. *Opp* REASSURE.

**frightened** *adj* afraid, aghast, alarmed, anxious, appalled, apprehensive, *inf* chicken, cowardly, craven, daunted, fearful, harrowed, horrified, horror-struck, panicky, panic-stricken, petrified, scared, shocked, terrified, terror-stricken, trembling, unnerved, upset, *inf* windy.

**frightening** *adj* alarming, appalling, blood-curdling, *inf* creepy, daunting, dire, dreadful, eerie, fearful, fearsome, formidable, ghostly, grim, hair-raising, horrifying, intimidating, petrifying, scary, sinister, spine-chilling, *inf* spooky, terrifying, traumatic, uncanny, unnerving, upsetting, weird, worrying. ▷ FRIGHTFUL.

**frightful** *adj* **1** awful, ghastly, grisly, gruesome, harrowing, hideous, horrible, horrid, horrific, macabre, shocking, terrible. ▷ FRIGHTENING. **2** ▷ BAD.

**fringe** *n* **1** borders, boundary, edge, limits, marches, margin, outskirts, perimeter, periphery. **2** border, edging, flounce, frill, gathering, ruffle, trimming, valance.

**frisky** *adj* active, animated, coltish, frolicsome, high-spirited, jaunty, lively, perky, playful, skittish, spirited, sprightly.

**frivolity** *n* childishness, facetiousness, flippancy, levity, light-heartedness, nonsense, playing about, silliness, triviality. ▷ FUN.

**frivolous** *adj* casual, childish, facetious, flighty, *inf* flip, flippant, foolish, inconsequential, insignificant, irresponsible, jocular, joking, minor, nugatory, paltry, petty, pointless, puerile, ridiculous, shallow, silly, stupid, superficial, trifling, trivial, trumpery, unimportant, unserious, vacuous, worthless. *Opp* SERIOUS.

**frock** *n* dress, gown, robe.

**frolic** *vb* caper, cavort, curvet, dance, frisk about, gambol, have fun, hop about, *inf* horse about, jump about, lark around, leap about, *inf* make whoopee, play about, prance, revel, rollick, romp, skip, skylark, sport.

**front** *adj* facing, first, foremost, leading, most advanced. • *n* **1** anterior, bow (*of ship*), façade, face, facing, forefront, foreground, frontage, head, nose, obverse, van, vanguard. **2** battle area, danger zone, front line. **3** *a brave front.* appearance, aspect, bearing, blind, *inf* cover-up, demeanour, disguise, expression, look, mask, pretence, show. *Opp* BACK.

**frontal** *adj* direct, facing, headon, oncoming, straight.

**frontier** *n* border, borderline, boundary, bounds, limit, marches, pale.

**froth** *n* bubbles, effervescence, foam, head (*on beer*), lather, scum, spume, suds.

**frown** *vb inf* give a dirty look, glare, glower, grimace, knit your brows, look sullen, lour, lower, scowl. **frown on** ▷ DISAPPROVE.

**fruit** *n* □ *apple, apricot, avocado, banana, berry, bilberry, blackberry, cherry, coconut, crabapple, cranberry, currant, damson, date, fig, gooseberry, grape, grapefruit, greengage, guava, hip, kiwi fruit, lemon, lichee, lime, litchi, loganberry, lychee, mango, medlar, melon, mulberry, nectarine, olive, orange, papaw, pawpaw, peach, pear, pineapple, plum, pomegranate, prune, quince, raisin, raspberry, satsuma, sloe, strawberry, sultana, tangerine, tomato, ugli.*

**fruitful** *adj* **1** abundant, bounteous, bountiful, copious, fecund, fertile, flourishing, lush, luxurious, plenteous, productive, profuse, prolific, rich. **2** advantageous, beneficial, effective, gainful, profitable, rewarding, successful, useful, well-spent, worthwhile. *Opp* FRUITLESS.

**fruitless** *adj* **1** barren, sterile, unfruitful, unproductive. **2** abortive, bootless, disappointing futile, ineffective, ineffectual, pointless, profitless, unavailing, unprofitable, unrewarding, unsuccessful, useless, vain. *Opp* FRUITFUL.

**frustrate** *vb* baffle, balk, baulk, block, check, disappoint, discourage, foil, halt, hamstring, hinder, impede, inhibit, nullify, prevent, *inf* scotch, stop, stymie, thwart. ▷ DEFEAT. *Opp* ENCOURAGE.

**frustrated** *adj* disappointed, embittered, loveless, lovesick, resentful, thwarted, unfulfilled, unsatisfied.

**fuel** *n* □ *anthracite, butane, charcoal, coal, coke, derv, diesel, electricity, gas, gasoline, kindling, logs, methylated spirit, nuclear fuel, oil, paraffin, peat, petrol, propane, tinder, wood.* • *vb* encourage, feed, inflame, keep going, nourish, put fuel on, stoke up, supply with fuel.

**fugitive** *adj* ▷ TRANSIENT. • *n* deserter, escapee, escaper, refugee, renegade, runaway.

**fulfil** *vb* **1** accomplish, achieve, bring about, bring off, carry off, carry out, complete, consummate, discharge, do, effect, effectuate, execute, implement, make come true, perform, realize. **2** *fulfil a need.* answer, comply with, conform to, meet, obey, respond to, satisfy.

**full** *adj* **1** brimming, bursting, *inf* chock-a-block, *inf* chock-full, congested, crammed, crowded, filled, jammed, *inf* jam-packed, loaded, overflowing, packed, replete, solid, stuffed, topped-up, well-filled, well-stocked, well-supplied. **2** *a full stomach.* gorged, sated, satiated, satisfied, well-fed. **3** *the full story.* complete, comprehensive, detailed, entire, exhaustive, plenary, thorough, total, unabridged, uncensored, uncut, unedited, unexpurgated, whole. **4** *full speed.* extreme, greatest, highest, maximum, top, utmost. **5** *a full figure.* ample, broad, buxom, fat, large, plump, rounded, voluptuous, well-built. **6** *a full skirt.* baggy, generous, voluminous, wide. *Opp* EMPTY, INCOMPLETE, SMALL.

**full-grown** *adj* adult, grown-up, mature, ready, ripe.

**fumble** *vb* grope at, feel, handle awkwardly, mishandle, stumble, touch clumsily.

**fume** *vb* emit fumes, smoke, smoulder. **fuming** ▷ ANGRY.

**fumes** *plur n* exhaust, fog, gases, pollution, smog, smoke, vapour.

**fun** *n* amusement, clowning, diversion, enjoyment, entertainment, festivity, *inf* fooling around, frolic, *inf* fun-and-games, gaiety, games, *inf* high jinks, high spirits, horseplay, jocularity, jokes, joking, *joc* jollification, jollity, laughter, merriment, merrymaking, mirth, pastimes, play, playfulness, pleasure, pranks, recreation, romp, *inf* skylarking, sport, teasing, tomfoolery. ▷ FRIVOLITY. **make fun of** ▷ MOCK.

**function** *n* **1** aim, purpose, *Fr* raison d'être, use. ▷ JOB. **2** *an official function.* affair, ceremony, *inf* do, event, occasion, party, reception. • *vb* act, behave, go, operate, perform, run, work.

**functional** *adj* functioning, practical, serviceable, useful, utilitarian, working. *Opp* DECORATIVE.

**fund** *n* cache, hoard, *inf* kitty, mine, pool, reserve, reservoir, stock, store, supply, treasure-house. **funds** capital, endowments, investments, reserves, resources, riches, savings, wealth. ▷ MONEY.

**fundamental** *adj* axiomatic, basic, cardinal, central, crucial, elementary, essential, important, key, main, necessary, primary, prime, principal, quintessential, rudimentary, underlying. *Opp* INESSENTIAL.

**funeral** *n* burial, cremation, entombment, exequies, interment, obsequies, Requiem Mass, wake.

**funereal** *adj* dark, depressing, dismal, gloomy, grave, mournful, sepulchral, solemn, sombre. ▷ SAD. *Opp* CHEERFUL.

**funnel** *n* chimney, smoke-stack. • *vb* channel, direct, filter, pour.

**funny** *adj* **1** absurd, amusing, comic, comical, crazy, *inf* daft, diverting, droll, eccentric, entertaining, facetious, farcical, foolish, grotesque, hilarious, humorous, *inf* hysterical, ironic, jocose, jocular, *inf* killing, laughable, ludicrous, mad, merry, nonsensical, preposterous, *inf* priceless, *inf* rich, ridiculous, risible, sarcastic, sardonic, satirical, *inf* side-splitting, silly, slapstick, uproarious, waggish, witty, zany. *Opp* SERIOUS. **2** ▷ PECULIAR.

**fur** *n* bristles, coat, down, fleece, hair, hide, pelt, skin, wool.

**furious** *adj* **1** boiling, enraged, fuming, incensed, infuriated, irate, livid, mad, raging, savage, wrathful. ▷ ANGRY. **2** *furious activity.* agitated, fierce, frantic, frenzied, intense, tempestuous, tumultuous, turbulent, violent, wild. *Opp* CALM.

**furnish** *vb* **1** decorate, equip, fit out, fit up, *inf* kit out. **2** *furnish information.* afford, give, grant, provide, supply.

**furniture** *n* antiques, chattels, effects, equipment, fitments, fittings, fixtures, furnishings, household goods, *inf* movables, possessions. □ *armchair, bed, bench, bookcase, bunk, bureau, cabinet, chair, chesterfield, chest of drawers, chiffonier, commode, cot, couch, cradle, cupboard, cushion, desk, divan, drawer, dresser, dressing-table, easel, fender, filing-cabinet, fireplace, mantelpiece, ottoman, overmantel, pelmet, pew, pouffe, rocking-chair, seat, settee, sideboard, sofa, stool, suite, table, trestle-table, wardrobe, workbench.*

**furrow** *n* channel, corrugation, crease, cut, ditch, drill, fissure, fluting, gash, groove, hollow, line, rut, score, scratch, track, trench, wrinkle.

**furrowed** *adj* **1** creased, crinkled, corrugated, fluted, grooved, ploughed, ribbed, ridged, rutted, scored. **2** *furrowed brow.* frowning, lined, worried, wrinkled. *Opp* SMOOTH.

**furry** *adj* bristly, downy, feathery, fleecy, fuzzy, hairy, woolly.

**further** *adj* accessory, additional, another, auxiliary, extra, fresh, more, new, other, spare, supplementary.

**furthermore** *adv* additionally, also, besides, moreover, too.

**furtive** *adj* clandestine, concealed, conspiratorial, covert, deceitful, disguised, hidden, mysterious, private, secret, secretive, shifty, sly, *inf* sneaky, stealthy, surreptitious, underhand, untrustworthy. ▷ CRAFTY. *Opp* BLATANT.

**fury** *n* ferocity, fierceness, force, intensity, madness, power, rage, savagery, tempestuousness, turbulence, vehemence, violence, wrath. ▷ ANGER.

**fuse** *vb* amalgamate, blend, coalesce, combine, commingle, compound, consolidate, join, meld, melt, merge, mix, solder, unite, weld.

**fusillade** *n* barrage, burst, firing, outburst, salvo, volley.

**fuss** *n* ▷ COMMOTION. • *vb* agitate, bother, complain, *inf* create, fidget, *inf* flap, *inf* get worked up, grumble, make a commotion, worry.

**fussy** *adj* **1** carping, choosy, difficult, discriminating, *inf* faddy, fastidious, *inf* finicky, hard to please, niggling, *inf* nit-picking, particular, *inf* pernickety, scrupulous, squeamish. **2** *fussy decorations.* Byzantine, complicated, detailed, elaborate, fancy, ornate, overdone, rococo.

**futile** *adj* abortive, absurd, barren, bootless, empty, foolish, forlorn, fruitless, hollow, impotent, ineffective, ineffectual, pointless, profitless, silly, sterile, unavailing, unproductive, unprofitable, unsuccessful, useless, vain, wasted, worthless. *Opp* FRUITFUL.

**future** *adj* approaching, awaited, coming, destined, expected, forthcoming, impending, intended, planned, prospective, subsequent, unborn. • *n* expectations, outlook, prospects, time to come, tomorrow. *Opp* PAST.

**fuzz** *n* down, floss, fluff, hair.

**fuzzy** *adj* **1** downy, feathery, fleecy, fluffy, frizzy, furry, linty, woolly. **2** bleary, blurred, cloudy, dim, faint, hazy, ill-defined, indistinct, misty, obscure, out of focus, shadowy, unclear, unfocused, vague. *Opp* CLEAR.

# G

**gadget** *n* apparatus, appliance, contraption, contrivance, device, implement, instrument, invention, machine, tool, utensil.

**gag** *n* ▷ JOKE. • *vb* check, curb, keep quiet, muffle, muzzle, prevent from speaking, quiet, silence, stifle, still, suppress.

**gaiety** *n* brightness, cheerfulness, colourfulness, delight, exhilaration, felicity, glee, happiness, high spirits, hilarity, jollity, joyfulness, joyousness, light-heartedness, liveliness, merriment, merrymaking, mirth.

**gain** *n* achievement, acquisition, advantage, asset, attainment, benefit, dividend, earnings, income, increase, proceeds, profit, return, revenue, winnings, yield. *Opp* LOSS. • *vb* **1** acquire, bring in, capture, collect, earn, garner, gather in, get, harvest, make, net, obtain, pick up, procure, profit, realize, reap, receive, win. *Opp* LOSE. **2** *gain your objective.* achieve, arrive at, attain, get to, reach, secure. *Opp* MISS. **gain on** approach, catch up with, close the gap, close with, go faster than, leave behind, overhaul, overtake.

**gainful** *adj* advantageous, beneficial, fruitful, lucrative, paid, productive, profitable, remunerative, rewarding, useful, worthwhile.

**gala** *n* carnival, celebration, fair, festival, festivity, fête, *inf* jamboree, party.

**gale** *n* blast, cyclone, hurricane, outburst, storm, tempest, tornado, typhoon, wind.

**gallant** *adj* attentive, chivalrous, courageous, courteous, courtly, dashing, fearless, gentlemanly, gracious, heroic, honourable, intrepid, magnanimous, noble, polite, valiant, well-bred. ▷ BRAVE. *Opp* VILLAINOUS.

**gallows** *n* gibbet, scaffold.

**gamble** *vb* back, bet, chance, draw lots, game, *inf* have a flutter, hazard, lay bets, risk money, speculate, stake money, *inf* take a chance, take risks, *inf* try your luck, venture, wager.

**game** *adj* ▷ BRAVE, WILLING. • *n* 1 amusement, diversion, entertainment, frolic, fun, jest, joke, *inf* lark, *inf* messing about, pastime, play, playing, recreation, romp, sport. 2 competition, contest, match, round, tournament. ▷ SPORT. 3 animals, game-birds, prey, quarry. **give the game away** ▷ REVEAL.

**gang** *n* band, crew, crowd, mob, pack, ring, team. ▷ GROUP. **gang together, gang up** ▷ COMBINE.

**gangster** *n* bandit, brigand, criminal, *inf* crook, desperado, gunman, hoodlum, hooligan, mafioso, mugger, racketeer, robber, ruffian, thug, tough.

**gaol** *n* borstal, cell, custody, dungeon, guardhouse, jail, *Amer* penitentiary, prison. • *vb* confine, detain, imprison, incarcerate, intern, *inf* send down, send to prison, *inf* shut away, shut up.

**gaoler** *n* guard, jailer, prison officer, *sl* screw, warder.

**gap** *n* 1 aperture, breach, break, cavity, chink, cleft, crack, cranny, crevice, gulf, hole, opening, rent, rift, rip, space, void. 2 breathing-space, discontinuity, hiatus, interlude, intermission, interruption, interval, lacuna, lapse, lull, pause, recess, respite, rest, suspension, wait. 3 *gap between political parties.* difference, disagreement, discrepancy, disparity, distance, divergence, division, incompatibility, inconsistency.

**gape** *vb* 1 open, part, split, yawn. 2 *inf* gawp, gaze, *inf* goggle, stare.

**garbage** *n* debris, detritus, junk, litter, muck, refuse, scrap, trash, waste. ▷ RUBBISH.

**garble** *vb* corrupt, distort, falsify, misconstrue, misquote, misrepresent, mutilate, pervert, slant, twist, warp. ▷ CONFUSE.

**garden** *n* allotment, patch, plot, yard. □ *arbour, bed, border, herbaceous border, lawn, orchard, patio, pergola, rock garden, rose garden, shrubbery, terrace, vegetable garden, walled garden, water garden, window-box.* **gardens** grounds, park.

**gardening** *n* cultivation, horticulture.

**garish** *adj* bright, Brummagem, cheap, crude, flamboyant, flashy, gaudy, harsh, loud, lurid, meretricious, ostentatious, raffish, showy, startling, tasteless, tawdry, vivid, vulgar. *Opp* DRAB, TASTEFUL.

**garment** *n* apparel, attire, clothing, costume, dress, garb, habit, outfit. ▷ CLOTHES.

**garrison** *n* 1 contingent, detachment, force, unit. 2 barracks, camp, citadel, fort, fortification, fortress, station, stronghold.

**gas** *n* exhalation, exhaust, fumes, miasma, vapour.

**gash** *vb* chop, cleave, cut, incise, lacerate, score, slash, slit, split, wound.

**gasp** *vb* blow, breathe with difficulty, choke, fight for breath, gulp, *inf* huff and puff, pant, puff, snort, wheeze. **gasping** ▷ BREATHLESS, THIRSTY.

**gate** *n* access, barrier, door, entrance, entry, exit, gateway, kissing-gate, opening, passage, *poet* portal, portcullis, turnstile, way in, way out, wicket, wicket-gate.

**gather** *vb* 1 accumulate, amass, assemble, bring together, build up, cluster, collect, come together, concentrate, congregate, convene, crowd, flock, forgather, get together, group, grow, heap up, herd, hoard, huddle together, marshal, mass, meet, mobilize, muster, rally, round up, pick up, pile up, stockpile, store up, swarm, throng. *Opp* DISPERSE. **2** *gather flowers*. cull, garner, glean, harvest, pick, pluck, reap. **3** *I gather he's ill*. assume, be led to believe, conclude, deduce, guess, infer, learn, surmise, understand.

**gathering** *n* assembly, conclave, congress, convention, convocation, function, *inf* get-together, meeting, party, rally, social. ▷ GROUP.

**gaudy** *adj* bright, Brummagem, cheap, crude, flamboyant, flashy, garish, harsh, loud, lurid, meretricious, ostentatious, raffish, showy, startling, tasteless, tawdry, vivid, vulgar. *Opp* DRAB, TASTEFUL.

**gauge** *n* **1** bench-mark, criterion, guide-line, measurement, norm, standard, test, yardstick. **2** capacity, dimensions, extent, measure, size, span, thickness, width. • *vb* ▷ ESTIMATE, MEASURE.

**gaunt** *adj* **1** bony, cadaverous, emaciated, haggard, hollow-eyed, lanky, lean, pinched, raw-boned, scraggy, scrawny, skeletal, starving, underweight, wasted away. ▷ THIN. *Opp* PLUMP. **2** *a gaunt ruin*. bare, bleak, desolate, dreary, forbidding, grim, stark, stern, unfriendly. *Opp* ATTRACTIVE.

**gawky** *adj* awkward, blundering, clumsy, gangling, gauche, gawky, inept, lumbering, maladroit, uncoordinated, ungainly, ungraceful, unskilful. *Opp* GRACEFUL.

**gay** *adj* **1** animated, bright, carefree, cheerful, colourful, festive, fun-loving, jolly, jovial, joyful, light-hearted, lively, merry, sparkling, sunny, vivacious. ▷ HAPPY. **2** ▷ HOMOSEXUAL.

**gaze** *vb* contemplate, gape, look, regard, stare, view, wonder (at).

**gear** *n* accessories, accoutrements, apparatus, appliances, baggage, belongings, equipment, *inf* get-up, harness, implements, instruments, kit, luggage, materials, paraphernalia, rig, stuff, tackle, things, tools, trappings. ▷ CLOTHES.

**gem** *n* gemstone, jewel, precious stone, *sl* sparkler.

**general** *adj* **1** accepted, accustomed, collective, common, communal, conventional, customary, everyday, familiar, habitual, normal, ordinary, popular, prevailing, prevalent, public, regular, *inf* run-of-the-mill, shared, typical, usual. **2** *general discussion*. across-the-board, all-embracing, blanket, broad-based, catholic, comprehensive, diversified, encyclopaedic, extensive, far-ranging, far-reaching, global, heterogeneous, hybrid, inclusive, sweeping, universal, wholesale, wide-ranging, widespread, worldwide. **3** *a general idea*. approximate, broad, ill-defined, imprecise, indefinite, inexact, in outline, loose, simplified, superficial, unclear, undefined, unspecific, vague. *Opp* SPECIFIC.

**generally** *adv* as a rule, broadly, chiefly, commonly, in the main, mainly, mostly, normally, on the whole, predominantly, principally, usually.

**generate** *vb* beget, breed, bring about, cause, create, engender, father, give rise to, make, originate, procreate, produce, propagate, sire, spawn, *inf* whip up.

**generosity** *n* bounty, largesse, liberality, munificence, philanthropy.

**generous** *adj* **1** benevolent, big-hearted, bounteous, bountiful, charitable, disinterested, forgiving, *inf* free, impartial, kind, liberal, magnanimous, munificent, noble, open, open-handed, philanthropic, public-spirited, unmercenary, unprejudiced, unselfish, unsparing, unstinting. **2** *generous gifts*. handsome, princely, undeserved, unearned, valuable. ▷ EXPENSIVE. **3** *generous portions*. abundant, ample, copious, lavish, plentiful, sizeable, substantial. ▷ BIG. *Opp* MEAN, SELFISH.

**genial** *adj* affable, agreeable, amiable, cheerful, convivial, cordial, easygoing, good-natured, happy, jolly, jovial, kindly, pleasant, relaxed, sociable, sunny, warm, warmhearted. ▷ FRIENDLY. *Opp* UNFRIENDLY.

**genitals** *n* genitalia, *inf* private parts, pudenda, sex organs.

**genius** *n* **1** ability, aptitude, bent, brains, brilliance, capability, flair, gift, intellect, intelligence, knack, talent, wit. **2** academic, *inf* egghead, expert, intellectual, *derog* know-all, mastermind, thinker, virtuoso.

**genteel** *adj derog* affected, chivalrous, courtly, gentlemanly, ladylike, mannered, overpolite, patrician, *inf* posh, refined, stylish, *inf* upper-crust. ▷ POLITE.

**gentle** *adj* **1** amiable, biddable, compassionate, docile, easygoing, good-tempered, harmless, humane, kind, kindly, lenient, loving, meek, merciful, mild, moderate, obedient, pacific, passive, peace-loving, pleasant, quiet, soft-hearted, sweet-tempered, sympathetic, tame, tender. **2** *gentle music*. low, muted, peaceful, reassuring, relaxing, soft, soothing. **3** *gentle wind*. balmy, delicate, faint, light, soft, warm. **4** *a gentle hint*. indirect, polite, subtle, tactful. **5** *a gentle hill*. easy, gradual, imperceptible, moderate, slight, steady. *Opp* HARSH, SEVERE.

**genuine** *adj* **1** actual, authentic, authenticated, *Lat* bona fide, legitimate, original, proper, *sl* pukka, real, sterling, veritable. **2** *genuine feelings*. candid, devout, earnest, frank, heartfelt, honest, sincere, true, unaffected, unfeigned. *Opp* FALSE.

**germ** *n* **1** basis, beginning, embryo, genesis, cause, nucleus, origin, root, seed, source, start. **2** bacterium, *inf* bug, microbe, micro-organism, virus.

**germinate** *vb* begin to grow, bud, develop, grow, root, shoot, spring up, sprout, start growing, take root.

**gesture** *n* action, flourish, gesticulation, indication, motion, movement, sign, signal. • *vb* gesticulate, indicate, motion, sign, signal. □ *beckon, bow, nod, point, salute, shake your head, shrug, smile, wave, wink.*

**get** *vb* **1** acquire, be given, bring, buy, come by, come in possession of, earn, fetch, gain, get hold of, inherit, *inf* land, *inf* lay hands on, obtain, pick up, procure, purchase, receive, retrieve, secure, take, win. **2** *get her by phone*. contact, get in touch with, reach, speak to. **3** *get a cold*. catch, come down with, contract, develop, fall ill with, suffer from. **4** *get a criminal*. apprehend, arrest, capture, catch, *inf* collar, *inf* nab, *sl* pinch, seize. **5** *get him to help*. cajole, cause, induce, influence, persuade, prevail on, *inf* twist someone's arm, wheedle. **6** *get tea.* cook, make ready, prepare. **7** *get what he means.* absorb, appreciate, apprehend, comprehend, fathom, follow, glean, grasp, know, take in, understand, work out. **8** *get what he says.* catch, distinguish, hear, make out. **9** *get somewhere.* arrive, come, go, journey, reach, travel. **10** *get cold.* become, grow, turn. **get across** ▷ COMMUNICATE. **get ahead** ▷ PROSPER. **get at** ▷ CRITICIZE. **get away** ▷ ESCAPE. **get down** ▷ DESCEND. **get in** ▷ ENTER. **get off** ▷ DESCEND. **get on** ▷ PROSPER. **get out** ▷ LEAVE. **get together** ▷ GATHER.

**getaway** *n* escape, flight, retreat.

**ghastly** *adj* appalling, awful, deathlike, dreadful, frightening, frightful, grim, grisly, gruesome, hideous, horrible, macabre, nasty, shocking, terrible, upsetting. ▷ UNPLEASANT.

**ghost** *n* apparition, banshee, *inf* bogey, *Ger* doppelgänger, ghoul, hallucination, illusion, phantasm, phantom, poltergeist, shade, shadow, spectre, spirit, *inf* spook, vision, visitant, wraith. **give up the ghost** ▷ DIE.

**ghostly** *adj* creepy, disembodied, eerie, frightening, illusory, phantasmal, scary, sinister, spectral, *inf* spooky, supernatural, uncanny, unearthly, weird, wraith-like.

**giant** *adj* ▷ GIGANTIC. • *n* colossus, Goliath, leviathan, monster, ogre, superhuman, titan, *inf* whopper.

**giddiness** *n* dizziness, faintness, unsteadiness, vertigo.

**giddy** *adj* dizzy, faint, light-headed, reeling, silly, spinning, unbalanced, unsteady, vertiginous.

**gift** *n* **1** benefaction, bonus, bounty, charity, contribution, donation, favour, *inf* give-away, grant, gratuity, *inf* hand-out, honorarium, largesse, offering, present, tip. **2** ability, aptitude, bent, capability, capacity, facility, flair, genius, knack, power, strength, talent.

**gifted** *adj* able, capable, expert, skilful, skilled, talented. ▷ CLEVER.

**gigantic** *adj* Brobdingnagian, colossal, elephantine, enormous, gargantuan, giant, herculean, huge, immense, *inf* jumbo, *inf* king-size, mammoth, massive, mighty, monstrous, prodigious, titanic, towering, vast. ▷ BIG. *Opp* SMALL.

**giggle** *vb* snicker, snigger, titter. ▷ LAUGH.

**gimcrack** *adj* cheap, *inf* cheap and nasty, flimsy, rubbishy, shoddy, tawdry, trashy, trumpery, useless, worthless.

**gimmick** *n* device, ploy, ruse, stratagem, stunt, subterfuge, trick.

**girder** *n* bar, beam, joist, rafter.

**girdle** *n* band, belt, corset, waistband. • *vb* ▷ SURROUND.

**girl** *n sl* bird, *old use* damsel, daughter, débutante, girlfriend, fiancé, hoyden, lass, *old use* maid, *old use* maiden, *inf* miss, schoolgirl, tomboy, virgin, *old use or sexist* wench. ▷ WOMAN.

**girth** *n* circumference, measurement round, perimeter.

**gist** *n* core, direction, drift, essence, general sense, main idea, meaning, nub, pith, point, quintessence, significance.

**give** *vb* **1** accord, allocate, allot, allow, apportion, assign, award, bestow, confer, contribute, deal out, *inf* dish out, distribute, *inf* dole out, donate, endow, entrust, *inf* fork out, furnish, give away, give out, grant, hand over, lend, let (someone) have, offer, pass over, pay, present, provide, ration out, render, share out, supply. **2** *give information*. deliver, display, express, impart, issue, notify, publish, put across, put into words, reveal, set out, show, tell, transmit. **3** *give a shout*. emit, let out, utter, voice. **4** *give medicine*. administer, dispense, dose with, impose, inflict, mete out, prescribe. **5** *give a party*. arrange, organize, provide, put on, run, set up. **6** *give trouble*. cause, create, engender, occasion. **7** *give under pressure*. be flexible, bend, buckle, collapse, distort, fail, fall apart, give way, warp, yield. *Opp* RECEIVE, TAKE. **give away** ▷ BETRAY. **give in** ▷ SURRENDER. **give off**, **give out** ▷ EMIT. **give up** ▷ ABANDON, SURRENDER.

**glad** *adj* **1** content, delighted, gratified, joyful, overjoyed, pleased. ▷ HAPPY. *Opp* GLOOMY. **2** *glad to help*. disposed, eager, inclined, keen, ready, willing. *Opp* RELUCTANT.

**glamorize** *vb* idealize, romanticize.

**glamorous** *adj* alluring, appealing, colourful, dazzling, enviable, exciting, exotic, fascinating, glittering, prestigious, romantic, smart, spectacular, wealthy. ▷ BEAUTIFUL.

**glamour** *n* allure, appeal, attraction, brilliance, charm, excitement, fascination, glitter, high-life, lustre, magic, romance. ▷ BEAUTY.

**glance** *vb* glimpse, have a quick look, peek, peep, scan, skim, *sl* take a dekko. ▷ LOOK.

**glare** *vb* **1** frown, *inf* give a nasty look, glower, *inf* look daggers, lour, lower, scowl, stare angrily. **2** blaze, dazzle, flare, reflect, shine. ▷ LIGHT. **glaring** ▷ BRIGHT.

**glass** *n* **1** crystal, glassware. **2** glazing, pane, plate-glass, window. **3** looking-glass, mirror, reflector. **4** beaker, drinking-glass, goblet, tumbler, wine-glass. **5** optical instrument. ▫ *binoculars, field-glasses, goggles, magnifying glass, microscope, opera-glasses, telescope, spyglass.* **glasses** *inf* specs, spectacles. ▫ *bifocals, contact-lenses, eyeglass, lorgnette, monocle, pince-nez, reading glasses, sun-glasses, trifocals.*

**glasshouse** *n* conservatory, greenhouse, hothouse, orangery, vinery.

**glassy** *adj* **1** glazed, gleaming, glossy, icy, polished, shining, shiny, smooth, vitreous. **2** *glassy stare*. ▷ EXPRESSIONLESS.

**glaze** *vb* burnish, enamel, gloss, lacquer, polish, shellac, shine, varnish.

**gleam** *vb* flash, glimmer, glint, glisten, glow, reflect, shine. ▷ LIGHT. **gleaming** ▷ BRIGHT.

**gleeful** *adj* cheerful, delighted, ecstatic, exuberant, exultant, gay, jovial, joyful, jubilant, overjoyed, pleased, rapturous, triumphant. ▷ HAPPY. *Opp* SAD.

**glib** *adj* articulate, facile, fast-talking, fluent, insincere, plausible, quick, ready, shallow, slick, smooth, smooth-tongued, suave, superficial, unctuous. ▷ TALKATIVE. *Opp* INARTICULATE, SINCERE.

**glide** *vb* coast, drift, float, fly, freewheel, glissade, hang, hover, move smoothly, sail, skate, ski, skid, skim, slide, slip, soar, stream.

**glimpse** *n* glance, look, peep, sight, *inf* squint, view. • *vb* discern, distinguish, espy, get a glimpse of, make out, notice, observe, see briefly, sight, spot, spy.

**glisten** *vb* flash, gleam, glimmer, glint, glitter, reflect, shine. ▷ LIGHT.

**glitter** *vb* coruscate, flash, scintillate, spark, sparkle, twinkle. ▷ LIGHT. **glittering** ▷ BRIGHT.

**gloat** *vb* boast, brag, *inf* crow, exult, glory, rejoice, *inf* rub it in, show off, triumph.

**global** *adj* broad, far-reaching, international, pandemic, total, universal, wide-ranging, worldwide. *Opp* LOCAL.

**globe** *n* **1** ball, globule, orb, sphere. **2** earth, planet, world.

**gloom** *n* blackness, cloudiness, darkness, dimness, dullness, dusk, murk, murkiness, obscurity, semi-darkness, shade, shadow, twilight. ▷ DEPRESSION.

**gloomy** *adj* **1** cheerless, cloudy, dark, depressing, dim, dingy, dismal, dreary, dull, glum, grim, heavy, joyless, murky, obscure, overcast, shadowy, shady, sombre. **2** *a gloomy mood.* depressed, downhearted, lugubrious, mournful, pessimistic, saturnine. ▷ SAD. *Opp* CHEERFUL.

**glorious** *adj* **1** celebrated, distinguished, eminent, famed, famous, heroic, illustrious, noble, noted, renowned, triumphant. **2** *glorious weather.* beautiful, bright, brilliant, dazzling, delightful, excellent, fine, gorgeous, grand, impressive, lovely, magnificent, majestic, marvellous, outstanding, pleasurable, resplendent, spectacular, splendid, *inf* super, superb, wonderful. *Opp* ORDINARY.

**glory** *n* **1** credit, distinction, eminence, fame, honour, *inf* kudos, praise, prestige, renown, repute, reputation, success, triumph. **2** *glory to God.* adoration, exaltation, glorification, gratitude, homage, praise, thanksgiving, veneration, worship. **3** *glory of sunrise.* brightness, brilliance, grandeur, magnificence, majesty, radiance, splendour, wonder. ▷ BEAUTY.

**gloss** *n* **1** brightness, brilliance, burnish, finish, glaze, gleam, lustre, polish, sheen, shine, varnish. **2** annotation, comment, definition, elucidation, exegesis, explanation, footnote, marginal note, note, paraphrase. • *vb* annotate, comment on, define, elucidate, explain, interpret, paraphrase. **gloss over** ▷ CONCEAL.

**glossary** *n* dictionary, phrase-book, vocabulary, word-list.

**glossy** *adj* bright, burnished, glassy, glazed, gleaming, glistening, lustrous, polished, reflective, shiny, silky, sleek, smooth, waxed. *Opp* DULL.

**glove** *n* gauntlet, mitt, mitten.

**glow** *n* **1** burning, fieriness, heat, incandescence, luminosity, lustre, phosphorescence, radiation, red-heat, redness. **2** ardour, blush, enthusiasm, fervour, flush, passion, rosiness, warmth. • *vb* blush, flush, gleam, incandesce, light up, phosphoresce, radiate heat, redden, smoulder, warm up. ▷ LIGHT.

**glower** *vb* frown, glare, lour, lower, scowl, stare angrily.

**glowing** *adj* **1** aglow, bright, hot, incandescent, lambent, luminous, phosphorescent, radiant, red, red-hot, white-hot. **2** *glowing praise.* complimentary, enthusiastic, fervent, passionate, warm.

**glue** *n* adhesive, cement, fixative, gum, paste, sealant, size, wallpaper-paste. • *vb* affix, bond, cement, fasten, fix, gum, paste, seal, stick.

**glum** *adj* cheerless, displeased, gloomy, grim, heavy, joyless, lugubrious, moody, mournful, *inf* out of sorts, saturnine, sullen. ▷ SAD. *Opp* CHEERFUL.

**glut** *n* abundance, excess, overabundance, overflow, overprovision, plenty, superfluity, surfeit, surplus. *Opp* SCARCITY.

**glutton** *n joc* good trencherman, gormandizer, gourmand, *inf* greedy-guts, guzzler, *inf* pig.

**gluttonous** *adj* gormandizing, greedy, *inf* hoggish, *inf* piggish, insatiable, ravenous, voracious.

**gnarled** *adj* bent, bumpy, contorted, crooked, distorted, knobbly, knotted, knotty, lumpy, rough, rugged, twisted, warped.

**gnaw** *vb* bite, chew, erode, wear away. ▷ EAT.

**go** *n* attempt, chance, *inf* crack, opportunity, *inf* shot, *inf* stab, try, turn. • *vb* **1** advance, begin, be off, commence, decamp, depart, disappear, embark, escape, get away, get going, get moving, get out, get under way, leave, make off, move, *inf* nip along, pass along, pass on, proceed, retire, retreat, run, set off, set out, *inf* shove off, start, take off, take your leave, vanish, *old use* wend your way, withdraw. ▷ RUN, TRAVEL, WALK. **2** die, fade, fail, give way. **3** extend, lead, reach, stretch. **4** *car won't go.* act, function, operate, perform, run, work. **5** *bomb went bang.* give off, make, produce, sound. **6** *Time goes slowly.* elapse, lapse, pass. **7** *go sour.* become, grow, turn. **8** *Milk goes in the fridge.* belong, feel at home, have a proper place, live. go away ▷ DEPART. **go down** ▷ DESCEND, SINK. **go in for** ▷ LIKE. **go into** ▷ INVESTIGATE. **go off** ▷ EXPLODE. **go on** ▷ CONTINUE. **go through** ▷ SUFFER. **go to** ▷ VISIT. **go together** ▷ MATCH. **go with** ▷ ACCOMPANY. **go without** ▷ ABSTAIN.

**goad** *vb* badger, *inf* chivvy, egg on, *inf* hassle, needle, prick, prod, prompt, spur, urge. ▷ STIMULATE.

**go-ahead** *adj* ambitious, enterprising, forward-looking, progressive, resourceful. • *n* approval, *inf* green light, permission, sanction, *inf* say-so, *inf* thumbs-up.

**goal** *n* aim, ambition, aspiration, design, end, ideal, intention, object, objective, purpose, target.

**gobble** *vb* bolt, devour, gulp, guzzle. ▷ EAT.

**go-between** *n* agent, broker, envoy, intermediary, liaison, mediator, messenger, middleman, negotiator. **act as go-between** ▷ MEDIATE.

**god, goddess** *ns* deity, divinity, godhead, spirit. **God** the Almighty, the Creator, the supreme being. **the gods** the immortals, the pantheon, the powers above.

**godsend** *n inf* bit of good luck, blessing, boon, gift, miracle, *inf* stroke of good fortune, windfall.

**golden** *adj* **1** aureate, gilded, gilt. **2** *golden hair.* blond, blonde, flaxen, yellow.

**good** *adj* **1** acceptable, admirable, agreeable, appropriate, approved of, commendable, delightful, enjoyable, esteemed, *inf* fabulous, fair, *inf* fantastic, fine, gratifying, happy, *inf* incredible, lovely, marvellous, nice, perfect, *inf* phenomenal, pleasant, pleasing, praiseworthy, proper, remarkable, right, satisfactory, *inf* sensational, sound, splendid, suitable, *inf* super, superb, useful, valid, valuable, wonderful, worthy. ▷ EXCELLENT. **2** *a good person*. angelic, benevolent, caring, charitable, chaste, considerate, decent, dependable, dutiful, ethical, friendly, helpful, holy, honest, honourable, humane, incorruptible, innocent, just, law-abiding, loyal, merciful, moral, noble, obedient, personable, pure, reliable, religious, righteous, saintly, sound, *inf* straight, thoughtful, true, trustworthy, upright, virtuous, well-behaved, well-mannered, worthy. ▷ KIND. **3** *a good worker*. able, accomplished, capable, conscientious, efficient, gifted, proficient, skilful, skilled, talented. ▷ CLEVER. **4** *good work*. careful, competent, correct, creditable, efficient, meritorious, neat, orderly, presentable, professional, thorough, well-done. **5** *good food*. beneficial, delicious, eatable, healthy, nourishing, nutritious, tasty, well-cooked, wholesome. **6** *a good book*. classic, exciting, great, interesting, readable, well-written. *Opp* BAD. **good-humoured** ▷ GOOD-TEMPERED. **good-looking** ▷ HANDSOME. **good-natured** ▷ GOOD-TEMPERED. **good person** *inf* angel, *inf* jewel, philanthropist, *inf* saint, Samaritan, worthy. **goods** **1** belongings, chattels, effects, possessions, property. **2** commodities, freight, load, mechandise, produce, stock, wares.

**goodbye** *n* farewell, departure, leave-taking, parting words, send-off, valediction. □ *adieu, adios, arrivederci, au revoir, auf Wiedersehen, bon voyage, ciao, cheerio, so long.*

**good-tempered** *adj* accommodating, amenable, amiable, benevolent, benign, cheerful, cheery, considerate, cooperative, cordial, friendly, genial, good-humoured, good-natured, helpful, in a good mood, obliging, patient, pleasant, relaxed, smiling, sympathetic, thoughtful, willing. ▷ KIND. *Opp* BAD-TEMPERED.

**gorge** *vb* be greedy, fill up, gormandize, guzzle, indulge yourself, *inf* make a pig of yourself, overeat, *inf* stuff yourself. ▷ EAT.

**gorgeous** *adj* colourful, dazzling, glorious, magnificent, resplendent, showy, splendid, sumptuous. ▷ BEAUTIFUL.

**gory** *adj* blood-stained, bloody, grisly, gruesome, sanguinary, savage.

**gospel** *n* creed, doctrine, good news, good tidings, message, religion, revelation, teaching, testament.

**gossip** *n* **1** casual talk, chatter, *inf* the grapevine, hearsay, prattle, rumour, scandal, small talk, *inf* tattle, *inf* tittle-tattle. **2** *inf* blab, busybody, chatterbox, *inf* Nosey Parker, rumourmonger, scandalmonger, tell-tale. • *vb inf* blab, chat, chatter, *inf* natter, prattle, spread scandal, *inf* tattle, tell tales, *inf* tittle-tattle. ▷ TALK.

**gouge** *vb* chisel, dig, gash, hollow, incise, scoop. ▷ CUT.

**gourmet** *n Fr* bon viveur, connoisseur, epicure, gastronome, *derog* gourmand.

**govern** *vb* **1** administer, be in charge of, command, conduct affairs, control, direct, guide, head, lead, look after, manage, oversee, preside over, reign, rule, run, steer, superintend, supervise. **2** *govern your anger*. bridle, check, control, curb, discipline, keep in check, keep under control, master, regulate, restrain, tame.

**government** *n* administration, authority, bureaucracy, conduct of state affairs, constitution, control, direction, domination, management, oversight, regime, regulation, rule, sovereignty, supervision, surveillance, sway. □ *commonwealth, democracy, dictatorship, empire, federation, kingdom, monarchy, oligarchy, republic.*

**gown** *n* dress, frock. ▷ CLOTHES.

**grab** *vb* appropriate, arrogate, *inf* bag, capture, catch, clutch, *inf* collar, commandeer, expropriate, get hold of, grasp, hold, *inf* nab, pluck, seize, snap up, snatch, usurp.

**grace** *n* **1** attractiveness, beauty, charm, ease, elegance, fluidity, gracefulness, loveliness, poise, refinement, softness, tastefulness. **2** *God's grace.* beneficence, benevolence, compassion, favour, forgiveness, goodness, graciousness, kindness, love, mercy. **3** *grace before meals.* blessing, prayer, thanksgiving.

**graceful** *adj* **1** agile, balletic, deft, dignified, easy, elegant, flowing, fluid, natural, nimble, pliant, slender, slim, smooth, supple, willowy. ▷ BEAUTIFUL. **2** *graceful compliments.* courteous, courtly, delicate, kind, polite, refined, suave, tactful, urbane. *Opp* GRACELESS.

**graceless** *adj* **1** awkward, clumsy, gangling, gawky, inelegant, maladroit, uncoordinated, ungainly. ▷ CLUMSY. *Opp* GRACEFUL. **2** *graceless manners.* boorish, gauche, inept, tactless, uncouth. ▷ RUDE.

**gracious** *adj* **1** affable, agreeable, civilized, cordial, courteous, dignified, elegant, friendly, good-natured, pleasant, polite, with grace. ▷ KIND. **2** clement, compassionate, forgiving, generous, indulgent, lenient, magnanimous, pitying, sympathetic. ▷ MERCIFUL. **3** *gracious living.* affluent, expensive, lavish, luxurious, opulent, self-indulgent, sumptuous.

**grade** *n* category, class, condition, degree, echelon, estate, level, mark, notch, point, position, quality, rank, rung, situation, standard, standing, status, step. • *vb* **1** arrange, categorize, classify, differentiate, group, organize, range, size, sort. **2** *grade students' work.* assess, evaluate, mark, rank, rate.

**gradient** *n* ascent, bank, declivity, hill, incline, rise, slope.

**gradual** *adj* continuous, easy, even, gentle, leisurely, moderate, regular, slow, steady, unhurried, unspectacular. *Opp* SUDDEN.

**graduate** *vb* **1** become a graduate, be successful, get a degree, pass, qualify. **2** *graduate a measuring-rod.* calibrate, divide into graded sections, gradate, mark off, mark with a scale.

**graft** *vb* implant, insert, join, splice.

**grain** *n* **1** atom, bit, crumb, fleck, fragment, granule, iota, jot, mite, molecule, morsel, mote, particle, scrap, seed, speck, trace. **2** ▷ CEREAL.

**grand** *adj* **1** aristocratic, august, dignified, eminent, glorious, great, important, imposing, impressive, lordly, magnificent, majestic, noble, opulent, palatial, regal, royal, splendid, stately, sumptuous, superb. ▷ BIG. **2** [*derog*] haughty, *inf* high-and-mighty, lofty, patronizing, pompous, posh, *inf* upper crust. ▷ GRANDIOSE. *Opp* MODEST.

**grandiloquent** *adj* bombastic, elaborate, florid, flowery, fustian, high-flown, inflated, melodramatic, ornate, poetic, pompous, rhetorical, turgid. ▷ GRANDIOSE. *Opp* SIMPLE.

**grandiose** *adj* affected, ambitious, exaggerated, extravagant, flamboyant, *inf* flashy, grand, highfalutin, ostentatious, overdone, *inf* over the top, pretentious, showy. ▷ GRANDILOQUENT. *Opp* MODEST.

**grant** *n* allocation, allowance, annuity, award, benefaction, bursary, concession, contribution, donation, endowment, expenses, gift, honorarium, investment, loan, pension, scholarship, sponsorship, subsidy, subvention. • *vb* **1** allocate, allot, allow, assign, award, bestow, confer, donate, give, pay, provide, supply. **2** *grant that I'm right.* accede, accept, acknowledge, admit, agree, concede, consent, vouchsafe.

**graph** *n* chart, column-graph, diagram, grid, pie chart, table.

**graphic** *adj* clear, descriptive, detailed, lifelike, lucid, photographic, plain, realistic, representational, vivid, well-drawn.

**grapple** *vb* clutch (at), grab, seize, tackle, wrestle. ▷ GRASP, FIGHT. **grapple with** *grapple with a problem.* attend to, come to grips with, contend with, cope with, deal with, engage with, get involved with, handle, *inf* have a go at, manage, try to solve.

**grasp** *vb* **1** catch, clasp, clutch, get hold of, *inf* get your hands on, grab, grapple with, grip, hang on to, hold, *inf* nab, seize, snatch, take hold of. **2** *grasp an idea.* appreciate, apprehend, comprehend, *inf* cotton on to, follow, *inf* get the drift of, get the hang of, get the point of, learn, master, realize, take in, understand. **grasping** ▷ GREEDY.

**grass** *n* downland, field, grassland, green, lawn, meadow, pasture, playing-field, prairie, savannah, steppe, *poet* sward, turf, veld. • *vb* ▷ INFORM.

**grate** *n* fireplace, hearth. • *vb* cut, grind, rasp, shred, triturate. **grate on** ▷ ANNOY. **grating** ▷ ANNOYING, HARSH.

**grateful** *adj* appreciative, beholden, gratified, indebted, obliged, thankful. *Opp* UNGRATEFUL.

**gratify** *vb* delight, fulfil, indulge, pander to, please, satisfy.

**gratis** *adj* complimentary, free, free of charge, gratuitous, without charge.

**gratitude** *n* appreciation, gratefulness, thankfulness, thanks.

**gratuitous** *adj* **1** ▷ GRATIS. **2** *gratuitous insults.* baseless, groundless, inappropriate, needless, unasked-for, uncalled-for, undeserved, unjustifiable, unmerited, unnecessary, unprovoked, unsolicited, unwarranted. *Opp* JUSTIFIABLE.

**gratuity** *n* bonus, *inf* perk, *Fr* pourboire, present, recompense, reward, tip.

**grave** *adj* **1** acute, critical, crucial, dangerous, important, *inf* life and death, major, momentous, perilous, pressing, serious, severe, significant, terminal (*illness*), threatening, urgent, vital, weighty, worrying. **2** *a grave offence.* criminal, indictable, punishable. **3** *a grave look.* dignified, earnest, grim, long-faced, pensive, sedate, serious, severe, sober, solemn, sombre, subdued, thoughtful, unsmiling. ▷ SAD. *Opp* CHEERFUL, TRIVIAL. • *n* barrow, burial-place, crypt, *inf* last resting-place, mausoleum, sepulchre, tomb, tumulus, vault. ▷ GRAVESTONE.

**gravel** *n* grit, pebbles, shingle, stones.

**gravestone** *n* headstone, memorial, monument, tombstone.

**graveyard** *n* burial-ground, cemetery, churchyard, necropolis.

**gravity** *n* **1** acuteness, danger, importance, magnitude, momentousness, seriousness, severity, significance, weightiness. **2** *behave with gravity*. ceremony, dignity, earnestness, *Lat* gravitas, pomp, reserve, sedateness, sobriety, solemnity. **3** *force of gravity*. attraction, gravitation, heaviness, ponderousness, pull, weight.

**graze** *n* abrasion, laceration, raw spot, scrape, scratch. ▷ WOUND.

**grease** *n* fat, lubrication, oil.

**greasy** *adj* **1** buttery, fatty, oily, slippery, slithery, smeary, waxy. **2** *greasy manner*. fawning, flattering, fulsome, grovelling, ingratiating, slick, *inf* smarmy, sycophantic, toadying, unctuous.

**great** *adj* **1** colossal, enormous, extensive, giant, gigantic, grand, huge, immense, large, massive, prodigious, *inf* tremendous, vast. ▷ BIG. **2** *great pain*. acute, considerable, excessive, extreme, intense, marked, pronounced. ▷ SEVERE. **3** *great events*. grand, imposing, large-scale, momentous, serious, significant, spectacular, weighty. ▷ IMPORTANT. **4** *great music*. brilliant, classic, *inf* fabulous, famous, *inf* fantastic, fine, first-rate, outstanding, wonderful. ▷ EXCELLENT. **5** *a great athlete*. able, celebrated, distinguished, eminent, gifted, notable, noted, prominent, renowned, talented, well-known. ▷ FAMOUS. **6** *a great friend*. chief, close, dedicated, devoted, faithful, fast, loyal, main, true, valued. **7** *a great reader*. active, ardent, assiduous, eager, enthusiastic, frequent, habitual, keen, passionate, zealous. **8** ▷ GOOD. *Opp* SMALL, UNIMPORTANT.

**greed** *n* **1** appetite, craving, gluttony, gormandizing, hunger, insatiability, intemperance, overeating, ravenousness, self-indulgence, voraciousness, voracity. **2** *greed for wealth*. acquisitiveness, avarice, covetousness, cupidity, desire, rapacity, self-interest. ▷ SELFISHNESS.

**greedy** *adj* **1** famished, gluttonous, gormandizing, *inf* hoggish, hungry, insatiable, intemperate, omnivorous, *inf* piggish, ravenous, self-indulgent, starving, voracious. *Opp* ABSTEMIOUS. **2** *greedy for wealth*. acquisitive, avaricious, avid, covetous, desirous, eager, grasping, materialistic, mean, mercenary, miserly, *inf* money-grubbing, rapacious, selfish. *Opp* UNSELFISH. **be greedy** ▷ GORGE. **greedy person** ▷ GLUTTON.

**green** *adj* **1** grassy, greenish, leafy, verdant. ◻ *emerald, grass-green, jade, khaki, lime, olive, pea-green, turquoise.* **2** ▷ IMMATURE.

**greenery** *n* foliage, leaves, plants, vegetation.

**greet** *vb* accost, acknowledge, address, give a greeting to, hail, receive, salute, *inf* say hello to, usher in, welcome.

**greeting** *n* salutation, reception, welcome. **greetings** compliments, congratulations, felicitations, good wishes, regards.

**grey** *adj* ashen, blackish, colourless, greying, grizzled, grizzly, hoary, leaden, livid, pearly, silver, silvery, slate-grey, smoky, sooty, whitish. ▷ GLOOMY.

**grid** *n* framework, grating, grille, lattice, network.

**grief** *n* affliction, anguish, dejection, depression, desolation, despondency, distress, heartache, heartbreak, melancholy, misery, mourning, pain, regret, remorse, sadness, sorrow, suffering, tragedy, unhappiness, woe, wretchedness. ▷ PAIN. *Opp* HAPPINESS. **come to grief** ▷ FAIL.

**grievance** *n* **1** calamity, damage, hardship, harm, indignity, injury, injustice. **2** allegation, *inf* bone to pick, charge, complaint, *inf* gripe, objection.

**grieve** *vb* **1** afflict, cause grief, depress, dismay, distress, hurt, pain, sadden, upset, wound. *Opp* PLEASE. **2** be in mourning, feel grief, *inf* eat your heart out, fret, lament, mope, mourn, suffer, wail, weep. *Opp* REJOICE.

**grim** *adj* alarming, appalling, awful, cruel, dire, dour, dreadful, fearsome, fierce, forbidding, formidable, frightening, frightful, frowning, ghastly, grisly, gruesome, harsh, hideous, horrible, *inf* horrid, inexorable, inflexible, joyless, louring, menacing, merciless, ominous, pitiless, relentless, ruthless, savage, severe, sinister, stark, stern, sullen, surly, terrible, threatening, unattractive, uncompromising, unfriendly, unpleasant, unrelenting, unsmiling, unyielding. ▷ GLOOMY. *Opp* CHEERFUL.

**grime** *n* dirt, dust, filth, grit, muck, scum, soot.

**grind** *vb* **1** abrade, comminute, crumble, crush, erode, granulate, grate, mill, pound, powder, pulverize, rasp, triturate. **2** file, polish, sand, sandpaper, scrape, sharpen, smooth, wear away, whet. **3** *grind your teeth.* gnash, grate, grit, rub together. **grind away** ▷ WORK. **grind down** ▷ OPPRESS.

**grip** *n* clasp, clutch, grasp, handclasp, hold, purchase, stranglehold. ▷ CONTROL. • *vb* **1** clasp, clutch, get a grip of, grab, grasp, hold, seize, take hold of. **2** *grip the imagination.* absorb, compel, engage, engross, enthral, entrance, fascinate, hypnotize, mesmerize, rivet, spellbind. **come to grips with** ▷ TACKLE.

**grisly** *adj* appalling, awful, bloody, disgusting, dreadful, fearful, frightful, ghastly, ghoulish, gory, grim, gruesome, hair-raising, hideous, horrible, *inf* horrid, horrifying, macabre, nauseating, repellent, repulsive, revolting, sickening, terrible.

**gristly** *adj* leathery, rubbery, tough, uneatable.

**gritty** *adj* abrasive, dusty, grainy, granular, gravelly, harsh, rasping, rough, sandy.

**groan** *vb* **1** cry out, lament, moan, sigh, wail, whimper, whine. **2** ▷ COMPLAIN.

**groom** *n* **1** ostler, stable-lad, stableman. **2** bridegroom, husband. • *vb* **1** brush, clean, make neat, neaten, preen, smarten up, spruce up, tidy, *inf* titivate. **2** *groom someone for a job.* coach, drill, educate, get ready, prepare, prime, train up, tutor.

**groove** *n* channel, cut, fluting, furrow, gouge, gutter, hollow, indentation, rut, score, scratch, slot, striation, track.

**grope** *vb* cast about, feel about, fish, flounder, fumble, search blindly.

**gross** *adj* **1** bloated, massive, obese, overweight, repellent, repulsive, revolting. ▷ FAT. **2** churlish, coarse, crude, rude, unrefined, unsophisticated, vulgar. **3** *gross injustice.* blatant, flagrant, glaring, manifest, monstrous, obvious, outrageous, shameful. **4** *gross income.* before tax, inclusive, overall, total, whole.

**grotesque** *adj* absurd, bizarre, curious, deformed, distorted, fantastic, freakish, gnarled, incongruous, ludicrous, macabre, malformed, misshapen, monstrous, outlandish, preposterous, queer, ridiculous, strange, surreal, twisted, ugly, unnatural, weird.

**ground** *n* **1** clay, dirt, earth, loam, mud, soil. **2** area, land, property, surroundings, terrain. **3** campus, estate, garden, park. **4** *sports ground.* arena, court, field, pitch, playground, playing-field, recreation ground, stadium. **5** *grounds for complaint.* argument, base, basis, case, cause, excuse, evidence, foundation, justification, motive, proof, rationale, reason. • *vb* **1** base, establish, found, set, settle. **2** coach, educate, instruct, prepare, teach, train, tutor. **3** beach, run ashore, shipwreck, strand, wreck.

**groundless** *adj* baseless, chimerical, false, gratuitous, hypothetical, illusory, imaginary, irrational, motiveless, needless, speculative, suppositional, uncalled for, unfounded, unjustifiable, unjustified, unproven, unreasonable, unsound, unsubstantiated, unsupported, unwarranted.

**group** *n* **1** [*people*] alliance, assemblage, assembly, association, band, bevy, body, brotherhood, *inf* bunch, cadre, cartel, caste, caucus, circle, clan, class, *derog* clique, club, cohort, colony, committee, community, company conclave, congregation, consortium, contingent, corps, coterie, coven, crew, crowd, delegation, faction, family, federation, force, fraternity, gang, gathering, group, guild, horde, host, knot, league, meeting, *derog* mob, multitude, number, organization, party, phalanx, picket, platoon, posse, *derog* rabble, ring, sect, *derog* shower, sisterhood, society, squad, squadron, swarm, team, throng, troop, troupe, union, unit **2** [*things, animals*] accumulation, agglomeration, assemblage, assortment, batch, battery (*guns*), brood (*chicks*), bunch, bundle, category, class, clump, cluster, clutch (*eggs*) collection, combination, conglomeration, constellation, convoy, covey (*birds*), fleet, flock, gaggle (*geese*), galaxy, grouping, heap, herd, hoard, host, litter, mass, pack, pile, pride (*lions*), school, set, shoal (*fish*), species. **3** ▷ MUSICIAN. • *vb* **1** arrange, assemble, assort, bracket together, bring together, categorize, classify, collect, deploy, gather, herd, marshal, order, organize, put together, set out, sort. **2** associate band, cluster, come together, congregate, crowd, flock, gather, get together, herd, make groups, swarm, team up, throng.

**grovel** *vb* abase yourself, be humble, cower, *inf* crawl, *inf* creep, cringe, demean yourself fawn, flatter, ingratiate yourself, *inf* kowtow, *inf* lick someone's boots, prostrate yourself, snivel, *inf* suck up, *inf* toady. **grovelling** ▷ OBSEQUIOUS.

**grow** *vb* **1** augment, become bigger, broaden, build up, burgeon, come to life, develop, emerge, enlarge, evolve, expand, extend, ill out, flourish, flower, germinate, improve, increase, lengthen, ive, make progress, mature, multiply, mushroom, progress, proliferate, prosper, put on growth, ripen, rise, shoot up, spread, spring up, sprout, survive, swell, thicken, thrive. **2** *grow roses.* cultivate, farm, help along, nurture, produce, propagate, raise. **3** *grow older.* become, get, turn.

**rown-up** *adj* adult, fully-grown, mature, well-developed.

**rowth** *n* **1** accretion, advance, augmentation, broadening, burgeoning, development, enlargement, evolution, expansion, extension, flowering, getting bigger, growing, improvement, increase, maturation, maturing, progress, proliferation, prosperity, spread, success. **2** crop, harvest, plants, produce, vegetation, yield. **3** cancer, cyst, excrescence, lump, swelling, tumour.

**rub** *n* **1** caterpillar, larva, maggot. **2** ▷ FOOD. • *vb* ▷ DIG.

**rudge** *n* ▷ RESENTMENT. • *vb* begrudge, covet, envy, resent.

**rudging** *adj* cautious, envious, guarded, half-hearted, hesitant, jealous, reluctant, resentful, secret, unenthusiastic, ungracious, unkind, unwilling. *Opp* ENTHUSIASTIC.

**ruelling** *adj* arduous, backbreaking, crippling, demanding, exhausting, fatiguing, laborious, punishing, severe, stiff, strenuous, taxing, tiring, tough, uphill, wearying. ▷ DIFFICULT. *Opp* EASY.

**ruesome** *adj* appalling, awful, bloody, disgusting, dreadful, fearful, fearsome, frightful, ghastly, ghoulish, gory, grim, grisly, hair-raising, hideous, horrible, *inf* horrid, horrific, horrifying, macabre, repellent, repugnant, revolting, shocking, sickening, terrible.

**gruff** *adj* **1** guttural, harsh, hoarse, husky, rasping, rough, throaty. **2** ▷ BAD-TEMPERED.

**grumble** *vb inf* beef, fuss, *inf* gripe, *inf* grouch, grouse, make a fuss, *inf* moan, object, protest, *inf* whinge. ▷ COMPLAIN.

**guarantee** *n* assurance, bond, oath, obligation, pledge, promise, surety, undertaking, warranty, word of honour. • *vb* **1** assure, certify, give a guarantee, pledge, promise, swear, undertake, vouch, vow. **2** ensure, make sure of, reserve, secure, stake a claim to.

**guard** *n* bodyguard, *inf* bouncer, custodian, escort, guardian, *sl* heavy, lookout, *sl* minder, patrol, picket, *sl* screw, security-guard, sentinel, sentry, warder, watchman. • *vb* be on guard over, care for, defend, keep safe, keep watch on, look after, mind, oversee, patrol, police, preserve, prevent from escaping, protect, safeguard, secure, shelter, shield, stand guard over, supervise, tend, watch, watch over. **on your guard** ▷ ALERT.

**guardian** *n* **1** adoptive parent, foster-parent. **2** champion, custodian, defender, keeper, preserver, protector, trustee, warden. ▷ GUARD.

**guess** *n* assumption, conjecture, estimate, feeling, *sl* guesstimate, guesswork, hunch, hypothesis, intuition, opinion, prediction, *inf* shot in the dark, speculation, supposition, surmise, suspicion, theory. • *vb* assume, conclude, conjecture, divine, estimate, expect, fancy, feel, have a hunch, have a theory, *inf* hazard a guess, hypothesize, imagine, intuit, judge, make a guess, postulate, predict, *inf* reckon, speculate, suppose, surmise, suspect, think likely, work out.

**guest** *n* 1 caller, company, visitor. 2 *hotel guests.* boarder, customer, lodger, patron, resident, tenant.

**guidance** *n* advice, briefing, counselling, direction, guidelines, guiding, help, instruction, leadership, management, *inf* spoon-feeding, *inf* taking by the hand, teaching, tips.

**guide** *n* 1 courier, escort, leader, navigator, pilot. 2 adviser, counsellor, director, guru, mentor. 3 atlas, directory, gazetteer, guidebook, handbook, *Lat* vade mecum. • *vb* 1 conduct, direct, escort, lead, manoeuvre, navigate, pilot, shepherd, show the way, steer, supervise, usher. 2 advise, brief, control, counsel, educate, give guidance to, govern, help along, influence, instruct, regulate, *inf* take by the hand, teach, train, tutor. *Opp* MISLEAD.

**guilt** *n* 1 blame, blameworthiness, criminality, culpability, fault, guiltiness, liability, responsibility, sinfulness, wickedness, wrongdoing. 2 *a look of guilt.* bad conscience, contriteness, contrition, dishonour, guilty feelings, penitence, regret, remorse, self-accusation, self-reproach, shame, sorrow. *Opp* INNOCENCE.

**guiltless** *adj* above suspicion, blameless, clear, faultless, free, honourable, immaculate, innocent, in the right, irreproachable, pure, sinless, untarnished, untroubled, virtuous. *Opp* GUILTY.

**guilty** *adj* 1 at fault, blameable, blameworthy, culpable, in the wrong, liable, reprehensible, responsible. 2 *a guilty look.* apologetic, ashamed, conscience-stricken, contrite, penitent, *inf* red-faced, regretful, remorseful, repentant, rueful, shamefaced, sheepish, sorry. *Opp* GUILTLESS, SHAMELESS.

**gullible** *adj* credulous, easily taken in, *inf* green, impressionable, inexperienced, innocent, naïve, suggestible, trusting, unsophisticated, unsuspecting, unwary. *Opp* WARY.

**gulp** *n* mouthful, swallow, *inf* swig. • *vb* 1 bolt down, gobble, swallow, *inf* wolf. ▷ EAT. 2 *inf* knock back, quaff, *inf* swig. ▷ DRINK. 3 *gulp back tears.* check, choke back, stifle, suppress.

**gumption** *n* cleverness, *inf* common sense, enterprise, initiative, judgement, *inf* nous, resourcefulness, sense, wisdom.

**gun** *n plur* artillery, firearm. □ *airgun, automatic, blunderbuss, cannon, machine-gun, mortar, musket, pistol, revolver, rifle, shot-gun,* plur *small arms, sub-machine-gun, tommy-gun.* **gun down** ▷ SHOOT.

**gunfire** *n* cannonade, cross-fire, firing, gunshots, salvo.

**gunman** *n* assassin, bandit, criminal, desperado, fighter, gangster, killer, murderer, sniper, terrorist.

**gurgle** *vb* babble, bubble, burble, ripple, purl, splash.

**gush** *n* burst, cascade, eruption, flood, flow, jet, outpouring, overflow, rush, spout, spurt, squirt, stream, tide, torrent. • *vb* 1 come in a gush, cascade, flood, flow freely, overflow, pour, run, rush, spout, spurt, squirt, stream, well up. 2 be enthusiastic, be sentimental, bubble over, fuss, *inf* go on, prattle on, talk on. **gushing** ▷ EFFUSIVE, SENTIMENTAL.

**gusto** *n* appetite, delight, enjoyment, enthusiasm, excitement, liveliness, pleasure, relish, satisfaction, spirit, verve, vigour, zest.

**gut** *vb* 1 clean, disembowel, draw, eviscerate, remove the guts of. 2 *gut a building.* clear, despoil, empty, loot, pillage, plunder, ransack, ravage, remove the contents of, sack, strip.

**guts** *plur n* 1 alimentary canal, belly, bowels, entrails, *inf* innards, insides, intestines, stomach, viscera. 2 ▷ COURAGE.

**gutter** *n* channel, conduit, ditch, drain, duct, guttering, sewer, sluice, trench, trough.

**gypsy** *n* nomad, Romany, traveller, wanderer.

**gyrate** *vb* circle, pirouette, revolve, rotate, spin, spiral, swivel, turn, twirl, wheel, whirl.

# H

**habit** *n* 1 convention, custom, pattern, policy, practice, routine, rule, usage, *old use* wont. 2 attitude, bent, disposition, inclination, manner, mannerism, penchant, predisposition, proclivity, propensity, quirk, tendency, way. 3 *bad habit.* addiction, compulsion, craving, dependence, fixation, obsession, vice.

**habitable** *adj* in good repair, inhabitable, liveable, usable. *Opp* UNINHABITABLE.

**habitual** *adj* 1 accustomed, common, conventional, customary, established, expected, familiar, fixed, frequent, natural, normal, ordinary, predictable, regular, ritual, routine, set, settled, standard, traditional, typical, usual, *old use* wonted. 2 addictive, besetting, chronic, established, ineradicable, ingrained, obsessive, persistent, recurrent. 3 *habitual smokers.* addicted, conditioned, confirmed, dependent, hardened, *inf* hooked, inveterate, persistent.

**hack** *vb* carve, chop, gash, hew, mangle, mutilate, slash. ▷ CUT.

**hackneyed** *adj* banal, clichéd, cliché-ridden, commonplace, conventional, *inf* corny, familiar, feeble, obvious, overused, pedestrian, platitudinous, predictable, stale, stereotyped, stock, threadbare, tired, trite, uninspired, unoriginal. *Opp* NEW.

**haggard** *adj inf* all skin and bone, careworn, drawn, emaciated, exhausted, gaunt, hollow-cheeked, hollow-eyed, pinched, run-down, scraggy, scrawny, shrunken, thin, tired out, ugly, unhealthy, wasted, weary, withered, worn out, *inf* worried to death. *Opp* HEALTHY.

**haggle** *vb* argue, bargain, barter, discuss terms, negotiate, quibble, wrangle. ▷ QUARREL.

**hail** *vb* 1 accost, address, call to, greet, signal to. 2 ▷ ACCLAIM.

**hair** *n* 1 beard, bristles, curls, fleece, fur, hank, locks, mane, *inf* mop, moustache, shock, tresses, whiskers. 2 coiffure, cut, haircut, *inf* hair-do, hairstyle, style. □ *bob, braid, bun, crew-cut, dreadlocks, fringe, Mohican,* inf *perm, permanent wave, pigtail, plait, pony-tail, quiff, ringlets, short back and sides, sideboards, sideburns, tonsure, topknot.* 3 *false hair.* hair-piece, toupee, wig.

**hairdresser** *n* barber, coiffeur, coiffeuse, hair-stylist.

**hairless** *adj* bald, bare, clean-shaven, naked, shaved, shaven, smooth. *Opp* HAIRY.

**hairy** *adj* bearded, bristly, downy, feathery, fleecy, furry, fuzzy, hirsute, long-haired, shaggy, stubbly, woolly. *Opp* HAIRLESS.

**half-hearted** *adj* apathetic, cool, easily distracted, feeble, indifferent, ineffective, lackadaisical, listless, lukewarm, nonchalant, passive, perfunctory, phlegmatic, uncaring, uncommitted, unconcerned, unenthusiastic, unreliable, wavering, weak, *inf* wishy-washy. *Opp* ENTHUSIASTIC.

**hall** *n* **1** auditorium, concert-hall, lecture room, theatre. **2** corridor, entrance-hall, foyer, hallway, lobby, passage, passageway, vestibule.

**hallowed** *adj* blessed, consecrated, dedicated, holy, honoured, revered, reverenced, sacred, sacrosanct, worshipped.

**hallucinate** *vb* day-dream, dream, fantasize, *inf* have a trip, have hallucinations, *inf* see things, see visions.

**hallucination** *n* apparition, chimera, day-dream, delusion, dream, fantasy, figment of the imagination, illusion, mirage, vision. ▷ GHOST.

**halt** *n* break, cessation, close, end, interruption, pause, standstill, stop, stoppage, termination. • *vb* **1** arrest, block, break off, cease, check, curb, end, impede, obstruct, stop, terminate. **2** come to a halt, come to rest, desist, discontinue, draw up, pull up, quit, stop, wait. *Opp* START. **halting** ▷ HESITANT, IRREGULAR.

**halve** *vb* bisect, cut by half, cut in half, decrease, divide into halves, lessen, reduce by half, share equally, split in two.

**hammer** *n* mallet, sledge-hammer. • *vb inf* bash, batter, beat, drive, knock, pound, smash, strike. ▷ DEFEAT, HIT.

**hamper** *vb* baulk, block, curb, curtail, delay, encumber, entangle, fetter, foil, frustrate, handicap, hinder, hold back, hold up, impede, inhibit, interfere with, obstruct, prevent, restrain, restrict, retard, shackle, slow down, thwart, trammel. *Opp* HELP.

**hand** *n* **1** fist, *sl* mitt, palm, *inf* paw. **2** *hand on a dial.* index, indicator, pointer. **3** [*old use*] *factory hands.* ▷ WORKER. • *vb* convey, deliver, give, offer, pass, present, submit. **at hand** ▷ HANDY. **give a hand** ▷ HELP. **hand down** ▷ BEQUEATH. **hand over** ▷ SURRENDER. **hand round** ▷ DISTRIBUTE. **lend a hand** ▷ HELP. **to hand** ▷ HANDY.

**handicap** *n* **1** barrier, burden, disadvantage, difficulty, drawback, encumbrance, hindrance, impediment, inconvenience, limitation, *inf* minus, nuisance, obstacle, problem, restraint, restriction, shortcoming, stumbling-block. *Opp* ADVANTAGE. **2** defect, disability, impairment. • *vb* be a handicap to, burden, check, curb, disable, disadvantage, encumber, hamper, hinder, hold back, impede, limit, restrain, restrict, retard, trammel. *Opp* HELP.

**handicapped** *adj* [*Some synonyms may cause offence*] autistic, bedridden, blind, crippled, deaf, disabled disadvantaged, dumb, dyslexic, incapacitated, invalid, lame, limbless, maimed, mute, paralysed, paraplegic, retarded, slow, spastic, unsighted. ▷ ILL.

**handiwork** *n* achievement, creation, doing, invention, production, responsibility, work.

**handle** *n* grip, haft, handgrip, helve, hilt, knob, stock (*of rifle*). • *vb* **1** caress, feel, finger, fondle, grasp, hold, *inf* maul, pat, *inf* paw, stroke, touch, treat. **2** *handle situations, people.* conduct, contend with, control, cope with, deal with, direct, guide, look after, manage, manipulate, tackle, treat. ▷ ORGANIZE. **3** *car handles well.* manoeuvre, operate, respond, steer, work. **4** *handle goods.* deal in, do trade in, market, sell, stock, touch, traffic in.

**andsome** *adj* **1** admirable, attractive, beautiful, comely, elegant, fair, fine-looking, good-looking, personable, tasteful. *Opp* UGLY. **2** *handsome gift.* big, bountiful, generous, goodly, gracious, large, liberal, magnanimous, munificent, sizeable, unselfish, valuable. *Opp* MEAN.

**andy** *adj* **1** convenient, easy to use, helpful, manageable, practical, serviceable, useful, well-designed, worth having. **2** *handy with tools.* adept, capable, clever, competent, practical, proficient, skilful. **3** *keep tools handy.* accessible, at hand, available, close at hand, easy to reach, get-at-able, nearby, reachable, ready, to hand. *Opp* AWKWARD, INACCESSIBLE.

**ang** *vb* **1** be suspended, dangle, depend, droop, flap, flop, sway, swing, trail down. **2** *hang washing.* attach, drape, fasten, fix, peg up, pin up, stick up, suspend. **3** *hang in the air.* drift, float, hover. **hang about** ▷ DAWDLE. **hang back** ▷ HESITATE. **hanging** ▷ PENDENT. **hangings** ▷ DRAPE. **hang on** ▷ WAIT. **hang on to** ▷ KEEP.

**ank** *n* coil, length, loop, piece, skein.

**anker** *vb* ache, covet, crave, desire, fancy, *inf* have a yen, hunger, itch, long, pine, thirst, want, wish, yearn.

**haphazard** *adj* accidental, adventitious, arbitrary, casual, chance, chaotic, confusing, disorderly, disorganized, fortuitous, *inf* higgledy-piggledy, *inf* hit-or-miss, illogical, irrational, random, serendipitous, unforeseen, unplanned, unstructured, unsystematic. *Opp* ORDERLY.

**happen** *vb* arise, befall, *old use* betide, chance, come about, crop up, emerge, follow, materialize, occur, result, take place, *inf* transpire, *inf* turn out. **happen on** ▷ FIND.

**happening** *n* accident, affair, chance, circumstance, episode, event, incident, occasion, occurrence, phenomenon.

**happiness** *n* bliss, cheer, cheerfulness, contentment, delight, ecstasy, elation, enjoyment, euphoria, exhilaration, exuberance, felicity, gaiety, gladness, glee, *inf* heaven, high spirits, joy, joyfulness, joyousness, jubilation, light-heartedness, merriment, pleasure, pride, rapture, well-being. *Opp* SADNESS.

**happy** *adj* **1** beatific, blessed, blissful, *poet* blithe, buoyant, cheerful, cheery, contented, delighted, ecstatic, elated, enraptured, euphoric, exhilarated, exuberant, exultant, felicitous, festive, gay, glad, gleeful, good-humoured, gratified, grinning, halcyon (*days*), *inf* heavenly, high-spirited, idyllic, jocose, jocular, jocund, joking, jolly, jovial, joyful, joyous, jubilant, laughing, light-hearted, lively, merry, *inf* on top of the world, overjoyed, *inf* over the moon, pleased, proud, radiant, rapturous, rejoicing, relaxed, satisfied, smiling, *inf* starry-eyed, sunny, thrilled, triumphant. *Opp* SAD. **2** *a happy accident.* advantageous, appropriate, apt, auspicious, beneficial, convenient, favourable, felicitous, fortuitous, fortunate, lucky, opportune, propitious, timely, welcome, well-timed.

**harangue** *n* diatribe, exhortation, lecture, *inf* pep talk, tirade. ▷ SPEECH. • *vb* chivvy, encourage, exhort, lecture, pontificate, preach, sermonize. ▷ SPEAK, TALK.

**harass** *vb* annoy, attack, badger, bait, bother, chivvy, disturb, harry, *inf* hassle, hound, irritate, molest, nag, persecute, pester, *inf* pick on, *inf* plague, torment, trouble, vex, worry.

**harassed** *adj inf* at the end of your tether, careworn, distraught, distressed, exhausted, frayed, pressured, strained, stressed, tired, weary, worn out.

**harbour** *n* anchorage, dock, haven, jetty, landing-stage, marina, mooring, pier, port, quay, safe haven, shelter, wharf. • *vb* **1** conceal, give asylum to, give refuge to, give sanctuary to, hide, protect, shelter, shield. **2** *harbour a grudge*. cherish, cling on to, hold on to, keep in mind, maintain, nurse, nurture, retain.

**hard** *adj* **1** adamantine, compact, compressed, dense, firm, flinty, frozen, hardened, impenetrable, impervious, inflexible, rigid, rocky, solid, solidified, steely, stiff, stony, unbreakable, unyielding. **2** *hard labour*. arduous, backbreaking, exhausting, fatiguing, formidable, gruelling, harsh, heavy, laborious, onerous, rigorous, severe, stiff, strenuous, taxing, tiring, tough, uphill, wearying. **3** *a hard problem*. baffling, complex, complicated, confusing, difficult, enigmatic, insoluble, intricate, involved, knotty, perplexing, puzzling, tangled, *inf* thorny. **4** *a hard heart*. callous, cold, cruel, *inf* hard-boiled, hard-hearted, harsh, heartless, hostile, inflexible, intolerant, merciless, obdurate, pitiless, ruthless, severe, stern, strict, unbending, unfeeling, unfriendly, unkind. **5** *a hard blow*. forceful, heavy, powerful, strong, violent. **6** *hard times*. austere, bad, calamitous, disagreeable, distressing, grim, intolerable, painful, unhappy, unpleasant. **7** *a hard worker*. assiduous, conscientious, devoted, indefatigable, industrious, keen, persistent, unflagging, untiring, zealous. *Opp* EASY, SOFT. **hard-headed** ▷ BUSINESSLIKE. **hard-hearted** ▷ CRUEL. **hard up** ▷ POOR. **hard-wearing** ▷ DURABLE.

**harden** *vb* bake, cake, clot, coagulate, congeal, freeze, gel, jell, ossify, petrify, reinforce, set, solidify, stiffen, strengthen, toughen. *Opp* SOFTEN.

**hardly** *adv* barely, faintly, only just, rarely, scarcely, seldom, with difficulty.

**hardship** *n* adversity, affliction, austerity, bad luck, deprivation, destitution, difficulty, distress, misery, misfortune, need, privation, suffering, *inf* trials and tribulations, trouble, unhappiness, want.

**hardware** *n* equipment, implements, instruments, ironmongery, machinery, tools.

**hardy** *adj* **1** durable, fit, healthy, hearty, resilient, robust, rugged, strong, sturdy, tough, vigorous. *Opp* TENDER. **2** ▷ BOLD.

**harm** *n* abuse, damage, detriment, disadvantage, disservice, havoc, hurt, inconvenience, injury, loss, mischief, misfortune, pain, unhappiness, *inf* upset, wrong. ▷ EVIL. • *vb* abuse, be harmful to, damage, hurt, ill-treat, impair, injure, ma

reat, misuse, ruin, spoil, wound. *pp* BENEFIT.
**armful** *adj* addictive, bad, baleul, damaging, dangerous, deadly, eleterious, destructive, detriental, disadvantageous, evil, atal, hurtful, injurious, lethal, align, negative, noxious, pernicious, poisonous, prejudicial, ruinus, unfavourable, unhealthy, npleasant, unwholesome. *pp* BENEFICIAL, HARMLESS.
**armless** *adj* acceptable, benign, entle, innocent, innocuous, inofensive, mild, non-addictive, nonoxic, safe, tame, unobjectionable. *pp* HARMFUL.
**armonious** *adj* **1** concordant, onsonant, *inf* easy on the ear, uphonious, harmonizing, melodius, musical, sweet-sounding, onal, tuneful. *Opp* DISCORDANT. *a harmonious meeting.* agreeble, amicable, compatible, congenal, congruous, cooperative, iendly, integrated, like-minded, ympathetic.
**armonize** *vb* agree, balance, be n harmony, blend, cooperate, cordinate, correspond, go together, atch, suit each other, tally, tone n.
**armony** *n* **1** assonance, concord, onsonance, euphony, tunefulness. accord, agreement, amity, balnce, compatibility, conformity, ongruence, cooperation, friendliess, goodwill, like-mindedness, eace, rapport, sympathy, getherness, understanding. *pp* DISCORD.
**arness** *n* equipment, *inf* gear, raps, tackle. • *vb* control, domestate, exploit, keep under control, ake use of, mobilize, tame, use, tilize.
**arsh** *adj* **1** abrasive, bristly, oarse, hairy, rough, scratchy. **2** *harsh sounds.* cacophonous, croaking, croaky, disagreeable, discordant, dissonant, grating, gravelly, grinding, gruff, guttural, hoarse, husky, irritating, jarring, rasping, raucous, rough, screeching, shrill, squawking, stertorous, strident, unpleasant. **3** *harsh colours,* light. bright, brilliant, dazzling, gaudy, glaring, lurid. **4** *harsh smell.* acrid, bitter, sour, unpleasant. **5** *harsh conditions.* arduous, austere, comfortless, difficult, hard, severe, stressful, tough. **6** *harsh criticism, treatment.* abusive, acerbic, bitter, blunt, brutal, cruel, Draconian, frank, hardhearted, hurtful, impolite, merciless, outspoken, pitiless, severe, sharp, stern, strict, uncivil, unforgiving, unkind, unrelenting, unsympathetic, untempered. *Opp* GENTLE.
**harvest** *n* crop, gathering-in, produce, reaping, return, yield. • *vb* bring in, collect, garner, gather, glean, mow, pick, reap, take in.
**hash** *n* **1** goulash, stew. **2** *inf* botch, confusion, farrago, *inf* hotchpotch, jumble, mess, *inf* mishmash, mixture. **make a hash of** ▷ BUNGLE.
**hassle** *n* altercation, argument, bother, confusion, difficulty, disagreement, disturbance, fighting, fuss, harassment, inconvenience, making difficulties, nuisance, persecution, problem, struggle, trouble, upset. • *vb* ▷ HARASS, QUARREL.
**haste** *n* dispatch, hurry, impetuosity, precipitateness, quickness, rashness, recklessness, rush, urgency. ▷ SPEED.
**hasty** *adj* **1** abrupt, fast, foolhardy, headlong, hot-headed, hurried, ill-considered, immediate,

impetuous, impulsive, incautious, instantaneous, *inf* pell-mell, precipitate, quick, rapid, rash, reckless, speedy, sudden, summary (*justice*), swift. **2** *hasty work*. brief, careless, cursory, hurried, ill-considered, perfunctory, rushed, short, slapdash, superficial, thoughtless, unthinking. *Opp* CAREFUL, SLOW.

**hat** *n* head-dress. □ *Balaclava, bearskin, beret, biretta, boater, bonnet, bowler, busby, cap, coronet, crash-helmet, crown, deerstalker, diadem, fez, fillet, headband, helmet, hood, mitre, skullcap, sombrero, sou'-wester, stetson, sun-hat, tiara, top hat, toque, trilby, turban, wig, wimple, yarmulke*.

**hatch** *vb* **1** brood, incubate. **2** conceive, concoct, contrive, *inf* cook up, design, devise, *inf* dream up, formulate, invent, plan, plot, scheme, think up.

**hate** *n* **1** ▷ HATRED. **2** *a pet hate*. abomination, aversion, *Fr* bête noir, dislike, loathing. • *vb* abhor, abominate, be averse to, be hostile to, be revolted by, *inf* can't bear, *inf* can't stand, deplore, despise, detest, dislike, execrate, fear, find intolerable, loathe, object to, recoil from, resent, scorn, shudder at. *Opp* LIKE, LOVE.

**hateful** *adj* abhorred, abhorrent, abominable, accursed, awful, contemptible, cursed, *inf* damnable, despicable, detestable, disgusting, distasteful, execrable, foul, hated, heinous, horrible, *inf* horrid, loathsome, nasty, nauseating, obnoxious, odious, offensive, repellent, repugnant, repulsive, revolting, vile. ▷ EVIL. *Opp* LOVABLE.

**hatred** *n* abhorrence, animosity, antagonism, antipathy, aversion, contempt, detestation, dislike, enmity, execration, hate, hostility, ill-will, intolerance, loathing, mi[santhropy], odium, repugnance, revulsion. *Opp* LOVE.

**haughty** *adj* arrogant, boastful, bumptious, cavalier, *inf* cocky, conceited, condescending, disdainful, egotistical, *inf* high-and-mighty, *inf* hoity-toity, impe[rious], lofty, lordly, offhand, patron[iz]ing, pompous, presumptuous, pr[e]tentious, proud, self-admiring, self-important, smug, snobbish, *inf* snooty, *inf* stuck-up, supercil[i]ous, superior, *inf* uppish, vain. *Opp* MODEST.

**haul** *vb* carry, cart, convey, drag, draw, heave, *inf* lug, move, pull, tow, trail, transport, tug.

**haunt** *vb* **1** frequent, *inf* hang around, keep returning to, loiter about, patronize, spend time at, visit regularly. **2** *haunt the mind*. beset, linger in, obsess, plague, prey on, torment.

**have** *vb* **1** be in possession of, keep, maintain, own, possess, us[e], utilize. **2** *house has six rooms*. co[m]prise, consist of, contain, embod[y], hold, include, incorporate, involve. **3** *have fun, illness*. be su[b]ject to, endure, enjoy, experience, feel, go through, know, live through, put up with, suffer, tol[er]ate, undergo. **4** *have presents*. accept, acquire, be given, gain, get, obtain, procure, receive. **5** *thieves had the lot*. *inf* get away with, remove, retain, secure, ste[al], take. **6** *have a snack*. consume, e[at], drink, partake of, swallow. **7** *ha[ve] a party*. arrange, hold, organize, prepare, set up. **8** *have guests*. be host to, cater for, entertain, put up. **have on** ▷ HOAX. **have to be** compelled to, be forced to, have obligation to, must, need to, oug[ht] to, should. **have up** ▷ ARREST.

**haven** *n* asylum, refuge, retreat, safety, sanctuary, shelter. ▷ HARBOUR.

**havoc** *n* carnage, chaos, confusion, damage, desolation, destruction, devastation, disorder, disruption, *inf* mayhem, *inf* rack and ruin, ruin, *inf* shambles, upset, waste, wreckage.

**hazard** *n* chance, danger, jeopardy, peril, risk, threat. • *vb* dare, gamble, jeopardize, risk, stake, take a chance with, venture.

**hazardous** *adj* chancy, dangerous, *inf* dicey, fraught with danger, parlous, perilous, precarious, risky, *inf* ticklish, *inf* tricky, uncertain, unpredictable, unsafe. *Opp* SAFE.

**haze** *n* cloud, film, fog, mist, steam, vapour.

**hazy** *adj* **1** blurred, blurry, clouded, cloudy, dim, faint, foggy, fuzzy, indefinite, milky, misty, obscure, unclear. **2** ▷ VAGUE. *Opp* CLEAR.

**head** *adj* ▷ CHIEF. • *n* **1** brain, cranium, skull. **2** *head for figures.* ability, brains, capacity, imagination, intelligence, intellect, mind, understanding. **3** *head of a mountain.* apex, crown, highest point, peak, summit, top, vertex. **4** boss, director, employer, leader, manager, ruler. ▷ CHIEF. **5** *head of a school.* headmaster, headmistress, head teacher, principal. **6** *head of a river.* ▷ SOURCE. • *vb* **1** be in charge of, command, control, direct, govern, guide, lead, manage, rule, run, superintend, supervise. **2** *head for home.* aim, go, make, *inf* make a beeline, point, set out, start, steer, turn. head off ▷ DEFLECT. **lose your head** ▷ PANIC. **off your head** ▷ MAD.

**heading** *n* caption, headline, rubric, title.

**headquarters** *n* administration, base, depot, head office, *inf* HQ, main office, *inf* nerve-centre.

**heal** *vb* **1** become healthy, get better, improve, knit, mend, recover, recuperate, unite. **2** cure, make better, minister to, nurse, rejuvenate, remedy, renew, restore, revitalize, tend, treat. **3** *heal differences.* patch up, put right, reconcile, repair, settle.

**health** *n* **1** condition, constitution, fettle, form, shape, trim. **2** *the picture of health.* fitness, robustness, soundness, strength, vigour, well-being.

**healthy** *adj* **1** active, blooming, fine, fit, flourishing, good, *inf* hale-and-hearty, hearty, *inf* in fine fettle, in good shape, lively, perky, robust, sound, strong, sturdy, vigorous, well. **2** bracing, health-giving, hygienic, invigorating, salubrious, sanitary, wholesome. *Opp* ILL, UNHEALTHY.

**heap** *n* accumulation, assemblage, bank, collection, hill, hoard, mass, mound, mountain, pile, stack. • *vb* accumulate, amass, bank up, collect, gather, hoard, mass, pile, stack, stockpile, store. **heaps** ▷ PLENTY.

**hear** *vb* **1** attend to, catch, *old use* hearken to, heed, listen to, overhear, pay attention to, pick up. **2** *hear evidence.* examine, investigate, judge, try. **3** *hear news.* be told, discover, find out, gather, get, *inf* get wind of, learn, receive.

**hearing** *n* case, inquest, inquiry, trial.

**heart** *n* **1** *sl* ticker. **2** centre, core, crux, essence, focus, hub, inside, kernel, marrow, middle, *sl* nitty-gritty, nub, nucleus, pith. **3** affection, compassion, concern, courage, feeling, goodness, humanity, kindness, love, pity, sensitivity, sympathy, tenderness, understanding, warmth.

**heartbreaking** *adj* bitter, distressing, grievous, heart-rending, pitiful, tragic.

**heartbroken** *adj* broken-hearted, dejected, desolate, despairing, dispirited, grieved, inconsolable, miserable, *inf* shattered. ▷ SAD.

**hearten** *vb* boost, cheer up, encourage, strengthen, uplift.

**heartless** *adj* callous, cold, icy, inhuman, pitiless, ruthless, steely, stony, unconcerned, unemotional, unkind, unsympathetic. ▷ CRUEL.

**hearty** *adj* **1** enthusiastic, exuberant, friendly, genuine, healthy, heartfelt, lively, positive, robust, sincere, spirited, strong, vigorous, warm. *Opp* HALF-HEARTED. **2** *a hearty dinner*. ▷ BIG.

**heat** *n* **1** calorific value, fever, fieriness, glow, hotness, incandescence, warmth. **2** closeness, heatwave, high temperature, hot weather, humidity, sultriness, torridity, warmth. **3** *heat of the moment*. anger, ardour, eagerness, enthusiasm, excitement, fervour, feverishness, fury, impetuosity, violence. ▷ PASSION. *Opp* COLD. • *vb* bake, blister, boil, burn, cook, *inf* frizzle, fry, grill, inflame, make hot, melt, reheat, roast, scald, scorch, simmer, sizzle, smoulder, steam, stew, swelter, toast, warm. *Opp* COOL. **heated** ▷ FERVENT, HOT.

**heath** *n* common land, moor, moorland, open country, waste land, wilderness.

**heathen** *adj* atheistic, barbaric, godless, idolatrous, infidel, irreligious, pagan, philistine, savage, unenlightened. • *n* atheist, barbarian, heretic, idolater, infidel, pagan, philistine, savage, sceptic, unbeliever.

**heave** *vb* **1** drag, draw, haul, hoist, lift, lug, move, pull, raise, tow, tug. **2** ▷ THROW. **heave into sight** ▷ APPEAR. **heave up** ▷ VOMIT.

**heaven** *n* **1** after-life, Elysium, eternal rest, the hereafter, the next world, nirvana, paradise. **2** bliss, contentment, delight, ecstasy, felicity, happiness, joy, perfection, pleasure, rapture, Utopia. *Opp* HELL.

**heavenly** *adj* angelic, beatific, beautiful, blissful, celestial, delightful, divine, exquisite, glorious, lovely, other-worldly, *inf* out of this world, saintly, spiritual, sublime, unearthly, wonderful.

**heavy** *adj* **1** bulky, burdensome, compact, concentrated, dense, hefty, immovable, large, leaden, massive, ponderous, unwieldy, weighty. ▷ BIG, FAT. **2** *heavy work*. arduous, demanding, difficult, hard, exhausting, laborious, onerous, strenuous, tough. **3** *heavy rain*. penetrating, pervasive, severe, torrential. **4** *a heavy crop*. abundant, copious, laden, loaded, profuse, thick. **5** *a heavy heart*. burdened, depressed, gloomy, miserable, sorrowful. ▷ SAD. **6** [*inf*] *a heavy lecture*. deep, dull, intellectual, intense, serious, tedious, wearisome. *Opp* LIGHT. **heavy-handed** ▷ CLUMSY. **heavy-hearted** ▷ SAD.

**hectic** *adj* animated, boisterous, brisk, bustling, busy, chaotic, excited, feverish, frantic, frenetic, frenzied, hurried, hyperactive, lively, mad, overactive, restless, riotous, rumbustious, *inf* rushed off your feet, turbulent, wild. *Opp* LEISURELY.

**hedge** *n* barrier, fence, hedgerow, screen. • *vb inf* beat about the bush, be evasive, equivocate, *inf* hum and haw, quibble, stall, temporize, waffle. **hedge in** ▷ ENCLOSE.

**hedonistic** *adj* epicurean, extravagant, intemperate, luxurious, pleasure-loving, self-indulgent, sensual, sybaritic, voluptuous. *Opp* PURITANICAL.

**heed** *vb* attend to, bear in mind, concern yourself about, consider, follow, keep to, listen to, mark, mind, note, notice, obey, observe, pay attention to, regard, take notice of. *Opp* DISREGARD.

**heedful** *adj* attentive, careful, concerned, considerate, mindful, observant, sympathetic, taking notice, vigilant, watchful. *Opp* HEEDLESS.

**heedless** *adj* blind, careless, deaf, inattentive, inconsiderate, neglectful, oblivious, reckless, regardless, thoughtless, uncaring, unconcerned, unmindful, unobservant, unsympathetic. *Opp* HEEDFUL.

**heel** *vb* careen, incline, lean, list, tilt, tip.

**hefty** *adj* beefy, brawny, bulky, burly, heavy, heavyweight, hulking, husky, large, massive, mighty, muscular, powerful, robust, rugged, solid, substantial, *inf* strapping, strong, tough. ▷ BIG. *Opp* SLIGHT.

**height** *n* **1** altitude, elevation, level, tallness, vertical measurement. **2** crag, fell, hill, mound, mountain, peak, prominence, ridge, summit, top. **3** *height of your career.* acme, apogee, climax, crest, culmination, extreme, high point, maximum, peak, pinnacle, zenith.

**heighten** *vb* add to, amplify, augment, boost, build up, elevate, enhance, improve, increase, intensify, lift up, magnify, make higher, maximize, raise, reinforce, sharpen, strengthen, supplement. *Opp* LOWER, REDUCE.

**hell** *n* **1** eternal punishment, Hades, infernal regions, lower regions, nether world, *sl* the other place, underworld. **2** ▷ MISERY. *Opp* HEAVEN.

**help** *n* advice, aid, assistance, avail, backing, benefit, boost, collaboration, contribution, cooperation, encouragement, friendship, guidance, moral support, patronage, relief, remedy, succour, support. *Opp* HINDRANCE. • *vb* **1** abet, advise, aid, aid and abet, assist, back, befriend, be helpful, boost, collaborate, contribute, cooperate, encourage, facilitate, forward, further the interests of, *inf* give a hand, *inf* lend a hand, profit, promote, prop up, *inf* rally round, serve, side with, *derog* spoonfeed, stand by, subsidize, succour, support, take pity on. *Opp* HINDER. **2** *linctus helps a cough.* alleviate, benefit, cure, ease, improve, lessen, make easier, relieve, remedy. **3** *can't help it.* ▷ AVOID, PREVENT.

**helper** *n* abettor, accessory, accomplice, ally, assistant, associate, collaborator, colleague, confederate, deputy, helpmate, *inf* henchman, partner, *inf* right-hand man, second, supporter, *inf* willing hands.

**helpful** *adj* **1** accommodating, benevolent, caring, considerate, constructive, cooperative, favourable, friendly, helping, kind, neighbourly, obliging, practical, supportive, sympathetic, thoughtful, willing. **2** *a helpful comment.*

advantageous, beneficial, informative, instructive, profitable, valuable, useful, worthwhile. **3** *a helpful tool.* convenient, easy to use, handy, manageable, practical, serviceable, useful, well designed, worth having. *Opp* UNHELPFUL, USELESS.

**helping** *adj* ▷ HELPFUL.
• *n* amount, *inf* dollop, plateful, portion, ration, serving, share.

**helpless** *adj* abandoned, crippled, defenceless, dependent, deserted, destitute, disabled, exposed, feeble, handicapped, impotent, incapable, in difficulties, infirm, lame, marooned, powerless, stranded, unprotected, vulnerable. *Opp* INDEPENDENT.

**herald** *n* **1** announcer, courier, messenger, town crier. **2** *herald of spring.* forerunner, harbinger, omen, precursor, sign.
• *vb* advertise, announce, indicate, make known, proclaim, promise, publicize. ▷ FORETELL.

**herd** *n* bunch, flock, mob, pack, swarm, throng. ▷ GROUP.
• *vb* assemble, collect, congregate, drive, gather, group together, round up, shepherd.

**hereditary** *adj* **1** ancestral, bequeathed, family, handed down, inherited, passed down, passed on, willed. **2** congenital, constitutional, genetic, inborn, inbred, inherent, inheritable, innate, native, natural, transmissible, transmittable.

**heresy** *n* blasphemy, dissent, idolatry, nonconformity, rebellion, *inf* stepping out of line, unorthodox ideas.

**heretic** *n* apostate, blasphemer, dissenter, free-thinker, iconoclast, nonconformist, rebel, renegade, unorthodox thinker.
*Opp* BELIEVER.

**heretical** *adj* apostate, atheistic, blasphemous, dissenting, freethinking, heathen, iconoclastic, idolatrous, impious, irreligious, nonconformist, pagan, rebellious, unorthodox. *Opp* ORTHODOX.

**heritage** *n* birthright, culture, history, inheritance, legacy, past, tradition.

**hermit** *n* anchoress, anchorite, eremite, monk, recluse, solitary.

**hero, heroine** *ns* champion, conqueror, daredevil, exemplar, ideal, idol, luminary, protagonist, star, superman, *inf* superstar, superwoman, victor, winner.

**heroic** *adj* adventurous, audacious, bold, brave, chivalrous, courageous, daring, dauntless, doughty, epic, fearless, gallant, herculean, intrepid, lion-hearted, noble, selfless, staunch, steadfast, stout-hearted, superhuman, unafraid, valiant, valorous.
*Opp* COWARDLY.

**hesitant** *adj* cautious, diffident, dithering, faltering, halfhearted, halting, hesitating, indecisive, irresolute, nervous, *inf* shilly-shallying, shy, stammering, stumbling, stuttering, tentative, timid, uncertain, uncommitted, undecided, underconfident, unsure, vacillating, wary, wavering.
*Opp* DECISIVE, FLUENT.

**hesitate** *vb* **1** be hesitant, be indecisive, *inf* be in two minds, delay, demur, *inf* dilly-dally, dither, equivocate, falter, halt, hang back, haver, *inf* hum and haw, pause, put it off, *inf* shilly-shally, shrink back, teeter, temporize, think twice, vacillate, wait, waver.
**2** stammer, stumble, stutter.

**hesitation** *n* caution, delay, diffidence, dithering, doubt, indecision, irresolution, nervousness, reluctance, *inf* shilly-shallying, uncertainty, vacillation, wavering.

**hidden** *adj* 1 camouflaged, concealed, covered, disguised, enclosed, invisible, obscured, out of sight, private, shrouded, *inf* under wraps, undetectable, unnoticeable, unseen, veiled. *Opp* VISIBLE. 2 *hidden meaning*. abstruse, arcane, coded, covert, cryptic, dark, esoteric, implicit, mysterious, mystical, obscure, occult, recondite, secret, unclear. *Opp* OBVIOUS.

**hide** *n* fur, leather, pelt, skin. • *vb* 1 blot out, bury, camouflage, cloak, conceal, cover, curtain, disguise, eclipse, enclose, mantle, mask, obscure, put away, put out of sight, screen, secrete, shelter, shroud, veil, wrap up. 2 *go into hiding*. *inf* go to ground, *inf* hole up, keep hidden, *inf* lie low, lurk, shut yourself away, take cover. 3 *hide facts*. censor, *inf* hush up, repress, silence, suppress, withhold.

**hideous** *adj* appalling, beastly, disgusting, dreadful, frightful, ghastly, grim, grisly, grotesque, gruesome, macabre, nauseous, odious, repellent, repulsive, revolting, shocking, sickening, terrible. ▷ UGLY. *Opp* BEAUTIFUL.

**hiding-place** *n* den, haven, hide, hideaway, *inf* hide-out, *inf* hidey-hole, lair, refuge, retreat, sanctuary.

**hierarchy** *n* grading, ladder, *inf* pecking-order, ranking, scale, sequence, series, social order, system.

**high** *adj* 1 elevated, extending upwards, high-rise, lofty, raised, soaring, tall, towering. 2 aristocratic, chief, distinguished, eminent, exalted, important, leading, powerful, prominent, royal, top, upper. 3 *high prices*. dear, excessive, exorbitant, expensive, extravagant, outrageous, *inf* steep, unreasonable. 4 *high winds*. exceptional, extreme, great, intense, *inf* stiff, stormy, strong. 5 *a high reputation*. favourable, good, noble, respected, virtuous. 6 *high sounds*. acute, high-pitched, penetrating, piercing, sharp, shrill, soprano, squeaky, treble. *Opp* LOW. **high-and-mighty** ▷ ARROGANT. **high-class** ▷ EXCELLENT. **high-handed** ▷ ARROGANT. **high-minded** ▷ MORAL. **high-powered** ▷ POWERFUL. **high-speed** ▷ FAST. **high-spirited** ▷ LIVELY.

**highbrow** *adj* 1 academic, bookish, brainy, cultured, intellectual, *derog* pretentious, sophisticated. 2 *highbrow books*. classical, cultural, deep, difficult, educational, improving, serious. *Opp* LOWBROW.

**highlight** *n* best moment, climax, high spot, peak, top point.

**hilarious** *adj* boisterous, cheerful, cheering, entertaining, jolly, jovial, lively, merry, mirthful, rollicking, side-splitting, uproarious. ▷ FUNNY.

**hill** *n* 1 elevation, eminence, foothill, height, hillock, hillside, hummock, knoll, mound, mount, mountain, peak, prominence, ridge, summit. □ *brae, down, fell, pike, stack, tor, wold.* 2 acclivity, ascent, declivity, drop, gradient, incline, ramp, rise, slope.

**hinder** *vb* arrest, bar, be a hindrance to, check, curb, delay, deter, endanger, frustrate, get in the way of, hamper, handicap, hit, hold back, hold up, impede, keep back, limit, obstruct, oppose, prevent, restrain, restrict, retard, sabotage, slow down, slow up, stand in the way of, stop, thwart. *Opp* HELP.

**hindrance** *n* bar, barrier, burden, check, curb, deterrent, difficulty, disadvantage, *inf* drag, drawback, encumbrance, handicap, hitch, impediment, inconvenience, limitation, obstacle, obstruction, restraint, restriction, snag, stumbling-block. *Opp* HELP.

**hinge** *n* articulation, joint, pivot. • *vb* depend, hang, rest, revolve, turn.

**hint** *n* **1** allusion, clue, idea, implication, indication, inkling, innuendo, insinuation, pointer, shadow, sign, suggestion, tip, *inf* tip-off. **2** *a hint of herbs.* dash, taste, tinge, touch, trace, undertone, whiff. • *vb* allude, give a hint, imply, indicate, insinuate, intimate, mention, suggest, tip off.

**hire** *vb* book, charter, employ, engage, lease, pay for the use of, rent, sign on, take on. **hire out** lease out, let, rent out, take payment for.

**hiss** *vb* buzz, fizz, purr, rustle, sizzle, whir, whizz.

**historic** *adj* celebrated, eminent, epoch-making, famed, famous, important, momentous, notable, outstanding, remarkable, renowned, significant, well-known. *Opp* INSIGNIFICANT.

**historical** *adj* actual, authentic, documented, factual, real, real-life, recorded, true, verifiable. *Opp* FICTITIOUS.

**history** *n* **1** antiquity, bygone days, heritage, historical events, the old days, the past. **2** annals, biography, chronicles, diaries, narratives, records.

**histrionic** *adj* actorish, dramatic, theatrical.

**hit** *n* **1** blow, bull's eye, collision, impact, shot, stroke. **2** success, triumph, *inf* winner. • *vb* **1** bang, bash, baste, batter, beat, belt, biff, birch, box, bludgeon, buffet, bump, butt, cane, cannon into, clap, clip, clobber, clock, clonk, clout, club, collide with, cosh, crack, crash into, cudgel, cuff, dash, deliver a blow, drive, elbow, flagellate, flail, flick, flip, flog, hammer, head, head-butt, impact, jab, jar, jog, kick, knee, knock, lam, lambaste, lash, nudge, pat, poke, pound, prod, pummel, punch, punt, putt, ram, rap, run into, scourge, slam, slap, slog, slosh, slug, smack, smash, smite, sock, spank, stab, strike, stub, swat, swipe, tan, tap, thrash, thump, thwack, wallop, whack, wham, whip. **2** *The slump hit sales.* affect, attack, bring disaster to, check, damage, do harm to, harm, have an effect on, hinder, hurt, make suffer, ruin. **hit back** ▷ RETALIATE. **hit on** ▷ DISCOVER.

**hoard** *n* accumulation, cache, collection, fund, heap, pile, reserve, stockpile, store, supply, treasure-trove. • *vb* accumulate, amass, assemble, collect, gather, keep, lay in, lay up, mass, pile up, put away, put by, save, stockpile, store, treasure. *Opp* SQUANDER, USE.

**hoarse** *adj* croaking, grating, gravelly, growling, gruff, harsh, husky, rasping, raucous, rough, throaty.

**hoax** *n* cheat, *inf* con, confidence trick, deception, fake, fraud, humbug, imposture, joke, *inf* leg-pull, practical joke, spoof, swindle, trick. • *vb* bluff, cheat, *inf* con, cozen, deceive, defraud, delude, dupe, fool, gull, *inf* have on, hoodwink, lead on, mislead, *inf* pull someone's leg, swindle, *inf* take for a ride, take in, trick. ▷ TEASE.

**hoaxer** *n inf* con-man, impostor, joker, practical joker, trickster. ▷ CHEAT.

**hobble** *vb* dodder, falter, limp, shuffle, stagger, stumble, totter. ▷ WALK.

**hobby** *n* amateur interest, avocation, diversion, interest, pastime, pursuit, recreation, relaxation, sideline.

**hoist** *n* block-and-tackle, crane, davit, jack, lift, pulley, tackle, winch, windlass. • *vb* elevate, heave, lift, pull up, raise, winch up.

**hold** *n* **1** clasp, clutch, foothold, grasp, grip, purchase, toehold. **2** *a hold over someone.* ascendancy, authority, control, dominance, influence, leverage, mastery, power, sway. • *vb* **1** bear, carry, catch, clasp, clench, cling to, clutch, cradle, embrace, enfold, grasp, grip, hang on to, have, hug, keep, possess, retain, seize, support, take. **2** *hold a suspect.* arrest, confine, coop up, detain, imprison, keep in custody, restrain. **3** *hold an opinion.* believe in, stick to, subscribe to, swear to. **4** *hold a pose.* continue, keep up, maintain, occupy, preserve, retain, sustain. **5** *hold a party.* celebrate, conduct, convene, have, organize. **6** *jug holds a litre.* contain, enclose, have a capacity of, include. **7** *My offer holds.* be unaltered, carry on, continue, endure, hold out, keep on, last, persist, remain unchanged, stay. **hold back** ▷ RESTRAIN. **hold forth** ▷ SPEAK, TALK. **hold out** ▷ OFFER, PERSIST. **hold over, hold up** ▷ DELAY. **hold-up** ▷ ROBBERY.

**hole** *n* **1** abyss, burrow, cave, cavern, cavity, chamber, chasm, crater, dent, depression, excavation, fault, fissure, hollow, indentation, niche, pit, pocket, pot-hole, recess, shaft, tunnel. **2** aperture, breach, break, chink, crack, cut, eyelet, fissure, gap, gash, leak, opening, orifice, perforation, puncture, rip, slit, slot, split, tear, vent.

**holiday** *n* bank holiday, break, day off, furlough, half-term, leave, recess, respite, rest, sabbatical, time off, vacation.

**holiness** *n* devotion, divinity, faith, godliness, piety, *derog* religiosity, sacredness, saintliness, *derog* sanctimoniousness, sanctity, venerability.

**hollow** *adj* **1** empty, unfilled, vacant, void. **2** cavernous, concave, deep, depressed, dimpled, indented, recessed, sunken. **3** *a hollow laugh, victory.* cynical, false, futile, insincere, insubstantial, meaningless, pointless, valueless, worthless. • *n* bowl, cave, cavern, cavity, concavity, crater, dent, depression, dimple, dint, dip, dish, excavation, furrow, hole, indentation, pit, trough. ▷ VALLEY. **hollow out** ▷ EXCAVATE.

**holocaust** *n* **1** conflagration, firestorm, inferno. **2** annihilation, bloodbath, destruction, devastation, extermination, genocide, massacre, pogrom.

**holy** *adj* **1** blessed, consecrated, dedicated, devoted, divine, hallowed, heavenly, revered, sacred, sacrosanct, venerable. **2** *holy pilgrims.* devout, faithful, God-fearing, godly, immaculate, *derog* pietistic, pious, prayerful, pure, religious, reverent, reverential, righteous, saintly, *derog* sanctimonious, sinless, unsullied. *Opp* IRRELIGIOUS.

**home** *n* **1** abode, accommodation, base, domicile, dwelling, dwelling-place, habitation, household, lodging, quarters, residence. ▷ HOUSE. **2** birthplace, native land. **3** *derog* institution. □ old use *almshouse, convalescent home,* *hospice, nursing-home,* old use *poorhouse, rest home, retirement home, retreat, shelter.*

**homeless** *adj* abandoned, destitute, dispossessed, down-and-out, evicted, exiled, forsaken, itinerant, nomadic, outcast, rootless, unhoused, vagrant, wandering. • *plur n* beggars, refugees, tramps, vagabonds, vagrants.

**homely** *adj* comfortable, congenial, cosy, easygoing, friendly, informal, intimate, modest, natural, relaxed, simple, unaffected, unassuming, unpretentious, unsophisticated. ▷ FAMILIAR. *Opp* FORMAL, SOPHISTICATED.

**homogeneous** *adj* akin, alike, comparable, compatible, consistent, identical, indistinguishable, matching, similar, uniform, unvarying. *Opp* DIFFERENT.

**homosexual** *adj inf* camp, gay, lesbian, *derog* queer.

**honest** *adj* above-board, blunt, candid, conscientious, direct, equitable, fair, forthright, frank, genuine, good, honourable, impartial, incorruptible, just, law-abiding, legal, legitimate, moral, *inf* on the level, open, outspoken, plain, principled, pure, reliable, respectable, scrupulous, sincere, square (*deal*), straight, straightforward, trustworthy, trusty, truthful, unbiased, unequivocal, unprejudiced, upright, veracious, virtuous. *Opp* DISHONEST.

**honesty** *n* **1** fairness, goodness, honour, integrity, morality, probity, rectitude, reliability, scrupulousness, sense of justice, trustworthiness, truthfulness, uprightness, veracity, virtue. *Opp* DECEIT. **2** bluntness, candour, directness, frankness, outspokenness, plainness, sincerity, straightforwardness.

**honorary** *adj* nominal, titular, unofficial, unpaid.

**honour** *n* **1** acclaim, accolade, compliment, credit, esteem, fame, good name, *inf* kudos, regard, renown, reputation, repute, respect, reverence, veneration. **2** distinction, duty, importance, pleasure, privilege. **3** *a sense of honour.* decency, dignity, honesty, integrity, loyalty, morality, nobility, principle, rectitude, righteousness, sincerity, uprightness, virtue. • *vb* acclaim, admire, applaud, celebrate, commemorate, commend, dignify, esteem, give credit to, glorify, pay homage to, pay respects to, pay tribute to, praise, remember, respect, revere, reverence, show respect to, sing the praises of, value, venerate, worship.

**honourable** *adj* admirable, chivalrous, creditable, decent, estimable, ethical, fair, good, high-minded, irreproachable, just, law-abiding, loyal, moral, noble, principled, proper, reputable, respectable, respected, righteous, sincere, *inf* straight, trustworthy, trusty, upright, venerable, virtuous, worthy. ▷ HONEST. *Opp* DISHONOURABLE.

**hoodwink** *vb* bluff, cheat, *inf* con, cozen, deceive, defraud, delude, dupe, fool, gull, *inf* have on, hoax, lead on, mislead, *inf* pull the wool over someone's eyes, swindle, *inf* take for a ride, take in, trick.

**hook** *n* barb, crook, peg. ▷ FASTENER. • *vb* **1** ▷ FASTEN. **2** *hook a fish.* capture, catch, take.

**hooligan** *n* bully, delinquent, hoodlum, lout, mugger, rough, ruffian, *inf* tearaway, thug, tough, trouble-maker, vandal, *inf* yob. ▷ CRIMINAL.

**hoop** *n* band, circle, girdle, loop, ring.

**hop** *vb* bound, caper, dance, jump, leap, limp, prance, skip, spring, vault.

**hope** *n* **1** ambition, aspiration, craving, day-dream, desire, dream, longing, wish, yearning. **2** *hope of better weather*. assumption, conviction, expectation, faith, likelihood, optimism, promise, prospect. • *vb inf* anticipate, aspire, be hopeful, believe, contemplate, count on, desire, expect, foresee, have faith, have hope, look forward (to), trust, wish. *Opp* DESPAIR.

**hopeful** *adj* **1** assured, confident, expectant, optimistic, positive, sanguine. **2** *hopeful signs*. auspicious, cheering, encouraging, favourable, heartening, promising, propitious, reassuring. *Opp* HOPELESS.

**hopefully** *adv* **1** confidently, expectantly, optimistically, with hope. **2** [*inf*] *Hopefully I'll be better tomorrow*. all being well, most likely, probably. [Many think this use of *hopefully* is wrong]

**hopeless** *adj* **1** defeatist, demoralized, despairing, desperate, disconsolate, fatalist, negative, pessimistic, resigned, wretched. **2** *a hopeless situation*. daunting, depressing, impossible, incurable, irremediable, irreparable, irreversible. **3** [*inf*] *He's hopeless!* feeble, inadequate, incompetent, inefficient, poor, useless, weak, worthless. *Opp* HOPEFUL.

**horde** *n* band, crowd, gang, mob, swarm, throng, tribe. ▷ GROUP.

**horizontal** *adj* even, flat, level, lying down, prone, prostrate, supine. *Opp* VERTICAL.

**horrible** *adj* awful, beastly, disagreeable, dreadful, ghastly, hateful, horrid, loathsome, macabre, nasty, objectionable, odious, offensive, revolting, terrible, unkind. ▷ HORRIFIC, UNPLEASANT. *Opp* PLEASANT.

**horrific** *adj* appalling, atrocious, blood-curdling, disgusting, dreadful, frightening, frightful, grisly, gruesome, hair-raising, harrowing, horrendous, horrifying, nauseating, shocking, sickening, spine-chilling, unacceptable, unnerving, unthinkable.

**horrify** *vb* alarm, appal, disgust, frighten, harrow, nauseate, outrage, scare, shock, sicken, stun, terrify, unnerve. **horrifying** ▷ HORRIFIC.

**horror** *n* **1** abhorrence, antipathy, aversion, detestation, disgust, dislike, dismay, distaste, dread, fear, hatred, loathing, panic, repugnance, revulsion, terror. **2** awfulness, frightfulness, ghastliness, gruesomeness, hideousness.

**horse** *n* bronco, carthorse, *old use* charger, cob, colt, filly, foal, *childish* gee-gee, gelding, hack, hunter, *old use* jade, mare, mount, mule, mustang, *inf* nag, *old use* palfrey, piebald, pony, racehorse, roan, skewbald, stallion, steed, warhorse.

**horseman, horsewoman** *ns* cavalryman, equestrian, jockey, rider.

**hospitable** *adj* cordial, courteous, generous, gracious, receptive, sociable, welcoming. ▷ FRIENDLY. *Opp* INHOSPITABLE.

**hospital** *n* clinic, convalescent home, dispensary, health centre, hospice, infirmary, medical centre, nursing home, sanatorium, sick bay.

**hospitality** *n* **1** accommodation, catering, entertainment. **2** cordiality, courtesy, friendliness, generosity, sociability, warmth, welcome.

**host** *n* **1** army, crowd, mob, multitude, swarm, throng, troop. ▷ GROUP. **2** ▷ COMPÈRE.

**hostage** *n* captive, pawn, prisoner, surety.

**hostile** *adj* **1** aggressive, antagonistic, antipathetic, attacking, averse, bellicose, belligerent, combative, confrontational, ill-disposed, inhospitable, inimical, malevolent, militant, opposed, oppressive, pugnacious, resentful, rival, unfriendly, unsympathetic, unwelcoming, warlike, warring. ▷ ANGRY. *Opp* FRIENDLY. **2** *hostile conditions*. adverse, contrary, opposing, unfavourable, unhelpful, unpropitious. ▷ BAD. *Opp* FAVOURABLE.

**hostility** *n* aggression, animosity, animus, antagonism, bad feeling, belligerence, confrontation, dissension, enmity, estrangement, friction, incompatibility, malevolence, malice, opposition, pugnacity, rancour, resentment, strife, unfriendliness. ▷ HATRED. *Opp* FRIENDSHIP. **hostilities** ▷ WAR.

**hot** *adj* **1** baking, blistering, boiling, burning, close, fiery, flaming, humid, oppressive, *inf* piping, red-hot, roasting, scalding, scorching, searing, sizzling, steamy, stifling, sultry, summery, sweltering, thermal, torrid, tropical, warm, white-hot. **2** *hot temper*. ardent, eager, emotional, excited, fervent, fervid, feverish, fierce, heated, hotheaded, impatient, impetuous, inflamed, intense, passionate, violent. **3** *hot taste*. acrid, biting, gingery, peppery, piquant, pungent, spicy, strong. *Opp* COLD, COOL. **hot-tempered** ▷ BAD-TEMPERED. **hot under the collar** ▷ ANGRY.

**hotel** *n* guest house, hostel, *joc* hostelry, inn, lodge, motel, pension. ▷ ACCOMMODATION.

**hound** *n* ▷ DOG. • *vb* annoy, badger, chase, harass, harry, hunt, nag, persecute, pester, pursue.

**house** *n old use* abode, domicile, dwelling, dwelling-place, habitation, home, homestead, household, place, residence. □ *apartment,* inf *back-to-back, bungalow, chalet, cottage, council house, croft, detached house, farmhouse, flat, grange, hovel, homestead, hut, igloo, lodge, maisonette, manor, manse, mansion, penthouse,* inf *prefab, public house, rectory,* inf *semi, semi-detached house, shack, shanty, terraced house, thatched house,* inf *two-up two-down, vicarage, villa.* • *vb* accommodate, billet, board, domicile, harbour, keep, lodge, place, *inf* put up, quarter, shelter, take in.

**household** *n* establishment, family, home, ménage, *inf* set-up.

**hovel** *n* cottage, *inf* dump, hole, hut, shack, shanty, shed.

**hover** *vb* **1** be suspended, drift, float, flutter, fly, hang, poise. **2** be indecisive, dally, dither, *inf* hang about, hang around, hesitate, linger, loiter, pause, vacillate, wait about, waver.

**howl** *vb* bay, bellow, cry, roar, shout, ululate, wail, yowl.

**hub** *n* axis, centre, core, focal point, focus, heart, middle, nucleus, pivot.

**huddle** *n* ▷ GROUP. • *vb* **1** cluster, converge, crowd, flock, gather, group, heap, herd, jam, jumble, pile, press, squeeze, swarm, throng. **2** cuddle, curl up, hug, nestle, snuggle.

**hue** *n* cast, complexion, dye, nuance, shade, tincture, tinge, tint, tone. ▷ COLOUR. **hue and cry** ▷ OUTCRY.

**hug** *vb* clasp, cling to, crush, cuddle, embrace, enfold, fold in your arms, hold close, huddle together, nestle together, nurse, snuggle against, squeeze.

**huge** *adj* **1** Brobdingnagian, colossal, elephantine, enormous, gargantuan, giant, gigantic, *inf* hulking, immense, imposing, impressive, *inf* jumbo, majestic, mammoth, massive, mighty, *inf* monster, monstrous, monumental, mountainous, prodigious, stupendous, titanic, towering, *inf* tremendous, vast, weighty, *inf* whopping. ▷ BIG. **2** *huge number*. ▷ INFINITE. *Opp* SMALL.

**hulk** *n* **1** body, carcass, frame, hull, shell, wreck. **2** *a clumsy hulk*. lout, lump, oaf.

**hulking** *adj* awkward, bulky, cumbersome, heavy, ungainly, unwieldy. ▷ BIG, CLUMSY.

**hull** *n* body, framework, structure.

**hum** *vb* buzz, drone, murmur, purr, sing, thrum, vibrate, whirr. **hum and haw** ▷ HESITATE.

**human** *adj* **1** anthropoid, hominid, hominoid, mortal. **2** *human feeling*. kind, rational, reasonable, sensible, sensitive, sympathetic, thoughtful. ▷ HUMANE. *Opp* INHUMAN. **human beings** folk, humanity, mankind, men and women, mortals, people.

**humane** *adj* altruistic, benevolent, charitable, civilized, compassionate, feeling, forgiving, good, human, humanitarian, kind-hearted, loving, magnanimous, merciful, philanthropic, pitying, refined, sympathetic, tender, understanding, unselfish, warm-hearted. ▷ KIND. *Opp* INHUMANE.

**humble** *adj* **1** deferential, docile, meek, modest, *derog* obsequious, polite, reserved, respectful, self-effacing, *derog* servile, submissive, subservient, *derog* sycophantic, unassertive, unassuming, unostentatious, unpresuming, unpretentious. *Opp* PROUD. **2** *humble birth*. base, commonplace, ignoble, inferior, insignificant, low, lowly, mean, obscure, ordinary, plebeian, poor, simple, undistinguished, unimportant, unprepossessing, unremarkable. • *vb* ▷ HUMILIATE.

**humid** *adj* clammy, damp, dank, moist, muggy, steamy, sticky, sultry, sweaty.

**humiliate** *vb* abase, abash, break, break someone's spirit, bring someone down, chagrin, chasten, crush, deflate, degrade, demean, discredit, disgrace, embarrass, humble, make someone ashamed, *inf* make someone eat humble pie, *inf* make someone feel small, mortify, *inf* put someone down, *inf* put someone in his/her place, shame, *inf* show someone up, *inf* take someone down a peg. **humiliating** ▷ SHAMEFUL.

**humiliation** *n* abasement, chagrin, degradation, discredit, disgrace, dishonour, embarrassment, ignominy, indignity, loss of face, mortification, obloquy, shame.

**humility** *n* deference, humbleness, lowliness, meekness, modesty, self-abasement, self-effacement, *derog* servility, shyness, unpretentiousness. *Opp* PRIDE.

**humorous** *adj* absurd, amusing, comic, comical, diverting, droll, entertaining, facetious, farcical, funny, hilarious, *inf* hysterical, ironic, jocose, jocular, *inf* killing, laughable, merry, *inf* priceless, risible, sarcastic, sardonic, satirical, *inf* side-splitting, slapstick, uproarious, waggish, whimsical, witty, zany. *Opp* SERIOUS.

**humour** *n* **1** absurdity, badinage, banter, comedy, drollness, facetiousness, fun, incongruity, irony, jesting, jocularity, jokes, joking, merriment, quips, raillery, repartee, satire, *inf* sense of fun, waggishness, wit, witticism, wittiness. **2** *in a good humour*. disposition, frame of mind, mood, spirits, state of mind, temper.

**hump** *n* bulge, bump, curve, growth, hunch, knob, lump, node, projection, protrusion, protuberance, swelling, tumescence. **2** *hump in the ground*. barrow, hillock, hummock, mound, rise, tumulus. • *vb* **1** arch, bend, crook, curl, curve, hunch, raise. **2** *hump a load*. drag, heave, hoist, lift, lug, raise, shoulder.

**hunch** *n* **1** ▷ HUMP. **2** feeling, guess, idea, impression, inkling, intuition, premonition, presentiment, suspicion. • *vb* arch, bend, crook, curl, curve, huddle, hump, raise, shrug.

**hunger** *n* **1** appetite, craving, greed, ravenousness, voracity. **2** deprivation, famine, lack of food, malnutrition, starvation, want. • *vb* ▷ DESIRE.

**hungry** *adj* aching, avid, covetous, craving, eager, emaciated, famished, famishing, greedy, longing, *inf* peckish, ravenous, starved, starving, underfed, undernourished, voracious.

**hunt** *n* chase, pursuit, quest, search. ▷ HUNTING. • *vb* **1** chase, course, dog, ferret, hound, poach, pursue, stalk, track, trail. **2** *hunt for lost property*. *inf* check out, enquire after, ferret out, look for, rummage, search for, seek, trace, track down.

**hunter** *n* huntsman, huntswoman, predator, stalker, trapper.

**hunting** *n* blood-sports, coursing, poaching, stalking, trapping.

**hurdle** *n* **1** barricade, barrier, fence, hedge, jump, obstacle, wall. **2** bar, check, complication, difficulty, handicap, hindrance, impediment, obstruction, problem, restraint, snag, stumbling block.

**hurl** *vb* cast, catapult, chuck, dash, fire, fling, heave, launch, *inf* let fly, pelt, pitch, project, propel, send, shy, sling, throw, toss.

**hurricane** *n* cyclone, storm, tempest, tornado, typhoon, whirlwind.

**hurry** *n* ▷ HASTE. • *vb* **1** *inf* belt, *inf* buck up, chase, dash, dispatch, *inf* fly, *inf* get a move on, hasten, hurtle, hustle, make haste, move quickly, rush, *inf* shift, speed, *inf* step on it, work faster. **2** *hurry a process*. accelerate, expedite, press on with, quicken, speed up. *Opp* DELAY. **hurried** ▷ HASTY.

**hurt** *vb* **1** ache, be painful, burn, pinch, smart, sting, suffer pain, throb, tingle. **2** *hurt physically*. abuse, afflict, agonize, bruise, cause pain to, cripple, cut, disable, injure, maim, misuse, mutilate, torture, wound. **3** *hurt mentally*. affect, aggrieve, be hurtful to, *inf* cut to the quick, depress, distress, grieve, humiliate, insult, offend, pain, sadden, torment, upset. **4** *hurt things*. damage, harm, impair, mar, ruin, sabotage, spoil.

**hurtful** *adj* biting, cruel, cutting, damaging, derogatory, detrimental, distressing, hard to bear, harmful, injurious, malicious, nasty, painful, sarcastic, scathing, spiteful, uncharitable, unkind, upsetting, vicious, wounding. *Opp* KIND.

**hurtle** *vb* charge, chase, dash, fly, plunge, race, rush, shoot, speed, tear.

**hush** *int* be quiet! be silent! *inf* hold your tongue! *sl* pipe down! *inf* shut up! • *vb* ▷ SILENCE. **hush up** ▷ SUPPRESS.

**hustle** *vb* **1** bustle, hasten, hurry, jostle, rush, scamper, scurry. **2** *hustled me away*. coerce, compel, force, push, shove, thrust.

**hut** *n* cabin, den, hovel, lean-to, shack, shanty, shed, shelter.

**hybrid** *n* amalgam, combination, composite, compound, cross, cross-breed, half-breed, mixture, mongrel.

**hygiene** *n* cleanliness, health, sanitariness, sanitation, wholesomeness.

**hygienic** *adj* aseptic, clean, disinfected, germ-free, healthy, pure, salubrious, sanitary, sterile, sterilized, unpolluted, wholesome. *Opp* UNHEALTHY.

**hypnotic** *adj* fascinating, irresistible, magnetic, mesmeric, mesmerizing, sleep-inducing, soothing, soporific, spellbinding.

**hypnotize** *vb* bewitch, captivate, cast a spell over, dominate, enchant, entrance, fascinate, gain power over, magnetize, mesmerize, *inf* put to sleep, spellbind, *inf* stupefy.

**hypocrisy** *n* cant, deceit, deception, double-dealing, double standards, double-talk, double-think, duplicity, falsity, *inf* humbug, inconsistency, insincerity.

**hypocritical** *adj* deceptive, double-dealing, double-faced, duplicitous, false, inconsistent, insincere, Pharisaical, *inf* phoney, self-deceiving, self-righteous, *inf* two-faced.

**hypothesis** *n* conjecture, guess, postulate, premise, proposition, speculation, supposition, theory, thesis.

**hypothetical** *adj* academic, alleged, assumed, conjectural, groundless, imaginary, presumed, putative, speculative, supposed, suppositional, theoretical, unreal.

**hysteria** *n* frenzy, hysterics, madness, mania, panic.

**hysterical** *adj* berserk, beside yourself, crazed, delirious, demented, distraught, frantic, frenzied, irrational, mad, overemotional, rabid, raving, uncontrollable, wild.

# I

**ice** *n* black ice, floe, frost, glacier, iceberg, icicle, rime.

**icy** *adj* **1** arctic, chilling, freezing, frosty, frozen, glacial, polar, Siberian. ▷ COLD. **2** *icy roads*. glassy, greasy, slippery, *inf* slippy.

**idea** *n* **1** abstraction, attitude, belief, concept, conception, conjecture, construct, conviction, doctrine, hypothesis, notion, opinion, philosophy, principle, sentiment, teaching, tenet, theory, thought, view. **2** *a bright idea*. brainwave, design, fancy, guess, inspiration, plan, proposal, scheme, suggestion. **3** *idea of a poem*. intention, meaning, point. **4** *idea of what to expect*. clue, guidelines, impression, inkling, intimation, model, pattern, perception, suspicion, vision.

**ideal** *adj* **1** best, classic, complete, excellent, faultless, model, optimum, perfect, supreme, unsurpassable. **2** *an ideal world*. chimerical, dream, hypothetical, illusory, imaginary, unattainable, unreal, Utopian, visionary. • *n* **1** acme, criterion, epitome, exemplar, model, paragon, pattern, standard. **2** ▷ PRINCIPLE.

**idealistic** *adj* high-minded, impractical, over-optimistic, quixotic, romantic, starry-eyed, unrealistic. *Opp* REALISTIC.

**idealize** *vb* apotheosize, deify, exalt, glamorize, glorify, *inf* put on a pedestal, romanticize. ▷ IDOLIZE.

**identical** *adj* alike, comparable, congruent, corresponding, duplicate, equal, equivalent, indistinguishable, interchangeable, like, matching, the same, similar, twin. *Opp* DIFFERENT.

**identifiable** *adj* detectable, discernible, distinctive, distinguishable, familiar, known, named, noticeable, perceptible, recognizable, unmistakable. *Opp* UNIDENTIFIABLE.

**identify** *vb* **1** distinguish, label, mark, name, pick out, pinpoint, *inf* put a name to, recognize, single out, specify, spot. **2** *identify an illness*. detect, diagnose, discover. **identify with** empathize with, feel for, *inf* put yourself in the shoes of, relate to, sympathize with.

**identity** *n* **1** *inf* ID, name. **2** character, distinctiveness, individuality, nature, particularity, personality, selfhood, singularity, uniqueness.

**ideology** *n* assumptions, beliefs, creed, convictions, ideas, philosophy, principles, tenets, theories, underlying attitudes.

**idiom** *n* argot, cant, choice of words, dialect, expression, jargon, language, manner of speaking, parlance, phrase, phraseology, phrasing, turn of phrase, usage.

**idiomatic** *adj* colloquial, natural, vernacular, well-phrased.

**idiosyncrasy** *n* characteristic, eccentricity, feature, habit, individuality, mannerism, oddity, peculiarity, quirk, trait.

**idiosyncratic** *adj* characteristic, distinctive, eccentric, individual, odd, peculiar, personal, quirky, singular, unique. *Opp* COMMON.

**idiot** *n* [*most synonyms inf*] ass, blockhead, bonehead, booby, chump, clot, cretin, dim-wit, dolt, dope, duffer, dumb-bell, dummy, dunce, dunderhead, fat-head, fool, half-wit, ignoramus, imbecile, moron, nincompoop, ninny, nitwit, simpleton, twerp, twit.

**idiotic** *adj* absurd, asinine, crazy, foolish, half-witted, imbecile, insane, irrational, mad, moronic, nonsensical, ridiculous, senseless. ▷ STUPID. *Opp* SENSIBLE.

**idle** *adj* **1** dormant, inactive, inoperative, in retirement, not working, redundant, retired, unemployed, unoccupied, unproductive, unused. **2** apathetic, good-for-nothing, indolent, lackadaisical, lazy, shiftless, slothful, slow, sluggish, torpid, uncommitted, work-shy. **3** *idle speculation*. casual, frivolous, futile, pointless, worthless. *Opp* BUSY. • *vb* be lazy, dawdle, do nothing, *inf* hang about, *inf* kill time, laze, loaf, loll, lounge about, *inf* mess about, potter, slack, stagnate, take it easy, vegetate. *Opp* WORK.

**idler** *n* *inf* good-for-nothing, *inf* layabout, *inf* lazybones, loafer, malingerer, shirker, *inf* skiver, slacker, sluggard, wastrel.

**idol** *n* 1 deity, effigy, fetish, god, graven image, icon, statue. 2 *pop idol.* celebrity, *inf* darling, favourite, hero, *inf* pin-up, star, *inf* superstar.

**idolize** *vb* adore, adulate, hero-worship, lionize, look up to, revere, reverence, venerate, worship. ▷ IDEALIZE.

**idyllic** *adj* Arcadian, bucolic, charming, delightful, happy, idealized, lovely, pastoral, peaceful, perfect, picturesque, rustic, unspoiled.

**ignite** *vb* burn, catch fire, fire, kindle, light, set alight, set on fire, spark off, touch off.

**ignoble** *adj* base, churlish, cowardly, despicable, disgraceful, dishonourable, infamous, low, mean, selfish, shabby, uncharitable, unchivalrous, unworthy. *Opp* NOBLE.

**ignorance** *n* inexperience, innocence, unawareness, unconsciousness, unfamiliarity. ▷ STUPIDITY. *Opp* KNOWLEDGE.

**ignorant** *adj* 1 ill-informed, innocent, lacking knowledge, oblivious, unacquainted, unaware, unconscious, unfamiliar (with), uninformed, unwitting. 2 benighted, *inf* clueless, illiterate, uncouth, uncultivated, uneducated, unenlightened, unlettered, unscholarly, unsophisticated. ▷ IMPOLITE, STUPID. *Opp* CLEVER, KNOWLEDGEABLE.

**ignore** *vb* disobey, disregard, leave out, miss out, neglect, omit, overlook, pass over, reject, *inf* shut your eyes to, skip, slight, snub, take no notice of, *inf* turn a blind eye to.

**ill** *adj* 1 ailing, bad, bedridden, bilious, *inf* dicky, diseased, feeble, frail, *inf* funny, *inf* groggy, indisposed, infected, infirm, invalid, nauseous, nauseated, *inf* off-colour, *inf* out of sorts, pasty, poorly, queasy, queer, *inf* seedy, sick, sickly, suffering, *inf* under the weather, unhealthy, unwell, valetudinarian, weak. *Opp* HEALTHY. 2 *ill effects.* bad, damaging, detrimental, evil, harmful, injurious, unfavourable, unfortunate, unlucky. *Opp* GOOD. • *plur n* the infirm, invalids, patients, the sick, sufferers, victims. **be ill** ail, languish, sicken. **ill-advised** ▷ MISGUIDED. **ill-bred** ▷ RUDE. **ill-fated** ▷ UNLUCKY. **ill-humoured** ▷ BAD-TEMPERED. **ill-mannered** ▷ RUDE. **ill-natured** ▷ UNKIND. **ill-omened** ▷ UNLUCKY. **ill-tempered** ▷ BAD-TEMPERED. **ill-treat** ▷ MISTREAT.

**illegal** *adj* actionable, against the law, banned, black-market, criminal, felonious, forbidden, illicit, invalid, irregular, outlawed, prohibited, proscribed, unauthorized, unconstitutional, unlawful, unlicensed, wrong, wrongful. ▷ ILLEGITIMATE. *Opp* LEGAL.

**illegible** *adj* indecipherable, indistinct, obscure, unclear, unreadable. *Opp* LEGIBLE.

**illegitimate** *adj* 1 against the rules, improper, inadmissible, incorrect, invalid, irregular, spurious, unauthorized, unjustifiable, unreasonable, unwarranted. ▷ ILLEGAL. 2 bastard, born out of wedlock, natural. *Opp* LEGITIMATE.

**illiterate** *adj* unable to read, uneducated, unlettered. ▷ IGNORANT. *Opp* LITERATE.

**illness** *n* abnormality, affliction, ailment, allergy, attack, blight, *inf* bug, complaint, condition, contagion, disability, disease, disorder, epidemic, fever, fit, health problem, indisposition, infection, infirmity, malady, malaise, pestilence, plague, sickness, *inf* trouble, *inf* turn, *inf* upset, weakness. ▷ WOUND.

**illogical** *adj* absurd, fallacious, inconsequential, inconsistent, invalid, irrational, senseless, unreasonable, unsound. ▷ SILLY. *Opp* LOGICAL.

**illuminate** *vb* **1** brighten, decorate with lights, light up, make brighter, reveal. **2** clarify, clear up, elucidate, enlighten, explain, explicate, throw light on.

**illusion** *n* **1** apparition, conjuring trick, day-dream, deception, delusion, dream, fancy, fantasy, figment of the imagination, hallucination, mirage. **2** *under an illusion.* error, false impression, misapprehension, misconception, mistake.

**illusory** *adj* chimerical, deceptive, deluding, delusive, fallacious, false, illusive, imagined, misleading, mistaken, sham, unreal, untrue. ▷ IMAGINARY. *Opp* REAL.

**illustrate** *vb* **1** demonstrate, elucidate, exemplify, explain, instance, show. **2** adorn, decorate, embellish, illuminate, ornament. **3** depict, draw pictures of, picture, portray.

**illustration** *n* **1** case in point, demonstration, example, exemplar, instance, sample, specimen. **2** decoration, depiction, diagram, drawing, figure, photograph, picture, sketch. ▷ IMAGE.

**image** *n* **1** imitation, likeness, projection, reflection, representation. ▷ PICTURE. **2** carving, effigy, figure, icon, idol, statue. **3** *the image of her mother*. counterpart, double, likeness, spitting-image, twin.

**imaginary** *adj* fabulous, fanciful, fictional, fictitious, hypothetical, imagined, insubstantial, invented, legendary, made-up, mythical, mythological, non-existent, supposed, unreal, visionary. ▷ ILLUSORY. *Opp* REAL.

**imagination** *n* artistry, creativity, fancy, ingenuity, insight, inspiration, inventiveness, *inf* mind's eye, originality, resourcefulness, sensitivity, thought, vision.

**imaginative** *adj* artistic, attractive, beautiful, clever, creative, fanciful, ingenious, innovative, inspired, inspiring, inventive, original, poetic, resourceful, sensitive, thoughtful, unusual, visionary, vivid. *Opp* UNIMAGINATIVE.

**imagine** *vb* **1** conceive, conjure up, *inf* cook up, create, dream up, envisage, fancy, fantasize, invent, make believe, make up, picture, pretend, see, think of, think up, visualize. **2** assume, believe, conjecture, guess, infer, judge, presume, suppose, surmise, suspect, think.

**imitate** *vb* **1** ape, burlesque, caricature, counterfeit, duplicate, echo, guy, mimic, parody, parrot, portray, reproduce, satirize, send up, simulate, *inf* take off, travesty. ▷ IMPERSONATE. **2** copy, emulate, follow, match, model yourself on.

**imitation** *adj* artificial, copied, counterfeit, dummy, ersatz, man-made, mock, model, *inf* phoney, reproduction, sham, simulated, synthetic. *Opp* REAL. • *n* **1** copying, duplication, emulation, mimicry, repetition. **2** *inf* clone, copy, counterfeit, dummy, duplicate, fake, forgery, impersonation, impression, likeness, *inf* mock-up, model, parody, reflection, replica, reproduction, sham, simulation, *inf* take-off, toy, travesty.

shortcoming, weakness. *Opp* PERFECTION.

**impermanent** *adj* changing, destructible, ephemeral, evanescent, fleeting, momentary, passing, shifting, short-lived, temporary, transient, transitory, unstable. ▷ CHANGEABLE. *Opp* PERMANENT.

**impersonal** *adj* aloof, businesslike, cold, cool, correct, detached, disinterested, dispassionate, distant, formal, hard, inhuman, mechanical, objective, official, remote, stiff, unapproachable, unemotional, unfriendly, unprejudiced, unsympathetic, without emotion, wooden. *Opp* FRIENDLY.

**impersonate** *vb* disguise yourself as, do impressions of, dress up as, masquerade as, mimic, pass yourself off as, portray, pose as, pretend to be, *inf* take off. ▷ IMITATE.

**impertinent** *adj* bold, brazen, cheeky, *inf* cocky, *inf* cool, discourteous, disrespectful, forward, fresh, impolite, impudent, insolent, insubordinate, insulting, irreverent, pert, saucy. ▷ RUDE. *Opp* RESPECTFUL.

**impervious** *adj* **1** hermetic, impenetrable, impermeable, non-porous, solid, waterproof, water-repellent, watertight. *Opp* POROUS. **2** ▷ RESISTANT.

**impetuous** *adj* abrupt, careless, eager, hasty, headlong, hot-headed, impulsive, incautious, off-hand, precipitate, quick, rash, reckless, speedy, spontaneous, *inf* spur-of-the-moment, *inf* tearing, thoughtless, unplanned, unpremeditated, unthinking, violent. *Opp* CAUTIOUS.

**impetus** *n* boost, drive, encouragement, energy, fillip, force, impulse, incentive, inspiration, momentum, motivation, power, push, spur, stimulation, stimulus, thrust.

**impiety** *n* blasphemy, godlessness, irreverence, profanity, sacrilege, sinfulness, ungodliness, unrighteousness, wickedness. *Opp* PIETY.

**impious** *adj* blasphemous, godless, irreligious, irreverent, profane, sacrilegious, sinful, unholy. ▷ WICKED. *Opp* PIOUS.

**implausible** *adj* doubtful, dubious, far-fetched, feeble, improbable, questionable, suspect, unconvincing, unlikely, unreasonable, weak. *Opp* PLAUSIBLE.

**implement** *n* apparatus, appliance, contrivance, device, gadget, instrument, mechanism, tool, utensil. • *vb* accomplish, achieve, bring about, carry out, effect, enforce, execute, fulfil, perform, put into effect, put into practice, realize, try out.

**implicate** *vb* associate, concern, connect, embroil, enmesh, ensnare, entangle, entrap, include, incriminate, inculpate, involve, show involvement in.

**implication** *n* **1** hidden meaning, hint, innuendo, insinua-tion, overtone, purport, significance. **2** *implication in crime.* association, connection, embroilment, entanglement, inclusion, involvement.

**implicit** *adj* **1** hinted at, implied, indirect, inherent, insinuated, tacit, understood, undeclared, unexpressed, unsaid, unspoken, unstated, unvoiced. *Opp* EXPLICIT. **2** *implicit faith.* ▷ ABSOLUTE.

**imply** *vb* **1** hint, indicate, insinuate, intimate, mean, point to, suggest. **2** ▷ SIGNIFY.

**impolite** *adj* discourteous, disrespectful, ill-bred, ill-mannered, uncivil, vulgar. ▷ RUDE. *Opp* POLITE.

**import** *vb* bring in, buy in, introduce, ship in. ▷ CONVEY.

**important** *adj* **1** basic, big, cardinal, central, chief, consequential, critical, epoch-making, essential, foremost, fundamental, grave, historic, key, main, major, momentous, newsworthy, noteworthy, once in a lifetime, outstanding, pressing, primary, principal, rare, salient, serious, signal, significant, strategic, substantial, urgent, valuable, vital, weighty. **2** celebrated, distinguished, eminent, famous, great, high-ranking, influential, known, leading, notable, noted, powerful, pre-eminent, prominent, renowned, top-level, well-known. *Opp* UNIMPORTANT. **be important** ▷ MATTER.

**importunate** *adj* demanding, impatient, insistent, persistent, pressing, relentless, urgent, unremitting.

**importune** *vb* badger, harass, hound, pester, plague, plead with, press, solicit, urge. ▷ ASK.

**impose** *vb* charge with, decree, dictate, enforce, exact, fix, foist, force, inflict, insist on, introduce, lay, levy, prescribe, set.

**impose on** ▷ BURDEN, EXPLOIT.

**imposing** ▷ IMPRESSIVE.

**impossible** *adj* hopeless, impracticable, impractical, inconceivable, insoluble, insuperable, insurmountable, *inf* not on, out of the question, unachievable, unattainable, unimaginable, unobtainable, unthinkable, unviable, unworkable. *Opp* POSSIBLE.

**impotent** *adj* debilitated, decrepit, emasculated, enervated, helpless, inadequate, incapable, incompetent, ineffective, ineffectual, inept, infirm, powerless, unable. ▷ WEAK. *Opp* POTENT.

**impracticable** *adj* not feasible, unachievable, unworkable, useless. ▷ IMPOSSIBLE. *Opp* PRACTICABLE.

**impractical** *adj* academic, idealistic, quixotic, romantic, theoretical, unrealistic, visionary. *Opp* PRACTICAL.

**imprecise** *adj* ambiguous, approximate, careless, estimated, fuzzy, guessed, hazy, ill-defined, inaccurate, inexact, inexplicit, loose, *inf* sloppy, undefined, unscientific, vague, *inf* waffly, *inf* woolly. *Opp* PRECISE.

**impregnable** *adj* impenetrable, invincible, inviolable, invulnerable, safe, secure, strong, unassailable, unconquerable. *Opp* VULNERABLE.

**impress** *vb* **1** affect, be memorable to, excite, influence, inspire, leave its mark on, move, persuade, *inf* stick in the mind of, stir, touch. **2** *impress a mark*. emboss, engrave, imprint, mark, print, stamp.

**impression** *n* **1** effect, impact, influence, mark. **2** belief, consciousness, fancy, feeling, hunch, idea, memory, notion, opinion, recollection, sense, suspicion, view. **3** dent, hollow, imprint, indentation, mark, print, stamp. **4** imitation, impersonation, mimicry, parody, *inf* take-off. **5** *impression of a book*. edition, printing, reprint.

**impressionable** *adj* easily influenced, gullible, inexperienced, naïve, persuadable, receptive, responsive, suggestible, susceptible.

**impressive** *adj* affecting, august, awe-inspiring, awesome, commanding, distinguished, evocative, exciting, formidable, grand, *derog* grandiose, great, imposing, magnificent, majestic, memorable, moving, powerful, redoubtable, remarkable, splendid, stately, stirring, striking, touching. ▷ BIG. *Opp* INSIGNIFICANT.

**imprison** *vb* cage, commit to prison, confine, detain, gaol, immure, incarcerate, intern, jail, keep in custody, keep under house arrest, *inf* keep under lock and key, lock away, lock up, *inf* put away, remand, *inf* send down, shut in, shut up. *Opp* FREE.

**imprisonment** *n* confinement, custody, detention, duress, gaol, house arrest, incarceration, internment, jail, remand, restraint.

**improbable** *adj* absurd, doubtful, dubious, far-fetched, *inf* hard to believe, implausible, incredible, preposterous, questionable, unbelievable, unconvincing, unexpected, unlikely. *Opp* PROBABLE.

**impromptu** *adj inf* ad-lib, extempore, extemporized, improvised, impulsive, made-up, offhand, *inf* off the cuff, *inf* off the top of your head, *inf* on the spur of the moment, spontaneous, unplanned, unpremeditated, unprepared, unrehearsed, unscripted. ▷ IMPULSIVE. *Opp* REHEARSED.

**improper** *adj* **1** ill-judged, ill-timed, inappropriate, incorrect, infelicitous, inopportune, irregular, mistaken, out of place, uncalled-for, unfit, unseemly, unsuitable, unwarranted. ▷ WRONG. **2** ▷ INDECENT. *Opp* PROPER.

**impropriety** *n* inappropriateness, incorrectness, indecency, indelicacy, infelicity, insensitivity, irregularity, rudeness, unseemliness. ▷ OBSCENITY. *Opp* PROPRIETY.

**improve** *vb* **1** advance, develop, get better, grow, increase, *inf* look up, move on, progress, *inf* take a turn for the better. **2** *improve after illness*. convalesce, *inf* pick up, rally, recover, recuperate, revive, strengthen, *inf* turn the corner. **3** *improve your ways*. ameliorate, amend, better, correct, enhance, enrich, make better, mend, polish (up), rectify, refine, reform, revise. **4** *improve a home*. decorate, extend, modernize, rebuild, recondition, refurbish, renovate, repair, touch up, update, upgrade. *Opp* WORSEN.

**improvement** *n* **1** advance, amelioration, betterment, correction, development, enhancement, gain, increase, progress, rally, recovery, reformation, upswing, upturn. **2** *home improvements*. alteration, extension, *inf* face-lift, modernization, modification, renovation.

**improvise** *vb* **1** *inf* ad-lib, concoct, contrive, devise, invent, make do, make up, *inf* throw together. **2** extemporize, perform impromptu, play by ear, vamp.

**impudent** *adj* audacious, bold, *inf* cheeky, disrespectful, forward, *inf* fresh, impertinent, insolent, pert, presumptuous, saucy. ▷ RUDE. *Opp* RESPECTFUL.

**impulse** *n* **1** drive, force, impetus, motive, pressure, push, stimulus, thrust. **2** caprice, desire, instinct, urge, whim.

**impulsive** *adj* automatic, emotional, hare-brained, hasty, headlong, hot-headed, impetuous, instinctive, intuitive, involuntary, madcap, precipitate, rash, reckless, *inf* snap, spontaneous, *inf* spur-of-the-moment, sudden, thoughtless, unconscious, unplanned, unpremeditated, unthinking, wild. ▷ IMPROMPTU. *Opp* DELIBERATE.

**impure** *adj* **1** adulterated, contaminated, defiled, foul, infected, polluted, tainted, unclean, unwholesome. ▷ DIRTY. **2** ▷ INDECENT.

**impurity** *n* contamination, defilement, infection, pollution, taint. ▷ DIRT.

**inaccessible** *adj* cut off, deserted, desolate, godforsaken, impassable, impenetrable, inconvenient, isolated, lonely, *inf* off the beaten track, outlying, out of reach, out-of-the-way, private, remote, solitary, unavailable, unfrequented, *inf* unget-at-able, unobtainable, unreachable, unusable. *Opp* ACCESSIBLE.

**inaccurate** *adj* erroneous, fallacious, false, faulty, flawed, imperfect, imprecise, incorrect, inexact, misleading, mistaken, unfaithful, unreliable, unsound, untrue, vague, wrong. *Opp* ACCURATE.

**inactive** *adj* asleep, dormant, hibernating, idle, immobile, inanimate, indolent, inert, languid, lazy, lethargic, out of action, passive, quiescent, quiet, sedentary, sleepy, slothful, slow, sluggish, somnolent, torpid, unemployed, unoccupied, vegetating. *Opp* ACTIVE.

**inadequate** *adj* deficient, disappointing, faulty, imperfect, incompetent, incomplete, ineffective, insufficient, limited, meagre, mean, niggardly, *inf* pathetic, scanty, scarce, *inf* skimpy, sparse, unacceptable, unsatisfactory, unsuitable. *Opp* ADEQUATE.

**inadvisable** *adj* foolish, ill-advised, imprudent, misguided, unwise. ▷ SILLY. *Opp* WISE.

**inanimate** *adj* cold, dead, dormant, immobile, inactive, insentient, lifeless, motionless, spiritless, unconscious. *Opp* ANIMATE.

**inappropriate** *adj* ill-judged, ill-suited, ill-timed, improper, inapplicable, inapposite, incompatible, incongruous, incorrect, inept, inopportune, irrelevant, out of place, tactless, tasteless, unbecoming, unbefitting, unfit, unseasonable, unseemly, unsuitable, unsuited, untimely, wrong. *Opp* APPROPRIATE.

**inarticulate** *adj* dumb, faltering, halting, hesitant, mumbling, mute, shy, silent, speechless, stammering, stuttering, tongue-tied, voiceless. ▷ INCOHERENT. *Opp* ARTICULATE.

**inattentive** *adj* absent-minded, abstracted, careless, day-dreaming, distracted, dreaming, drifting, heedless, *inf* in a world of your own, lacking concentration, negligent, preoccupied, rambling, remiss, slack, unobservant, vague, wandering, wool-gathering. *Opp* ATTENTIVE.

**inaudible** *adj* imperceptible, mumbled, quiet, silenced, silent, stifled, undetectable, undistinguishable, unheard. ▷ FAINT. *Opp* AUDIBLE.

**incapable** *adj* **1** clumsy, helpless, impotent, inadequate, incompetent, ineffective, ineffectual, inept, powerless, stupid, unable, unfit, unqualified, useless, weak. *Opp* CAPABLE. **2** ▷ DRUNK.

**incentive** *n* bait, *inf* carrot, encouragement, enticement, impetus, incitement, inducement, lure, motivation, reward, stimulus, *inf* sweetener.

**incessant** *adj* ceaseless, chronic, constant, continual, continuous, endless, eternal, everlasting, interminable, never-ending, non-stop, perennial, permanent, perpetual, persistent, relentless, unbroken, unceasing, unending, unremitting. *Opp* INTERMITTENT, TEMPORARY.

**incident** *n* **1** affair, circumstance, episode, event, fact, happening, occasion, occurrence, proceeding. **2** *a nasty incident.* accident, confrontation, disturbance, fight, scene, upset. ▷ COMMOTION.

**incidental** *adj* accidental, adventitious, attendant, casual, chance, fortuitous, inessential, minor, odd, random, secondary, serendipitous, subordinate, subsidiary, unplanned. *Opp* ESSENTIAL.

**incipient** *adj* beginning, developing, early, embryonic, growing, new, rudimentary, starting.

**incisive** *adj* acute, clear, concise, cutting, decisive, direct, penetrating, percipient, precise, sharp, telling, trenchant. *Opp* VAGUE.

**incite** *vb* awaken, encourage, excite, fire, foment, inflame, inspire, prompt, provoke, rouse, spur on, stimulate, stir, urge, whip up, work up.

**inclination** *n* affection, bent, bias, disposition, fondness, habit, instinct, leaning, liking, partiality, penchant, predilection, predisposition, preference, proclivity, propensity, readiness, tendency, trend, willingness. ▷ DESIRE.

**incline** *n* acclivity, ascent, declivity, descent, drop, grade, gradient, hill, pitch, ramp, rise, slope. • *vb* angle, ascend, bank, bend, bow, descend, drop, gravitate, lean, rise, slant, slope, tend, tilt, tip, veer. **inclined (to)** ▷ LIABLE.

**include** *vb* **1** add in, blend in, combine, comprehend, comprise, consist of, contain, embody, embrace, encompass, incorporate, involve, make room for, mix, subsume, take in. **2** *The price includes tea.* allow for, cover, take into account. *Opp* EXCLUDE.

**incoherent** *adj* confused, disconnected, disjointed, disordered, disorganized, garbled, illogical, incomprehensible, inconsistent, irrational, jumbled, mixed up, muddled, rambling, scrambled, unclear, unconnected, unstructured, unsystematic. ▷ INARTICULATE. *Opp* COHERENT.

**incombustible** *adj* fireproof, fire-resistant, flameproof, nonflammable. *Opp* COMBUSTIBLE.

**income** *n* earnings, gain, interest, pay, pension, proceeds, profits, receipts, return, revenue, salary, takings, wages. *Opp* EXPENSE.

**incoming** *adj* **1** approaching, arriving, entering, coming, landing, new, next, returning. **2** *incoming tide.* flowing, rising. *Opp* OUTGOING.

**incompatible** *adj* antipathetic, at variance, clashing, conflicting, contradictory, contrasting, different, discordant, discrepant, incongruous, inconsistent, irreconcilable, mismatched, opposed, unsuited. *Opp* COMPATIBLE.

**incompetent** *adj* **1** bungling, clumsy, feckless, gauche, helpless, *inf* hopeless, incapable, ineffective, ineffectual, inefficient, inexperienced, maladroit, unfit, unqualified, unskilled, untrained. **2** bungled, inadequate, inexpert, unacceptable, unsatisfactory, unskilful, useless. *Opp* COMPETENT.

**incomplete** *adj* abbreviated, abridged, *inf* bitty, deficient, edited, expurgated, faulty, fragmentary, imperfect, insufficient, partial, selective, shortened, sketchy, unfinished, unpolished, wanting. *Opp* COMPLETE.

**incomprehensible** *adj* abstruse, arcane, baffling, beyond comprehension, cryptic, deep, enigmatic, esoteric, illegible, impenetrable, indecipherable, meaningless, mysterious, mystifying, obscure, opaque, *inf* over my head, perplexing, puzzling, recondite, strange, too difficult, unclear, unfathomable, unintelligible. *Opp* COMPREHENSIBLE.

**inconceivable** *adj* implausible, impossible to understand, incredible, *inf* mind-boggling, staggering, unbelievable, undreamed-of, unimaginable, unthinkable. *Opp* CREDIBLE.

**inconclusive** *adj* ambiguous, equivocal, indecisive, indefinite, interrogative, open, open-ended, questionable, uncertain, unconvincing, unresolved, *inf* up in the air. *Opp* CONCLUSIVE.

**incongruous** *adj* clashing, conflicting, contrasting, discordant, ill-matched, ill-suited, inappropriate, incompatible, inconsistent, irreconcilable, odd, out of keeping, out of place, surprising, uncoordinated, unsuited. ▷ ABSURD. *Opp* COMPATIBLE.

**inconsiderate** *adj* careless, cruel, heedless, insensitive, intolerant, negligent, rude, self-centred, selfish, tactless, thoughtless, uncaring, unconcerned, unfriendly, ungracious, unhelpful, unkind, unsympathetic, unthinking. *Opp* CONSIDERATE.

**inconsistent** *adj* capricious, changeable, erratic, fickle, inconstant, patchy, unpredictable, unreliable, unstable, *inf* up-and-down, variable. ▷ INCOMPATIBLE. *Opp* CONSISTENT.

**inconspicuous** *adj* camouflaged, concealed, discreet, hidden, insignificant, in the background, invisible, modest, ordinary, out of sight, plain, restrained, retiring, self-effacing, small, unassuming, unobtrusive, unostentatious. *Opp* CONSPICUOUS.

**inconvenience** *n* annoyance, bother, discomfort, disruption, drawback, encumbrance, hindrance, impediment, irritation, nuisance, trouble. • *vb* annoy, bother, discommode, disturb, incommode, irk, irritate, *inf* put out, trouble.

**inconvenient** *adj* annoying, awkward, bothersome, cumbersome, difficult, embarrassing, ill-timed, inopportune, irksome, irritating, tiresome, troublesome, unsuitable, untimely, untoward, unwieldy. *Opp* CONVENIENT.

**incorporate** *vb* admit, combine, comprehend, comprise, consist of, contain, embody, embrace, encompass, include, involve, mix in, subsume, take in, take into account, unite. *Opp* EXCLUDE.

**incorrect** *adj* erroneous, fallacious, false, faulty, imprecise, improper, inaccurate, inexact, mendacious, misinformed, misleading, mistaken, specious, untrue. *Opp* CORRECT.

**incorrigible** *adj* confirmed, *inf* dyed-in-the-wool, habitual, hardened, *inf* hopeless, impenitent, incurable, inveterate, irredeemable, obdurate, shameless, unalterable, unreformable, unrepentant. ▷ WICKED.

**incorruptible** *adj* **1** honest, honourable, just, moral, sound, *inf* straight, true, trustworthy, unbribable, upright. *Opp* CORRUPT. **2** ▷ EVERLASTING.

**increase** *n* addition, amplification, augmentation, boost, build-up, crescendo, development, enlargement, escalation, expansion, extension, gain, growth, increment, inflation, intensification, proliferation, rise, spread, upsurge, upturn. • *vb* 1 add to, advance, amplify, augment, boost, broaden, build up, develop, enlarge, expand, extend, improve, lengthen, magnify, make bigger, maximize, multiply, prolong, put up, raise, *inf* step up, strengthen, stretch, swell, widen. 2 escalate, gain, get bigger, grow, intensify, proliferate, *inf* snowball, spread, wax. *Opp* DECREASE.

**incredible** *adj* beyond belief, far-fetched, implausible, impossible, improbable, inconceivable, miraculous, surprising, unbelievable, unconvincing, unimaginable, unlikely, untenable, unthinkable. ▷ EXTRAORDINARY. *Opp* CREDIBLE.

**incredulous** *adj* disbelieving, distrustful, doubtful, dubious, mistrustful, questioning, sceptical, suspicious, unbelieving, uncertain, unconvinced. *Opp* CREDULOUS.

**incriminate** *vb* accuse, blame, charge, embroil, implicate, inculpate, indict, involve, *inf* point the finger at. *Opp* EXCUSE.

**incur** *vb* earn, expose yourself to, get, lay yourself open to, provoke, run up, suffer.

**incurable** *adj* 1 fatal, hopeless, inoperable, irremediable, irreparable, terminal, untreatable. *Opp* CURABLE. 2 ▷ INCORRIGIBLE.

**indebted** *adj old use* beholden, bound, grateful, obliged, thankful, under an obligation.

**indecent** *adj inf* blue, coarse, crude, dirty, immodest, impolite, improper, impure, indelicate, insensitive, naughty, obscene, offensive, risqué, rude, *inf* sexy, *inf* smutty, suggestive, titillating, unprintable, unrepeatable, unsuitable, vulgar. ▷ INDECOROUS. *Opp* DECENT.

**indecisive** *adj* doubtful, equivocal, evasive, *inf* in two minds, irresolute, undecided. ▷ HESITANT, INDEFINITE. *Opp* DECISIVE. **be indecisive** ▷ HESITATE.

**indecorous** *adj* churlish, ill-bred, inappropriate, *inf* in bad taste, tasteless, unbecoming, uncouth, undignified, unseemly, vulgar. ▷ INDECENT. *Opp* DECOROUS.

**indefensible** *adj* incredible, insupportable, unjustifiable, unpardonable, unreasonable, unsound, untenable, vulnerable, weak. ▷ WRONG.

**indefinite** *adj* ambiguous, blurred, confused, dim, general, ill-defined, imprecise, indeterminate, inexact, inexplicit, *inf* leaving it open, neutral, obscure, uncertain, unclear, unsettled, unspecific, unspecified, unsure, vague. ▷ INDECISIVE. *Opp* DEFINITE.

**indelible** *adj* fast, fixed, indestructible, ineffaceable, ineradicable, ingrained, lasting, unfading, unforgettable. ▷ PERMANENT.

**indentation** *n* cut, dent, depression, dimple, dip, furrow, groove, hollow, indent, mark, nick, notch, pit, recess, score, serration, toothmark, zigzag.

**independence** *n* 1 autonomy, freedom, individualism, liberty, nonconformity, self-confidence, self-reliance, self-sufficiency. 2 autarchy, home rule, self-determination, self-government, self-rule, sovereignty.

**independent** *adj* 1 carefree, *inf* footloose, free, freethinking, individualistic, nonconformist, non-partisan, open-minded, private, self-confident, self-reliant, separate, spontaneous, unbeholden, unbiased, uncommitted, unconventional, unprejudiced, untrammelled, without ties. **2** autonomous, liberated, neutral, non-aligned, self-determining, self-governing, sovereign.

**indescribable** *adj* beyond words, indefinable, inexpressible, stunning, unspeakable, unutterable.

**indestructible** *adj* durable, enduring, eternal, everlasting, immortal, imperishable, ineradicable, lasting, permanent, shatterproof, solid, strong, tough, toughened, unbreakable.

**index** *n* **1** catalogue, directory, guide, key, register, table (*of contents*). **2** ▷ INDICATOR.

**indicate** *vb* announce, betoken, convey, communicate, denote, describe, designate, display, evidence, express, give an indication (*of*), give notice of, imply, intimate, make known, manifest, mean, notify, point out, register, reveal, say, show, signal, signify, specify, spell, stand for, suggest, symbolize, warn.

**indication** *n* augury, clue, evidence, forewarning, hint, inkling, intimation, omen, portent, sign, signal, suggestion, symptom, token, warning.

**indicator** *n* clock, dial, display, gauge, index, instrument, marker, meter, needle, pointer, screen, sign, signal.

**indifferent** *adj* **1** aloof, apathetic, blasé, bored, casual, cold, cool, detached, disinterested, dispassionate, distant, half-hearted, impassive, incurious, insouciant, neutral, nonchalant, not bothered, uncaring, unconcerned, unemotional, unenthusiastic, unexcited, unimpressed, uninterested, uninvolved, unmoved. ▷ IMPARTIAL. *Opp* ENTHUSIASTIC. **2** commonplace, fair, mediocre, middling, moderate, *inf* nothing to write home about, *inf* poorish, undistinguished, unexciting. ▷ ORDINARY. *Opp* EXCELLENT.

**indigestion** *n* dyspepsia, flatulence, heartburn.

**indignant** *adj inf* aerated, annoyed, cross, disgruntled, exasperated, furious, heated, infuriated, *inf* in high dudgeon, irate, irked, irritated, livid, mad, *inf* miffed, *inf* peeved, piqued, provoked, *inf* put out, riled, sore, upset, vexed. ▷ ANGRY.

**indirect** *adj* **1** *inf* all round the houses, ambagious, bendy, circuitous, devious, erratic, long, meandering, oblique, rambling, roundabout, roving, tortuous, twisting, winding, zigzag. **2** *an indirect insult.* ambiguous, backhanded, circumlocutory, disguised, equivocal, euphemistic, evasive, implicit, implied, oblique. *Opp* DIRECT.

**indiscreet** *adj* careless, foolish, ill-advised, ill-considered, ill-judged, impolite, impolitic, incautious, injudicious, insensitive, tactless, undiplomatic, unguarded, unthinking, unwise. *Opp* DISCREET.

**indiscriminate** *adj* aimless, careless, casual, confused, desultory, general, haphazard, *inf* hit or miss, imperceptive, miscellaneous mixed, promiscuous, random, uncritical, undifferentiated, undiscerning, undiscriminating, uninformed, unplanned, unselective, unsystematic, wholesale. *Opp* SELECTIVE.

**indispensable** *adj* basic, central, compulsory, crucial, essential, imperative, important, key, mandatory, necessary, needed, obligatory, required, requisite, vital. *Opp* UNNECESSARY.

**indisputable** *adj* absolute, accepted, acknowledged, axiomatic, beyond doubt, certain, clear, definite, evident, incontestable, incontrovertible, indubitable, irrefutable, positive, proved, proven, self-evident, sure, unanswerable, unarguable, undeniable, undisputed, undoubted, unimpeachable, unquestionable. *Opp* DEBATABLE.

**indistinct** *adj* **1** bleary, blurred, confused, dim, dull, faint, fuzzy, hazy, ill-defined, indefinite, misty, obscure, shadowy, unclear, vague. **2** deadened, muffled, mumbled, muted, slurred, unintelligible, woolly. *Opp* DISTINCT.

**indistinguishable** *adj* alike, identical, interchangeable, the same, twin, undifferentiated. *Opp* DIFFERENT.

**individual** *adj* characteristic, different, distinct, distinctive, exclusive, idiosyncratic, individualistic, particular, peculiar, personal, private, separate, singular, special, specific, unique. *Opp* COLLECTIVE, GENERAL. • *n* ▷ PERSON.

**indoctrinate** *vb* brainwash, implant, instruct, re-educate, train. ▷ TEACH.

**induce** *vb* **1** coax, encourage, incite, influence, inspire, motivate, persuade, press, prevail on, stimulate, sway, *inf* talk into, tempt, urge. *Opp* DISCOURAGE. **2** *induce a fever*. bring on, cause, effect, engender, generate, give rise to, lead to, occasion, produce, provoke.

**inducement** *n* attraction, bait, bribe, encouragement, enticement, incentive, spur, stimulus, *inf* sweetener.

**indulge** *vb* be indulgent to, cosset, favour, give in to, gratify, humour, mollycoddle, pamper, pander to, spoil, *inf* spoonfeed, treat. *Opp* DEPRIVE. **indulge in** ▷ ENJOY. **indulge yourself** be self-indulgent, drink too much, eat too much, give in to temptation, overdo it, overeat, spoil yourself, succumb, yield.

**indulgent** *adj* compliant, easygoing, fond, forbearing, forgiving, genial, kind, lenient, liberal, overgenerous, patient, permissive, tolerant. *Opp* STRICT.

**industrious** *adj* assiduous, busy, conscientious, diligent, dynamic, earnest, energetic, enterprising, hard-working, involved, keen, laborious, persistent, pertinacious, productive, sedulous, tireless, unflagging, untiring, zealous. *Opp* LAZY.

**industry** *n* **1** business, commerce, manufacturing, production, trade. **2** activity, application, commitment, determination, diligence, dynamism, effort, energy, enterprise, industriousness, keenness, labour, perseverance, persistence, sedulousness, tirelessness, toil, zeal. ▷ WORK. *Opp* LAZINESS.

**inedible** *adj* bad for you, harmful, indigestible, nauseating, *inf* off, poisonous, rotten, tough, uneatable, unpalatable, unwholesome. *Opp* EDIBLE.

**ineffective** *adj* **1** fruitless, futile, *inf* hopeless, inept, unconvincing, unproductive, unsuccessful, useless, vain, worthless. **2** disorganized, feckless, feeble, idle, impotent, inadequate, incapable, incompetent, ineffectual, inefficient, powerless, shiftless, unenterprising, weak. *Opp* EFFECTIVE.

**inefficient** *adj* **1** extravagant, prodigal, uneconomic, wasteful. **2** ▷ INEFFECTIVE. *Opp* EFFICIENT.

**inelegant** *adj* awkward, clumsy, crude, gauche, graceless, inartistic, rough, uncouth, ungainly, unpolished, unskilful, unsophisticated, unstylish. ▷ UGLY. *Opp* ELEGANT.

**ineligible** *adj* disqualified, inappropriate, *inf* out of the running, *inf* ruled out, unacceptable, unauthorized, unfit, unqualified, unsuitable, unworthy. *Opp* ELIGIBLE.

**inept** *adj* **1** awkward, bumbling, bungling, clumsy, gauche, incompetent, inexpert, maladroit, unskilful, unskilled. **2** ▷ INAPPROPRIATE.

**inequality** *n* contrast, difference, discrepancy, disparity, dissimilarity, imbalance, incongruity, prejudice. *Opp* EQUALITY.

**inert** *adj* apathetic, dormant, idle, immobile, inactive, inanimate, lifeless, passive, quiescent, quiet, slow, sluggish, static, stationary, still, supine, torpid. *Opp* LIVELY.

**inertia** *n* apathy, deadness, idleness, immobility, inactivity, indolence, lassitude, laziness, lethargy, listlessness, numbness, passivity, sluggishness, torpor. *Opp* LIVELINESS.

**inessential** *adj* dispensable, expendable, minor, needless, non-essential, optional, ornamental, secondary, spare, superfluous, unimportant, unnecessary. *Opp* ESSENTIAL.

**inevitable** *adj* assured, *inf* bound to happen, certain, destined, fated, ineluctable, inescapable, inexorable, ordained, predictable, sure, unavoidable. ▷ RELENTLESS.

**inexcusable** *adj* ▷ UNFORGIVABLE.

**inexpensive** *adj* ▷ CHEAP.

**inexperienced** *adj inf* born yesterday, callow, *inf* green, immature, inexpert, innocent, naïve, new, probationary, raw, unaccustomed, unfledged, uninitiated, unskilled, unsophisticated, untried, *inf* wet behind the ears, young. *Opp* EXPERT.

**inexplicable** *adj* baffling, bewildering, confusing, enigmatic, incomprehensible, inscrutable, insoluble, mysterious, mystifying, perplexing, puzzling, strange, unaccountable, unexplainable, unfathomable, unsolvable. *Opp* STRAIGHTFORWARD.

**infallible** *adj* certain, dependable, faultless, foolproof, impeccable, perfect, reliable, sound, sure, trustworthy, unbeatable, unerring, unfailing. *Opp* FALLIBLE.

**infamous** *adj* disgraceful, disreputable, ill-famed, notorious, outrageous, well-known. ▷ WICKED.

**infant** *n* baby, *inf* toddler, *inf* tot. ▷ CHILD.

**infantile** *adj* [*derog*] adolescent, babyish, childish, immature, juvenile, puerile. ▷ SILLY. *Opp* MATURE.

**infatuated** *adj* besotted, charmed, enchanted, *inf* head over heels, in love, obsessed, *inf* smitten.

**infatuation** *n inf* crush, obsession, passion. ▷ LOVE.

**infect** *vb* **1** blight, contaminate, defile, poison, pollute, spoil, taint. **2** affect, influence, inspire, touch. **infected** ▷ SEPTIC.

**infection** *n* blight, contagion, contamination, epidemic, pestilence, pollution, virus. ▷ ILLNESS.

**infectious** *adj* catching, communicable, contagious, spreading, transmissible, transmittable.

**infer** *vb* assume, conclude, deduce, derive, draw a conclusion, extrapolate, gather, guess, reach the conclusion, surmise, understand, work out.

**inferior** *adj* **1** humble, junior, lesser, lower, lowly, mean, menial, secondary, second-class, servile, subordinate, subsidiary, unimportant. **2** cheap, indifferent, mediocre, poor, shoddy, tawdry, *inf* tinny. *Opp* SUPERIOR. • *n* ▷ SUBORDINATE.

**infertile** *adj* barren, sterile, unfruitful, unproductive.

**infest** *vb* infiltrate, overrun, pervade, plague. **infested** alive, crawling, swarming, teeming, verminous.

**infidelity** *n* **1** adultery, unfaithfulness. **2** ▷ DISLOYALTY.

**infiltrate** *vb* enter secretly, insinuate, intrude, penetrate, spy on.

**infinite** *adj* astronomical, big, boundless, countless, endless, eternal, everlasting, immeasurable, immense, incalculable, indeterminate, inestimable, inexhaustible, innumerable, interminable, limitless, multitudinous, never-ending, numberless, perpetual, uncountable, undefined, unending, unfathomable, unlimited, unnumbered, untold. ▷ HUGE. *Opp* FINITE.

**infinity** *n* endlessness, eternity, infinite distance, infinite quantity, infinitude, perpetuity, space.

**infirm** *adj* bedridden, crippled, elderly, feeble, frail, lame, old, poorly, senile, sickly, unwell. ▷ ILL, WEAK. *Opp* HEALTHY.

**inflame** *vb* arouse, encourage, excite, fire, foment, goad, ignite, incense, incite, kindle, madden, provoke, rouse, stimulate, stir up, work up. ▷ ANGER. *Opp* COOL. **inflamed** ▷ PASSIONATE, SEPTIC.

**inflammable** *adj* burnable, combustible, flammable, volatile. *Opp* INCOMBUSTIBLE.

**inflammation** *n* abscess, boil, infection, irritation, redness, sore, soreness, swelling.

**inflate** *vb* **1** blow up, dilate, distend, enlarge, puff up, pump up, swell. **2** ▷ EXAGGERATE.

**inflexible** *adj* **1** adamantine, firm, hard, hardened, immovable, rigid, solid, stiff, unbending, unyielding. **2** adamant, entrenched, fixed, immutable, inexorable, intractable, intransigent, obdurate, obstinate, *inf* pig-headed, refractory, resolute, rigorous, strict, stubborn, unalterable, unchangeable, uncompromising, unhelpful. *Opp* FLEXIBLE.

**inflict** *vb* administer, apply, deal out, enforce, force, impose, mete out, perpetrate, wreak.

**influence** *n* ascendancy, authority, control, direction, dominance, effect, guidance, hold, impact, leverage, power, pressure, pull, sway, weight. • *vb* **1** affect, bias, change, control, direct, dominate, exert influence on, guide, impinge on, impress, manipulate, modify, motivate, move, persuade, prejudice, prompt, put pressure on, stir, sway. **2** *influence a judge.* bribe, corrupt, lead astray, suborn, tempt.

**influential** *adj* authoritative, compelling, controlling, convincing, dominant, effective, far-reaching, forceful, guiding, important, inspiring, leading, moving, persuasive, potent, powerful, prestigious, significant, strong, telling, weighty. *Opp* UNIMPORTANT.

**influx** *n* flood, flow, inflow, inundation, invasion, rush, stream.

**inform** *vb* **1** advise, apprise, brief, communicate to, enlighten, *inf* fill in, give information to, instruct, leak, notify, *inf* put in the picture, teach, tell, *inf* tip off. **2** *inf* blab, give information, *sl* grass, *sl* peach, *sl* rat, *inf* sneak, *inf* split on, *inf* tell, *inf* tell tales. **inform against** ▷ BETRAY. **informed** ▷ KNOWLEDGEABLE.

**informal** *adj* **1** approachable, casual, comfortable, cosy, easy, easy-going, everyday, familiar, free and easy, friendly, homely, natural, ordinary, relaxed, simple, unceremonious, unofficial, unpretentious, unsophisticated. **2** *informal language.* chatty, colloquial, personal, slangy, vernacular. **3** *an informal design.* asymmetrical, flexible, fluid, intuitive, irregular, spontaneous. *Opp* FORMAL.

**information** *n* **1** announcement, briefing, bulletin, communication, enlightenment, instruction, message, news, report, statement, *old use* tidings, *inf* tip-off, word. **2** data, database, dossier, evidence, facts, intelligence, knowledge, statistics.

**informative** *adj* communicative, edifying, educational, enlightening, factual, giving information, helpful, illuminating, instructive, meaningful, revealing, useful. *Opp* MEANINGLESS.

**informer** *n sl* grass, informant, spy, *sl* stool-pigeon, *inf* tell-tale, traitor.

**infrequent** *adj* exceptional, intermittent, irregular, occasional, *inf* once in a blue moon, rare, spasmodic, uncommon, unusual. *Opp* FREQUENT.

**infringe** *vb* breach, break, contravene, defy, disobey, disregard, flout, ignore, overstep, sin against, transgress, violate.

**ingenious** *adj* adroit, artful, astute, brilliant, clever, complex, crafty, creative, cunning, deft, imaginative, inspired, intelligent, intricate, inventive, neat, original, resourceful, shrewd, skilful, *inf* smart, subtle, talented. *Opp* UNIMAGINATIVE.

**ingenuous** *adj* artless, childlike, frank, guileless, honest, innocent, naïve, open, plain, simple, sincere trusting, unaffected, uncomplicated, unsophisticated. *Opp* SOPHISTICATED.

**ingredient** *n* component, constituent, element, factor, *plur* makings, part.

**inhabit** *vb old use* abide in, colonize, dwell in, live in, make your home in, occupy, people, populate, possess, reside in, settle in, set up home in.

**inhabitable** *adj* habitable, in good repair, liveable, usable. *Opp* UNINHABITABLE.

**inhabitant** *n* citizen, *old use* denizen, dweller, inmate, native, occupant, occupier, *plur* population, resident, settler, tenant, *plur* townsfolk, *plur* townspeople.

**inherent** *adj* built-in, congenital, essential, fundamental, hereditary, immanent, inborn, inbred, indwelling, ingrained, intrinsic, native, natural.

**inherit** *vb* be the inheritor of, be left, *inf* come into, receive as an inheritance, succeed to. **inherited** ▷ HEREDITARY.

**inheritance** *n* bequest, birthright, estate, fortune, heritage, legacy, patrimony.

**inhibit** *vb* bridle, check, control, curb, discourage, frustrate, hinder, hold back, prevent, quell, repress, restrain. **inhibited** ▷ REPRESSED, SHY.

**inhibition** *n* **1** bar, barrier, check, constraint, curb, impediment, interference, restraint, stricture. **2** blockage, diffidence, *inf* hang-up, repression, reserve, self-consciousness, shyness.

**inhospitable** *adj* antisocial, reclusive, reserved, solitary, standoffish, unkind, unsociable, unwelcoming. ▷ UNFRIENDLY. **2** bleak, cold, comfortless, desolate, grim, hostile, lonely. *Opp* HOSPITABLE.

**inhuman** *adj* animal, barbaric, barbarous, bestial, bloodthirsty, brutish, diabolical, fiendish, merciless, pitiless, ruthless, savage, unnatural, vicious. ▷ INHUMANE. *Opp* HUMAN.

**inhumane** *adj* cold-hearted, cruel, hard, hard-hearted, heartless, inconsiderate, insensitive, uncaring, uncharitable, uncivilized, unfeeling, unkind, unsympathetic. ▷ INHUMAN. *Opp* HUMANE.

**initial** *adj* beginning, commencing, earliest, first, inaugural, incipient, introductory, opening, original, primary, starting. *Opp* FINAL.

**initiate** *vb* activate, actuate, begin, commence, enter upon, get going, get under way, inaugurate, instigate, institute, introduce, launch, originate, set going, set in motion, set up, start, take the initiative, trigger.

**initiative** *n* ambition, drive, dynamism, enterprise, *inf* get-up-and-go, inventiveness, lead, leadership, originality, resourcefulness. **take the initiative** ▷ INITIATE.

**injection** *n inf* fix, inoculation, *inf* jab, vaccination.

**injure** *vb* break, crush, cut, damage, deface, disfigure, harm, hurt, ill-treat, impair, mar, ruin, spoil, vandalize. ▷ WOUND.

**injurious** *adj* **1** damaging, deleterious, destructive, detrimental, harmful, insalubrious, painful, ruinous. **2** ▷ ABUSIVE.

**injury** *n* damage, harm, hurt, mischief. ▷ WOUND.

**injustice** *n* bias, bigotry, discrimination, dishonesty, favouritism, illegality, inequality, inequity, one-sidedness, oppression, partiality, partisanship, prejudice, unfairness, unlawfulness, wrong, wrongness. *Opp* JUSTICE.

**inn** *n old use* hostelry, hotel, *inf* local, pub, tavern.

**inner** *adj* central, concealed, hidden, innermost, inside, interior, internal, intimate, inward, mental, middle, private, secret. *Opp* OUTER.

**innocence** *n* **1** goodness, honesty, incorruptibility, purity, righteousness, sinlessness, virtue. **2** [*derog*] gullibility, inexperience, naïvety, simple-mindedness.

**innocent** *adj* **1** above suspicion, angelic, blameless, chaste, childlike, faultless, free from blame, guiltless, harmless, honest, immaculate, incorrupt, inoffensive, pure, righteous, sinless, spotless, untainted, virginal, virtuous. *Opp* CORRUPT, GUILTY. **2** artless, childlike, credulous, *inf* green, guileless, gullible, inexperienced, ingenuous, naïve, simple, simple-minded, trusting, unsophisticated.

**innovation** *n* change, departure, invention, new feature, novelty, reform, revolution.

**innovator** *n* discoverer, experimenter, inventor, pioneer, reformer, revolutionary.

**innumerable** *adj* countless, many, numberless, uncountable, untold. ▷ INFINITE.

**inquest** *n* hearing. ▷ INQUIRY.

**inquire** *vb* ask, explore, investigate, seek information, *inf* probe, search, survey. ▷ ENQUIRE.

**inquiry** *n* cross-examination, examination, inquest, inquisition, interrogation, investigation, poll, *inf* post-mortem, *inf* probe, referendum, review, study, survey.

**inquisitive** *adj* curious, impertinent, indiscreet, inquiring, interfering, intrusive, investigative, meddlesome, meddling, *inf* nosy, probing, prying, questioning, sceptical, searching, *inf* snooping, spying. **be inquisitive** ▷ PRY.

**insane** *adj inf* crazy, deranged, lunatic, *inf* mental, psychotic, unbalanced, unhinged. ▷ MAD. *Opp* SANE.

**inscription** *n* dedication, engraving, epigraph, superscription, writing.

**insect** *n inf* bug, *inf* creepy-crawly. □ *ant, aphid, bee, beetle, blackfly, butterfly, cicada, cockchafer, cockroach, crane-fly, cricket, daddy-long-legs, damselfly, dragonfly, earwig, firefly, fly, glow-worm, gnat, grasshopper, hornet, ladybird, locust, mantis, mayfly, midge, mosquito, moth, sawfly, termite, tsetse (fly), wasp, weevil.*

**insecure** *adj* **1** dangerous, flimsy, loose, precarious, rickety, rocky, shaky, unsafe, unsound, unstable, unsteady, unsupported, weak, wobbly. **2** *an insecure feeling*. anxious, apprehensive, defenceless, exposed, open, uncertain, under-confident, unprotected, vulnerable, worried. *Opp* SECURE.

**insensible** *adj* anaesthetized, benumbed, *inf* dead to the world, inert, insensate, insentient, knocked out, numb, *inf* out, senseless, unaware, unconscious. *Opp* CONSCIOUS.

**insensitive** *adj* **1** anaesthetized, dead, numb, unresponsive. **2** boorish, callous, crass, cruel, imperceptive, obtuse, tactless, *inf* thick-skinned, thoughtless, uncaring, unfeeling, unsympathetic. *Opp* SENSITIVE.

**inseparable** *adj* always together, attached, indissoluble, indivisible, integral.

**insert** *vb* drive in, embed, implant, intercalate, interject, interleave, interpolate, interpose, introduce, place in, *inf* pop in, push in, put in, stick in, tuck in.

**inside** *adj* central, indoor, inner, innermost, interior, internal. • *n* bowels, centre, contents, core, heart, indoors, interior, lining, middle. *Opp* OUTSIDE. **insides** ▷ ENTRAILS.

**insidious** *adj* creeping, deceptive, furtive, pervasive, secretive, stealthy, subtle, surreptitious, treacherous, underhand. ▷ CRAFTY.

**insignificant** *adj* forgettable, inconsiderable, irrelevant, insubstantial, lightweight, meaningless, minor, negligible, paltry, small, trifling, trivial, undistinguished, unimpressive, valueless, worthless, unimportant. *Opp* SIGNIFICANT.

**insincere** *adj* artful, crafty, deceitful, deceptive, devious, dishonest, disingenuous, dissembling, false, feigned, flattering, *inf* foxy, hollow, hypocritical, lying, *inf* mealy-mouthed, mendacious, perfidious, *inf* phoney, pretended, *inf* put on, *inf* smarmy, sycophantic, treacherous, *inf* two-faced, untrue, untruthful, wily. *Opp* SINCERE.

**insist** *vb* **1** assert, asseverate, aver, avow, declare, emphasize, hold, maintain, state, stress, swear, take an oath, vow. **2** assert yourself, be assertive, command, persist, *inf* put your foot down, stand firm, *inf* stick to your guns. **insist on** ▷ DEMAND.

**insistent** *adj* assertive, demanding, dogged, emphatic, firm, forceful, importunate, inexorable, obstinate, peremptory, persistent, relentless, repeated, resolute, stubborn, unrelenting, unremitting, urgent.

**insolence** *n* arrogance, boldness, *inf* cheek, defiance, disrespect, effrontery, impertinence, impudence, incivility, insubordination, *inf* lip, presumptuousness, rudeness, *inf* sauce.

**insolent** *adj* arrogant, audacious, bold, brazen, *inf* cheeky, contemptuous, defiant, disdainful, disrespectful, forward, *inf* fresh, impertinent, impolite, impudent, insubordinate, insulting, offensive, pert, presumptuous, saucy, shameless, sneering, uncivil. ▷ RUDE. *Opp* POLITE.

**insoluble** *adj* baffling, enigmatic, incomprehensible, inexplicable, mystifying, puzzling, strange, unaccountable, unanswerable, unfathomable, unsolvable. *Opp* SOLUBLE.

**insolvent** *adj* bankrupt, *inf* bust, failed, ruined. ▷ POOR.

**inspect** *vb* check, examine, *sl* give it the once over, investigate, peruse, pore over, scan, scrutinize, study, survey, vet.

**inspection** *n* check, check-up, examination, *inf* going-over, investigation, review, scrutiny, survey.

**inspector** *n* controller, examiner, investigator, official, scrutineer, superintendent, supervisor, tester.

**inspiration** *n* **1** creativity, genius, imagination, muse. **2** enthusiasm, impulse, incitement, influence, motivation, prompting, spur, stimulation, stimulus. **3** *a sudden inspiration.* brainwave, idea, insight, revelation, thought.

**inspire** *vb* activate, animate, arouse, awaken, *inf* egg on, encourage, energize, enthuse, fire, galvanize, influence, inspirit, instigate, kindle, motivate, prompt, provoke, quicken, reassure, set off, spark off, spur, stimulate, stir, support.

**instability** *n* capriciousness, change, changeableness, fickleness, fluctuation, flux, impermanence, inconstancy, insecurity, mutability, precariousness, shakiness, transience, uncertainty, unpredictability, unreliability, unsteadiness, *inf* ups-and-downs, vacillation, variability, variations, weakness. *Opp* STABILITY.

**install** *vb* ensconce, establish, fit, fix, instate, introduce, place, plant, position, put in, settle, set up, situate, station. *Opp* REMOVE.

**instalment** *n* **1** payment, rent, rental. **2** chapter, episode, part.

**instance** *n* case, example, exemplar, illustration, occurrence, precedent, sample.

**instant** *adj* direct, fast, immediate, instantaneous, on-the-spot, prompt, quick, rapid, speedy, split-second, swift, unhesitating, urgent. • *n* flash, *inf* jiffy, moment, point of time, second, split second, *inf* tick, *inf* trice, *inf* twinkling.

**instigate** *vb* activate, begin, be the instigator of, bring about, cause, encourage, foment, generate, incite, initiate, inspire, kindle, prompt, provoke, set up, start, stimulate, stir up, urge, *inf* whip up.

**instigator** *n* agitator, fomenter, inciter, initiator, inspirer, leader, mischief-maker, provoker, ringleader, troublemaker.

**instil** *vb inf* din into, imbue, implant, inculcate, indoctrinate, infuse, ingrain, inject, insinuate, introduce.

**instinct** *n* bent, faculty, feel, feeling, guesswork, hunch, impulse, inclination, instinctive urge, intuition, presentiment, propensity, sixth-sense, the subconscious, tendency, urge.

**instinctive** *adj* automatic, congenital, constitutional, *inf* gut, impulsive, inborn, inbred, inherent, innate, instinctual, intuitive, involuntary, irrational, mechanical, native, natural, reflex, spontaneous, subconscious, unconscious, unreasoning, unthinking, visceral. *Opp* DELIBERATE.

**institute** *n* ▷ INSTITUTION. • *vb* begin, create, establish, fix up, found, inaugurate, initiate, introduce, launch, open, organize, originate, pioneer, set up, start.

**institution** *n* **1** creation, establishing, formation, founding, inauguration, inception, initiation, introduction, launching, opening, setting-up. **2** academy, asylum, college, establishment, foundation, home, hospital, institute, organization, school, *inf* set-up. **3** convention, custom, habit, practice, ritual, routine, rule, tradition.

**instruct** *vb* **1** coach, drill, educate, indoctrinate, inform, lecture, prepare, school, teach, train, tutor. **2** authorize, brief, charge, command, direct, enjoin, give the order, order, require, tell.

**instruction** *n* **1** briefing, coaching, demonstration, drill, education, guidance, indoctrination, lecture, lesson, schooling, teaching, training, tuition, tutorial, tutoring. **2** authorization, brief, charge, command, direction, directive, order, requisition.

**instructive** *adj* didactic, edifying, educational, enlightening, helpful, illuminating, improving, informational, informative, instructional, revealing.

**instructor** *n* adviser, coach, trainer, tutor. ▷ TEACHER.

**instrument** *n* apparatus, appliance, contraption, device, equipment, gadget, implement, machine, mechanism, tool, utensil. **musical instrument** □ *accordion, bagpipes, banjo, bassoon, bugle, castanets, celesta, cello, clarinet, clavichord, clavier, concertina, cor anglais, cornet, cymbals, double-bass, drum, dulcimer, euphonium, fiddle, fife, flugelhorn, flute, fortepiano, French horn, glockenspiel, gong, guitar, harmonica, harmonium, harp, harpsichord, horn, hurdy-gurdy, kettledrum, keyboard, lute, lyre, mouth-organ, oboe, organ, piano, piccolo, pipes, recorder, saxophone, sitar, spinet, synthesizer, tambourine, timpani, triangle, trombone, trumpet, tuba, tubular bells, ukulele, viol, viola, violin, virginals, xylophone, zither.*

**instrumental** *adj* active, advantageous, beneficial, contributory, helpful, influential, supportive, useful, valuable.

**insubordinate** *adj* defiant, disobedient, insurgent, mutinous, rebellious, riotous, seditious, undisciplined, unruly. ▷ IMPERTINENT. *Opp* OBEDIENT.

**insufficient** *adj* deficient, disappointing, inadequate, incomplete, little, meagre, mean, niggardly, *inf* pathetic, poor, scanty, scarce, short, skimpy, sparse, unsatisfactory. *Opp* EXCESSIVE, SUFFICIENT.

**insular** *adj* closed, cut-off, isolated, limited, narrow, narrow-minded, parochial, provincial, remote, separated. *Opp* BROADMINDED, COSMOPOLITAN.

**insulate** *vb* **1** cocoon, cover, cushion, enclose, isolate, lag, protect, shield, surround, wrap up. **2** cut off, detach, isolate, keep apart, quarantine, segregate, separate.

**insult** *n* abuse, affront, aspersion, *inf* cheek, contumely, defamation, impudence, indignity, insulting behaviour, libel, *inf* put-down, rudeness, slander, slight, slur, snub. • *vb* abuse, affront, be rude to, *inf* call names, *inf* cock a snook at, defame, dishonour, disparage, libel, mock, offend, outrage, patronize, revile, slander, slang, slight, sneer at, snub, *inf* thumb your nose at, vilify. *Opp* COMPLIMENT. **insulting** ▷ RUDE.

**insuperable** *adj* insurmountable, overwhelming, unconquerable. ▷ IMPOSSIBLE.

**insurance** *n* assurance, cover, indemnification, indemnity, policy, protection, security.

**insure** *vb* cover yourself, indemnify, protect, take out insurance.

**intact** *adj* complete, entire, integral, solid, sound, unbroken, undamaged, whole. ▷ PERFECT.

**intangible** *adj* abstract, airy, disembodied, elusive, ethereal, evanescent, fleeting, impalpable, imperceptible, imponderable, incorporeal, indefinite, insubstantial, invisible, shadowy, unreal, vague. *Opp* TANGIBLE.

**integral** *adj* **1** basic, constituent, essential, fundamental, indispensable, intrinsic, irreplaceable, necessary, requisite. *Opp* INESSENTIAL. **2** *an integral unit.* attached, complete, full, indivisible, whole. *Opp* SEPARATE.

**integrate** *vb* amalgamate, assemble, blend, bring together, coalesce, combine, consolidate, desegregate, fuse, harmonize, join, knit, merge, mix, put together, unify, unite, weld. *Opp* SEPARATE.

**integrity** *n* **1** decency, fidelity, goodness, honesty, honour, incorruptibility, loyalty, morality, principle, probity, rectitude, reliability, righteousness, sincerity, trustworthiness, uprightness, veracity, virtue. **2** ▷ UNITY.

**intellect** *n inf* brains, cleverness, genius, mind, rationality, reason, sense, understanding, wisdom, *old use* wit. ▷ INTELLIGENCE.

**intellectual** *adj* **1** academic, *inf* bookish, cerebral, cultured, educated, scholarly, studious, thinking, thoughtful. ▷ INTELLIGENT. **2** cultural, deep, difficult, educational, highbrow, improving, thought-provoking. • *n* academic, *inf* egghead, genius, highbrow, intellectual person, *inf* mastermind, *inf* one of the intelligentsia, savant, thinker.

**intelligence** *n* **1** ability, acumen, alertness, astuteness, brainpower, *inf* brains, brightness, brilliance, capacity, cleverness, discernment, genius, *inf* grey matter, insight, intellect, judgement, keenness, mind, *inf* nous, perceptiveness, perspicaciousness, perspicacity, quickness, reason, sagacity, sense, sharpness, shrewdness, understanding, wisdom, wit, wits. **2** data, facts, information, knowledge, *inf* low-down, news, notification, report, *inf* tip-off, warning. **3** espionage, secret service, spying.

**intelligent** *adj* able, acute, alert, astute, brainy, bright, brilliant, *inf* canny, clever, discerning, educated, intellectual, knowing, penetrating, perceptive, percipient, perspicacious, profound, quick, ratiocinative, rational, reasonable, sagacious, sensible, sharp, shrewd, *inf* smart, thinking, thoughtful, trenchant, wise, *inf* with it, witty. *Opp* STUPID.

**intelligible** *adj* clear, comprehensible, decipherable, fathomable, legible, logical, lucid, meaningful, plain, straightforward, unambiguous, understandable. *Opp* INCOMPREHENSIBLE.

**intend** *vb* aim, aspire, contemplate, design, determine, have in mind, mean, plan, plot, propose, purpose, resolve, scheme.

**intense** *adj* **1** ardent, burning, consuming, deep, eager, earnest, emotional, fanatical, fervent, fervid, impassioned, passionate, powerful, profound, serious, strong, towering, vehement, violent, zealous. *Opp* COOL, HALF-HEARTED. **2** *intense pain.* acute, agonizing, excruciating, extreme, fierce, great, harsh, keen, severe, sharp. *Opp* SLIGHT.

**intensify** *vb* add to, aggravate, augment, become greater, boost, build up, deepen, emphasize, escalate, fire, focus, fuel, heighten, *inf* hot up, increase, magnify, make greater, quicken, raise, redouble, reinforce, sharpen, *inf* step up, strengthen, whet. *Opp* REDUCE.

**intensive** *adj inf* all-out, comprehensive, concentrated, detailed, exhaustive, high-powered, thorough, unremitting.

**intent** *adj* absorbed, attentive, bent, committed, concentrated, concentrating, determined, eager, engrossed, enthusiastic, firm, focused, keen, occupied, preoccupied, resolute, set, steadfast, watchful, zealous. *Opp* CASUAL.
• *n* ▷ INTENTION.

**intention** *n* aim, ambition, design, end, goal, intent, object, objective, plan, point, purpose, target.

**intentional** *adj* calculated, conscious, contrived, deliberate, designed, intended, knowing, planned, pre-arranged, preconceived, premeditated, prepared, studied, wilful. *Opp* ACCIDENTAL.

**intercept** *vb* ambush, arrest, block, catch, check, cut off, deflect, head off, impede, interrupt, obstruct, stop, thwart, trap.

**intercourse** *n* **1** communication, conversation, dealings, interaction, traffic. **2** *sexual intercourse.* carnal knowledge, coition, coitus, congress, copulation, intimacy, love-making, mating, rape, sex, union.

**interest** *n* **1** attention, attentiveness, care, commitment, concern, curiosity, involvement, notice, regard, scrutiny. **2** *of no interest.* consequence, importance, moment, note, significance, value. **3** *leisure interests.* activity, diversion, hobby, pastime, preoccupation, pursuit, relaxation.
• *vb* absorb, appeal to, arouse the curiosity of, attract, capture the imagination of, captivate, concern, divert, enchant, engage, engross, entertain, enthral, fascinate, intrigue, involve, occupy, preoccupy, stimulate, *inf* turn on. ▷ EXCITE. *Opp* BORE.

**interested** *adj* **1** absorbed, attentive, curious, engrossed, enthusiastic, immersed, intent, involved, keen, occupied, preoccupied, rapt, responsive, riveted. *Opp* UNINTERESTED. **2** concerned, involved, partial. ▷ BIASED. *Opp* DISINTERESTED.

**interesting** *adj* absorbing, appealing, attractive, challenging, compelling, curious, engaging, engrossing, entertaining, enthralling, fascinating, gripping, imaginative, important, intriguing, inviting, original, piquant, *often ironic* riveting, spellbinding, unpredictable, unusual, varied. *Opp* BORING.

**interfere** *vb* be a busybody, butt in, interrupt, intervene, intrude, meddle, obtrude, *inf* poke your nose in, pry, snoop, *inf* stick your oar in, tamper. **interfere with** ▷ OBSTRUCT. **interfering** ▷ NOSY.

**interim** *adj* half-time, halfway, provisional, stopgap, temporary.

**interior** *adj* ▷ INTERNAL. • *n* centre, core, depths, heart, inside, middle, nucleus.

**interlude** *n* entràcte, intermezzo, intermission. ▷ INTERVAL.

**intermediary** *n* agent, ambassador, arbiter, arbitrator, broker, go-between, mediator, middleman, negotiator, referee, spokesperson, umpire.

**intermediate** *adj* average, *inf* betwixt and between, halfway, intermediary, intervening, mean, medial, median, middle, midway, *inf* neither one thing nor the other, neutral, *inf* sitting on the fence, transitional.

**intermittent** *adj* broken, discontinuous, erratic, fitful, irregular, occasional, *inf* on and off, periodic, random, recurrent, spasmodic, sporadic. *Opp* CONTINUOUS.

**internal** *adj* **1** inner, inside, interior. *Opp* EXTERNAL. **2** confidential, hidden, intimate, inward, personal, private, secret, undisclosed.

**international** *adj* cosmopolitan, global, intercontinental, universal, worldwide.

**interpret** *vb* clarify, clear up, construe, decipher, decode, define, elucidate, explain, explicate, expound, gloss, make clear, make sense of, paraphrase, render, rephrase, reword, simplify, sort out, translate, understand, unravel, work out.

**interpretation** *n* clarification, definition, elucidation, explanation, gloss, paraphrase, reading, rendering, translation, understanding, version.

**interrogation** *n* cross-examination, debriefing, examination, grilling, inquisition, questioning, *inf* third degree.

**interrogative** *adj* asking, inquiring, inquisitive, interrogatory, investigatory, questioning.

**interrupt** *vb* **1** *inf* barge in, break in, butt in, *inf* chime in, *inf* chip in, cut in, disrupt, disturb, heckle, hold up, interfere, intervene, intrude, punctuate, obstruct, spoil. **2** break off, call a halt to, cut off, cut short, discontinue, halt, stop, suspend, terminate.

**interruption** *n* break, check, disruption, division, gap, halt, hiatus, interference, intrusion, stop, suspension. ▷ INTERVAL.

**intersect** *vb* bisect each other, converge, criss-cross, cross, divide, meet, pass across each other.

**interval** *n* **1** adjournment, break, *inf* breather, breathing-space, delay, distance, gap, hiatus, interruption, lapse, lull, opening, pause, recess, respite, rest, space, void, wait. **2** entràcte, interlude, intermezzo, intermission.

**intervene** *vb* **1** come between, elapse, happen, occur, pass. **2** arbitrate, butt in, intercede, interfere, interpose, interrupt, intrude, mediate, *inf* step in.

**interview** *n* appraisal, audience, duologue, formal discussion, meeting, questioning, selection procedure, vetting. • *vb* appraise, ask questions, evaluate, examine, interrogate, question, sound out, vet.

**interweave** *vb* criss-cross, entwine, interlace, intertwine, knit, tangle, weave together.

**intestines** *plur n* bowels, entrails, innards, insides, offal.

**intimate** *adj* **1** affectionate, close, familiar, informal, loving, sexual. ▷ FRIENDLY. **2** *intimate details.* confidential, detailed, exhaustive, personal, private, secret. • *n* ▷ FRIEND. • *vb* ▷ INDICATE.

**intimidate** *vb* alarm, browbeat, bully, coerce, cow, daunt, dismay, frighten, hector, make afraid, menace, overawe, persecute, petrify, scare, terrify, terrorize, threaten, tyrannize.

**intolerable** *adj* excruciating, impossible, insufferable, insupportable, unacceptable, unbearable, unendurable. *Opp* TOLERABLE.

**intolerant** *adj* biased, bigoted, chauvinistic, classist, discriminatory, dogmatic, illiberal, narrow-minded, one-sided, opinionated, prejudiced, racist, sexist, uncharitable, unsympathetic, xenophobic. *Opp* TOLERANT.

**intonation** *n* accent, delivery, inflection, modulation, pronunciation, sound, speech pattern, tone.

**intoxicate** *vb* addle, inebriate, make drunk, stupefy. **intoxicated** ▷ DRUNK, EXCITED. **intoxicating** ▷ ALCOHOLIC, EXCITING.

**intricate** *adj* complex, complicated, convoluted, delicate, detailed, elaborate, entangled, fancy, *inf* fiddly, involved, *inf* knotty, labyrinthine, ornate, sophisticated, tangled, tortuous. *Opp* SIMPLE.

**intrigue** *n* ▷ PLOT. • *vb* **1** appeal to, arouse the curiosity of, attract, beguile, captivate, capture the interest of, engage, engross, excite the curiosity of, fascinate, interest, stimulate, *inf* turn on. *Opp* BORE.

**intrinsic** *adj* basic, essential, fundamental, immanent, inborn, inbred, in-built, inherent, native, natural, proper, real.

**introduce** *vb* **1** acquaint, make known, present. **2** announce, give an introduction to, lead into, preface. **3** add, advance, bring in, bring out, broach, create, establish, inaugurate, initiate, inject, insert, interpose, launch, make available, offer, phase in, pioneer, put forward, set up, start, suggest, usher in. ▷ BEGIN.

**introduction** *n* foreword, *inf* intro, *inf* lead-in, opening, overture, preamble, preface, prelude, prologue. ▷ BEGINNING.

**introductory** *adj* basic, early, first, fundamental, inaugural, initial, opening, prefatory, preliminary, preparatory, starting. *Opp* FINAL.

**introverted** *adj* contemplative, introspective, inward-looking, meditative, pensive, quiet, reserved, retiring, self-contained, shy, thoughtful, unsociable, withdrawn. *Opp* EXTROVERTED.

**intrude** *vb* break in, butt in, eavesdrop, encroach, gatecrash, interfere, interpose, interrupt, intervene, join uninvited, obtrude, *inf* snoop.

**ntruder** *n* **1** eavesdropper, gate-crasher, infiltrator, interloper, *inf* uninvited guest. **2** burglar, housebreaker, invader, prowler, raider, robber, *inf* snooper, thief, trespasser.

**ntuition** *n* insight, perceptiveness, percipience. ▷ INSTINCT.

**nvade** *vb* descend on, encroach on, enter, impinge on, infest, infringe, march into, occupy, overrun, penetrate, raid, subdue, violate. ▷ ATTACK.

**nvalid** *adj* **1** null and void, out-of-date, unacceptable, unusable, void, worthless. **2** fallacious, false, illogical, incorrect, irrational, spurious, unconvincing, unfounded, unreasonable, unscientific, unsound, untenable, untrue, wrong. *Opp* VALID. **3** ▷ ILL.
• *n* cripple, incurable, patient, sufferer, valetudinarian.

**nvaluable** *adj* incalculable, inestimable, irreplaceable, precious, priceless, useful. ▷ VALUABLE. *Opp* WORTHLESS.

**nvariable** *adj* certain, changeless, constant, eternal, even, immutable, inflexible, permanent, predictable, regular, reliable, rigid, solid, stable, steady, unalterable, unchangeable, unchanging, unfailing, uniform, unvarying, unwavering. *Opp* VARIABLE.

**nvasion** *n* **1** encroachment, incursion, infiltration, inroad, intrusion, onslaught, raid, violation. ▷ ATTACK. **2** colony, flood, horde, infestation, spate, stream, swarm, throng.

**nvasive** *adj* burgeoning, increasing, mushrooming, profuse, proliferating, relentless, unstoppable.

**invent** *vb* coin, conceive, concoct, construct, contrive, *inf* cook up, create, design, devise, discover, *inf* dream up, fabricate, formulate, *inf* hit upon, imagine, improvise, make up, originate, plan, put together, think up, trump up.

**invention** *n* **1** brainchild, coinage, contrivance, creation, design, discovery. **2** contraption, device, gadget. **3** deceit, fabrication, falsehood, fantasy, fiction, figment, lie. **4** ▷ INVENTIVENESS.

**inventive** *adj* clever, creative, enterprising, fertile, imaginative, ingenious, innovative, inspired, original, resourceful. *Opp* BANAL.

**inventiveness** *n* creativity, genius, imagination, ingenuity, inspiration, invention, originality, resourcefulness.

**inventor** *n* architect, author, *inf* boffin, creator, designer, discoverer, maker, originator.

**inverse** *adj* opposite, reversed, transposed.

**invert** *vb* capsize, overturn, reverse, transpose, turn upside down, upset.

**invest** *vb* **1** buy stocks and shares, play the market, speculate. **2** lay out, put to work, *inf* sink, use profitably, venture. **invest in** ▷ BUY.

**investigate** *vb* analyse, consider, enquire about, examine, explore, follow up, gather evidence about, *inf* go into, inquire into, look into, probe, research, scrutinize, sift (*evidence*), study, *inf* suss out, weigh up.

**investigation** *n* enquiry, examination, inquiry, inquisition, inspection, *inf* post-mortem, *inf* probe, quest, research, review, scrutiny, search, study, survey.

**invidious** *adj* discriminatory, objectionable, offensive, undesirable, unfair, unjust, unwarranted.

**invigorating** *adj* bracing, enlivening, exhilarating, fresh, healthful, health-giving, healthy, refreshing, rejuvenating, revitalizing, salubrious, stimulating, tonic, vitalizing. *Opp* EXHAUSTING.

**invincible** *adj* impregnable, indestructible, indomitable, insuperable, invulnerable, strong, unassailable, unbeatable, unconquerable, unstoppable.

**invisible** *adj* camouflaged, concealed, covered, disguised, hidden, imperceptible, inconspicuous, obscured, out of sight, secret, undetectable, unnoticeable, unnoticed, unseen. *Opp* VISIBLE.

**invite** *vb* **1** ask, encourage, request, summon, urge. **2** attract, entice, solicit, tempt. **inviting** ▷ ATTRACTIVE.

**invoice** *n* account, bill, list, statement.

**invoke** *vb* appeal to, call for, cry out for, entreat, implore, pray for, solicit, supplicate.

**involuntary** *adj* automatic, conditioned, impulsive, instinctive, mechanical, reflex, spontaneous, unconscious, uncontrollable, unintentional, unthinking, unwitting. *Opp* DELIBERATE.

**involve** *vb* **1** comprise, contain, embrace, entail, hold, include, incorporate, take in. **2** affect, concern, interest, touch. **3** *involve in crime.* embroil, implicate, include, incriminate, inculpate, *inf* mix up. **involved** ▷ BUSY, COMPLEX.

**involvement** *n* **1** activity, interest, participation. **2** association, complicity, entanglement, partnership.

**ironic** *adj* derisive, double-edged, ironical, mocking, sarcastic, satirical, wry.

**irony** *n* double meaning, mockery, paradox, sarcasm, satire.

**irrational** *adj* absurd, arbitrary, biased, crazy, emotional, emotive, illogical, insane, mad, nonsensical, prejudiced, senseless, subjective, surreal, unconvincing, unintelligent, unreasonable, unreasoning, unsound, unthinking, wild. ▷ SILLY. *Opp* RATIONAL.

**irregular** *adj* **1** erratic, fitful, fluctuating, halting, haphazard, intermittent, occasional, random, spasmodic, sporadic, unequal, unpredictable, unpunctual, variable, varying, wavering. **2** abnormal, anomalous, eccentric, exceptional, extraordinary, illegal, improper, odd, peculiar, quirky, unconventional, unofficial, unplanned, unscheduled, unusual. **3** *irregular surface.* broken, bumpy, jagged, lumpy, patchy, pitted, ragged, rough, uneven, up and down. *Opp* REGULAR.

**irrelevant** *adj inf* beside the point, extraneous, immaterial, impertinent, inapplicable, inappoite, inappropriate, inessential, mal apropos, *inf* neither here nor there, pointless, unconnected, unnecessary, unrelated. *Opp* RELEVANT.

**irreligious** *adj* agnostic, atheistic, godless, heathen, humanist, impious, irreverent, pagan, sinful, uncommitted, unconverted, ungodly, unrighteous, wicked. *Opp* RELIGIOUS.

**irreparable** *adj* hopeless, incurable, irrecoverable, irremediable, irretrievable, irreversible, lasting, permanent, unalterable.

**irreplaceable** *adj* inimitable, priceless, unique. ▷ RARE.

**irrepressible** *adj* boisterous, bouncy, *inf* bubbling, buoyant, ebullient, resilient, uncontrollable, ungovernable, uninhibited, unmanageable, unrestrainable, unstoppable, vigorous. ▷ LIVELY. *Opp* LETHARGIC.

**irresistible** *adj* compelling, inescapable, inexorable, irrepressible, not to be denied, overpowering, overriding, overwhelming, persuasive, powerful, relentless, seductive, strong, unavoidable, uncontrollable. *Opp* WEAK.

**irresolute** *adj* doubtful, fickle, flexible, *inf* hedging your bets, indecisive, open to compromise, tentative, uncertain, undecided, vacillating, wavering, weak, weak-willed. ▷ HESITANT. *Opp* RESOLUTE.

**irresponsible** *adj* antisocial, careless, conscienceless, devil-may-care, feckless, immature, immoral, inconsiderate, negligent, rash, reckless, selfish, shiftless, thoughtless, unethical, unreliable, unthinking, untrustworthy, wild. *Opp* RESPONSIBLE.

**irreverent** *adj* blasphemous, disrespectful, impious, irreligious, profane, sacrilegious, ungodly, unholy. ▷ RUDE. *Opp* REVERENT.

**irrevocable** *adj* binding, final, fixed, hard and fast, immutable, irreparable, irretrievable, irreversible, permanent, settled, unalterable, unchangeable.

**irrigate** *vb* flood, inundate, supply water to, water.

**irritable** *adj* bad-tempered, cantankerous, choleric, crabby, cross, crotchety, crusty, curmudgeonly, dyspeptic, easily annoyed, edgy, fractious, grumpy, ill-humoured, ill-tempered, impatient, irascible, oversensitive, peevish, pettish, petulant, *inf* prickly, querulous, *inf* ratty, short-tempered, snappy, testy, tetchy, touchy, waspish. ▷ ANGRY. *Opp* EVEN-TEMPERED.

**irritate** *vb* **1** cause irritation, itch, rub, tickle, tingle. **2** ▷ ANNOY.

**island** *n plur* archipelago, atoll, coral reef, isle, islet.

**isolate** *vb* cloister, cordon off, cut off, detach, exclude, insulate, keep apart, place apart, quarantine, seclude, segregate, separate, sequester, set apart, shut off, shut out, single out. **isolated** ▷ SOLITARY.

**issue** *n* **1** affair, argument, controversy, dispute, matter, point, problem, question, subject, topic. **2** conclusion, consequence, effect, end, impact, outcome, *inf* payoff, repercussions, result, upshot. **3** *issue of a magazine.* copy, edition, instalment, number, printing, publication, version. • *vb* **1** appear, come out, emerge, erupt, flow out, gush, leak, rise, spring. **2** announce, bring out, broadcast, circulate, declare, disseminate, distribute, give out, make public, print, produce, promulgate, publicize, publish, put out, release, send out, supply.

**itch** *n* **1** irritation, prickling, tickle, tingling. **2** ache, craving, desire, hankering, hunger, impatience, impulse, longing, need, restlessness, thirst, urge, wish, yearning, *inf* yen. • *vb* **1** be irritated, prickle, tickle, tingle. **2** ▷ DESIRE.

**item** *n* **1** article, bit, component, contribution, entry, ingredient, lot, matter, object, particular, thing. **2** *item in a newspaper.* account, article, feature, notice, piece, report.

# J

**jab** *vb* dig, elbow, nudge, poke, prod, stab, thrust. ▷ HIT.

**jacket** *n* casing, cover, covering, envelope, folder, sheath, skin, wrapper, wrapping. ▷ COAT.

**jaded** *adj* **1** ▷ WEARY. **2** bored, *inf* fed up, gorged, listless, sated, satiated, *inf* sick and tired, surfeited. *Opp* LIVELY.

**jagged** *adj* angular, barbed, broken, chipped, denticulate, indented, irregular, notched, ragged, rough, serrated, sharp, snagged, spiky, toothed, uneven, zigzag. *Opp* SMOOTH.

**jail, jailer** *ns* ▷ GAOL, GAOLER.

**jam** *n* **1** blockage, bottleneck, congestion, crush, obstruction, press, squeeze, stoppage, throng. ▷ CROWD. **2** difficulty, dilemma, embarrassment, *inf* fix, *inf* hole, *inf* hot water, *inf* pickle, plight, predicament, quandary, tight corner, trouble. **3** conserve, jelly, marmalade, preserve. • *vb* **1** block, *inf* bung up, clog, congest, cram, crowd, crush, fill, force, pack, obstruct, overcrowd, ram, squash, squeeze, stop up, stuff. **2** prop, stick, wedge.

**jar** *n* amphora, carafe, container, crock, ewer, glass, jug, mug, pitcher, pot, receptacle, urn, vessel. ▷ BOTTLE. *vb* **1** jerk, jog, jolt, *inf* rattle, shake, shock, upset. **2** *That noise jars on me.* grate, grind, *inf* jangle. ▷ ANNOY. **jarring** ▷ HARSH.

**jargon** *n* argot, cant, creole, dialect, idiom, language, patois, slang, vernacular.

**jaunt** *n* excursion, expedition, ou[ting], tour, trip. ▷ JOURNEY.

**jaunty** *adj* alert, breezy, brisk, buoyant, bright, carefree, *inf* cheeky, debonair, frisky, lively, perky, spirited, sprightly. ▷ HAPPY.

**jazzy** *adj* **1** animated, rhythmic, spirited, swinging, syncopated, vivacious. ▷ LIVELY. *Opp* SEDATE. **2** *jazzy colours.* bold, clashing, co[n]trasting, flashy, gaudy, loud.

**jealous** *adj* **1** bitter, covetous, envious, *inf* green-eyed, *inf* green with envy, grudging, jaundiced, resentful. **2** *jealous of your reputation.* careful, possessive, protective, vigilant, watchful.

**jeer** *vb* barrack, boo, chaff, deride, disapprove, gibe, heckle, hiss, *inf* knock, laugh, make fun (of), mock, scoff, sneer, taunt, *inf* twit. ▷ RIDICULE. *Opp* CHEER.

**jeopardize** *vb* endanger, gamble, imperil, menace, put at risk, risk, threaten, venture.

**jerk** *vb* jar, jiggle, jog, jolt, lurch, move jerkily, move suddenly, pluck, pull, *inf* rattle, shake, tug, tweak, twist, twitch, wrench, *inf* yank.

**jerky** *adj* bouncy, bumpy, convul[s]ive, erratic, fitful, jolting, jumpy, rough, shaky, spasmodic, *inf* stopping and starting, twitchy, uncontrolled, uneven. *Opp* STEAD[Y].

**jest** *n, vb* ▷ JOKE.

**jester** *n* buffoon, clown, comedia[n], comic, fool, joker.

**jet** *adj* ▷ BLACK. • *n* **1** flow, fountain, gush, rush, spout, spray, spurt, squirt, stream. **2** nozzle, sprinkler.

**jetty** *n* breakwater, groyne, landing-stage, mole, pier, quay, whar[f].

**jewel** *n* brilliant, gem, gemstone, ornament, precious stone, *inf* rock, *inf* sparkler. ▷ JEWELLERY. □ *amber, cairngorm, carnelian, coral, diamond, emerald, garnet, ivory, jade, jasper, jet, lapis lazuli, moonstone, onyx, opal, pearl, rhinestone, ruby, sapphire, topaz, turquoise.*

**jeweller** *n* goldsmith, silversmith.

**jewellery** *n* gems, jewels, ornaments, treasure, *inf* sparklers. □*bangle, beads, bracelet, brooch, chain, charm, clasp, cuff-links, earring, locket, necklace, pendant, pin, ring, signet ring, tie-pin, watch, watch-chain.*

**jilt** *vb* abandon, break with, desert, *inf* ditch, drop, *inf* dump, forsake, *inf* give someone the brush-off, leave behind, *inf* leave in the lurch, renounce, repudiate, *inf* throw over, *inf* wash your hands of.

**jingle** *n* **1** doggerel, rhyme, song, tune, verse. **2** chinking, clinking, jangling, ringing, tinkling, tintinnabulation. • *vb* chime, chink, clink, jangle, ring, tinkle. ▷ SOUND.

**job** *n* **1** activity, assignment, charge, chore, duty, errand, function, housework, mission, operation, project, pursuit, responsibility, role, stint, task, undertaking, work. **2** appointment, business, calling, career, craft, employment, livelihood, living, métier, occupation, position, post, profession, sinecure, trade, vocation.

**jobless** *adj* out of work, redundant, unemployed, unwaged.

**jocular** *adj* cheerful, gay, glad, gleeful, happy, jocund, jokey, joking, jolly, jovial, joyous, jubilant, merry, overjoyed, rejoicing. *Opp* SAD, SERIOUS.

**jog** *vb* **1** bounce, jar, jerk, joggle, jolt, knock, nudge, shake. ▷ HIT. **2** *jog the memory*. activate, arouse, prompt, refresh, remind, set off, stimulate, stir. **3** *jog round the park*. exercise, lope, run, trot.

**join** *n* connection, joint, knot, link, mend, seam. • *vb* **1** add, amalgamate, attach, combine, connect, couple, dock, dovetail, fit, fix, juxtapose, knit, link, marry, merge, put together, splice, tack on, unite, yoke. ▷ FASTEN. *Opp* SEPARATE. **2** abut, adjoin, border on, come together, converge, meet, touch, verge on. **3** *join a crowd*. accompany, associate with, follow, go with, *inf* latch on to, tag along with, team up with. **4** *join a club*. affiliate with, become a member of, enlist in, enrol in, participate in, register for, sign up for, subscribe to, volunteer for. *Opp* LEAVE.

**joint** *adj* collaborative, collective, combined, common, communal, concerted, cooperative, corporate, general, mutual, shared, united. *Opp* SEPARATE. • *n* articulation, connection, hinge, junction, union.

**joist** *n* beam, girder, rafter.

**joke** *n* *inf* crack, funny story, *inf* gag, *old use* jape, jest, laugh, *inf* one-liner, pleasantry, pun, quip, wisecrack, witticism. • *vb* banter, be facetious, clown, fool about, have a laugh, jest, make jokes, quip, tease.

**jolly** *adj* cheerful, delighted, gay, glad, gleeful, grinning, high-spirited, jocose, jocular, jocund, joking, jovial, joyful, joyous, jubilant, laughing, merry, playful, rejoicing, rosy-faced, smiling, sportive. ▷ HAPPY. *Opp* SAD.

**jolt** *vb* **1** bounce, bump, jar, jerk, jog, shake, twitch. ▷ HIT. **2** *jolted me into action*. astonish, disturb, nonplus, shake up, shock, startle, stun, surprise.

**jostle** *vb* crowd in on, hustle, press, push, shove.

**jot** *vb* **jot down** note, scribble, take down. ▷ WRITE.

**journal** *n* **1** gazette, magazine, monthly, newsletter, newspaper, paper, periodical, publication, review, weekly. **2** account, annals, chronicle, diary, dossier, history, log, memoir, record, scrapbook.

**journalist** *n* broadcaster, columnist, contributor, correspondent, *derog* hack, *inf* newshound, newspaperman, newspaperwoman, pressman, reporter, writer.

**journey** *n* excursion, expedition, itinerary, jaunt, mission, odyssey, outing, peregrination, progress, route, tour, transition, travelling, trip, wandering. □ *cruise, drive, flight, hike, joy-ride, pilgrimage, ramble, ride, run, safari, sail, sea crossing, sea passage, trek, voyage, walk.* • *vb* ▷ TRAVEL.

**joy** *n* bliss, cheer, cheerfulness, delight, ecstasy, elation, euphoria, exaltation, exhilaration, exultation, felicity, gaiety, gladness, glee, gratification, happiness, high spirits, hilarity, jocularity, joviality, joyfulness, joyousness, jubilation, light-heartedness, merriment, mirth, pleasure, rapture, rejoicing, triumph. *Opp* SORROW.

**joyful** *adj* buoyant, cheerful, delighted, ecstatic, elated, euphoric, exhilarated, exultant, gay, glad, gleeful, jocund, jolly, jovial, joyous, jubilant, light-hearted, merry, overjoyed, pleased, rapturous, rejoicing, triumphant. ▷ HAPPY. *Opp* SAD.

**jubilee** *n* anniversary, celebration, commemoration, festival.

**judge** *n* **1** *sl* beak, justice, magistrate. **2** adjudicator, arbiter, arbitrator, moderator, referee, umpire. **3** *judge of wine.* authority, connoisseur, critic, expert, reviewer. • *vb* **1** condemn, convict, examine, pass judgement on, pronounce judgement on, punish, sentence, try. **2** adjudicate, mediate, moderate, referee, umpire. **3** believe, conclude, consider, decide, decree, deem, determine, estimate, gauge, guess, reckon, rule, suppose. **4** *judge others.* appraise, assess, criticize, evaluate, give your opinion of, rate, rebuke, scold, sit in judgement on, size up, weigh.

**judgement** *n* **1** arbitration, award, conclusion, conviction, decision, decree, *old use* doom, finding, outcome, penalty, punishment, result, ruling, verdict. **2** *use your judgement.* acumen, common sense, discernment, discretion, discrimination, expertise, good sense, reason, wisdom. ▷ INTELLIGENCE. **3** *in my judgement.* assessment, belief, estimation, evaluation, idea, impression, mind, notion, opinion, point of view, valuation.

**judicial** *adj* **1** forensic, legal, official. **2** ▷ JUDICIOUS.

**judicious** *adj* appropriate, astute, careful, circumspect, considered, diplomatic, discerning, discreet, discriminating, enlightened, expedient, judicial, politic, prudent, sage, sensible, shrewd, sober, thoughtful, well-advised, well-judged. ▷ WISE.

**jug** *n* bottle, carafe, container, decanter, ewer, flagon, flask, jar, pitcher, vessel.

**juggle** *vb* alter, *inf* cook, *inf* doctor, falsify, *inf* fix, manipulate, misrepresent, move about, rearrange, rig, tamper.

**uice** *n* drink, extract, fluid, liquid, ap.

**uicy** *adj* lush, moist, soft, *nf* squelchy, succulent, wet. *Opp* DRY.

**umble** *n* chaos, clutter, confuion, disarray, disorder, farrago, otchpotch, mess, muddle, tangle. • *vb* confuse, disarrange, disorganze, *inf* mess up, mingle, mix up, nuddle, shuffle, tangle. *Opp* ARRANGE.

**ump** *n* **1** bounce, bound, hop, eap, pounce, skip, spring, vault. **2** ditch, fence, gap, gate, hurdle, bstacle. **3** *a jump in prices.* ▷ RISE. • *vb* **1** bounce, bound, aper, dance, frisk, frolic, gambol, op, leap, pounce, prance, skip, pring. **2** *jump a fence.* clear, urdle, vault. **3** *jump in surprise.* linch, recoil, start, wince. **jump n** ▷ ATTACK. **make someone ump** ▷ STARTLE.

**unction** *n* confluence, connecion, corner, crossroads, interhange, intersection, joining, juncure, *inf* link-up, meeting, points, T-junction, union.

**ungle** *n* forest, rain-forest, tangle, ndergrowth, woods.

**unior** *adj* inferior, lesser, lower, ninor, secondary, subordinate, ubsidiary, younger. *Opp* SENIOR.

**unk** *n* clutter, debris, flotsamnd-jetsam, garbage, litter, lumer, oddments, odds and ends, efuse, rubbish, rummage, scrap, rash, waste. • *vb* ▷ DISCARD.

**ust** *adj* apt, deserved, equitable, thical, even-handed, fair, fairninded, honest, impartial, justiied, lawful, legal, legitimate, merted, neutral, proper, reasonable, ightful, right-minded, unbiased, nprejudiced, upright. ▷ MORAL. *Opp* UNJUST.

**justice** *n* **1** equity, even-handedness, fair-mindedness, fair play, impartiality, integrity, legality, neutrality, objectivity, right. ▷ MORALITY. **2** the law, legal proceedings, the police, punishment, retribution, vengeance.

**justifiable** *adj* acceptable, allowable, defensible, excusable, forgivable, justified, lawful, legitimate, pardonable, permissible, reasonable, understandable, warranted. *Opp* UNJUSTIFIABLE.

**justify** *vb* condone, defend, exculpate, excuse, exonerate, explain, explain away, forgive, legitimate, legitimize, pardon, rationalize, substantiate, support, sustain, uphold, validate, vindicate, warrant.

**jut** *vb* beetle, extend, overhang, poke out, project, protrude, stick out. *Opp* RECEDE.

**juvenile** *adj* **1** babyish, childish, immature, infantile, puerile, unsophisticated. **2** adolescent, *inf* teenage, underage, young, youthful. *Opp* MATURE.

# K

**keen** *adj* **1** active, ambitious, anxious, ardent, assiduous, avid, bright, clever, committed, dedicated, devoted, diligent, eager, enthusiastic, fervent, fervid, industrious, intelligent, intense, intent, interested, motivated, passionate, quick, zealous. **2** *a keen knife.* knife-edged, piercing, razor-sharp, sharp, sharpened. **3** *a keen wit.* acerbic, acid, acute, biting, clever, cutting, discerning, incisive, lively, mordant, observant, pungent, rapier-like, sarcastic, satirical, scathing, shrewd, sophisticated, stinging. **4** *keen eyesight*. acute, clear, fine, perceptive, sensitive. **5** *a keen wind*. bitter, cold, extreme, icy, intense, penetrating, severe. **6** *keen prices*. competitive, low, rock-bottom. *Opp* APATHETIC, DULL.

**keep** *vb* **1** accumulate, amass, conserve, guard, hang on to, hoard, hold, maintain, preserve, protect, put aside, put away, retain, safeguard, save, store, stow away, withhold. **2** *keep going*. carry on, continue, do again and again, do for a long time, keep on, persevere in, persist in. **3** *keep left*. remain, stay. **4** *keep a family*. be responsible for, care for, cherish, feed, finance, foster, guard, have charge of, look after, maintain, manage, mind, own, pay for, protect, provide for, subsidize, support, take charge of, tend, watch over. **5** *keep a birthday*. celebrate, commemorate, mark, observe, solemnize. **6** *food keeps in the fridge*. be preserved, be usable, last, survive, stay fresh, stay good. **7** *won't keep you*. block, check, confine, curb, delay, detain, deter, get in the way of, hamper, hinder, hold up, impede, imprison, obstruct, prevent, restrain, retard. **keep still** ▷ STAY. **keep to** ▷ FOLLOW, OBEY. **keep up** ▷ PROLONG, SUSTAIN.

**keeper** *n* caretaker, curator, custodian, gaoler, guard, guardian, *inf* minder, warden, warder.

**kernel** *n* centre, core, essence, heart, middle, nub, pith.

**key** *n* **1** answer, clarification, clue, explanation, indicator, pointer, secret, solution. **2** *key to a map*. glossary, guide, index, legend.

**keyboard** *n* □ *accordion, celesta, clavichord, clavier, fortepiano, harmonium, harpsichord, organ, piano,* old use *pianoforte, spinet, synthesizer, virginals*.

**kick** *vb* boot, heel, punt. ▷ HIT.

**kidnap** *vb* abduct, capture, carr off, run away with, seize, snatch

**kill** *vb* annihilate, assassinate, b the killer of, *sl* bump off, butche cull, decimate, destroy, *inf* dispatch, *inf* do away with, *sl* do in, execute, exterminate, *inf* finish off, *inf* knock off, liqui ate, martyr, massacre, murder, put down, put to death, put to sleep, slaughter, slay, *inf* snuff o take life, *Amer sl* waste. □ *behea brain, choke, crucify, decapitate, disembowel, drown, electrocute, eviscerate, garrotte, gas, guillotir hang, knife, lynch, poison, pole-a shoot, smother, stab, starve, stifle stone, strangle, suffocate, throttle*

**killer** *n* assassin, butcher, cutthroat, destroyer, executioner, exterminator, gunman, *sl* hit ma murderer, slayer.

**killing** *n* annihilation, assassination, bloodbath, bloodshed, butchery, carnage, decimation, destruction, elimination, eradication, euthanasia, execution, extermination, extinction, fratricide, genocide, homicide, infanticide, liquidation, manslaughter, martyrdom, massacre, matricide murder, parricide, patricide, pogrom, regicide, slaughter, sororicide, suicide, unlawful killing, uxoricide.

**kin** *n* clan, family, *inf* folks, kindred, kith and kin, relations, relatives.

**kind** *adj* accommodating, affable affectionate, agreeable, altruisti amenable, amiable, amicable, approachable, attentive, avuncular, beneficent, benevolent, benign, bountiful, brotherly, car ng, charitable, comforting, compassionate, congenial, considerate, ordial, courteous, encouraging, atherly, favourable, friendly, generous, genial, gentle, good-natured, good-tempered, gracious, elpful, hospitable, humane, umanitarian, indulgent, kind-earted, kindly, lenient, loving, nerciful, mild, motherly, neighourly, nice, obliging, patient, philnthropic, pleasant, polite, publicpirited, sensitive, sisterly, oft-hearted, sweet, sympathetic, actful, tender, tender-hearted, houghtful, tolerant, understandng, unselfish, warm, warm-earted, well-intentioned, well-neaning, well-meant. *Opp* UNKIND. ● *n* brand, breed, category, class, lescription, family, form, genre, genus, make, manner, nature, peruasion, race, set, sort, species, style, type, variety.

**indle** *vb* **1** burn, fire, ignite, ight, set alight, set fire to, spark off. **2** ▷ AROUSE.

**ing** *n* **1** crowned head, His Majesty, monarch, ruler, sovereign. **2** ▷ CHIEF.

**ingdom** *n* country, empire, land, nonarchy, realm.

**ink** *n* **1** bend, coil, crimp, rinkle, curl, knot, loop, tangle, wist, wave. **2** ▷ QUIRK.

**iosk** *n* booth, stall. □ *bookstall, ews-stand, telephone-box.*

**iss** *vb* caress, embrace, *sl* neck, sculate, *old use* spoon. ▷ TOUCH.

**it** *n* accoutrements, apparatus, appurtenances, baggage, effects, equipment, gear, *joc* impedimenta, mplements, luggage, outfit, paraphernalia, rig, supplies, tackle, ools, tools of the trade, utensils.

**itchen** *n* cookhouse, galley, kit-henette, scullery.

**knack** *n* ability, adroitness, aptitude, art, bent, dexterity, expertise, facility, flair, genius, gift, habit, intuition, *inf* know-how, skill, talent, trick, *inf* way.

**knapsack** *n* backpack, haversack, rucksack.

**knead** *vb* manipulate, massage, pound, press, pummel, squeeze, work.

**kneel** *vb* bend, bow, crouch, fall to your knees, genuflect, stoop.

**knickers** *n old use* bloomers, boxer-shorts, briefs, drawers, panties, pants, shorts, trunks, underpants.

**knife** *n* blade. □ *butter-knife, carving-knife, clasp-knife, cleaver, dagger, flick-knife, machete, penknife, pocket-knife, scalpel, sheath knife.* ● *vb* cut, pierce, slash, stab. ▷ KILL, WOUND.

**knit** *vb* **1** crochet, weave. **2** bind, combine, connect, fasten, heal, interlace, interweave, join, knot, link, marry, mend, tie, unite. **knit your brow** ▷ FROWN.

**knob** *n* boss, bulge, bump, handle, lump, projection, protrusion, protuberance, stud, swelling.

**knock** *vb* **1** bang, *inf* bash, buffet, bump, pound, rap, smack, *old use* smite, strike, tap, thump. ▷ HIT. **2** ▷ CRITICIZE. **knock down** ▷ DEMOLISH. **knock off** ▷ CEASE. **knock out** ▷ STUN.

**knot** *n* **1** bond, bow, ligature, tangle, tie. **2** ▷ GROUP. ● *vb* bind, do up, entangle, entwine, join, knit, lash, link, tether, tie, unite. ▷ FASTEN. *Opp* UNTIE.

**know** *vb* **1** be certain, have confidence, have no doubt. **2** *know facts.* be cognizant of, be familiar with, be knowledgeable about, comprehend, discern, have experience of, have in mind, realize, remember, understand. **3** *know a person.* be acquainted with, be a friend of, be friends with. **4** differentiate, distinguish, identify, make out, perceive, recognize, see.

**know-all** *n* expert, pundit, *inf* show-off, wiseacre.

**knowing** *adj* astute, clever, conspiratorial, cunning, discerning, experienced, expressive, meaningful, perceptive, shrewd. ▷ CUNNING, KNOWLEDGEABLE. *Opp* INNOCENT.

**knowledge** *n* **1** data, facts, information, intelligence, *sl* low-down. **2** acquaintance, awareness, background, cognition, competence, consciousness, education, erudition, experience, expertise, familiarity, grasp, insight, *inf* know-how, learning, lore, memory, science, scholarship, skill, sophistication, technique, training. *Opp* IGNORANCE.

**knowledgeable** *adj Fr* au fait, aware, cognizant, conversant, educated, enlightened, erudite, experienced, expert, familiar (with), *sl* genned up, informed, *inf* in the know, learned, scholarly, versed (in), well-informed. *Opp* IGNORANT.

# L

**label** *n* brand, docket, hallmark, identification, imprint, logo, marker, sticker, tag, ticket, trademark. • *vb* brand, call, categorize, class, classify, define, describe, docket, identify, mark, name, pigeon-hole, stamp, tag.

**laborious** *adj* **1** arduous, backbreaking, difficult, exhausting, fatiguing, gruelling, hard, heavy, herculean, onerous, stiff, strenuous, taxing, tiresome, tough, uphill, wearisome, wearying. *Opp* EASY. **2** *a laborious style.* art ficial, contrived, forced, heavy, laboured, overdone, overworked, pedestrian, ponderous, strained, unnatural. *Opp* FLUENT.

**labour** *n* **1** *inf* donkey-work, drudgery, effort, exertion, industry, navvying, *inf* pains, slavery, strain, *inf* sweat, toil, work. **2** employees, *old use* hands, wag earners, workers, workforce. **3** childbirth, contractions, delivery, labour pains, parturition, *o use* travail. • *vb* drudge, exert yo self, navvy, *inf* slave, strain, strive, struggle, *inf* sweat, toil, travail. ▷ WORK. laboured ▷ LABORIOUS.

**labourer** *n* blue-collar worker, employee, *old use* hand, manual worker, *inf* navvy, wage-earner, worker.

**labour-saving** *adj* convenient, handy, helpful, time-saving.

**labyrinth** *n* complex, jungle, maze, network, tangle.

**lace** *n* **1** filigree, mesh, net, netting, openwork, tatting, web. **2** bootlace, cord, shoelace, string thong. • *vb* ▷ FASTEN.

**lacerate** *vb* claw, cut, gash, graz mangle, rip, scrape, scratch, sla tear. ▷ WOUND.

**lack** *n* absence, dearth, deficienc deprivation, famine, insufficien need, paucity, privation, scarcit shortage, want. *Opp* PLENTY. • *vb* be lacking in, be short of, b without, miss, need, require, want.

**lacking** *adj* defective, deficient, inadequate, insufficient, short, unsatisfactory, weak. ▷ STUPID.

**laden** *adj* burdened, *inf* chock-full, fraught, full, hampered, loaded, oppressed, piled high, weighed down.

**lady** *n* **1** wife, woman. ▷ FEMALE. **2** aristocrat, peeress. ▷ TITLE.

**ladylike** *adj* aristocratic, courtly, cultured, dainty, decorous, elegant, genteel, modest, noble, polished, posh, prim and proper, *derog* prissy, refined, respectable, well-born, well-bred. ▷ POLITE.

**lag** *vb* **1** be slow, *inf* bring up the rear, come last, dally, dawdle, delay, drop behind, fall behind, go too slow, hang about, hang back, idle, linger, loiter, saunter, straggle, trail. **2** *lag pipes.* insulate, wrap up.

**lair** *n* den, hide-out, hiding-place, refuge, resting-place, retreat, shelter.

**lake** *n* boating-lake, lagoon, lido, *Scot* loch, mere, pool, pond, reservoir, sea, tarn, water.

**lame** *adj* **1** crippled, disabled, halting, handicapped, hobbled, hobbling, incapacitated, limping, maimed, spavined. **2** *a lame leg.* dragging, game, *inf* gammy, injured, stiff. **3** *a lame excuse.* feeble, flimsy, inadequate, poor, tame, thin, unconvincing, weak. • *vb* cripple, disable, hobble, incapacitate, maim. **be lame** ▷ LIMP.

**lament** *n* dirge, elegy, lamentation, moaning, monody, mourning, requiem, threnody. • *vb* bemoan, bewail, complain, cry, deplore, express your sorrow, grieve, keen, mourn, regret, shed tears, sorrow, wail, weep.

**lamentable** *adj* deplorable, regrettable, unfortunate, unhappy. ▷ SAD.

**lamentation** *n* complaints, crying, grief, grieving, lamenting, moaning, mourning, regrets, sobbing, tears, wailing, weeping.

**lamp** *n* bulb, fluorescent lamp, headlamp, lantern, standard lamp, street light, torch. ▷ LIGHT.

**land** *n* **1** coast, ground, landfall, shore, *joc* terra firma. **2** *lie of the land.* geography, landscape, terrain, topography. **3** country, fatherland, homeland, motherland, nation, region, state, territory. **4** farmland, earth, soil. **5** *land you own.* estate, grounds, property. • *vb* **1** alight, arrive, berth, come ashore, come to rest, disembark, dismount, dock, end a journey, get down, go ashore, light, reach landfall, settle, touch down. **2** *land yourself a job.* ▷ GET.

**landing** *n* **1** docking, re-entry, return, splashdown, touchdown. **2** alighting, arrival, deplaning, disembarkation. **3** ▷ LANDING-STAGE.

**landing-stage** *n* berth, dock, harbour, jetty, landing, pier, quay, wharf.

**landlady, landlord** *ns* **1** host, hostess, hotelier, *old use* innkeeper, licensee, publican, restaurateur. **2** landowner, lessor, letter, manager, manageress, owner, proprietor.

**landmark** *n* **1** feature, guidepost, high point, identification, visible feature. **2** milestone, new era, turning-point, watershed.

**landscape** *n* aspect, countryside, outlook, panorama, prospect, rural scene, scene, scenery, view, vista.

**language** *n* **1** parlance, speech, tongue. □ *argot, cant, colloquialism, dialect, formal language, idiolect, idiom, informal language, jargon, journalese, lingua franca,* inf *lingo, patois, register, slang, vernacular.* **2** linguistics. □ *etymology, lexicography,* *orthography, philology, phonetics, psycholinguistics, semantics, semiotics, sociolinguistics.* **3** *computer language.* code, system of signs.

**languid** *adj* apathetic, *inf* droopy, feeble, inactive, inert, lackadaisical, lazy, lethargic, slow, sluggish, torpid, unenthusiastic, weak. *Opp* ENERGETIC.

**languish** *vb* decline, flag, lose momentum, mope, pine, slow down, stagnate, suffer, sulk, waste away, weaken, wither. *Opp* FLOURISH.

**lank** *adj* **1** drooping, lifeless, limp, long, straight, thin. **2** ▷ LANKY.

**lanky** *adj* angular, awkward, bony, gangling, gaunt, lank, lean, long, scraggy, scrawny, skinny, tall, thin, ungraceful, weedy. *Opp* GRACEFUL, STURDY.

**lap** *n* **1** knees, thighs. **2** circle, circuit, course, orbit, revolution. • *vb* ▷ DRINK.

**lapse** *n* **1** backsliding, blunder, decline, error, failing, fault, flaw, mistake, omission, relapse, shortcoming, slip, *inf* slip-up, temporary failure, weakness. **2** *lapse of time.* break, gap, hiatus, *inf* hold-up, intermission, interruption, interval, lacuna, lull, pause. • *vb* **1** decline, deteriorate, diminish, drop, fall, sink, slide, slip, slump, subside. **2** *My membership lapsed.* become invalid, expire, finish, run out, stop, terminate.

**large** *adj* above average, abundant, ample, big, bold, broad, bulky, burly, capacious, colossal, commodious, considerable, copious, elephantine, enormous, extensive, fat, formidable, gargantuan, generous, giant, gigantic, grand, great, heavy, hefty, high, huge, *inf* hulking, immense, immeasurable, impressive, incalculable, infinite, *inf* jumbo, *inf* king-sized, largish, lofty, long, mammoth, massive, mighty, monstrous, monumental, mountainous, outsize, overgrown, oversized, prodigious, *inf* roomy, sizeable, spacious, substantial, swingeing (*increase*), tall, thick, *inf* thumping, *inf* tidy (*sum*), titanic, towering, *inf* tremendous, vast, voluminous, weighty, *inf* whacking, *inf* whopping, wide. *Opp* SMALL.

**larva** *n* caterpillar, grub, maggot.

**lash** *n* ▷ WHIP. • *vb* **1** beat, birch, cane, flail, flog, scourge, strike, thrash, whip. ▷ HIT. **2** ▷ CRITICIZE.

**last** *adj* closing, concluding, final, furthest, hindmost, latest, most recent, rearmost, terminal, terminating, ultimate. *Opp* FIRST. • *vb* carry on, continue, endure, hold, hold out, keep on, linger, live, persist, remain, stay, survive, *inf* wear well. *Opp* DIE, FINISH. **lasting** ▷ PERMANENT.

**late** *adj* **1** behindhand, belated, delayed, dilatory, overdue, slow, tardy, unpunctual. **2** *a late edition.* current, last, new, recent, up-to-date. **3** *the late king.* dead, deceased, departed, ex-, former, past, previous.

**latent** *adj* dormant, hidden, invisible, potential, undeveloped, undiscovered.

**latitude** *n inf* elbow-room, freedom, leeway, liberty, room, scope, space.

**latter** *adj* closing, concluding, last, last-mentioned, later, recent, second. *Opp* FORMER.

**lattice** *n* criss-cross, framework, grid, mesh, trellis.

**laugh** *vb* beam, be amused, burst into laughter, chortle, chuckle, *sl* fall about, giggle, grin, guffaw, roar with laughter, simper, smile, smirk, sneer, snicker, snigger, *inf* split your sides, titter. **laugh at** ▷ RIDICULE.

**laughable** *adj* absurd, derisory, ludicrous, preposterous, ridiculous. ▷ FUNNY.

**laughing-stock** *n* butt, figure of fun, victim.

**laughter** *n* chuckling, giggling, guffawing, hilarity, *inf* hysterics, laughing, laughs, merriment, mirth, snickering, sniggering, tittering. ▷ RIDICULE.

**launch** *vb* **1** begin, embark on, establish, float, found, inaugurate, initiate, open, organize, set in motion, set off, set up, start. **2** blast off, catapult, dispatch, fire, propel, send off, set off, shoot.

**lavatory** *n* bathroom, cloakroom, convenience, *inf* Gents, *inf* Ladies, latrine, *inf* loo, *inf* men's room, *childish* potty, *old use* privy, public convenience, toilet, urinal, water-closet, WC, *inf* women's room.

**lavish** *adj* **1** abundant, bountiful, copious, exuberant, free, generous, liberal, luxuriant, luxurious, munificent, opulent, plentiful, profuse, sumptuous, unselfish, unsparing, unstinting. **2** excessive, extravagant, improvident, prodigal, self-indulgent, wasteful. *Opp* ECONOMICAL.

**law** *n* **1** act, bill [= *draft law*], bylaw, commandment, decree, directive, edict, injunction, mandate, measure, order, ordinance, pronouncement, regulation, rule, statute. **2** *laws of science.* axiom, formula, postulate, principle, proposition, theory. **3** *laws of decency.* code, convention, practice. **4** *court of law.* justice, litigation.

**law-abiding** *adj* compliant, decent, disciplined, good, honest, obedient, orderly, peaceable, peaceful, respectable, well-behaved. *Opp* LAWLESS.

**lawful** *adj* allowable, allowed, authorized, constitutional, documented, just, justifiable, legal, legitimate, permissible, permitted, prescribed, proper, recognized, regular, right, rightful, valid. *Opp* ILLEGAL.

**lawless** *adj* anarchic, anarchical, badly-behaved, chaotic, disobedient, disorderly, illdisciplined, insubordinate, mutinous, rebellious, riotous, rowdy, seditious, turbulent, uncontrolled, undisciplined, ungoverned, unregulated, unrestrained, unruly, wild. ▷ WICKED. *Opp* LAW-ABIDING.

**lawlessness** *n* anarchy, chaos, disobedience, disorder, mob-rule, rebellion, rioting. *Opp* ORDER.

**lawyer** *n* advocate, barrister, counsel, legal representative, member of the bar, solicitor.

**lax** *adj* careless, casual, easygoing, flexible, indulgent, lenient, loose, neglectful, negligent, permissive, relaxed, remiss, slack, slipshod, unreliable, vague. *Opp* STRICT.

**laxative** *n* aperient, enema, purgative.

**lay** *vb* **1** apply, arrange, deposit, leave, place, position, put down, rest, set down, set out, spread. **2** *lay foundations.* build, construct, establish. **3** *lay the blame on someone.* ascribe, assign, attribute, burden, charge, impose, plant, *inf* saddle. **4** *lay plans.* concoct, create, design, organize, plan, set up. **lay bare** ▷ REVEAL. **lay bets** ▷ GAMBLE. **lay by** ▷ STORE. **lay down the law** ▷ DICTATE. **lay in** ▷ STORE. **lay into** ▷ ATTACK. **lay low** ▷ DEFEAT. **lay off something**

▷ CEASE. **lay someone off** ▷ DISMISS. **lay to rest** ▷ BURY. **lay up** ▷ STORE. **lay waste** ▷ DESTROY.

**layer** *n* **1** coat, coating, covering, film, sheet, skin, surface, thickness. **2** *layer of rock.* seam, stratum, substratum. *in layers* laminated, layered, sandwiched, stratified.

**layman** *n* **1** amateur, non-specialist, untrained person. *Opp* PROFESSIONAL. **2** [*church*] lay-person, member of the congregation, parishioner, unordained person. *Opp* CLERGYMAN.

**laze** *vb* be lazy, do nothing, idle, lie about, loaf, lounge, relax, sit about, unwind.

**laziness** *n* dilatoriness, idleness, inactivity, indolence, lethargy, loafing, lounging about, shiftlessness, slackness, sloth, slowness, sluggishness, torpor. *Opp* INDUSTRY.

**lazy** *adj* **1** dilatory, easily pleased, easygoing, idle, inactive, indolent, languid, lethargic, listless, shiftless, *inf* skiving, slack, slothful, slow, sluggish, torpid, unenterprising, work-shy. **2** peaceful, quiet, relaxing. *Opp* ENERGETIC, INDUSTRIOUS. **be lazy** ▷ LAZE. **lazy person** ▷ SLACKER.

**lead** *n* **1** direction, example, guidance, leadership, model, pattern, precedent. **2** *lead on a crime.* clue, hint, line, tip, tip-off. **3** *lead in a race.* first place, front, spearhead, van, vanguard. **4** *lead in a play.* chief part, hero, heroine, principal, protagonist, starring role, title role. **5** cable, flex, wire. **6** *dog's lead.* chain, leash, strap. • *vb* **1** conduct, draw, escort, guide, influence, pilot, prompt, show the way, steer, usher. **2** be in charge of, captain, command, direct, govern, head, manage, preside over, rule, *inf* skipper, superintend, supervise. **3** be first, be in front, be in the lead, go first, head the field. **4** *lead the field.* beat, defeat, excel, outdo, outstrip, precede, surpass, vanquish. *Opp* FOLLOW. **lead astray** ▷ MISLEAD. **leading** ▷ CHIEF, INFLUENTIAL. **lead off** ▷ BEGIN.

**leader** *n* **1** ayatollah, boss, captain, chieftain, commander, conductor, courier, demagogue, director, figure-head, godfather, guide, head, patriarch, premier, prime minister, principal, ringleader, superior, *inf* supremo. ▷ CHIEF, RULER. **2** *leader in a newspaper.* editorial, leading article.

**leaf** *n* **1** blade, foliage, frond, greenery. **2** folio, page, sheet.

**leaflet** *n* advertisement, bill, booklet, brochure, circular, flyer, folder, handbill, handout, notice, pamphlet.

**league** *n* alliance, association, coalition, confederation, *derog* conspiracy, federation, fraternity, guild, society, union. ▷ GROUP. **be in league with** ▷ CONSPIRE.

**leak** *n* **1** discharge, drip, emission, escape, exudation, leakage, oozing, seepage, trickle. **2** aperture, break, chink, crack, crevice, cut, fissure, flaw, hole, opening, perforation, puncture, rent, split, tear. **3** *security leak.* disclosure, revelation. • *vb* **1** discharge, drip, escape, exude, ooze, percolate, seep, spill, trickle. **2** *leak secrets.* disclose, divulge, give away, let out, let slip, *inf* let the cat out of the bag about, make known, pass on, reveal, *inf* spill the beans about.

**leaky** *adj* cracked, dripping, holed, perforated, punctured.

**lean** *adj* angular, bony, emaciated, gangling, gaunt, hungry-looking, lanky, long, rangy, scraggy, scrawny, skinny, slender, slim, spare, thin, weedy, wiry. *Opp* FAT. • *vb* 1 bank, careen, heel over, incline, keel over, list, slant, slope, tilt, tip. 2 loll, prop yourself up, recline, rest, support yourself.

**leaning** *n* bent, bias, favouritism, inclination, instinct, liking, partiality, penchant, predilection, preference, propensity, readiness, taste, tendency, trend.

**leap** *vb* 1 bound, clear (*a fence*), hop over, hurdle, jump, leapfrog, skip over, spring, vault. 2 caper, cavort, dance, frisk, frolic, gambol, hop, prance, romp. 3 *leap on someone.* ambush, attack, pounce.

**learn** *vb* acquire, ascertain, assimilate, become aware of, become proficient in, be taught, *inf* catch on, commit to memory, discover, find out, gain, gain understanding of, gather, grasp, master, memorize, *inf* mug up, pick up, remember, study, *inf* swot up. **learned** ▷ ACADEMIC, EDUCATED.

**learner** *n* apprentice, beginner, cadet, initiate, L-driver, novice, pupil, scholar, starter, student, trainee, tiro.

**learning** *n* culture, education, erudition, information, knowledge, lore, scholarship, wisdom.

**lease** *n* agreement, contract. • *vb* charter, hire out, let, rent out, sublet.

**least** *adj* fewest, lowest, minimum, negligible, poorest, slightest, smallest, tiniest.

**leather** *n* chamois, hide, skin, suede.

**leave** *n* 1 authorization, consent, dispensation, liberty, licence, permission, sanction. 2 *leave from work.* absence, free time, furlough, holiday, recess, sabbatical, time off, vacation. • *vb* 1 *inf* be off, *inf* check out, decamp, depart, disappear, *sl* do a bunk, escape, exeunt [*they go out*], exit [*he/she goes out*], get away, get out, go away, go out, *sl* hop it, *inf* pull out, *sl* push off, retire, retreat, run away, say goodbye, set off, *inf* take off, take your leave, withdraw. 2 abandon, desert, evacuate, forsake, vacate. 3 *leave your job.* *inf* chuck in, *inf* drop out of, give up, quit, relinquish, renounce, resign from, retire from, *inf* walk out of, *inf* wash your hands of. 4 *leave it there.* allow to stay, *inf* let alone, let be. 5 *left it here.* deposit, place, position, put down, set down. 6 *left it somewhere.* forget, lose, mislay. 7 *leave it to you.* assign, cede, consign, entrust, refer, relinquish. 8 *leave in a will.* bequeath, hand down, will. **leave off** ▷ STOP. **leave out** ▷ OMIT.

**lecture** *n* 1 address, discourse, disquisition, instruction, lesson, paper, speech, talk, treatise. 2 *lecture on bad manners.* diatribe, harangue, sermon. ▷ REPRIMAND. • *vb* 1 be a lecturer, teach. 2 discourse, give a lecture, harangue, *inf* hold forth, pontificate, preach, sermonize, speak, talk formally. 3 ▷ REPRIMAND.

**lecturer** *n* don, fellow, instructor, professor, speaker, teacher, tutor.

**ledge** *n* mantel, overhang, projection, ridge, shelf, sill, step, window-sill.

**left** *adj, n* 1 left-hand, port [= *left facing bow of ship*], sinistral. 2 *left wing in politics.* communist, Labour, leftist, liberal, Marxist, progressive, radical, *derog* red, revolutionary, socialist. *Opp* RIGHT.

**leg** *n* 1 limb, member, *inf* peg, *inf* pin, shank. □ *ankle, calf, foot, hock, knee, shin, thigh.* 2 brace, column, pillar, prop, support, upright. 3 *leg of a journey.* lap, length, part, section, stage, stretch. **pull someone's leg** ▷ HOAX.

**legacy** *n* bequest, endowment, estate, inheritance.

**legal** *adj* 1 above-board, acceptable, admissible, allowable, allowed, authorized, constitutional, just, lawful, legalized, legitimate, licensed, licit, permitted, permissible, proper, regular, right, rightful, valid. *Opp* ILEGAL. 2 *legal proceedings.* forensic, judicial, judiciary.

**legalize** *vb* allow, authorize, legitimate, legitimize, license, make legal, normalize, permit, regularize, validate. *Opp* BAN.

**legend** *n* epic, folk-tale, myth, saga, tradition. ▷ STORY.

**legendary** *adj* 1 apocryphal, epic, fabled, fabulous, fictional, fictitious, imaginary, invented, made-up, mythical, non-existent, story-book, traditional. 2 *a legendary name.* ▷ FAMOUS.

**legible** *adj* clear, decipherable, distinct, intelligible, neat, plain, readable, understandable. *Opp* ILLEGIBLE.

**legitimate** *adj* 1 authentic, genuine, proper, real, regular, true. ▷ LEGAL. 2 *a legitimate deception.* ethical, just, justifiable, moral, proper, reasonable, right. *Opp* ILLEGITIMATE.
• *vb* ▷ LEGALIZE.

**leisure** *n* breathing-space, ease, holiday, liberty, opportunity, quiet, recreation, relaxation, relief, repose, respite, rest, spare time, time off.

**leisurely** *adj* easy, gentle, lingering, peaceful, relaxed, relaxing, restful, unhurried. ▷ SLOW. *Opp* BRISK.

**lend** *vb* advance, loan. *Opp* BORROW.

**length** *n* 1 distance, extent, footage, measure, measurement, mileage, reach, size, span, stretch. 2 duration, period, stretch, term.

**lengthen** *vb* continue, drag out, draw out, enlarge, elongate, expand, extend, get longer, increase, make longer, *inf* pad out, prolong, protract, pull out, stretch. *Opp* SHORTEN.

**lenient** *adj* charitable, easygoing, forbearing, forgiving, gentle, humane, indulgent, merciful, mild, permissive, soft, soft-hearted, sparing, tolerant. ▷ KIND. *Opp* STRICT.

**less** *adj* fewer, reduced, shorter, smaller. *Opp* MORE.

**lessen** *vb* 1 assuage, cut, deaden, decrease, ease, lighten, lower, make less, minimize, mitigate, reduce, relieve, tone down. 2 abate, become less, decline, decrease, die away, diminish, dwindle, ease off, let up, moderate, slacken, subside, tail off, weaken. *Opp* INCREASE.

**lesson** *n* 1 class, drill, instruction laboratory, lecture, practical, seminar, session, task, teaching, tutorial, workshop. 2 *a moral lesson.* admonition, example, moral, warning.

**let** *vb* 1 agree to, allow to, authorize to, consent to, enable to, give permission to, license to, permit to, sanction to. 2 *let a house.* charter, contract out, hire, lease, rent. **let alone, let be** ▷ LEAVE. **let go, let loose** ▷ LIBERATE. **let off** ▷ FIRE. **let out** ▷ LIBERATE. **let**

**someone off** ▷ ACQUIT. **let up** ▷ LESSEN.

**letdown** *n* anti-climax, disappointment, disillusionment, *inf* wash-out.

**lethal** *adj* deadly, fatal, mortal, poisonous.

**lethargic** *adj* apathetic, comatose, dull, heavy, inactive, indifferent, indolent, languid, lazy, listless, phlegmatic, sleepy, slow, slothful, sluggish, torpid. ▷ WEARY. *Opp* ENERGETIC.

**lethargy** *n* apathy, idleness, inactivity, indolence, inertia, laziness, listlessness, slothfulness, slowness, sluggishness, torpor, weariness. *Opp* ENERGY.

**letter** *n* **1** character, consonant, sign, symbol, vowel. **2** *old use* billet-doux, card, communication, dispatch, epistle, message, missive, note, postcard. **letters** correspondence, junk mail, mail, post.

**level** *adj* **1** even, flat, flush, horizontal, plane, regular, smooth, straight, true, uniform. **2** horizontal. **3** *level scores*. balanced, even, equal, matching, *inf* neck-and-neck, the same. *Opp* UNEVEN. • *n* **1** altitude, depth, elevation, height, value. **2** degree, echelon, grade, plane, position, rank, *inf* rung on the ladder, stage, standard, standing, status. **3** *level in a building*. floor, storey. • *vb* **1** even out, flatten, rake, smooth. **2** bulldoze, demolish, destroy, devastate, knock down, lay low, raze, tear down, wreck. **level-headed** ▷ SENSIBLE.

**lever** *vb* force, prise, wrench.

**liable** *adj* **1** accountable, answerable, blameworthy, responsible. **2** *liable to fall over*. apt, disposed, inclined, in the habit of, likely, minded, predisposed, prone, ready, susceptible, tempted, vulnerable, willing.

**liaison** *n* **1** communication, contact, cooperation, liaising, linkage, links, mediation, relationship, tie. **2** ▷ AFFAIR.

**liar** *n* deceiver, false witness, *inf* fibber, perjurer, *inf* story-teller.

**libel** *n* calumny, defamation, denigration, insult, lie, misrepresentation, obloquy, scandal, slander, slur, smear, vilification. • *vb* blacken the name of, calumniate, defame, denigrate, disparage, malign, misrepresent, slander, slur, smear, write lies about, traduce, vilify.

**libellous** *adj* calumnious, cruel, damaging, defamatory, disparaging, false, hurtful, insulting, lying, malicious, mendacious, scurrilous, slanderous, untrue, vicious.

**liberal** *adj* **1** abundant, ample, bounteous, bountiful, copious, free, generous, lavish, munificent, open-handed, plentiful, unstinting. **2** *liberal attitudes*. big-hearted, broad-minded, charitable, easy-going, enlightened, fair-minded, humanitarian, indulgent, impartial, latitudinarian, lenient, magnanimous, open-minded, permissive, philanthropic, tolerant, unbiased, unbigoted, unopinionated, unprejudiced, unselfish. *Opp* NARROW-MINDED. **3** *liberal politics*. progressive, radical, reformist. *Opp* CONSERVATIVE.

**liberalize** *vb* broaden, ease, enlarge, make more liberal, moderate, open up, relax, soften, widen.

**liberate** *vb* deliver, discharge, disenthral, emancipate, enfranchise, free, let go, let loose, let out, loose, manumit, ransom, release, rescue, save, set free, untie. *Opp* CAPTURE, SUBJUGATE.

**liberty** *n* autonomy, emancipation, independence, liberation, release, self-determination, self-rule. ▷ FREEDOM. **at liberty** ▷ FREE.

**licence** *n* **1** certificate, credentials, document, papers, permit, warrant. **2** ▷ FREEDOM.

**license** *vb* **1** allow, approve, authorize, certify, commission, empower, entitle, give a licence to, permit, sanction, validate. **2** buy a licence for, make legal.

**lid** *n* cap, cover, covering, top.

**lie** *n* deceit, dishonesty, disinformation, fabrication, falsehood, falsification, *inf* fib, fiction, invention, misrepresentation, prevarication, untruth, *inf* whopper. *Opp* TRUTH. ● *vb* **1** *inf* be economical with the truth, bluff, commit perjury, deceive, falsify the facts, *inf* fib, perjure yourself, prevaricate, tell lies. **2** be horizontal, be prone, be prostrate, be recumbent, be supine, lean back, lounge, recline, repose, rest, sprawl, stretch out. **3** *The house lies in a valley.* be, be found, be located, be situated, exist. **lie low** ▷ HIDE.

**life** *n* **1** being, existence, living. **2** activity, animation, dash, élan, energy, enthusiasm, exuberance, *inf* go, liveliness, soul, sparkle, spirit, sprightliness, verve, vigour, vitality, vivacity, zest. **3** autobiography, biography, memoir, story.

**lifeless** *adj* **1** comatose, dead, deceased, inanimate, inert, insensate, insensible, killed, motionless, unconscious. **2** *lifeless desert.* arid, bare, barren, desolate, empty, sterile, uninhabited, waste. **3** *a lifeless performance.* apathetic, boring, dull, flat, heavy, lacklustre, lethargic, slow, torpid, unexciting, wooden. *Opp* LIVELY, LIVING.

**lifelike** *adj* authentic, convincing, faithful, graphic, natural, photographic, realistic, true-to-life, vivid. *Opp* UNREALISTIC.

**lift** *n* elevator, hoist. ● *vb* **1** buoy up, carry, elevate, heave up, hoist, jack up, pick up, pull up, raise, rear. **2** ascend, fly, lift off, rise, soar. **3** boost, cheer, enhance, improve, promote. **4** ▷ STEAL.

**light** *adj* **1** lightweight, portable, underweight, weightless. *Opp* HEAVY. **2** bright, illuminated, lit-up, well-lit. *Opp* DARK. **3** *light work.* ▷ EASY. **4** *a light wind.* ▷ GENTLE. **5** *a light touch.* ▷ DELICATE. **6** *light colours.* ▷ PALE. **7** *a light heart.* ▷ CHEERFUL. **8** *light traffic.* ▷ SPARSE. ● *n* **1** beam, blaze, brightness, brilliance, effulgence, flare, flash, fluorescence, glare, gleam, glint, glitter, glow, halo, illumination, incandescence, luminosity, lustre, phosphorescence, radiance, ray, reflection, scintillation, shine, sparkle, twinkle. □ *candlelight, daylight, firelight, gaslight, moonlight, starlight, sunlight, torchlight, twilight.* □ *arc light, beacon, bulb, candelabra, candle, chandelier, electric light, flare, floodlight, fluorescent lamp, headlamp, headlight, lamp, lantern, laser, lighthouse, lightship, neon light, pilot light, searchlight, spotlight, standard lamp, street light, strobe, stroboscope, taper, torch, traffic lights.* ● *vb* **1** fire, ignite, kindle, put a match to, set alight, set fire to, switch on. *Opp* EXTINGUISH. **2** ▷ LIGHTEN. **bring to light** ▷ DISCOVER. **give light, reflect light** be bright, be luminous, be phosphorescent, blaze, blink, burn, coruscate, dazzle, flash, flicker, glare, gleam, glimmer, glint, glisten, glitter, glow, radiate, reflect, scintillate, shimmer, shine, spark, sparkle, twinkle. **light-headed** ▷ DIZZY. **light-hearted** ▷ CHEERFUL. **light up** ▷ LIGHTEN. **shed light on** ▷ EXPLAIN.

**lighten** *vb* **1** cast light on, floodlight, illuminate, irradiate, light up, shed light on, shine on. **2** *The sky lightened.* become lighter, brighten, cheer up, clear. **3** ▷ LESSEN.

**lighthouse** *n* beacon, light, lightship, warning-light.

**like** *adj* akin to, analogous to, close to, cognate with, comparable to, congruent with, corresponding to, equal to, equivalent to, identical to, parallel to, similar to. • *n* liking, partiality, predilection, preference. • *vb* admire, approve of, appreciate, be attracted to, be fond of, be interested in, be keen on, be partial to, be pleased by, delight in, enjoy, find pleasant, *sl* go for, *inf* go in for, have a high regard for, *inf* have a weakness for, prefer, relish, revel in, take pleasure in, *inf* take to, welcome. ▷ LOVE. *Opp* HATE.

**likeable** *adj* admirable, attractive, charming, congenial, endearing, interesting, lovable, nice, personable, pleasant, pleasing. ▷ FRIENDLY. *Opp* HATEFUL.

**likelihood** *n* chance, hope, possibility, probablility, prospect.

**likely** *adj* **1** anticipated, expected, feasible, foreseeable, plausible, possible, predictable, probable, reasonable, unsurprising. **2** *a likely candidate.* able, acceptable, appropriate, convincing, credible, favourite, fitting, hopeful, promising, qualified, suitable, *inf* tipped to win. **3** *likely to come.* apt, disposed, inclined, liable, prone, ready, tempted, willing. *Opp* UNLIKELY.

**liken** *vb* compare, equate, juxtapose, match.

**likeness** *n* **1** affinity, analogy, compatibility, congruity, correspondence, resemblance, similarity. *Opp* DIFFERENCE. **2** copy, depiction, drawing, duplicate, facsimile, image, model, picture, portrait, replica, representation, reproduction, study.

**liking** *n* affection, affinity, appetite, eye, fondness, inclination, partiality, penchant, predilection, predisposition, preference, propensity, *inf* soft spot, taste, weakness. ▷ LOVE. *Opp* HATRED.

**limb** *n* appendage, member, offshoot, projection. □ *arm, bough, branch, flipper, foreleg, forelimb, leg, wing.*

**limber** *vb* **limber up** exercise, get ready, loosen up, prepare, warm up.

**limbo** *n* **in limbo** abandoned, forgotten, in abeyance, left out, neglected, neither one thing nor the other, *inf* on hold, *inf* on the back burner, unattached.

**limit** *n* **1** border, boundary, bounds, brink, confines, demarcation line, edge, end, extent, extreme point, frontier, perimeter. **2** ceiling, check, curb, cut-off point, deadline, inhibition, limitation, maximum, restraint, restriction, stop, threshold. • *vb* bridle, check, circumscribe, confine, control, curb, define, fix, hold in check, put a limit on, ration, restrain, restrict. **limited** ▷ FINITE, INADEQUATE.

**limitation** *n* **1** ▷ LIMIT. **2** defect, deficiency, fault, inadequacy, shortcoming, weakness.

**limitless** *adj* boundless, countless, endless, everlasting, immeasurable, incalculable, inexhaustible, infinite, innumersable, never-ending, numberless, perpetual, renewable, unbounded, unconfined, unending, unimaginable, unlimited, unrestricted. ▷ VAST. *Opp* FINITE.

**limp** *adj inf* bendy, drooping, flabby, flaccid, flexible, *inf* floppy, lax, loose, pliable, sagging, slack, soft, weak, wilting, yielding. ▷ WEARY. *Opp* RIGID. • *vb* be lame, falter, hobble, hop, stagger, totter.

**line** *n* **1** band, borderline, boundary, contour, contour line, dash, mark, streak, striation, strip, stripe, stroke, trail. **2** corrugation, crease, *inf* crow's feet, fold, furrow, groove, score, wrinkle. **3** cable, cord, flex, hawser, lead, rope, string, thread, wire. **4** chain, column, cordon, crocodile, file, procession, queue, rank, row, series. **5** *railway line.* branch, main line, route, service, track. • *vb* **1** mark with lines, rule, score, streak, striate, underline. **2** *line the street.* border, edge, fringe. **3** *line a garment.* cover the inside, insert a lining, pad, reinforce. **line up** ▷ ALIGN, QUEUE.

**linger** *vb* continue, dally, dawdle, delay, dither, endure, hang about, hover, idle, lag, last, loiter, pause, persist, procrastinate, remain, *inf* shilly-shally, stay, stay behind, *inf* stick around, survive, temporize, wait about. *Opp* HURRY.

**lining** *n* inner coat, inner layer, interfacing, liner, padding.

**link** *n* **1** bond, connection, connector, coupling, join, joint, linkage, tie, yoke. ▷ FASTENER. **2** affiliation, alliance, association, communication, interdependence, liaison, partnership, relationship, *inf* tie-up, twinning, union. • *vb* **1** amalgamate, associate, attach, compare, concatenate, connect, couple, interlink, join, juxtapose, make a link, merge, network, relate, see a link, twin, unite, yoke. ▷ FASTEN.

**lip** *n* brim, brink, edge, rim.

**liquefy** *vb* become liquid, dissolve, liquidize, melt, run, thaw. *Opp* SOLIDIFY.

**liquid** *adj* aqueous, flowing, fluid, liquefied, melted, molten, running, *inf* runny, sloppy, *inf* sloshy, thin, watery, wet. *Opp* SOLID. • *n* fluid, juice, liquor, solution, stock.

**liquidate** *vb* annihilate, destroy, *inf* do away with, *inf* get rid of, remove, silence, wipe out. ▷ KILL.

**liquor** *n* **1** alcohol, *sl* booze, *sl* hard stuff, intoxicants, *sl* shorts, spirits, strong drink. **2** ▷ LIQUID.

**list** *n* catalogue, column, directory, file, index, inventory, listing, register, roll, roster, rota, schedule, shopping-list, table. • *vb* **1** catalogue, enumerate, file, index, itemize, make a list of, note, record, register, tabulate, write down. **2** bank, careen, heel, incline, keel over, lean, slant, slope, tilt, tip.

**listen** *vb* attend, concentrate, eavesdrop, *old use* hark, hear, heed, *inf* keep your ears open, lend an ear, overhear, pay attention, take notice.

**listless** *adj* apathetic, enervated, feeble, heavy, languid, lazy, lethargic, lifeless, phlegmatic, sluggish, tired, torpid, unenthusiastic, uninterested, weak. ▷ WEARY. *Opp* LIVELY.

**literal** *adj* close, exact, faithful, matter of fact, plain, prosaic, strict, unimaginative, verbatim, word for word.

**literary** *adj* 1 cultured, educated, erudite, imaginative, learned, refined, scholarly, well-read, widely-read. 2 *literary style*. ornate, *derog* pedantic, poetic, polished, rhetorical, *derog* self-conscious, sophisticated, stylish.

**literate** *adj* 1 educated, wellread. 2 accurate, correct, readable, well-written.

**literature** *n* books, brochures, circulars, creative writing, handbills, handouts, leaflets, pamphlets, papers, writings. □ *autobiography, biography, comedy, crime fiction, criticism, drama, epic, essay, fantasy, fiction, folktale, journalism, myth and legend, novels, parody, poetry, propaganda, prose, romance, satire, science fiction, tragedy*. ▷ WRITING.

**lithe** *adj* agile, flexible, limber, lissom, loose-jointed, pliable, pliant, supple. *Opp* STIFF.

**litter** *n* bits and pieces, clutter, debris, fragments, garbage, jumble, junk, mess, odds and ends, refuse, rubbish, trash, waste. • *vb* clutter, fill with litter, make untidy, *inf* mess up, scatter, strew.

**little** *adj* 1 *inf* baby, bantam, compact, concise, diminutive, *inf* dinky, dwarf, exiguous, fine, fractional, infinitesimal, lean, lilliputian, microscopic, midget, *inf* mini, miniature, minuscule, minute, narrow, petite (*woman*), *inf* pint-sized, *inf* pocket-sized, *inf* poky, portable, pygmy, short, slender, slight, small, *inf* teeny, thin, tiny, toy, undergrown, undersized, *inf* wee, *inf* weeny. *Opp* BIG. 2 *little food*. inadequate, insufficient, meagre, mean, *inf* measly, miserly, modest, niggardly, parsimonious, *inf* piddling, scanty, skimpy, stingy, ungenerous, unsatisfactory. 3 *of little importance*. inconsequential, insignificant, minor, negligible, nugatory, slim (*chance*), slight, trifling, trivial, unimportant.

**live** *adj* 1 ▷ LIVING. 2 *a live fire*. ▷ ALIGHT. 3 *a live issue*. contemporary, current, important, pressing, relevant, topical, vital. *Opp* DEAD. • *vb* 1 breathe, continue, endure, exist, flourish, function, last, persist, remain, stay alive, survive. *Opp* DIE. 2 be accommodated, dwell, lodge, reside, room, stay. 3 *live on £20 a week*. fare, *inf* get along, keep going, make a living, pay the bills, subsist. **live in** ▷ INHABIT. **live on** ▷ EAT.

**liveliness** *n* activity, animation, boisterousness, bustle, dynamism, energy, enthusiasm, exuberance, *inf* go, gusto, high spirits, spirit, sprightliness, verve, vigour, vitality, vivacity, zeal. *Opp* APATHY.

**lively** *adj* active, agile, alert, animated, boisterous, bubbly, bustling, busy, cheerful, colourful, dashing, eager, energetic, enthusiastic, exciting, expressive, exuberant, frisky, gay, high-spirited, irrepressible, jaunty, jazzy, jolly, merry, nimble, *inf* perky, playful, quick, spirited, sprightly, stimulating, strong, vigorous, vital, vivacious, vivid, *inf* zippy. ▷ HAPPY. *Opp* APATHETIC.

**livestock** *n* cattle, farm animals.

**living** *adj* active, actual, alive, animate, breathing, existing, extant, flourishing, functioning, live, living, *old use* quick, sentient, surviving, vigorous, vital. ▷ LIVELY. *Opp* DEAD, EXTINCT. • *n* income, livelihood, occupation, subsistence, way of life.

**load** *n* 1 burden, cargo, consignment, freight, lading, lorry-load, shipment, van-load. 2 *load of responsibility*. *inf* albatross, anxiety, care, *inf* cross, encumbrance, *inf* millstone, onus, trouble, weight, worry. • *vb* 1 burden, encumber, fill, heap, overwhelm, pack, pile, ply, saddle, stack, stow, weigh down. 2 *load a gun.* charge, prime. **loaded** ▷ BIASED, LADEN, WEALTHY.

**loafer** *n* idler, *inf* good-for-nothing, layabout, *inf* lazybones, lounger, shirker, *sl* skiver, vagrant, wastrel.

**loan** *n* advance, credit, mortgage. • *vb* advance, allow, credit, lend.

**loathe** *vb* abhor, abominate, be averse to, be revolted by, despise, detest, dislike, execrate, find intolerable, hate, object to, recoil from, resent, scorn, shudder at. *Opp* LOVE.

**lobby** *n* 1 ante-room, corridor, entrance hall, entry, foyer, hall, hallway, porch, reception, vestibule. 2 *environmental lobby.* campaign, campaigners, pressure-group, supporters. • *vb* persuade, petition, pressurize, try to influence, urge.

**local** *adj* 1 adjacent, adjoining, nearby, neighbouring, serving the locality. 2 *local politics.* community, limited, narrow, neighbourhood, parochial, particular, provincial, regional. *Opp* GENERAL, NATIONAL. • *n* 1 inhabitant, resident, townsman, townswoman. 2 ▷ PUB.

**locality** *n* area, catchment area, community, district, location, neighbourhood, parish, region, residential area, town, vicinity, zone.

**localize** *vb* concentrate, confine, contain, enclose, keep within bounds, limit, narrow down, pin down, restrict. *Opp* SPREAD.

**locate** *vb* 1 come across, detect, discover, find, identify, *inf* lay your hands on, *inf* run to earth, search out, track down, unearth. 2 build, establish, find a place for, found, place, position, put, set up, site, situate, station.

**location** *n* 1 locale, locality, place, point, position, setting, site, situation, spot, venue, whereabouts. 2 *film locations.* background, scene, setting.

**lock** *n* bar, bolt, catch, clasp, fastening, hasp, latch, padlock. • *vb* bolt, close, fasten, padlock, seal, secure, shut. **lock away** ▷ IMPRISON. **lock out** ▷ EXCLUDE. **lock up** ▷ IMPRISON.

**lodge** *n* ▷ HOUSE. • *vb* 1 accommodate, billet, board, house, *inf* put up. 2 abide, dwell, live, *inf* put up, reside, stay, stop. 3 *lodge a complaint.* enter, file, make formally, put on record, record, register, submit.

**lodger** *n* boarder, guest, inmate, paying guest, resident, tenant.

**lodgings** *n* accommodation, apartment, billet, boarding-house, *inf* digs, lodging-house, *sl* pad, quarters, rooms, shelter, *sl* squat, temporary home.

**lofty** *adj* 1 elevated, high, imposing, majestic, noble, soaring, tall, towering. 2 ▷ ARROGANT.

**log** *n* 1 timber, wood. 2 account, diary, journal, record.

**logic** *n* clarity, deduction, intelligence, logical thinking, ratiocination, rationality, reasonableness, reasoning, sense, validity.

**logical** *adj* clear, cogent, coherent consistent, deductive, intelligent, methodical, rational, reasonable, sensible, sound, *inf* step-by-step, structured, systematic, valid, well-reasoned, well-thought-out, wise. *Opp* ILLOGICAL.

**oiter** *vb* be slow, dally, dawdle, hang back, linger, *inf* loaf about, *inf* mess about, skulk, *inf* stand about, straggle.

**one** *adj* isolated, separate, single, solitary, solo, unaccompanied. ▷ LONELY.

**onely** *adj* 1 abandoned, alone, forlorn, forsaken, friendless, lonesome, loveless, neglected, outcast, reclusive, retiring, solitary, unsociable, withdrawn. ▷ SAD. 2 *inf* cut off, deserted, desolate, distant, faraway, isolated, *inf* off the beaten track, out of the way, remote, secluded, unfrequented, uninhabited.

**ong** *adj* big, drawn out, elongated, endless, extended, extensive, great, interminable, large, lasting, lengthy, longish, prolonged, protracted, slow, stretched, sustained, time-consuming, unending. • *vb* crave, hanker, have a longing (for), hunger, *inf* itch, pine, thirst, wish, yearn. **long for** ▷ DESIRE. **long-lasting, long-lived** ▷ PERMANENT. **long-standing** ▷ OLD. **long-suffering** ▷ PATIENT. **long-winded** ▷ TEDIOUS.

**onging** *n* appetite, craving, desire, hankering, hunger, *inf* itch, need, thirst, urge, wish, yearning, *inf* yen.

**ook** *n* 1 gaze, glance, glimpse, observation, peek, peep, sight, *inf* squint, view. 2 air, appearance, aspect, attractiveness, bearing, beauty, complexion, countenance, demeanour, expression, face, looks, manner, mien. • *vb* 1 behold, *inf* cast your eye, consider, contemplate, examine, eye, gape, *inf* gawp, gaze, glance, glimpse, goggle, have a look, inspect, observe, ogle, pay attention (to), peek, peep, peer, read, regard, scan, scrutinize, see, skim through, squint, stare, study, survey, *sl* take a dekko, take note (of), view, watch. 2 *The house looks south.* face, overlook. 3 *look pleased.* appear, seem. **look after** ▷ TEND. **look down on** ▷ DESPISE. **look for** ▷ SEEK. **look into** ▷ INVESTIGATE. **look out** ▷ BEWARE. **look up to** ▷ ADMIRE.

**look-out** *n* guard, sentinel, sentry, watchman.

**loom** *vb* arise, appear, dominate, emerge, hover, impend, materialize, menace, rise, stand out, stick up, take shape, threaten, tower.

**loop** *n* bend, bow, circle, coil, curl, eye, hoop, kink, noose, ring, turn, twist, whorl. • *vb* bend, coil, curl, entwine, make a loop, turn, twist, wind.

**loophole** *n* escape, *inf* get-out, *inf* let-out, outlet, way out.

**loose** *adj* 1 detachable, detached, disconnected, independent, insecure, loosened, movable, moving, scattered, shaky, unattached, unconnected, unfastened, unsteady, wobbly. 2 *loose animals.* at large, at liberty, escaped, free, free-range, released, roaming, uncaged, unconfined, unfettered, unrestricted, untied. 3 *loose hair.* dangling, hanging, spread out, straggling, trailing. 4 *loose clothing.* baggy, *inf* floppy, loose-fitting, slack, unbuttoned. 5 *loose thinking.* broad, careless, casual, diffuse, general, ill-defined, illogical, imprecise, inexact, informal, lax, rambling, rough, *inf* sloppy, unscientific, unstructured, vague. *Opp* PRECISE, SECURE, TIGHT. 6 ▷ IMMORAL. • *vb* ▷ FREE, LOOSEN.

**loosen** *vb* 1 ease off, free, let go, loose, make loose, relax, release, separate, slacken, unfasten, unloose, untie. ▷ UNDO. 2 become loose, come adrift, open up. *Opp* TIGHTEN.

**loot** *n* booty, contraband, haul, *inf* ill-gotten gains, plunder, prize, spoils, *inf* swag, takings. • *vb* despoil, pillage, plunder, raid, ransack, ravage, rifle, rob, sack, steal from.

**lopsided** *adj* askew, asymmetrical, awry, *inf* cockeyed, crooked, one-sided, tilting, unbalanced, unequal, uneven.

**lord** *n* aristocrat, noble, peer. □ *baron, count, duke, earl,* old use *thane, viscount.* ▷ RULER.

**lose** *vb* **1** be deprived of, cease to have, drop, find yourself without, forfeit, forget, leave (somewhere), mislay, misplace, miss, part with, stray from. *Opp* FIND. **2** admit defeat, be defeated, capitulate, *inf* come to grief, fail, get beaten, get thrashed, succumb, suffer defeat. *Opp* WIN. **3** *lose your chance.* fritter, let slip, squander, waste. **4** *lose pursuers.* escape from, evade, get rid of, give the slip, leave behind, outrun, shake off, throw off. **losing** ▷ UNSUCCESSFUL.

**loser** *n* the defeated, runner-up, the vanquished. *Opp* WINNER.

**loss** *n* bereavement, damage, defeat, deficit, depletion, deprivation, destruction, diminution, disappearance, erosion, failure, forfeiture, impairment, privation, reduction, sacrifice. *Opp* GAIN. **losses** casualties, deaths, death toll, fatalities.

**lost** *adj* **1** abandoned, departed, destroyed, disappeared, extinct, forgotten, gone, irrecoverable, irretrievable, left behind, mislaid, misplaced, missing, strayed, untraceable, vanished. **2** absorbed, day-dreaming, dreamy, distracted, engrossed, preoccupied, rapt. **3** corrupt, damned, fallen. ▷ WICKED.

**lot** *n a lot in a sale.* ▷ ITEM. **a lot of, lots of** ▷ PLENTY. **draw lots** ▷ GAMBLE. **the lot** all (of), everything, the whole thing, *inf* the works.

**lotion** *n* balm, cream, embrocation, liniment, ointment, pomade, salve, unguent.

**lottery** *n* **1** raffle, sweepstake. **2** gamble, speculation, venture.

**loud** *adj* **1** audible, blaring, booming, clamorous, clarion (*call*), deafening, ear-splitting, echoing, fortissimo, high, noisy, penetrating, piercing, raucous, resounding, reverberant, reverberating, roaring, rowdy, shrieking, shrill, sonorous, stentorian, strident, thundering, thunderous, vociferous. **2** *loud colours.* ▷ GAUDY. *Opp* QUIET.

**lounge** *n* drawing-room, front room, living-room, parlour, salon, sitting-room. • *vb* be idle, be lazy, dawdle, hang about, idle, *inf* kill time, laze, loaf, lie around, loiter, *inf* loll about, *inf* mess about, *inf* mooch about, relax, *inf* skive, slouch, slump, sprawl, stand about, take it easy, vegetate, wast[e] time.

**lout** *n* boor, churl, rude person, oaf, *inf* yob.

**lovable** *adj* adorable, appealing, attractive, charming, cuddly, *inf* cute, *inf* darling, dear, enchanting, endearing, engaging, fetching, likeable, lovely, pleasing, taking, winning, winsome. *Opp* HATEFUL.

**love** *n* **1** admiration, adoration, adulation, affection, ardour, attachment, attraction, desire, devotion, fancy, fervour, fondness, infatuation, liking, passion, regard, tenderness, warmth. ▷ FRIENDSHIP. 2 beloved, darling, dear, dearest, loved one. ▷ LOVER. • *vb* 1 admire, adore, be charmed by, be fond of, be infatuated by, be in love with, care for, cherish, desire, dote on, fancy, *inf* have a crush on, have a passion for, idolize, lose your heart to, lust after, treasure, value, want, worship. ▷ LIKE. *Opp* HATE. **in love** besotted, devoted, enamoured, fond, *inf* head over heels, infatuated. **love affair** affair, amour, courtship, intrigue, liaison, relationship, romance. **make love** be intimate, *inf* canoodle, caress, copulate, court, cuddle, embrace, flirt, fornicate, have intercourse, have sex, kiss, mate, *sl* neck, *inf* pet, philander, *old use* spoon, woo. ▷ SEX.

**loved** *adj* beloved, cherished, darling, dear, dearest, esteemed, favourite, precious, treasured, valued, wanted.

**loveless** *adj* cold, frigid, heartless, passionless, undemonstrative, unfeeling, unloving, unresponsive. *Opp* LOVING. ▷ UNLOVED.

**lovely** *adj* appealing, attractive, charming, delightful, enjoyable, fine, good, nice, pleasant, pretty, sweet. ▷ BEAUTIFUL. *Opp* NASTY.

**lover** *n* admirer, boyfriend, companion, concubine, fiancé(e), *old use* follower, friend, gigolo, girlfriend, *inf* intended, mate, mistress, *old use* paramour, suitor, sweetheart, *sl* toy boy, valentine.

**lovesick** *adj* frustrated, languishing, lovelorn, pining.

**loving** *adj* admiring, adoring, affectionate, amorous, ardent, attached, brotherly, caring, close, concerned, dear, demonstrative, devoted, doting, fatherly, fond, friendly, inseparable, kind, maternal, motherly, passionate, paternal, protective, sisterly, tender, warm. ▷ FRIENDLY, SEXY. *Opp* LOVELESS.

**low** *adj* 1 flat, low-lying, sunken. 2 *low trees*. little, short, squat, stumpy, stunted. 3 *low status*. abject, base, degraded, humble, inferior, junior, lesser, lower, lowly, menial, miserable, modest, servile. 4 *low behaviour*. churlish, coarse, common, cowardly, crude, *old use* dastardly, disreputable, ignoble, mean, nasty, vulgar, wicked. ▷ IMMORAL. 5 *low sounds*. gentle, indistinct, muffled, murmurous, muted, pianissimo, quiet, soft, subdued, whispered. 6 *low notes*. bass, deep, reverberant. *Opp* HIGH. **in low spirits** ▷ SAD. **low point** ▷ NADIR.

**lowbrow** *adj* accessible, easy, ordinary, pop, popular, *derog* rubbishy, simple, straightforward, *derog* trashy, *derog* uncultured, undemanding, unpretentious, unsophisticated. *Opp* HIGHBROW.

**lower** *vb* 1 dip, drop, haul down, let down, take down. 2 *lower prices*. bring down, cut, decrease, discount, lessen, mark down, reduce, *inf* slash. 3 *lower the volume*. abate, diminish, quieten, tone down, turn down. 4 *lower yourself*. abase, belittle, debase, degrade, demean, discredit, disgrace, humble, humiliate, stoop. *Opp* RAISE.

**lowly** *adj* base, humble, insignificant, little-known, low, low-born, meek, modest, obscure, unimportant. ▷ ORDINARY. *Opp* EMINENT.

**loyal** *adj* committed, constant, dedicated, dependable, devoted, dutiful, faithful, honest, patriotic, reliable, sincere, stable, staunch, steadfast, steady, true, trustworthy, trusty, unswerving, unwavering. *Opp* DISLOYAL.

**loyalty** *n* allegiance, constancy, dedication, dependability, devotion, duty, faithfulness, fealty, fidelity, honesty, patriotism, reliability, staunchness, steadfastness, trustworthiness. *Opp* DISLOYALTY.

**lubricate** *vb* grease, oil.

**luck** *n* **1** accident, chance, coincidence, destiny, fate, *inf* fluke, fortune, serendipity. **2** *wished her luck.* *inf* break, good fortune, happiness, prosperity, success.

**lucky** *adj* **1** accidental, appropriate, chance, *inf* fluky, fortuitous, opportune, providential, timely, unintentional, unplanned, welcome. **2** blessed, favoured, fortunate, successful. ▷ HAPPY. **3** *lucky number.* advantageous, auspicious. *Opp* UNLUCKY.

**luggage** *n* baggage, belongings, *inf* gear, impedimenta, paraphernalia, *inf* things. □ *bag, basket, box, brief-case, case, chest, hamper, handbag, hand luggage, haversack, holdall, knapsack, pannier,* old use *portmanteau, purse, rucksack, satchel, suitcase, trunk, wallet.*

**lukewarm** *adj* **1** room temperature, tepid, warm. **2** apathetic, cool, half-hearted, indifferent, unenthusiastic.

**lull** *n* break, calm, delay, gap, halt, hiatus, interlude, interruption, interval, lapse, *inf* let-up, pause, respite, rest, silence. • *vb* calm, hush, pacify, quell, quieten, soothe, subdue, tranquillize.

**lumber** *n* **1** beams, boards, planks, timber, wood. **2** bits and pieces, clutter, jumble, junk, litter, odds and ends, rubbish, trash, *inf* white elephants. • *vb* **1** blunder, move clumsily, shamble, trudge. **2** ▷ BURDEN.

**luminous** *adj* bright, glowing, luminescent, lustrous, phosphorescent, radiant, refulgent, shining. ▷ LIGHT.

**lump** *n* **1** ball, bar, bit, block, cake, chunk, clod, clot, cube, *inf* dollop, gob, gobbet, hunk, ingot, mass, nugget, piece, slab, wad, wedge, *inf* wodge. **2** boil, bulge, bump, carbuncle, cyst, excrescence, growth, hump, knob, node, nodule, protrusion, protuberance, spot, swelling, tumescence, tumour. • *vb* **lump together** ▷ COMBINE.

**lunacy** *n* delirium, dementia, derangement, frenzy, hysteria, illogicality, insanity, madness, mania, psychosis, unreason. ▷ STUPIDITY.

**lunatic** *adj* ▷ MAD. • *n* *inf* crackpot, *inf* crank, *inf* loony, madman, madwoman, maniac, *inf* mental case, *inf* nutcase, *inf* nutter, psychopath, psychotic.

**lunge** *vb* **1** jab, stab, strike, thrust. **2** charge, dash, dive, lurch, plunge, pounce, rush, spring, throw yourself.

**lurch** *vb* heave, lean, list, lunge, pitch, plunge, reel, roll, stagger, stumble, sway, totter, wallow. **leave in the lurch** ▷ ABANDON.

**lure** *vb* allure, attract, bait, charm, coax, decoy, draw, entice, induce, inveigle, invite, lead on, persuade, seduce, tempt.

**lurid** *adj* **1** bright, gaudy, glaring, glowing, striking, vivid. **2** ▷ SENSATIONAL.

**lurk** *vb* crouch, hide, lie in wait, lie low, prowl, skulk, steal.

**luscious** *adj* appetizing, delectable, delicious, juicy, mouthwatering, rich, succulent, sweet, tasty.

**lust** *n* **1** carnality, concupiscence, desire, lasciviousness, lechery, libido, licentiousness, passion, sensuality, sexuality. **2** appetite, craving, greed, hunger, itch, longing.

**lustful** *adj* carnal, concupiscent, erotic, lascivious, lecherous, lewd, libidinous, licentious, on heat, passionate, *sl* randy, salacious, sensual, *sl* turned on. ▷ SEXY.

**lustrous** *adj* burnished, glazed, gleaming, glossy, metallic, polished, reflective, sheeny, shiny.

**luxuriant** *adj* **1** abundant, ample, copious, dense, exuberant, fertile, flourishing, green, lush, opulent, plenteous, plentiful, profuse, prolific, rank, rich, teeming, thick, thriving, verdant. **2** ▷ ORNATE. *Opp* SPARSE.

**luxurious** *adj* comfortable, costly, expensive, extravagant, grand, hedonistic, lavish, lush, magnificent, opulent, pampered, *inf* plush, *inf* posh, rich, *inf* ritzy, self-indulgent, splendid, sumptuous, sybaritic, voluptuous. *Opp* SPARTAN.

**luxury** *n* affluence, comfort, ease, enjoyment, extravagance, grandeur, hedonism, high living, indulgence, magnificence, opulence, pleasure, relaxation, self-indulgence, splendour, sumptuousness, voluptuousness.

**lying** *adj* crooked, deceitful, deceptive, dishonest, double-dealing, duplicitous, false, hypocritical, inaccurate, insincere, mendacious, misleading, perfidious, unreliable, untrustworthy, untruthful. *Opp* TRUTHFUL. • *n* deceit, deception, dishonesty, duplicity, falsehood, *inf* fibbing, hypocrisy, mendacity, perfidy, perjury, prevarication.

**lyrical** *adj* emotional, expressive, impassioned, inspired, melodious, musical, poetic, rapturous, rhapsodic, song-like, sweet, tuneful. *Opp* PROSAIC.

# M

**macabre** *adj* eerie, fearsome, frightful, ghoulish, grim, grisly, grotesque, gruesome, morbid, *inf* sick, unhealthy, weird.

**machine** *n* appliance, contraption, contrivance, device, engine, gadget, implement, instrument, mechanism, motor, robot, tool. ▷ MACHINERY.

**machinery** *n* **1** apparatus, equipment, gear, machines, plant, **2** constitution, method, organization, procedure, structure, system.

**mackintosh** *n* anorak, cape, mac, sou'wester, waterproof. ▷ COAT.

**mad** *adj* **1** *inf* batty, berserk, *inf* bonkers, *inf* certified, *inf* crackers, crazed, crazy, *inf* daft, delirious, demented, deranged, disordered, distracted, *inf* dotty, eccentric, fanatical, frantic, frenzied, hysterical, insane, irrational, *inf* loony, lunatic, maniacal, manic, *inf* mental, moonstruck, *Lat* non compos mentis, *inf* nutty, *inf* off your head, *inf* off your rocker, *inf* out of your mind, possessed, *inf* potty, psychotic, *inf* queer in the head, *inf* round the bend, *inf* round the twist, *inf* screwy, *inf* touched, unbalanced, unhinged, unstable, *inf* up the pole, wild. *Opp* SANE. **2** *a mad comedy*. ▷ ABSURD. **3** ▷ ANGRY. **4** ▷ ENTHUSIASTIC.

**madden** *vb* anger, craze, derange, *inf* drive crazy, enrage, exasperate, excite, incense, inflame, infuriate, irritate, make you mad, *inf* make your blood boil, *inf* make you see red, provoke, *inf* send you round the bend, unhinge, vex.

**madman, madwoman** *ns* *inf* crackpot, *inf* crank, eccentric, *inf* loony, lunatic, maniac, *inf* mental case, *inf* nutcase, *inf* nutter, psychopath, psychotic.

**madness** *n* delirium, dementia, derangement, eccentricity, folly, frenzy, hysteria, illogicality, insanity, lunacy, mania, mental illness, psychosis, unreason. ▷ STUPIDITY.

**magazine** *n* **1** comic, journal, monthly, newspaper, pamphlet, paper, periodical, publication, quarterly, weekly. **2** *magazine of weapons.* ammunition dump, armoury, arsenal, storehouse.

**magic** *adj* **1** conjuring, miraculous, mystic, necromantic, supernatural. **2** bewitching, charming, enchanting, entrancing, magical, spellbinding. • *n* **1** black magic, charm, enchantment, hocus-pocus, incantation, *inf* mumbo-jumbo, necromancy, occultism, sorcery, spell, voodoo, witchcraft, witchery, wizardry. **2** conjuring, illusion, legerdemain, sleight of hand, trickery, tricks.

**magician** *n* conjuror, enchanter, enchantress, illusionist, magus, necromancer, sorcerer, *old use* warlock, witch, witch-doctor, wizard.

**magnetic** *adj* alluring, attractive, bewitching, captivating, charismatic, charming, compelling, engaging, enthralling, entrancing, fascinating, hypnotic, inviting, irresistible, seductive, spellbinding. *Opp* REPULSIVE.

**magnetism** *n* allure, appeal, attractiveness, charisma, charm, drawing power, fascination, irresistibility, lure, power, pull, seductiveness.

**magnificent** *adj* awe-inspiring, beautiful, distinguished, excellent, fine, glorious, gorgeous, grand, grandiose, great, imposing, impressive, majestic, marvellous, noble, opulent, *inf* posh, regal, resplendent, rich, spectacular, splendid, stately, sumptuous, superb, wonderful. *Opp* ORDINARY.

**magnify** *vb* **1** amplify, augment, *inf* blow up, enlarge, expand, increase, intensify, make larger. *Opp* SHRINK. **2** *magnify difficulties.* *inf* blow up out of proportion, dramatize, exaggerate, heighten, inflate, make too much of, maximize, overdo, overestimate, overstate. *Opp* MINIMIZE.

**magnitude** *n* bigness, enormousness, extent, greatness, immensity, importance, size.

**mail** *n* correspondence, letters, parcels, post. • *vb* dispatch, forward, post, send.

**maim** *vb* cripple, disable, hamstring, handicap, incapacitate, lame, mutilate. ▷ WOUND.

**main** *adj* basic, biggest, cardinal, central, chief, critical, crucial, dominant, dominating, essential, first, foremost, fundamental, greatest, largest, leading, major, most important, outstanding, paramount, predominant, pre-eminent, prevailing, primary, prime, principal, special, strongest, supreme, top, vital. *Opp* MINOR.

**mainly** *adv* above all, as a rule, chiefly, especially, essentially, first and foremost, generally, in the main, largely, mostly, normally, on the whole, predominantly, primarily, principally, usually.

**maintain** *vb* **1** carry on, continue, hold to, keep going, keep up, perpetuate, persevere in, persist in, preserve, retain, stick to, sustain. **2** *maintain a car.* care for, keep in good condition, look after, service, take care of. **3** *maintain a family.* feed, keep, pay for, provide for, stand by, support. **4** *maintain your innocence.* affirm, allege, argue, assert, aver, claim, contend, declare, defend, insist, proclaim, profess, state, uphold.

**maintenance** *n* **1** care, conservation, looking after, preservation, repairs, servicing, upkeep. **2** alimony, allowance, contribution, subsistence, support.

**majestic** *adj* august, awe-inspiring, awesome, dignified, distinguished, elevated, exalted, glorious, grand, grandiose, imperial, imposing, impressive, kingly, lofty, lordly, magisterial, magnificent, monumental, noble, pompous, princely, queenly, regal, royal, splendid, stately, striking, sublime.

**majesty** *n* awesomeness, dignity, glory, grandeur, kingliness, loftiness, magnificence, nobility, pomp, royalty, splendour, stateliness, sublimity.

**major** *adj* bigger, chief, considerable, extensive, greater, important, key, larger, leading, outstanding, principal, serious, significant. ▷ MAIN. *Opp* MINOR.

**majority** *n* **1** *inf* best part, *inf* better part, bulk, greater number, mass, preponderance. **2** adulthood, coming of age, manhood, maturity, womanhood. **be in the majority** ▷ DOMINATE.

**make** *n* brand, kind, model, sort, type, variety. • *vb* **1** assemble, beget, bring about, build, compose, constitute, construct, contrive, craft, create, devise, do, engender, erect, execute, fabricate, fashion, forge, form, frame, generate, invent, make up, manufacture, mass-produce, originate, produce, put together, think up. **2** *make dinner.* concoct, cook, *inf* fix, prepare. **3** *make clothes.* knit, *inf* run up, sew, weave. **4** *make an effigy.* carve, cast, model, mould, sculpt, shape. **5** *make a speech.* deliver, pronounce, utter. ▷ SPEAK. **6** *made her chairperson.* appoint, elect, nominate, ordain. **7** *make P into B.* alter, change, convert, modify, transform, turn. **8** *make a fortune.* earn, gain, get, obtain, receive. **9** *make a good employee.* become, change into, grow into, turn into. **10** *make your objective.* accomplish, achieve, arrive at, attain, catch, get to, reach, win. **11** *2 + 2 makes 4.* add up to, amount to, come to, total. **12** *make rules.* agree, arrange, codify, establish, decide on, draw up, fix, write. **13** *make her happy.* cause to become, render. **14** *make trouble.* bring about, carry out, cause, give rise to, provoke, result in. **15** *make them obey.* coerce, compel, constrain, force, induce, oblige, order, pressurize, prevail on, require. **make amends** ▷ COMPENSATE. **make believe** ▷ IMAGINE. **make fun of** ▷ RIDICULE. **make good** ▷ PROSPER. **make love** ▷ LOVE. **make off** ▷ DEPART. **make off with** ▷ STEAL. **make out** ▷ UNDERSTAND. **make up** ▷ INVENT. **make up for** ▷ COMPENSATE. **make up your mind** ▷ DECIDE.

**make-believe** *adj* fanciful, feigned, imaginary, made-up, mock, *inf* pretend, pretended, sham, simulated, unreal.

• *n* dream, fantasy, play-acting, pretence, self-deception, unreality.

**maker** *n* architect, author, builder, creator, director, manufacturer, originator, producer.

**makeshift** *adj* emergency, improvised, provisional, stopgap, temporary.

**maladjusted** *adj* disturbed, muddled, neurotic, unbalanced.

**male** *adj* manly, masculine, virile. • *n* □ *bachelor,* inf *bloke, boy, boyfriend,* inf *bridegroom, brother, chap,* inf *codger, father, fellow, gentleman, groom,* inf *guy, husband, lad, lover, man, son,* inf *squire, uncle, widower.* □ *buck, bull, cock, dog, ram, stallion, tom(cat).* *Opp* FEMALE.

**malefactor** *n* delinquent, lawbreaker, offender, villain, wrongdoer. ▷ CRIMINAL.

**malice** *n* animosity, *inf* bitchiness, bitterness, *inf* cattiness, enmity, hatred, hostility, ill-will, malevolence, maliciousness, malignity, nastiness, rancour, spite, spitefulness, vengefulness, venom, viciousness, vindictiveness.

**malicious** *adj inf* bitchy, bitter, *inf* catty, evil, evil-minded, hateful, ill-natured, malevolent, malignant, mischievous, nasty, rancorous, revengeful, sly, spiteful, vengeful, venomous, vicious, villainous, vindictive, wicked. *Opp* KIND.

**malignant** *adj* dangerous, deadly, destructive, fatal, harmful, injurious, life-threatening, poisonous, *inf* terminal, uncontrollable, virulent. ▷ MALICIOUS.

**malleable** *adj* ductile, plastic, pliable, soft, tractable, workable. *Opp* BRITTLE.

**malnutrition** *n* famine, hunger, starvation, undernourishment.

**man** *n* **1** [*either sex*] ▷ MANKIND. **2** ▷ MALE. • *vb* cover, crew, provide staff for, staff.

**manage** *vb* **1** administer, be in charge of, be manager of, command, conduct, control, direct, dominate, govern, head, lead, look after, mastermind, operate, organize, oversee, preside over, regulate, rule, run, superintend, supervise, take care of, take control of, take over. **2** *manage your jobs.* accomplish, achieve, bring about, carry out, contend with, cope with, deal with, do, finish, get through, handle, manipulate, muddle through, perform, sort out, succeed in, undertake. **3** *Can you manage?* cope, fend for yourself, *inf* make it, scrape by, shift for yourself, succeed, survive. **4** *can manage £10.* afford, spare.

**manageable** *adj* **1** acceptable, convenient, easy to manage, handy, neat, reasonable. *Opp* AWKWARD. **2** amenable, compliant, controllable, disciplined, docile, governable, submissive, tame, tractable. ▷ OBEDIENT. *Opp* DISOBEDIENT.

**manager, manageress** *ns* administrator, *inf* boss, chief, controller, director, executive, foreman, forewoman, governor, head, organizer, overseer, proprietor, ruler, superintendent, supervisor. ▷ CHIEF.

**mandatory** *adj* compulsory, essential, necessary, needed, obligatory, required, requisite.

**mangle** *vb* butcher, cripple, crush, cut, damage, deform, disfigure, hack, injure, lacerate, maim, maul, mutilate, ruin, spoil, squash, tear, wound.

**mangy** *adj* dirty, filthy, motheaten, nasty, scabby, scruffy, shabby, slovenly, squalid, *inf* tatty, unkempt, wretched.

**manhandle** *vb* **1** carry, haul, heave, hump, lift, manoeuvre, move, pull, push. **2** abuse, batter, *inf* beat up, ill-treat, knock about, maltreat, mistreat, misuse, *inf* rough up, treat roughly.

**mania** *n* craving, craze, enthusiasm, fad, fetish, frenzy, infatuation, obsession, passion, preoccupation, rage. ▷ MADNESS.

**maniac** *n* lunatic, psychopath, psychotic. ▷ MADMAN.

**manifest** *adj* apparent, blatant, clear, conspicuous, discernible, evident, explicit, glaring, noticeable, obvious, patent, plain, recognizable, undisguised, visible. • *vb* ▷ SHOW.

**manifesto** *n* declaration, policy statement.

**manipulate** *vb* **1** feel, massage, rub, stimulate. **2** *manipulate people.* control, direct, engineer, exploit, guide, handle, influence, manage, manoeuvre, orchestrate, steer.

**mankind** *n Lat* homo sapiens, human beings, humanity, humankind, the human race, man, men and women, mortals, people.

**manly** *adj* chivalrous, gallant, heroic, *inf* macho, male, mannish, masculine, strong, swashbuckling, vigorous, virile. ▷ BRAVE. *Opp* EFFEMINATE.

**man-made** *adj* artificial, imitation, manufactured, mass-produced, processed, simulated, synthetic, unnatural. *Opp* NATURAL.

**manner** *n* **1** approach, fashion, means, method, mode, procedure, process, style, technique, way. **2** air, aspect, attitude, bearing, behaviour, character, conduct, demeanour, deportment, disposition, look, mien. **3** *all manner of things.* genre, kind, sort, type, variety.

**manners** behaviour, breeding, civility, conduct, courtesy, decorum, etiquette, gentility, politeness, protocol, refinement, social graces.

**mannerism** *n* characteristic, habit, idiosyncrasy, peculiarity, quirk, trait.

**manoeuvre** *n* device, dodge, gambit, intrigue, move, operation, plan, plot, ploy, ruse, scheme, stratagem, strategy, tactics, trick. • *vb* contrive, engineer, guide, jockey, manipulate, move, navigate, negotiate, pilot, steer.

**manoeuvres** army exercise, operation, training.

**mansion** *n* castle, château, manor, manor-house, palace, stately home, villa. ▷ HOUSE.

**mantle** *n* cape, cloak, covering, hood, shawl, shroud, wrap. • *vb* ▷ COVER.

**manufacture** *vb* assemble, build, construct, create, fabricate, make, mass-produce, prefabricate, process, put together, *inf* turn out. **manufactured** ▷ MAN-MADE.

**manufacturer** *n* factory-owner, industrialist, maker, producer.

**manure** *n* compost, dung, fertilizer, *inf* muck.

**manuscript** *n* document, papers, script. ▷ BOOK.

**many** *adj* abundant, assorted, copious, countless, diverse, frequent, innumerable, multifarious, myriad, numberless, numerous, profuse, *inf* umpteen, uncountable, untold, varied, various. *Opp* FEW.

**map** *n* chart, diagram, plan.

**mar** *vb* blight, blot, damage, deface, disfigure, harm, hurt, impair, ruin, spoil, stain, taint, tarnish, wreck.

**marauder** *n* bandit, buccaneer, invader, pirate, plunderer, raider.

**march** *n* cortège, demonstration, march-past, parade, procession, progress. • *vb* file, pace, parade, step, stride, troop. ▷ WALK.

**margin** *n* **1** border, boundary, brink, edge, frieze, perimeter, periphery, rim, side, verge. **2** allowance, latitude, leeway, room, scope, space.

**marginal** *adj* borderline, doubtful, insignificant, minimal, negligible, peripheral, slight, small, unimportant.

**marital** *adj* conjugal, matrimonial, nuptial.

**mark** *n* **1** blemish, blot, blotch, dent, dot, fingermark, *plur* graffiti, impression, line, marking, pockmark, print, scar, scratch, scribble, smear, smudge, smut, *inf* splotch, spot, stain, *plur* stigmata, streak, trace, vestige. **2** *mark of breeding*. characteristic, feature, indication, indicator, marker, token. **3** *identifying mark*. badge, brand, device, emblem, fingerprint, hallmark, label, seal, sign, stamp, standard, symbol, trademark. • *vb* **1** blemish, blot, brand, bruise, cut, damage, deface, dent, dirty, disfigure, draw on, make a mark on, mar, scar, scratch, scrawl over, scribble on, smudge, spot, stain, stamp, streak, tattoo, write on. **2** *mark pupils' work*. appraise, assess, correct, evaluate, grade, tick. **3** *mark my words*. attend to, heed, listen to, mind, note, notice, observe, pay attention to, take note of, take seriously, *inf* take to heart, watch.

**market** *n* auction, bazaar, exchange, fair, marketplace, sale. ▷ SHOP. • *vb* advertise, deal in, make available, merchandise, peddle, promote, put on the market, retail, sell, *inf* tout, trade, trade in, try to sell, vend.

**marksman** *n* crack shot, gunman sharpshooter, sniper.

**maroon** *vb* abandon, cast away, desert, forsake, isolate, leave, put ashore, strand.

**marriage** *n* **1** matrimony, partnership, union, wedlock. **2** nuptials, union, wedding. □ *bigamy, monogamy, polygamy*.

**marriageable** *adj* adult, mature, nubile.

**marry** *vb* espouse, *inf* get hitched, *inf* get spliced, join in matrimony, *inf* tie the knot, unite, wed.

**marsh** *n* bog, fen, marshland, morass, mud, mudflats, quagmire, quicksands, saltings, saltmarsh, *old use* slough, swamp, wetland.

**marshal** *vb* arrange, assemble, collect, deploy, draw up, gather, group, line up, muster, organize, set out.

**martial** *adj* aggressive, bellicose, belligerent, militant, military, pugnacious, soldierly, warlike. *Opp* PEACEABLE.

**marvel** *n* miracle, phenomenon, wonder. • *vb* **marvel at** admire, applaud, be amazed by, be astonished by, be surprised by, gape at, praise, wonder at.

**marvellous** *adj* admirable, amazing, astonishing, astounding, breathtaking, excellent, exceptional, extraordinary, *inf* fabulous, *inf* fantastic, glorious, impressive, incredible, magnificent, miraculous, out of the ordinary, *inf* out of this world, phenomenal, praiseworthy, prodigious, remarkable, *inf* sensational, spectacular, splendid, staggering, stunning, stupendous, *inf* super, superb, surprising, *inf* terrific, unbelievable, wonderful, wondrous. *Opp* ORDINARY.

**masculine** *adj* boyish, *inf* butch, gentlemanly, heroic, *inf* macho, male, manly, mannish, muscular, powerful, strong, vigorous, virile. *Opp* FEMININE.

**mash** *vb* beat, crush, grind, mangle, pound, pulp, pulverize, smash, squash.

**mask** *n* camouflage, cloak, cover, cover-up, disguise, façade, front, guise, screen, shield, veil, visor. • *vb* blot out, camouflage, cloak, conceal, cover, disguise, hide, obscure, screen, shield, shroud, veil.

**masonry** *n* bricks, brickwork, stone, stonework.

**mass** *adj* comprehensive, general, large-scale, popular, universal, wholesale, widespread. • *n* **1** accumulation, agglomeration, aggregation, body, bulk, *inf* chunk, collection, concretion, conglomeration, *inf* dollop, heap, hoard, *inf* hunk, *inf* load, lot, lump, mound, mountain, pile, profusion, quantity, stack, volume. **2** ▷ GROUP. • *vb* accumulate, aggregate, amass, assemble, collect, congregate, convene, flock together, gather, marshal, meet, mobilize, muster, pile up, rally.

**massacre** *vb* annihilate, slaughter. ▷ KILL.

**massage** *vb* knead, manipulate, rub.

**mast** *n* aerial, flagpole, maypole, pylon, transmitter.

**master** *n* **1** *inf* boss, employer, governor, keeper, lord, overseer, owner, person in charge, proprietor, ruler, taskmaster. ▷ CHIEF. **2** captain, skipper. **3** *master of an art. inf* ace, authority, expert, genius, mastermind, maestro, virtuoso. **4** ▷ TEACHER. • *vb* **1** become expert in, *inf* get off by heart, *inf* get the hang of, grasp, know, learn, understand. **2** break in, bridle, check, conquer, control, curb, defeat, dominate, *inf* get the better of, govern, manage, overcome, overpower, quell, regulate, repress, rule, subdue, subjugate, suppress, tame, triumph over, vanquish.

**masterly** *adj* accomplished, adroit, consummate, dexterous, excellent, expert, masterful, matchless, practiced, proficient, skilful, skilled, unsurpassable.

**mastermind** *n* architect, brains, conceiver, contriver, creator, engineer, expert, genius, intellectual, inventor, manager, originator, planner, prime mover. • *vb* carry through, conceive, contrive, devise, direct, engineer, execute, manage, organize, originate, plan, plot. ▷ MANAGE.

**masterpiece** *n* best work, *Fr* chef-d'oeuvre, classic, *inf* hit, *Lat* magnum opus, masterwork, *Fr* pièce de résistance.

**match** *n* **1** bout, competition, contest, duel, game, test match, tie, tournament, tourney. **2** *met my match*. complement, counterpart, double, equal, equivalent, twin. **3** *a good match*. combination, fit, pair, similarity. **4** *a love match*. alliance, friendship, marriage, partnership, relationship, union. • *vb* **1** agree, accord, be compatible, be equivalent, be the same, be similar, blend, coincide, combine, compare, coordinate, correspond, fit, *inf* go together, harmonize, suit, tally, tie in, tone in. *Opp* CONTRAST. **2** ally, combine, fit, join, link up, marry, mate, pair off, pair up, put together, team up. *Opp* SPLIT. **matching** ▷ SIMILAR.

**mate** *n* **1** *inf* better half, companion, consort, helpmeet, husband, partner, spouse, wife. ▷ FRIEND. **2** assistant, associate, collaborator, colleague, helper. • *vb* become partners, copulate, couple, have intercourse, *inf* have sex, join, marry, *inf* pair up, unite, wed.

**material** *adj* concrete, corporeal, palpable, physical, solid, substantial, tangible. • *n* **1** fabric, stuff, textile. ▷ CLOTH. **2** components, constituents, content, data, facts, ideas, information, matter, notes, resources, statistics, stuff, subject matter, substance, supplies, things.

**materialize** *vb* appear, become visible, emerge, occur, take shape, *inf* turn up.

**mathematics** *n* mathematical science, *inf* maths, number work. □ *addition, algebra, arithmetic, calculus, division, geometry, multiplication, operational research, statistics, subtraction, trigonometry.*

**matted** *adj* knotted, tangled, uncombed, unkempt. ▷ DISHEVELLED.

**matter** *n* **1** body, material, stuff, substance. **2** discharge, pus, suppuration. **3** *a matter of life and death.* affair, business, concern, episode, event, fact, incident, issue, occurrence, question, situation, subject, thing, topic. **4** *What's the matter?* difficulty, problem, trouble, upset, worry. • *vb* be important, be of consequence, be significant, count, make a difference, mean something, signify. **matter-of-fact** ▷ PROSAIC.

**mature** *adj* **1** adult, advanced, experienced, full-grown, grown-up, nubile, of age, perfect, sophisticated, well-developed. **2** mellow, ready, ripe, seasoned. *Opp* IMMATURE. • *vb* age, come to fruition, develop, grow up, mellow, reach maturity, ripen.

**maturity** *n* adulthood, completion, majority, mellowness, perfection, readiness, ripeness.

**maul** *vb* claw, injure, *inf* knock about, lacerate, mangle, manhandle, mutilate, paw, savage, treat roughly, wound.

**maximize** *vb* **1** add to, augment, build up, increase, make the most of. **2** inflate, magnify, overdo, overstate. ▷ EXAGGERATE. *Opp* MINIMIZE.

**maximum** *adj* biggest, extreme, full, fullest, greatest, highest, largest, maximal, most, peak, supreme, top, topmost, utmost, uttermost. • *n* apex, ceiling, climax, extreme, highest point, peak, pinnacle, top, upper limit, zenith. *Opp* MINIMUM.

**maybe** *adv* conceivably, perhaps, possibly.

**maze** *n* complex, confusion, convolution, labyrinth, network, tangle, web.

**meadow** *n* field, *old use* mead, paddock, pasture.

**meagre** *adj* deficient, inadequate, insufficient, lean, mean, paltry, poor, puny, scanty, skimpy, slight, sparse, thin, unsatisfying. ▷ SMALL. *Opp* GENEROUS.

**meal** *n inf* blow-out, *old use* collation, repast, *inf* spread. □ *banquet, barbecue, breakfast, buffet, dinner,* inf *elevenses, feast, high tea, lunch, luncheon, picnic, snack,*

upper, take-away, tea, tea-break,
ld use *tiffin*.

**ean** *adj* 1 beggarly, *f* cheese-paring, close, close-sted, illiberal, *inf* mingy, miserly, iggardly, parsimonious, *f* penny-pinching, selfish, spar-ig, stingy, *inf* tight, tight-fisted, ngenerous. 2 *a mean disposition.* allous, churlish, contemptible, ruel, despicable, hard-hearted, noble, ill-tempered, malicious, asty, shabby, shameful, small-inded, *inf* sneaky, spiteful, ncharitable, unkind, vicious. 3 *a ean dwelling.* base, common, umble, inferior, insignificant, w, lowly, miserable, poor, nabby, squalid, wretched. *pp* GENEROUS, VALUABLE. • *vb* augur, betoken, communicate, onnote, convey, denote, *inf* drive t, express, foretell, *inf* get over, erald, hint at, imply, indicate, timate, portend, presage, refer , represent, say, show, signal, ignify, specify, spell out, stand r, suggest, symbolize. 2 *I mean to ucceed.* aim, desire, have in mind, ope, intend, plan, propose, pur-ose, want, wish. 3 *The job means ng hours.* entail, involve, neces-itate.

**eander** *vb* ramble, rove, snake, vist and turn, wander, wind, zig-ag. **meandering** ▷ TWISTY.

**eaning** *n* connotation, content, efinition, denotation, drift, xplanation, force, gist, idea, nplication, import, importance, nterpretation, message, point, urport, purpose, relevance, ense, significance, signification, ubstance, thrust, value.

**eaningful** *adj* deep, eloquent, xpressive, meaning, pointed, pos-ive, pregnant, relevant, serious, gnificant, suggestive, telling, tell-tale, weighty, worthwhile. *Opp* MEANINGLESS.

**meaningless** *adj* 1 absurd, coded, incomprehensible, incoherent, inconsequential, irrelevant, nonsensical, pointless, senseless. 2 *meaningless compliments.* empty, flattering, hollow, insincere, shallow, silly, sycophantic, vacuous, worthless. *Opp* MEANINGFUL.

**means** *n* 1 ability, capacity, channel, course, fashion, machinery, manner, medium, method, mode, process, way. 2 *private means.* ▷ WEALTH.

**measurable** *adj* appreciable, considerable, perceptible, quantifiable, reasonable, significant. *Opp* NEGLIGIBLE.

**measure** *n* 1 allocation, allowance, amount, amplitude, extent, magnitude, portion, quantity, quota, range, ration, scope, size, unit. ▷ MEASUREMENT. 2 criterion, *inf* litmus test, standard, test, touchstone, yardstick. 3 *measures to curb crime.* act, action, bill, control, course of action, expedient, law, means, procedure, step. • *vb* assess, calculate, calibrate, compute, count, determine, estimate, gauge, judge, mark out, meter, plumb (*depth*), quantify, rank, rate, reckon, survey, take measurements of, weigh. **measure out** ▷ DISPENSE.

**measurement** *n* amount, calculation, dimensions, extent, figure, mensuration, size. ▷ MEASURE. □ *area, breadth, bulk, capacity, depth, distance, height, length, mass, speed, time, volume, weight, width.* □ *acreage, footage, mileage, tonnage.*

**meat** *n* flesh. ▷ FOOD. □ *bacon, beef, chicken, game, gammon, ham, lamb, mutton, pork, poultry, turkey, veal, venison.* □ *brawn, breast,* *burger, brisket, chine, chops, chuck, cutlet, fillet, flank, hamburger, leg, loin, mince, offal, paté, potted meat, rib, rissole, rump, sausage, scrag, shoulder, silverside, sirloin, spare-rib, steak, topside, tripe.*

**mechanic** *n* engineer, technician.

**mechanical** *adj* **1** automated, automatic, machine-driven, technological. **2** cold, habitual, impersonal, inhuman, instinctive, lifeless, matter-of-fact, perfunctory, reflex, routine, soulless, unconscious, unemotional, unfeeling, unimaginative, uninspired, unthinking. *Opp* HUMAN.

**mechanize** *vb* automate, bring up to date, computerize, equip with machines, modernize.

**medal** *n* award, decoration, honour, medallion, prize, reward, trophy.

**medallist** *n* champion, victor, winner.

**meddle** *vb inf* be a busybody, butt in, interfere, *inf* poke your nose in, pry, snoop, tamper.

**mediate** *vb* act as go-between, act as mediator, arbitrate, intercede, liaise, negotiate.

**mediator** *n* arbiter, arbitrator, broker, conciliator, go-between, intercessor, intermediary, judge, liaison officer, middleman, moderator, negotiator, peacemaker, referee, umpire.

**medicinal** *adj* curative, healing, medical, remedial, restorative, therapeutic.

**medicine** *n* **1** healing, surgery, therapeutics, therapy, treatment. **2** cure, dose, drug, medicament, medication, nostrum, panacea, *old use* physic, prescription, remedy, treatment. □ *anaesthetic, antibiotic, antidote, antiseptic, aspirin, capsule, gargle, herbal remedy, i[c]ine, inhaler, linctus, lotion, lozenge, narcotic, ointment, painkill[er], pastille, penicillin, pill, sedative, suppository, tablet, tonic, tranqui[l]lizer.*

**mediocre** *adj* amateurish, average, *inf* common-or-garden, commonplace, everyday, fair, indiffe[r]ent, inferior, medium, middling, moderate, ordinary, passable, pe[d]estrian, poorish, *inf* run-of-the-mill, second-rate, *inf* so-so, undistinguished, unexceptional, unexciting, uninspire[d], unremarkable, weakish. *Opp* OUTSTANDING.

**meditate** *vb* be lost in thought, brood, cerebrate, chew over, cog[it]ate, consider, contemplate, delib[er]ate, mull things over, muse, ponder, pray, reflect, ruminate, think, turn over.

**meditation** *n* cerebration, contemplation, deliberation, musin[g], prayer, reflection, rumination, thought, yoga.

**meditative** *adj* brooding, conte[m]plative, pensive, prayerful, rapt, reflective, ruminative, thoughtf[ul]

**medium** *adj* average, intermediate, mean, medial, median, mi[d], middle, middling, mid-sized, mi[d]way, moderate, normal, ordinar[y], standard, usual. • *n* **1** average, centre, compromise, mean, middle, midpoint, norm. **2** agenc[y], approach, channel, form, means method, mode, vehicle, way. **3** clairvoyant, seer, spiritualist. **the media, mass media** ▷ COMMUNICATION.

**meek** *adj* acquiescent, complian[t], deferential, docile, forbearing, gentle, humble, long-suffering, lowly, mild, modest, non-militan[t], obedient, patient, peaceable, qui[et], resigned, retiring, self-effacing,

hy, soft, spineless, submissive, ame, timid, tractable, unambiıous, unassuming, unprotesting, ṿeak, *inf* wimpish.
)*pp* AGGRESSIVE.

**ıeet** *vb* **1** *inf* bump into, chance ıpon, collide with, come across, onfront, contact, encounter, face, ıappen on, have a meeting with, un across, run into, see. **2** be ntroduced to, make the acquaintnce of. **3** come and fetch, greet, *nf* pick up, rendezvous with, welome. **4** assemble, collect, come ɔgether, congregate, convene, for;ather, gather, have a meeting, ıuster, rally, rendezvous. **5** *The nds don't meet.* abut, adjoin, come ɔgether, connect, converge, cross, ntersect, join, link up, merge, ɔuch, unite. **6** *meet a request.* .cquiesce in, agree to, answer, omply with, deal with, fulfil, *ıf* measure up to, observe, pay, atisfy, settle, take care of. **7** *meet lifficulties.* encounter, endure, xperience, go through, suffer, ındergo.

**ıeeting** *n* **1** assembly, gathering, *ıf* get-together, *inf* powwow. ı *audience, board, briefing, cabıet, caucus, committee, conclave, onference, congregation, congress, onvention, council, discussion roup, forum, prayer meeting, ally, seminar, service, synod.* ı appointment, assignation, date, ngagement, *inf* get-together, ren.ezvous, *old use* tryst. **3** *chance ıeeting.* confrontation, contact, ncounter. **4** *meeting of lines, oads.* confluence (*of rivers*), con'ergence, crossing, crossroads, ntersection, joining, junction, '-junction, union.

**ıelancholy** *adj* cheerless, .ejected, depressed, depressing, espondent, disconsolate, dismal, dispirited, dispiriting, *inf* down, down-hearted, forlorn, gloomy, glum, joyless, lifeless, low, lugubrious, melancholic, miserable, moody, morose, mournful, sombre, sorrowful, unhappy, woebegone, woeful. ▷ SAD.
*Opp* CHEERFUL. • *n* ▷ SADNESS.

**mellow** *adj* **1** mature, rich, ripe, smooth, sweet. **2** *mellow mood.* agreeable, amiable, comforting, cordial, genial, gentle, happy, kindly, peaceful, pleasant, reassuring, soft, subdued, warm.
*Opp* HARSH. • *vb* age, develop, improve with age, mature, ripen, soften, sweeten.

**melodious** *adj* dulcet, *inf* easy on the ear, euphonious, harmonious, lyrical, mellifluous, melodic, sweet, tuneful.

**melodramatic** *adj* emotional, exaggerated, *inf* hammy, histrionic, overdone, overdrawn, *inf* over the top, sensationalized, sentimental, theatrical.

**melody** *n* air, song, strain, subject, theme, tune.

**melt** *vb* deliquesce, dissolve, liquefy, soften, thaw, unfreeze.
**melt away** ▷ DISAPPEAR.

**member** *n* associate, colleague, fellow, life-member, paid-up member.

**memorable** *adj* catchy (*tune*), distinguished, extraordinary, haunting, historic, impressive, indelible, ineradicable, never-to-be-forgotten, notable, outstanding, remarkable, striking, unforgettable.

**memorial** *n* cairn, cenotaph, gravestone, headstone, monument, plaque, reminder, statue, tablet, tomb.

**memorize** *n* commit to memory, *inf* get off by heart, learn, learn by rote, learn parrot-fashion, remember, retain.

**memory** *n* **1** ability to remember, recall, retention. **2** impression, recollection, reminder, reminiscence, souvenir. **3** *memory of the dead.* fame, honour, name, remembrance, reputation, respect.

**menace** *n* danger, peril, threat, warning. • *vb* alarm, bully, cow, intimidate, terrify, terrorize, threaten. ▷ FRIGHTEN.

**mend** *vb* **1** fix, patch up, put right, rectify, remedy, renew, renovate, repair, restore. □ *beat out, darn, patch, replace parts, sew up, solder, stitch up, touch up, weld.* **2** *mend your ways.* ameliorate, amend, correct, cure, improve, make better, reform, revise. **3** *mend after illness.* convalesce, get better, heal, improve, recover, recuperate.

**menial** *adj* base, boring, common, degrading, demeaning, humble, inferior, insignificant, low, lowly, servile, slavish, subservient, unskilled, unworthy.
• *n inf* dogsbody, lackey, minion, slave, underling. ▷ SERVANT.

**mental** *adj* **1** abstract, cerebral, cognitive, conceptual, intellectual, rational, theoretical. **2** *mental illness.* emotional, psychological, subjective, temperamental. ▷ MAD.

**mentality** *n* attitude, bent, character, disposition, frame of mind, inclination, *inf* make-up, outlook, personality, predisposition, propensity, psychology, set, temperament, way of thinking.

**mention** *vb* acknowledge, allude to, animadvert on, bring up, broach, cite, comment on, disclose, draw attention to, enumerate, hint at, *inf* let drop, let out, make known, make mention, name, note, observe, pay tribute to, point out, quote, recognize, refer to, remark, report, reveal, say, speak about, touch on, write about.

**mercenary** *adj* acquisitive, avaricious, covetous, grasping, greedy, *inf* money-mad, venal.
• *n* ▷ FIGHTER.

**merchandise** *n* commodities, goods, items for sale, produce, products, stock. • *vb* ▷ ADVERTISE.

**merchant** *n* broker, dealer, distributor, retailer, salesman, seller, shopkeeper, stockist, supplier, trader, tradesman, tradeswoman, vendor, wholesaler.

**merciful** *adj* beneficent, benevolent, charitable, clement, compassionate, forbearing, forgiving, generous, gracious, humane, humanitarian, indulgent, kind, kind-hearted, kindly, lenient, liberal, magnanimous, mild, pitying, *inf* soft, soft-hearted, sympathetic, tender-hearted, tolerant.
*Opp* MERCILESS.

**merciless** *adj* barbaric, barbarous, brutal, callous, cold, cruel, cut-throat, hard, hard-hearted, harsh, heartless, indifferent, inexorable, inflexible, inhuman, inhumane, intolerant, malevolent, pitiless, relentless, remorseless, rigorous, ruthless, savage, severe, stern, stony-hearted, strict, tyrannical, unbending, unfeeling, unforgiving, unkind, unmerciful, unrelenting, unremitting, vicious.
*Opp* MERCIFUL.

**mercy** *n* beneficence, benignity, charity, clemency, compassion, feeling, forbearance, forgiveness, generosity, grace, humaneness, humanity, indulgence, kind-heartedness, kindness, leniency, love, pity, quarter, sympathy, understanding.

**merge** *vb* **1** amalgamate, blend, coalesce, combine, come together, confederate, consolidate, fuse, integrate, join together, link up, mingle, mix, pool, put together, unite. **2** *motorways merge.* converge, join, meet. *Opp* SEPARATE.

**merit** *n* credit, distinction, excellence, good, goodness, importance, quality, strength, talent, value, virtue, worth, worthiness. • *vb* be entitled to, be worthy of, deserve, earn, have a right to, incur, justify, rate, warrant.

**meritorious** *adj* admirable, commendable, estimable, exemplary, honourable, laudable, praiseworthy, worthy.

**merriment** *n* amusement, cheerfulness, conviviality, exuberance, gaiety, glee, good cheer, high spirits, hilarity, jocularity, joking, jollity, joviality, *inf* larking about, laughter, levity, light-heartedness, liveliness, mirth, vivacity. ▷ MERRYMAKING.

**merry** *adj* bright, *inf* bubbly, carefree, cheerful, cheery, *inf* chirpy, convivial, festive, fun-loving, gay, glad, hilarious, jocular, jolly, jovial, joyful, joyous, lighthearted, lively, mirthful, rollicking, spirited, vivacious. ▷ HAPPY. *Opp* SERIOUS.

**merrymaking** *n* carousing, celebration, conviviality, festivity, frolic, fun, *inf* fun and games, *inf* jollification, *inf* junketing, merriment, revelry, roistering, sociability, *old use* wassailing. ▷ PARTY.

**mesh** *n* grid, lace, lacework, lattice, lattice-work, net, netting, network, reticulation, screen, sieve, tangle, tracery, trellis, web, webbing.

**mess** *n* **1** chaos, clutter, confusion, dirt, disarray, disorder, hotchpotch, jumble, litter, *inf* mishmash, muddle, *inf* shambles, tangle, untidiness. ▷ CONFUSION, DIRT. **2** *made a mess of it.* *inf* botch, failure, *inf* hash, *inf* mix-up. **3** *got into a mess.* difficulty, dilemma, *inf* fix, *inf* jam, *inf* pickle, plight, predicament, problem, trouble. • *vb* **mess about** amuse yourself, loaf, loiter, lounge about, *inf* monkey about, *inf* muck about, *inf* play about. **make a mess of** ▷ BUNGLE, MUDDLE. **mess up** ▷ MUDDLE. **mess up a job** ▷ BUNGLE.

**message** *n* announcement, bulletin, cable, communication, communiqué, dispatch, information, intelligence, letter, memo, memorandum, missive, news, note, notice, report, statement, *old use* tidings.

**messenger** *n* bearer, carrier, courier, dispatch-rider, emissary, envoy, errand-boy, errand-girl, gobetween, harbinger, herald, intermediary, legate, Mercury, messenger-boy, messenger-girl, nuncio, postman, runner.

**messy** *adj* blowzy, careless, chaotic, cluttered, dirty, dishevelled, disorderly, filthy, grubby, mucky, muddled, *inf* shambolic, slapdash, sloppy, slovenly, unkempt, untidy. *Opp* NEAT.

**metallic** *adj* **1** gleaming, lustrous, shiny. **2** clanking, clinking, ringing.

**metaphorical** *adj* allegorical, figurative, non-literal, symbolic. *Opp* LITERAL.

**method** *n* **1** approach, fashion, *inf* knack, manner, means, methodology, mode, *Lat* modus operandi, plan, procedure, process, programme, recipe, scheme, style, technique, trick, way. **2** arrangement, design, discipline, neatness, order, orderliness, organization, pattern, routine, structure, system.

**methodical** *adj* businesslike, careful, deliberate, disciplined, logical, meticulous, neat, ordered, orderly, organized, painstaking, precise, rational, regular, routine, structured, systematic, tidy. *Opp* DISORGANIZED.

**meticulous** *adj* accurate, exact, exacting, fastidious, *inf* finicky, painstaking, particular, perfectionist, precise, punctilious, scrupulous, thorough. *Opp* CARELESS.

**microbe** *n* bacillus, bacterium, *inf* bug, germ, micro-organism, virus.

**middle** *adj* central, centre, half-way, inner, inside, intermediate, intervening, mean, medial, median, mid, middle-of-the-road, midway, neutral. • *n* bull's eye, centre, core, crown (*of road*), focus, half-way point, heart, hub, inside, middle position, midpoint, midst, nucleus.

**middling** *adj* average, fair, *inf* fair to middling, indifferent, mediocre, moderate, modest, *inf* nothing to write home about, ordinary, passable, run-of-the-mill, *inf* so-so, unremarkable. *Opp* OUTSTANDING.

**might** *n* capability, capacity, energy, force, muscle, potency, power, strength, superiority, vigour.

**mighty** *adj* brawny, dominant, doughty, energetic, enormous, forceful, great, hefty, muscular, potent, powerful, robust, *inf* strapping, strong, sturdy, vigorous, weighty. ▷ BIG. *Opp* WEAK.

**migrate** *vb* emigrate, go, immigrate, move, relocate, resettle, settle, travel.

**mild** *adj* **1** affable, amiable, conciliatory, docile, easygoing, equable, forbearing, forgiving, gentle, good-tempered, harmless, indulgent, inoffensive, kind, kindly, lenient, meek, merciful, modest, non-violent, pacific, peaceable, placid, quiet, *inf* soft, soft-hearted, submissive, sympathetic, tractable, unassuming, understanding, yielding. **2** *mild weather*. balmy, calm, clement, fair, peaceful, pleasant, serene, temperate, warm. **3** *a mild illness*. insignificant, minor, modest, slight, trivial, unimportant. **4** *mild flavour*. bland, delicate, faint, mellow, soothing, subtle. *Opp* SEVERE, STRONG.

**mildness** *n* affability, amiability, clemency, docility, forbearance, gentleness, kindness, leniency, moderation, placidity, softness, sympathy, tenderness. *Opp* ASPERITY.

**militant** *adj* active, aggressive, assertive, attacking, combative, fierce, hostile, positive, pugnacious. *Opp* PASSIVE. • *n* activist, extremist, *inf* hawk, partisan.

**militaristic** *adj* bellicose, belligerent, combative, fond of fighting, hawkish, hostile, pugnacious, warlike. *Opp* PEACEABLE.

**military** *adj* armed, belligerent, combatant, enlisted, fighting, martial, uniformed, warlike. *Opp* CIVIL.

**militate** *vb* militate against cancel out, counter, counteract, countervail, discourage, hinder, oppose, prevent, resist.

**milk** *vb* bleed, drain, exploit, extract, tap, wring.

**milky** *adj* chalky, cloudy, misty, opaque, whitish. *Opp* CLEAR.

**mill** *n* **1** factory, foundry, plant, shop, works, workshop. **2** crusher, grinder, quern, watermill, windmill. • *vb* crush, granulate, grate, grind, pound, powder, pulverize.

**ill about** move aimlessly,
ethe, swarm, throng, wander.

**imic** *n* caricaturist, imitator,
npersonator, impressionist.
*vb* ape, caricature, copy, do
npressions of, echo, imitate,
npersonate, lampoon, look like,
*f* make fun of, mirror, mock, par-
dy, parrot, pretend to be, repro-
uce, ridicule, satirize, simulate,
ound like, *inf* take off.

**ind** *n* 1 astuteness, brain, brain-
ower, brains, cleverness, *inf* grey
atter, head, insight, intellect,
ntelligence, judgement, memory,
ental power, perception, psyche,
ationality, reason, reasoning,
emembrance, sagacity, sapience,
ense, shrewdness, thinking,
nderstanding, wisdom, wit, wits.
attitude, belief, bias, disposition,
umour, inclination, intention,
pinion, outlook, persuasion, plan,
oint of view, position, view, view-
oint, way of thinking, wishes.
*vb* 1 attend to, care for, guard,
eep an eye on, look after, take
are, take charge of, watch.
*mind the warning*. be careful
bout, beware of, heed, listen to,
ok out for, mark, note, obey, pay
ttention to, remember, take
otice of, watch out for. 3 *won't
ind if he's late*. be annoyed, be
othered, be offended, be resent-
ul, bother, care, complain, disap-
rove, grumble, object, take
ffence, worry. **be in two minds**
HESITATE. **make up your mind**
DECIDE. **out of your mind**
MAD.

**indful** *adj* alert, attentive,
ware, conscious, heedful, *inf* on
e lookout, vigilant, watchful.
CAREFUL. *Opp* CARELESS.

**indless** *adj* brainless, fatuous,
iotic, obtuse, senseless, thick,
oughtless, unintelligent,
unthinking, witless. ▷ STUPID.
*Opp* INTELLIGENT.

**mine** *n* **1** coalfield, colliery,
excavation, opencast mine, pit,
quarry, shaft, tunnel, working.
**2** *mine of information*. fund, repository, source, store, storehouse,
supply, treasury, vein, wealth.
• *vb* dig, excavate, extract, quarry,
remove, scoop out, unearth.

**mineral** *n* metal, ore, rock.

**mingle** *vb* amalgamate, associate,
blend, circulate, combine, commingle, get together, fraternize,
*inf* hobnob, intermingle, intermix,
join, merge, mix, move about,
*inf* rub shoulders, socialize, unite.

**miniature** *adj* baby, diminutive,
dwarf, pocket, pygmy, reduced,
scaled-down, small-scale, tiny, toy.
▷ SMALL.

**minimal** *adj* least, minimum, negligible, nominal, slightest, smallest, token.

**minimize** *vb* **1** cut down,
decrease, diminish, lessen, pare,
prune, reduce. **2** *minimize problems*. belittle, decry, devalue,
depreciate, gloss over, make light
of, play down, underestimate,
undervalue. *Opp* MAXIMIZE.

**minimum** *adj* bottom, least, lit-
tlest, lowest, minimal, minutest,
nominal, *inf* rock bottom, slightest, smallest. • *n* least, lowest, minimum amount, minimum quant-
ity, nadir. *Opp* MAXIMUM.

**minister** *n* ▷ CLERGYMAN, OFFI-
CIAL. • *vb* **minister to** aid, assist,
attend to, care for, help, look after,
nurse, see to, support, wait on.

**minor** *adj* inconsequential,
inferior, insignificant, lesser,
little, negligible, petty, secondary,
smaller, subordinate, subsidiary,
trivial, unimportant. ▷ SMALL.

*Opp* MAJOR. • *n* ▷ ADOLESCENT, CHILD.

**minstrel** *n* balladeer, bard, entertainer, jongleur, musician, singer, troubadour.

**mint** *adj* brand-new, first-class, fresh, immaculate, new, perfect, unblemished, unmarked, unused. • *n* fortune, heap, *inf* packet, pile, stack, unlimited supply, vast amount. • *vb* cast, coin, forge, make, manufacture, produce, stamp out, strike.

**minute** *adj* diminutive, dwarf, infinitesimal, insignificant, lilliputian, microscopic, *inf* mini, miniature, minuscule, *inf* pint-sized, pocket, pygmy, tiny. ▷ SMALL.

**minutes** *plur n* log, notes, proceedings, record, résumé, summary, transactions.

**miracle** *n* marvel, miraculous event, mystery, wonder.

**miraculous** *adj* abnormal, extraordinary, incredible, inexplicable, magic, magical, mysterious, paranormal, phenomenal, preternatural, remarkable, supernatural, unaccountable, unbelievable, unexplainable. ▷ MARVELLOUS.

**mirage** *n* delusion, hallucination, illusion, vision.

**mire** *n* bog, fen, marsh, morass, mud, ooze, quagmire, quicksand, slime, *old use* slough, swamp. ▷ DIRT.

**mirror** *n* glass, looking-glass, reflector, speculum. • *vb* echo, reflect, repeat, send back.

**misadventure** *n* accident, calamity, catastrophe, disaster, ill fortune, mischance, misfortune, mishap.

**misanthropic** *adj* anti-social, cynical, mean, nasty, surly, unfriendly, unpleasant, unsociable. *Opp* PHILANTHROPIC.

**misappropriate** *vb* defalcate, embezzle, expropriate, peculate. ▷ STEAL.

**misbehave** *vb* be a nuisance, be bad, behave badly, be mischievous, *inf* blot your copybook, *inf* carry on, commit an offence, default, disobey, do wrong, err, fool about, make mischief, *inf* mess about, *inf* muck about, offend, *inf* play about, play up, *inf* raise Cain, sin, transgress.

**misbehaviour** *n* badness, delinquency, disobedience, disorderliness, horseplay, indiscipline, insubordination, mischief, mischief-making, misconduct, misdemeanour, naughtiness, rowdyism, rudeness, sin, vandalism, wrongdoing.

**miscalculate** *vb* *inf* boob, err, *inf* get it wrong, go wrong, make mistake, miscount, misjudge, misread, overestimate, overrate, overvalue, *inf* slip up, underestimate, underrate.

**miscarriage** *n* **1** abortion, premature birth, termination of pregnancy. **2** *miscarriage of justice.* breakdown, collapse, defeat, error, failure, perversion.

**miscarry** *vb* **1** abort, *inf* lose a baby, suffer a miscarriage. **2** break down, *inf* come to grief, come to nothing, fail, fall through, founder, go wrong, misfire. *Opp* SUCCEED.

**miscellaneous** *adj* assorted, different, *old use* divers, diverse, heterogeneous, manifold, mixed, motley, multifarious, sundry, varied, various.

**miscellany** *n* assortment, diversity, gallimaufry, hotchpotch, jumble, medley, mélange, *inf* mixed bag, mixture, pot-pourri, *inf* ragbag, variety.

**mischief** *n* **1** devilment, devilry, escapade, impishness, misbehaviour, misconduct, *inf* monkey business, naughtiness, playfulness, prank, rascality, roguishness, scrape, *inf* shenanigans, trouble. **2** damage, difficulty, evil, harm, hurt, injury, misfortune, trouble.

**mischievous** *adj* annoying, badly behaved, boisterous, disobedient, elvish, fractious, frolicsome, full of mischief, impish, lively, naughty, playful, *inf* puckish, rascally, roguish, sportive, uncontrollable, *inf* up to no good. ▷ WICKED. *Opp* WELL-BEHAVED.

**miser** *n* hoarder, miserly person, niggard, *inf* Scrooge, *inf* skinflint. *Opp* SPENDTHRIFT.

**miserable** *adj* **1** broken-hearted, crestfallen, *inf* cut up, dejected, depressed, desolate, despairing, despondent, disappointed, disconsolate, dismayed, dispirited, distressed, doleful, *inf* down, downcast, downhearted, forlorn, friendless, gloomy, glum, grief-stricken, heartbroken, hopeless, in low spirits, *inf* in the doldrums, *inf* in the dumps, joyless, lachrymose, languishing, lonely, low, melancholy, mournful, moping, sad, sorrowful, suicidal, tearful, uneasy, unfortunate, unhappy, unlucky, woebegone, woeful, wretched. **2** churlish, cross, disagreeable, discontented, *inf* grumpy, ill-natured, mean, miserly, morose, pessimistic, sour, sulky, sullen, surly, taciturn, unfriendly, unhelpful, unsociable. **3** *miserable living conditions.* abject, awful, bad, deplorable, destitute, disgraceful, distressing, heart-breaking, hopeless, impoverished, inadequate, inhuman, lamentable, pathetic, pitiable, pitiful, poor, shameful, sordid, soul-destroying, squalid, vile, uncivilized, uncomfortable, vile, worthless, wretched. **4** *miserable weather.* cheerless, damp, depressing, dismal, dreary, grey, inclement, sunless, unpleasant, wet. *Opp* HAPPY, PLEASANT.

**miserly** *adj* avaricious, *inf* cheese-paring, *inf* close, *inf* close-fisted, covetous, economical, grasping, greedy, mean, mercenary, mingy, niggardly, parsimonious, penny-pinching, penurious, sparing, stingy, *inf* tight, *inf* tight-fisted. *Opp* GENEROUS.

**misery** *n* **1** angst, anguish, anxiety, bitterness, dejection, depression, despair, desperation, despondency, discomfort, distress, dolour, gloom, grief, heartache, heartbreak, *inf* hell, hopelessness, melancholy, sadness, sorrow, suffering, unhappiness, woe, wretchedness. *Opp* HAPPINESS. **2** adversity, affliction, deprivation, destitution, hardship, indigence, misfortune, need, oppression, penury, poverty, privation, squalor, suffering, *inf* trials and tribulations, tribulation, trouble, want, wretchedness.

**misfire** *vb* abort, fail, fall through, *inf* flop, founder, go wrong, miscarry. *Opp* SUCCEED.

**misfortune** *n* accident, adversity, affliction, bad luck, blow, calamity, catastrophe, contretemps, curse, disappointment, disaster, evil, hard luck, hardship, ill-luck, misadventure, mischance, mishap, reverse, setback, tragedy, trouble, vicissitude.

**misguided** *adj* erroneous, foolish, ill-advised, ill-judged, inappropriate, incorrect, inexact, misinformed, misjudged, misled, mistaken, unfounded, unjust, unsound, unwise. ▷ WRONG.

**misjudge** *vb* get wrong, guess wrongly, *inf* jump to the wrong conclusion, make a mistake, misinterpret, overestimate, overvalue, underestimate, undervalue. ▷ MISCALCULATE.

**mislay** *vb* lose, mislocate, misplace, put in the wrong place.

**mislead** *vb* bluff, confuse, delude, fool, give misleading information to, give a wrong impression to, lead astray, *inf* lead up the garden path, lie to, misdirect, misguide, misinform, outwit, *inf* take for a ride, take in, *inf* throw off the scent, trick. ▷ DECEIVE. **misleading** ▷ DECEPTIVE, PUZZLING.

**miss** *vb* **1** absent yourself from, avoid, be absent from, be too late for, dodge, escape, evade, fail to keep, forget, forgo, let go, lose, play truant from, *inf* skip, *inf* skive off. **2** *miss a target*. be wide of, fail to hit, fall short of. **3** *miss absent friends*. feel nostalgia for, grieve for, lament, long for, need, pine for, want, yearn for. **miss out** ▷ OMIT.

**misshapen** *adj* awry, bent, contorted, corkscrew, crippled, crooked, crumpled, deformed, disfigured, distorted, gnarled, grotesque, knotted, malformed, monstrous, screwed up, tangled, twisted, twisty, ugly, warped. *Opp* PERFECT.

**missile** *n* projectile. □ *arrow, ballistic missile, bomb, boomerang, brickbat, bullet, dart, grenade, guided missile, rocket, shell, shot, torpedo*. ▷ WEAPON.

**missing** *adj* absent, disappeared, lost, mislaid, *inf* skiving, straying, truant, unaccounted-for. *Opp* PRESENT.

**mission** *n* **1** delegation, deputation, expedition, exploration, journey, sortie, task-force, voyage. **2** *mission in life*. aim, assignment, calling, commitment, duty, function, goal, job, life's work, métier, objective, occupation, profession, purpose, quest, undertaking, vocation. **3** *evangelical mission*. campaign, crusade, holy war.

**missionary** *n* campaigner, crusader, evangelist, minister, preacher, proselytizer.

**mist** *n* **1** cloud, drizzle, fog, haze, smog, vapour. **2** condensation, film, steam.

**mistake** *n inf* bloomer, blunder, *inf* boob, *inf* botch, *inf* clanger, erratum, error, false step, fault, *Fr* faux pas, gaffe, *inf* howler, inaccuracy, indiscretion, lapse, misapprehension, miscalculation, misconception, misjudgement, misprint, misspelling, misunderstanding, omission, oversight, slip, slip-up, solecism, wrong move. • *vb* confuse, *inf* get the wrong end of the stick, get wrong, misconstrue, misinterpret, misjudge, misread, misunderstand, mix up, *inf* take the wrong way.

**mistaken** *adj* erroneous, distorted, false, faulty, ill-judged, inaccurate, inappropriate, incorrect, inexact, misguided, misinformed, unfounded, unjust, unsound. ▷ WRONG. *Opp* CORRECT.

**mistimed** *adj* badly timed, early, inconvenient, inopportune, late, unseasonable, untimely. *Opp* OPPORTUNE.

**mistreat** *vb* abuse, batter, damage, harm, hurt, ill-treat, ill-use, injure, *inf* knock about, maltreat, manhandle, misuse, molest, treat roughly.

**mistress** *n* [*mostly old use*] **1** chief, head, keeper, owner, person in charge, proprietor. **2** ▷ TEACHER. **3** ▷ LOVER.

**mistrust** *n* apprehension, *inf* chariness, distrust, doubt, misgiving, reservation, scepticism, suspicion, uncertainty, unsureness, wariness. • *vb* be sceptical about, be suspicious of, be wary of, disbelieve, distrust, doubt, fear, have doubts about, have misgivings about, have reservations about, question, suspect. *Opp* TRUST.

**misty** *adj* bleary, blurred, blurry, clouded, cloudy, dim, faint, foggy, fuzzy, hazy, indistinct, murky, obscure, opaque, shadowy, smoky, steamy, unclear, vague. *Opp* CLEAR.

**misunderstand** *vb inf* get the wrong end of the stick, get wrong, misapprehend, miscalculate, misconceive, misconstrue, mishear, misinterpret, misjudge, misread, miss the point of, mistake, mistranslate. *Opp* UNDERSTAND.

**misunderstanding** *n* **1** error, failure of understanding, false impression, misapprehension, miscalculation, misconception, misconstruction, misinterpretation, misjudgement, misreading, mistake, *inf* mix up, wrong idea. **2** argument, *inf* contretemps, controversy, difference of opinion, disagreement, discord, dispute. ▷ QUARREL.

**misuse** *n* abuse, careless use, corruption, ill-treatment, ill-use, maltreatment, misapplication, misappropriation, mishandling, mistreatment, perversion. • *vb* **1** damage, harm, mishandle, treat carelessly. **2** *misuse an animal.* abuse, batter, damage, harm, hurt, ill-treat, ill-use, injure, *inf* knock about, maltreat, manhandle, mistreat, molest, treat roughly. **3** *misuse funds.* fritter away, misappropriate, squander, use wrongly, waste.

**mitigate** *vb* abate, allay, alleviate, decrease, ease, extenuate, lessen, lighten, make milder, moderate, palliate, qualify, reduce, relieve, soften, *inf* take the edge off, temper, tone down. *Opp* AGGRAVATE.

**mix** *n* amalgam, assortment, blend, combination, compound, range, variety. • *vb* **1** alloy, amalgamate, blend, coalesce, combine, commingle, compound, confuse, diffuse, emulsify, fuse, homogenize, integrate, intermingle, join, jumble up, make a mixture, meld, merge, mingle, mix up, muddle, put together, shuffle, stir together, unite. *Opp* SEPARATE. **2** *mix with people.* ▷ SOCIALIZE.

**mixed** *adj* **1** assorted, different, diverse, heterogeneous, miscellaneous, varied, various. **2** *mixed with other things.* adulterated, alloyed, diluted, impure. **3** *mixed ingredients.* amalgamated, combined, composite, hybrid, integrated, joint, mongrel, united. **4** *mixed feelings.* ambiguous, ambivalent, confused, equivocal, muddled, uncertain.

**mixture** *n* **1** alloy, amalgam, amalgamation, association, assortment, blend, collection, combination, composite, compound, concoction, conglomeration, emulsion, farrago, fusion, gallimaufry, *inf* hotchpotch, intermingling, jumble, medley, mélange, merger, mess, mingling, miscellany, *inf* mishmash, mix, *inf* motley collection, pastiche, potpourri, selection, suspension, synthesis, variety. **2** cross-breed, half-caste, hybrid, mongrel.

**moan** *n* complaint, grievance, lament, lamentation. • *vb* **1** complain, grieve, *inf* grouse, grumble, lament. **2** cry, groan, keen, sigh, ululate, wail, weep, whimper, whine. ▷ SOUND.

**mob** *n inf* bunch, crowd, gang, herd, horde, host, multitude, pack, press, rabble, riot, *inf* shower, swarm, throng. ▷ GROUP. • *vb* besiege, crowd round, hem in, jostle, surround, swarm round, throng round.

**mobile** *adj* **1** itinerant, motorized, movable, portable, transportable, travelling, unfixed. **2** able to move, active, agile, independent, moving, nimble, *inf* on the go, *inf* up and about. **3** *mobile features.* animated, changeable, changing, expressive, flexible, fluid, plastic, shifting. *Opp* IMMOVABLE.

**mobilize** *vb* activate, assemble, call up, conscript, enlist, enrol, gather, get together, levy, marshal, muster, organize, rally, stir up, summon.

**mock** *adj* artificial, counterfeit, ersatz, fake, false, imitation, make-believe, man-made, *inf* pretend, simulated, sham, substitute. • *vb* decry, deride, disparage, flout, gibe at, insult, jeer at, lampoon, laugh at, make fun of, make sport of, parody, poke fun at, ridicule, satirize, scoff at, scorn, *inf* send up, sneer at, tantalize, taunt, tease, travesty. ▷ MIMIC.

**mockery** *n* derision, insults, jeering, laughter, ridicule, scorn. □ *burlesque, caricature, lampoon, parody, sarcasm, satire,* inf *send-up,* inf *spoof,* inf *take-off, travesty.*

**mocking** *adj* contemptuous, derisive, disparaging, disrespectful, insulting, irreverent, jeering, rude, sarcastic, satirical, scornful, taunting, teasing, uncomplimentary, unkind. *Opp* RESPECTFUL.

**mode** *n* **1** approach, configuration, manner, medium, method, *Lat* modus operandi, procedure, set-up, system, technique, way. **2** ▷ FASHION.

**model** *adj* **1** imitation, miniature, scaled-down, toy. **2** *model pupil.* exemplary, ideal, perfect, unequalled. • *n* **1** archetype, copy, dummy, effigy, facsimile, image, imitation, likeness, miniature, *inf* mock-up, paradigm, prototype, replica, representation, scale model, toy. **2** *model of excellence.* byword, epitome, example, exemplar, ideal, nonpareil, paragon, pattern, standard, yardstick. **3** *artist's model.* poser, sitter, subject. **4** *latest model.* brand, design, kind, mark, type, version. **5** *fashion model.* mannequin. • *vb* carve, fashion, form, make, mould, sculpt, shape. **model yourself on** ▷ IMITATE.

**moderate** *adj* **1** average, balanced, calm, cautious, commonsensical, cool, deliberate, fair, judicious, medium, middle, *inf* middle-of-the-road, middling, modest, normal, ordinary, rational, reasonable, respectable, sensible, sober, steady, temperate, unexceptional, usual. *Opp* EXTREME. **2** *moderate winds.* gentle, light, mild. • *vb* **1** abate, become less extreme, decline, decrease, die down, ease off, subside. **2** blunt, calm, check, curb, dull, ease, keep down, lessen, make less extreme, mitigate, modify, modulate, mollify, reduce, regulate, restrain, slacken, subdue, temper, tone down.

**moderately** *adv* comparatively, fairly, passably, *inf* pretty, quite, rather, reasonably, somewhat, to some extent.

**moderation** *n* balance, caution, common sense, fairness, reasonableness, restraint, reticence, sobriety, temperance.

**modern** *adj* advanced, avant-garde, contemporary, current, fashionable, forward-looking, fresh, futuristic, in vogue, latest, modish, new, newfangled, novel, present, present-day, progressive, recent, stylish, *inf* trendy, up-to-date, up-to-the-minute, *inf* with it. *Opp* OLD.

**modernize** *vb* bring up-to-date, *inf* do up, improve, make modern, rebuild, redesign, redo, refurbish, regenerate, rejuvenate, renovate, revamp, update.

**modest** *adj* **1** diffident, humble, inconspicuous, lowly, meek, plain, quiet, reserved, restrained, reticent, retiring, self-effacing, simple, unassuming, unobtrusive, unostentatious, unpretentious. *Opp* CONCEITED. **2** bashful, chaste, coy, decent, demure, discreet, proper, seemly, self-conscious, shamefaced, shy, simple. **3** *a modest income.* limited, medium, middling, moderate, normal, ordinary, reasonable, unexceptional. *Opp* EXCESSIVE.

**modesty** *n* **1** humbleness, humility, lowliness, meekness, reserve, restraint, reticence, self-effacement, simplicity. *Opp* OSTENTATION. **2** *modesty about undressing.* bashfulness, coyness, decency, demureness, discretion, propriety, seemliness, self-consciousness, shame, shyness.

**modify** *vb* adapt, adjust, alter, amend, change, convert, improve, reconstruct, redesign, remake, remodel, reorganize, revise, reword, rework, transform, vary. ▷ MODERATE.

**modulate** *vb* adjust, balance, change key, change the tone, lower the tone, moderate, regulate, soften, tone down.

**moist** *adj* affected by moisture, clammy, damp, dank, dewy, humid, misty, *inf* muggy, rainy, *inf* runny, steamy, watery, wettish. ▷ WET. *Opp* DRY.

**moisten** *vb* damp, dampen, humidify, make moist, moisturize, soak, spray, wet. *Opp* DRY.

**moisture** *n* condensation, damp, dampness, dankness, dew, humidity, liquid, precipitation, spray, steam, vapour, water, wet, wetness.

**molest** *vb* abuse, accost, annoy, assault, attack, badger, bother, disturb, harass, harry, hassle, hector, ill-treat, interfere with, irk, irritate, manhandle, mistreat, *inf* needle, persecute, pester, plague, set on, tease, torment, vex, worry.

**molten** *adj* fluid, liquid, liquefied, melted, soft.

**moment** *n* **1** flash, instant, *inf* jiffy, minute, second, split second, *inf* tick, *inf* trice, *inf* twinkling of an eye, *inf* two shakes. **2** *a historic moment.* hour, juncture, occasion, opportunity, point in time, stage, time.

**momentary** *adj* brief, ephemeral, evanescent, fleeting, fugitive, hasty, passing, quick, short, short-lived, temporary, transient, transitory. *Opp* PERMANENT.

**momentous** *adj* consequential, critical, crucial, decisive, epoch-making, fateful, grave, historic, important, portentous, serious, significant, weighty. *Opp* UNIMPORTANT.

**monarch** *n* crowned head, emperor, empress, king, potentate, queen, ruler, tsar.

**monarchy** *n* empire, domain, kingdom, realm.

**money** *n* affluence, arrears, assets, bank-notes, *inf* bread, capital, cash, change, cheque, coin, copper, credit card, credit transfer, currency, damages, debt, dividend, *inf* dough, dowry, earnings, endowment, estate, expenditure, finance, fortune, fund, grant, income, interest, investment, legal tender, loan, *inf* lolly, *old use* lucre, mortgage, *inf* nest-egg, notes, outgoings, patrimony, pay, penny, pension, pocket-money, proceeds, profit, *inf* the ready, remittance, resources, revenue, riches, salary, savings, silver, sterling, takings, tax, traveller's cheque, wage, wealth, *inf* the wherewithal, winnings.

**mongrel** *n* cross-breed, cur, half-breed, hybrid, mixed breed.

**monitor** *n* **1** detector, guardian, prefect, supervisor, watchdog. **2** *TV monitor*. screen, set, television, TV, VDU, visual display unit. • *vb* audit, check, examine, *inf* keep an eye on, oversee, record, supervise, trace, track, watch.

**monk** *n* brother, friar, hermit.

**monkey** *n* ape, primate, simian. □ *baboon, chimpanzee, gibbon, gorilla, marmoset, orang-utan.*

**monopolize** *vb* control, *inf* corner the market in, dominate, have a monopoly of, *inf* hog, keep for yourself, own, shut others out of, take over. *Opp* SHARE.

**monotonous** *adj* boring, colourless, dreary, dull, featureless, flat, level, repetitious, repetitive, soporific, tedious, tiresome, tiring, toneless, unchanging, uneventful, unexciting, uniform, uninteresting, unvarying, wearisome. *Opp* INTERESTING.

**monster** *n* abortion, beast, bogeyman, brute, demon, devil, fiend, freak, giant, horror, monstrosity, monstrous creature, mutant, ogre, troll.

**monstrous** *adj* **1** colossal, elephantine, enormous, gargantuan, giant, gigantic, great, huge, hulking, immense, *inf* jumbo, mammoth, mighty, prodigious, titanic, towering, tremendous, vast. ▷ BIG **2** *a monstrous crime*. abhorrent, atrocious, awful, beastly, brutal, cruel, devilish, dreadful, disgusting, evil, ghoulish, grisly, gross, gruesome, heinous, hideous, *inf* horrendous, horrible, horrific, horrifying, inhuman, nightmarish, obscene, outrageous, repulsive, shocking, terrible, ugly, villainous, wicked. ▷ EVIL.

**monument** *n* cairn, cenotaph, cross, gravestone, headstone, mausoleum, memorial, obelisk, pillar, prehistoric remains, relic, reminder, shrine, tomb, tombstone.

**monumental** *adj* **1** awe-inspiring, awesome, classic, enduring, epoch-making, grand, historic, impressive, lasting, large-scale, major, memorable, unforgettable. ▷ BIG. **2** *a monumental plaque*. commemorative, memorial.

**mood** *n* **1** attitude, disposition, frame of mind, humour, inclination, nature, spirit, state of mind, temper, vein. **2** atmosphere, feeling, tone. ▷ ANGRY, HAPPY, SAD, etc. **in the mood** ▷ READY.

**moody** *adj* abrupt, bad-tempered, cantankerous, capricious, changeable, crabby, cross, crotchety, depressed, depressive, disgruntled, erratic, fickle, gloomy, grumpy, *inf* huffy, ill-humoured, inconstant, irritable, melancholy, mercurial, miserable, morose, peevish, petulant, *inf* short, short-tempered, snappy, sulky, sullen, temperamental, testy, *inf* touchy, unpredictable, unreliable, unstable, volatile. ▷ SAD.

**moor** *n* fell, heath, moorland, wasteland. • *vb* anchor, berth, dock, make fast, secure, tie up. ▷ FASTEN.

**mope** *vb* be sad, brood, despair, grieve, languish, *inf* moon, pine, sulk.

**moral** *adj* **1** blameless, chaste, decent, ethical, good, highminded, honest, honourable, incorruptible, innocent, irreproachable, just, law-abiding, noble, principled, proper, pure, respectable, responsible, right, righteous, sinless, trustworthy, truthful, upright, upstanding, virtuous. **2** *a moral tale.* allegorical, cautionary, didactic, moralistic, moralizing. *Opp* IMMORAL. • *n* lesson, maxim, meaning, message, point, precept, principle, teaching. **morals** ▷ MORALITY. **moral tale** allegory, cautionary tale, fable, parable.

**morale** *n* attitude, cheerfulness, confidence, *Fr* esprit de corps, *inf* heart, mood, self-confidence, self-esteem, spirit, state of mind.

**morality** *n* behaviour, conduct, decency, ethics, ethos, fairness, goodness, honesty, ideals, integrity, justice, morals, principles, propriety, rectitude, righteousness, rightness, scruples, standards, uprightness, virtue.

**moralize** *vb* lecture, philosophize, pontificate, preach, sermonize.

**morbid** *adj* black (*humour*), brooding, dejected, depressed, ghoulish, gloomy, grim, grotesque, gruesome, lugubrious, macabre, melancholy, monstrous, morose, pathological, pessimistic, *inf* sick, sombre, unhappy, unhealthy, unpleasant, unwholesome. *Opp* CHEERFUL.

**more** *adj* added, additional, extra, further, increased, longer, new, other, renewed, supplementary. *Opp* LESS.

**moreover** *adv* also, as well, besides, further, furthermore, in addition, *old use* to boot, too.

**morose** *adj* bad-tempered, churlish, depressed, gloomy, glum, grim, humourless, ill-natured, melancholy, moody, mournful, pessimistic, saturnine, sour, sulky, sullen, surly, taciturn, unhappy, unsociable. ▷ SAD. *Opp* CHEERFUL.

**morsel** *n* bite, crumb, fragment, gobbet, mouthful, nibble, piece, sample, scrap, small amount, soupçon, spoonful, taste, titbit. ▷ BIT.

**mortal** *adj* **1** ephemeral, human, passing, temporal, transient. *Opp* IMMORTAL. **2** *mortal sickness.* deadly, fatal, lethal, terminal. **3** *mortal enemies.* deadly, implacable, irreconcilable, remorseless, sworn, unrelenting. • *n* creature, human being, man, person, soul, woman.

**mortality** *n* **1** corruptibility, humanity, impermanence, transience. **2** *infant mortality.* death-rate, dying, fatalities, loss of life.

**mortify** *vb* abash, chagrin, chasten, *inf* crush, deflate, embarrass, humble, humiliate, *inf* put down, shame.

**mostly** *adv* chiefly, commonly, generally, largely, mainly, normally, predominantly, primarily, principally, typically, usually.

**moth-eaten** *adj* antiquated, decrepit, holey, mangy, ragged, shabby, *inf* tatty. ▷ OLD.

**mother** *n old use* dam, *inf* ma, *inf* mamma, *old use* mater, *inf* mum, *inf* mummy, parent. • *vb* care for, cherish, coddle, comfort, cuddle, fuss over, indulge, look after, love, nourish, nurse, nurture, pamper, protect, spoil, take care of.

**motherly** *adj* caring, kind, maternal, protective. ▷ LOVING.

**motif** *n* decoration, design, device, figure, idea, leitmotif, ornament, pattern, symbol, theme.

**motion** *n* action, activity, agitation, change, commotion, development, evolution, move, movement, progress, rise and fall, shift, stir, stirring, to and fro, travel, travelling, trend. • *vb* ▷ GESTURE.

**motionless** *adj* at rest, calm, frozen, immobile, inanimate, inert, lifeless, paralysed, peaceful, resting, stagnant, static, stationary, still, stock-still, unmoving. *Opp* MOVING.

**motivate** *vb* activate, actuate, arouse, cause, drive, egg on, encourage, excite, galvanize, goad, incite, induce, influence, inspire, instigate, move, occasion, persuade, prompt, provoke, push, rouse, spur, stimulate, stir, urge.

**motive** *n* aim, ambition, cause, drive, encouragement, end, enticement, grounds, impulse, incentive, incitement, inducement, inspiration, instigation, intention, lure, motivation, object, provocation, purpose, push, rationale, reason, spur, stimulation, stimulus, thinking.

**motor** *n* ▷ ENGINE, VEHICLE. • *vb* drive, go by car. ▷ TRAVEL.

**mottled** *adj* blotchy, brindled, dappled, flecked, freckled, marbled, patchy, spattered, speckled, spotted, spotty, streaked, streaky, variegated.

**motto** *n* adage, aphorism, catchphrase, maxim, precept, proverb, rule, saw, saying, slogan.

**mould** *n* blight, fungus, growth, mildew. • *vb* cast, fashion, forge, form, model, *inf* sculpt, shape, stamp, work.

**mouldy** *adj* carious, damp, decaying, decomposing, fusty, mildewed, mouldering, musty, putrefying, rotten, stale.

**mound** *n* bank, dune, elevation, heap, hill, hillock, hummock, hump, knoll, pile, stack, tumulus.

**mount** *n* ▷ MOUNTAIN. • *vb* **1** ascend, clamber up, climb, fly up, go up, rise, rocket upwards, scale, shoot up, soar. *Opp* DESCEND. **2** *mount a horse.* get astride, get on, jump onto. **3** *savings mount.* accumulate, build up, escalate, expand, get bigger, grow, increase, intensify, multiply, pile up, swell. *Opp* DECREASE. **4** *mount a picture.* display, exhibit, frame, install, prepare, put in place, put on, set up.

**mountain** *n* alp, arête, *Scot* ben, elevation, eminence, height, hill, mound, mount, peak, prominence, range, ridge, sierra, summit, tor, volcano.

**mountainous** *adj* alpine, craggy, daunting, formidable, high, hilly, precipitous, rocky, rugged, steep, towering. ▷ BIG.

**mourn** *vb* bemoan, bewail, fret, go into mourning, grieve, keen, lament, mope, pine, regret, wail, weep. *Opp* REJOICE.

**mournful** *adj* dismal, distressed, distressing, doleful, funereal, gloomy, grief-stricken, grieving, heartbreaking, heartbroken, lamenting, lugubrious, melancholy, plaintive, plangent, sad, sorrowful, tearful, tragic, unhappy, woeful. *Opp* CHEERFUL.

**mouth** *n* 1 *inf* chops, *sl* gob, jaws, *sl* kisser, lips, maw, muzzle, palate. 2 *mouth of cave*. aperture, door, doorway, entrance, exit, gate, gateway, inlet, opening, orifice, outlet, vent, way in. 3 *mouth of a river*. delta, estuary, outflow. • *vb* articulate, enunciate, form, pronounce. ▷ SAY.

**mouthful** *n* bite, gobbet, gulp, morsel, sip, spoonful, swallow, taste.

**movable** *adj* adjustable, changeable, detachable, floating, mobile, portable, transferable, transportable, unfixed, variable. *Opp* IMMOVABLE.

**move** *n* 1 act, action, deed, device, dodge, gambit, manoeuvre, measure, movement, ploy, ruse, step, stratagem, *inf* tack, tactic. 2 *a career move*. change, changeover, relocation, shift, transfer. 3 *your move*. chance, go, opportunity, turn. • *vb* 1 *move about*. be agitated, be astir, budge, change places, change position, fidget, flap, roll, shake, shift, stir, swing, toss, tremble, turn, twist, twitch, wag, *inf* waggle, wave, *inf* wiggle. 2 *move along*. cruise, fly, jog, journey, make headway, make progress, march, pass, proceed, travel, walk. 3 *move quickly*. bolt, *inf* bowl along, canter, career, dash, dart, flit, flounce, fly, gallop, hasten, hurry, hurtle, hustle, *inf* nip, race, run, rush, shoot, speed, stampede, streak, sweep along, sweep past, *inf* tear, *inf* zip, *inf* zoom. 4 *move slowly*. amble, crawl, dawdle, drift, stroll. 5 *move gracefully*. dance, flow, glide, skate, skim, slide, slip, sweep. 6 *move awkwardly*. dodder, falter, flounder, lumber, lurch, pitch, shuffle, stagger, stumble, sway, totter, trip, trundle. 7 *move stealthily*. crawl, creep, edge, slink, slither. 8 *move things*. carry, export, import, shift, ship, relocate, transfer, transplant, transport, transpose. 9 *moved him to act*. encourage, impel, influence, inspire, persuade, prompt, stimulate, urge. 10 *move the crowd's feelings*. affect, arouse, enrage, fire, impassion, rouse, stir, touch. 11 *moved to improve the situation*. act, do something, make a move, take action. **move away** ▷ DEPART. **move back** ▷ RETREAT. **move down** ▷ DESCEND. **move in** ▷ ENTER. **move round** ▷ CIRCULATE, ROTATE. **move towards** ▷ APPROACH. **move up** ▷ ASCEND.

**movement** *n* 1 action, activity, migration, motion, shifting, stirring. ▷ GESTURE, MOVE. 2 *movement towards green issues*. change, development, drift, evolution, progress, shift, swing, tendency, trend. 3 *a political movement*. campaign, crusade, drive, faction, group, organization, party. 4 *military movements*. exercise, operation.

**movie** *n* film, *inf* flick, motion picture.

**moving** *adj* 1 active, alive, astir, dynamic, flowing, going, in motion, mobile, movable, on the move, travelling, under way. *Opp* MOTIONLESS. 2 *a moving tale*. affecting, emotional, emotive, exciting, heart-rending, heartwarming, inspirational, inspiring, pathetic, poignant, spine-tingling, stirring, *inf* tear-jerking, thrilling, touching.

**mow** *vb* clip, cut, scythe, shear, trim.

**muck** *n* dirt, droppings, dung, excrement, faeces, filth, grime, *inf* gunge, manure, mess, mire, mud, ooze, ordure, rubbish, scum, sewage, slime, sludge.

**mucky** *adj* dirty, filthy, foul, grimy, grubby, messy, muddy, scummy, slimy, soiled, sordid, squalid. *Opp* CLEAN.

**mud** *n* clay, dirt, mire, muck, ooze, silt, slime, sludge, slurry, soil.

**muddle** *n* chaos, clutter, confusion, disorder, *inf* hotchpotch, jumble, mess, *inf* mishmash, *inf* mix up, *inf* shambles, tangle, untidiness. • *vb* **1** bemuse, bewilder, confound, confuse, disorient, disorientate, mislead, perplex, puzzle. *Opp* CLARIFY. **2** disarrange, disorder, disorganize, entangle, *inf* foul up, jumble, make a mess of, *inf* mess up, mix up, scramble, shuffle, tangle. *Opp* TIDY.

**muddy** *adj* **1** caked, dirty, filthy, messy, mucky, soiled. **2** *muddy water.* cloudy, impure, misty, opaque. **3** *muddy ground.* boggy, marshy, sloppy, sodden, soft, spongy, waterlogged, wet. *Opp* CLEAN, FIRM.

**muffle** *vb* **1** cloak, conceal, cover, enclose, enfold, envelop, shroud, swathe, wrap up. **2** *muffle noise.* damp, dampen, deaden, disguise, dull, hush, mask, mute, quieten, silence, soften, stifle, still, suppress, tone down.

**muffled** *adj* damped, deadened, dull, fuzzy, indistinct, muted, silenced, stifled, suppressed, unclear, woolly. *Opp* CLEAR.

**mug** *n* beaker, cup, *old use* flagon, pot, tankard. • *vb* assault, beat up, jump on, molest, rob, set on, steal from. ▷ ATTACK. **mug up** ▷ LEARN.

**mugger** *n* attacker, hooligan, robber, ruffian, thief, thug. ▷ CRIMINAL.

**mugging** *n* attack, robbery, street crime. ▷ CRIME.

**muggy** *adj* clammy, close, damp, humid, moist, oppressive, steamy, sticky, stuffy, sultry, warm.

**multiple** *adj* complex, compound, double, many, numerous, plural, quadruple, quintuple, triple.

**multiplicity** *n* abundance, array, complex, diversity, number, plurality, profusion, variety.

**multiply** *vb* **1** double, quadruple, quintuple, *inf* times, triple. **2** become numerous, breed, increase, proliferate, propagate, reproduce, spread.

**multitude** *n* crowd, host, large number, legion, lots, mass, myriad, swarm, throng. ▷ GROUP.

**mumble** *vb* be inarticulate, murmur, mutter, speak indistinctly, swallow your words.

**munch** *vb* bite, chew, champ, chomp, crunch, eat, gnaw, masticate.

**mundane** *adj* banal, common, commonplace, down-to-earth, dull, everyday, familiar, human, material, physical, practical, quotidian, routine, temporal, worldly. ▷ ORDINARY. *Opp* EXTRAORDINARY, SPIRITUAL.

**municipal** *adj* borough, city, civic, community, district, local, public, town, urban.

**murder** *n* assassination, fratricide, genocide, homicide, infanticide, killing, manslaughter, matricide, parricide, patricide, regicide, sororicide, unlawful killing, uxoricide. • *vb* ▷ KILL.

**murderer** *n* assassin, *inf* butcher, cutthroat, gunman, homicide, killer, slayer.

**murderous** *adj* barbarous, bloodthirsty, bloody, brutal, cruel, dangerous, deadly, fell, ferocious, fierce, homicidal, inhuman, pitiless, ruthless, savage, vicious, violent.

**murky** *adj* clouded, cloudy, dark, dim, dismal, dreary, dull, foggy, funereal, gloomy, grey, misty, muddy, obscure, overcast, shadowy, sombre. *Opp* CLEAR.

**murmur** *n* background noise, buzz, drone, grumble, hum, mutter, rumble, susurration, undertone, whisper. • *vb* drone, hum, moan, mumble, mutter, rumble, speak in an undertone, whisper. ▷ GRUMBLE, TALK.

**muscular** *adj* athletic, *inf* beefy, brawny, broad-shouldered, burly, hefty, *inf* hulking, husky, powerful, powerfully built, robust, sinewy, *inf* strapping, strong, sturdy, tough, well-built, well-developed, wiry. *Opp* WEAK.

**muse** *vb* cogitate, consider, contemplate, deliberate, meditate, mull over, ponder, reflect, ruminate, study, think.

**mushy** *adj* pulpy, spongy, squashy. ▷ SOFT.

**music** *n* harmony. □ *blues, chamber music, choral music, classical music, dance music, disco music, folk, instrumental music, jazz, orchestral music, plain-song, pop, ragtime, reggae, rock, soul, swing.* □ *anthem, ballad, cadenza, calypso, canon, cantata, canticle, carol, chant, concerto, dance, dirge, duet, étude, fanfare, fugue, hymn, improvisation, intermezzo, lullaby, march, musical, nocturne, nonet, octet, opera, operetta, oratorio, overture, prelude, quartet, quintet, rhapsody, rondo, scherzo, sea shanty, septet, sextet, sonata, song, spiritual, symphony, toccata, trio.*

**musical** *adj* euphonious, harmonious, lyrical, melodious, pleasant, sweet-sounding, tuneful. **musical instrument** ▷ INSTRUMENT.

**musician** *n* composer, music-maker, performer, player, singer. □ *accompanist, bass, bugler, cellist, clarinettist, conductor, contralto, drummer, fiddler, flautist, guitarist, harpist, instrumentalist, maestro, minstrel, oboist, organist, percussionist, pianist, piper, soloist, soprano, tenor, timpanist, treble, trombonist, trumpeter, violinist, virtuoso, vocalist.* **musicians** □ *band, choir, chorus, consort, duet, duo, ensemble, group, nonet, octet, orchestra, quartet, quintet, septet, sextet, trio.*

**muster** *vb* assemble, call together, collect, come together, convene, convoke, gather, get together, group, marshal, mobilize, rally, round up, summon.

**musty** *adj* airless, damp, dank, fusty, mildewed, mildewy, mouldy, smelly, stale, stuffy, unventilated.

**mutant** *n* abortion, anomaly, deviant, freak, monster, monstrosity, sport, variant.

**mutation** *n* alteration, deviance, evolution, metamorphosis, modification, transfiguration, transformation, transmutation, variation. ▷ CHANGE.

**mute** *adj* dumb, quiet, silent, speechless, tacit, taciturn, tight-lipped, tongue-tied, voiceless. • *vb* damp, dampen, deaden, dull, hush, make quieter, mask, muffle, quieten, silence, soften, stifle, still, suppress, tone down.

**mutilate** *vb* cripple, damage, deface, disable, disfigure, dismember, injure, lame, maim, mangle, mar, spoil, vandalize, wound.

**mutinous** *adj* contumacious, defiant, disobedient, insubordinate, insurgent, insurrectionary, rebellious, refractory, revolutionary, seditious, subversive, ungovernable, unmanageable, unruly. *Opp* OBEDIENT.

**mutiny** *n* defiance, disobedience, insubordination, insurgency, insurrection, rebellion, revolt, revolution, sedition, subversion, unruliness, uprising. • *vb* agitate, be mutinous, disobey, rebel, revolt, rise up, strike.

**mutter** *vb* drone, grumble, mumble, murmur, speak in an undertone, whisper. ▷ GRUMBLE, TALK.

**mutual** *adj* common, interactive, joint, reciprocal, reciprocated, requited, shared.

**muzzle** *n* jaws, mouth, nose, snout. • *vb* censor, gag, restrain, silence, stifle, suppress.

**mysterious** *adj* arcane, baffling, bewildering, bizarre, confusing, cryptic, curious, dark, enigmatic, incomprehensible, inexplicable, inscrutable, insoluble, magical, miraculous, mystical, mystifying, obscure, perplexing, puzzling, recondite, secret, strange, uncanny, unexplained, unfathomable, unknown, weird. *Opp* STRAIGHTFORWARD.

**mystery** *n* conundrum, enigma, miracle, problem, puzzle, question, riddle, secret.

**mystical** *adj* abnormal, arcane, cabalistic, ineffable, metaphysical, mysterious, occult, other-worldly, preternatural, religious, spiritual, supernatural. *Opp* MUNDANE.

**mystify** *vb* baffle, *inf* bamboozle, *inf* beat, bewilder, confound, confuse, *inf* flummox, fool, hoax, perplex, puzzle, *inf* stump.

**myth** *n* **1** allegory, fable, legend, mythology, symbolism. **2** fabrication, falsehood, fiction, invention, make-believe, pretence, untruth.

**mythical** *adj* **1** allegorical, fabled, fabulous, legendary, mythic, mythological, poetic, symbolic. **2** false, fanciful, fictional, imaginary, invented, make-believe, nonexistent, pretended, unreal. *Opp* REAL.

# N

**nadir** *n* bottom, depths, low point, zero. *Opp* ZENITH.

**nag** *n* ▷ HORSE. • *vb* annoy, badger, chivvy, find fault with, goad, *inf* go on at, harass, hector, *inf* henpeck, keep complaining, pester, *inf* plague, scold, worry.

**nail** *n* pin, spike, stud, tack. ▷ FASTEN.

**naïve** *adj* artless, *inf* born yesterday, candid, childlike, credulous, *inf* green, guileless, gullible, inexperienced, ingenuous, innocent, open, simple, simple-minded, stupid, trustful, trusting, unsophisticated, unsuspecting, unwary. *Opp* ARTFUL.

**naked** *adj* bare, denuded, disrobed, exposed, in the nude, nude, stark-naked, stripped, unclothed, unconcealed, uncovered, undraped, undressed.

**name** *n* **1** alias, appellation, Christian name, first name, forename, given name, *inf* handle, identity, nickname, nom de plume, pen name, personal name, pseudonym, sobriquet, surname, title. **2** denomination, designation, epithet, term. • *vb* **1** baptize, call, christen, dub, style. **2** *name a book.* entitle, label. **3** *named him man of the match.* appoint, choose, commission, delegate, designate, elect, nominate, select, single out, specify. **named** ▷ SPECIFIC.

**nameless** *adj* **1** anonymous, incognito, unheard-of, unidentified, unnamed, unsung. **2** *nameless horrors.* dreadful, horrible, indescribable, inexpressible, shocking, unmentionable, unspeakable, unutterable.

**nap** *n* catnap, doze, *inf* forty winks, rest, *inf* shut-eye, siesta, sleep, snooze.

**narrate** *vb* chronicle, describe, detail, recount, rehearse, relate, repeat, report, retail, tell, unfold.

**narration** *n* commentary, reading, recital, recitation, relation, storytelling, telling, voiceover.

**narrative** *n* account, chronicle, description, history, report, story, tale, *inf* yarn.

**narrator** *n* author, chronicler, raconteur, reporter, storyteller.

**narrow** *adj* attenuated, close, confined, constricted, constricting, cramped, enclosed, fine, limited, restricted, slender, slim, thin, tight. *Opp* WIDE.

**narrow-minded** *adj* biased, bigoted, conservative, conventional, hidebound, illiberal, inflexible, insular, intolerant, narrow, old-fashioned, parochial, petty, prejudiced, prim, prudish, puritanical, reactionary, rigid, small-minded, straitlaced, *inf* stuffy. *Opp* BROAD-MINDED.

**nasty** *adj* [*Nasty* refers to anything you do not like. The range of synonyms is almost limitless: we give only a selection here.] bad, beastly, dangerous, difficult, dirty, disagreeable, disgusting, distasteful, foul, hateful, horrible, loathsome, *sl* lousy, objectionable, obnoxious, obscene, *inf* off-putting, repulsive, revolting, severe, sickening, unkind, unpleasant. *Opp* NICE.

**nation** *n* civilization, community, country, domain, land, people, population, power, race, realm, society, state, superpower.

**national** *adj* **1** ethnic, popular, racial. **2** *a national emergency.* countrywide, general, nationwide, state, widespread. • *n* citizen, inhabitant, native, resident, subject.

**nationalism** *n* chauvinism, jingoism, loyalty, patriotism, xenophobia.

**native** *adj* **1** aboriginal, indigenous, local, original. **2** *native wit.* congenital, hereditary, inborn, inbred, inherent, inherited, innate, mother (*wit*), natural. • *n* aborigine, life-long resident.

**natural** *adj* **1** common, everyday, habitual, normal, ordinary, predictable, regular, routine, standard, typical, usual. **2** *natural feelings.* healthy, hereditary, human, inborn, inherited, innate, instinctive, intuitive, kind, maternal, native, paternal, proper, right. **3** *a natural smile.* artless, authentic, candid, genuine, guileless, sincere, spontaneous, unaffected, unpretentious, unselfconscious, unstudied. **4** *natural resources.* crude (*oil*), raw, unadulterated, unprocessed, unrefined. **5** *a natural leader.* born, congenital, untaught. *Opp* UNNATURAL.

**nature** *n* 1 countryside, creation, ecology, environment, natural history, scenery, wildlife. 2 attributes, character, complexion, constitution, disposition, essence, humour, make-up, manner, personality, properties, quality, temperament, traits. 3 category, description, kind, sort, species, type, variety.

**naughty** *adj* 1 bad, badly-behaved, bad-mannered, boisterous, contrary, defiant, delinquent, disobedient, disorderly, disruptive, fractious, headstrong, impish, impolite, incorrigible, insubordinate, intractable, misbehaved, mischievous, obstinate, obstreperous, perverse, playful, puckish, rascally, rebellious, refractory, roguish, rude, self-willed, stubborn, troublesome, uncontrollable, undisciplined, ungovernable, unmanageable, unruly, wayward, wicked, wild, wilful. 2 [*inf*] cheeky, improper, ribald, risqué, shocking, *inf* smutty, vulgar. ▷ OBSCENE. *Opp* POLITE, WELL BEHAVED.

**nauseate** *vb* disgust, offend, repel, revolt, sicken.

**nauseous** *adj* disgusting, foul, loathsome, nauseating, offensive, repulsive, revolting, sickening, stomach-turning.

**nautical** *adj* marine, maritime, naval, seafaring, seagoing, yachting.

**navigate** *vb* captain, direct, drive, guide, handle, manoeuvre, map-read, pilot, sail, skipper, steer.

**navy** *n* armada, convoy, fleet, flotilla.

**near** *adj* 1 abutting, adjacent, adjoining, bordering, close, connected, contiguous, immediate, nearby, neighbouring, next-door. 2 *Christmas is near.* approaching, coming, forthcoming, imminent, impending, looming, *inf* round the corner. 3 *near friends.* close, dear, familiar, intimate, related. *Opp* DISTANT.

**nearly** *adv* about, all but, almost, approaching, approximately, around, as good as, close to, just about, not quite, practically, roughly, virtually.

**neat** *adj* 1 adroit, clean, dainty, deft, dexterous, elegant, *inf* natty, orderly, organized, pretty, *inf* shipshape, smart, *inf* spick and span, spruce, straight, systematic, tidy, trim, uncluttered, well-kept. 2 accurate, expert, methodical, meticulous, precise, skilful. 3 *neat alcohol.* pure, *inf* straight, unadulterated, undiluted. *Opp* CLUMSY, UNTIDY.

**necessary** *adj* compulsory, destined, essential, fated, imperative, important, indispensable, ineluctable, inescapable, inevitable, inexorable, mandatory, needed, needful, obligatory, predestined, required, requisite, unavoidable, vital. *Opp* UNNECESSARY.

**necessity** *n* 1 compulsion, essential, inevitability, *inf* must, need, obligation, prerequisite, requirement, requisite, *Lat* sine qua non. 2 beggary, destitution, hardship, indigence, need, penury, poverty, privation, shortage, suffering, want.

**need** *n* call, demand, lack, requirement, want. ▷ NECESSITY. • *vb* be short of, call for, crave, demand, depend on, lack, miss, rely on, require, want.

**needless** *adj* excessive, gratuitous, pointless, redundant, superfluous, unnecessary.

**needy** *adj* badly off, destitute, *inf* hard up, impecunious, impoverished, indigent, necessitous, penurious, poverty-stricken, underpaid. ▷ POOR.

**negate** *vb* annul, cancel out, deny, gainsay, invalidate, nullify, oppose.

**negative** *adj* adversarial, antagonistic, *inf* anti, contradictory, destructive, disagreeing, dissenting, grudging, nullifying, obstructive, opposing, pessimistic, uncooperative, unenthusiastic, unresponsive, unwilling.
• *n* denial, no, refusal, rejection, veto. *Opp* POSITIVE.

**neglect** *n* carelessness, dereliction of duty, disregard, inadvertence, inattention, indifference, negligence, oversight, slackness.
• *vb* abandon, be remiss about, disregard, forget, ignore, leave alone, let slide, lose sight of, miss, omit, overlook, pay no attention to, shirk, skip. **neglected** ▷ DERELICT.

**negligent** *adj* careless, forgetful, heedless, inattentive, inconsiderate, indifferent, irresponsible, lax, offhand, reckless, remiss, slack, sloppy, slovenly, thoughtless, uncaring, unthinking. *Opp* CAREFUL.

**negligible** *adj* imperceptible, inconsequential, inconsiderable, insignificant, minor, nugatory, paltry, petty, slight, small, tiny, trifling, trivial, unimportant. *Opp* CONSIDERABLE.

**negotiate** *vb* arbitrate, bargain, come to terms, confer, deal, discuss terms, haggle, intercede, make arrangements, mediate, parley, transact.

**negotiation** *n* arbitration, bargaining, conciliation, debate, diplomacy, discussion, mediation, parleying, transaction.

**negotiator** *n* agent, ambassador, arbitrator, broker, conciliator, diplomat, go-between, intercessor, intermediary, mediator, middleman.

**neighbourhood** *n* area, community, district, environs, locality, place, purlieus, quarter, region, surroundings, vicinity, zone.

**neighbouring** *adj* adjacent, adjoining, attached, bordering, close, closest, connecting, contiguous, near, nearby, nearest, next-door, surrounding.

**neighbourly** *adj* civil, considerate, friendly, helpful, kind, sociable, thoughtful, well-disposed.

**nerve** *n* coolness, determination, firmness, fortitude, resolution, resolve, will-power. ▷ COURAGE.

**nervous** *adj* afraid, agitated, anxious, apprehensive, disturbed, edgy, excitable, fearful, fidgety, flustered, fretful, highly-strung, ill-at-ease, *inf* in a tizzy, insecure, *inf* jittery, *inf* jumpy, *inf* nervy, neurotic, on edge, *inf* on tenterhooks, *inf* rattled, restive, restless, ruffled, shaky, shy, strained, tense, timid, *inf* touchy, *inf* twitchy, uneasy, unnerved, unsettled, *inf* uptight, worried. ▷ FRIGHTENED. *Opp* CALM.

**nestle** *vb* cuddle, curl up, huddle, nuzzle, snuggle.

**net** *n* lace, lattice-work, mesh, netting, network, web. • *vb* **1** catch, capture, enmesh, ensnare, trammel, trap. **2** *net £200 a week.* accumulate, bring in, clear, earn, get, make, realize, receive, *inf* take home.

**network** *n* **1** *inf* criss-cross, grid, labyrinth, lattice, maze, mesh, net, netting, tangle, tracery, web. **2** complex, organization, system.

**neurosis** *n* abnormality, anxiety, depression, mental condition, obsession, phobia.

**neurotic** *adj* anxious, distraught, disturbed, irrational, maladjusted, nervous, obsessive, overwrought, unbalanced, unstable.

**neuter** *adj* ambiguous, ambivalent, asexual, indeterminate, uncertain. • *vb* castrate, *inf* doctor, emasculate, geld, spay, sterilize.

**neutral** *adj* **1** detached, disinterested, dispassionate, fair, impartial, indifferent, non-aligned, non-belligerent, non-partisan, objective, unaffiliated, unaligned, unbiased, uncommitted, uninvolved, unprejudiced. *Opp* BIASED. **2** *neutral colours.* characterless, colourless, dull, drab, indefinite, indeterminate, intermediate, neither one thing nor the other, pale, vague. *Opp* DISTINCTIVE.

**neutralize** *vb* annul, cancel out, compensate for, counteract, counterbalance, invalidate, make ineffective, make up for, negate, nullify, offset, wipe out.

**new** *adj* **1** brand-new, clean, different, fresh, mint, strange, unfamiliar, unheard of, untried, unused. **2** *new ideas.* advanced, contemporary, current, different, fashionable, latest, modern, modernistic, newfangled, novel, original, recent, revolutionary, *inf* trendy, up-to-date. **3** *new data.* added, additional, changed, extra, further, just arrived, supplementary, unexpected, unknown. *Opp* OLD.

**newcomer** *n* alien, arrival, immigrant, new boy, new girl, outsider, settler, stranger.

**news** *n* account, advice, announcement, bulletin, communication, communiqué, dispatch, headlines, information, intelligence, *inf* the latest, message, newscast, newsflash, newsletter, notice, press-release, proclamation, report, rumour, statement, *old use* tidings, word.

**newspaper** *n inf* daily, gazette, journal, paper, periodical, *inf* rag, tabloid.

**next** *adj* **1** adjacent, adjoining, closest, nearest, neighbouring, next-door. **2** *the next moment.* following, soonest, subsequent, succeeding.

**nice** *adj* **1** accurate, careful, delicate, discriminating, exact, fine, hair-splitting, meticulous, precise, punctilious, scrupulous, subtle. **2** *nice manners.* dainty, elegant, fastidious, fussy, particular, *inf* pernickety, polished, refined, well-mannered. **3** [*inf:* In this sense, nice refers to anything which you like. The range of synonyms is almost limitless: we give only a selection here.] acceptable, agreeable, amiable, attractive, beautiful, delicious, delightful, friendly, good, gratifying, kind, likeable, pleasant, satisfactory, welcome. *Opp* NASTY.

**niche** *n* alcove, corner, hollow, nook, recess.

**nickname** *n* alias, sobriquet.

**niggardly** *adj* mean, miserly, parsimonious, stingy.

**nimble** *adj* acrobatic, active, adroit, agile, brisk, deft, dextrous, limber, lithe, lively, *inf* nippy, quick-moving, sprightly, spry, swift. *Opp* CLUMSY.

**nip** *vb* bite, clip, pinch, snag, snap at, squeeze.

**nobility** *n* **1** dignity, glory, grandeur, greatness, high-mindedness, integrity, magnanimity, morality, nobleness, uprightness, virtue, worthiness. **2** *the nobility.* aristocracy, élite, gentry, nobles, peerage, the ruling classes, *inf* the upper crust.

**noble** *adj* 1 aristocratic, *inf* blue-blooded, courtly, distinguished, élite, gentle, high-born, high-ranking, patrician, princely, royal, thoroughbred, titled, upper-class. 2 *noble deeds*. brave, chivalrous, courageous, gallant, glorious, heroic. 3 *noble thoughts*. elevated, high-flown, honourable, lofty, magnanimous, moral, upright, virtuous, worthy. 4 *noble music*. dignified, elegant, grand, great, imposing, impressive, magnificent, majestic, splendid, stately. *Opp* BASE, COMMON. • *n* aristocrat, gentleman, gentlewoman, grandee, lady, lord, nobleman, noblewoman, patrician, peer, peeress.

**nod** *vb* bend, bob, bow. **nod off** ▷ SLEEP.

**noise** *n inf* babel, bawling, bedlam, blare, cacophony, *inf* caterwauling, clamour, clangour, clatter, commotion, din, discord, *inf* fracas, hubbub, *inf* hullabaloo, outcry, pandemonium, *inf* racket, row, *inf* rumpus, screaming, screeching, shouting, shrieking, tumult, uproar, yelling. ▷ SOUND. *Opp* SILENCE.

**noiseless** *adj* inaudible, mute, muted, quiet, silent, soft, soundless, still. *Opp* NOISY.

**noisy** *adj* blaring, boisterous, booming, cacophonous, chattering, clamorous, deafening, discordant, dissonant, ear-splitting, fortissimo, harsh, loud, raucous, resounding, reverberating, rowdy, screaming, screeching, shrieking, shrill, strident, talkative, thunderous, tumultuous, unmusical, uproarious, vociferous. *Opp* NOISELESS.

**nomadic** *adj* itinerant, peripatetic, roving, travelling, vagrant, wandering, wayfaring.

**nominal** *adj* 1 formal, in name only, ostensible, self-styled, *inf* so-called, supposed, theoretical, titular. 2 *a nominal sum*. insignificant, minimal, minor, small, token.

**nominate** *vb* appoint, choose, designate, elect, name, propose, put forward, *inf* put up, recommend, select, specify.

**non-existent** *adj* chimerical, fictional, fictitious, hypothetical, imaginary, imagined, legendary, made-up, mythical, unreal. *Opp* REAL.

**nonplus** *vb* amaze, baffle, confound, disconcert, dumbfound, flummox, perplex, puzzle, render speechless.

**nonsense** *n* 1 [Most synonyms *inf*] balderdash, bilge, boloney, bosh, bunk, bunkum, claptrap, codswallop, double-Dutch, drivel, eyewash, fiddlesticks, foolishness, gibberish, gobbledegook, mumbo jumbo, piffle, poppycock, rot, rubbish, silliness, stuff and nonsense, stupidity, tommy-rot, trash, tripe, twaddle. 2 *The plan was a nonsense*. absurdity, inanity, mistake, nonsensical idea.

**nonsensical** *adj* absurd, asinine, crazy, *inf* daft, fatuous, foolish, idiotic, illogical, impractical, inane, incomprehensible, irrational, laughable, ludicrous, mad, meaningless, preposterous, ridiculous, senseless, stupid, unreasonable. ▷ SILLY. *Opp* SENSIBLE.

**non-stop** *adj* ceaseless, constant, continual, continuous, endless, eternal, incessant, interminable, perpetual, persistent, *inf* round-the-clock, steady, unbroken, unending, uninterrupted, unremitting.

**norm** *n* criterion, measure, model, pattern, rule, standard, type, yardstick.

**normal** *adj* **1** accepted, accustomed, average, common, commonplace, conventional, customary, established, everyday, familiar, general, habitual, natural, ordinary, orthodox, predictable, prosaic, quotidian, regular, routine, *inf* run-of-the-mill, standard, typical, universal, unsurprising, usual. **2** *a normal person.* balanced, healthy, rational, reasonable, sane, stable, *inf* straight, well-adjusted. *Opp* ABNORMAL.

**normalize** *vb* legalize, regularize, regulate, return to normal.

**nose** *n* **1** nostrils, proboscis, snout. **2** *nose of a boat.* bow, front, prow. • *vb* enter cautiously, insinuate yourself, intrude, nudge your way, penetrate, probe, push, shove. **nose about** ▷ PRY.

**nostalgia** *n* longing, memory, pining, regret, reminiscence, sentiment, sentimentality, yearning.

**nostalgic** *adj* emotional, maudlin, regretful, romantic, sentimental, wistful, yearning.

**nosy** *adj* curious, eavesdropping, inquisitive, interfering, meddlesome, prying.

**notable** *adj* celebrated, conspicuous, distinctive, distinguished, eminent, evident, extraordinary, famous, illustrious, important, impressive, memorable, noted, noteworthy, noticeable, obvious, outstanding, pre-eminent, prominent, rare, remarkable, renowned, singular, striking, uncommon, unforgettable, unusual, well-known. *Opp* ORDINARY.

**note** *n* **1** billet-doux, chit, communication, correspondence, epistle, jotting, letter, *inf* memo, memorandum, message, postcard. **2** annotation, comment, cross-reference, explanation, footnote, gloss, jotting, marginal note. **3** *a note in your voice.* feeling, quality, sound, tone. **4** *a £5 note.* banknote, bill, currency, draft. • *vb* **1** enter, jot down, record, scribble, write down. **2** *note mentally.* detect, discern, discover, feel, find, heed, mark, mind, notice, observe, pay attention to, remark, register, see, spy, take note of. **noted** ▷ FAMOUS.

**noteworthy** *adj* exceptional, extraordinary, rare, remarkable, uncommon, unique, unusual. *Opp* ORDINARY.

**nothing** *n cricket* duck, *tennis* love, *old use* naught, *football* nil, nought, zero, *sl* zilch.

**notice** *n* **1** advertisement, announcement, handbill, handout, intimation, leaflet, message, note, notification, placard, poster, sign, warning. **2** attention, awareness, cognizance, consciousness, heed, note, regard. • *vb* be aware, detect, discern, discover, feel, find, heed, make out, mark, mind, note, observe, pay attention to, perceive, register, remark, see, spy, take note. **give notice** ▷ NOTIFY, WARN.

**noticeable** *adj* appreciable, audible, clear, clear-cut, considerable, conspicuous, detectable, discernible, distinct, distinguishable, manifest, marked, measurable, notable, observable, obtrusive, obvious, overt, palpable, perceivable, perceptible, plain, prominent, pronounced, salient, significant, striking, unconcealed, unmistakable, visible. *Opp* IMPERCEPTIBLE.

**notify** *vb* acquaint, advise, alert, announce, apprise, give notice, inform, make known, proclaim, publish, report, tell, warn.

**notion** *n* apprehension, belief, concept, conception, fancy, hypothesis, idea, impression, *inf* inkling, opinion, sentiment, theory, thought, understanding, view.

**notorious** *adj* disgraceful, disreputable, flagrant, ill-famed, infamous, outrageous, overt, patent, scandalous, shocking, talked about, undisguised, undisputed, well-known. ▷ FAMOUS, WICKED.

**nourish** *vb* feed, maintain, nurse, nurture, provide for, strengthen, support, sustain. **nourishing** ▷ NUTRITIOUS.

**nourishment** *n* diet, food, goodness, nutrient, nutriment, nutrition, sustenance, *old use* victuals.

**novel** *adj* different, fresh, imaginative, innovative, new, odd, original, rare, singular, startling, strange, surprising, uncommon, unconventional, unfamiliar, untested, unusual. *Opp* FAMILIAR.
• *n* best-seller, *inf* blockbuster, fiction, novelette, novella, romance, story. ▷ WRITING.

**novelty** *n* **1** freshness, newness, oddity, originality, strangeness, surprise, unfamiliarity, uniqueness. **2** bauble, curiosity, gimmick, knick-knack, ornament, souvenir, trifle, trinket.

**novice** *n* amateur, apprentice, beginner, *inf* greenhorn, inexperienced person, initiate, learner, probationer, tiro, trainee.

**now** *adv* at once, at present, here and now, immediately, instantly, just now, nowadays, promptly, *inf* right now, straight away, today.

**noxious** *adj* corrosive, foul, harmful, nasty, noisome, objectionable, poisonous, polluting, sulphureous, sulphurous, unwholesome.

**nub** *n* centre, cord, crux, essence, gist, heart, kernel, nucleus, pith, point.

**nucleus** *n* centre, core, heart, kernel, middle.

**nude** *adj* bare, disrobed, exposed, in the nude, naked, stark-naked, stripped, unclothed, uncovered, undressed.

**nudge** *vb* bump, dig, elbow, hit, jab, jog, jolt, poke, prod, push, shove, touch.

**nuisance** *n* annoyance, bother, burden, inconvenience, irritant, irritation, *inf* pain, pest, plague, trouble, vexation, worry.

**nullify** *vb* abolish, annul, cancel, do away with, invalidate, negate, neutralize, quash, repeal, rescind, revoke, stultify.

**numb** *adj* anaesthetized, *inf* asleep, benumbed, cold, dead, deadened, frozen, immobile, insensible, insensitive, paralysed, senseless, suffering from pins and needles. *Opp* SENSITIVE.
• *vb* anaesthetize, benumb, deaden, desensitize, drug, dull, freeze, immobilize, make numb, paralyse, stun, stupefy.

**number** *n* **1** digit, figure, integer, numeral, unit. **2** aggregate, amount, *inf* bunch, collection, crowd, multitude, quantity, sum, total. ▷ GROUP. **3** *musical number*. item, piece, song. **4** *a number of a magazine*. copy, edition, impression, issue, printing, publication.
• *vb* add up to, total, work out at. ▷ COUNT.

**numerous** *adj* abundant, copious, countless, endless, incalculable, infinite, innumerable, many, multitudinous, myriad, numberless, plentiful, several, uncountable, untold. *Opp* FEW.

**nun** *n* abbess, mother-superior, novice, prioress, sister.

**nurse** *n* **1** district-nurse, *old use* matron, sister. **2** nanny, nursemaid. • *vb* **1** care for, look after, minister to, nurture, tend, treat. **2** breast-feed, feed, suckle, wet-nurse. **3** cherish, coddle, cradle, cuddle, dandle, hold, hug, mother, pamper.

**nursery** *n* **1** crèche, kindergarten, nursery school. **2** garden centre, market garden.

**nurture** *vb* bring up, cultivate, educate, feed, look after, nourish, nurse, rear, tend, train.

**nut** *n* kernel. □ *almond, brazil, cashew, chestnut, cob-nut, coconut, filbert, hazel, peanut, pecan, pistachio, walnut.*

**nutrient** *n* fertilizer, goodness, nourishment.

**nutriment** *n* food, goodness, nourishment, nutrition, sustenance.

**nutritious** *adj* alimentary, beneficial, good for you, health-giving, healthy, nourishing, sustaining, wholesome.

# O

**oasis** *n* **1** spring, watering-hole, well. **2** asylum, haven, refuge, resort, retreat, safe harbour, sanctuary.

**oath** *n* **1** assurance, avowal, guarantee, pledge, promise, undertaking, vow, word of honour. **2** blasphemy, curse, exclamation, expletive, *inf* four-letter word, imprecation, malediction, obscenity, profanity, swearword.

**obedient** *adj* acquiescent, amenable, biddable, compliant, conformable, deferential, disciplined, docile, duteous, dutiful, law-abiding, manageable, submissive, subservient, tamed, tractable, well-behaved, well-trained. *Opp* DISOBEDIENT.

**obese** *adj* corpulent, gross, overweight. ▷ FAT.

**obey** *vb* abide by, accept, acquiesce in, act in accordance with, adhere to, agree to, be obedient to, be ruled by, bow to, carry out, comply with, conform to, defer to, do what you are told, execute, follow, fulfil, give in to, heed, honour, implement, keep to, mind, observe, perform, *inf* stick to, submit to, take orders from. *Opp* DISOBEY.

**object** *n* **1** article, body, entity, item, thing. **2** aim, end, goal, intent, intention, objective, point, purpose, reason. **3** *object of ridicule.* butt, destination, target. • *vb* argue, be opposed, carp, cavil, complain, demur, disapprove, dispute, dissent, expostulate, *sl* grouse, grumble, make an objection, *inf* mind, *inf* moan, oppose, protest, quibble, raise objections, raise questions, remonstrate, take a stand, take exception. *Opp* ACCEPT, AGREE.

**objection** *n* argument, cavil, challenge, complaint, demur, demurral, disapproval, exception, opposition, outcry, protest, query, question, quibble, refusal, remonstration.

**objectionable** *adj* abhorrent, detestable, disagreeable, disgusting, dislikeable, displeasing, distasteful, foul, hateful, insufferable, intolerable, loathsome, nasty, nauseating, noisome, obnoxious, odious, offensive, *inf* off-putting, repellent, repugnant, repulsive, revolting, sickening, unacceptable, undesirable, unwanted. ▷ UNPLEASANT. *Opp* ACCEPTABLE.

**objective** *adj* 1 detached, disinterested, dispassionate, factual, impartial, impersonal, neutral, open-minded, outward-looking, rational, scientific, unbiased, uncoloured, unemotional, unprejudiced. 2 *objective evidence*. empirical, existing, observable, real. *Opp* SUBJECTIVE. • *n* aim, ambition, aspiration, design, destination, end, goal, hope, intent, intention, object, point, purpose, target.

**obligation** *n* commitment, compulsion, constraint, contract, duty, liability, need, requirement, responsibility. ▷ PROMISE. *Opp* OPTION.

**obligatory** *adj* binding, compulsory, essential, mandatory, necessary, required, requisite, unavoidable. *Opp* OPTIONAL.

**oblige** *vb* 1 coerce, compel, constrain, force, make, require. 2 *Please oblige me*. accommodate, gratify, indulge, please. **obliged** ▷ BOUND, GRATEFUL. **obliging** ▷ HELPFUL, POLITE.

**oblique** *adj* 1 angled, askew, aslant, canted, declining, diagonal, inclined, leaning, listing, raked, rising, skewed, slanted, slanting, slantwise, sloping, tilted. 2 *an oblique insult*. backhanded, circuitous, circumlocutory, devious, implicit, implied, indirect, roundabout. ▷ EVASIVE. *Opp* DIRECT.

**obliterate** *vb* blot out, cancel, cover over, delete, destroy, efface, eliminate, eradicate, erase, expunge, extirpate, leave no trace of, rub out, wipe out.

**oblivion** *n* 1 anonymity, darkness, disregard, extinction, limbo, neglect, obscurity. 2 amnesia, coma, forgetfulness, ignorance, insensibility, obliviousness, unawareness, unconsciousness.

**oblivious** *adj* forgetful, heedless, ignorant, insensible, insensitive, unacquainted, unaware, unconscious, unfeeling, uninformed, unmindful, unresponsive. *Opp* AWARE.

**obscene** *adj* abominable, bawdy, *inf* blue, coarse, corrupting, crude, debauched, degenerate, depraved, dirty, disgusting, distasteful, filthy, foul, foul-mouthed, gross, immodest, immoral, improper, impure, indecent, indecorous, indelicate, *inf* kinky, lecherous, lewd, loathsome, nasty, *inf* off-colour, offensive, outrageous, perverted, pornographic, prurient, repulsive, ribald, risqué, rude, salacious, scatological, scurrilous, shameful, shameless, shocking, *inf* sick, smutty, suggestive, unchaste, vile, vulgar. ▷ OBJECTIONABLE, SEXY. *Opp* DECENT.

**obscenity** *n* abomination, blasphemy, coarseness, dirtiness, evil, filth, foulness, grossness, immorality, impropriety, indecency, lewdness, licentiousness, offensiveness, outrage, perversion, pornography, profanity, scurrility, vileness. ▷ SWEARWORD.

**obscure** *adj* 1 blurred, clouded, concealed, covered, dark, dim, faint, foggy, hazy, hidden, inconspicuous, indefinite, indistinct, masked, misty, murky, nebulous, secret, shadowy, shady, shrouded, unclear, unlit, unrecognizable, vague, veiled. *Opp* CLEAR. 2 *an obscure joke*. arcane, baffling, complex, cryptic, delphic, enigmatic, esoteric, incomprehensible, mystifying, perplexing, puzzling, recherché, recondite, strange. *Opp* OBVIOUS. **3** *an obscure poet.* forgotten, minor, undistinguished, unfamiliar, unheard of, unimportant, unknown, unnoticed. *Opp* FAMOUS. • *vb* block out, blur, cloak, cloud, conceal, cover, darken, disguise, eclipse, envelop, hide, make obscure, mask, obfuscate, overshadow, screen, shade, shroud, veil. *Opp* CLARIFY.

**obsequious** *adj* abject, *inf* bootlicking, crawling, cringing, deferential, effusive, fawning, flattering, fulsome, *inf* greasy, grovelling, ingratiating, insincere, mealy-mouthed, menial, *inf* oily, servile, *inf* smarmy, submissive, subservient, sycophantic, unctuous. **be obsequious** ▷ GROVEL.

**observant** *adj* alert, astute, attentive, aware, careful, eagle-eyed, heedful, mindful, on the lookout, *inf* on the qui vive, perceptive, percipient, quick, sharp-eyed, shrewd, vigilant, watchful, with eyes peeled. *Opp* INATTENTIVE.

**observation** *n* **1** attention (to), examination, inspection, monitoring, scrutiny, study, surveillance, viewing, watching. **2** comment, note, opinion, reaction, reflection, remark, response, sentiment, statement, thought, utterance.

**observe** *vb* **1** consider, contemplate, detect, discern, examine, *inf* keep an eye on, look at, monitor, note, notice, perceive, regard, scrutinize, see, spot, spy, stare at, study, view, watch, witness. **2** *observe rules.* abide by, adhere to, comply with, conform to, follow, heed, honour, keep, obey, pay attention to, respect. **3** *observe Easter.* celebrate, commemorate, keep, mark, recognize, remember, solemnize. **4** *observed that it was badly acted.* animadvert (on), comment, declare, explain, make an observation, mention, reflect, remark, say, state.

**observer** *n* beholder, bystander, commentator, eyewitness, looker-on, onlooker, spectator, viewer, watcher, witness.

**obsess** *vb* become an obsession with, bedevil, consume, control, dominate, grip, haunt, monopolize, plague, possess, preoccupy, rule, take hold of.

**obsession** *n* addiction, *inf* bee in your bonnet, conviction, fetish, fixation, *inf* hang-up, *inf* hobby-horse, *Fr* idée fixe, infatuation, mania, passion, phobia, preoccupation, *sl* thing.

**obsessive** *adj* addictive, compulsive, consuming, controlling, dominating, haunting, passionate.

**obsolescent** *adj* ageing, aging, declining, dying out, fading, going out of use, losing popularity, moribund, *inf* on the way out, waning.

**obsolete** *adj* anachronistic, antiquated, antique, archaic, dated, dead, discarded, disused, extinct, old-fashioned, *inf* old hat, outdated, out-of-date, outmoded, passé, primitive, superannuated, superseded, unfashionable. ▷ OLD. *Opp* CURRENT.

**obstacle** *n* bar, barricade, barrier, block, blockage, catch, check, difficulty, hindrance, hurdle, impediment, obstruction, problem, restriction, snag, *inf* stumbling-block.

**obstinate** *adj* adamant, *sl* bloody-minded, defiant, determined, dogged, firm, headstrong, immovable, inflexible, intractable, intransigent, *inf* mulish, obdurate, persistent, pertinacious, perverse, *inf* pig-headed, refractory, resolute, rigid, self-willed, single-minded, *inf* stiff-necked, stubborn, tenacious, uncooperative, unreasonable, unyielding, wilful, wrong-headed. *Opp* AMENABLE.

**obstreperous** *adj* awkward, boisterous, disorderly, irrepressible, naughty, rough, rowdy, *inf* stroppy, turbulent, uncontrollable, undisciplined, unmanageable, unruly, vociferous, wild. ▷ NOISY. *Opp* WELL-BEHAVED.

**obstruct** *vb* arrest, bar, block, bring to a standstill, check, curb, delay, deter, frustrate, halt, hamper, hinder, hold up, impede, inhibit, interfere with, interrupt, occlude, prevent, restrict, retard, slow down, stand in the way of, *inf* stonewall, stop, *inf* stymie, thwart. *Opp* HELP.

**obtain** *vb* **1** acquire, attain, be given, bring, buy, capture, come by, come into possession of, earn, elicit, enlist (*help*), extort, extract, find, gain, get, get hold of, *inf* lay your hands on, *inf* pick up, procure, purchase, receive, secure, seize, take possession of, win. **2** *rules still obtain*. apply, be in force, be in use, be relevant, be valid, exist, prevail, stand.

**obtrusive** *adj* blatant, conspicuous, forward, importunate, inescapable, interfering, intrusive, meddling, meddlesome, noticeable, out of place, prominent, unwanted, unwelcome. ▷ OBVIOUS. *Opp* INCONSPICUOUS.

**obtuse** *adj* dense, dull, imperceptive, slow, slow-witted. ▷ STUPID. *Opp* CLEVER.

**obviate** *vb* avert, forestall, make unnecessary, preclude, prevent, remove, take away.

**obvious** *adj* apparent, bald, blatant, clear, clear-cut, conspicuous, distinct, evident, eye-catching, flagrant, glaring, gross, inescapable, intrusive, manifest, notable, noticeable, obtrusive, open, overt, palpable, patent, perceptible, plain, prominent, pronounced, recognizable, self-evident, self-explanatory, straightforward, unconcealed, undisguised, undisputed, unmistakable, visible. *Opp* HIDDEN, OBSCURE.

**occasion** *n* **1** chance, circumstance, moment, occurrence, opportunity, time. **2** *no occasion for rudeness*. call, cause, excuse, grounds, justification, need, reason. **3** *a happy occasion*. affair, celebration, ceremony, event, function, *inf* get-together, happening, incident, occurrence, party.

**occasional** *adj* casual, desultory, fitful, infrequent, intermittent, irregular, odd, *inf* once in a while, periodic, random, rare, scattered, spasmodic, sporadic, uncommon, unpredictable. *Opp* FREQUENT, REGULAR.

**occult** *adj* ▷ SUPERNATURAL. • *n* black arts, black magic, cabbalism, diabolism, occultism, sorcery, the supernatural, witchcraft.

**occupant** *n* denizen, householder, incumbent, inhabitant, lessee, lodger, occupier, owner, resident, tenant.

**occupation** *n* **1** incumbency, lease, occupancy, possession, residency, tenancy, tenure, use. **2** appropriation, colonization, conquest, invasion, oppression, seizure, subjection, subjugation, suzerainty, *inf* takeover, usurpation. **3** appointment, business, calling, career, employment, job, *inf* line, métier, position, post, profession, situation, trade, vocation, work. **4** *leisure occupation*. activity, diversion, entertainment, hobby, interest, pastime, pursuit, recreation.

**occupy** *vb* **1** dwell in, inhabit, live in, move into, reside in, take up residence in, tenant. **2** *occupy space.* fill, take up, use, utilize. **3** capture, colonize, conquer, garrison, invade, overrun, possess, subjugate, take over, take possession of. **4** *occupy your time.* absorb, busy, divert, engage, engross, involve, preoccupy. **occupied** ▷ BUSY.

**occur** *vb* appear, arise, befall, be found, chance, come about, come into being, *old use* come to pass, *inf* crop up, develop, exist, happen, manifest itself, materialize, *inf* show up, take place, *inf* transpire, *inf* turn out, *inf* turn up.

**occurrence** *n* affair, case, circumstance, development, event, happening, incident, manifestation, matter, occasion, phenomenon, proceeding.

**odd** *adj* **1** *odd numbers.* uneven. *Opp* EVEN. **2** *an odd sock.* extra, left over, *inf* one-off, remaining, single, spare, superfluous, surplus, unmatched, unused. **3** *odd jobs.* casual, irregular, miscellaneous, occasional, part-time, random, sundry, varied, various. **4** *odd behaviour.* abnormal, anomalous, atypical, bizarre, *inf* cranky, curious, deviant, different, eccentric, exceptional, extraordinary, freak, funny, idiosyncratic, incongruous, inexplicable, *inf* kinky, outlandish, out of the ordinary, peculiar, puzzling, queer, rare, singular, strange, uncharacteristic, uncommon, unconventional, unexpected, unusual, weird. *Opp* NORMAL.

**oddments** *plur n* bits, bits and pieces, fragments, *inf* junk, leftovers, litter, odds and ends, offcuts, remnants, scraps, shreds, unwanted pieces.

**odious** *adj* detestable, execrable, loathsome, offensive, repugnant, repulsive. ▷ HATEFUL.

**odorous** *adj* fragrant, odoriferous, perfumed, scented. ▷ SMELLING.

**odour** *n* aroma, bouquet, fragrance, nose, redolence, scent, smell, stench, *inf* stink.

**odourless** *adj* deodorized, unscented. *Opp* ODOROUS.

**offence** *n* **1** breach, crime, fault, felony, infringement, lapse, malefaction, misdeed, misdemeanour, outrage, peccadillo, sin, transgression, trespass, violation, wrong, wrongdoing. **2** anger, annoyance, disgust, displeasure, hard feelings, indignation, irritation, pique, resentment, *inf* upset. **give offence** ▷ OFFEND.

**offend** *vb* **1** affront, anger, annoy, cause offence, chagrin, disgust, displease, embarrass, give offence, hurt your feelings, insult, irritate, make angry, *inf* miff, outrage, pain, provoke, *inf* put your back up, revolt, rile, sicken, slight, snub, upset, vex. **2** *offend against the law.* do wrong, transgress, violate. **be offended** be annoyed, *inf* take umbrage.

**offender** *n* criminal, culprit, delinquent, evil-doer, guilty party, lawbreaker, malefactor, miscreant, outlaw, sinner, transgressor, wrongdoer.

**offensive** *adj* **1** abusive, annoying, antisocial, coarse, detestable, disagreeable, disgusting, displeasing, disrespectful, embarrassing, foul, impolite, improper, indecent, insulting, loathsome, nasty, nauseating, nauseous, noxious, objectionable, obnoxious, *inf* off-putting, repugnant, revolting, rude, sickening, unpleasant, unsavoury, vile, vulgar. ▷ OBSCENE. *Opp* PLEASANT.

2 *offensive action*. aggressive, antagonistic, attacking, belligerent, hostile, threatening, warlike. *Opp* PEACEABLE. • *n* ▷ ATTACK.

**offer** *n* bid, proposal, proposition, suggestion, tender. • *vb* 1 bid, extend, give the opportunity of, hold out, make an offer of, make available, proffer, put forward, put up, suggest. 2 *offer to help*. come forward, propose, *inf* show willing, volunteer.

**offering** *n* contribution, donation, gift, oblation, offertory, present, sacrifice.

**offhand** *adj* 1 abrupt, aloof, careless, cavalier, cool, curt, offhanded, perfunctory, unceremonious, uncooperative, uninterested. ▷ CASUAL. 2 ▷ IMPROMPTU.

**office** *n* 1 bureau, room, workplace, workroom. 2 appointment, assignment, commission, duty, function, job, occupation, place, position,post, responsibility, role, situation, work.

**officer** *n* 1 adjutant, aide-de-camp, CO, commandant, commanding officer. 2 *police officer*. constable, PC, policeman, policewoman, WPC. 3 ▷ OFFICIAL.

**official** *adj* accredited, approved, authentic, authoritative, authorized, bona fide, certified, formal, lawful, legal, legitimate, licensed, organized, proper, recognized, true, trustworthy, valid. ▷ FORMAL. • *n* administrator, agent, appointee, authorized person, bureaucrat, dignitary, executive, functionary, mandarin, officer, organizer, representative, responsible person. □ *bailiff, captain, chief, clerk of court, commander, commissioner, consul, customs officer, director, elder (*of church*), equerry, governor, manager, marshal, mayor, mayoress, minister, monitor, ombudsman, overseer, prefect, president, principal, proctor, proprietor, registrar, sheriff, steward, superintendent, supervisor, usher.*

**officiate** *vb* adjudicate, be in charge, be responsible, chair (*a meeting*), conduct, have authority, manage, preside, referee, *inf* run (*a meeting*), umpire.

**officious** *adj inf* bossy, bumptious, *inf* cocky, dictatorial, forward, impertinent, interfering, meddlesome, meddling, overzealous, *inf* pushy, self-appointed, self-important.

**offset** *vb* cancel out, compensate for, counteract, counterbalance, make amends for, make good, make up for, redress. ▷ BALANCE.

**offshoot** *n* branch, by-product, derivative, development, *inf* spin-off, subsidiary product.

**offspring** *n* [*sing*] baby, child, descendant, heir, successor. [*plur*] brood, family, fry, issue, litter, progeny, *old use* seed, spawn, young.

**often** *adv* again and again, *inf* all the time, commonly, constantly, continually, frequently, generally, habitually, many times, regularly, repeatedly, time after time, time and again, usually.

**oil** *vb* grease, lubricate.

**oily** *adj* 1 buttery, fat, fatty, greasy, oleaginous. 2 *an oily manner*. ▷ OBSEQUIOUS.

**ointment** *n* balm, cream, embrocation, emollient, liniment, lotion, paste, salve, unguent.

**old** *adj* 1 ancient, antediluvian, antiquated, antique, crumbling, decayed, decaying, decrepit, dilapidated, early, historic, medieval, obsolete, primitive, quaint, ruined, superannuated, time-worn, venerable, veteran, vintage.

▷ OLD-FASHIONED. **2** *old times.* bygone, classical, forgotten, former, (*time*) immemorial, *old use* olden, past, prehistoric, previous, primeval, primitive, primordial, remote. **3** *old people.* advanced in years, aged, *inf* doddery, elderly, geriatric, *inf* getting on, grey-haired, hoary, *inf* in your dotage, long-lived, oldish, *inf* past it, senile. **4** *old customs.* age-old, enduring, established, lasting, long-standing, time-honoured, traditional, well-established. **5** *old clothes.* moth-eaten, ragged, scruffy, shabby, threadbare, worn, worn-out. **6** *old bread.* dry, stale. **7** *old tickets.* cancelled, expired, invalid, used. **8** *an old hand.* experienced, expert, familiar, mature, practised, skilled, veteran. *Opp* NEW, YOUNG. **old age** *inf* declining years, decrepitude, *inf* dotage, senility. **old person** centenarian, *inf* fogey, *inf* fogy, nonagenarian, octogenarian, pensioner, septuagenarian.

**old-fashioned** *adj* anachronistic, antiquated, archaic, backward-looking, conventional, dated, fusty, hackneyed, narrow-minded, obsolete, old, *inf* old hat, outdated, out-of-date, out-of-touch, outmoded, passé, pedantic, prim, proper, prudish, reactionary, time-honoured, traditional, unfashionable. *Opp* MODERN. **old-fashioned person** *inf* fogey, *inf* fogy, *inf* fuddy-duddy, pedant, reactionary, *inf* square.

**omen** *n* augury, auspice, foreboding, forewarning, harbinger, indication, portent, premonition, presage, prognostication, sign, token, warning, *inf* writing on the wall.

**ominous** *adj* baleful, dire, fateful, forbidding, foreboding, grim, ill-omened, ill-starred, inauspicious, lowering, menacing, portentous, prophetic, sinister, threatening, unfavourable, unlucky, unpromising, unpropitious, warning. *Opp* AUSPICIOUS.

**omission** *n* **1** deletion, elimination, exception, excision, exclusion. **2** failure, gap, neglect, negligence, oversight, shortcoming.

**omit** *vb* **1** cross out, cut, dispense with, drop, edit out, eliminate, erase, except, exclude, ignore, jump, leave out, miss out, overlook, pass over, reject, skip, strike out. **2** fail, forget, neglect.

**omnipotent** *adj* all-powerful, almighty, invincible, supreme, unconquerable.

**oncoming** *adj* advancing, approaching, facing, looming, nearing.

**onerous** *adj* burdensome, demanding, heavy, laborious, taxing. ▷ DIFFICULT.

**one-sided** *adj* **1** biased, bigoted, partial, partisan, prejudiced. **2** *one-sided game.* ill-matched, unbalanced, unequal, uneven.

**onlooker** *n* bystander, eye-witness, looker-on, observer, spectator, watcher, witness.

**only** *adj* lone, one, single, sole, solitary, unique. • *adv* barely, exclusively, just, merely, simply, solely.

**ooze** *vb* bleed, discharge, emit, exude, leak, secrete, seep, weep.

**opaque** *adj* cloudy, dark, dim, dull, filmy, hazy, impenetrable, muddy, murky, obscure, turbid, unclear. *Opp* CLEAR.

**open** *adj* **1** agape, ajar, gaping, unbolted, unfastened, unlocked, unsealed, unwrapped, wide, wide-open, yawning. **2** accessible, available, exposed, free, public, revealed, unenclosed, unprotected, unrestricted. **3** *open space*. bare, broad, clear, empty, extensive, spacious, treeless, uncrowded, undefended, unfenced, unobstructed, vacant. **4** *open arms*. extended, outstretched, spread out, unfolded. **5** *open nature*. artless, candid, communicative, flexible, frank, generous, guileless, honest, innocent, magnanimous, open-minded, responsive, sincere, straightforward, transparent, uninhibited. **6** *open defiance*. apparent, barefaced, blatant, conspicuous, downright, evident, flagrant, obvious, outspoken, overt, plain, unconcealed, undisguised, visible. **7** *an open question*. arguable, debatable, moot, problematical, unanswered, undecided, unresolved, unsettled. *Opp* CLOSED, HIDDEN. • *vb* **1** unbar, unblock, unbolt, unclose, uncork, undo, unfasten, unfold, unfurl, unlatch, unlock, unroll, unseal, untie, unwrap. **2** become open, gape, yawn. **3** *open proceedings*. activate, begin, commence, establish, *inf* get going, inaugurate, initiate, *inf* kick off, launch, set in motion, set up, start. *Opp* CLOSE.

**opening** *adj* first, inaugural, initial, introductory. *Opp* FINAL. • *n* **1** aperture, breach, break, chink, cleft, crack, crevice, cut, door, doorway, fissure, gap, gash, gate, gateway, hatch, hole, leak, mouth, orifice, outlet, rent, rift, slit, slot, space, split, tear, vent. **2** beginning, birth, commencement, dawn, inauguration, inception, initiation, launch, outset, start. **3** *a business opening*. *inf* break, chance, opportunity, way in.

**operate** *vb* **1** act, function, go, perform, run, work. **2** *operate a machine*. control, deal with, drive, handle, manage, use, work. **3** *operate on a patient*. do an operation, perform surgery.

**operation** *n* **1** control, direction, function, functioning, management, operating, performance, running, working. **2** action, activity, business, campaign, effort, enterprise, exercise, manoeuvre, movement, procedure, proceeding, process, project, transaction, undertaking, venture. **3** [*medical*] biopsy, surgery, transplant.

**operational** *adj* functioning, going, in operation, in use, in working order, operating, operative, running, *inf* up and running, usable, working.

**operative** *adj* **1** ▷ OPERATIONAL. **2** *the operative word*. crucial, important, key, principal, relevant, significant. • *n* ▷ WORKER.

**opinion** *n* assessment, attitude, belief, comment, conclusion, conjecture, conviction, estimate, feeling, guess, idea, impression, judgement, notion, perception, point of view, sentiment, theory, thought, view, viewpoint, way of thinking.

**opponent** *n* adversary, antagonist, challenger, competitor, contender, contestant, enemy, foe, opposer, opposition, rival. *Opp* ALLY.

**opportune** *adj* advantageous, appropriate, auspicious, beneficial, convenient, favourable, felicitous, fortunate, good, happy, lucky, propitious, right, suitable, timely, well-timed. *Opp* INCONVENIENT.

**opportunity** *n inf* break, chance, moment, occasion, opening, possibility, time.

**oppose** *vb* argue with, attack, be at variance with, be opposed to, challenge, combat, compete against, confront, contend with, contest, contradict, controvert, counter, counterattack, defy, disagree with, disapprove of, dissent from, face, fight, object to, obstruct, *inf* pit your wits against, quarrel with, resist, rival, stand up to, *inf* take a stand against, take issue with, withstand. *Opp* SUPPORT. **opposed** ▷ HOSTILE, OPPOSITE.

**opposite** *adj* **1** antithetical, conflicting, contradictory, contrasting, converse, different, hostile, incompatible, inconsistent, opposed, opposing, rival. **2** contrary, reverse. **3** *your opposite* number. corresponding, equivalent, facing, matching, similar. • *n* antithesis, contrary, converse, reverse.

**opposition** *n* antagonism, antipathy, competition, defiance, disapproval, enmity, hostility, objection, resistance, scepticism, unfriendliness. ▷ OPPONENT. *Opp* SUPPORT.

**oppress** *vb* abuse, afflict, burden, crush, depress, encumber, enslave, exploit, grind down, harass, intimidate, keep under, maltreat, overburden, persecute, pressurize, *inf* ride roughshod over, subdue, subjugate, terrorize, *inf* trample on, tyrannize, weigh down.

**oppressed** *adj* browbeaten, downtrodden, enslaved, exploited, misused, persecuted, subjugated, tyrannized.

**oppression** *n* abuse, despotism, enslavement, exploitation, harassment, injustice, maltreatment, persecution, pressure, subjection, subjugation, suppression, tyranny.

**oppressive** *adj* **1** brutal, cruel, despotic, harsh, repressive, tyrannical, undemocratic, unjust. **2** airless, close, heavy, hot, humid, muggy, stifling, stuffy, suffocating, sultry.

**optimism** *n* buoyancy, cheerfulness, confidence, hope, idealism, positiveness. *Opp* PESSIMISM.

**optimistic** *adj* buoyant, cheerful, confident, expectant, hopeful, idealistic, *inf* looking on the bright side, positive, sanguine. *Opp* PESSIMISTIC.

**optimum** *adj* best, finest, first-class, first-rate, highest, ideal, maximum, most favourable, perfect, prime, superlative, top.

**option** *n* alternative, chance, choice, election, possibility, selection.

**optional** *adj* avoidable, discretionary, dispensable, elective, inessential, possible, unnecessary, unforced, voluntary. *Opp* COMPULSORY.

**oral** *adj* by mouth, said, spoken, unwritten, uttered, verbal, vocal, voiced.

**oratory** *n* declamation, eloquence, enunciation, fluency, *inf* gift of the gab, grandiloquence, magniloquence, rhetoric, speaking, speech making.

**orbit** *n* circuit, course, path, revolution, trajectory. • *vb* circle, encircle, go round, travel round.

**orbital** *adj* circular, encircling.

**orchestrate** *vb* **1** arrange, compose. **2** ▷ ORGANIZE.

**ordeal** *n* affliction, anguish, difficulty, distress, hardship, misery, *inf* nightmare, pain, suffering, test, torture, trial, tribulation, trouble.

**order** *n* **1** arrangement, array, classification, codification, disposition, lay-out, *inf* line-up, neatness, organization, pattern, progression, sequence, series, succession, system, tidiness. **2** calm, control, discipline, good behaviour, government, harmony, law and order, obedience, orderliness, peace, peacefulness, quiet, rule. **3** *social orders.* caste, category, class, degree, group, hierarchy, kind, level, rank, sort, status. **4** *in good order.* condition, repair, state. **5** *orders to be obeyed.* command, decree, direction, directive, edict, fiat, injunction, instruction, law, mandate, ordinance, regulation, requirement, rule. **6** *an order for goods.* application, booking, commission, demand, mandate, request, requisition, reservation. **7** *religious orders.* association, brotherhood, community, fraternity, group, guild, lodge, sect, sisterhood, society, sodality, sorority. *Opp* DISORDER. • *vb* **1** arrange, categorize, classify, codify, lay out, organize, put in order, sort out, tidy up. **2** *old use* bid, charge, command, compel, decree, demand, direct, enjoin, instruct, ordain, require, tell. **3** apply for, ask for, book, reserve, requisition, send away for.

**orderly** *adj* **1** careful, methodical, neat, organized, regular, symmetrical, systematic, tidy, well-arranged, well-organized, well-prepared. *Opp* CONFUSED, DISORGANIZED. **2** civilized, controlled, decorous, disciplined, law-abiding, peaceable, polite, restrained, well-behaved, well-mannered. *Opp* UNDISCIPLINED.

**ordinary** *adj* accustomed, average, common, *inf* common or garden, commonplace, conventional, customary, established, everyday, fair, familiar, habitual, humble, *inf* humdrum, indifferent, mediocre, medium, middling, moderate, modest, mundane, nondescript, normal, orthodox, passable, pedestrian, plain, prosaic, quotidian, reasonable, regular, routine, *inf* run-of-the-mill, satisfactory, simple, *inf* so-so, standard, stock, traditional, typical, undistinguished, unexceptional, unexciting, unimpressive, uninspired, uninteresting, unpretentious, unremarkable, unsurprising, usual, well-known, workaday. *Opp* EXTRAORDINARY.

**organic** *adj* **1** animate, biological, growing, live, living, natural. **2** *an organic whole.* coherent, coordinated, evolving, integral, integrated, methodical, organized, structured, systematic.

**organism** *n* animal, being, cell, creature, living thing, plant.

**organization** *n* **1** arrangement, categorization, classification, codification, composition, coordination, logistics, organizing, planning, regimentation, *inf* running, structuring. **2** *a business organization.* alliance, association, body, business, club, combine, company, concern, confederation, conglomerate, consortium, corporation, federation, firm, group, institute, institution, league, network, *inf* outfit, party, society, syndicate, union.

**organize** *vb* **1** arrange, catalogue, categorize, classify, codify, compose, coordinate, group, order, *inf* pigeon-hole, put in order, rearrange, regiment, *inf* run, sort, sort out, structure, systematize, tabulate, tidy up. *Opp* JUMBLE. **2** build, coordinate, create, deal with, establish, make arrangements for, manage, mobilize, orchestrate, plan, put together, run, *inf* see to, set up, take care of. **organized** ▷ OFFICIAL, SYSTEMATIC.

**orgy** *n* Bacchanalia, *inf* binge, debauch, *inf* fling, party, *inf* rave-up, revel, revelry, Saturnalia, *inf* spree.

**orient** *vb* acclimatize, accommodate, accustom, adapt, adjust, condition, familiarize, orientate, position.

**oriental** *adj* Asiatic, eastern, far-eastern.

**origin** *n* **1** base, basis, beginning, birth, cause, commencement, cradle, creation, dawn, derivation, foundation, fount, fountainhead, genesis, inauguration, inception, launch, outset, provenance, root, source, start, well-spring. *Opp* END. **2** *humble origins.* ancestry, background, descent, extraction, family, genealogy, heritage, lineage, parentage, pedigree, start in life, stock.

**original** *adj* **1** aboriginal, archetypal, earliest, first, initial, native, primal, primitive, primordial. **2** *original antiques.* actual, authentic, genuine, real, true, unique. **3** *original ideas.* creative, first-hand, fresh, imaginative, ingenious, innovative, inspired, inventive, new, novel, resourceful, thoughtful, unconventional, unfamiliar, unique, unusual. *Opp* HACKNEYED.

**originate** *vb* **1** arise, be born, be derived, be descended, begin, come, commence, crop up, derive, emanate, emerge, issue, proceed, spring up, start, stem. **2** beget, be the inventor of, bring about, coin, conceive, create, design, discover, engender, found, give birth to, inaugurate, initiate, inspire, institute, introduce, invent, launch, mastermind, pioneer, produce, think up.

**ornament** *n* accessory, adornment, bauble, beautification, decoration, embellishment, embroidery, enhancement, filigree, frill, frippery, garnish, gewgaw, ornamentation, tracery, trimming, trinket. ▷ JEWELLERY. • *vb* adorn, beautify, deck, decorate, dress up, elaborate, embellish, emblazon, emboss, embroider, enhance, festoon, garnish, prettify, trim.

**ornamental** *adj* attractive, decorative, fancy, flashy, pretty, showy.

**ornate** *adj* arabesque, baroque, *inf* busy, decorated, elaborate, fancy, florid, flamboyant, flowery, fussy, luxuriant, ornamented, overdone, pretentious, rococo. *Opp* PLAIN.

**orphan** *n* foundling, stray, waif.

**orthodox** *adj* accepted, accustomed, approved, authorized, common, conformist, conservative, conventional, customary, established, mainstream, normal, official, ordinary, prevailing, recognized, regular, standard, traditional, usual, well-established. *Opp* UNCONVENTIONAL.

**ostensible** *adj* alleged, apparent, outward, pretended, professed, *inf* put-on, reputed, specious, supposed, visible. *Opp* REAL.

**ostentation** *n* affectation, display, exhibitionism, flamboyance, *inf* flashiness, flaunting, parade, pretention, pretentiousness, self-advertisement, show, showing-off, *inf* swank. *Opp* MODESTY.

**ostentatious** *adj* flamboyant, *inf* flashy, pretentious, showy, *inf* swanky, vainglorious. ▷ BOASTFUL. *Opp* MODEST.

**ostracize** *vb* avoid, banish, *inf* black, blackball, blacklist, boycott, cast out, cold-shoulder, *inf* cut, *inf* cut dead, exclude, excommunicate, expel, isolate, reject, *inf* send to Coventry, shun, shut out, snub. *Opp* BEFRIEND.

**oust** *vb* banish, drive out, eject, expel, *inf* kick out, remove, replace, *inf* sack, supplant, take over from, unseat.

**outbreak** *n* epidemic, *inf* flare-up, plague, rash, upsurge.

**outburst** *n* attack, effusion, eruption, explosion, fit, flood, outbreak, outpouring, paroxysm, rush, spasm, surge, upsurge.

**outcast** *n* castaway, displaced person, exile, leper, outlaw, outsider, pariah, refugee, reject, untouchable.

**outcome** *n* conclusion, consequence, effect, end-product, result, sequel, upshot.

**outcry** *n* cry of disapproval, dissent, hue and cry, objection, opposition, protest, protestation, remonstrance.

**outdo** *vb* beat, defeat, exceed, excel, *inf* get the better of, outbid, outdistance, outrun, outshine, outstrip, outweigh, overcome, surpass, top, trump.

**outdoor** *adj* alfresco, open-air, out of doors, outside.

**outer** *adj* **1** exterior, external, outside, outward, superficial, surface. **2** distant, further, outlying, peripheral, remote. *Opp* INNER.

**outfit** *n* **1** accoutrements, attire, costume, ensemble, equipment, garb, *inf* gear, *inf* get-up, *inf* rig, suit, trappings, *inf* turn-out. **2** ▷ ORGANIZATION.

**outgoing** *adj* **1** *outgoing president.* departing, emeritus, ex-, former, last, leaving, past, retiring. **2** *outgoing tide.* ebbing, falling, retreating. **3** ▷ SOCIABLE. *Opp* INCOMING. **outgoings** ▷ EXPENSE.

**outing** *n* excursion, expedition, jaunt, picnic, ride, tour, trip.

**outlast** *vb* outlive, survive.

**outlaw** *n* bandit, brigand, criminal, deserter, desperado, fugitive, highwayman, marauder, outcast, renegade, robber. • *vb* ban, exclude, forbid, prohibit, proscribe. ▷ BANISH.

**outlet** *n* **1** channel, discharge, duct, egress, escape route, exit, mouth, opening, orifice, safety valve, vent, way out. **2** ▷ SHOP.

**outline** *n* **1** abstract, *inf* bare bones, diagram, digest, draft, framework, plan, précis, résumé, *inf* rough idea, *inf* rundown, scenario, skeleton, sketch, summary, synopsis, thumbnail sketch. **2** contour, figure, form, profile, shadow, shape, silhouette. • *vb* delineate, draft, give the outline, give the gist of, plan out, précis, rough out, sketch out, summarize.

**outlook** *n* **1** aspect, panorama, scene, sight, vantage point, view, vista. **2** *your mental outlook.* angle, attitude, frame of mind, opinion, perspective, point of view, position, slant, standpoint, viewpoint. **3** *the weather outlook.* expectations, forecast, *inf* look-out, prediction, prognosis, prospect.

**outlying** *adj* distant, far-away, far-flung, far-off, outer, outermost, remote. *Opp* CENTRAL.

**output** *n* achievement, crop, harvest, production, productivity, result, yield.

**outrage** *n* **1** atrocity, crime, *inf* disgrace, enormity, indignity, outrageous act, scandal, *inf* sensation, violation. **2** anger, bitterness, disgust, fury, horror, indignation, resentment, revulsion, shock, wrath. • *vb* ▷ ANGER.

**outrageous** *adj* 1 abominable, atrocious, barbaric, beastly, bestial, criminal, cruel, disgraceful, disgusting, execrable, infamous, iniquitous, monstrous, nefarious, notorious, offensive, preposterous, revolting, scandalous, shocking, unspeakable, unthinkable, vile, villainous, wicked. 2 *outrageous prices.* excessive, extortionate, extravagant, immoderate, unreasonable. *Opp* REASONABLE.

**outside** *adj* 1 exterior, external, facing, outer, outward, superficial, surface, visible. 2 *outside interference.* alien, extraneous, foreign. 3 *outside chance.* ▷ REMOTE.
• *n* appearance, case, casing, exterior, façade, face, front, look, shell, skin, surface.

**outsider** *n* alien, foreigner, *inf* gatecrasher, guest, immigrant, interloper, intruder, invader, newcomer, non-member, non-resident, outcast, stranger, trespasser, visitor.

**outskirts** *plur n* borders, edge, environs, fringe, margin, outer areas, periphery, purlieus, suburbs. *Opp* CENTRE.

**outspoken** *adj* blunt, candid, direct, explicit, forthright, frank, plain-spoken, tactless, unambiguous, undiplomatic, unequivocal, unreserved. ▷ HONEST. *Opp* EVASIVE.

**outstanding** *adj* 1 above the rest, celebrated, conspicuous, distinguished, dominant, eminent, excellent, exceptional, extraordinary, first-class, first-rate, great, important, impressive, memorable, notable, noteworthy, noticeable, predominant, pre-eminent, prominent, remarkable, singular, special, striking, superior, top rank, unrivalled. ▷ FAMOUS. *Opp* ORDINARY. 2 ▷ OVERDUE.

**outward** *adj* apparent, evident, exterior, external, manifest, noticeable, observable, obvious, ostensible, outer, outside, superficial, surface, visible.

**outwit** *vb* deceive, dupe, fool, *inf* get the better of, gull, hoax, hoodwink, make a fool of, outfox, outmanoeuvre, *inf* outsmart, *inf* put one over on, *inf* take in, trick. ▷ CHEAT.

**oval** *adj* egg-shaped, ellipsoidal, elliptical, oviform, ovoid.

**ovation** *n* acclaim, acclamation, applause, cheering, plaudits, praise.

**overcast** *adj* black, clouded, cloudy, dark, dismal, dull, gloomy, grey, leaden, lowering, murky, sombre, starless, stormy, sunless, threatening. *Opp* CLOUDLESS.

**overcoat** *n* greatcoat, mackintosh, top-coat, trench-coat.

**overcome** *adj* at a loss, beaten, *inf* bowled over, *inf* done in, exhausted, overwhelmed, prostrate, speechless.
• *vb* ▷ OVERTHROW.

**overcrowded** *adj* congested, crammed, crawling, filled to capacity, full, jammed, *inf* jampacked, overloaded, packed.

**overdue** *adj* 1 belated, delayed, late, slow, tardy, unpunctual. *Opp* EARLY. 2 *overdue bills.* due, outstanding, owing, unpaid, unresolved, unsettled.

**overeat** *vb* be greedy, eat too much, feast, gorge, gormandize, *inf* guzzle, indulge yourself, *inf* make a pig of yourself, overindulge, *inf* stuff yourself.

**overflow** *vb* brim over, flood, pour over, run over, spill, well up.

**overgrown** *adj* 1 outsize, oversized. ▷ BIG. 2 *overgrown garden.* overrun, rank, tangled, uncut, unkempt, untidy, untrimmed, unweeded, weedy, wild.

**overhang** *vb* beetle, bulge, jut, project, protrude, stick out.

**overhaul** *vb* 1 check over, examine, *inf* fix, inspect, mend, rebuild, recondition, refurbish, renovate, repair, restore, service. 2 ▷ OVERTAKE.

**overhead** *adj* aerial, elevated, high, overhanging, raised, upper.

**overlook** *vb* 1 fail to notice, forget, leave out, miss, neglect, omit. 2 condone, disregard, excuse, gloss over, ignore, let pass, make allowances for, pardon, pass over, pay no attention to, *inf* shut your eyes to, *inf* turn a blind eye to, *inf* write off. 3 *overlook a lake.* face, front, have a view of, look at, look down on, look on to.

**overpower** *vb* ▷ OVERTHROW.

**overpowering** *adj* compelling, consuming, inescapable, insupportable, irrepressible, irresistible, overriding, overwhelming, powerful, strong, unbearable, uncontrollable, unendurable.

**oversee** *vb* administer, be in charge of, control, direct, invigilate, *inf* keep an eye on, preside over, superintend, supervise, watch over.

**oversight** *n* 1 carelessness, dereliction of duty, error, failure, fault, mistake, omission. 2 *oversight of a job.* administration, control, direction, management, supervision, surveillance.

**overstate** *vb inf* blow up out of proportion, embroider, exaggerate, magnify, make too much of, maximize, overemphasize, overstress.

**overt** *adj* apparent, blatant, clear, evident, manifest, obvious, open, patent, plain, unconcealed, undisguised, visible. *Opp* SECRET.

**overtake** *vb* catch up with, gain on, leave behind, outdistance, outpace, outstrip, overhaul, pass.

**overthrow** *n* conquest, defeat, destruction, mastery, rout, subjugation, suppression, unseating. • *vb* beat, bring down, conquer, crush, deal with, defeat, depose, dethrone, get the better of, *inf* lick, master, oust, overcome, overpower, overturn, overwhelm, rout, *inf* send packing, subdue, *inf* topple, triumph over, unseat, vanquish, win against.

**overtone** *n* association, connotation, hint, implication, innuendo, reverberation, suggestion, undertone.

**overturn** *vb* 1 capsize, flip, invert, keel over, knock over, spill, tip over, topple, turn over, *inf* turn turtle, turn upside down, up-end, upset. 2 ▷ OVERTHROW.

**overwhelm** *vb* 1 engulf, flood, immerse, inundate, submerge, swamp. 2 ▷ OVERTHROW. **overwhelming** ▷ OVERPOWERING.

**owe** *vb* be in debt, have debts.

**owing** *adj* due, outstanding, overdue, owed, payable, unpaid, unsettled. **owing to** because of, caused by, on account of, resulting from, thanks to, through.

**own** *vb* be the owner of, have, hold, possess. **own up** ▷ CONFESS.

**owner** *n* freeholder, holder, landlady, landlord, possessor, proprietor.

# P

**pace** *n* **1** step, stride. **2** *a fast pace.* gait, *inf* lick, movement, quickness, rate, speed, tempo, velocity. • *vb* ▷ WALK.

**pacify** *vb* appease, assuage, calm, conciliate, humour, mollify, placate, quell, quieten, soothe, subdue, tame, tranquillize. *Opp* ANGER.

**pack** *n* **1** bale, box, bundle, package, packet, parcel. **2** backpack, duffel bag, haversack, kitbag, knapsack, rucksack. **3** ▷ GROUP. • *vb* **1** bundle (up), fill, load, package, parcel up, put, put together, store, stow, wrap up. **2** compress, cram, crowd, jam, overcrowd, press, ram, squeeze, stuff, tamp down, wedge. **pack off** ▷ DISMISS. **pack up** ▷ FINISH.

**pact** *n* agreement, alliance, armistice, arrangement, bargain, compact, concord, concordat, contract, covenant, deal, *Fr* entente, league, peace, settlement, treaty, truce, understanding.

**pad** *n* **1** cushion, filler, hassock, kneeler, padding, pillow, stuffing, wad. **2** jotter, memo pad, notebook, stationery, writing pad. • *vb* cushion, fill, line, pack, protect, stuff, upholster, wad. **pad out** ▷ EXTEND.

**padding** *n* **1** filling, protection, upholstery, stuffing, wadding. **2** prolixity, verbiage, verbosity, *inf* waffle, wordiness.

**paddle** *n* oar, scull. • *vb* **1** propel, row, scull. **2** dabble, splash about, wade.

**paddock** *n* enclosure, field, meadow, pasture.

**pagan** *adj* atheistic, godless, heathen, idolatrous, infidel, irreligious, polytheistic, unchristian. • *n* atheist, heathen, infidel, savage, unbeliever.

**page** *n* **1** folio, leaf, recto, sheet, side, verso. **2** errand-boy, messenger, page-boy.

**pageant** *n* ceremony, display, extravaganza, parade, procession, show, spectacle, tableau.

**pageantry** *n* ceremony, display, formality, grandeur, magnificence, pomp, ritual, show, spectacle, splendour.

**pain** *n* ache, aching, affliction, agony, anguish, cramp, crick, discomfort, distress, headache, hurt, irritation, ordeal, pang, smart, smarting, soreness, spasm, stab, sting, suffering, tenderness, throb, throes, toothache, torment, torture, twinge. • *vb* ▷ HURT.

**painful** *adj* **1** aching, *inf* achy, agonizing, burning, cruel, excruciating, *old use* grievous, hard to bear, hurting, inflamed, piercing, raw, severe, sharp, smarting, sore, *inf* splitting (*head*), stabbing, stinging, tender, throbbing. **2** distressing, harrowing, hurtful, laborious, *inf* traumatic, trying, unpleasant, upsetting, vexing. **3** *a painful decision.* difficult, hard, troublesome, uncongenial. *Opp* PAINLESS. **be painful** ▷ HURT.

**painkiller** *n* anaesthetic, analgesic, anodyne, palliative, sedative.

**painless** *adj* comfortable, easy, effortless, pain-free, simple, trouble-free, undemanding. *Opp* PAINFUL.

**paint** *n* colour, colouring, dye, pigment, stain, tint. □ *distemper, emulsion, enamel, gloss paint, lacquer, matt paint, oil-colour, oil-paint, oils, pastel, primer, tempera, under-*

coat, varnish, water-colour, whitewash. • *vb* **1** apply paint to, coat, colour, cover, daub, decorate, dye, enamel, gild, lacquer, redecorate, stain, tint, touch up, varnish, whitewash. **2** delineate, depict, describe, picture, portray, represent.

**painter** *n* artist, decorator, illustrator, miniaturist.

**painting** *n* fresco, landscape, miniature, mural, oil-painting, portrait, still-life, water-colour.

**pair** *n* brace, couple, duet, duo, mates, partners, partnership, set of two, twins, twosome. • *vb* **pair off, pair up** couple, double up, find a partner, get together, join up, *inf* make a twosome, match up, *inf* pal up, team up.

**palace** *n* castle, château, mansion, official residence, stately home. ▷ HOUSE.

**palatable** *adj* acceptable, agreeable, appetizing, easy to take, eatable, edible, nice to eat, pleasant, tasty. *Opp* UNPALATABLE.

**palatial** *adj* aristocratic, grand, large-scale, luxurious, majestic, opulent, *inf* posh, splendid, stately, up-market.

**pale** *adj* **1** anaemic, ashen, blanched, bloodless, cadaverous, colourless, corpse-like, *inf* deathly, drained, etiolated, ghastly, ghostly, ill-looking, pallid, pasty, *inf* peaky, sallow, sickly, unhealthy, wan, *inf* washed-out, *inf* whey-faced, white, whitish. **2** *pale colours*. bleached, dim, faded, faint, light, pastel, subtle, weak. *Opp* BRIGHT. • *vb* become pale, blanch, blench, dim, etiolate, fade, lighten, lose colour, whiten.

**pall** *n* cloth, mantle, shroud, veil. ▷ COVERING. • *vb* become boring, become uninteresting, cloy, irritate, jade, sate, satiate, weary.

**palliative** *adj* alleviating, calming, reassuring, sedative, soothing. • *n* painkiller, sedative, tranquillizer.

**palpable** *adj* apparent, corporeal, evident, manifest, obvious, patent, physical, real, solid, substantial, tangible, touchable, visible. *Opp* INTANGIBLE.

**palpitate** *vb* beat, flutter, pound, pulsate, quiver, shiver, throb, tremble, vibrate.

**paltry** *adj* contemptible, inconsequential, insignificant, petty, *inf* piddling, pitiable, puny, trifling, unimportant, worthless. ▷ SMALL. *Opp* IMPORTANT.

**pamper** *vb* coddle, cosset, humour, indulge, mollycoddle, overindulge, pet, spoil, spoonfeed.

**pamphlet** *n* booklet, brochure, bulletin, catalogue, circular, flyer, folder, handbill, handout, leaflet, notice, tract.

**pan** *n* container, utensil. □ *billycan, casserole, frying-pan, pot, saucepan, skillet*. • *vb* ▷ CRITICIZE.

**panache** *n* animation, brio, confidence, dash, élan, energy, enthusiasm, flair, flamboyance, flourish, savoir-faire, self-assurance, spirit, style, swagger, verve, zest.

**pandemonium** *n* babel, bedlam, chaos, confusion, hubbub, noise, rumpus, turmoil, uproar. ▷ COMMOTION.

**pander** *n* go-between, *inf* pimp, procurer. • *vb* **pander to** bow to, cater for, fulfil, gratify, humour, indulge, please, provide, satisfy.

**pane** *n* glass, light, panel, sheet of glass, window.

**panel** *n* **1** insert, pane, panelling, rectangle, *plur* wainscot. **2** committee, group, jury, team.

**panic** *n* alarm, consternation, *inf* flap, horror, hysteria, stampede, terror. • *vb* become panic-stricken, *inf* fall apart, *inf* flap, *inf* go to pieces, *inf* lose your head, *inf* lose your nerve, overreact, stampede. ▷ FEAR.

**panic-stricken** *adj* alarmed, *inf* beside yourself, disorientated, frantic, frenzied, horrified, hysterical, *inf* in a cold sweat, *inf* in a tizzy, jumpy, overexcited, panicky, panic-struck, terror-stricken, undisciplined, unnerved, worked-up. ▷ FRIGHTENED. *Opp* CALM.

**panorama** *n* landscape, perspective, prospect, scene, view, vista.

**panoramic** *adj* commanding, extensive, scenic, sweeping, wide.

**pant** *vb* blow, breathe quickly, gasp, *inf* huff and puff, puff, wheeze. **panting** ▷ BREATHLESS.

**pants** *n* **1** *old use* bloomers, boxer shorts, briefs, camiknickers, drawers, knickers, panties, pantihose, shorts, *inf* smalls, trunks, underpants, *inf* undies, Y-fronts. **2** ▷ TROUSERS.

**paper** *n* **1** folio, leaf, sheet. □ *A4 (A1, A2, etc), card, cardboard, cartridge paper, foolscap, manila, notepaper, papyrus, parchment, postcard, quarto, stationery, tissue-paper, toilet paper, tracing-paper, vellum, wallpaper, wrapping-paper, writing-paper.* **2** certificate, credentials, deed, document, form, *inf* ID, identification, licence, record. **3** *the daily paper.* *inf* daily, journal, newspaper, *inf* rag, tabloid. **4** *an academic paper.* article, discourse, dissertation, essay, monograph, thesis, treatise.

**parable** *n* allegory, exemplum, fable, moral tale. ▷ WRITING.

**parade** *n* cavalcade, ceremony, column, cortge, display, file, march-past, motorcade, pageant, procession, review, show, spectacle. • *vb* **1** assemble, file past, form up, line up, make a procession, march past, present yourself, process. ▷ WALK. **2** ▷ DISPLAY.

**paradise** *n* Eden, Elysium, heaven, Shangri-La, Utopia.

**paradox** *n* absurdity, anomaly, contradiction, incongruity, inconsistency, self-contradiction.

**paradoxical** *adj* absurd, anomalous, conflicting, contradictory, illogical, improbable, incongruous, self-contradictory.

**parallel** *adj* **1** equidistant. **2** *parallel events.* analogous, cognate, contemporary, corresponding, equivalent, like, matching, similar. • *n* **1** analogue, counterpart, equal, likeness, match. **2** analogy, comparison, correspondence, equivalence, kinship, resemblance, similarity. • *vb* be parallel to, be parallel with, compare with, correspond to, duplicate, echo, equate with, keep pace with, match, remind you of, run alongside.

**paralyse** *vb* anaesthetize, cripple, deactivate, deaden, desensitize, disable, freeze, halt, immobilize, incapacitate, lame, numb, petrify, stop, stun.

**paralysed** *adj* crippled, dead, desensitized, disabled, handicapped, immobile, immovable, incapacitated, lame, numb, palsied, paralytic, paraplegic, rigid, unusable, useless.

**paralysis** *n* deadness, immobility, numbness, palsy, paraplegia.

**paraphernalia** *plur n* accessories, apparatus, baggage, belongings, chattels, *inf* clobber, effects, equipment, gear, impedimenta, materials, *inf* odds and ends, possessions, property, *inf* rig, stuff, tackle, things, trappings.

**paraphrase** *vb* explain, interpret, put into other words, rephrase, restate, reword, rewrite, translate.

**parcel** *n* bale, box, bundle, carton, case, pack, package, packet. **parcel out** ▷ DIVIDE. **parcel up** ▷ PACK.

**parch** *vb* bake, burn, dehydrate, desiccate, dry, scorch, shrivel, wither. **parched** ▷ DRY, THIRSTY.

**pardon** *n* absolution, amnesty, condonation, discharge, exculpation, exoneration, forgiveness, indulgence, mercy, release, reprieve. • *vb* absolve, condone, exculpate, excuse, exonerate, forgive, free, grant pardon, let off, overlook, release, remit, reprieve, set free, spare.

**pardonable** *adj* allowable, condonable, excusable, forgivable, justifiable, minor, negligible, petty, understandable, venial (*sin*). *Opp* UNFORGIVABLE.

**parent** *n* begetter, father, guardian, mother, procreator, progenitor.

**parentage** *n* ancestry, birth, descent, extraction, family, line, lineage, pedigree, stock.

**park** *n* common, gardens, green, recreation ground. □ *amusement park, arboretum, botanical gardens, car-park, estate, national park, nature reserve, parkland, reserve, theme park*. • *vb* deposit, leave, place, position, put, station, store. **park yourself** ▷ SETTLE.

**parliament** *n* assembly, conclave, congress, convocation, council, diet, government, legislature, lower house, senate, upper house.

**parody** *n* burlesque, caricature, distortion, imitation, lampoon, mimicry, satire, *inf* send-up, *inf* spoof, *inf* take-off, travesty. • *vb* ape, burlesque, caricature, guy, imitate, lampoon, mimic, satirize, *inf* send up, *inf* take off, travesty. ▷ RIDICULE.

**parry** *vb* avert, block, deflect, evade, fend off, push away, repel, repulse, stave off, ward off.

**part** *n* **1** bit, branch, component, constituent, department, division, element, fraction, fragment, ingredient, parcel, particle, percentage, piece, portion, ramification, scrap, section, sector, segment, shard, share, single item, subdivision, unit. **2** department, faction, party, section, subdivision, unit. **3** *part of a book*. chapter, episode. **4** *part of a town*. area, district, neighbourhood, quarter, region, sector, vicinity. **5** *part of the body*. limb, member, organ. **6** *part in a play*. cameo, character, role. • *vb* **1** cut off, detach, disconnect, divide, pull apart, separate, sever, split, sunder. *Opp* JOIN. **2** break away, depart, go away, leave, part company, quit, say goodbye, separate, split up, take leave, withdraw. *Opp* MEET. **part with** ▷ RELINQUISH. **take part** ▷ PARTICIPATE.

**partial** *adj* **1** imperfect, incomplete, limited, qualified, unfinished. *Opp* COMPLETE. **2** *partial judge*. biased, one-sided, partisan, prejudiced, unfair. *Opp* IMPARTIAL. **be partial to** ▷ LIKE.

**participate** *vb* assist, be active, be involved, contribute, cooperate, engage, enter, help, join in, partake, share, take part.

**participation** *n* activity, assistance, complicity, contribution, cooperation, engagement, involvement, partnership, sharing.

**particle** *n* 1 bit, crumb, dot, drop, fragment, grain, hint, iota, jot, mite, morsel, *old use* mote, piece, scintilla, scrap, shred, sliver, *inf* smidgen, speck, trace. 2 atom, electron, molecule, neutron.

**particular** *adj* 1 distint, idiosyncratic, individual, peculiar, personal, singular, specific, uncommon, unique, unmistakable. 2 *particular with detail.* exact, nice, painstaking, precise, rigorous, scrupulous, thorough. 3 *gave particular pleasure.* especial, exceptional, important, marked, notable, noteworthy, outstanding, significant, special, unusual. 4 *particular about food.* choosy, critical, discriminating, fastidious, finical, finicky, fussy, meticulous, nice, *inf* pernickety, selective. *Opp* GENERAL, EASYGOING. **particulars** circumstances, details, facts, information, *sl* low-down.

**parting** *n* departure, farewell, going away, leave-taking, leaving, saying goodbye, separation, splitting up, valediction.

**partisan** *adj* biased, bigoted, blinkered, devoted, factional, fanatical, narrow-minded, one-sided, partial, prejudiced, sectarian, unfair. *Opp* IMPARTIAL. • *n* adherent, devotee, fanatic, follower, freedom fighter, guerrilla, resistance fighter, supporter, underground fighter, zealot.

**partition** *n* 1 break-up, division, separation, splitting up. 2 barrier, panel, room-divider, screen, wall. • *vb* cut up, divide, parcel out, separate off, share out, split up, subdivide.

**partner** *n* 1 accessory, accomplice, ally, assistant, associate, *inf* bedfellow, collaborator, colleague, companion, comrade, confederate, helper, *inf* mate, *sl* sidekick. 2 consort, husband, mate, spouse, wife.

**partnership** *n* 1 affiliation, alliance, association, combination, company, confederation, cooperative, syndicate. 2 collaboration, complicity, cooperation. 3 marriage, relationship, union.

**party** *n* 1 celebration, *inf* do, festivity, function, gathering, *inf* get-together, *inf* jollification, *inf* knees-up, merrymaking, *inf* rave-up, *inf* shindig, social gathering. □ *ball, banquet, barbecue, ceilidh, dance,* inf *disco, discothque, feast,* inf *hen-party, house-warming, orgy, picnic, reception, reunion,* inf *stag-party, tea-party, wedding.* 2 *a political party.* alliance, association, bloc, cabal, *inf* camp, caucus, clique, coalition, faction, junta, league, sect, side. ▷ GROUP.

**pass** *n* 1 canyon, col, cut, defile, gap, gorge, gully, opening, passage, ravine, valley, way through. 2 *identity pass.* authority, authorization, clearance, *inf* ID, licence, passport, permission, permit, safe-conduct, ticket, warrant. • *vb* 1 go beyond, go by, move on, move past, outstrip, overhaul, overtake, proceed, progress, *inf* thread your way. 2 *time passes.* disappear, elapse, fade, go away, lapse, tick by, vanish. 3 *pass drinks.* circulate, deal out, deliver, give, hand over, offer, present, share, submit, supply, transfer. 4 *pass a resolution.* agree, approve, authorize, confirm, decree, enact, establish, ordain, pronounce, ratify, validate. 5 *I pass! inf* give in, opt out, say nothing, waive your rights. **pass away** ▷ DIE. **pass on**

▷ TRANSFER. **pass out** ▷ FAINT. **pass over** ▷ IGNORE.

**passable** *adj* **1** acceptable, adequate, admissible, allowable, all right, fair, indifferent, mediocre, middling, moderate, not bad, ordinary, satisfactory, *inf* so-so, tolerable. *Opp* UNACCEPTABLE. **2** clear, navigable, open, traversable, unblocked, unobstructed, usable. *Opp* IMPASSABLE.

**passage** *n* **1** corridor, entrance, hall, hallway, lobby, passageway, vestibule. **2** *passage of time.* advance, flow, lapse, march, movement, moving on, passing, progress, progression, transition. **3** *sea passage.* crossing, cruise, voyage. ▷ JOURNEY. **4** *through passage.* pass, route, thoroughfare, tunnel, way through. ▷ OPENING. **5** *passage from a book.* citation, episode, excerpt, extract, paragraph, part, piece, portion, quotation, scene, section, selection.

**passenger** *n* commuter, rider, traveller, voyager.

**passer-by** *n* bystander, onlooker, witness.

**passion** *n* appetite, ardour, avidity, avidness, commitment, craving, craze, desire, drive, eagerness, emotion, enthusiasm, fanaticism, fervency, fervour, fire, flame, frenzy, greed, heat, hunger, infatuation, intensity, keenness, love, lust, mania, obsession, strong feeling, suffering, thirst, urge, urgency, vehemence, zeal, zest.

**passionate** *adj* ardent, aroused, avid, burning, committed, eager, emotional, enthusiastic, excited, fanatical, fervent, fiery, frenzied, greedy, heated, hot, hungry, impassioned, infatuated, inflamed, intense, lustful, manic, obsessive, roused, sexy, strong, urgent, vehement, violent, worked up, zealous. *Opp* APATHETIC.

**passive** *adj* apathetic, complaisant, compliant, deferential, docile, impassive, inert, inactive, long-suffering, malleable, non-violent, patient, phlegmatic, pliable, quiescent, receptive, resigned, sheepish, submissive, supine, tame, tractable, unassertive, unmoved, unresisting, yielding. ▷ CALM. *Opp* ACTIVE.

**past** *adj* bygone, dead, *inf* dead and buried, earlier, ended, finished, forgotten, former, gone, historical, late, olden (*days*), *inf* over and done with, previous, recent, sometime. • *n* antiquity, days gone by, former times, history, old days, olden days, past times. *Opp* FUTURE.

**paste** *n* **1** adhesive, fixative, glue, gum. **2** pâté, spread. • *vb* fix, glue, stick. ▷ FASTEN.

**pastiche** *n* blend, composite, compound, *inf* hotchpotch, mess, miscellany, mixture, *inf* motley collection, patchwork, selection.

**pastime** *n* activity, amusement, avocation, distraction, diversion, entertainment, fun, game, hobby, leisure activity, occupation, play, recreation, relaxation, sport.

**pastoral** *adj* **1** agrarian, agricultural, Arcadian, bucolic, country, farming, idyllic, outdoor, provincial, rural, rustic. ▷ PEACEFUL. *Opp* URBAN. **2** *pastoral duties.* clerical, ecclesiastical, parochial, ministerial, priestly.

**pasture** *n* field, grassland, grazing, mead, meadow, paddock, pasturage.

**pat** *vb* caress, dab, slap, stroke, tap. ▷ TOUCH.

**patch** *n* darn, mend, piece, reinforcement, repair. • *vb* cover, darn, fix, mend, reinforce, repair, sew up, stitch up.

**patchy** *adj inf* bitty, blotchy, changing, dappled, erratic, inconsistent, irregular, speckled, spotty, uneven, unpredictable, variable, varied, varying. *Opp* UNIFORM.

**patent** *adj* apparent, blatant, clear, conspicuous, evident, flagrant, manifest, obvious, open, plain, self-evident, transparent, undisguised, visible.

**path** *n* **1** alley, bridle-path, bridleway, esplanade, footpath, footway, pathway, pavement, *Amer* sidewalk, towpath, track, trail, walk, walkway, way. ▷ ROAD. **2** approach, course, direction, flight path, orbit, route, trajectory, way.

**pathetic** *adj* **1** affecting, distressing, emotional, emotive, heartbreaking, heart-rending, lamentable, moving, piteous, pitiable, pitiful, plaintive, poignant, stirring, touching, tragic. ▷ SAD. **2** ▷ INADEQUATE.

**pathos** *n* emotion, feeling, pity, poignancy, sadness, tragedy.

**patience** *n* **1** calmness, composure, endurance, equanimity, forbearance, fortitude, leniency, long-suffering, resignation, restraint, self-control, serenity, stoicism, toleration, *inf* unflappability. **2** *work with patience.* assiduity, determination, diligence, doggedness, endurance, firmness, perseverance, persistence, pertinacity, *inf* stickability, tenacity.

**patient** *adj* **1** accommodating, acquiescent, calm, compliant, composed, docile, easygoing, even-tempered, forbearing, forgiving, lenient, long-suffering, mild, philosophical, quiet, resigned, self-possessed, serene, stoical, submissive, tolerant, uncomplaining. **2** *a patient worker.* assiduous, determined, diligent, dogged, persevering, persistent, steady, tenacious, unhurried, untiring. *Opp* IMPATIENT. • *n* case, invalid, outpatient, sufferer.

**patriot** *n derog* chauvinist, loyalist, nationalist, *derog* xenophobe.

**patriotic** *adj derog* chauvinistic, *derog* jingoistic, loyal, nationalistic, *derog* xenophobic.

**patriotism** *n derog* chauvinism, *derog* jingoism, loyalty, nationalism, *derog* xenophobia.

**patrol** *n* **1** beat, guard, policing, sentry-duty, surveillance, vigilance, watch. **2** guard, lookout, patrolman, sentinel, sentry, watchman. • *vb* be on patrol, defend, guard, inspect, keep a lookout, make the rounds, police, protect, stand guard, tour, walk the beat, watch over.

**patron** *n* **1** advocate, *inf* angel, backer, benefactor, champion, defender, helper, philanthropist, promoter, sponsor, subscriber, supporter. **2** *patron of a shop.* client, customer, frequenter, *inf* regular, shopper.

**patronage** *n* backing, business, custom, help, sponsorship, support, trade.

**patronize** *vb* **1** back, be a patron of, bring trade to, buy from, deal with, encourage, frequent, give patronage to, shop at, support. **2** *be patronizing towards.* humiliate, *inf* look down on, *inf* look down your nose at, *inf* put down, talk down to. **patronizing** ▷ SUPERIOR.

**pattern** *n* **1** arrangement, decoration, design, device, figuration, figure, motif, ornamentation, sequence, shape, system, tessellation. **2** archetype, criterion, xample, exemplar, guide, ideal, nodel, norm, original, paragon, recedent, prototype, sample, specimen, standard, yardstick.

**ause** *n* break, *inf* breather, breathing space, caesura, check, delay, gap, halt, hesitation, hiatus, hold-up, interlude, intermission, interruption, interval, lacuna, apse, *inf* let-up, lull, moratorium, espite, rest, standstill, stop, stoppage, suspension, wait. • *vb* break off, delay, falter, halt, hang back, have a pause, hesitate, hold, mark ime, rest, stop, *inf* take a break, *nf* take a breather, wait.

**ave** *vb* asphalt, concrete, cover with paving, flag, *old use* macadamize, *inf* make up, surace, tarmac, tile. **pave the way** > PREPARE.

**avement** *n* footpath, *Amer* sidewalk. ▷ PATH.

**ay** *n* cash in hand, compensation, dividend, earnings, emoluments, ee, gain, honorarium, income, money, payment, profit, recompense, reimbursement, remittance, return, salary, settlement, tipend, take-home pay, wages. *vb* 1 *inf* cough up, *inf* fork out, give, grant, hand over, proffer, ecompense, remunerate, requite, pend, *inf* stump up. 2 *pay debts.* bear the cost of, clear, compensate, *inf* foot, honour, indemnify, meet, pay back, pay off, pay up, efund, reimburse, repay, settle. *crime doesn't pay.* avail, benefit, be profitable, pay off, produce esults, prove worthwhile, yield a eturn. 4 *pay for mistakes.* be punshed, suffer. ▷ ATONE. **pay back** > RETALIATE.

**ayment** *n* advance, alimony, llowance, charge, commission, ompensation, contribution, cost, leposit, disbursement, donation, expenditure, fare, fee, figure, fine, instalment, loan, outgoings, outlay, pocket-money, premium, price, ransom, rate, remittance, reward, royalty, *inf* sub, subscription, subsistence, supplement, surcharge, tip, toll, wage. *Opp* INCOME.

**peace** *n* 1 accord, agreement, amity, conciliation, concord, friendliness, harmony, order. *Opp* CONFLICT. 2 alliance, armistice, cease-fire, pact, treaty, truce. *Opp* WAR. 3 *peace of mind.* calm, calmness, peace and quiet, peacefulness, placidity, quiet, repose, serenity, silence, stillness, tranquillity. *Opp* ANXIETY.

**peaceable** *adj* amicable, civil, conciliatory, cooperative, friendly, gentle, harmonious, inoffensive, mild, non-violent, pacific, peace-loving, placid, temperate, understanding. *Opp* QUARRELSOME.

**peaceful** *adj* balmy, calm, easy, gentle, pacific, placid, pleasant, quiet, relaxing, restful, serene, slow-moving, soothing, still, tranquil, undisturbed, unruffled, untroubled. *Opp* NOISY, STORMY.

**peacemaker** *n* adjudicator, appeaser, arbitrator, conciliator, diplomat, intercessor, intermediary, mediator, reconciler, referee, umpire.

**peak** *n* 1 apex, brow, cap, crest, crown, eminence, hill, mountain, pinnacle, point, ridge, summit, tip, top. 2 *peak of your career.* acme, apogee, climax, consummation, crisis, crown, culmination, height, highest point, zenith.

**peal** *n* carillon, chime, chiming, clangour, knell, reverberation, ringing, tintinnabulation, toll. • *vb* chime, clang, resonate, ring, ring the changes, sound, toll.

**peasant** *n* [*derog*] boor, bumpkin, churl, oaf, rustic, serf, swain, village idiot, yokel.

**pebbles** *plur n* cobbles, gravel, stones.

**peculiar** *adj* **1** aberrant, abnormal, anomalous, atypical, bizarre, curious, deviant, eccentric, exceptional, freakish, funny, odd, offbeat, outlandish, out of the ordinary, quaint, queer, quirky, surprising, strange, uncommon, unconventional, unusual, weird. **2** *your peculiar style.* characteristic, different, distinctive, identifiable, idiosyncratic, individual, natural, particular, personal, private, singular, special, unique, unmistakable. *Opp* COMMON, ORDINARY.

**peculiarity** *n* abnormality, characteristic, difference, distinctiveness, eccentricity, foible, idiosyncrasy, individuality, mannerism, oddity, outlandishness, quirk, singularity, speciality, trait, uniqueness.

**pedantic** *adj* **1** academic, bookish, donnish, dry, formal, humourless, learned, old-fashioned, pompous, scholarly, schoolmasterly, stiff, stilted, *inf* stuffy. **2** *inf* by the book, doctrinaire, exact, fastidious, fussy, inflexible, *inf* nit-picking, precise, punctilious, strict, unimaginative. *Opp* INFORMAL, LAX.

**peddle** *vb inf* flog, hawk, market, *inf* push, sell, traffic in, vend.

**pedestrian** *adj* **1** pedestrianized, traffic-free. **2** banal, boring, dreary, dull, commonplace, flat-footed, lifeless, mundane, prosaic, run-of-the-mill, tedious, unimaginative, uninteresting. ▷ ORDINARY. • *n inf* foot-slogger, foot-traveller, stroller, walker.

**pedigree** *adj* pure-bred, thoroug bred. • *n* ancestry, blood, descen extraction, family, family histor genealogy, line, lineage, parentage, roots, stock, strain.

**pedlar** *n old use* chapman, *inf* cheapjack, *old use* colporteur door-to-door salesman, hawker, *inf* pusher, seller, street-trader, trafficker, vendor.

**peel** *n* coating, rind, skin. • *vb* denude, flay, hull, pare, ski strip. ▷ UNDRESS.

**peep** *vb* **1** glance, have a look, peek, squint. ▷ LOOK. **2** ▷ SHOW.

**peer** *n* aristocrat, grandee, noble nobleman, noblewoman, patricia titled person. □ *baron, baroness, countess, duchess, duke, earl, lad lord, marchioness, marquis, viscount, viscountess.* • *vb* have a look, look earnestly, spy, squint. ▷ LOOK. **peers 1** aristocracy, nobility, peerage. **2** colleagues, compeers, confrères, equals, fellows, peer-group.

**peevish** *adj* cantankerous, chur ish, crabby, crusty, curmudgeonly, grumpy, illhumoured, irr able, petulant, querulous, testy, touchy, waspish. ▷ BAD-TEMPERE

**peg** *n* bolt, dowel, pin, rod, stick, thole-pin. • *vb* ▷ FASTEN.

**pelt** *n* coat, fur, hide, skin. • *vb* assail, bombard, shower, strafe. ▷ THROW.

**pen** *n* **1** coop, *Amer* corral, enclo ure, fold, hutch, pound. **2** ball-point, biro, felt-tip, founta pen, *old use* quill.

**penalize** *vb* discipline, fine, impose a penalty on, punish.

**penalty** *n* fine, forfeit, price. ▷ PUNISHMENT. **pay the penalty** ▷ ATONE.

**penance** *n* amends, atonement, contrition, penitence, punishment, reparation. **do penance** ▷ ATONE.

**pendent** *adj* dangling, hanging, loose, pendulous, suspended, swaying, swinging, trailing.

**pending** *adj* about to happen, forthcoming, *inf* hanging fire, imminent, impending, *inf* in the offing, undecided, waiting.

**penetrate** *vb* **1** bore through, break through, drill into, enter, get into, get through, infiltrate, lance, make a hole, perforate, pierce, probe, puncture, stab, stick in. **2** *damp penetrates.* filter through, impregnate, percolate through, permeate, pervade, seep into, suffuse.

**penitent** *adj* apologetic, conscience-stricken, contrite, regretful, remorseful, repentant, rueful, shamefaced, sorry. *Opp* UNREPENTANT.

**pennon** *n* banner, flag, pennant, standard, streamer.

**pension** *n* annuity, benefit, old age pension, superannuation.

**pensive** *adj* brooding, cogitating, contemplative, day-dreaming, *inf* far-away, *inf* in a brown study, lost in thought, meditative, reflective, ruminative, thoughtful.

**penury** *n* beggary, destitution, impoverishment, indigence, lack, need, poverty, scarcity, want.

**people** *n* **1** folk, human beings, humanity, humans, individuals, ladies and gentlemen, mankind, men and women, mortals, persons. **2** citizenry, citizens, community, electorate, *inf* grass roots, *Gr* hoi polloi, nation, *inf* the plebs, populace, population, the public, society, subjects. **3** *your own people.* clan, family, kinsmen, kith and kin, nation, race, relations, relatives, tribe. • *vb* colonize, fill, inhabit, occupy, overrun, populate, settle.

**perceive** *vb* **1** become aware of, catch sight of, descry, detect, discern, discover, distinguish, espy, glimpse, hear, identify, make out, notice, note, observe, recognize, see, spot. **2** appreciate, apprehend, comprehend, deduce, feel, *inf* figure out, gather, grasp, infer, know, realize, sense, understand.

**perceptible** *adj* appreciable, audible, detectable, discernible, distinct, distinguishable, evident, identifiable, manifest, marked, notable, noticeable, observable, obvious, palpable, perceivable, plain, recognizable, unmistakable, visible. *Opp* IMPERCEPTIBLE.

**perception** *n* appreciation, apprehension, awareness, cognition, comprehension, consciousness, discernment, insight, instinct, intuition, knowledge, observation, perspective, realization, recognition, sensation, sense, understanding, view.

**perceptive** *adj* acute, alert, astute, attentive, aware, clever, discerning, discriminating, observant, penetrating, percipient, perspicacious, quick, responsive, sensitive, sharp, sharp-eyed, shrewd, sympathetic, understanding. ▷ INTELLIGENT.

**perch** *n* rest, resting-place, roost. • *vb* balance, rest, roost, settle, sit.

**percussion** *n* □ *bell, castanets, celesta, chime bar, cymbal, glockenspiel, gong, kettledrum, maracas, rattle, tambourine, timpani, triangle, tubular bells, vibraphone, whip, wood block, xylophone.* ▷ DRUM.

**perdition** *n* damnation, doom, downfall, hell, hellfire, ruin, ruination.

**perfect** *adj* 1 absolute, complete, completed, consummate, excellent, exemplary, faultless, finished, flawless, ideal, immaculate, incomparable, matchless, mint, superlative, unbeatable, undamaged, unexceptionable, unqualified, whole. 2 blameless, irreproachable, pure, sinless, spotless, unimpeachable. 3 accurate, authentic, correct, exact, faithful, immaculate, impeccable, precise, tailor-made, true. *Opp* IMPERFECT. • *vb* bring to fruition, bring to perfection, carry through, complete, consummate, effect, execute, finish, fulfil, make perfect, realize, *inf* see through.

**perfection** *n* 1 beauty, completeness, excellence, faultlessness, flawlessness, ideal, precision, purity, wholeness. *Opp* IMPERFECTION. 2 *the perfection of a plan.* accomplishment, achievement, completion, consummation, end, fruition, fulfilment, realization.

**perforate** *vb* bore through, drill, penetrate, pierce, prick, punch, puncture, riddle.

**perform** *vb* 1 accomplish, achieve, bring about, carry on, carry out, commit, complete, discharge, dispatch, do, effect, execute, finish, fulfil, *inf* pull off. 2 behave, function, go, operate, run, work. 3 *perform on stage.* act, appear, dance, feature, figure, take part. 4 *perform a play, song.* enact, mount, present, play, produce, put on, render, represent, serenade, sing, stage.

**performance** *n* 1 accomplishment, achievement, carrying out, completion, doing, execution, fulfilment. 2 act, behaviour, conduct, deception, exhibition, exploit, feat, play-acting, pretence. 3 *stage performance.* acting, début, impersonation, interpretation, play, playing, portrayal, presentation, production, rendition, representation. □ *concert, dress rehearsal, first night, last night, matinée, première, preview, rehearsal, show, sketch, turn.*

**performer** *n* actor, actress, artist, artiste, player, singer, star, *inf* superstar, thespian, trouper. ▷ ENTERTAINER.

**perfume** *n* 1 aroma, bouquet, fragrance, odour, scent. ▷ SMELL. 2 after-shave, eau de Cologne, scent, toilet water.

**perfunctory** *adj* apathetic, automatic, brief, cursory, dutiful, fleeting, half-hearted, hurried, inattentive, indifferent, mechanical, offhand, routine, superficial, uncaring, unenthusiastic, uninterested, uninvolved, unthinking. *Opp* ENTHUSIASTIC.

**perhaps** *adv* conceivably, maybe *old use* peradventure, *old use* perchance, possibly.

**peril** *n* danger, hazard, insecurity, jeopardy, risk, susceptibility, threat, vulnerability.

**perilous** *adj* dangerous, hazardous, insecure, risky, uncertain, unsafe, vulnerable. *Opp* SAFE.

**perimeter** *n* border, borderline, boundary, bounds, circumference, confines, edge, fringe, frontier, limit, margin, periphery, verge.

**period** *n* 1 duration, interval, phase, season, session, span, spell, stage, stint, stretch, term, while. 2 aeon, age, epoch, era. ▷ TIME.

**periodic** *adj* cyclical, intermittent, occasional, recurrent, repeated, spasmodic, sporadic.

**peripheral** *adj* 1 distant, on the perimeter, outer, outermost, outlying. 2 borderline, incidental, inessential, irrelevant, marginal, minor, nonessential, secondary, tangential, unimportant, unnecessary. *Opp* CENTRAL.

**perish** *vb* **1** be destroyed, be killed, die, expire, fall, lose your life, meet your death, pass away. **2** crumble away, decay, decompose, disintegrate, go bad, rot.

**perishable** *adj* biodegradable, destructible, liable to perish, unstable. *Opp* PERMANENT.

**perjury** *n* bearing false witness, lying, mendacity.

**permanent** *adj* abiding, ceaseless, changeless, chronic, constant, continual, continuous, durable, endless, enduring, eternal, everlasting, fixed, immutable, incessant, incurable, indestructible, indissoluble, ineradicable, interminable, invariable, irreparable, irreversible, lasting, lifelong, long-lasting, never-ending, non-stop, ongoing, perennial, perpetual, persistent, stable, steady, unalterable, unceasing, unchanging, undying, unending. *Opp* TEMPORARY.

**permeate** *vb* diffuse, filter through, flow through, impregnate, infiltrate, penetrate, percolate, pervade, saturate, soak through, spread through.

**permissible** *adj* acceptable, admissible, allowable, allowed, excusable, lawful, legal, legitimate, licit, permitted, proper, right, sanctioned, tolerable, valid, venial (*sin*). *Opp* UNACCEPTABLE.

**permission** *n* agreement, approbation, approval, aquiescence, assent, authority, authorization, consent, dispensation, franchise, *inf* go-ahead, *inf* green light, leave, licence, *inf* rubber stamp, sanction, seal of approval, stamp of approval, support. ▷ PERMIT.

**permissive** *adj* aquiescent, consenting, easygoing, indulgent, latitudinarian, lenient, liberal, libertarian, tolerant.

**permit** *n* authority, authorization, certification, charter, licence, order, pass, passport, ticket, visa, warrant. • *vb* admit, agree to, allow, approve of, authorize, consent to, endorse, enfranchise, give an opportunity for, give permission for, give your blessing to, legalize, license, make possible, sanction, *old use* suffer, support, tolerate.

**perpendicular** *adj* at right angles, erect, plumb, straight up and down, upright, vertical.

**perpetual** *adj* abiding, ageless, ceaseless, chronic, constant, continual, continuous, endless, enduring, eternal, everlasting, frequent, immortal, immutable, incessant, incurable, indestructible, ineradicable, interminable, invariable, lasting, long-lasting, never-ending, non-stop, ongoing, perennial, permanent, persistent, protracted, recurrent, recurring, repeated, *old use* sempiternal, timeless, unceasing, unchanging, undying, unending, unfailing, unremitting. *Opp* TEMPORARY.

**perpetuate** *vb* continue, eternalize, eternize, extend, immortalize, keep going, maintain, make permanent, preserve.

**perplex** *vb* baffle, *inf* bamboozle, befuddle, bewilder, confound, confuse, disconcert, distract, dumbfound, muddle, mystify, nonplus, puzzle, *inf* stump, *inf* throw, worry.

**perquisite** *n* benefit, bonus, *inf* consideration, emolument, extra, fringe benefit, gratuity, *inf* perk, tip.

**persecute** *vb* abuse, afflict, annoy, badger, bother, bully, discriminate against, harass, hector, hound, ill-treat, intimidate, maltreat, martyr, molest, oppress, pester, *inf* put the screws on, suppress, terrorize, torment, torture, trouble, tyrannize, victimize, worry.

**persist** *vb* be diligent, be steadfast, carry on, continue, endure, go on, *inf* hang on, hold out, *inf* keep at it, keep going, *inf* keep it up, keep on, last, linger, persevere, *inf* plug away, remain, *inf* soldier on, stand firm, stay, *inf* stick at it. *Opp* CEASE.

**persistent** *adj* **1** ceaseless, chronic, constant, continual, continuous, endless, eternal, everlasting, incessant, interminable, lasting, long-lasting, never-ending, obstinate, permanent, perpetual, persisting, recurrent, recurring, remaining, repeated, unending, unrelenting, unrelieved, unremitting. *Opp* BRIEF, INTERMITTENT. **2** assiduous, determined, dogged, hard-working, indefatigable, patient, persevering, pertinacious, relentless, resolute, steadfast, steady, stubborn, tenacious, tireless, unflagging, untiring, unwavering, zealous. *Opp* LAZY.

**person** *n* adolescent, adult, baby, being, *inf* body, character, child, *inf* customer, figure, human, human being, individual, infant, mortal, personage, soul, *inf* type, woman. ▷ MAN, PEOPLE, WOMAN.

**persona** *n* character, exterior, façade, guise, identity, image, part, personality, role, self-image.

**personal** *adj* **1** distinct, distinctive, exclusive, idiosyncratic, individual, inimitable, particular, peculiar, private, special, unique, your own. *Opp* GENERAL. **2** *a personal appearance.* actual, in person, in the flesh, live, physical. **3** *personal letters.* confidential, friendly, individual, informal, intimate, private, secret. *Opp* PUBLIC. **4** *personal friends.* bosom, close, dear, familiar, intimate, known. **5** *personal remarks.* belittling, critical, derogatory, disparaging, insulting, offensive, pejorative, rude, slighting, unfriendly. **6** *personal knowledge.* direct, empirical, experiential, firsthand.

**personality** *n* **1** attractiveness, character, charisma, charm, disposition, identity, individuality, magnetism, *inf* make-up, nature, persona, psyche, temperament. **2** *inf* big name, celebrity, idol, luminary, name, public figure, star, superstar.

**personification** *n* allegorical representation, embodiment, epitome, human likeness, incarnation, living image, manifestation.

**personify** *vb* allegorize, embody, epitomize, exemplify, give human shape to, incarnate, manifest, personalize, represent, stand for, symbolize, typify.

**personnel** *n* employees, manpower, people, staff, workforce, workers.

**perspective** *n* angle, approach, attitude, outlook, point of view, position, prospect, slant, standpoint, view, viewpoint.

**persuade** *vb* bring round, cajole, coax, convert, convince, entice, exhort, importune, induce, influence, inveigle, press, prevail upon, prompt, talk into, tempt, urge, use persuasion, wheedle (into), win over. *Opp* DISSUADE.

**persuasion** *n* **1** argument, blandishment, brainwashing, cajolery, coaxing, conditioning, enticement, exhortation, inducement, persuad-

ng, propaganda, reasoning.
: affiliation, belief, conviction, :reed, denomination, faith, reli- ;ion, sect.

**ersuasive** *adj* cogent, compel- ing, conclusive, convincing, cred- ble, effective, efficacious, elo- quent, forceful, influential, logical, plausible, potent, reasonable, sound, strong, telling, unarguable, valid, watertight.
*Opp* UNCONVINCING.

**ertain** *vb* appertain, apply, be rel- evant, have bearing, have refer- ence, have relevance, refer. **per- tain to** affect, concern.

**ertinent** *adj* apposite, appropri- ate, apropos, apt, fitting, germane, relevant, suitable.
*Opp* IRRELEVANT.

**erturb** *vb* agitate, alarm, bother, confuse, discompose, discomfit, disconcert, disquiet, distress, dis- turb, fluster, frighten, make anxious, ruffle, scare, shake, trouble, unnerve, unsettle, upset, vex, worry. *Opp* REASSURE.

**eruse** *vb* examine, inspect, look over, read, run your eye over, scan, scrutinize, study.

**ervade** *vb* affect, diffuse, fill, fil- ter through, flow through, impreg- nate, penetrate, percolate, per- meate, saturate, spread through, suffuse.

**ervasive** *adj* general, inescap- able, insidious, omnipresent, pen- etrating, permeating, pervading, prevalent, rife, ubiquitous, univer- sal, widespread.

**erverse** *adj* adamant, contradict- ory, contrary, disobedient, frac- tious, headstrong, illogical, inap- propriate, inflexible, intractable, intransigent, obdurate, obstinate, peevish, *inf* pig-headed, rebellious, refractory, self-willed, stubborn, tiresome, uncooperative, unhelp- ful, unreasonable, wayward, wil- ful, wrong-headed.
*Opp* REASONABLE.

**perversion** *n* **1** corruption, distor- tion, falsification, misrepresenta- tion, misuse, twisting.
**2** aberration, abnormality, deprav- ity, deviance, deviation, immoral- ity, impropriety, *inf* kinkiness, per- versity, unnaturalness, vice, wickedness.

**pervert** *n* debauchee, degenerate, deviant, perverted person, proflig- ate. • *vb* **1** bend, deflect, distort, divert, falsify, misrepresent, per- jure, subvert, twist, undermine.
**2** *pervert a witness.* bribe, corrupt, lead astray.

**perverted** *adj* abnormal, amoral, bad, corrupt, debauched, degener- ate, depraved, deviant, dissolute, eccentric, evil, immoral, improper, *inf* kinky, profligate, sick, twisted, unnatural, unprin- cipled, warped, wicked, wrong.
▷ OBSCENE. *Opp* NATURAL.

**pessimism** *n* cynicism, despair, despondency, fatalism, gloom, hopelessness, negativeness, resig- nation, unhappiness.
*Opp* OPTIMISM.

**pessimistic** *adj* bleak, cynical, defeatist, despairing, despondent, fatalistic, gloomy, hopeless, melan- choly, morbid, negative, resigned, unhappy. ▷ SAD. *Opp* OPTIMISTIC.

**pest** *n* **1** annoyance, bane, bother, curse, irritation, nuisance, *inf* pain in the neck, *inf* thorn in your flesh, trial, vexation.
**2** *inf* bug, *inf* creepy-crawly, insect, parasite, *plur* vermin.

**pester** *n* annoy, badger, bait, besiege, bother, *inf* get under some- one's skin, harass, harry, *inf* hassle, irritate, molest, nag, nettle, plague, provoke, torment, trouble, worry.

**pestilence** *n* blight, curse, epidemic, pandemic, plague, scourge. ▷ ILLNESS.

**pet** *n inf* apple of your eye, darling, favourite, idol. • *vb* caress, cuddle, fondle, kiss, nuzzle, pat, stroke. ▷ TOUCH.

**petition** *n* appeal, application, entreaty, list of signatures, plea, request, solicitation, suit, supplication. • *vb* appeal to, call upon, deliver a petition to, entreat, importune, solicit, sue, supplicate. ▷ ASK.

**petty** *adj* **1** inconsequential, insignificant, minor, niggling, small, trivial, trifling. ▷ UNIMPORTANT. *Opp* IMPORTANT. **2** *petty complaints.* grudging, mean, nit-picking, small-minded, ungenerous. *Opp* GENEROUS.

**phase** *n* development, period, season, spell, stage, state, step. ▷ TIME. **phase in** ▷ INTRODUCE. **phase out** ▷ FINISH.

**phenomenal** *adj* amazing, astonishing, astounding, exceptional, extraordinary, *inf* fantastic, incredible, marvellous, *inf* mind-boggling, miraculous, notable, outstanding, prodigious, rare, remarkable, *inf* sensational, singular, staggering, stunning, unbelievable, unorthodox, unusual, *inf* wonderful. *Opp* ORDINARY.

**phenomenon** *n* **1** circumstance, event, experience, fact, happening, incident, occasion, occurrence, sight. **2** *an unusual phenomenon.* curiosity, marvel, miracle, phenomenal person, phenomenal thing, prodigy, rarity, sensation, spectacle, wonder.

**philanthropic** *adj* altruistic, beneficent, benevolent, bountiful, caring, charitable, generous, humane, humanitarian, magnanimous, munificent, public-spirite ungrudging. ▷ KIND. *Opp* MISANTHROPIC.

**philanthropist** *n* altruist, benefactor, donor, giver, *inf* Good Samaritan, humanitarian, patron, provider, sponsor.

**philistine** *adj* boorish, ignorant, lowbrow, materialistic, uncivilized, uncultivated, uncultured, unenlightened, unlettered, vulgar

**philosopher** *n* sage, student of philosophy, thinker.

**philosophical** *adj* **1** abstract, academic, analytical, erudite, esoteric, ideological, impractical, intellectual, learned, logical, metaphysical, rational, reasoned, scholarly, theoretical, thoughtful, wise **2** calm, collected, composed, detached, equable, imperturbable judicious, patient, reasonable, resigned, serene, sober, stoical, unemotional, unruffled. *Opp* EMOTIONAL.

**philosophize** *vb* analyse, moralize, pontificate, preach, rationalize, reason, sermonize, theorize, think things out.

**philosophy** *n* **1** epistemology, ideology, logic, metaphysics, rationalism, thinking. **2** *philosophy of life.* attitude, convictions, outlook, set of beliefs, tenets, values, viewpoint, wisdom

**phlegmatic** *adj* apathetic, cold, cool, frigid, impassive, imperturbable, indifferent, lethargic, passive, placid, slow, sluggish, stoical stolid, torpid, undemonstrative, unemotional, unenthusiastic, unfeeling, uninvolved, unresponsive. *Opp* EXCITABLE.

**phobia** *n* anxiety, aversion, dislike, dread, *inf* hang-up, hatred, horror, loathing, neurosis, obsession, repugnance, revulsion. ▷ FEAR. □ *agoraphobia (*open space*), arachnophobia (*spiders*), claustrophobia (*enclosed space*), xenophobia (*foreigners*).*

**phone** *vb* call, dial, *inf* give a buzz, ring, telephone.

**phoney** *adj* affected, artificial, assumed, bogus, cheating, contrived, counterfeit, deceitful, ersatz, factitious, fake, faked, false, fictitious, fraudulent, hypocritical, imitation, insincere, mock, pretended, *inf* pseudo, *inf* put-on, *inf* put-up, sham, spurious, synthetic, trick, unreal. *Opp* REAL.

**photocopy** *vb* copy, duplicate, photostat, print off, reproduce, *inf* run off.

**photograph** *n* enlargement, exposure, negative, *inf* photo, picture, plate, positive, print, shot, slide, *inf* snap, snapshot, transparency. • *vb* film, shoot, snap, take a photograph of.

**photographic** *adj* **1** accurate, exact, faithful, graphic, lifelike, naturalistic, realistic, representational, true to life. **2** *photographic memory.* pictorial, retentive, visual.

**phrase** *n* clause, expression. ▷ SAYING. • *vb* ▷ SAY.

**phraseology** *n* diction, expression, idiom, language, parlance, phrasing, style, turn of phrase, wording.

**physical** *adj* actual, bodily, carnal, concrete, corporal, corporeal, earthly, fleshly, incarnate, material, mortal, palpable, physiological, real, solid, substantial, tangible. *Opp* INTANGIBLE, SPIRITUAL.

**physician** *n* consultant, doctor, general practitioner, *inf* GP, *inf* medic, medical practitioner, specialist.

**physiological** *adj* anatomical, bodily, physical. *Opp* PSYCHOLOGICAL.

**physique** *n* body, build, figure, form, frame, muscles, physical condition, shape.

**pick** *n* **1** choice, election, option, preference, selection. **2** best, cream, élite, favourite, flower, pride. • *vb* **1** cast (*actor*), choose, decide on, elect, fix on, make a choice of, name, nominate, opt for, prefer, select, settle on, single out, vote for. **2** *pick flowers.* collect, cull, cut, gather, harvest, pluck, pull off, take. **pick on** ▷ BULLY. **pick up** ▷ GET, IMPROVE.

**pictorial** *adj* diagrammatic, graphic, illustrated, realistic, representational, vivid.

**picture** *n* **1** delineation, depiction, image, likeness, outline, portrayal, profile, representation. □ *abstract, cameo, caricature, cartoon, collage, design, doodle, drawing, engraving, etching, fresco,* plur *graffiti,* plur *graphics, icon, identikit, illustration, landscape, montage, mosaic, mural, oil-painting, old master, painting, photofit, photograph, pin-up, plate, portrait, print, reproduction, self-portrait, silhouette, sketch, slide,* inf *snap, snapshot, still life, transfer, transparency, triptych,* Fr *trompe l'oeil, video, vignette.* **2** film, movie, moving picture, video. ▷ FILM. • *vb* **1** caricature, delineate, depict, display, doodle, draw, engrave, etch, evoke, film, illustrate, outline, paint, photograph, portray, print, represent, show, sketch, video. **2** *picture the future.* conceive, describe, dream up, envisage, envision, fancy, imagine, see in your mind's eye, think up, visualize.

**picturesque** *adj* **1** attractive, charming, colourful, idyllic, lovely, pleasant, pretty, quaint, scenic, *inf* story-book. ▷ BEAUTIFUL. *Opp* UNATTRACTIVE. **2** *picturesque language.* colourful, descriptive, expressive, graphic, imaginative, poetic, vivid. *Opp* PROSAIC.

**pie** *n* flan, pasty, patty, quiche, tart, tartlet, turnover, vol-au-vent.

**piece** *n* **1** bar, bit, bite, block, chip, chunk, crumb, division, *inf* dollop, fraction, fragment, grain, helping, hunk, length, lump, morsel, part, particle, portion, quantity, remnant, sample, scrap, section, segment, shard, share, shred, slab, slice, sliver, snippet, speck, stick, tablet, *inf* titbit, wedge. **2** component, constituent, element, spare part, unit. **3** *piece of music, work.* article, composition, example, instance, item, number, passage, specimen, work. ▷ MUSIC, WRITING. **piece together** ▷ ASSEMBLE.

**pied** *adj* dappled, flecked, mottled, particoloured, patchy, piebald, spotted, variegated.

**pier** *n* **1** breakwater, jetty, landing-stage, quay, wharf. ▷ DOCK. **2** buttress, column, pile, pillar, post, support, upright.

**pierce** *vb* bayonet, bore through, cut, drill, enter, go through, impale, jab, lance, make a hole in, penetrate, perforate, poke through, prick, punch, puncture, riddle, skewer, spear, spike, spit, stab, stick into, thrust into, transfix, tunnel through, wound. **piercing** ▷ SHARP.

**piety** *n* dedication, devotion, devotedness, devoutness, faith, godliness, holiness, piousness, religion, *derog* religiosity, saintliness, sanctity. *Opp* IMPIETY.

**pig** *n* boar, hog, *inf* piggy, piglet, runt, sow, swine.

**pile** *n* **1** abundance, accumulation, agglomeration, collection, concentration, conglomeration, deposit, heap, hoard, *inf* load, mass, mound, *inf* mountain, plethora, quantity, stack, stockpile, supply, *inf* tons. **2** column, pier, post, support, upright. • *vb* accumulate, amass, assemble, bring together, build up, collect, concentrate, deposit, gather, heap, hoard, load, mass, stack up, stockpile, store.

**pilfer** *vb inf* filch, *inf* pinch, rob, shoplift. ▷ STEAL.

**pilgrim** *n* crusader, *old use* palmer. ▷ TRAVELLER.

**pill** *n* bolus, capsule, lozenge, pastille, pellet, pilule, tablet. ▷ MEDICINE.

**pillage** *n* buccaneering, depredation, despoliation, devastation, looting, marauding, piracy, plunder, plundering, ransacking, rape, rapine, robbery, robbing, sacking, stealing, stripping. • *vb* despoil, devastate, loot, maraud, plunder, raid, ransack, ravage, raze, rob, sack, steal, strip, vandalize.

**pillar** *n* baluster, caryatid, column, pier, pilaster, pile, post, prop, shaft, stanchion, support, upright.

**pilot** *n* **1** airman, *old use* aviator, captain, flier. **2** coxswain, helmsman, navigator, steersman. • *vb* conduct, convey, direct, drive, fly, guide, lead, navigate, shepherd, steer.

**pimple** *n* blackhead, boil, eruption, pustule, spot, swelling, *sl* zit. **pimples** acne, rash.

**pin** *n old use* bodkin, bolt, brooch, clip, dowel, drawing-pin, hatpin, nail, peg, rivet, safety-pin, spike, staple, thole, tiepin. • *vb* clip, nail, pierce, staple, tack, transfix. ▷ FASTEN.

**pinch** *vb* **1** crush, grip, hurt, nip, press, squeeze, tweak. **2** ▷ STEAL.

**pine** *vb* mope, mourn, sicken, waste away. **pine for** ▷ WANT.

**pinnacle** *n* **1** acme, apex, cap, climax, consummation, crest, crown, crowning point, height, highest point, peak, summit, top, zenith. **2** *pinnacle on a roof.* spire, steeple, turret.

**pioneer** *n* **1** colonist, discoverer, explorer, frontiersman, frontierswoman, pathfinder, settler, trail-blazer. **2** innovator, inventor, originator, pace-maker, trend-setter. • *vb* begin, *inf* bring out, create, develop, discover, establish, experiment with, found, inaugurate, initiate, institute, introduce, invent, launch, open up, originate, set up, start.

**pious** *adj* **1** dedicated, devoted, devout, faithful, god-fearing, godly, good, holy, moral, religious, reverent, reverential, saintly, sincere, spiritual, virtuous. *Opp* IMPIOUS. **2** [*derog*] *inf* goody-goody, *inf* holier-than-thou, hypocritical, insincere, mealy-mouthed, pietistic, Pharisaical, sanctimonious, self-righteous, self-satisfied, *inf* smarmy, unctuous. *Opp* SINCERE.

**pip** *n* **1** pit, seed, stone. **2** mark, spot, star. **3** bleep, blip, sound, stroke.

**pipe** *n* conduit, channel, duct, hose, hydrant, line, main, pipeline, piping, tube. • *vb* **1** carry along a pipe, carry along a wire, channel, convey, deliver, supply, transmit. **2** *pipe a tune.* blow, play, sound, *inf* tootle, whistle. **pipe up** ▷ SPEAK. **piping** ▷ HOT, SHRILL.

**piquant** *adj* **1** appetizing, pungent, salty, sharp, spicy, tangy, tart, tasty. *Opp* BLAND. **2** *a piquant notion.* arresting, exciting, interesting, provocative, stimulating. *Opp* BANAL.

**pirate** *n* buccaneer, *old use* corsair, marauder, privateer, sea rover. ▷ THIEF. • *vb* ▷ PLAGIARIZE.

**pit** *n* **1** abyss, chasm, crater, depression, ditch, excavation, hole, hollow, pothole, rut, trench, well. **2** coal-mine, colliery, mine, mineshaft, quarry, shaft, working.

**pitch** *n* **1** bitumen, tar. **2** *pitch of a roof.* angle, gradient, incline, slope, steepness, tilt. **3** *musical pitch.* frequency, tuning. **4** *soccer pitch.* arena, ground, playing-field, stadium. • *vb* **1** erect, put up, raise, set up. **2** *pitch stones.* bowl, *inf* bung, cast, *inf* chuck, fire, fling, heave, hurl, launch, lob, sling, throw, toss. **3** *pitch into the water.* dive, drop, fall headlong, plunge, plummet, *inf* take a nosedive, topple. **pitch about** ▷ TOSS. **pitch in** ▷ COOPERATE. **pitch into** ▷ ATTACK.

**piteous** *adj* affecting, distressing, heartbreaking, heart-rending, lamentable, miserable, moving, pathetic, pitiable, pitiful, plaintive, poignant, touching, woeful, wretched. ▷ SAD.

**pitfall** *n* catch, danger, difficulty, hazard, peril, snag, trap.

**pitiful** *adj* **1** abject, contemptible, deplorable, hopeless, inadequate, incompetent, insignificant, laughable, mean, *inf* miserable, *inf* pathetic, pitiable, ridiculous, sorry, trifling, unimportant, useless, worthless. *Opp* ADMIRABLE. **2** ▷ PITEOUS.

**pitiless** *adj* bloodthirsty, brutal, callous, cruel, ferocious, hard, heartless, inexorable, inhuman, merciless, relentless, ruthless, sadistic, unfeeling, unrelenting, unrelieved, unremitting, unsympathetic. *Opp* MERCIFUL.

**pitted** *adj* dented, eaten away, eroded, *inf* holey, marked, pockmarked, rough, scarred, uneven. *Opp* SMOOTH.

**pity** *n* charity, clemency, commiseration, compassion, condolence, feeling, forbearance, forgiveness, grace, humanity, kindness, leniency, love, mercy, regret, *old use* ruth, softness, sympathy, tenderness, understanding, warmth. *Opp* CRUELTY. • *vb inf* bleed for, commiserate with, *inf* feel for, feel sorry for, show pity for, sympathize with, weep for.

**pivot** *n* axis, axle, centre, fulcrum, gudgeon, hinge, hub, pin, point of balance, spindle, swivel. • *vb* hinge, revolve, rotate, spin, swivel, turn, twirl, whirl.

**placard** *n* advert, advertisement, bill, notice, poster, sign.

**placate** *vb* appease, calm, conciliate, humour, mollify, pacify, soothe.

**place** *n* **1** area, country, district, locale, location, locality, locus, neighbourhood, part, point, position, quarter, region, scene, setting, site, situation, *inf* spot, town, venue, vicinity, *inf* whereabouts. **2** *a place in society.* condition, degree, estate, function, grade, job, mission, niche, office, position, rank, role, standing, station, status. **3** *a place to live.* ▷ HOUSE. **4** *a place to sit.* ▷ SEAT. • *vb* **1** deposit, dispose, *inf* dump, lay, leave, locate, pinpoint, plant, position, put down, rest, set down, set out, settle, situate, stand, station, *inf* stick. **2** arrange, categorize, class, classify, grade, order, position, put in order, rank, sort. **3** *can't place it.* identify, put a name to, put into context, recognize, remember.

**placid** *adj* **1** collected, composed, cool, equable, even-tempered, imperturbable, level-headed, mild, phlegmatic, restful, sensible, stable, steady, unexcitable. **2** calm, motionless, peaceful, quiet, tranquil, unruffled, untroubled. *Opp* EXCITABLE, STORMY.

**plagiarize** *vb* appropriate, borrow, copy, *inf* crib, imitate, infringe copyright, *inf* lift, pirate, purloin, reproduce. ▷ STEAL.

**plague** *n* **1** affliction, bane, blight, calamity, contagion, epidemic, infection, outbreak, pandemic, pestilence. ▷ ILLNESS. **2** infestation, invasion, nuisance, scourge, swarm, visitation. • *vb* afflict, annoy, be a nuisance to, bother, distress, disturb, harass, harry, hound, irritate, molest, *inf* nag, persecute, pester, torment, torture, trouble, vex, worry.

**plain** *adj* **1** apparent, audible, certain, clear, comprehensible, definite, distinct, evident, intelligible, legible, lucid, manifest, obvious, patent, simple, transparent, unambiguous, understandable, unmistakable, visible, well-defined. *Opp* OBSCURE. **2** *plain speech.* basic, blunt, candid, direct, downright, explicit, forthright, frank, honest, informative, outspoken, plain-spoken, prosaic, sincere, straightforward, unequivocal, unvarnished. **3** *plain living.* austere, drab, everyday, frugal, homely, modest, ordinary, simple, Spartan, stark, unadorned, unattractive, undecorated, unexciting, unprepossessing, unpretentious, unremarkable, workaday. *Opp* SOPHISTICATED. • *n* grassland, pasture, pampas, prairie, savannah, steppe, tundra, veld.

**plaintive** *adj* doleful, melancholy, mournful, plangent, sorrowful, wistful. ▷ SAD.

**plan** *n* **1** *inf* bird's-eye view, blueprint, chart, design, diagram, drawing, layout, map, representation, sketch-map. **2** *a plan of action.* aim, course of action, design, formula, idea, intention, method, plot, policy, procedure, programme, project, proposal, proposition, *inf* scenario, scheme, strategy, system. • *vb* **1** arrange, concoct, contrive, design, devise, draw up a plan, formulate, invent, map out, *inf* mastermind, organize, outline, plot, prepare, scheme, think out, work out. **2** *I plan to go away.* aim, conspire, contemplate, envisage, expect, intend, mean, propose, think of. **planned** ▷ DELIBERATE.

**plane** *adj* even, flat, flush, level, smooth, uniform. • *n* **1** flat surface, level, surface. **2** ▷ AIRCRAFT.

**planet** *n* globe, orb, satellite, sphere, world. □ *Earth, Jupiter, Mars, Mercury, Neptune, Pluto, Saturn, Uranus, Venus.*

**plank** *n* beam, board, planking, timber.

**planning** *n* arrangement, design, drafting, forethought, organization, preparation, setting up, thinking out.

**plant** *n* greenery, growth, undergrowth, vegetation. □ *annual, bulb, cactus, cereal, climber, fern, flower, fungus, grass, herb, lichen, moss, perennial, shrub, tree, vegetable, vine, water-plant, weed.* ▷ FLOWER, TREE, VEGETABLE. **2** *a manufacturing plant.* factory, foundry, mill, shop, works, workshop. **3** *industrial plant.* apparatus, equipment, machinery, machines. • *vb* **1** bed out, set out, sow, transplant. **2** locate, place, position, put, situate, station.

**plaster** *n* **1** mortar, stucco. **2** dressing, sticking-plaster. • *vb* apply, bedaub, coat, cover, daub, smear, spread.

**plastic** *adj* ductile, flexible, malleable, pliable, shapable, soft, supple, workable. □ *bakelite, celluloid, polystyrene, polythene, polyurethane, polyvinyl, PVC, vinyl.*

**plate** *n* **1** *old use* charger, dinner-plate, dish, platter, salver, side-plate, soup-plate, *old use* trencher. **2** lamina, lamination, layer, leaf, pane, panel, sheet, slab, stratum. **3** *plates in a book.* illustration, *inf* photo, photograph, picture, print. **4** *a dental plate.* dentures, false teeth. • *vb* anodize, coat, cover, electroplate, galvanize (*with zinc*), gild (*with gold*).

**platform** **1** dais, podium, rostrum, stage, stand. **2** *political platform.* ▷ POLICY.

**platitude** *n* banality, cliché, commonplace, truism.

**plausible** *adj* **1** acceptable, believable, conceivable, credible, imaginable, likely, logical, persuasive, possible, probable, rational, reasonable, sensible, tenable, thinkable. *Opp* IMPLAUSIBLE. **2** deceptive, glib, meretricious, misleading, specious, smooth, sophistical.

**play** *n* **1** amusement, diversion, entertainment, frivolity, fun, *inf* fun and games, *inf* horseplay, joking, make-believe, merrymaking, playing, pretending, recreation, revelry, *inf* skylarking, sport. **2** *play in moving parts.* flexibility, freedom, freedom of movement, *inf* give, latitude, leeway, looseness, movement, tolerance. 3 ▷ DRAMA. • *vb* 1 amuse yourself, caper, cavort, disport yourself, enjoy yourself, fool about, frisk, frolic, gambol, *inf* have a good time, have fun, *inf* mess about, romp, sport. 2 *play a game.* join in, participate, take part. 3 *play an opponent.* challenge, compete against, oppose, rival, take on, vie with. 4 *play a role.* act, depict, impersonate, perform, portray, pretend to be, represent, take the part of. 5 *play an instrument.* make music on, perform on, strum. 6 *play the radio.* have on, listen to, operate, put on, switch on. play about ▷ MISBEHAVE. **play along, play ball** ▷ COOPERATE. **play down** ▷ MINIMIZE. **play for time** ▷ DELAY. **play it by ear** ▷ IMPROVISE. **play up** ▷ MISBEHAVE. **play up to** ▷ FLATTER.

**player** *n* 1 athlete, competitor, contestant, participant, sportsman, sportswoman. 2 actor, actress, artiste, entertainer, instrumentalist, musician, performer, soloist, Thespian, trouper. ▷ ENTERTAINER, MUSICIAN.

**playful** *adj* active, cheerful, coltish, facetious, flirtatious, frisky, frolicsome, fun-loving, good-natured, high-spirited, humorous, impish, jesting, *inf* jokey, joking, kittenish, light-hearted, lively, mischievous, puckish, roguish, skittish, spirited, sportive, sprightly, teasing, *inf* tongue-in-cheek, vivacious, waggish. *Opp* SERIOUS.

**plea** *n* 1 appeal, entreaty, invocation, petition, prayer, request, solicitation, suit, supplication. 2 argument, excuse, explanation, justification, pretext, reason.

**plead** *vb* 1 appeal, ask, beg, beseech, cry out, demand, entreat, implore, importune, petition, request, seek, solicit, supplicate. 2 allege, argue, assert, aver, declare, maintain, reason, swear.

**pleasant** *adj* acceptable, affable, agreeable, amiable, approachable, attractive, balmy, beautiful, charming, cheerful, congenial, decent, delicious, delightful, enjoyable, entertaining, excellent, fine, friendly, genial, gentle, good, gratifying, *inf* heavenly, hospitable, kind, likeable, lovely, mellow, mild, nice, palatable, peaceful, pleasing, pleasurable, pretty, reassuring, relaxed, satisfying, soothing, sympathetic, warm, welcome, welcoming. *Opp* ANNOYING, UNPLEASANT.

**please** *vb* 1 amuse, cheer up, content, delight, divert, entertain, give pleasure to, gladden, gratify, humour, make happy, satisfy, suit. 2 *Do what you please.* ▷ WANT.

**pleasing** ▷ PLEASANT.

**pleased** *adj inf* chuffed, *derog* complacent, contented, delighted, elated, euphoric, glad, grateful, gratified, *sl* over the moon, satisfied, thankful, thrilled. ▷ HAPPY. *Opp* ANNOYED.

**pleasure** *n* 1 bliss, comfort, contentment, delight, ecstasy, enjoyment, euphoria, fulfilment, gladness, gratification, happiness, joy, rapture, satisfaction, solace. 2 amusement, diversion, entertainment, fun, luxury, recreation, self-indulgence.

**pleat** *n* crease, flute, fold, gather, tuck.

**plebiscite** *n* ballot, poll, referendum, vote.

**pledge** *n* 1 assurance, covenant, guarantee, oath, pact, promise, undertaking, vow, warranty, word. 2 *a pledge left at a pawnbroker's.* bail, bond, collateral, deposit, pawn, security, surety. • *vb* agree, commit yourself, contract, give your word, guarantee, promise, swear, undertake, vouch, vouchsafe, vow.

**plenary** *adj* full, general, open.

**plentiful** *adj* abounding, abundant, ample, bounteous, bountiful, bristling, bumper (*crop*), copious, generous, inexhaustible, lavish, liberal, overflowing, plenteous, profuse, prolific. *Opp* SCARCE. **be plentiful** ▷ ABOUND.

**plenty** *n* abundance, adequacy, affluence, cornucopia, excess, fertility, flood, fruitfulness, glut, *inf* heaps, *inf* lashings, *inf* loads, a lot, *inf* lots, *inf* masses, much, more than enough, *inf* oceans, *inf* oodles, *inf* piles, plenitude, plentifulness, plethora, prodigality, profusion, prosperity, quantities, *inf* stacks, sufficiency, superabundance, surfeit, surplus, *inf* tons, wealth. *Opp* SCARCITY.

**pliable** *adj* **1** bendable, *inf* bendy, ductile, flexible, plastic, pliant, springy, supple. **2** *a pliable character.* adaptable, compliant, docile, easily influenced, easily led, easily persuaded, impressionable, manageable, persuadable, receptive, responsive, succeptible, suggestible, tractable, yielding.

**plod** *vb* **1** slog, tramp, trudge. ▷ WALK. **2** drudge, grind on, labour, persevere, *inf* peg away, *inf* plug away, toil. ▷ WORK.

**plot** *n* **1** acreage, allotment, area, estate, garden, lot, parcel, patch, smallholding, tract. **2** *a plot of a novel.* chain of events, narrative, organization, outline, scenario, story, story-line, thread. **3** *a subversive plot.* cabal, conspiracy, intrigue, machination, plan, scheme. • *vb* **1** chart, compute, draw, map out, mark, outline, plan, project. **2** *plot to rob a bank.* collude, conspire, have designs, intrigue, machinate, scheme. **3** *plot a crime.* arrange, *inf* brew, conceive, concoct, *inf* cook up, design, devise, dream up, hatch.

**pluck** *n* ▷ COURAGE. *vb* **1** collect, gather, harvest, pick, pull off, remove. **2** grab, jerk, pull, seize, snatch, tear away, tweak, yank. **3** *pluck a chicken.* denude, remove feathers from, strip. **4** *pluck a violin.* play pizzicato, strum, twang.

**plug** *n* **1** bung, cork, stopper. **2** ▷ ADVERTISEMENT. • *vb* **1** block up, *inf* bung up, close, cork, fill, jam, seal, stop up, stuff up. **2** advertise, commend, mention frequently, promote, publicize, puff, recommend. **plug away** ▷ WORK.

**plumb** *adv* **1** accurately, *inf* dead, exactly, precisely, *inf* slap. **2** perpendicularly, vertically. • *vb* fathom, measure, penetrate, probe, sound.

**plumbing** *n* heating system, pipes, water-supply.

**plume** *n* feather, *plur* plumage, quill.

**plump** *adj* ample, buxom, chubby, dumpy, overweight, podgy, portly, pudgy, *inf* roly-poly, rotund, round, squat, stout, tubby, *inf* well-upholstered. ▷ FAT. *Opp* THIN. **plump for** ▷ CHOOSE.

**plunder** *n* booty, contraband, loot, pickings, pillage, prize, spoils, swag, takings. • *vb* capture, despoil, devastate, lay waste, loot, maraud, pillage, raid, ransack, ravage, rifle, rob, sack, seize, spoil, steal from, strip, vandalize.

**plunge** *vb* **1** descend, dip, dive, drop, engulf, fall, fall headlong, hurtle, immerse, jump, leap, lower, nosedive, pitch, plummet, sink, submerge, swoop, tumble. 2 force, push, stick, thrust.

**poach** *vb* 1 hunt, steal. 2 ▷ COOK.

**pocket** *n* bag, container, pouch, receptacle. • *vb* ▷ TAKE.

**pod** *n* case, hull, shell.

**poem** *n inf* ditty, *inf* jingle, piece of poetry, rhyme, verse. □ *ballad, ballade, doggerel, eclogue, elegy, epic, epithalamium, haiku, idyll, lay, limerick, lyric, nursery-rhyme, pastoral, ode, sonnet, vers libre.* ▷ VERSE.

**poet** *n* bard, lyricist, minstrel, poetaster, rhymer, rhymester, sonneteer, versifier. ▷ WRITER.

**poetic** *adj* emotive, *derog* flowery, imaginative, lyrical, metrical, musical, poetical. *Opp* PROSAIC.

**poignant** *adj* affecting, distressing, heartbreaking, heartfelt, heart-rending, moving, painful, pathetic, piquant, piteous, pitiful, stirring, tender, touching, upsetting. ▷ SAD.

**point** *n* 1 apex, peak, prong, sharp end, spike, spur, tine, tip. 2 *a point in space.* location, place, position, site, situation, spot. 3 *a point in time.* instant, juncture, moment, second, stage, time. 4 *decimal point.* dot, full stop, mark, speck, spot. 5 *the point of an argument.* aim, burden, crux, drift, end, essence, gist, goal, heart, import, intention, meaning, motive, nub, object, objective, pith, purpose, quiddity, relevance, significance, subject, substance, theme, thrust, use, usefulness. 6 *points to raise.* aspect, detail, idea, item, matter, particular, question, thought, topic. 7 *good points in her character.* attribute, characteristic, facet, feature, peculiarity, property, quality, trait. • *vb* 1 call attention to, direct attention to, draw attention to, indicate, point out, show, signal. 2 aim, direct, guide, lead, steer. **pointed** ▷ SHARP. **to the point** ▷ RELEVANT.

**pointer** *n* arrow, hand (*of clock*), indicator.

**pointless** *adj* aimless, fatuous, fruitless, futile, inane, ineffective, senseless, silly, unproductive, useless, vain, worthless. ▷ STUPID.

**poise** *n* aplomb, assurance, balance, calmness, composure, coolness, dignity, equanimity, equilibrium, equipoise, imperturbability, presence, sang-froid, self-confidence, self-control, self-possession, serenity, steadiness. • *vb* balance, be poised, hover, keep in balance, support, suspend.

**poised** *adj* 1 balanced, hovering, in equilibrium, standing, steady, teetering, wavering. 2 *poised to begin.* keyed up, prepared, ready, set, standing by, waiting. 3 *a poised performer.* assured, calm, composed, cool, cool-headed, dignified, self-confident, self-possessed, serene, suave, *inf* unflappable, unruffled, urbane.

**poison** *n* bane, toxin, venom. • *vb* 1 adulterate, contaminate, infect, pollute, taint. ▷ KILL. 2 *poison the mind.* corrupt, defile, deprave, envenom, pervert, prejudice, subvert, warp. **poisoned** ▷ DIRTY, POISONOUS.

**poisonous** *adj* deadly, fatal, infectious, lethal, mephitic, miasmic, mortal, noxious, poisoned, septic, toxic, venomous, virulent.

**poke** *vb* butt, dig, elbow, goad, jab, jog, nudge, prod, stab, stick, thrust. ▷ HIT. **poke about** ▷ SEARCH. **poke fun at**

▷ RIDICULE. **poke out** ▷ PROTRUDE.

**poky** *adj* confined, cramped, inconvenient, restrictive, uncomfortable. ▷ SMALL. *Opp* SPACIOUS.

**polar** *adj* antarctic, arctic, freezing, glacial, icy, Siberian. ▷ COLD.

**polarize** *vb* diverge, divide, move to opposite positions, separate, split.

**pole** *n* **1** bar, beanpole, column, flag-pole, mast, post, rod, shaft, spar, staff, stake, standard, stick, stilt, upright. **2** *opposite poles.* end, extreme, limit. **poles apart** ▷ DIFFERENT.

**police** *n* sl the Bill, constabulary, *sl* the fuzz, *inf* the law, police force, policemen. • *vb* control, guard, keep in order, keep the peace, monitor, oversee, patrol, protect, provide a police presence, supervise, watch over.

**policeman, policewoman** *ns* *inf* bobby, constable, *sl* cop, *sl* copper, detective, *Fr* gendarme, inspector, officer, PC, police constable, *sl* rozzer, woman police constable, WPC.

**policy** *n* **1** approach, code of conduct, custom, guidelines, *inf* line, method, practice, principles, procedure, protocol, regulations, rules, stance, strategy, tactics. **2** intentions, manifesto, plan of action, platform, programme, proposals.

**polish** *n* **1** brightness, brilliance, finish, glaze, gleam, gloss, lustre, sheen, shine, smoothness, sparkle. **2** beeswax, French polish, oil, shellac, varnish, wax. **3** *His manners show polish.* *inf* class, elegance, finesse, grace, refinement, sophistication, style, suavity, urbanity. • *vb* brighten, brush up, buff up, burnish, French-polish, gloss, rub down, rub up, shine, smooth, wax. **polish off** ▷ FINISH. **polish up** ▷ IMPROVE.

**polished** *adj* **1** bright, burnished, glassy, gleaming, glossy, lustrous, shining, shiny. **2** *polished manners.* civilized, *inf* classy, cultivated, cultured, debonair, elegant, expert, faultless, fine, finished, flawless, genteel, gracious, impeccable, perfect, perfected, polite, *inf* posh, refined, soigné(e), sophisticated, suave, urbane. *Opp* ROUGH.

**polite** *adj* agreeable, attentive, chivalrous, civil, considerate, correct, courteous, courtly, cultivated, deferential, diplomatic, discreet, euphemistic, formal, gallant, genteel, gentlemanly, gracious, ladylike, obliging, polished, proper, respectful, tactful, thoughtful, well-bred, well-mannered, well-spoken. *Opp* RUDE.

**political** *adj* **1** administrative, civil, diplomatic, governmental, legislative, parliamentary, state. **2** activist, factional, militant, partisan, party-political. □ *anarchist, capitalist, communist, conservative, democrat, fascist, Labour, leftist, left-wing, liberal, Marxist, moderate, monarchist, nationalist, Nazi, parliamentarian, radical, republican, revolutionary, rightist, right-wing, socialist, Tory,* old use *Whig.*

**politics** *n* diplomacy, government, political affairs, political science, public affairs, statecraft, statesmanship. □ *anarchy, capitalism, communism, democracy, dictatorship, martial law, monarchy, oligarchy, republic.*

**poll** *n* **1** ballot, election, vote. **2** canvass, census, plebiscite, referendum, survey. • *vb* ballot, canvass, question, sample. survey.

**pollute** *vb* adulterate, befoul, blight, contaminate, corrupt, defile, dirty, foul, infect, poison, soil, taint.

**pomp** *n* brilliance, ceremonial, ceremony, display, formality, glory, grandeur, magnificence, ostentation, pageantry, ritual, show, solemnity, spectacle, splendour.

**pompous** *adj* affected, arrogant, bombastic, conceited, grandiloquent, grandiose, haughty, *inf* highfalutin, imperious, long-winded, magisterial, ornate, ostentatious, overbearing, pedantic, pontifical, posh, pretentious, self-important, sententious, showy, smug, snobbish, *inf* snooty, *inf* stuck-up, *inf* stuffy, supercilious, turgid, vain, vainglorious. ▷ PROUD. *Opp* MODEST.

**ponderous** *adj* **1** awkward, bulky, burdensome, cumbersome, heavy, hefty, huge, massive, unwieldy, weighty. *Opp* LIGHT. **2** *a ponderous style.* dreary, dull, elephantine, heavy-handed, humourless, inflated, laboured, lifeless, long-winded, overdone, pedestrian, plodding, prolix, slow, stilted, stodgy, tedious, tiresome, verbose, *inf* windy. *Opp* LIVELY.

**pool** *n* lagoon, lake, mere, oasis, paddling-pool, pond, puddle, swimming-pool, tarn. • *vb* ▷ COMBINE.

**poor** *adj* **1** badly off, bankrupt, beggarly, *inf* broke, deprived, destitute, disadvantaged, *inf* down-and-out, *inf* hard up, homeless, impecunious, impoverished, in debt, indigent, insolvent, necessitous, needy, *inf* on your uppers, penniless, penurious, poverty-stricken, *sl* skint, straitened, underpaid, underprivileged. **2** *poor soil.* barren, exhausted, infertile, sterile, unfruitful, unproductive. **3** *a poor salary.* inadequate, insufficient, low, meagre, mean, scanty, small, sparse, unprofitable, unrewarding. **4** *poor in health. inf* below par, poorly. ▷ ILL. **5** *poor quality.* amateurish, bad, cheap, defective, deficient, disappointing, faulty, imperfect, inferior, low-grade, mediocre, paltry, second-rate, shoddy, substandard, unacceptable, unsatisfactory, useless, worthless. **6** *poor child!* forlorn, hapless, ill-fated, luckless, miserable, pathetic, pitiable, sad, unfortunate, unhappy, unlucky, wretched. *Opp* GOOD, LARGE, LUCKY, RICH. • *plur n* beggars, the destitute, down-and-outs, the homeless, paupers, tramps, the underprivileged, vagrants, wretches.

**populace** *n* commonalty, *derog Gk* hoi polloi, masses, people, public, *derog* rabble, *derog* riff-raff.

**popular** *adj* **1** accepted, acclaimed, *inf* all the rage, approved, celebrated, famous, fashionable, favoured, favourite, *inf* in, in demand, liked, lionized, loved, renowned, sought-after, *inf* trendy, well-known, well-liked, well-received. *Opp* UNPOPULAR. **2** *popular opinion.* average, common, conventional, current, democratic, general, of the people, ordinary, predominant, prevailing, representative, standard, universal.

**popularize** *vb* **1** make popular, promote, spread. **2** *popularize classics.* make easy, simplify, *derog* tart up.

**populate** *vb* colonize, dwell in, fill, inhabit, live in, occupy, overrun, people, reside in, settle.

**population** *n* citizenry, citizens, community, denizens, folk, inhabitants, natives, occupants, people, populace, public, residents.

**opulous** *adj* crowded, full, heav-ily populated, jammed, over-crowded, overpopulated, packed, swarming, teeming.

**orch** *n* doorway, entrance, lobby, portico.

**ore** *vb* **pore over** examine, go over, peruse, read, scrutinize, study.

**ornographic** *adj* arousing, *inf* blue, erotic, explicit, exploitative, sexual, sexy, titillating. > OBSCENE.

**orous** *adj* absorbent, cellular, holey, penetrable, permeable, pervious, spongy. *Opp* IMPERVIOUS.

**ort** *n* anchorage, dock, dockyard, harbour, haven, marina, mooring, sea-port.

**ortable** *adj* compact, convenient, easy to carry, handy, light, lightweight, manageable, mobile, movable, pocket, pocket-sized, small, transportable. *Opp* UNWIELDY.

**orter** *n* **1** caretaker, concierge, door-keeper, doorman, gatekeeper, janitor, security-guard, watchman. **2** baggage-handler, bearer, carrier.

**ortion** *n* allocation, allowance, bit, chunk, division, fraction, fragment, helping, hunk, measure, part, percentage, piece, quantity, quota, ration, scrap, section, segment, serving, share, slice, sliver, subdivision, wedge. **portion out** > SHARE.

**ortrait** *n* depiction, image, likeness, picture, portrayal, profile, representation, self-portrait. > PICTURE.

**ortray** *vb* **1** delineate, depict, describe, evoke, illustrate, paint, picture, represent, show. **2** ▷ IMPERSONATE.

**ose** *n* **1** attitude, position, posture, stance. **2** act, affectation, attitudinizing, façade, masquerade, pretence. • *vb* **1** keep still, model, sit, strike a pose. **2** attitudinize, *inf* be a poser, be a poseur, posture, *inf* put on airs, show off. **3** *pose a question.* advance, ask, broach, posit, postulate, present, put forward, submit, suggest. **pose as** ▷ IMPERSONATE.

**poser** *n* **1** dilemma, enigma, problem, puzzle, question, riddle. **2** ▷ POSEUR.

**poseur** *n* attitudinizer, exhibitionist, fraud, impostor, masquerader, *inf* phoney, *inf* poser, pretender, *inf* show-off.

**posh** *adj inf* classy, elegant, fashionable, formal, grand, lavish, luxurious, ostentation, rich, showy, smart, snobbish, stylish, sumptuous, *inf* swanky, *inf* swish.

**position** *n* **1** locality, location, locus, place, placement, point, reference, site, situation, spot, whereabouts. **2** *an awkward position.* circumstances, condition, predicament, situation, state. **3** *position of the body.* angle, pose, posture, stance. **4** *intellectual position.* assertion, attitude, contention, hypothesis, opinion, outlook, perspective, principle, proposition, standpoint, thesis, view, viewpoint. **5** *position in a firm.* appointment, degree, employment, function, grade, job, level, niche, occupation, place, post, rank, role, standing, station, status, title. • *vb* arrange, deploy, dispose, fix, locate, place, put, settle, site, situate, stand, station.

**positive** *adj* **1** affirmative, assured, categorical, certain, clear, conclusive, confident, convinced, decided, definite, emphatic, explicit, firm, incontestable, incontrovertible, irrefutable, real, sure, undeniable, unequivocal. **2** *positive advice.* beneficial, constructive, helpful, optimistic, practical, useful, worthwhile. *Opp* NEGATIVE.

**possess** *vb* **1** be in possession of, enjoy, have, hold, own. **2** be gifted with, embody, embrace, include. **3** *possess territory*. acquire, control, dominate, govern, invade, occupy, rule, seize, take over. **4** *possess a person*. bewitch, captivate, cast a spell over, charm, enthral, haunt, hypnotize, obsess.

**possessions** *plur n* assets, belongings, chattels, effects, estate, fortune, goods, property, riches, things, wealth, worldly goods.

**possessive** *adj* clinging, dominating, domineering, jealous, overbearing, proprietorial, protective, selfish. ▷ GREEDY.

**possibility** *n* capability, chance, danger, feasibility, likelihood, odds, opportunity, plausibility, potential, potentiality, practicality, probability, risk.

**possible** *adj* achievable, admissible, attainable, conceivable, credible, *inf* doable, feasible, imaginable, likely, obtainable, *inf* on, plausible, potential, practicable, practical, probable, prospective, realizable, reasonable, tenable, thinkable, viable, workable. *Opp* IMPOSSIBLE.

**possibly** *adv* God willing, *inf* hopefully, if possible, maybe, *old use* peradventure, *old use* perchance, perhaps.

**post** *n* **1** baluster, bollard, brace, capstan, column, gate-post, leg, newel, pale, paling, picket, pier, pile, pillar, pole, prop, pylon, shaft, stake, stanchion, standard, starting-post, strut, support, upright, winning-post. **2** *a sentry's post*. location, place, point, position, station. **3** *post in a firm*. appointment, assignment, employment, function, job, occupation, office, place, position, situation, task, work. **4** airmail, cards, delivery, junk mail, letters, mail, packets, parcels, postcards. • *vb* **1** advertise, announce, display, pin up, proclaim, promulgate, publicize, publish, put up, stick up. **2** *post a letter*. dispatch, mail, send, transmit. **3** *post a sentry*. appoint, assign, locate, place, position, set, situate, station.

**poster** *n* advertisement, announcement, bill, broadsheet, circular, display, flyer, notice, placard, sign.

**posterity** *n* descendants, future generations, heirs, issue, offspring, progeny, successors.

**postpone** *vb* adjourn, defer, delay, extend, hold over, keep in abeyance, lay aside, put back, put off, *inf* put on ice, *inf* put on the back burner, *inf* shelve, stay, suspend, temporize.

**postscript** *n* addendum, addition, afterthought, codicil (*to will*), epilogue, *inf* PS.

**postulate** *vb* assume, hypothesize, posit, propose, suppose, theorize.

**posture** *n* **1** appearance, bearing, carriage, deportment, pose, position, stance. **2** ▷ ATTITUDE.

**posy** *n* bouquet, bunch of flowers, buttonhole, corsage, nosegay, spray.

**pot** *n* basin, bowl, casserole, cauldron, container, crock, crucible, dish, jar, pan, saucepan, stewpot, teapot, urn, vessel.

**potent** *adj* **1** effective, forceful, formidable, influential, intoxicating (*drink*), mighty, overpowering, overwhelming, powerful, puissant, strong, vigorous. ▷ STRONG. **2** *a potent argument*. ▷ PERSUASIVE. *Opp* WEAK.

**potential** *adj* **1** aspiring, budding, embryonic, future, *inf* hopeful, intending, latent, likely, possible, probable, promising, prospective, *inf* would-be. **2** *potential disaster.* imminent, impending, looming, threatening. • *n* aptitude, capability, capacity, possibility, resources.

**potion** *n* brew, concoction, decoction, dose, draught, drink, drug, elixir, liquid, medicine, mixture, philtre, potation, tonic.

**potter** *vb* dabble, do odd jobs, fiddle about, loiter, mess about, tinker, work.

**pottery** *n* ceramics, china, crockery, crocks, earthenware, porcelain, stoneware, terracotta.

**pouch** *n* bag, pocket, purse, reticule, sack, wallet.

**poultry** *n* □ *bantam, chicken, duck, fowl, goose, guinea-fowl, hen, pullet, turkey.*

**pounce** *vb* ambush, attack, drop on, jump on, leap on, seize, snatch, spring at, strike, swoop down on, take by surprise.

**pound** *n* compound, corral, enclosure, pen. • *vb* batter, beat, crush, grind, hammer, knead, mash, powder, pulp, pulverize, smash. ▷ HIT.

**pour** *vb* **1** cascade, course, discharge, disgorge, flood, flow, gush, run, spew, spill, spout, spurt, stream. **2** *pour wine.* decant, empty, serve, tip.

**poverty** *n* **1** beggary, bankruptcy, debt, destitution, hardship, impecuniousness, indigence, insolvency, necessity, need, penury, privation, want. **2** *a poverty of talent.* absence, dearth, insufficiency, lack, paucity, scarcity, shortage. *Opp* WEALTH.

**powder** *n* dust, particles, talc. • *vb* **1** atomize, comminute, crush, granulate, grind, pound, pulverize, reduce to powder. **2** besprinkle, coat, cover with powder, dredge, dust, sprinkle.

**powdered** *adj* **1** ▷ POWDERY. **2** dehydrated, dried, freeze-dried.

**powdery** *adj* chalky, crumbly, crushed, disintegrating, dry, dusty, fine, friable, granular, granulated, ground, loose, powdered, pulverized, sandy. *Opp* SOLID, WET.

**power** *n* **1** ability, capability, capacity, competence, drive, energy, faculty, force, might, muscle, potential, skill, talent, vigour. **2** *power to arrest.* authority, privilege, right. **3** *power of a tyrant.* ascendancy, *inf* clout, command, control, dominance, domination, dominion, influence, mastery, omnipotence, oppression, potency, rule, sovereignty, supremacy, sway. ▷ STRENGTH. *Opp* WEAKNESS.

**powerful** *adj* authoritative, cogent, commanding, compelling, consuming, convincing, dominant, dynamic, effective, effectual, energetic, forceful, high-powered, influential, invincible, irresistible, mighty, muscular, omnipotent, overpowering, overwhelming, persuasive, potent, sovereign, vigorous, weighty. ▷ STRONG. *Opp* POWERLESS.

**powerless** *adj* defenceless, disabled, feeble, helpless, impotent, incapable, incapacitated, ineffective, ineffectual, paralysed, unable, unfit. ▷ WEAK. *Opp* POWERFUL.

**practicable** *adj* achievable, attainable, *inf* doable, feasible, performable, possible, practical, realistic, sensible, viable, workable. *Opp* IMPRACTICABLE.

**practical** *adj* **1** applied, empirical, experimental. **2** businesslike, capable, competent, down-to-earth, efficient, expert, hard-headed, matter-of-fact, *inf* no-nonsense, pragmatic, proficient, realistic, sensible, skilled. **3** *a practical tool.* convenient, functional, handy, usable, useful, utilitarian. **4** ▷ PRACTICABLE. *Opp* IMPRACTICAL, THEORETICAL.

**practical joke** ▷ TRICK.

**practically** *adv* almost, close to, just about, nearly, to all intents and purposes, virtually.

**practice** *n* **1** action, actuality, application, doing, effect, operation, reality, use. **2** *inf* dummy-run, exercise, practising, preparation, rehearsal, *inf* run-through, training, *inf* try-out, *inf* work-out. **3** *common practice.* convention, custom, habit, modus operandi, routine, tradition, way, wont. **4** *a doctor's practice.* business, office, work.

**practise** *vb* **1** do exercises, drill, exercise, prepare, rehearse, train, warm up, *inf* work out. **2** *practise what you preach.* apply, carry out, do, engage in, follow, make a practice of, perform, put into practice.

**praise** *n* **1** acclaim, acclamation, accolade, admiration, adulation, applause, approbation, approval, commendation, compliment, congratulation, encomium, eulogy, homage, honour, ovation, panegyric, plaudits, testimonial, thanks, tribute. **2** *praise to God.* adoration, devotion, glorification, worship. • *vb* **1** acclaim, admire, applaud, cheer, clap, commend, compliment, congratulate, *inf* crack up, eulogize, extol, give a good review of, marvel at, offer praise to, pay tribute to, *inf* rave about, recommend, *inf* say nice things about, show approval of. *Opp* CRITICIZE. **2** *praise God.* adore, exalt, glorify, honour, laud, magnify, worship. *Opp* CURSE.

**praiseworthy** *adj* admirable, commendable, creditable, deserving, laudable, meritorious, worthy. ▷ GOOD. *Opp* BAD.

**pram** *n* baby-carriage, *old use* perambulator, push-chair.

**prance** *vb* bound, caper, cavort, dance, frisk, frolic, gambol, hop, jig about, jump, leap, play, romp, skip, spring.

**prattle** *vb* babble, blather, chatter, gabble, maunder, *inf* rattle on, *inf* witter on.

**pray** *vb* beseech, call upon, invoke, say prayers, supplicate. ▷ ASK.

**prayer** *n* collect, devotion, entreaty, invocation, litany, meditation, petition, praise, supplication.

**prayer-book** *n* breviary, missal.

**preach** *vb* **1** deliver a sermon, evangelize, expound, proselytize, spread the Gospel. **2** expatiate, give moral advice, harangue, *inf* lay down the law, lecture, moralize, pontificate, sermonize.

**preacher** *n* cleric, crusader, divine, ecclesiastic, evangelist, minister, missionary, moralist, pastor, revivalist. ▷ CLERGYMAN.

**prearranged** *adj* arranged beforehand, fixed, planned, predetermined, prepared, rehearsed, thought out. *Opp* SPONTANEOUS.

**precarious** *adj* dangerous, *inf* dicey, *inf* dodgy, dubious, hazardous, insecure, perilous, risky, rocky, shaky, slippery, treacherous, uncertain, unreliable, unsafe, unstable, unsteady, vulnerable, wobbly. *Opp* SAFE.

**precaution** *n* anticipation, defence, insurance, preventive measure, protection, provision, safeguard, safety measure.

**precede** *vb* be in front of, come before, go ahead, go before, go in front, herald, introduce, lead, lead into, pave the way for, preface, prefix, start, usher in. *Opp* FOLLOW.

**precious** *adj* **1** costly, expensive, invaluable, irreplaceable, priceless, valuable. *Opp* WORTHLESS. **2** adored, beloved, darling, dear, loved, prized, treasured, valued, venerated.

**precipice** *n* bluff, cliff, crag, drop, escarpment, precipitous face, rock.

**precipitate** *adj* breakneck, hasty, headlong, meteoric, premature. ▷ QUICK. • *vb* accelerate, advance, bring on, cause, encourage, expedite, further, hasten, hurry, incite, induce, instigate, occasion, provoke, spark off, trigger off.

**precipitation** *n* □ *dew, downpour, drizzle, hail, rain, rainfall, shower, sleet, snow, snowfall.*

**precipitous** *adj* abrupt, perpendicular, sharp, sheer, steep, vertical.

**precise** *adj* **1** accurate, clear-cut, correct, defined, definite, distinct, exact, explicit, fixed, measured, right, specific, unambiguous, unequivocal, well-defined. *Opp* IMPRECISE. **2** *precise work.* careful, critical, exacting, fastidious, faultless, finicky, flawless, meticulous, nice, perfect, punctilious, rigorous, scrupulous. *Opp* CARELESS.

**preclude** *vb* avert, avoid, bar, debar, exclude, forestall, frustrate, impede, make impossible, obviate, pre-empt, prevent, prohibit, rule out, thwart.

**precocious** *adj* advanced, forward, gifted, mature, quick. ▷ CLEVER. *Opp* BACKWARD.

**preconception** *n* assumption, bias, expectation, preconceived idea, predisposition, prejudgement, prejudice, presupposition.

**predatory** *adj* acquisitive, avaricious, covetous, extortionate, greedy, hunting, marauding, pillaging, plundering, preying, rapacious, ravenous, voracious.

**predecessor** *n* ancestor, antecedent, forebear, forefather, forerunner, precursor.

**predetermined** *adj* **1** fated, destined, doomed, ordained, predestined. **2** agreed, prearranged, preplanned, recognized, *inf* set up.

**predicament** *n* crisis, difficulty, dilemma, embarrassment, emergency, *inf* fix, impasse, *inf* jam, *inf* mess, *inf* pickle, plight, problem, quandary, situation, state.

**predict** *vb* augur, forebode, forecast, foresee, foreshadow, foretell, foretoken, forewarn, hint, intimate, presage, prognosticate, prophesy, tell fortunes.

**predictable** *adj* anticipated, certain, expected, foreseeable, foreseen, likely, *inf* on the cards, probable, sure, unsurprising. *Opp* UNPREDICTABLE.

**predominant** *adj* ascendant, chief, dominating, leading, main, preponderant, prevailing, prevalent, primary, ruling, sovereign.

**predominate** *vb* be in the majority, control, dominate, *inf* have the upper hand, hold sway, lead, outnumber, outweigh, preponderate, prevail, reign, rule.

**pre-eminent** *adj* distinguished, eminent, excellent, incomparable, matchless, outstanding, peerless, supreme, unrivalled, unsurpassed.

**pre-empt** *vb* anticipate, appropriate, arrogate, expropriate, forestall, seize, take over.

**preface** *n* exordium, foreword, introduction, *inf* lead-in, overture, preamble, prelude, proem, prolegomenon, prologue. • *vb* begin, introduce, lead into, open, precede, prefix, start.

**prefer** *vb* advocate, *inf* back, be partial to, choose, fancy, favour, *inf* go for, incline towards, like, like better, opt for, pick out, *inf* plump for, *inf* put your money on, recommend, select, single out, think preferable, vote for, want.

**preferable** *adj* advantageous, better, better-liked, chosen, desirable, favoured, likely, nicer, preferred, recommended, wanted. *Opp* OBJECTIONABLE.

**preference** *n* **1** choice, fancy, favourite, liking, option, pick, selection, wish. **2** favouritism, inclination, partiality, predilection, prejudice, proclivity.

**preferential** *adj* advantageous, better, biased, favourable, favoured, privileged, showing favouritism, special, superior.

**pregnant** *adj* **1** carrying a child, expectant, *inf* expecting, gestating, gravid, parturient, *old use* with child. **2** *pregnant remark.* ▷ MEANINGFUL.

**prejudice** *n* bias, bigotry, chauvinism, discrimination, dogmatism, fanaticism, favouritism, intolerance, jingoism, leaning, narrow-mindedness, partiality, partisanship, predilection, predisposition, prejudgement, racialism, racism, sexism, unfairness, xenophobia. *Opp* TOLERANCE. • *vb* **1** bias, colour, incline, influence, interfere with, make prejudiced, predispose, sway. **2** *prejudice your chances.* damage, harm, injure, ruin, spoil, undermine.

**prejudiced** *adj* biased, bigoted, chauvinist, discriminatory, illiberal, intolerant, jaundiced, jingoistic, leading (*question*), loaded, narrow-minded, one-sided, parochial, partial, partisan, racist, sexist, tendentious, unfair, xenophobic. *Opp* IMPARTIAL. **prejudiced person** bigot, chauvinist, fanatic, racist, sexist, zealot.

**prejudicial** *adj* damaging, deleterious, detrimental, disadvantageous, harmful, inimical, injurious, unfavourable.

**preliminary** *adj* advance, earliest, early, experimental, exploratory, first, inaugural, initial, introductory, opening, prefatory, preparatory, qualifying, tentative, trial. • *n* ▷ PRELUDE.

**prelude** *n* beginning, *inf* curtain-raiser, exordium, introduction, opener, opening, overture, preamble, precursor, preface, preliminary, preparation, proem, prolegomenon, prologue, start, starter, *inf* warm-up. *Opp* CONCLUSION, POSTSCRIPT.

**premature** *adj* abortive, before time, early, hasty, ill-timed, precipitate, *inf* previous, too early, too soon, undeveloped, untimely. *Opp* LATE.

**premeditated** *adj* calculated, conscious, considered, contrived, deliberate, intended, intentional, planned, prearranged, preconceived, predetermined, preplanned, studied, wilful. *Opp* SPONTANEOUS.

**premiss** *n* assertion, assumption, basis, grounds, hypothesis, proposition, supposition, theorem.

**premonition** *n* anxiety, fear, foreboding, forewarning, *inf* funny feeling, *inf* hunch, indication, intuition, misgiving, omen, portent, presentiment, suspicion, warning, worry.

**preoccupied** *adj* **1** absorbed, engaged, engrossed, immersed, interested, involved, obsessed, sunk, taken up, wrapped up. **2** absent-minded, abstracted, daydreaming, distracted, faraway, inattentive, lost in thought, musing, pensive, pondering, rapt, reflecting, thoughtful.

**preparation** *n* arrangements, briefing, *inf* gearing up, getting ready, groundwork, making provision, measures, organization, plans, practice, preparing, setting up, spadework, training.

**prepare** *vb* **1** arrange, cook, devise, *inf* do what's necessary, *inf* fix up, get ready, make arrangements, make ready, organize, pave the way, plan, process, set up. ▷ MAKE. **2** *prepare for exams.* *inf* cram, practise, revise, study, *inf* swot. **3** *prepare pupils for exams.* brief, coach, educate, equip, instruct, rehearse, teach, train, tutor. **prepared** ▷ PRE-ARRANGED, READY. **prepare yourself** be prepared, be ready, brace yourself, discipline yourself, fortify yourself, steel yourself.

**preposterous** *adj* bizarre, excessive, extreme, grotesque, monstrous, outrageous, surreal, unreasonable, unthinkable. ▷ ABSURD.

**prerequisite** *adj* compulsory, essential, indispensable, mandatory, necessary, obligatory, prescribed, required, requisite, specified, stipulated. *Opp* OPTIONAL. • *n* condition, essential, necessity, precondition, proviso, qualification, requirement, requisite, *Lat* sine qua non, stipulation.

**prescribe** *n* advise, assign, command, demand, dictate, direct, fix, impose, instruct, lay down, ordain, order, recommend, require, specify, stipulate, suggest.

**presence** *n* **1** attendance, closeness, companionship, company, nearness, propinquity, proximity, society. **2** air, appearance, aura, bearing, comportment, demeanour, impressiveness, mien, personality, poise, selfassurance, self-possession.

**present** *adj* **1** adjacent, at hand, close, here, in attendance, nearby. **2** contemporary, current, existing, extant, present-day, up-to-date. • *n* **1** *inf* here and now, today. **2** alms, bonus, bounty, charity, contribution, donation, endowment, gift, grant, gratuity, handout, offering, tip. • *vb* **1** award, bestow, confer, dispense, distribute, donate, give, hand over, offer. **2** *present evidence.* adduce, bring forward, demonstrate, display, exhibit, furnish, proffer, put forward, reveal, set out, show, submit. **3** *present a guest.* announce, introduce, make known. **4** *present a play.* act, bring out, perform, put on, stage. **present yourself** ▷ ATTEND, REPORT.

**presentable** *adj* acceptable, adequate, all right, clean, decent, decorous, fit to be seen, good enough, neat, passable, proper, respectable, satisfactory, suitable, tidy, tolerable, *inf* up to scratch, worthy.

**presently** *adv old use* anon, before long, by and by, *inf* in a jiffy, shortly, soon.

**preserve** *n* **1** conserve, jam, jelly, marmalade. **2** *wildlife preserve.* reservation, reserve, sanctuary. • *vb* care for, conserve, defend, guard, keep, lay up, look after, maintain, perpetuate, protect, retain, safeguard, save, secure, stockpile, store, support, sustain, uphold, watch over. *Opp* DESTROY. **2** *preserve food.* bottle, can, chill, cure, dehydrate, dry, freeze, freeze-dry, irradiate, jam, pickle, refrigerate, salt, tin. **3** *preserve a corpse.* embalm, mummify.

**preside** *vb* be in charge, chair, officiate, take charge, take the chair. **preside over** ▷ GOVERN.

**press** *n* newspapers, magazines, the media. • *vb* **1** apply pressure to, compress, condense, cram, crowd, crush, depress, force, gather, *inf* jam, push, shove, squash, squeeze, subject to pressure. **2** *press laundry.* flatten, iron, smooth. **3** *press someone to stay.* ask, beg, bully, coerce, constrain, dragoon, entreat, exhort, implore, importune, induce, *inf* lean on, persuade, pressure, pressurize, put pressure on, request, require, urge. **pressing** ▷ URGENT.

**pressure** *n* **1** burden, compression, force, heaviness, load, might, power, stress, weight. **2** *pressure of modern life.* adversity, affliction, constraints, demands, difficulties, exigencies, *inf* hassle, hurry, oppression, problems, strain, stress, urgency. • *vb* ask, beg, bully, coerce, constrain, dragoon, entreat, exhort, implore, importune, induce, *inf* lean on, persuade, press, pressurize, put pressure on, request, require, urge.

**prestige** *n* cachet, celebrity, credit, distinction, eminence, esteem, fame, glory, good name, honour, importance, influence, *inf* kudos, regard, renown, reputation, respect, standing, stature, status.

**prestigious** *adj* acclaimed, august, celebrated, creditable, distinguished, eminent, esteemed, estimable, famed, famous, highly-regarded, high-ranking, honourable, honoured, important, influential, pre-eminent, renowned, reputable, respected, significant, wellknown. *Opp* INSIGNIFICANT.

**presume** *vb* **1** assume, believe, conjecture, gather, guess, hypothesize, imagine, infer, postulate, suppose, surmise, suspect, *inf* take for granted, *inf* take it, think. **2** *He presumed to correct me.* be presumptuous enough, dare, have the effrontery, make bold, take the liberty, venture.

**presumptuous** *adj* arrogant, bold, brazen, *inf* cheeky, conceited, forward, impertinent, impudent, insolent, overconfident, *inf* pushy, shameless, unauthorized, unwarranted. ▷ PROUD.

**pretence** *n* act, acting, affectation, appearance, artifice, camouflage, charade, counterfeiting, deception, disguise, display, dissembling, dissimulation, excuse, façade, falsification, feigning, feint, fiction, front, guise, hoax, *inf* humbug, hypocrisy, insincerity, invention, lying, make-believe, masquerade, pose, posing, posturing, pretext, ruse, sham, show, simulation, subterfuge, trickery, wile. ▷ DECEIT.

**pretend** *vb* **1** act, affect, allege, behave insincerely, bluff, counterfeit, deceive, disguise, dissemble, dissimulate, fake, feign, fool, hoax, hoodwink, imitate, impersonate, *inf* kid, lie, *inf* make out, mislead, play-act, play a part, perform, pose, posture, profess, purport, put on an act, sham, simulate, take someone in, trick. **2** ▷ IMAGINE. **3** ▷ CLAIM.

**pretender** *n* aspirant, claimant, rival, suitor.

**pretentious** *adj* affected, *inf* arty, conceited, exaggerated, extravagant, grandiose, *inf* highfalutin, immodest, inflated, ostentatious, overblown, *inf* over the top, pompous, showy, *inf* snobbish, superficial. *Opp* UNPRETENTIOUS.

**pretext** *n* cloak, cover, disguise, excuse, pretence.

**pretty** *adj* appealing, attractive, *inf* bonny, charming, *inf* cute, dainty, delicate, *inf* easy on the eye, fetching, good-looking, lovely, nice, pleasing, *derog* pretty-pretty, winsome. ▷ BEAUTIFUL. *Opp* UGLY. • *adv* fairly, moderately, quite, rather, reasonably, somewhat, tolerably. *Opp* VERY.

**prevail** *vb* be prevalent, hold sway, predominate, preponderate, succeed, triumph, *inf* win the day. ▷ WIN. **prevailing** ▷ PREVALENT.

**prevalent** *adj* accepted, ascendant, chief, common, commonest, current, customary, dominant, dominating, effectual, established, extensive, familiar, fashionable, general, governing, influential, main, mainstream, normal, ordinary, orthodox, pervasive, popular, powerful, predominant, prevailing, principal, ruling, ubiquitous, universal, usual, widespread. *Opp* UNUSUAL.

**prevaricate** *vb inf* beat about the bush, be evasive, cavil, deceive, equivocate, *inf* fib, hedge, lie, mislead, quibble, temporize.

**prevent** *vb* anticipate, avert, avoid, baffle, bar, block, check, control, curb, deter, fend off, foil, forestall, frustrate, hamper, *inf* head off, *inf* help (*can't help it*), hinder, impede, inhibit, inoculate against, intercept, *inf* nip in the bud, obstruct, obviate, preclude, pre-empt, prohibit, *inf* put a stop to, restrain, save, stave off, stop, take precautions against, thwart, ward off. ▷ FORBID. *Opp* ENCOURAGE.

**preventive** *adj* anticipatory, counteractive, deterrent, obstructive, precautionary, pre-emptive, preventative.

**previous** *adj* **1** above-mentioned, aforementioned, aforesaid, antecedent, earlier, erstwhile, foregoing, former, past, preceding, prior. *Opp* SUBSEQUENT. **2** ▷ PREMATURE.

**prey** *n* kill, quarry, victim. • *vb* **prey on** eat, feed on, hunt, kill, live off. ▷ EXPLOIT.

**price** *n* **1** amount, charge, cost, *inf* damage, expenditure, expense, fare, fee, figure, outlay, payment, rate, sum, terms, toll, value, worth. **2** *Give me a price.* estimate, offer, quotation, valuation. ▷ PAYMENT. **pay the price for** ▷ ATONE.

**priceless** *adj* **1** costly, dear, expensive, incalculable, inestimable, invaluable, irreplaceable, precious, *inf* pricey, rare, valuable. *Opp* WORTHLESS. **2** ▷ FUNNY.

**prick** *vb* **1** bore into, jab, lance, perforate, pierce, punch, puncture, riddle, stab, sting. **2** ▷ STIMULATE.

**prickle** *n* **1** barb, bristle, bur, needle, spike, spine, thorn. **2** irritation, itch, pricking, prickling, tingle, tingling. • *vb* irritate, itch, make your skin crawl, scratch, sting, tingle.

**prickly** *adj* **1** barbed, bristly, rough, scratchy, sharp, spiky, spiny, stubbly, thorny, unshaven. *Opp* SMOOTH. **2** ▷ IRRITABLE.

**pride** *n* **1** *Fr* amour propre, gratification, happiness, honour, pleasure, satisfaction, self-respect, self-satisfaction. ▷ DIGNITY. **2** *her pride and joy.* jewel, treasure, treasured possession. 3 *pride before a fall.* arrogance, being proud, *inf* big-headedness, boastfulness, conceit, egotism, haughtiness, hubris, megalomania, narcissism, overconfidence, presumption, self-admiration, self-esteem, self-importance, self-love, smugness, snobbery, snobbishness, vainglory, vanity. *Opp* HUMILITY.

**priest** *n* confessor, Druid, lama, minister, preacher. ▷ CLERGYMAN.

**priggish** *adj* conservative, fussy, *inf* goody-goody, haughty, moralistic, prudish, self-righteous, sententious, stiff-necked, *inf* stuffy. ▷ PRIM.

**prim** *adj* demure, fastidious, formal, inhibited, narrow-minded, precise, *inf* prissy, proper, prudish, *inf* starchy, strait-laced. *Opp* BROADMINDED.

**primal** *adj* 1 early, earliest, first, original, primeval, primitive, primordial. 2 ▷ PRIMARY.

**primarily** *adv* basically, chiefly, especially, essentially, firstly, fundamentally, generally, mainly, mostly, particularly, predominantly, pre-eminently, principally.

**primary** *adj* basic, cardinal, chief, dominant, first, foremost, fundamental, greatest, important, initial, leading, main, major, outstanding, paramount, predominant, pre-eminent, primal, prime, principal, supreme, top.

**prime** *adj* 1 best, first-class, first-rate, foremost, select, superior, top, top-quality. ▷ EXCELLENT. 2 ▷ PRIMARY. • *vb* get ready, prepare.

**primitive** *adj* 1 aboriginal, ancient, barbarian, early, first, prehistoric, primeval, savage, uncivilized, uncultivated. 2 *primitive technology.* antediluvian, backward, basic, *inf* behind the times, crude, elementary, obsolete, rough, rudimentary, simple, simplistic, undeveloped. ▷ OLD. 3 *primitive art.* childlike, crude, naïve, unpolished, unrefined, unsophisticated. *Opp* ADVANCED, SOPHISTICATED.

**principal** *adj* basic, cardinal, chief, dominant, dominating, first, foremost, fundamental, greatest, highest, important, key, leading, main, major, outstanding, paramount, preeminent, predominant, prevailing, primary, prime, starring, supreme, top. • *n* 1 ▷ CHIEF. 2 *the principal in a play.* diva, hero, heroine, lead, leading role, prima ballerina, protagonist, star.

**principle** *n* 1 assumption, axiom, belief, creed, criterion, doctrine, dogma, ethic, idea, ideal, maxim, notion, precept, proposition, rule, standard, teaching, tenet, truism, truth, values. 2 *a person of principle.* conscience, high-mindedness, honesty, honour, ideals, integrity, morality, probity, scruples, standards, uprightness, virtue. **principles** basics, elements, essentials, fundamentals, laws, philosophy, theory.

**print** *n* 1 impression, imprint, indentation, mark, stamp. 2 characters, fount, lettering, letters, printing, text, type, typeface. 3 copy, duplicate, engraving, etching, facsimile, linocut, lithograph, monoprint, photograph, reproduction, silk screen, woodcut. ▷ PICTURE. • *vb* 1 copy, impress, imprint, issue, publish, run off, stamp. 2 ▷ WRITE.

**prior** *adj* earlier, erstwhile, former, late, old, onetime, previous.

**priority** *n* first place, greater importance, precedence, preference, prerogative, right-of-way, seniority, superiority, urgency.

**prise** *vb* force, lever, prize, wrench.

**prison** *n old use* approved school, *old use* Borstal, cell, *sl* clink, custody, detention centre, dungeon, gaol, guardhouse, house of correction, jail, *inf* lock-up, *Amer* penitentiary, oubliette, reformatory, *sl* stir, youth custody centre. ▷ CAPTIVITY.

**prisoner** *n* captive, convict, detainee, *inf* gaolbird, hostage, inmate, internee, lifer, *inf* old lag, *inf* trusty.

**privacy** *n* concealment, isolation, monasticism, quietness, retirement, retreat, seclusion, secrecy, solitude.

**private** *adj* **1** exclusive, individual, particular, personal, privately owned, reserved. **2** classified, confidential, *inf* hush-hush, *inf* off the record, restricted, secret, top secret, undisclosed. **3** *a private meeting*. clandestine, closed, covert, intimate, surreptitious. **4** *a private hideaway*. concealed, hidden, inaccessible, isolated, little-known, quiet, secluded, sequestered, solitary, unknown, withdrawn. *Opp* PUBLIC.

**privilege** *n* advantage, benefit, concession, entitlement, exemption, freedom, immunity, licence, prerogative, right.

**privileged** *adj* **1** advantaged, authorized, élite, entitled, favoured, honoured, immune, licensed, powerful, protected, sanctioned, special, superior. **2** ▷ WEALTHY.

**prize** *n* accolade, award, jackpot, *inf* purse, reward, trophy, winnings. • *vb* **1** appreciate, approve of, cherish, esteem, hold dear, like, rate highly, regard, revere, treasure, value. **2** ▷ PRISE.

**probable** *adj* believable, convincing, credible, expected, feasible, likely, *inf* odds-on, plausible, possible, predictable, presumed, undoubted, unquestioned. *Opp* IMPROBABLE.

**probationer** *n* apprentice, beginner, inexperienced worker, learner, novice, tiro.

**probe** *n* enquiry, examination, exploration, inquiry, investigation, research, scrutiny, study. • *vb* **1** delve, dig, penetrate, plumb, poke, prod. **2** *probe a problem*. examine, explore, go into, inquire into, investigate, look into, research into, scrutinize, study.

**problem** *n* **1** brain-teaser, conundrum, enigma, mystery, *inf* poser, puzzle, question, riddle. **2** *a worrying problem*. burden, *inf* can of worms, complication, difficulty, dilemma, dispute, *inf* facer, *inf* headache, *inf* hornet's nest, predicament, quandary, set-back, snag, trouble, worry.

**problematic** *adj* complicated, controversial, debatable, difficult, disputed, doubtful, enigmatic, hard to deal with, *inf* iffy, intractable, moot (*point*), problematical, puzzling, questionable, sensitive, taxing, *inf* tricky, uncertain, unsettling, worrying. *Opp* STRAIGHTFORWARD.

**procedure** *n* approach, conduct, course of action, *inf* drill, formula, method, methodology, *Lat* modus operandi, plan of action, policy, practice, process, routine, scheme, strategy, system, technique, way.

**proceed** *vb* **1** advance, carry on, continue, follow, forge ahead, *inf* get going, go ahead, go on, make headway, make progress, move along, move forward, *inf* press on, progress. **2** arise, be derived, begin, develop, emerge, grow, originate, spring up, start. ▷ RESULT.

**proceedings** *plur n* 1 events, *inf* goings-on, happenings, things. 2 *legal proceedings.* action, lawsuit, procedure, process. 3 *proceedings of a meeting.* annals, business, dealings, *inf* doings, matters, minutes, records, report, transactions.

**proceeds** *plur n* earnings, gain, gate, income, profit, receipts, returns, revenue, takings. ▷ MONEY.

**process** *n* 1 function, method, operation, procedure, proceeding, system, technique. 2 *process of ageing.* course, development, evolution, experience, progression. • *vb* 1 alter, change, convert, deal with, make usable, manage, modify, organize, prepare, refine, transform, treat. 2 ▷ PARADE.

**procession** *n* cavalcade, chain, column, cortège, file, line, march, march-past, motorcade, pageant, parade, sequence, string, succession, train.

**proclaim** *vb* 1 announce, advertise, assert, declare, give out, make known, profess, promulgate, pronounce, publish. 2 ▷ DECREE.

**procrastinate** *vb* be dilatory, be indecisive, dally, defer a decision, delay, *inf* dilly-dally, dither, *inf* drag your feet, equivocate, evade the issue, hesitate, *inf* hum and haw, pause, *inf* play for time, postpone, put things off, *inf* shilly-shally, stall, temporize, vacillate, waver.

**procure** *vb* acquire, buy, come by, find, get, *inf* get hold of, *inf* lay your hands on, obtain, *inf* pick up, purchase, requisition.

**prod** *vb* dig, elbow, goad, jab, nudge, poke, push, urge on. ▷ HIT, URGE.

**prodigal** *adj* excessive, extravagant, immoderate, improvident, irresponsible, lavish, profligate, reckless, self-indulgent, wasteful. *Opp* THRIFTY.

**prodigy** *n* curiosity, freak, genius, marvel, miracle, phenomenon, rarity, sensation, talent, virtuoso, *inf* whizz kid, wonder, *Ger* Wunderkind.

**produce** *n* crop, harvest, output, yield. ▷ PRODUCT. • *vb* 1 assemble, bring out, cause, compose, conjure up, construct, create, cultivate, develop, fabricate, form, generate, give rise to, grow, initiate, invent, make, manufacture, originate, provoke, result in, supply, think up, turn out, yield. 2 *produce evidence.* advance, bring out, disclose, display, exhibit, furnish, introduce, offer, present, provide, put forward, reveal, show, supply, throw up. 3 *produce children.* bear, beget, breed, give birth to, raise, rear. 4 *produce a play.* direct, mount, present, put on, stage.

**product** *n* 1 artefact, by-product, commodity, end-product, goods, merchandise, output, produce, production. 2 consequence, effect, fruit, issue, outcome, result, upshot, yield.

**productive** *adj* 1 beneficial, busy, constructive, creative, effective, efficient, gainful (*employment*), inventive, profitable, profitmaking, remunerative, rewarding, useful, valuable, worthwhile. 2 *a productive garden.* abundant, bounteous, bountiful, fecund, fertile, fruitful, lush, prolific, vigorous. *Opp* UNPRODUCTIVE.

**profess** *vb* **1** affirm, announce, assert, asseverate, aver, confess, confirm, declare, maintain, state, vow. **2** *profess to be an expert.* allege, claim, make out, pretend, purport.

**profession** *n* **1** business, calling, career, craft, employment, job, line of work, métier, occupation, trade, vocation, work. **2** *profession of love.* acknowledgement, affirmation, announcement, assertion, avowal, confession, declaration, statement, testimony.

**professional** *adj* **1** able, authorized, educated, experienced, expert, knowledgeable, licensed, official, proficient, qualified, skilled, trained. *Opp* AMATEUR. **2** full-time, paid. **3** *professional work.* businesslike, competent, conscientious, efficient, masterly, proper, skilful, thorough, well-done. *Opp* UNPROFESSIONAL. • *n* expert, professional player, professional worker.

**proficient** *adj* able, accomplished, adept, capable, competent, efficient, expert, gifted, professional, skilled, talented. *Opp* INCOMPETENT.

**profile** *n* **1** contour, outline, shape, side view, silhouette. **2** *personal profile.* account, biography, *Lat* curriculum vitae, sketch, study.

**profit** *n* advantage, benefit, excess, gain, interest, proceeds, return, revenue, surplus, yield. • *vb* **1** advance, avail, benefit, further the interests of, pay, serve. ▷ HELP. **2** *profit from a sale.* capitalize (on), *inf* cash in, earn money, gain, *inf* make a killing, make a profit, make money. **profit by, profit from** ▷ EXPLOIT.

**profitable** *adj* advantageous, beneficial, commercial, enriching, fruitful, gainful, lucrative, money-making, paying, productive, profit-making, remunerative, rewarding, useful, valuable, well-paid, worthwhile. *Opp* UNPROFITABLE.

**profiteer** *n* black-marketeer, exploiter, extortionist, racketeer. • *vb* exploit, extort, fleece, overcharge.

**profligate** *adj* **1** abandoned, debauched, degenerate, depraved, dissolute, immoral, libertine, licentious, loose, perverted, promiscuous, sinful, sybaritic, unprincipled, wanton. ▷ WICKED. **2** extravagant, prodigal, reckless, spendthrift, wasteful.

**profound** *adj* **1** deep, heartfelt, intense, sincere. **2** *a profound discussion.* abstruse, arcane, erudite, esoteric, imponderable, informed, intellectual, knowledgeable, learned, penetrating, philosophical, recondite, sagacious, scholarly, serious, thoughtful, wise. **3** *profound silence.* absolute, complete, extreme, fundamental, perfect, thorough, total, unqualified. *Opp* SUPERFICIAL.

**profuse** *adj* abundant, ample, bountiful, copious, extravagant, exuberant, generous, lavish, luxuriant, plentiful, productive, prolific, superabundant, thriving, unsparing, unstinting. *Opp* MEAN, SPARSE.

**programme** *n* **1** agenda, bill of fare, calendar, curriculum, *inf* line-up, listing, menu, plan, routine, prospectus, schedule, scheme, syllabus, timetable. **2** *a TV programme.* broadcast, performance, presentation, production, show, transmission.

**progress** *n* **1** advance, breakthrough, development, evolution, forward movement, furtherance, gain, growth, headway, improvement, march (*of time*), maturation, progression, *inf* step forward. 2 journey, route, travels, way. 3 *progress in a career*. advancement, betterment, elevation, promotion, rise, *inf* step up. • *vb* advance, *inf* come on, develop, *inf* forge ahead, go forward, go on, make headway, make progress, move forward, press forward, press on, proceed, prosper. ▷ IMPROVE. *Opp* REGRESS, STAGNATE.

**progression** *n* **1** ▷ PROGRESS. **2** chain, concatenation, course, flow, order, row, sequence, series, string, succession.

**progressive** *adj* **1** accelerating, advancing, continuing, continuous, developing, escalating, gradual, growing, increasing, ongoing, steady. **2** *progressive ideas*. advanced, avant-garde, contemporary, dynamic, enterprising, forward-looking, *inf* go-ahead, modernistic, radical, reformist, revisionist, revolutionary, up-to-date. *Opp* CONSERVATIVE.

**prohibit** *vb* ban, bar, block, censor, check, *inf* cut out, debar, disallow, exclude, foil, forbid, hinder, impede, inhibit, interdict, make illegal, outlaw, place an embargo on, preclude, prevent, proscribe, restrict, rule out, shut out, stop, taboo, veto. *Opp* ALLOW.

**prohibitive** *adj* discouraging, excessive, exorbitant, impossible, *inf* out of reach, out of the question, unreasonable, unthinkable.

**project** *n* activity, assignment, contract, design, enterprise, idea, job, piece of research, plan, programme, proposal, scheme, task, undertaking, venture. • *vb* **1** concoct, contrive, design, devise, invent, plan, propose, scheme, think up. **2** beetle, bulge, extend, jut out, overhang, protrude, stand out, stick out. **3** *project into space*. cast, *inf* chuck, fling, hurl, launch, lob, propel, shoot, throw. **4** *project light*. cast, flash, shine, throw out. **5** *project future profits*. estimate, forecast, predict.

**proliferate** *vb* burgeon, flourish, grow, increase, multiply, mushroom, reproduce, thrive.

**prolific** *adj* **1** abundant, bounteous, bountiful, copious, fruitful, numerous, plenteous, profuse, rich. **2** *a prolific writer*. creative, fertile, productive. *Opp* UNPRODUCTIVE.

**prolong** *vb* delay, *inf* drag out, draw out, elongate, extend, increase, keep up, lengthen, make longer, *inf* pad out, protract, *inf* spin out, stretch out. *Opp* SHORTEN.

**prominent** *adj* **1** conspicuous, discernible, distinguishable, evident, eye-catching, large, notable, noticeable, obtrusive, obvious, pronounced, recognizable, salient, significant, striking. *Opp* INCONSPICUOUS. **2** bulging, jutting out, projecting, protruding, protuberant, sticking out. **3** celebrated, distinguished, eminent, familiar, foremost, illustrious, important, leading, major, much-publicized, noted, outstanding, public, renowned. ▷ FAMOUS. *Opp* UNKNOWN.

**promiscuous** *adj* casual, haphazard, indiscriminate, irresponsible, non-selective, random, undiscriminating. ▷ IMMORAL. *Opp* MORAL.

**promise** *n* **1** assurance, commitment, compact, contract, covenant, guarantee, oath, pledge, undertaking, vow, word, word of honour. **2** *actor with promise*. capability, expectation(s), latent ability, potential, promising qualities, talent. • *vb* 1 agree, assure, commit yourself, consent, contract, engage, give a promise, give your word, guarantee, pledge, swear, take an oath, undertake, vow. 2 *The clouds promise rain.* augur, *old use* betoken, forebode, foretell, hint at, indicate, presage, prophesy, show signs of, suggest.

**promising** *adj* auspicious, budding, encouraging, favourable, hopeful, likely, optimistic, propitious, talented, *inf* up-and-coming.

**promontory** *n* cape, foreland, headland, peninsula, point, projection, ridge, spit, spur.

**promote** *vb* 1 advance, elevate, exalt, give promotion, move up, prefer, raise, upgrade. 2 *promote a product.* advertise, back, boost, champion, encourage, endorse, further, help, make known, market, patronize, *inf* plug, popularize, publicize, *inf* push, recommend, sell, speak for, sponsor, support. ▷ HELP.

**promoter** *n* backer, champion, patron, sponsor, supporter.

**promotion** *n* 1 advancement, elevation, preferment, rise, upgrading. 2 *promotion of a product.* advertising, backing, encouragement, furtherance, marketing, publicity, recommendation, selling, sponsorship.

**prompt** *adj* eager, efficient, expeditious, immediate, instantaneous, on time, punctual, timely, unhesitating, willing. ▷ QUICK. *Opp* UNPUNCTUAL. • *n* cue, line, reminder. • *vb* advise, coax, egg on, encourage, exhort, help, incite, influence, inspire, jog the memory, motivate, nudge, persuade, prod, provoke, remind, rouse, spur, stimulate, urge.

**prone** *adj* 1 face down, flat, horizontal, lying, on your front, prostrate, stretched out. *Opp* SUPINE. 2 *prone to colds.* apt, disposed, given, inclined, liable, likely, predisposed, subject, susceptible, tending, vulnerable. *Opp* IMMUNE.

**prong** *n* point, spike, spur, tine.

**pronounce** *vb* 1 articulate, aspirate, enunciate, express, put into words, say, sound, speak, utter, vocalize, voice. 2 *pronounce judgement.* announce, assert, asseverate, declare, decree, judge, make known, proclaim, state. ▷ SPEAK.

**pronounced** *adj* clear, conspicuous, decided, definite, distinct, evident, inescapable, marked, noticeable, obvious, prominent, recognizable, striking, unambiguous, undisguised, unmistakable, well-defined.

**pronunciation** *n* accent, articulation, delivery, diction, elocution, enunciation, inflection, intonation, modulation, speech.

**proof** *n* 1 authentication, certification, confirmation, corroboration, demonstration, evidence, facts, grounds, substantiation, testimony, validation, verification. 2 *the proof of the pudding.* criterion, judgement, measure, test, trial.

**prop** *n* brace, buttress, crutch, post, stay, strut, support, truss, upright. • *vb* 1 bolster, brace, buttress, hold up, reinforce, shore up, support, sustain. 2 lean, rest, stand.

**propaganda** *n* advertising, brainwashing, disinformation, indoctrination, persuasion, publicity.

**propagate** *vb* 1 breed, generate, increase, multiply, produce, proliferate, reproduce. 2 *propagate ideas.* circulate, disseminate, pass on, promote, promulgate, publish, spread, transmit. **3** *propagate plants*. grow from seed, layer, sow, take cuttings.

**propel** *vb* drive, force, impel, launch, move, *inf* pitchfork, push, send, set in motion, shoot, spur, thrust, urge.

**propeller** *n* rotor, screw, vane.

**proper** *adj* **1** becoming, conventional, decent, decorous, delicate, dignified, formal, genteel, gentlemanly, grave, in good taste, ladylike, modest, polite, *derog* prim, *derog* prudish, respectable, sedate, seemly, serious, solemn, tactful, tasteful. **2** acceptable, accepted, advisable, apposite, appropriate, apropos, apt, deserved, fair, fitting, just, lawful, legal, normal, orthodox, rational, sensible, suitable, unexceptionable, usual, valid. **3** *the proper time*. accurate, correct, exact, precise, right. **4** *the proper place*. allocated, distinctive, individual, own, particular, reserved, separate, special, unique. *Opp* IMPROPER.

**property** *n* **1** assets, belongings, capital, chattels, effects, fortune, *inf* gear, goods, holdings, patrimony, possessions, resources, riches, wealth. **2** acreage, buildings, estate, land, premises. **3** attribute, characteristic, feature, hallmark, idiosyncrasy, oddity, peculiarity, quality, quirk, trait.

**prophecy** *n* augury, crystal-gazing, divination, forecast, foretelling, fortune-telling, oracle, prediction, prognosis, prognostication, vaticination.

**prophesy** *vb* augur, bode, divine, forecast, foresee, foreshadow, foretell, portend, predict, presage, prognosticate, promise, vaticinate.

**prophet** *n* clairvoyant, forecaster, fortune-teller, oracle, seer, sibyl, soothsayer.

**prophetic** *adj* apocalyptic, far-seeing, oracular, predictive, prescient, prognostic, prophesying, sibylline.

**propitious** *adj* advantageous, auspicious, favourable, fortunate, happy, lucky, opportune, promising, providential, rosy, timely, well-timed.

**proportion** *n* **1** balance, comparison, correlation, correspondence, distribution, equivalence, ratio, statistical relationship. **2** allocation, fraction, part, percentage, piece, quota, ration, section, share. ▷ NUMBER, QUANTITY. **proportions** dimensions, extent, magnitude, measurements, size, volume.

**proportional** *adj* analogous, balanced, commensurate, comparable, corresponding, equitable, in proportion, just, proportionate, relative, symmetrical. *Opp* DISPROPORTIONATE.

**proposal** *n* bid, declaration, draft, motion, offer, plan, project, proposition, recommendation, scheme, statement, suggestion, tender.

**propose** *vb* **1** advance, ask for, *inf* come up with, present, propound, put forward, recommend, submit, suggest. **2** aim, have in mind, intend, mean, offer, plan, purpose. **3** *propose a candidate*. nominate, put forward, put up, sponsor.

**propriety** *n* appropriateness, aptness, correctness, courtesy, decency, decorum, delicacy, dignity, etiquette, fairness, fitness, formality, gentility, good form, good manners, gravity, justice, modesty, politeness, *derog* prudishness, refinement, respectability, sedateness, seemliness, sensitivity, suitability, tact, tastefulness. *Opp* IMPROPRIETY.

**prosaic** *adj* **1** clear, direct, down to earth, factual, matter-of-fact, plain, simple, straightforward, to the point, unadorned, understandable, unemotional, unsentimental, unvarnished. **2** [*derog*] characterless, clichéd, commonplace, dry, dull, flat, hackneyed, lifeless, monotonous, mundane, pedestrian, prosy, routine, stereotyped, trite, unfeeling, unimaginative, uninspired, uninspiring, unpoetic, unromantic. ▷ ORDINARY. *Opp* POETIC.

**prosecute** *vb* **1** accuse, arraign, bring an action against, bring to trial, charge, indict, institute legal proceedings against, prefer charges against, put on trial, sue, take legal proceedings against, take to court. **2** ▷ PURSUE.

**prospect** *n* **1** aspect, landscape, outlook, panorama, perspective, scene, seascape, sight, spectacle, view, vista. **2** *prospect of fine weather.* anticipation, chance, expectation, hope, likelihood, opportunity, possibility, probability, promise. • *vb* explore, quest, search, survey.

**prospective** *adj* anticipated, approaching, awaited, coming, expected, forthcoming, future, imminent, impending, intended, likely, looked-for, negotiable, pending, possible, potential, probable.

**prospectus** *n* announcement, brochure, catalogue, leaflet, manifesto, pamphlet, programme, scheme, syllabus.

**prosper** *vb* become prosperous, be successful, *inf* boom, burgeon, develop, do well, fare well, flourish, *inf* get ahead, *inf* get on, *inf* go from strength to strength, grow, *inf* make good, *inf* make your fortune, profit, progress, strengthen, succeed, thrive. *Opp* FAIL.

**prosperity** *n* affluence, *inf* bonanza, *inf* boom, good fortune, growth, opulence, plenty, profitability, riches, success, wealth.

**prosperous** *adj* affluent, *inf* blooming, *inf* booming, buoyant, expanding, flourishing, fruitful, healthy, moneyed, money-making, productive, profitable, prospering, rich, successful, thriving, vigorous, wealthy, *inf* well-heeled, well-off, well-to-do. *Opp* UNSUCCESSFUL.

**prostitute** *n old use* bawd, call girl, *old use* camp follower, *old use* courtesan, *old use* harlot, *inf* hooker, streetwalker, *old use* strumpet, *inf* tart, toy boy, trollop, whore. • *vb* cheapen, debase, degrade, demean, devalue, lower, misuse.

**prostrate** *adj* ▷ OVERCOME, PRONE. • *vb prostrate yourself* abase yourself, bow, kneel, kowtow, lie flat, submit. ▷ GROVEL.

**protagonist** *n* chief actor, contender, contestant, hero, heroine, lead, leading figure, principal, title role.

**protect** *vb* care for, cherish, conserve, defend, escort, guard, harbour, insulate, keep, keep safe, look after, mind, preserve, provide cover for, safeguard, screen, secure, shield, stand up for, support, take care of, tend, watch over. *Opp* ENDANGER, NEGLECT.

**protection** *n* **1** care, conservation, custody, defence, guardianship, patronage, preservation, safekeeping, safety, security, tutelage. **2** barrier, buffer, bulwark, cloak, cover, guard, insulation, screen, shelter, shield.

**protective** *adj* 1 fireproof, insulating, preservative, protecting, sheltering, shielding, waterproof. 2 *protective parents.* careful, defensive, heedful, jealous, paternalistic, possessive, solicitous, vigilant, watchful.

**protector** *n* benefactor, bodyguard, champion, defender, guard, guardian, *sl* minder, patron.

**protest** *n* 1 complaint, cry of disapproval, demur, demurral, dissent, exception, grievance, *inf* gripe, *inf* grouse, grumble, objection, opposition, outcry, protestation, remonstrance. 2 *inf* demo, demonstration, march, rally. • *vb* 1 appeal, argue, challenge a decision, complain, cry out, expostulate, express disapproval, fulminate, *inf* gripe, *inf* grouse, grumble, make a protest, *inf* moan, object, remonstrate, take exception. 2 demonstrate, *inf* hold a demo, march. 3 *protest your innocence.* affirm, assert, asseverate, aver, declare, insist on, profess, swear.

**protracted** *adj* endless, extended, interminable, long-drawn-out, long-winded, never-ending, prolonged, spun-out. ▷ LONG. *Opp* SHORT.

**protrude** *vb* balloon, bulge, extend, jut out, overhang, poke out, project, stand out, stick out, stick up, swell.

**protruding** *adj* bulbous, bulging, distended, gibbous, humped, jutting, overhanging, projecting, prominent, protuberant, swollen, tumescent.

**proud** *adj* 1 appreciative, delighted, glad, gratified, happy, honoured, pleased, satisfied. 2 *a proud bearing.* brave, dignified, independent, self-respecting. 3 *a proud history.* august, distinguished, glorious, great, honourable, illustrious, noble, reputable, respected, splendid, worthy. 4 [*derog*] arrogant, *inf* big-headed, boastful, bumptious, *inf* cocky, *inf* cocksure, conceited, disdainful, egocentric, egotistical, grand, haughty, *inf* high and mighty, immodest, lordly, narcissistic, self-centred, self-important, self-satisfied, smug, snobbish, *inf* snooty, *inf* stuck-up, supercilious, *inf* swollen-headed, *inf* toffee-nosed, vain, vainglorious. *Opp* MODEST.

**provable** *adj* demonstrable, verifiable. *Opp* UNPROVABLE.

**prove** *vb* ascertain, assay, attest, authenticate, *inf* bear out, certify, check, confirm, corroborate, demonstrate, establish, explain, justify, show to be true, substantiate, test, verify. *Opp* DISPROVE.

**proven** *adj* accepted, proved, reliable, tried and tested, trustworthy, undoubted, unquestionable, valid, verified. *Opp* DOUBTFUL, THEORETICAL.

**proverb** *n* adage, maxim, *old use* saw. ▷ SAYING.

**proverbial** *adj* aphoristic, axiomatic, clichéd, conventional, customary, famous, legendary, time-honoured, traditional, well-known.

**provide** *vb* afford, allot, allow, arrange for, cater, contribute, donate, endow, equip, *inf* fix up with, *inf* fork out, furnish, give, grant, lay on, lend, make provision, offer, present, produce, purvey, spare, stock, supply, yield.

**providence** *n* destiny, divine intervention, fate, fortune, karma, kismet.

**provident** *adj* careful, economical, far-sighted, forward-looking, frugal, judicious, prudent, thrifty.

**providential** *adj* felicitous, fortunate, happy, lucky, opportune, timely.

**provincial** *adj* **1** local, regional. *Opp* NATIONAL. **2** [*derog*] backward, boorish, bucolic, insular, narrow-minded, parochial, rural, rustic, small-minded, uncultivated, uncultured, unsophisticated. *Opp* COSMOPOLITAN.

**provisional** *adj* conditional, interim, stopgap, temporary, tentative, transitional. *Opp* DEFINITIVE, PERMANENT.

**provisions** *plur n* food, foodstuff, groceries, provender, rations, requirements, stocks, stores, subsistence, supplies, *old use* victuals.

**proviso** *n* condition, exception, limitation, provision, qualification, requirement, restriction, rider, stipulation.

**provocation** *n inf* aggravation, cause, challenge, grievance, grounds, incentive, incitement, inducement, justification, motivation, motive, reason, stimulus, taunts, teasing.

**provocative** *adj* **1** alluring, arousing, erotic, pornographic, *inf* raunchy, seductive, sensual, sensuous, *inf* sexy, tantalizing, tempting. **2** *inf* aggravating, annoying, infuriating, irksome, irritating, maddening, provoking, teasing, vexing.

**provoke** *vb* **1** activate, arouse, awaken, bring about, call forth, cause, elicit, encourage, excite, foment, generate, give rise to, induce, initiate, inspire, instigate, kindle, motivate, promote, prompt, spark off, start, stimulate, stir up, urge on, work up. **2** *inf* aggravate, anger, annoy, enrage, exasperate, gall, *inf* get on your nerves, goad, incense, incite, inflame, infuriate, insult, irk, irritate, madden, offend, outrage, pique, rile, rouse, tease, torment, upset, vex, *inf* wind up, worry. *Opp* PACIFY.

**prowess** *n* **1** ability, adeptness, adroitness, aptitude, cleverness, competence, dexterity, excellence, expertise, genius, mastery, proficiency, skill, talent. **2** *prowess in battle.* boldness, bravery, courage, daring, doughtiness, gallantry, heroism, mettle, spirit, valour.

**prowl** *vb* creep, lurk, roam, rove, skulk, slink, sneak, steal. ▷ WALK.

**proximity** *n* **1** closeness, nearness, propinquity. **2** locality, neighbourhood, vicinity.

**prudent** *adj* advisable, careful, cautious, circumspect, discreet, economical, far-sighted, frugal, judicious, politic, proper, provident, reasonable, sagacious, sage, sensible, shrewd, thoughtful, thrifty, vigilant, watchful, wise. *Opp* UNWISE.

**prudish** *adj* decorous, easily shocked, illiberal, intolerant, narrow-minded, old-fashioned, *inf* old-maidish, priggish, prim, *inf* prissy, proper, puritanical, rigid, shockable, strait-laced, strict. *Opp* BROAD-MINDED.

**prune** *vb* clip, cut back, lop, pare down, trim. ▷ CUT.

**pry** *vb* be curious, be inquisitive, be nosy, delve, *inf* ferret, inquire, interfere, intrude, investigate, meddle, *inf* nose about, peer, poke about, *inf* poke your nose in, search, *inf* snoop, *inf* stick your nose in. **prying** ▷ INQUISITIVE.

**pseudonym** *n* alias, assumed name, false name, incognito, nickname, *Fr* nom de plume, pen-name, sobriquet, stage name.

**psychic** *adj* clairvoyant, extra-sensory, magical, mental, meta-physical, mystic, occult, preternat-ural, psychical, spiritual, supernatural, telepathic. • *n* astrologer, clairvoyant, crystal-gazer, fortune-teller, medium, mind-reader, spiritualist, telepath-ist.

**psychological** *adj* cerebral, emo-tional, mental, subconscious, sub-jective, subliminal, unconscious. *Opp* PHYSIOLOGICAL.

**pub** *n old use* alehouse, bar, *sl* boozer, cocktail lounge, *old use* hostelry, inn, *inf* local, public house, saloon, tavern, wine bar.

**puberty** *n* adolescence, growing-up, juvenescence, pubes-cence, sexual maturity, *inf* teens.

**public** *adj* **1** accessible, available, common, familiar, free, known, open, shared, unconcealed, unres-tricted, visible, well-known. **2** *public support.* civic, civil, col-lective, communal, community, democratic, general, majority, national, popular, social, univer-sal. **3** *a public figure.* ▷ PROMINENT. *Opp* PRIVATE. • *n the public.* citizens, the community, the country, the nation, people, the populace, society, voters.

**publication** *n* **1** appearance, issu-ing, printing, production. ▷ BOOK, MAGAZINE. **2** advertising, announcement, broadcasting, declaration, disclosure, dissemina-tion, proclamation, promulgation, publicizing, reporting.

**publicity** *n* **1** attention, *inf* ballyhoo, fame, *inf* hype, lime-light, notoriety. **2** advertising, mar-keting, promotion. ▷ ADVERTISEMENT.

**publicize** *vb* advertise, *sl* hype, *inf* plug, promote, *old use* puff. ▷ PUBLISH.

**publish** *vb* **1** bring out, circulate issue, make available, print, pro-duce, put on sale, release. **2** *publish secrets.* advertise, announce, break the news about, broadcast, communicate, declare, disclose, disseminate, divulge, issue a statement about, *inf* leak, make known, make public, pro-claim, promulgate, publicize, *inf* put about, report, reveal, spread.

**pucker** *vb* compress, contract, crease, crinkle, draw together, purse, screw up, squeeze, tighten, wrinkle.

**puerile** *adj* babyish, boyish, child ish, immature, infantile, juvenile. ▷ SILLY.

**puff** *n* **1** blast, blow, breath, draught, flurry, gust, whiff, wind. **2** *a puff of smoke.* cloud, wisp. • *vb* **1** blow, breathe heavily, gasp, huff, pant, wheeze. **2** *puff at a cigar. inf* drag, draw, inhale, pull, smoke, suck. **3** *sails puffed by the wind.* balloon, billow, distend, enlarge, inflate, rise, swell.

**pugnacious** *adj* aggressive, antag onistic, argumentative, bellicose, belligerent, combative, conten-tious, disputatious, excitable, frac tious, hostile, hottempered, litigi-ous, militant, unfriendly, warlike ▷ QUARRELSOME. *Opp* PEACEABLE.

**pull** *vb* **1** drag, draw, haul, lug, tow, trail. *Opp* PUSH. **2** jerk, tug, pluck, rip, wrench, *inf* yank. **3** *pull a tooth.* extract, pull out, remove, take out. **pull off** ▷ DETACH. **pull out** ▷ WITHDRAW. **pull round** ▷ RECOVER. **pull someone's leg** ▷ TEASE. **pull through** ▷ RECOVER. **pull together** ▷ COOPERATE. **pull up** ▷ HALT.

**pulp** *n* mash, mush, paste, pap, puré. • *vb* crush, liquidize, mash, pound, pulverize, purée, smash, squash.

**pulsate** *vb* beat, drum, oscillate, palpitate, pound, pulse, quiver, reverberate, throb, tick, vibrate.

**pulse** *n* beat, drumming, oscillation, pounding, pulsation, rhythm, throb, ticking, vibration.

**pump** *vb* drain, draw off, empty, force, raise, siphon. **pump up** blow up, fill, inflate.

**punch** *vb* **1** beat, *sl* biff, box, *inf* clout, cuff, jab, poke, prod, pummel, slog, *sl* slug, *sl* sock, strike, thump. ▷ HIT. **2** ▷ PIERCE.

**punctual** *adj* in good time, *inf* on the dot, on time, prompt. *Opp* UNPUNCTUAL.

**punctuate** *vb* **1** insert punctuation, point. **2** *punctuated by applause.* break, interrupt, intersperse, *inf* pepper.

**punctuation** *n* marks, points, stops. □ *accent, apostrophe, asterisk, bracket, caret, cedilla, colon, comma, dash, exclamation mark, full stop, hyphen, question mark, quotation marks, speech marks, semicolon.*

**puncture** *n* blow-out, burst, burst tyre, *inf* flat, flat tyre, hole, leak, opening, perforation, pin-prick, rupture. • *vb* deflate, go through, let down, penetrate, perforate, pierce, prick, rupture.

**pungent** *adj* **1** aromatic, hot, peppery, piquant, seasoned, sharp, spicy, strong, tangy. **2** acid, acrid, astringent, caustic, *inf* chemically, harsh, sour, stinging. **3** *pungent criticism.* biting, bitter, incisive, mordant, sarcastic, scathing, trenchant.

**punish** *vb* castigate, chasten, chastise, correct, discipline, exact retribution from, impose punishment on, inflict punishment on, *inf* make an example of, pay back, penalize, *inf* rap over the knuckles, scold, *inf* teach someone a lesson.

**punishment** *n* chastisement, correction, discipline, forfeit, imposition, *inf* just deserts, penalty, punitive measure, retribution, revenge, sentence. □ *banishment, beating, the birch,* old use *Borstal, the cane, capital punishment, cashiering, confiscation of property, corporal punishment, detention, excommunication, execution, exile, fine, flogging, gaol,* inf *hiding, imprisonment, jail, keelhauling, lashing, pillory, prison, probation, scourging, spanking, the stocks, torture, whipping.*

**punitive** *adj* disciplinary, penal, retaliatory, retributive, revengeful, vindictive.

**puny** *adj* diminutive, dwarf, feeble, frail, sickly, stunted, underdeveloped, undernourished, undersized. ▷ SMALL. *Opp* LARGE, STRONG.

**pupil** *n* apprentice, beginner, disciple, follower, learner, novice, protégé(e), scholar, schoolboy, schoolchild, schoolgirl, student, tiro.

**puppet** *n* doll, dummy, finger-puppet, glove-puppet, hand-puppet, marionette, string-puppet.

**purchase** *n* **1** acquisition, *inf* buy (*a good buy*), investment. **2** grasp, grip, hold, leverage, support. • *vb* acquire, buy, get, invest in, obtain, pay for, procure, secure.

**pure** *adj* **1** authentic, genuine, neat, real, solid, sterling, straight, unadulterated, unalloyed, undiluted. **2** *pure food.* eatable, germ-free, hygienic, natural, pasteurized, uncontaminated, untainted, wholesome. **3** *pure water.* clean, clear, distilled, drinkable, fresh, potable, sterile, unpolluted. **4** *pure in morals.* blameless, chaste, decent, good, impeccable, innocent, irreproachable, maidenly, modest, moral, proper, sinless, stainless, virginal, virtuous. **5** *pure genius.* absolute, complete, downright, *inf* out-and-out, perfect, sheer, thorough, total, true, unmitigated, unqualified, utter. **6** *pure science.* abstract, academic, conjectural, conceptual, hypothetical, speculative, theoretical. *Opp* IMPURE, PRACTICAL.

**purgative** *n* aperient, cathartic, enema, laxative, purge.

**purge** *vb* **1** clean out, cleanse, clear, depurate, empty, purify, wash out. **2** *purge your opponents.* eject, eliminate, eradicate, expel, get rid of, liquidate, oust, remove, root out.

**purify** *vb* clarify, clean, cleanse, decontaminate, depurate, disinfect, distil, filter, fumigate, make pure, purge, refine, sanitize, sterilize.

**puritan** *n* fanatic, *derog* killjoy, moralist, *derog* prude, zealot.

**puritanical** *adj* ascetic, austere, moralistic, narrow-minded, pietistic, prim, proper, prudish, rigid, self-denying, self-disciplined, severe, stern, stiff-necked, strait-laced, strict, temperate, unbending, uncompromising. *Opp* HEDONISTIC.

**purpose** *n* **1** aim, ambition, aspiration, design, end, goal, hope, intent, intention, motivation, motive, object, objective, outcome, plan, point, rationale, result, target, wish. **2** determination, devotion, drive, firmness, persistence, resolution, resolve, steadfastness, tenacity, will, zeal. **3** *purpose of a tool.* advantage, application, benefit, good (*what's the good of it?*), point, practicality, use, usefulness, utility, value. • *vb* ▷ INTEND.

**purposeful** *adj* calculated, decided, decisive, deliberate, determined, devoted, firm, persistent, positive, resolute, steadfast, *derog* stubborn, tenacious, unwavering, wilful, zealous. ▷ INTENTIONAL. *Opp* HESITANT.

**purposeless** *adj* aimless, bootless, empty, gratuitous, meaningless, pointless, senseless, unnecessary, useless, vacuous, wanton. *Opp* MEANINGFUL, USEFUL.

**purposely** *adv* consciously, deliberately, intentionally, knowingly, on purpose, wilfully.

**purse** *n* bag, handbag, moneybag, pocketbook, pouch, wallet.

**pursue** *vb* **1** chase, follow, go after, go in pursuit of, harry, hound, hunt, keep up with, run after, shadow, stalk, *inf* tail, trace, track down, trail. **2** aim for, aspire to, be committed to, carry on, conduct, continue, dedicate yourself to, engage in, follow up, *inf* go for, persevere in, persist in, proceed with, prosecute, *inf* stick with, strive for, try for. **3** *pursue truth.* inquire into, investigate, quest after, search for, seek.

**pursuit** *n* **1** chase, chasing, following, harrying, *inf* hue and cry, hunt, hunting, pursuing, shadowing, stalking, tracking down, trail. **2** *leisure pursuits.* activity, employment, enthusiasm, hobby, interest, obsession, occupation, pastime, pleasure, speciality, specialization.

**push** *vb* **1** advance, drive, force, hustle, impel, jostle, move, nudge, poke, press, prod, propel, set in motion, shove, thrust. **2** *push a button.* depress, press. **3** *push into* *a space.* compress, cram, crowd, crush, insert, jam, pack, put, ram, squash, squeeze. **4** *push someone to act.* browbeat, bully, coerce, compel, constrain, dragoon, encourage, force, hurry, importune, incite, induce, influence, *inf* lean on, motivate, nag, persuade, pressurize, prompt, put pressure on, spur, stimulate, urge. **5** *push a new product.* advertise, boost, make known, market, *inf* plug, promote, publicize. *Opp* PULL. **push around** ▷ BULLY. **push off** ▷ DEPART. **push on** ▷ ADVANCE.

**put** *vb* **1** arrange, assign, commit, consign, deploy, deposit, dispose, fix, hang, lay, leave, locate, park, place, *inf* plonk, position, rest, set down, settle, situate, stand, station. **2** *put a question.* express, formulate, frame, phrase, say, state, utter, voice, word, write. **3** *put a proposal.* advance, bring forward, offer, outline, present, propose, submit, suggest, tender. **4** *put blame on someone.* attach, attribute, cast, fix, impose, inflict, lay, *inf* pin. **put across** ▷ COMMUNICATE. **put back** ▷ RETURN. **put by** ▷ SAVE. **put down** ▷ KILL, SUPPRESS. **put in** ▷ INSERT, INSTALL. **put off** ▷ POSTPONE. **put out** ▷ EJECT, EXTINGUISH. **put over** ▷ COMMUNICATE. **put right** ▷ REPAIR. **put someone up** ▷ ACCOMMODATE. **put up** ▷ RAISE. **put your foot down** ▷ INSIST. **put your foot in it** ▷ BLUNDER.

**putative** *adj* alleged, assumed, conjectural, presumed, reputed, rumoured, supposed, suppositious.

**putrefy** *vb* decay, decompose, go bad, go off, moulder, rot, spoil.

**putrid** *adj* bad, corrupt, decaying, decomposing, fetid, foul, mouldy, putrefying, rotten, rotting, spoilt.

**puzzle** *n inf* brain-teaser, conundrum, difficulty, dilemma, enigma, mystery, paradox, *inf* poser, problem, quandary, question, riddle. • *vb* baffle, bewilder, confuse, confound, *inf* floor, *inf* flummox, mystify, nonplus, perplex, set thinking, *inf* stump, *inf* stymie, worry. **puzzle out** ▷ SOLVE. **puzzle over** ▷ CONSIDER.

**puzzling** *adj* ambiguous, baffling, bewildering, confusing, cryptic, enigmatic, impenetrable, inexplicable, insoluble, *inf* mind-boggling, mysterious, mystifying, perplexing, strange, unaccountable, unanswerable, unfathomable, worrying. *Opp* STRAIGHTFORWARD.

**pygmy** *adj* dwarf, lilliputian, midget, tiny. ▷ SMALL.

---

# Q

**quadrangle** *n* cloisters, courtyard, enclosure, *inf* quad, yard.

**quagmire** *n* bog, fen, marsh, mire, morass, mud, quicksand, *old use* slough, swamp.

**quail** *vb* back away, be apprehensive, blench, cower, cringe, falter, flinch, quake, recoil, show fear, shrink, tremble, wince.

**quaint** *adj* antiquated, antique, charming, curious, eccentric, fanciful, fantastic, odd, offbeat, old-fashioned, old-world, outlandish, peculiar, picturesque, strange, *inf* twee, unconventional, unexpected, unfamiliar, unusual, whimsical.

**quake** *vb* convulse, heave, move, quaver, quiver, rock, shake, shiver, shudder, stagger, sway, tremble, vibrate, wobble.

**qualification** *n* **1** ability, aptitude, capability, capacity, certification, competence, eligibility, experience, fitness, *inf* know-how, knowledge, proficiency, quality, skill, suitability, training. **2** certificate, degree, diploma, doctorate, first degree, Master's degree, matriculation. **3** *agree without qualification.* caveat, condition, exception, limitation, modification, proviso, reservation, restriction.

**qualified** *adj* **1** able, capable, certificated, competent, equipped, experienced, expert, fit, practised, professional, proficient, skilled, trained, well-informed. *Opp* UNSKILLED. **2** *qualified applicants.* appropriate, eligible, suitable. **3** *qualified praise.* cautious, conditional, equivocal, guarded, half-hearted, limited, modified, provisional, reserved, restricted. *Opp* UNCONDITIONAL.

**qualify** *vb* **1** authorize, empower, entitle, equip, fit, make eligible, permit, sanction. **2** become eligible, get through, *inf* make the grade, meet requirements, pass. **3** *qualify your praise.* abate, lessen, limit, mitigate, moderate, modulate, restrain, restrict, soften, temper, weaken.

**quality** *n* **1** calibre, class, condition, excellence, grade, rank, sort, standard, status, value, worth. **2** *personal qualities.* attribute, characteristic, distinction, feature, mark, peculiarity, property, trait.

**quandary** *n* *inf* catch-22, *inf* cleft stick, confusion, difficulty, dilemma, perplexity, plight, predicament, uncertainty.

**quantity** *n* aggregate, amount, bulk, consignment, dosage, dose, expanse, extent, length, load, lot, magnitude, mass, measurement, number, part, portion, proportion, quantum, sum, total, volume, weight. ▷ MEASURE.

**quarrel** *n* altercation, argument, bickering, clash, conflict, confrontation, contention, controversy, debate, difference, disagreement, discord, disharmony, dispute, dissension, division, feud, *inf* hassle, misunderstanding, row, *inf* ructions, rupture, *inf* scene, schism, *inf* slanging match, split, squabble, strife, *inf* tiff, vendetta, wrangle. • *vb* argue, *inf* be at loggerheads, *inf* be at odds, bicker, clash, conflict, contend, *inf* cross swords, differ, disagree, dispute, dissent, *inf* fall out, feud, haggle, misunderstand one another, *inf* row, squabble, wrangle. ▷ FIGHT. **quarrel with** ▷ DISPUTE.

**quarrelsome** *adj* aggressive, angry, argumentative, bad-tempered, cantankerous, choleric, contentious, contrary, cross, defiant, disagreeable, dyspeptic, explosive, fractious, impatient, irascible, irritable, petulant, peevish, querulous, quick-tempered, *inf* stroppy, testy, truculent, volatile, unfriendly. ▷ PUGNACIOUS. *Opp* PEACEABLE.

**quarry** *n* **1** game, kill, object, prey, victim. **2** excavation, mine, pit, working. • *vb* dig out, excavate, extract, mine.

**quarter** *n* area, district, division, locality, neighbourhood, part, region, section, sector, territory, vicinity, zone. • *vb* accommodate, billet, board, house, lodge, *inf* put up, shelter, station. **quarters** abode, accommodation, barracks, billet, domicile, dwelling-place, home, housing, living quarters, lodgings, residence, rooms, shelter.

**quash** *vb* **1** abolish, annul, cancel, invalidate, overrule, overthrow, reject, rescind, reverse, revoke. **2** ▷ QUELL.

**quaver** *vb* falter, fluctuate, oscillate, pulsate, quake, quiver, shake, shiver, shudder, tremble, vibrate, waver.

**quay** *n* berth, dock, harbour, jetty, landing-stage, pier, wharf.

**queasy** *adj* bilious, *inf* green, *inf* groggy, nauseated, nauseous, *inf* poorly, *inf* queer, sick, unwell. ▷ ILL.

**queer** *adj* **1** aberrant, abnormal, anomalous, atypical, bizarre, curious, different, eerie, exceptional, extraordinary, *inf* fishy, freakish, *inf* funny, incongruous, inexplicable, irrational, mysterious, odd, offbeat, outlandish, peculiar, puzzling, quaint, remarkable, *inf* rum, singular, strange, unaccountable, uncanny, uncommon, unconventional, unexpected, unnatural, unorthodox, unusual, weird. **2** *inf* cranky, deviant, eccentric, questionable, *inf* shady (*customer*), *inf* shifty, suspect, suspicious. ▷ MAD. *Opp* NORMAL. **3** ▷ ILL. **4** ▷ HOMOSEXUAL.

**quell** *vb* **1** crush, overcome, put down, quash, repress, subdue, suppress. **2** *quell fears.* allay, alleviate, calm, mitigate, moderate, mollify, pacify, soothe, tranquillize.

**quench** **1** allay, appease, cool, sate, satisfy, slake. **2** *quench a fire.* damp down, douse, extinguish, put out, smother, snuff out, stifle, suppress.

**quest** *n* crusade, expedition, exploration, hunt, mission, pilgrimage, pursuit, search, voyage of discovery. • *vb* **quest after** ▷ SEEK.

**question** *n* **1** *inf* brain-teaser, conundrum, demand, enquiry, inquiry, *inf* poser, query, request, riddle. **2** *an unresolved question.* argument, controversy, debate, difficulty, dispute, doubt, misgiving, mystery, objection, problem, puzzle, uncertainty. • *vb* **1** ask, catechize, cross-examine, cross-question, debrief, enquire of, examine, *inf* grill, inquire of, interrogate, interview, probe, *inf* pump, quiz. **2** *question a decision.* argue over, be sceptical about, call into question, cast doubt upon, challenge, dispute, doubt, enquire about, impugn, inquire about, object to, oppose, quarrel with, query.

**questionable** *adj* arguable, borderline, debatable, disputable, doubtful, dubious, *inf* iffy, moot, problematical, *inf* shady (*customer*), suspect, suspicious, uncertain, unclear, unprovable, unreliable.

**questionnaire** *n* catechism, opinion poll, question sheet, quiz, survey, test.

**queue** *n* chain, concatenation, column, *inf* crocodile, file, line, line-up, procession, row, string, succession, tail-back, train. • *vb* fall in, form a queue, line up, wait in a queue.

**quibble** *n* ▷ OBJECTION. • *vb inf* bandy words, be evasive, carp, cavil, equivocate, *inf* nit-pick, object, pettifog, *inf* split hairs, wrangle.

**quick** *adj* **1** breakneck, brisk, expeditious, express, fast, *old use* fleet, headlong, high-speed, *inf* nippy, precipitate, rapid, *inf* smart (*pace*), *inf* spanking, speedy, swift. **2** *a quick reaction.* adroit, agile, animated, brisk, deft, dexterous, energetic, lively, nimble, spirited, spry, vigorous. **3** *a quick response.* abrupt, early, hasty, hurried, immediate, instant, instantaneous, perfunctory, precipitate, prompt, punctual, ready, sudden, summary, unhesitating. **4** *a quick mind.* acute, alert, apt, astute, bright, clever, intelligent, perceptive, quick-witted, sharp, shrewd, smart. *Opp* SLOW. **5** *a quick rest.* brief, fleeting, momentary, passing, perfunctory, short, short-lived, temporary, transitory. **6** [*old use*] *the quick and the dead.* ▷ ALIVE. *Opp* SLOW.

**quicken** *vb* **1** accelerate, expedite, hasten, hurry, go faster, speed up. **2** ▷ AROUSE.

**quiet** *adj* **1** inaudible, noiseless, silent, soundless. **2** *quiet music.* hushed, low, pianissimo, soft, *It* sotto voce. **3** *a quiet person.* composed, contemplative, contented, gentle, introverted, meditative, meek, mild, modest, peaceable, reserved, retiring, shy, taciturn, thoughtful, uncommunicative, unforthcoming, unsociable, withdrawn. **4** *a quiet life.* cloistered, sheltered, tranquil, unadventurous, unexciting, untroubled. **5** *a quiet place.* isolated, lonely, peaceful, private, secluded, sequestered, undisturbed, unfrequented. **6** *quiet weather.* calm, motionless, placid, restful, serene, still. *Opp* BUSY, NOISY, RESTLESS.

**quieten** *vb* **1** calm, compose, hush, lull, pacify, sedate, soothe, subdue, tranquillize. **2** deaden, dull, muffle, mute, reduce the volume of, silence, soften, stifle, suppress, tone down.

**quirk** *n* aberration, caprice, crotchet, eccentricity, idiosyncrasy, kink, oddity, peculiarity, trick, whim.

**quit** *vb* **1** abandon, decamp from, depart from, desert, exit from, forsake, go away from, leave, walk out (on), withdraw. **2** abdicate, discontinue, drop, give up, leave, *inf* pack it in, relinquish, renounce, repudiate, resign from, retire from, withdraw from. **3** [*inf*] *Quit pushing!* cease, desist from, discontinue, leave off, stop.

**quite** *adv* [NB: the two senses are almost opposite.] **1** *Yes, I've quite finished.* absolutely, altogether, completely, entirely, perfectly, thoroughly, totally, unreservedly, utterly, wholly. **2** *quite good, but not perfect.* comparatively, fairly, moderately, *inf* pretty, rather, relatively, somewhat, to some extent.

**quits** *adj* equal, even, level, repaid, revenged, square.

**quiver** *vb* flicker, fluctuate, flutter, oscillate, palpitate, pulsate, quake, quaver, shake, shiver, shudder, tremble, vibrate, wobble.

**quixotic** *adj* fanciful, foolhardy, idealistic, impracticable, impractical, romantic, *inf* starry-eyed, unrealistic, unrealizable, unselfish, Utopian, visionary. *Opp* REALISTIC.

**quiz** *n* competition, exam, examination, questioning, questionnaire, quiz-game, test. • *vb* ▷ QUESTION.

**quizzical** *adj* amused, comical, curious, intrigued, perplexed, puzzled, queer, questioning.

**quota** *n* allocation, allowance, apportionment, assignment, *inf* cut, part, portion, proportion, ration, share.

**quotation** *n* **1** allusion, citation, *inf* clip, cutting, excerpt, extract, passage, piece, reference, selection. **2** estimate, price, tender, valuation.

**quote** *vb* **1** cite, instance, mention, produce a quotation from, refer to, repeat, reproduce. **2** *quote a price.* estimate, tender.

# R

**rabble** *n* crowd, gang, herd, *Gk* hoi polloi, horde, mob, *inf* riffraff, swarm, throng. ▷ GROUP.

**race** *n* **1** breed, clan, ethnic group, family, folk, genus, kind, lineage, nation, people, species, stock, tribe, variety. **2** chase, competition, contention, contest, heat, rivalry. □ *cross-country, greyhound race, horse-race, hurdles, marathon, motor-race, regatta, relay, road-race, rowing, scramble, speedway, sprint, steeple-chase, stock-car race, swimming, track event.* • *vb* **1** *I'll race you!* compete with, contest with, have a race with, try to beat. **2** *race along.* career, dash, *inf* fly, gallop, hasten, hurry, move fast, run, rush, speed, sprint, *inf* tear, *inf* zip, *inf* zoom.

**racetrack** *n* cinder-track, circuit, dog-track, lap, racecourse.

**racial** *adj* ethnic, folk, genetic, national, tribal.

**racism** *n* bias, bigotry, chauvinism, discrimination, intolerance, prejudice, racialism, xenophobia. □ *anti-Semitism, apartheid.*

**racist** *adj* biased, bigoted, chauvinist, discriminatory, intolerant, prejudiced, racialist, xenophobic. □ *anti-Semitic.*

**rack** *n* frame, framework, holder, scaffold, scaffolding, shelf, stand, support. • *vb* ▷ TORTURE.

**radiant** *adj* **1** beaming, bright, brilliant, effulgent, gleaming, glorious, glowing, incandescent, luminous, phosphorescent, refulgent, shining. **2** *The bride was radiant.* ▷ HAPPY.

**radiate** *vb* beam, diffuse, emanate, emit, give off, gleam, glow, send out, shed, shine, spread, transmit.

**radical** *adj* **1** basic, cardinal, deep-seated, elementary, essential, fundamental, primary, principal, profound. **2** complete, comprehensive, drastic, entire, exhaustive, thorough, thoroughgoing. **3** *radical politics.* extreme, extremist, fanatical, far-reaching, revolutionary, *derog* subversive. *Opp* MODERATE, SUPERFICIAL.

**radio** *n* CB, *sl* ghettoblaster, portable, receiver, set, *inf* transistor, transmitter, walkie-talkie, *old use* wireless. • *vb* broadcast, send out, transmit.

**rafter** *n* beam, girder, joist.

**rage** *n* ▷ ANGER. • *vb* be angry, boil, fume, go berserk, lose control, rave, *inf* see red, seethe, storm.

**ragged** *adj* **1** chafed, frayed, in ribbons, old, patched, patchy, ravelled, rent, ripped, rough, rough-edged, shabby, shaggy, tattered, tatty, threadbare, torn, unkempt, unravelled, untidy, worn out. **2** *ragged line.* denticulated, disorganized, erratic, irregular, jagged, serrated, uneven, zigzag.

**rags** *plur n* bits and pieces, cloths, fragments, old clothes, remnants, ribbons, scraps, shreds, tatters.

**raid** *n* assault, attack, blitz, foray, incursion, inroad, invasion, onslaught, sally, sortie, strike, surprise attack, swoop. • *vb* 1 assault, attack, descend on, invade, pounce on, rush, storm, swoop down on. 2 loot, maraud, pillage, plunder, ransack, rifle, rob, sack, steal from, strip.

**raider** *n* attacker, brigand, invader, looter, marauder, outlaw, pillager, pirate, plunderer, ransacker, robber, rustler, thief.

**railway** *n* line, permanent way, *Amer* railroad, rails, track. □ *branch line, cable railway, funicular, light railway, main line, metro, mineral line, monorail, mountain railway, narrow gauge, rack-and-pinion, rapid transit system, siding, standard gauge, tramway, tube, underground.* ▷ TRAIN.

**rain** *n* cloudburst, deluge, downpour, drizzle, precipitation, raindrops, rainfall, rainstorm, shower, squall. • *vb inf* bucket down, drizzle, pelt, pour, *inf* rain cats and dogs, spit, teem.

**rainy** *adj* damp, drizzly, showery, wet.

**raise** *vb* 1 elevate, heave up, hoist, hold up, jack up, lift, loft, pick up, put up, rear. 2 *raise prices.* augment, boost, increase, inflate, put up, *inf* up. 3 *raise to a higher rank.* exalt, prefer, promote, upgrade. 4 *raise a monument.* build, construct, create, erect, set up. 5 *raise hopes.* activate, arouse, awaken, build up, buoy up, encourage, engender, enlarge, excite, foment, foster, heighten, incite, kindle, motivate, provoke, rouse, stimulate, uplift. 6 *raise animals, children, crops.* breed, bring up, care for, cultivate, educate, farm, grow, look after, nurture, produce, propagate, rear. 7 *raise money.* amass, collect, get, make, receive, solicit. 8 *raise questions.* advance, bring up, broach, express, instigate, introduce, mention, moot, originate, pose, present, put forward, suggest. *Opp* LOWER, REDUCE. **raise from the dead** ▷ RESURRECT. **raise the alarm** ▷ WARN.

**rally** *n* 1 assembly, *inf* demo, demonstration, gathering, march, mass meeting, protest. 2 ▷ COMPETITION. • *vb* 1 assemble, convene, get together, marshal, muster, organize, round up, summon. 2 come together, reassemble, reform, regroup. 3 *rally after illness.* ▷ RECOVER.

**ram** *vb* 1 bump, butt, collide with, crash into, slam into, smash into, strike. ▷ HIT. 2 compress, cram, crowd, crush, drive, force, jam, pack, press, push, squash, squeeze, tamp down, wedge.

**ramble** *n* hike, tramp, trek, walk. • *vb* 1 hike, range, roam, rove, tramp, trek, stroll, wander. ▷ WALK. 2 digress, drift, *inf* lose the thread, maunder, *inf* rabbit on, *inf* rattle on, talk aimlessly, wander, *inf* witter on.

**rambling** *adj* 1 circuitous, indirect, labyrinthine, meandering, roundabout, tortuous, twisting, wandering, winding, zigzag. *Opp* DIRECT. 2 aimless, circumlocutory, confused, diffuse, digressive, disconnected, discursive, disjointed, illogical, incoherent, jumbled, muddled, periphrastic, unstructured, verbose, wordy. *Opp* COHERENT. 3 *a rambling house.* asymmetrical, extensive, irregular, large, sprawling, straggling, straggly. *Opp* COMPACT.

**ramification** *n* branch, by-product, complication, consequence, division, effect, extension, implication, offshoot, result, subdivision, upshot.

**ramp** *n* acclivity, gradient, incline, rise, slope.

**rampage** *n* frenzy, riot, tumult, uproar, vandalism, violence. • *vb* behave violently, go berserk, go wild, lose control, race about, run amok, run riot, rush about, storm about. **on the rampage** ▷ WILD.

**ramshackle** *adj* broken-down, crumbling, decrepit, derelict, dilapidated, flimsy, jerry-built, rickety, ruined, run-down, shaky, tottering, tumbledown, unsafe, unstable, unsteady. *Opp* SOLID.

**random** *adj* accidental, adventitious, aimless, arbitrary, casual, chance, fortuitous, haphazard, *inf* hit-or-miss, indiscriminate, irregular, serendipitous, stray, unconsidered, unplanned, unpremeditated, unspecific, unsystematic. *Opp* DELIBERATE, SYSTEMATIC.

**range** *n* **1** area, compass, distance, extent, field, gamut, limit, orbit, radius, reach, scope, span, spectrum, sphere, spread, sweep. **2** *a wide range of goods.* diversity, selection, variety. **3** *range of mountains.* chain, file, line, rank, row, series, string, tier. • *vb* **1** differ, extend, fluctuate, go, reach, run the gamut, spread, stretch, vary. **2** ▷ RANK. **3** ▷ ROAM.

**rank** *adj* **1** *rank growth.* ▷ ABUNDANT. **2** *rank smell.* ▷ SMELLING. • *n* **1** column, file, formation, line, order, queue, row, series, tier. **2** birth, blood, caste, class, condition, degree, echelon, estate, grade, level, position, standing, station, status, stratum, title. • *vb* arrange, array, assort, categorize, class, classify, grade, graduate, line up, order, organize, range, rate, set out in order, sort.

**ransack** *vb* **1** comb, explore, go through, rake through, rummage through, scour, search, *inf* turn upside down. **2** *ransack a shop.* despoil, loot, pillage, plunder, raid, ravage, rob, sack, strip, wreck.

**ransom** *n* payment, *inf* payoff, price, redemption. • *vb* buy the release of, deliver, redeem.

**rap** *vb* **1** knock, strike, tap. ▷ HIT. **2** ▷ CRITICIZE.

**rape** *n* **1** assault, sexual attack. **2** ▷ PILLAGE. • *vb* assault, defile, deflower, dishonour, force yourself on, *inf* have your way with, *old use* ravish, violate.

**rapid** *adj* alacritous, breakneck, brisk, expeditious, express, fast, *old use* fleet, hasty, headlong, high-speed, hurried, immediate, impetuous, instant, instantaneous, *inf* lightning, *inf* nippy, precipitate, prompt, quick, smooth, speedy, swift, unchecked, uninterrupted. *Opp* SLOW.

**rapids** *plur n* cataract, current, waterfall, white water.

**rapture** *n* bliss, delight, ecstasy, elation, euphoria, exaltation, happiness, joy, pleasure, thrill, transport.

**rare** *adj* abnormal, atypical, curious, exceptional, extraordinary, *inf* few and far between, infrequent, irreplaceable, limited, occasional, odd, out of the ordinary, peculiar, scarce, singular, special, strange, surprising, uncommon, unfamiliar, unusual. *Opp* COMMON.

**rascal** *n* blackguard, *old use* bounder, devil, good-fornothing, imp, knave, mischiefmaker, miscreant, ne'er-do-well, rapscallion, rogue, *inf* scally-wag, scamp, scoundrel, troublemaker, villain, wastrel. ▷ CRIMINAL.

**rash** *adj* careless, foolhardy, harebrained, hasty, headlong, headstrong, heedless, hotheaded, hurried, ill-advised, ill-considered, impetuous, imprudent, impulsive, incautious, indiscreet, injudicious, madcap, precipitate, reckless, risky, thoughtless, unthinking, wild. *Opp* CAREFUL. • *n* **1** efflorescence, eruption, spots. **2** *a rash of thefts.* ▷ OUTBREAK.

**rasp** *vb* **1** abrade, file, grate, rub, scrape. **2** *rasp orders.* croak, screech, speak hoarsely. ▷ SPEAK. **rasping** ▷ HARSH.

**rate** *n* **1** gait, pace, speed, tempo, velocity. **2** amount, charge, cost, fare, fee, figure, payment, price, scale, tariff, wage. • *vb* **1** appraise, assess, class, classify, compute, consider, estimate, evaluate, gauge, grade, judge, measure, prize, put a price on, rank, reckon, regard, value, weigh. **2** *rate a prize.* be worthy of, deserve, merit. **3** ▷ REPRIMAND.

**rather** *adv* **1** fairly, moderately, *inf* pretty, quite, relatively, slightly, somewhat. **2** *would rather have tea than coffee.* more willingly, preferably, sooner.

**ratify** *vb* approve, authorize, confirm, endorse, sanction, sign, validate, verify.

**rating** *n* classification, evaluation, grade, grading, mark, order, placing, ranking.

**ratio** *n* balance, correlation, correspondence, fraction, percentage, proportion, relationship.

**ration** *n* allocation, allotment, allowance, amount, helping, measure, percentage, portion, quota, share. • *vb* allocate, allot, apportion, conserve, control, distribute fairly, dole out, give out, limit, parcel out, restrict, share equally. **rations** food, necessaries, necessities, provisions, stores, supplies.

**rational** *adj* balanced, clear-headed, commonsense, enlightened, intelligent, judicious, logical, lucid, normal, ratiocinative, reasonable, reasoned, reasoning, sane, sensible, sound, thoughtful, wise. *Opp* IRRATIONAL.

**rationale** *n* argument, case, cause, excuse, explanation, grounds, justification, logical basis, principle, reason, reasoning, theory, vindication.

**rationalize** *vb* **1** account for, be rational about, elucidate, excuse, explain, justify, make rational, provide a rationale for, ratiocinate, think through, vindicate. **2** ▷ REORGANIZE.

**rattle** *vb* **1** clatter, vibrate. **2** agitate, jar, joggle, *inf* jiggle about, jolt, shake about. **3** [*inf*] *rattled him by booing.* alarm, discomfit, discompose, disconcert, disturb, fluster, frighten, make nervous, put off, unnerve, upset, worry. **rattle off** ▷ RECITE. **rattle on** ▷ RAMBLE, TALK.

**raucous** *adj* ear-splitting, harsh, husky, grating, jarring, noisy, rasping, rough, screeching, shrill, squawking, strident.

**ravage** *vb* damage, despoil, destroy, devastate, lay waste, loot, pillage, plunder, raid, ransack, ruin, sack, spoil, wreak havoc on, wreck.

**rave** *vb* **1** be angry, fulminate, fume, rage, rant, roar, storm, thunder. **2** be enthusiastic, enthuse, *inf* go into raptures, *inf* gush, rhapsodize.

**avenous** *adj* famished, hungry, nsatiable, ravening, starved, starving, voracious. ▷ GREEDY.

**avish** *vb* **1** bewitch, captivate, capture, charm, delight, enchant, entrance, spellbind, transport. **2** ▷ RAPE. **ravishing** ▷ BEAUTIFUL.

**aw** *adj* **1** fresh, rare (*steak*), uncooked, underdone, unprepared, wet (*fish*). **2** *raw materials.* crude, natural, unprocessed, unrefined, untreated. **3** *raw recruits. inf* green, ignorant, immature, inexperienced, innocent, new, unseasoned, untrained, untried. **4** *raw skin.* bloody, chafed, grazed, inflamed, painful, red, rough, scraped, scratched, sensitive, sore, tender, vulnerable. **5** *raw wind.* ▷ COLD.

**ay** *n* **1** bar, beam, laser, pencil, shaft, streak, stream. **2** *a ray of hope.* flicker, gleam, glimmer, hint, indication, scintilla, sign, trace.

**aze** *vb* bulldoze, demolish, destroy, flatten, level, tear down.

**azor** *n* □ *cut-throat razor, disposable razor, electric razor, safety razor.*

**each** *n* compass, distance, orbit, range, scope, sphere. • *vb* **1** achieve, arrive at, attain, come to, get hold of, get to, go as far as, grasp, *inf* make, take, touch. **2** *reach for the salt.* put out your hand, stretch, try to get. **3** *reach me by phone.* communicate with, contact, get in touch with. **reach out** ▷ EXTEND.

**eact** *vb* act, answer, behave, conduct yourself, reciprocate, reply, respond, retaliate, retort, take revenge. **react to** ▷ COUNTER.

**eaction** *n* answer, backlash, *inf* come-back, counter, countermove, effect, feedback, parry, reciprocation, reflex, rejoinder, reply, reprisal, response, retaliation, retort, revenge, riposte.

**reactionary** *adj* conservative, die-hard, old-fashioned, rightist, right-wing, *inf* stick-in-the-mud, traditionalist, unprogressive. *Opp* PROGRESSIVE.

**read** *vb* **1** devour, *inf* dip into, glance at, interpret, look over, peruse, pore over, review, scan, skim, study. **2** *can't read the handwriting.* decipher, decode, interpret, make out, understand.

**readable** *adj* **1** absorbing, compulsive, easy, engaging, enjoyable, entertaining, gripping, interesting, stimulating, well-written. *Opp* BORING. **2** clear, comprehensible, decipherable, distinct, intelligible, legible, neat, plain, understandable. *Opp* ILLEGIBLE.

**readily** *adv* cheerfully, eagerly, easily, effortlessly, freely, gladly, happily, promptly, quickly, ungrudgingly, unhesitatingly, voluntarily, willingly.

**ready** *adj* **1** accessible, *inf* all set, arranged, at hand, available, complete, convenient, done, finalized, finished, fit, obtainable, prepared, primed, ripe, set, set up, waiting. **2** *ready to help.* agreeable, consenting, content, disposed, eager, equipped, *inf* game, glad, inclined, in the mood, keen, *inf* keyed up, liable, likely, minded, of a mind, open, organized, pleased, poised, predisposed, primed, *inf* psyched up, raring (*to go*), trained, willing. **3** *ready wit.* acute, adroit, alert, apt, facile, immediate, prompt, quick, quick-witted, rapid, sharp, smart, speedy. *Opp* SLOW, UNPREPARED.

**real** *adj* **1** actual, authentic, certain, corporeal, everyday, existing, factual, genuine, material, natural, ordinary, palpable, physical, pure, realistic, tangible, visible. **2** authenticated, *Lat* bona fide, legal, legitimate, official, valid, verifiable. **3** *real friends.* dependable, positive, reliable, sound, true, trustworthy, worthy. **4** *real grief.* earnest, heartfelt, honest, sincere, truthful, unaffected, undoubted, unfeigned, unquestionable. *Opp* FALSE.

**realism** *n* **1** authenticity, fidelity, naturalism, verisimilitude. **2** *realism in business.* clear-sightedness, common sense, objectivity, practicality, pragmatism.

**realistic** *adj* **1** businesslike, clear-sighted, commonsense, down-to-earth, feasible, hardheaded, *inf* hard-nosed, levelheaded, logical, matter-of-fact, *inf* no-nonsense, objective, possible, practicable, practical, pragmatic, rational, sensible, tough, unemotional, unsentimental, viable, workable. **2** *realistic pictures.* authentic, convincing, faithful, graphic, lifelike, natural, recognizable, representational, true-to-life, truthful, vivid. **3** *realistic prices.* acceptable, adequate, fair, genuine, justifiable, moderate, reasonable. *Opp* UNREALISTIC.

**reality** *n* actuality, authenticity, certainty, empirical knowledge, experience, fact, life, *inf* nitty-gritty, real life, the real world, truth, verity. *Opp* FANTASY.

**realize** *vb* **1** accept, appreciate, apprehend, be aware of, become conscious of, *inf* catch on to, comprehend, conceive of, *inf* cotton on to, grasp, know, perceive, recognize, see, sense, *inf* twig, understand, *inf* wake up to. **2** *realize an ambition.* accomplish, achieve, bring about, complete, effect, effectuate, fulfil, implement, make a reality of, obtain, perform, put into effect. **3** *realize a price.* *inf* bring in, *inf* clear, earn, fetch, make, net, obtain, produce.

**realm** *n* country, domain, empire kingdom, monarchy, principality

**reap** *vb* **1** cut, garner, gather in, glean, harvest, mow. **2** *reap a reward.* acquire, bring in, collect, get, obtain, receive, win.

**rear** *adj* back, end, hind, hinder, hindmost, last, rearmost. *Opp* FRONT. • *n* **1** back, end, sterr (*of ship*), tail-end. **2** ▷ BUTTOCKS. • *vb* **1** breed, bring up, care for, cultivate, educate, feed, look afteı nurse, nurture, produce, raise, train. **2** *rear your head.* elevate, hold up, lift, raise, uplift. **3** ▷ BUILD.

**rearrange** *vb* change round, regroup, reorganize, switch roun swop round, transpose. ▷ CHANG

**rearrangement** *n* anagram, reo ganization, transposition. ▷ CHANGE.

**reason** *n* **1** apology, argument, case, cause, defense, excuse, explanation, grounds, incentive, justification, motive, occasion, pr text, rationale, vindication. **2** brains, common sense, *inf* gumption, intelligence, judgement, logic, mind, *inf* nous, persp cacity, rationality, reasonablenes sanity, sense, understanding, wis dom, wit. ▷ REASONING. **3** *reason for living.* aim, goal, intention, motivation, motive, object, object ive, point, purpose, spur, stimulus. • *vb* **1** act rationally, calculate, cerebrate, conclude, conside deduce, estimate, figure out, hypc thesize, infer, intellectualize, judge, *inf* put two and two together, ratiocinate, resolve, the orize, think, use your head, work out. **2** *I reasoned with her.* argue, debate, discuss, expostulate, remonstrate.

**reasonable** *adj* **1** calm, helpful, honest, intelligent, rational, realistic, sane, sensible, sincere, sober, thinking, thoughtful, unemotional, wise. **2** *reasonable argument.* arguable, believable, credible, defensible, justifiable, logical, plausible, practical, reasoned, sound, tenable, viable, well-thought-out. **3** *reasonable prices.* acceptable, appropriate, average, cheap, competitive, conservative, fair, inexpensive, moderate, ordinary, proper, right, suitable, tolerable, unexceptionable. *Opp* IRRATIONAL.

**reasoning** *n* analysis, argument, case, *derog* casuistry, cerebration, deduction, dialectic, hypothesis, line of thought, logic, proof, rationalization, *derog* sophistry, theorizing, thinking.

**reassure** *vb* assure, bolster, buoy up, calm, cheer, comfort, encourage, give confidence to, hearten, *inf* set someone's mind at rest, support, uplift. *Opp* ALARM, THREATEN. **reassuring** ▷ SOOTHING, SUPPORTIVE.

**rebel** *adj* ▷ REBELLIOUS. • *n* anarchist, apostate, dissenter, freedom fighter, heretic, iconoclast, insurgent, malcontent, maverick, mutineer, nonconformist, recusant, resistance fighter, revolutionary, schismatic. • *vb* be a rebel, disobey, dissent, fight, *inf* kick over the traces, mutiny, refuse to obey, revolt, rise up, *inf* run riot, *inf* take a stand. *Opp* CONFORM. **rebel against** ▷ DEFY.

**rebellion** *n* contumacy, defiance, disobedience, insubordination, insurgency, insurrection, mutiny, rebelliousness, resistance, revolt, revolution, rising, schism, sedition, uprising.

**rebellious** *adj inf* bolshie, breakaway, contumacious, defiant, difficult, disaffected, disloyal, disobedient, incorrigible, insubordinate, insurgent, intractable, malcontent, mutinous, obstinate, quarrelsome, rebel, recalcitrant, refractory, resistant, revolting, revolutionary, seditious, uncontrollable, ungovernable, unmanageable, unruly, wild. *Opp* OBEDIENT.

**rebirth** *n* reawakening, regeneration, renaissance, renewal, resurgence, resurrection, return, revival.

**rebound** *vb inf* backfire, *inf* boomerang, bounce, misfire, recoil, ricochet, spring back.

**rebuff** *n inf* brush-off, check, discouragement, refusal, rejection, slight, snub. • *vb* cold-shoulder, decline, discourage, refuse, reject, repulse, slight, snub, spurn, turn down.

**rebuild** *n* reassemble, reconstruct, recreate, redevelop, refashion, regenerate, remake. ▷ RECONDITION.

**rebuke** *vb* admonish, castigate, censure, chide, reprehend, reproach, reprove, scold, upbraid. ▷ REPRIMAND.

**recall** *vb* **1** bring back, call in, summon, withdraw. **2** ▷ REMEMBER.

**recede** *vb* abate, decline, dwindle, ebb, fall back, go back, lessen, regress, retire, retreat, return, shrink back, sink, slacken, subside, wane, withdraw.

**receipt** *n* **1** account, acknowledgement, bill, proof of purchase, sales slip, ticket. **2** *receipt of goods.* acceptance, delivery, reception, receipts gains, gate, income, proceeds, profits, return, takings.

**receive** *vb* 1 accept, acquire, be given, be paid, be sent, collect, come by, come into, derive, earn, gain, get, gross, inherit, make, net, obtain, take. 2 *receive an injury.* bear, be subjected to, endure, experience, meet with, suffer, sustain, undergo. 3 *receive visitors.* accommodate, admit, entertain, greet, let in, meet, show in, welcome. *Opp* GIVE.

**recent** *adj* brand-new, contemporary, current, fresh, just out, latest, modern, new, novel, present-day, up-to-date, young. *Opp* OLD.

**reception** *n* 1 greeting, response, welcome. 2 ▷ PARTY.

**receptive** *adj* amenable, favourable, flexible, interested, open, open-minded, responsive, susceptible, sympathetic, tractable, welcoming, well-disposed. *Opp* RESISTANT.

**recess** *n* 1 alcove, apse, bay, cavity, corner, cranny, hollow, indentation, niche, nook. 2 adjournment, break, *inf* breather, breathing-space, interlude, intermission, interval, respite, rest, time off.

**recession** *n* decline, depression, downturn, slump.

**recipe** *n* directions, formula, instructions, method, plan, prescription, procedure, technique.

**reciprocal** *adj* corresponding, exchanged, joint, mutual, requited, returned, shared.

**reciprocate** *vb* exchange, give the same in return, match, repay, requite, return.

**recital** *n* 1 concert, performance, programme. 2 *recital of events.* account, description, narration, narrative, recounting, rehearsal, relation, repetition, story, telling. ▷ RECITATION.

**recitation** *n* declaiming, declamation, delivery, monologue, narration, performance, presentation, reading, *old use* rendition, speaking, telling.

**recite** *vb* articulate, declaim, deliver, narrate, perform, present quote, *inf* rattle off, recount, reel off, rehearse, relate, repeat, speak tell.

**reckless** *adj* 1 brash, careless, *inf* crazy, daredevil, *inf* devil-may-care, foolhardy, harebrained, *inf* harum-scarum, hasty, heedless impetuous, imprudent, impulsive, inattentive, incautious, indiscreet, injudicious, irresponsible, *inf* mad madcap, negligent, rash, thoughtless, unconsidered, unwise, wild. *Opp* CAREFUL. 2 *reckless criminals.* dangerous, desperate, hardened, violent.

**reckon** *vb* 1 add up, appraise, assess, calculate, compute, count, enumerate, estimate, evaluate, figure out, gauge, number, tally, total, value, work out. 2 ▷ THINK.

**reclaim** *vb* 1 get back, *inf* put in for, recapture, recover, regain. 2 *reclaim derelict land.* make usable, redeem, regenerate, reinstate, rescue, restore, salvage, save.

**recline** *vb* lean back, lie, loll, lounge, repose, rest, sprawl, stretch out.

**recluse** *n* anchoress, anchorite, hermit, loner, monk, nun, solitary.

**recognizable** *adj* detectable, distinctive, distinguishable, identifiable, known, noticeable, perceptible, undisguised, unmistakable, visible.

**recognize** *vb* **1** detect, diagnose, discern, distinguish, identify, know, name, notice, perceive, pick out, place (*can't place him*), *inf* put a name to, recall, recollect, remember, see, spot. **2** *recognize your faults.* accept, acknowledge, admit to, appreciate, be aware of, concede, confess, grant, realize, understand. **3** *recognize someone's rights.* approve of, *inf* back, endorse, legitimize, ratify, sanction, support, validate.

**recoil** *vb* blench, draw back, falter, flinch, jerk back, jump, quail, shrink, shy away, start, wince. ▷ REBOUND.

**recollect** *vb* hark back to, recall, reminisce about, summon up, think back to. ▷ REMEMBER.

**recommend** *vb* **1** advise, advocate, counsel, exhort, prescribe, propose, put forward, suggest, urge. **2** applaud, approve of, *inf* back, commend, favour, *inf* plug, praise, *inf* push, *inf* put in a good word for, speak well of, support, vouch for. ▷ ADVERTISE.

**recommendation** *n* advice, advocacy, approbation, approval, *inf* backing, commendation, counsel, favourable mention, reference, seal of approval, support, testimonial.

**reconcile** *vb* bring together, *old use* conciliate, harmonize, make friendly again, placate, reunite, settle differences between. **be reconciled to** accept, adjust to, resign yourself to, submit to.

**recondition** *vb* make good, overhaul, rebuild, renew, renovate, repair, restore.

**reconnaissance** *n* examination, exploration, inspection, investigation, observation, *inf* recce, reconnoitring, scouting, spying, survey.

**reconnoitre** *vb inf* case, *inf* check out, examine, explore, gather intelligence (about), inspect, investigate, patrol, scout, scrutinize, spy, survey, *sl* suss out.

**reconsider** *vb* be converted, change your mind, come round, reappraise, reassess, re-examine, rethink, review your position, think better of.

**reconstruct** *vb* act out, mock up, recreate, rerun. ▷ REBUILD.

**record** *n* **1** account, annals, archives, catalogue, chronicle, diary, documentation, dossier, file, journal, log, memorandum, minutes, narrative, note, register, report, transactions. **2** best performance, best time. **3** ▷ RECORDING. • *vb* **1** chronicle, document, enter, inscribe, list, log, minute, note, put down, register, set down, take down, transcribe, write down. **2** *record on tape.* keep, preserve, tape, tape-record, video.

**recording** *n* performance, release. □ *album, audio-tape, cassette, CD, compact disc, digital recording, disc, long-playing record, LP, mono recording, record, single, stereo recording, tape, tape-recording, tele-recording, video, video-cassette, video disc, videotape.*

**record-player** *n* CD player, gramophone, midi system, *old use* phonograph, record deck, turntable.

**recount** *vb* communicate, describe, detail, impart, narrate, recite, relate, report, tell, unfold.

**recover** *vb* **1** find, get back, get compensation for, make good, make up for, recapture, reclaim, recoup, regain, repossess, restore, retrieve, salvage, trace, track down, win back. **2** *inf* be on the mend, come round, convalesce, *inf* get back on your feet, get better, heal, improve, mend, *inf* pull round, *inf* pull through, rally, recuperate, regain your strength, revive, survive, *inf* take a turn for the better.

**recovery** *n* **1** recapture, reclamation, repossession, restoration, retrieval, salvage, salvaging. **2** *recovery from illness.* advance, convalescence, cure, deliverance, healing, improvement, progress, rally, recuperation, revival, upturn.

**recreation** *n* amusement, distraction, diversion, enjoyment, entertainment, fun, games, hobby, leisure, pastime, play, pleasure, refreshment, relaxation, sport.

**recrimination** *n* accusation, *inf* come-back, counter-attack, counter-charge, reprisal, retaliation, retort.

**recruit** *n* apprentice, beginner, conscript, *inf* greenhorn, initiate, learner, neophyte, *inf* new boy, new girl, novice, tiro, trainee. *Opp* VETERAN. • *vb* advertise for, conscript, draft in, engage, enlist, enrol, *old use* impress, mobilize, muster, register, sign on, sign up, take on.

**rectify** *vb* amend, correct, cure, *inf* fix, make good, put right, repair, revise.

**recumbent** *adj* flat, flat on your back, horizontal, lying down, prone, reclining, stretched out, supine. *Opp* UPRIGHT.

**recuperate** *vb* convalesce, get better, heal, improve, mend, rally, regain strength, revive. ▷ RECOVER.

**recur** *vb* be repeated, come back again, happen again, persist, reappear, repeat, return.

**recurrent** *adj* chronic, cyclical, frequent, intermittent, iterative, periodic, persistent, recurring, regular, repeated, repetitive, returning. ▷ CONTINUAL.

**recycle** *vb* reclaim, recover, retrieve, reuse, salvage, use again.

**red** *adj* bloodshot, blushing, embarrassed, fiery, flaming, florid, flushed, glowing, inflamed, rosy, rubicund, ruddy. □ *auburn, blood-red, brick-red, cardinal, carmine, carroty, cerise, cherry, chestnut, crimson, damask, flame-coloured, foxy, magenta, maroon, orange, pink, rose, roseate, ruby, scarlet, titian, vermilion, wine-coloured.* **red herring** ▷ DECOY.

**redden** *vb* blush, colour, flush, glow.

**redeem** *vb* buy back, cash in, exchange for cash, reclaim, recover, re-purchase, trade in, win back. ▷ LIBERATE. **redeem yourself** ▷ ATONE.

**redolent** *adj* **1** aromatic, fragrant, perfumed, scented, smelling. **2** *redolent of the past.* reminiscent, suggestive.

**reduce** *vb* **1** abate, abbreviate, abridge, clip, compress, curtail, cut, cut back, cut down, decimate, decrease, detract from, devalue, dilute, diminish, *inf* dock (*wages*), *inf* ease up on, halve, impair, lessen, limit, lower, make less, minimize, moderate, narrow, prune, shorten, shrink, simplify, *inf* slash, slim down, tone down, trim, truncate, weaken, whittle. **2** become less, contract, dwindle, shrink. **3** *reduce a liquid.* concentrate, condense, thicken. **4** *reduce to rubble.* break up, destroy, grind, pulp, pulverize, triturate. **5** *reduce to poverty.* degrade, demote, downgrade, humble, impoverish, move down, put down, ruin. *Opp* INCREASE, RAISE.

**reduction** *n* **1** contraction, curtailment, *inf* cutback, deceleration (*of speed*), decimation, decline, decrease, diminution, drop, impairment, lessening, limitation, loss, moderation, narrowing, remission, shortening, shrinkage, weakening. **2** *reduction in price*. concession, cut, depreciation, devaluation, discount, rebate, refund. *Opp* INCREASE.

**redundant** *adj* excessive, inessential, non-essential, superfluous, supernumerary, surplus, too many, unnecessary, unneeded, unwanted. *Opp* NECESSARY.

**reek** *n* stench, stink. ▷ SMELL.

**reel** *n* bobbin, spool. • *vb* falter, lurch, pitch, rock, roll, spin, stagger, stumble, sway, totter, waver, whirl, wobble. **reel off** ▷ RECITE.

**refer** *vb* **refer to 1** allude to, bring up, cite, comment on, draw attention to, make reference to, mention, name, point to, quote, speak of, specify, touch on. **2** *refer one person to another*. direct to, guide to, hand over to, pass on to, recommend to, send to. **3** *refer to the dictionary*. consult, go to, look up, resort to, study, turn to.

**referee** *n* adjudicator, arbiter, arbitrator, judge, mediator, umpire.

**reference** *n* **1** allusion, citation, example, illustration, instance, intimation, mention, note, quotation, referral, remark. **2** endorsement, recommendation, testimonial.

**refill** *vb* fill up, refuel, renew, replenish, top up.

**refine** *vb* **1** clarify, cleanse, clear, decontaminate, distil, process, purify, treat. **2** *refine manners*. civilize, cultivate, improve, perfect, polish.

**refined** *adj* **1** aristocratic, civilized, courteous, courtly, cultivated, cultured, delicate, dignified, discerning, discriminating, educated, elegant, fastidious, genteel, gentlemanly, gracious, ladylike, nice, polished, polite, *inf* posh, precise, *derog* pretentious, *derog* prissy, sensitive, sophisticated, stylish, subtle, tasteful, *inf* upper-crust, urbane, well-bred, well brought-up. *Opp* RUDE. **2** *refined oil*. clarified, distilled, processed, purified, treated. *Opp* CRUDE.

**refinement** *n* **1** breeding, *inf* class, courtesy, cultivation, delicacy, discernment, discrimination, elegance, finesse, gentility, graciousness, polish, *derog* pretentiousness, sensitivity, sophistication, style, subtlety, taste, urbanity. **2** *refinements in design*. alteration, change, enhancement, improvement, modification, perfection.

**reflect** *vb* **1** echo, mirror, return, send back, shine back, throw back. **2** brood, cerebrate, *inf* chew things over, consider, contemplate, deliberate, meditate, ponder, remind yourself, reminisce, ruminate. ▷ THINK. **3** *Her success reflects her hard work*. bear witness to, correspond to, demonstrate, evidence, exhibit, illustrate, indicate, match, point to, reveal, show.

**reflection** *n* **1** echo, image, likeness. **2** *reflection of hard work*. demonstration, evidence, indication, manifestation, result. **3** *no reflection on you*. aspersion, censure, criticism, discredit, imputation, reproach, shame, slur. **4** *time for reflection*. cerebration, cogitation, contemplation, deliberation, meditation, pondering, rumination, self-examination, study, thinking, thought.

**reflective** *adj* **1** glittering, lustrous, reflecting, shiny, silvery. **2** ▷ THOUGHTFUL.

**reform** *vb* **1** ameliorate, amend, become better, better, change, convert, correct, improve, make better, mend, put right, reconstruct, rectify, remodel, reorganize, save. **2** *reform a system.* purge, reconstitute, regenerate, revolutionize.

**refrain** *vb* **refrain from** abstain from, avoid, cease, desist from, do without, eschew, forbear, leave off, *inf* quit, renounce, stop.

**refresh** *vb* **1** cool, energize, enliven, fortify, freshen, invigorate, *inf* perk up, quench the thirst of, reanimate, rejuvenate, renew, restore, resuscitate, revitalize, revive, slake (*thirst*). **2** *refresh the memory.* activate, awaken, jog, remind, prod, prompt, stimulate.

**refreshing** *adj* **1** bracing, cool, enlivening, exhilarating, inspiriting, invigorating, restorative, reviving, stimulating, thirst-quenching, tingling, tonic. *Opp* EXHAUSTING. **2** *a refreshing change.* different, fresh, interesting, new, novel, original, unexpected, unfamiliar, unforeseen, unpredictable, welcome. *Opp* BORING.

**refreshments** *n* drinks, eatables, *inf* eats, *inf* nibbles, snack. ▷ DRINK, FOOD.

**refrigerate** *vb* chill, cool, freeze, ice, keep cold.

**refuge** *n* asylum, *inf* bolt-hole, cover, harbour, haven, *inf* hideaway, hideout, *inf* hidey-hole, hiding-place, protection, retreat, safety, sanctuary, security, shelter, stronghold.

**refugee** *n* displaced person, émigré, exile, fugitive, outcast, runaway.

**refund** *n* rebate, repayment. • *vb* give back, pay back, recoup, reimburse, repay, return.

**refusal** *n inf* brush-off, denial, disagreement, disapproval, rebuff, rejection, veto. *Opp* ACCEPTANCE.

**refuse** *n* detritus, dirt, garbage, junk, litter, rubbish, trash, waste. • *vb* baulk at, decline, deny, disallow, *inf* jib at, *inf* pass up, rebuff, reject, repudiate, say no to, spurn, turn down, veto, withhold. *Opp* ACCEPT, GRANT.

**refute** *vb* counter, discredit, disprove, negate, prove wrong, rebut.

**regain** *vb* be reunited with, find, get back, recapture, reclaim, recoup, recover, repossess, retake, retrieve, return to, win back.

**regal** *adj derog* haughty, imperial, kingly, lordly, majestic, noble, palatial, *derog* pompous, princely, queenly, royal, stately. ▷ SPLENDID.

**regard** *n* **1** gaze, look, scrutiny, stare. **2** attention, care, concern, consideration, deference, heed, notice, reference, respect, sympathy, thought. **3** admiration, affection, appreciation, approbation, approval, deference, esteem, favour, honour, love, respect, reverence, veneration. • *vb* **1** behold, contemplate, eye, gaze at, keep an eye on, look at, note, observe, scrutinize, stare at, view, watch. **2** *regarded me as a liability.* account, consider, deem, esteem, judge, look upon, perceive, rate, reckon, respect, think of, value, view, weigh up.

**regarding** *prep* about, apropos, concerning, connected with, involving, on the subject of, pertaining to, *inf* re, respecting, with reference to, with regard to.

**regardless** *adj* **regardless of** careless about, despite, heedless of, indifferent to, neglectful of, notwithstanding, unconcerned about, unmindful of.

**regime** *n* administration, control, discipline, government, leadership, management, order, reign, rule, system.

**regiment** *vb* arrange, control, discipline, organize, regulate, systematize.

**region** *n* area, country, department, district, division, expanse, land, locality, neighbourhood, part, place, province, quarter, sector, territory, tract, vicinity, zone.

**register** *n* archives, catalogue, chronicle, diary, directory, file, index, inventory, journal, ledger, list, record, roll, tally. • *vb* **1** enlist, enrol, enter your name, join, sign on. **2** *register a complaint.* catalogue, enter, list, log, make official, minute, present, record, set down, submit, write down. **3** *register emotion.* betray, display, divulge, express, indicate, manifest, reflect, reveal, show. **4** *register in a hotel. inf* check in, sign in. **5** *register what someone says.* keep in mind, make a note of, mark, notice, take account of.

**regress** *vb* backslide, degenerate, deteriorate, fall back, go back, move backwards, retreat, retrogress, revert, slip back. *Opp* PROGRESS.

**regret** *n* **1** bad conscience, compunction, contrition, guilt, penitence, pricking of conscience, remorse, repentance, self-accusation, self-condemnation, self-reproach, shame. **2** disappointment, grief, sadness, sorrow, sympathy. • *vb* accuse yourself, bemoan, be regretful, be sad, bewail, deplore, deprecate, feel remorse, grieve (about), lament, mourn, repent (of), reproach yourself, rue, weep (over).

**regretful** *adj* apologetic, ashamed, conscience-stricken, contrite, disappointed, guilty, penitent, remorseful, repentant, rueful, sorry. ▷ SAD. *Opp* UNREPENTANT.

**regrettable** *adj* deplorable, disappointing, distressing, lamentable, reprehensible, sad, shameful, undesirable, unfortunate, unhappy, unlucky, unwanted, upsetting, woeful, wrong.

**regular** *adj* **1** consistent, constant, equal, even, fixed, measured, ordered, predictable, recurring, repeated, rhythmic, steady, symmetrical, systematic, uniform, unvarying. □ *daily, hourly, monthly, weekly, yearly.* **2** *a regular procedure.* accustomed, common, commonplace, conventional, customary, established, everyday, familiar, frequent, habitual, known, normal, official, ordinary, orthodox, prevailing, proper, routine, scheduled, standard, traditional, typical, usual. **3** *a regular supporter.* dependable, faithful, reliable. *Opp* IRREGULAR. • *n inf* faithful, frequenter, habitué, regular customer, patron.

**regulate** *vb* **1** administer, conduct, control, direct, govern, manage, monitor, order, organize, oversee, restrict, supervise. **2** *regulate temperature.* adjust, alter, balance, change, get right, moderate, modify, set, vary.

**regulation** *n* by-law, commandment, decree, dictate, directive, edict, law, order, ordinance, requirement, restriction, rule, ruling, statute.

**rehearsal** *n* dress rehearsal, *inf* dry run, exercise, practice, preparation, *inf* read-through, *inf* run-through, *inf* try-out.

**rehearse** *vb* drill, go over, practise, prepare, *inf* run over, *inf* run through, try out.

**rehearsed** *adj* calculated, practised, pre-arranged, premeditated, prepared, scripted, studied, thought out. *Opp* IMPROMPTU.

**reign** *n* administration, ascendancy, command, empire, government, jurisdiction, kingdom, monarchy, power, rule, sovereignty. • *vb* be king, be on the throne, be queen, command, govern, have power, hold sway, rule, *inf* wear the crown.

**reincarnation** *n* rebirth, return to life, transmigration.

**reinforce** *vb* **1** back up, bolster, buttress, fortify, give strength to, hold up, prop up, stay, stiffen, srrengthen, support, toughen. **2** *reinforce an army*. add to, assist, augment, help, increase the size of, provide reinforcements for, supplement.

**reinforcements** *plur n* additional troops, auxiliaries, back-up, help, reserves, support.

**reinstate** *vb* recall, rehabilitate, restore, take back, welcome back. *Opp* DISMISS.

**reject** *vb* **1** cast off, discard, discount, dismiss, eliminate, exclude, jettison, *inf* junk, put aside, scrap, send back, throw away, throw out. **2** *reject friends*. disown, *inf* drop, *inf* give someone the cold shoulder, jilt, rebuff, renounce, repel, repudiate, repulse, *inf* send packing, shun, spurn, turn your back on. **3** *reject an invitation*. brush aside, decline, refuse, say no to, turn down, veto. *Opp* ACCEPT, ADOPT.

**rejoice** *vb* be happy, celebrate, delight, exult, glory, revel, triumph. *Opp* GRIEVE.

**relapse** *n* degeneration, deterioration, recurrence (*of illness*), regression, reversion, *inf* set-back, worsening. • *vb* backslide, degenerate, deteriorate, fall back, have a relapse, lapse, regress, retreat, revert, sink back, slip back, weaken.

**relate** *vb* **1** communicate, describe, detail, divulge, impart, make known, narrate, present, recite, recount, rehearse, report, reveal, tell. **2** ally, associate, compare, connect, consider together, coordinate, correlate, couple, join, link. **relate to 1** appertain to, apply to, bear upon, be relevant to, concern, *inf* go with, pertain to, refer to. **2** *relate to other people*. be friends with, empathize with, fraternize with, handle, have a relationship with, identify with, socialize with, understand.

**related** *adj* affiliated, akin, allied, associated, cognate, comparable, connected, consanguineous, interconnected, interdependent, interrelated, joined, joint, linked, mutual, parallel, reciprocal, relative, similar, twin. ▷ RELEVANT. *Opp* UNRELATED.

**relation** *n* **1** *old use* kinsman, *old use* kinswoman, *plur* kith and kin member of the family, relative. ▷ FAMILY. **2** *relation of a story*. ▷ NARRATION.

**relationship** *n* **1** affiliation, affinity, association, attachment, bond, closeness, connection, consanguinity, correlation, correspondence, interconnection, interdependence, kinship, link, parallel, pertinence, rapport, ratio tie, understanding. ▷ SIMILARITY. *Opp* CONTRAST. **2** affair, *inf* intrigue, *inf* liaison, love affair, romance, sexual relations. ▷ FRIENDSHIP.

**relative** *adj* ▷ RELATED, RELEVANT. **relative to** commensurate (with), comparative, proportional, proportionate. *Opp* UNRELATED. • *n* ▷ RELATION.

**relax** *vb* **1** be easy, be relaxed, calm down, cool down, feel at home, *inf* let go, *inf* put your feet up, rest, *inf* slow down, *inf* take it easy, unbend, unwind. *Opp* TENSION. **2** *relax your vigilance.* abate, curb, decrease, diminish, ease off, lessen, loosen, mitigate, moderate, reduce, release, relieve, slacken, soften, temper, *inf* tone down, unclench, unfasten, weaken. *Opp* INCREASE.

**relaxation** *n* **1** ease, informality, loosening up, relaxing, repose, rest, unwinding. ▷ RECREATION. *Opp* TENSION. **2** abatement, alleviation, diminution, lessening, *inf* let-up, mitigation, moderation, remission, slackening, weakening. *Opp* INCREASE.

**relaxed** *adj derog* blasé, calm, carefree, casual, comfortable, contented, cool, cosy, easygoing, *inf* free and easy, friendly, good-humoured, happy, *inf* happy-go-lucky, informal, insouciant, *inf* laid-back, *derog* lax, leisurely, light-hearted, nonchalant, peaceful, reassuring, restful, serene, *derog* slack, tranquil, unconcerned, unhurried, untroubled. *Opp* TENSE.

**relay** *n* **1** shift, turn. **2** *live relay.* broadcast, programme, transmission. • *vb* broadcast, communicate, pass on, send out, spread, televise, transmit.

**release** *vb* **1** acquit, allow out, deliver, discharge, dismiss, emancipate, excuse, exonerate, free, let go, let loose, let off, liberate, loose, pardon, rescue, save, set free, set loose, unchain, unfasten, unfetter, unleash, unshackle, untie. *Opp* DETAIN. **2** fire off, launch, let fly, let off, send off. **3** *release information.* circulate, disseminate, distribute, issue, make available, present, publish, put out, send out, unveil.

**relegate** *vb* **1** consign to a lower position, demote, downgrade, put down. **2** banish, dispatch, exile.

**relent** *vb* acquiesce, become more lenient, be merciful, capitulate, give in, give way, relax, show pity, soften, weaken, yield.

**relentless** *adj* **1** dogged, fierce, hard-hearted, implacable, incessant, inexorable, intransigent, merciless, obdurate, obstinate, pitiless, remorseless, ruthless, uncompromising, unfeeling, unforgiving, unmerciful, unyielding. ▷ CRUEL. **2** unceasing, unrelieved, unstoppable, unyielding. ▷ CONTINUAL.

**relevant** *adj* appertaining, applicable, apposite, appropriate, apropos, apt, connected, essential, fitting, germane, linked, material, pertinent, proper, related, relative, significant, suitable, suited, to the point. *Opp* IRRELEVANT.

**reliable** *adj* certain, conscientious, consistent, constant, dependable, devoted, efficient, faithful, honest, infallible, loyal, predictable, proven, punctilious, regular, reputable, responsible, safe, solid, sound, stable, staunch, steady, sure, trusted, trustworthy, trusty, unchanging, unfailing. *Opp* UNRELIABLE.

**relic** *n* heirloom, heritage, inheritance, keepsake, memento, remains, reminder, remnant, souvenir, survival, token, vestige.

**relief** *n* abatement, aid, alleviation, assistance, assuagement, comfort, cure, deliverance, diversion, ease, easement, help, *inf* let-up, mitigation, palliation, relaxation, release, remedy, remission, respite, rest.

**relieve** *vb* abate, alleviate, anaesthetize, assuage, bring relief to, calm, comfort, console, cure, diminish, disburden, disencumber, dull, ease, lessen, lift, lighten, make less, mitigate, moderate, palliate, reduce, relax, release, rescue, soften, soothe, unburden. ▷ HELP. *Opp* INTENSIFY.

**religion** *n* **1** belief, creed, divinity, doctrine, dogma, *derog* pietism, theology. **2** creed, cult, denomination, faith, persuasion, sect. □ *Buddhism, Christianity, Hinduism, Islam, Judaism, Sikhism, Taoism, Zen.*

**religious** *adj* **1** devotional, divine, holy, sacramental, sacred, scriptural, theological. *Opp* SECULAR. **2** church-going, committed, dedicated, devout, God-fearing, godly, *derog* pietistic, pious, *derog* religiose, reverent, righteous, saintly, *derog* sanctimonious, spiritual. *Opp* IRRELIGIOUS. **3** *religious wars.* bigoted, doctrinal, fanatical, sectarian, schismatic.

**relinquish** *vb* concede, give in, hand over, part with, submit, surrender, yield.

**relish** *n* **1** appetite, delight, enjoyment, enthusiasm, gusto, pleasure, zest. **2** flavour, piquancy, savour, tang, taste. • *vb* appreciate, delight in, enjoy, like, love, revel in, savour, take pleasure in.

**reluctant** *adj* averse, disinclined, grudging, hesitant, loath, unenthusiastic, unwilling. *Opp* EAGER.

**rely** *vb* **rely on** *inf* bank on, count on, depend on, have confidence in, lean on, put your faith in, *inf* swear by, trust.

**remain** *vb old use* abide, be left, carry on, continue, endure, keep on, linger, live on, persevere, persist, stay, *inf* stay put, survive, tarry, wait. **remaining** ▷ RESIDUAL.

**remainder** *n* balance, excess, extra, remnant, residue, residuum, rest, surplus. ▷ REMAINS.

**remains** *plur n* **1** crumbs, debris, detritus, dregs, fragments, *inf* leftovers, oddments, *inf* odds and ends, offcuts, remainder, remnants, residue, rubble, ruins, scraps, traces, vestiges, wreckage. **2** *historic remains.* heirloom, heritage, inheritance, keepsake, memento, monument, relic, reminder, souvenir, survival. **3** *human remains.* ashes, body, bones, carcass, corpse.

**remake** *vb* piece together, rebuild, reconstitute, reconstruct, redo. ▷ RENEW.

**remark** *n* comment, mention, observation, opinion, reflection, statement, thought, utterance, word. • *vb* **1** assert, comment, declare, mention, note, observe, pass comment, reflect, say, state. **2** heed, mark, notice, observe, perceive, see, take note of.

**remarkable** *adj* amazing, astonishing, astounding, conspicuous, curious, different, distinguished, exceptional, extraordinary, important, impressive, marvellous, memorable, notable, noteworthy, odd, out-of-the-ordinary, outstanding, peculiar, phenomenal, prominent, signal, significant, singular, special, strange, striking, surprising, *inf* terrific, *inf* tremendous, uncommon, unforgettable, unusual, wonderful. *Opp* ORDINARY.

**remedy** *n inf* answer, antidote, corrective, countermeasure, cure, cure-all, drug, elixir, medicament, medication, medicine, nostrum, palliative, panacea, prescription, redress, relief, restorative, solution, therapy, treatment.
• *vb* alleviate, *inf* ameliorate, answer, control, correct, counteract, *inf* fix, heal, help, mend, mitigate, palliate, put right, rectify, redress, relieve, repair, solve, treat. ▷ CURE.

**remember** *vb* **1** be mindful of, have a memory of, have in mind, keep in mind, recognize. **2** learn, memorize, retain. **3** *remember old times.* be nostalgic about, hark back to, recall, recollect, reminisce about, review, summon up, tell stories about, think back to. **4** *remember Christmas.* celebrate, commemorate, observe. *Opp* FORGET.

**remind** *vb* give a reminder to, jog the memory of, nudge, prompt.

**reminder** *n* **1** aide-mémoire, cue, hint, *inf* memo, memorandum, mnemonic, note, *inf* nudge, prompt, *inf* shopping list. **2** heirloom, inheritance, keepsake, memento, relic, souvenir, survival.

**reminisce** *vb* be nostalgic, hark back, look back, recall, remember, review, tell stories, think back.

**reminiscence** *n* account, anecdote, memoir, memory, recollection, remembrance.

**reminiscent** *adj* evocative, nostalgic, recalling, redolent, suggestive.

**remiss** *adj* careless, dilatory, forgetful, irresponsible, lax, negligent, slack, thoughtless. *Opp* CAREFUL.

**remit** *vb* **1** *remit a debt.* cancel, let off, settle. **2** abate, decrease, ease off, lessen, relax, slacken. **3** dispatch, forward, send, transmit. ▷ PAY.

**remittance** *n* allowance, fee, payment.

**remnants** *plur n* bits, fragments, *inf* leftovers, oddments, offcuts, residue, scraps, traces, vestiges. ▷ REMAINS.

**remodel** *vb* ▷ RENEW.

**remorse** *n* bad conscience, compunction, contrition, grief, guilt, mortification, pangs of conscience, penitence, pricking of conscience, regret, repentance, sadness, self-accusation, self-reproach, shame, sorrow.

**remorseful** *adj* ashamed, conscience-stricken, contrite, grief-stricken, guilt-ridden, guilty, penitent, regretful, repentant, rueful, sorry. *Opp* UNREPENTANT.

**remorseless** *adj* dogged, implacable, inexorable, intransigent, merciless, obdurate, pitiless, relentless, ruthless, uncompromising, unforgiving, unkind, unmerciful, unremitting. ▷ CRUEL.

**remote** *adj* **1** alien, cut off, desolate, distant, far-away, foreign, God-forsaken, hard to find, inaccessible, isolated, lonely, outlying, out of reach, out of reach, out of the way, secluded, solitary, unfamiliar, unfrequented, *inf* unget-at-able, unreachable. *Opp* CLOSE. **2** *a remote chance.* doubtful, implausible, improbable, negligible, outside, poor, slender, slight, small, unlikely. *Opp* SURE. **3** *a remote manner.* abstracted, aloof, cold, cool, detached, haughty, preoccupied, reserved, standoffish, uninvolved, withdrawn. *Opp* FRIENDLY.

**removal** *n* **1** relocation, removing, taking away, transfer, transportation. **2** elimination, eradication, extermination, liquidation, purge, purging. ▷ KILLING. **3** *removal from a job.* deposition, dethronement, dislodgement, dismissal, displacement, ejection, expulsion, *inf* firing, making redundant, ousting, redundancy, *inf* sacking, transference, unseating. **4** *removal of teeth.* drawing, extraction, pulling, taking out, withdrawal.

**remove** *vb* **1** abolish, abstract, amputate (*limb*), banish, clear away, cut off, cut out, delete, depose, detach, disconnect, dismiss, dispense with, displace, dispose of, do away with, eject, eliminate, eradicate, erase, evict, excise, exile, expel, expunge, *inf* fire, *inf* get rid of, *inf* kick out, kill, oust, purge, root out, rub out, *inf* sack, send away, separate, strike out, sweep away, take out, throw out, turn out, undo, unfasten, uproot, wash off, wipe (*tape-recording*), wipe out. **2** *remove furniture.* carry away, convey, move, take away, transfer, transport. **3** *remove a tooth.* draw out, extract, pull out, take out. **4** *remove clothes.* doff (*a hat*), peel off, strip off, take off.

**rend** *vb* cleave, lacerate, pull apart, rip, rupture, shred, split, tear.

**render** *vb* **1** cede, deliver, furnish, give, hand over, offer, present, proffer, provide, surrender, tender, yield. **2** *render a song.* execute, interpret, perform, play, produce, sing. **3** *rendered me speechless.* cause to be, make.

**rendezvous** *n* appointment, assignation, date, engagement, meeting, meeting-place, *old use* tryst.

**renegade** *n* apostate, backslider, defector, deserter, fugitive, heretic, mutineer, outlaw, rebel, runaway, traitor, turncoat.

**renege** *vb* **renege on** abjure, abrogate, *inf* back out of, break, default on, fail to keep, go back on, *sl* rat on, repudiate, *sl* welsh on.

**renew** *vb* **1** bring up to date, *inf* do up, *inf* give a face-lift to, improve, mend, modernize, overhaul, recondition, reconstitute, recreate, redecorate, redesign, redevelop, redo, refit, refresh, refurbish, regenerate, reintroduce rejuvenate, remake, remodel, renovate, repaint, repair, replace, replenish, restore, resume, resurrect, revamp, revitalize, revive, touch up, transform, update. **2** *renew an activity.* come back to, pick up again, recommence, restart, resume, return to. **3** *renew vows.* confirm, reaffirm, reiterate, repeat, restate.

**renounce** *vb* **1** abandon, abjure, abstain from, declare your opposition to, deny, desert, discard, disown, eschew, forgo, forsake, forswear, give up, reject, repudiate, spurn. **2** *renounce the throne.* abdicate, *inf* quit, relinquish, resign, surrender.

**renovate** *vb* ▷ RENEW.

**renovation** *n* improvement, modernization, overhaul, reconditioning, redevelopment, refit, refurbishment, renewal, repair, restoration, transformation, updating.

**renowned** *adj* celebrated, distinguished, eminent, illustrious, noted, prominent, well-known. ▷ FAMOUS.

**rent** *n* 1 fee, hire, instalment, payment, rental. 2 *a rent in a garment.* ▷ SPLIT. • *vb* charter, farm out, hire, lease, let.

**reorganize** *vb* rationalize, rearrange, re-deploy, reshuffle, restructure.

**repair** *vb* 1 *inf* fix, mend, overhaul, patch up, put right, rectify, refit, restore, service. ▷ RENEW. 2 darn, patch, sew up.

**repay** *vb* 1 compensate, give back, pay back, recompense, refund, reimburse, remunerate, settle. 2 avenge, get even, *inf* get your own back, reciprocate, requite, retaliate, return, revenge.

**repeal** *vb* abolish, abrogate, annul, cancel, invalidate, nullify, rescind, reverse, revoke.

**repeat** *vb* 1 do again, duplicate, redo, rehearse, replay, replicate, reproduce, re-run, show again. 2 echo, quote, recapitulate, re-echo, regurgitate, reiterate, restate, retell, say again.

**repel** *vb* 1 check, drive away, fend off, fight off, hold off, *inf* keep at bay, parry, push away, rebuff, repulse, resist, ward off, withstand. 2 *repel water.* be impermeable to, exclude, keep out, reject. 3 *cruelty repels us.* alienate, be repellent to, disgust, nauseate, offend, *inf* put off, revolt, sicken, *inf* turn off. *Opp* ATTRACT.

**repellent** *adj* 1 impermeable, impervious, resistant, unsusceptible. 2 ▷ REPULSIVE.

**repent** *vb* bemoan, be repentant about, bewail, feel repentance for, lament, regret, reproach yourself for, rue.

**repentance** *n* contrition, guilt, penitence, regret, remorse, self-accusation, self-reproach, shame, sorrow.

**repentant** *adj* apologetic, ashamed, conscience-stricken, contrite, grief-stricken, guilt-ridden, guilty, penitent, regretful, remorseful, rueful, sorry. *Opp* UNREPENTANT.

**repertory** *n* collection, repertoire, repository, reserve, stock, store, supply.

**repetitive** *adj* boring, incessant, iterative, monotonous, recurrent, repeated, repeating, repetitious, tautologous, tedious, unchanging, unvaried. ▷ CONTINUAL.

**replace** *vb* 1 make good, put back, reinstate, restore, return. 2 be a replacement for, come after, follow, oust, succeed, supersede, supplant, take over from, take the place of. ▷ DEPUTIZE. 3 *replace worn parts.* change, renew, substitute.

**replacement** *n inf* fill-in, proxy, stand-in, substitute, successor, understudy.

**replenish** *vb* fill up, refill, renew, restock, top up.

**replete** *adj inf* bursting, crammed, gorged, *inf* jam-packed, overloaded, sated, stuffed. ▷ FULL.

**replica** *n inf* carbon copy, clone, copy, duplicate, facsimile, imitation, likeness, model, reconstruction, reproduction.

**reply** *n* acknowledgement, answer, *inf* come-back, reaction, rejoinder, response, retort, riposte. • *vb* answer, give a reply, react, rejoin, respond. **reply to** ▷ ACKNOWLEDGE, COUNTER.

**report** *n* 1 account, announcement, article, communication, communiqué, description, dispatch, narrative, news, record, statement, story, *inf* write-up. **2** backfire, bang, blast, boom, crack, detonation, discharge, explosion, noise. • *vb* **1** announce, broadcast, circulate, communicate, declare, describe, disclose, divulge, document, give an account of, notify, present a report on, proclaim, publish, put out, record, recount, reveal, state, tell. **2** *report for duty.* announce yourself, check in, clock in, introduce yourself, make yourself known, present yourself, sign in. **3** *report someone to the police.* complain about, denounce, inform against, *inf* tell on.

**reporter** *n* columnist, commentator, correspondent, journalist, newscaster, newsman, newswoman, newspaperman, newspaperwoman, news presenter, photojournalist.

**repose** *n* calm, calmness, comfort, ease, inactivity, peace, peacefulness, poise, quiescence, quiet, quietness, relaxation, respite, rest, serenity, stasis, stillness, tranquillity. ▷ SLEEP. *Opp* ACTIVITY.

**reprehensible** *adj* blameworthy, culpable, deplorable, disgraceful, immoral, objectionable, regrettable, remiss, shameful, unworthy, wicked. ▷ GUILTY. *Opp* INNOCENT.

**represent** *vb* **1** act out, assume the guise of, be an example of, embody, enact, epitomize, exemplify, exhibit, express, illustrate, impersonate, incarnate, masquerade as, personify, pose as, present, pretend to be, stand for, symbolize, typify. **2** characterize, define, delineate, depict, describe, draw, paint, picture, portray, reflect, show, sketch. **3** act for, speak for, stand up for.

**representation** *n* depiction, figure, icon, image, imitation, likeness, model, picture, portrait, portrayal, resemblance, semblance, statue.

**representative** *adj* **1** archetypal, average, characteristic, illustrative, normal, typical. *Opp* ABNORMAL. **2** *representative government.* chosen, democratic, elected, elective, popular. *Opp* TOTALITARIAN. • *n* **1** delegate, deputy, proxy, spokesman, spokeswoman, stand-in, substitute. **2** agent, *inf* rep, salesman, salesperson, saleswoman, *inf* traveller. **3** ambassador, consul, diplomat, emissary, envoy, legate. **4** *Amer* congressman, councillor, Member of Parliament, MP, ombudsman.

**repress** *vb* **1** control, crush, curb, keep down, limit, oppress, overcome, put down, quell, restrain, subdue, subjugate. **2** *repress emotion. inf* bottle up, frustrate, inhibit, stifle, suppress.

**repressed** *adj* **1** cold, frigid, frustrated, inhibited, neurotic, *inf* prim and proper, tense, unbalanced, undemonstrative, *inf* uptight. *Opp* UNINHIBITED. **2** *repressed emotion. inf* bottled up, hidden, latent, subconscious, suppressed, unconscious, unfulfilled.

**repression** *n* **1** authoritarianism, censorship, coercion, control, despotism, dictatorship, oppression, restraint, subjugation, totalitarianism, tyranny. **2** *repression of emotion. inf* bottling up, frustration, inhibition, suffocation, suppression.

**repressive** *adj* authoritarian, autocratic, brutal, coercive, cruel, despotic, dictatorial, fascist, harsh, illiberal, oppressive, restricting, severe, totalitarian, tyrannical, undemocratic, unenlightened. *Opp* LIBERAL.

**reprieve** *n* amnesty, pardon, postponement, respite, stay of execution. • *vb* commute a sentence, forgive, let off, pardon, postpone execution, set free, spare.

**reprimand** *n* admonition, castigation, censure, condemnation, criticism, *inf* dressing-down, *inf* going-over, *inf* lecture, lesson, *inf* rap on the knuckles, rebuke, remonstration, reproach, reproof, scolding, *inf* slap on the wrist, *inf* slating, *inf* talking-to, *inf* telling-off, *inf* ticking-off, upbraiding, *inf* wigging. • *vb* admonish, berate, blame, *inf* carpet, castigate, censure, chide, condemn, correct, criticize, disapprove of, *inf* dress down, find fault with, *inf* haul over the coals, *inf* lecture, *inf* rap, rate, *inf* read the riot act to, rebuke, reprehend, reproach, reprove, scold, *inf* slate, *inf* take to task, *inf* teach a lesson, *inf* tell off, *inf* tick off, upbraid. *Opp* PRAISE.

**reprisal** *n* counter-attack, getting even, redress, repayment, retaliation, retribution, revenge, vengeance.

**reproach** *n* blame, disapproval, disgrace, scorn. • *vb* censure, criticize, scold, show disapproval of, upbraid. ▷ REPRIMAND. *Opp* PRAISE.

**reproachful** *adj* admonitory, censorious, condemnatory, critical, disapproving, disparaging, reproving, scornful, withering.

**reproduce** *vb* **1** copy, counterfeit, duplicate, forge, imitate, mimic, photocopy, print, redo, reissue, reprint, simulate. ▷ REPEAT. **2** beget young, breed, increase, multiply, procreate, produce offspring, propagate, regenerate, spawn.

**reproduction** *n* **1** breeding, cloning, increase, multiplying, procreation, proliferation, propagation, spawning. **2** *inf* carbon copy, clone, copy, duplicate, facsimile, fake, forgery, imitation, likeness, print, replica.

**repudiate** *vb* **1** deny, disagree with, dispute, rebuff, refute, reject, scorn, turn down. *Opp* ACKNOWLEDGE. **2** *repudiate an agreement.* abrogate, discard, disown, go back on, recant, renounce, rescind, retract, reverse, revoke.

**repugnant** *adj* ▷ REPULSIVE.

**repulsive** *adj* abhorrent, abominable, beastly, disagreeable, disgusting, distasteful, distressing, foul, gross, hateful, hideous, loathsome, nasty, nauseating, nauseous, objectionable, obnoxious, odious, offensive, *inf* off-putting, repellent, repugnant, revolting, *inf* sick, sickening, unattractive, unpalatable, unpleasant, unsavoury, unsightly, vile. ▷ UGLY. *Opp* ATTRACTIVE.

**reputable** *adj* creditable, dependable, esteemed, famous, good, highly regarded, honourable, honoured, prestigious, reliable, respectable, respected, trustworthy, unimpeachable, *inf* up-market, well-thought-of, worthy. *Opp* DISREPUTABLE.

**reputation** *n* character, fame, name, prestige, recognition, renown, repute, standing, stature, status.

**reputed** *adj* alleged, assumed, believed, considered, deemed, famed, judged, purported, reckoned, regarded, rumoured, said, supposed, thought.

**request** *n* appeal, application, call, demand, entreaty, petition, plea, prayer, question, requisition, solicitation, suit, supplication. • *vb* adjure, appeal, apply (for), ask, beg, beseech, call for, claim, demand, desire, entreat, implore, importune, invite, petition, pray for, require, requisition, seek, solicit, supplicate.

**require** *vb* **1** be missing, be short of, depend on, lack, need, want. **2** *require a response.* call for, coerce, command, compel, direct, force, insist, instruct, make, oblige, order, put pressure on. ▷ REQUEST. **required** ▷ REQUISITE.

**requirement** *n* condition, demand, essential, necessity, need, precondition, prerequisite, provision, proviso, qualification, *Lat* sine qua non, stipulation.

**requisite** *adj* compulsory, essential, imperative, indispensable, mandatory, necessary, needed, obligatory, prescribed, required, set, stipulated. *Opp* OPTIONAL.

**requisition** *n* application, authorization, demand, mandate, order, request, voucher. • *vb* **1** demand, order, *inf* put in for, request. **2** appropriate, commandeer, confiscate, expropriate, occupy, seize, take over, take possession of.

**rescue** *n* deliverance, emancipation, freeing, liberation, recovery, release, relief, salvage. • *vb* **1** deliver, emancipate, extricate, free, let go, liberate, loose, ransom, release, save, set free. **2** get back, recover, retrieve, salvage.

**research** *n* analysis, enquiry, examination, experimentation, exploration, fact-finding, inquiry, investigation, *inf* probe, scrutiny, searching, study. • *vb inf* check out, *inf* delve into, experiment, investigate, *inf* probe, search, study.

**resemblance** *n* affinity, closeness, coincidence, comparability, comparison, conformity, congruity, correspondence, equivalence, likeness, similarity, similitude.

**resemble** *vb* approximate to, bear resemblance to, be similar to, compare with, look like, mirror, sound like, *inf* take after.

**resent** *vb* begrudge, be resentful about, dislike, envy, feel bitter about, grudge, grumble at, object to, *inf* take exception to, *inf* take umbrage at.

**resentful** *adj* aggrieved, annoyed, begrudging, bitter, disgruntled, displeased, embittered, envious, grudging, hurt, indignant, irked, jaundiced, jealous, malicious, offended, *inf* peeved, *inf* put out, spiteful, unfriendly, ungenerous, upset, vexed, vindictive. ▷ ANGRY.

**resentment** *n* animosity, bitterness, discontent, envy, grudge, hatred, hurt, ill-will, indignation, irritation, jealousy, malevolence, malice, pique, rancour, spite, unfriendliness, vexation, vindictiveness. ▷ ANGER.

**reservation** *n* **1** condition, doubt, hedging, hesitation, misgiving, proviso, qualification, qualm, reluctance, reticence, scepticism, scruple. **2** *hotel reservation.* appointment, booking. **3** *a wildlife reservation.* ▷ RESERVE.

**reserve** *n* **1** cache, fund, hoard, *inf* nest-egg, reservoir, savings, stock, stockpile, store, supply. **2** *inf* back-up, deputy, *plur* reinforcements, replacement, stand-by, *inf* stand-in, substitute, understudy. **3** *a wildlife reserve.* enclave, game park, preserve, protected area, reservation, safari-park, sanctuary. **4** aloofness, caution, modesty, quietness, reluctance, reticence, self-consciousness, self-effacement, shyness, *derog* stand-offishness, taciturnity, timidity. • *vb* **1** earmark, hoard, hold back, keep, keep back, preserve, put aside, retain, save, set aside, stockpile, store up. **2** *reserve seats. inf* bag, book, order, pay for. **reserved** ▷ RETICENT.

**reside** *vb* **reside in** dwell in, inhabit, live in, lodge in, occupy, settle in.

**residence** *n old use* abode, address, domicile, dwelling, dwelling-place, habitation, home, quarters, seat. ▷ HOUSE.

**resident** *adj* in residence, livingin, permanent, remaining, staying. • *n* citizen, denizen, dweller, householder, houseowner, inhabitant, *inf* local, native.

**residual** *adj* abiding, continuing, left over, outstanding, persisting, remaining, surviving, unconsumed, unused.

**resign** *vb* abandon, abdicate, *sl* chuck in, forsake, give up, leave, quit, relinquish, renounce, retire, stand down, step down, surrender, vacate. **resigned** ▷ PATIENT. **resign yourself to** ▷ ACCEPT.

**resilient** *adj* **1** bouncy, elastic, firm, plastic, pliable, rubbery, springy, supple. *Opp* BRITTLE. **2** *a resilient person.* adaptable, buoyant, irrepressible, strong, tough, unstoppable. *Opp* VULNERABLE.

**resist** *vb* avoid, be resistant to, check, confront, counteract, defy, face up to, hinder, *inf* hold out against, *inf* hold your ground against, impede, inhibit, keep at bay, oppose, prevent, rebuff, refuse, stand up to, withstand. ▷ FIGHT. *Opp* ASSIST, YIELD.

**resistant** *adj* defiant, hostile, intransigent, invulnerable, obstinate, opposed, refractory, stubborn, uncooperative, unresponsive, unyielding. **resistant to** against, impervious to, invulnerable to, opposed to, proof against, repellent of, unaffected by, unsusceptible to, unyielding to. *Opp* SUSCEPTIBLE.

**resolute** *adj* adamant, bold, committed, constant, courageous, decided, decisive, determined, dogged, firm, immovable, immutable, indefatigable, *derog* inflexible, *derog* obstinate, persevering, persistent, pertinacious, relentless, resolved, single-minded, staunch, steadfast, strong-minded, strong-willed, *derog* stubborn, tireless, unbending, undaunted, unflinching, unshakable, unswerving, untiring, unwavering. *Opp* IRRESOLUTE.

**resolution** *n* **1** boldness, commitment, constancy, determination, devotion, doggedness, firmness, *derog* obstinacy, perseverance, persistence, pertinacity, purposefulness, resolve, single-mindedness, staunchness, steadfastness, *derog* stubbornness, tenacity, will-power. ▷ COURAGE. **2** commitment, oath, pledge, promise, undertaking, vow. **3** *resolution at a meeting.* decision, motion, proposal, proposition, statement. **4** *resolution of a problem.* answer, denouement, disentangling, resolving, settlement, solution, sorting out.

**resolve** *n* ▷ RESOLUTION. • *vb* **1** agree, conclude, decide formally, determine, elect, fix, make a decision, make up your mind, opt, pass a resolution, settle, undertake, vote. 2 *resolve a problem.* answer, clear up, disentangle, figure out, settle, solve, sort out, work out.

**resonant** *adj* booming, echoing, full, pulsating, resounding, reverberant, reverberating, rich, ringing, sonorous, thunderous, vibrant, vibrating.

**resort** *n* **1** alternative, course of action, expedient, option, recourse, refuge, remedy, reserve. **2** *a seaside resort.* holiday town, retreat, spa, *old use* watering-place. • *vb* **resort to** adopt, *inf* fall back on, have recourse to, make use of, turn to, use. **2** frequent, go to, *inf* hang out in, haunt, invade, patronize, visit.

**resound** *vb* boom, echo, pulsate, resonate, reverberate, ring, rumble, thunder, vibrate.

**resounding** ▷ RESONANT.

**resourceful** *adj* clever, creative, enterprising, imaginative, ingenious, innovative, inspired, inventive, original, skilful, *inf* smart, talented. *Opp* SHIFTLESS.

**resources** *plur n* **1** assets, capital, funds, possessions, property, reserves, riches, wealth. ▷ MONEY. **2** *natural resources.* materials, raw materials.

**respect** *n* **1** admiration, appreciation, awe, consideration, courtesy, deference, esteem, homage, honour, liking, love, politeness, regard, reverence, tribute, veneration. **2** *perfect in every respect.* aspect, attribute, characteristic, detail, element, facet, feature, particular, point, property, quality, trait, way. • *vb* admire, appreciate, be polite to, defer to, esteem, have high regard for, honour, look up to, pay homage to, revere, reverence, show respect to, think well of, value, venerate. *Opp* DESPISE.

**respectable** *adj* **1** decent, gente honest, honourable, law-abiding, refined, respected, unimpeachabl upright, worthy. **2** *respectable clothes.* chaste, clean, decorous, dignified, modest, presentable, proper, seemly. *Opp* DISREPUTABLE. **3** *a respectable sum.* ▷ CONSIDERABLE.

**respectful** *adj* admiring, civil, considerate, cordial, courteous, deferential, dutiful, gentlemanly, gracious, humble, ladylike, obliging, polite, proper, reverent, reverential, *derog* servile, subservient, thoughtful, well-mannered. *Opp* DISRESPECTFUL.

**respective** *adj* individual, own, particular, personal, relevant, separate, several, special, specific.

**respite** *n* break, *inf* breather, delay, hiatus, holiday, intermission, interruption, interval, *inf* let-up, lull, pause, recess, relaxation, relief, remission, rest time off, time out, vacation.

**resplendent** *adj* brilliant, dazzling, glittering, shining, splendi ▷ BRIGHT.

**respond** *vb* **respond to** **1** acknowledge, answer, counter, give a response to, react to, reciprocate, reply to. **2** *respond to nee* ▷ SYMPATHIZE.

**response** *n* acknowledgement, answer, *inf* comeback, counter, counterblast, feedback, reaction, rejoinder, reply, retort, riposte.

**responsible** *adj* **1** at fault, culpable, guilty, liable, to blame. **2** *a responsible person.* accountable, answerable, concerned, conscientious, creditable, dependable, diligent, dutiful, ethical, honest, in charge, law-abiding, loyal, matur moral, reliable, sensible, sober, steady, thinking, thoughtful, trustworthy, unselfish. *Opp* IRRESPON

IBLE. **3** *a responsible job.* burdensome, decision-making, executive, *inf* front-line, important, managerial, *inf* top. *Opp* MENIAL.

**responsive** *adj* alert, alive, aware, impressionable, interested, open, perceptive, receptive, sensitive, sharp, sympathetic, warm-hearted, wideawake, willing. *Opp* UNINTERESTED.

**rest** *n* **1** break, *inf* breather, holiday, breathing-space, comfort, ease, hiatus, holiday, idleness, inactivity, indolence, interlude, intermission, interval, leisure, *inf* let-up, *inf* lie-down, *inf* loafing, lull, nap, pause, quiet, recess, relaxation, relief, remission, repose, respite, siesta, tea-break, time off, vacation. ▷ SLEEP. **2** base, brace, bracket, holder, prop, stand, support, trestle, tripod. **3** ▷ REMAINDER. • *vb* **1** be still, doze, have a rest, idle, laze, lie back, lie down, lounge, nod off, *inf* put your feet up, recline, relax, snooze, *inf* take a nap, *inf* take it easy, unwind. ▷ SLEEP. **2** lean, place, position, prop, set, stand, support. **3** *It all rests on the weather.* depend, hang, hinge, rely, turn. **come to rest** ▷ HALT.

**restaurant** *n* eating-place. □ *bistro, brasserie, buffet, café, cafeteria, canteen, carvery, diner, dining-room, grill, refectory, snack-bar, steak-house.*

**restful** *adj* calm, calming, comfortable, leisurely, peaceful, quiet, relaxed, relaxing, reposeful, soothing, still, tranquil, undisturbed, unhurried, untroubled. *Opp* EXHAUSTING.

**restless** *adj* **1** agitated, anxious, edgy, excitable, fidgety, highly-strung, impatient, *inf* jittery, jumpy, nervous, *inf* on tenterhooks, restive, skittish, uneasy, worked up, worried. ▷ ACTIVE. **2** *a restless night.* disturbed, interrupted, sleepless, *inf* tossing and turning, troubled, uncomfortable, unsettled. *Opp* RESTFUL.

**restore** *vb* **1** bring back, give back, make restitution, put back, reinstate, replace, return. **2** *restore antiques.* clean, *inf* do up, fix, *inf* make good, mend, rebuild, recondition, reconstruct, refurbish, renew, renovate, repair, touch up. **3** *restore good relations.* re-establish, rehabilitate, reinstate, reintroduce, rekindle, revive. **4** *restore to health.* cure, nurse, rejuvenate, resuscitate, revitalize.

**restrain** *vb* **1** check, control, curb, govern, hold back, inhibit, keep back, keep under control, limit, regulate, rein in, repress, restrict, stifle, stop, strait-jacket, subdue, suppress. **2** arrest, bridle, confine, detain, fetter, handcuff, harness, imprison, incarcerate, jail, *inf* keep under lock and key, lock up, manacle, muzzle, pinion, tie up. **restrained** ▷ CALM, DISCREET.

**restrict** *vb* circumscribe, confine, control, cramp, delimit, enclose, impede, imprison, inhibit, keep within bounds, limit, regulate, shut. ▷ RESTRAIN. *Opp* FREE.

**restriction** *n* ban, check, constraint, control, curb, curfew, inhibition, limit, limitation, proviso, qualification, regulation, restraint, rule, stipulation.

**result** *n* **1** conclusion, consequence, effect, end-product, fruit, issue, outcome, repercussion, sequel, upshot. **2** *result of a trial.* decision, judgement, verdict. **3** *result of a sum.* answer, product, score, total. • *vb* arise, be produced, come about, develop, emanate, emerge, ensue, eventuate, follow, happen, issue, occur, proceed, spring, stem, take place, turn out. **result in** ▷ CAUSE.

**resume** *vb* begin again, carry on, continue, *inf* pick up the threads, proceed, recommence, reconvene, re-open, restart.

**resumption** *n* continuation, recommencement, re-opening, *inf* restart.

**resurrect** *vb* breathe new life into, bring back, raise (from the dead), reawaken, restore, resuscitate, revitalize, revive. ▷ RENEW.

**retain** *vb* **1** *inf* hang on to, hold, hold back, keep, keep control of, maintain, preserve, reserve, save. *Opp* LOSE. **2** *retain moisture.* absorb, soak up. **3** *retain facts.* keep in mind, learn, memorize, remember. *Opp* FORGET.

**retaliate** *vb* avenge yourself, be revenged, counter-attack, exact retribution, *inf* get even, *inf* get your own back, *inf* give tit for tat, hit back, pay back, repay, revenge yourself, seek retribution, *inf* settle a score, strike back, *inf* take an eye for an eye, take revenge, wreak vengeance.

**retaliation** *n* counter-attack, reprisal, retribution, revenge, vengeance.

**retard** *vb* check, handicap, hinder, hold back, hold up, impede, obstruct, postpone, put back, set back, slow down. ▷ DELAY.

**retarded** ▷ BACK-WARD.

**reticent** *adj* aloof, *derog* anti-social, bashful, cautious, *derog* cold, cool, demure, diffident, discreet, distant, modest, quiet, remote, reserved, restrained, retiring, secretive, self-conscious, self-effacing, shy, silent, *derog* standoffish, taciturn, timid, uncommunicative, unemotional, undemonstrative, unforthcoming, unresponsive, unsociable, withdrawn. *Opp* DEMONSTRATIVE.

**retinue** *n* attendants, company, entourage, followers, *inf* hangers-on, servants, suite, train.

**retire** *vb* **1** give up, leave, quit, resign. **2** *retire from society.* become reclusive, cloister yourself, go away, go into retreat, retreat from the world, sequester yourself, withdraw. **3** aestivate, go to bed, hibernate, *sl* hit the hay. ▷ SLEEP.

**retort** *n* answer, *inf* comeback, rebuttal, rejoinder, reply, response, retaliation, riposte. • *vb* answer, counter, react, rejoin, reply, respond, retaliate, return.

**retract** *vb* **1** draw in, pull back, pull in. **2** abandon, cancel, disclaim, disown, forswear, *inf* have second thoughts about, recant, renounce, repeal, repudiate, rescind, reverse, revoke, withdraw.

**retreat** *n* **1** departure, escape, evacuation, exit, flight, retirement, withdrawal. **2** *a secluded retreat.* asylum, den, haven, *inf* hideaway, hideout, hiding-place, refuge, resort, sanctuary, shelter. • *vb* **1** back away, back down, climb down, decamp, depart, evacuate, fall back, flee, give ground, go away, leave, move back, pull back, retire, *inf* run away, take flight, *inf* take to your heels, *inf* turn tail, withdraw. **2** *the floods retreated.* ebb, flow back, recede, shrink back. *Opp* ADVANCE.

**retribution** *n* compensation, *Lat* quid pro quo, recompense, redress, reprisal, retaliation, revenge, *old use* satisfaction, vengeance. *Opp* FORGIVENESS.

**retrieve** *vb* bring back, come back with, fetch back, find, get back, make up for, recapture, reclaim, recoup, recover, regain, repossess, rescue, restore, return, salvage, save, take back, trace, track down.

**retrograde** *adj* backward, negative, regressive, retreating, retrogressive, reverse.

**retrospective** *adj* backward-looking, looking back, looking behind, nostalgic, with hindsight.

**return** *n* **1** advent, arrival, homecoming, reappearance, re-entry. **2** *return to normality*. re-establishment (of), regression, reversion. **3** *return of a problem* recrudescence, recurrence, re-emergence, repetition. **4** *return of stolen goods*. replacing, restitution, restoration, retrieval. **5** *return on an investment*. benefit, earnings, gain, income, interest, proceeds, profit, yield. • *vb* **1** backtrack, come back, do a U-turn, double back, go back, reassemble, reconvene, re-enter, regress, retrace your steps, revert, turn back. **2** put back, readdress, repatriate, replace, restore, send back. **3** *return money*. exchange, give back, refund, reimburse, repay. **4** *return a verdict*. *inf* come up with, deliver, give, proffer, report. **5** *The problem returned*. *inf* crop up again, happen again, reappear, recur, resurface.

**reveal** *vb* announce, bare, betray, bring to light, communicate, confess, declare, denude, dig up, disclose, display, divulge, exhibit, expose, *inf* give the game away, lay bare, leak, *inf* let on, *inf* let out, *inf* let slip, make known, open, proclaim, produce, publish, show, show up, *inf* spill the beans about, *inf* take the wraps off, tell, uncover, undress, unearth, unfold, unmask, unveil. *Opp* HIDE.

**revel** *n* carnival, festival, fête, *inf* jamboree, *inf* rave-up, *inf* spree. ▷ REVELRY. • *vb* carouse, celebrate, *inf* have a spree, have fun, indulge in revelry, *inf* live it up, make merry, *inf* paint the town red. **revel in** ▷ ENJOY.

**revelation** *n* admission, announcement, communiqué, confession, declaration, disclosure, discovery, exposé, exposure, information, *inf* leak, news, proclamation, publication, revealing, unmasking, unveiling.

**revelry** *n* carousing, celebration, conviviality, debauchery, festivity, fun, gaiety, *inf* high jinks, jollification, jollity, *inf* junketing, *inf* living it up, merry-making, revelling, revels, roistering, *inf* spree. ▷ PARTY.

**revenge** *n* reprisal, retaliation, retribution, spitefulness, vengeance, vindictiveness. • *vb* avenge, repay. **be revenged** ▷ RETALIATE.

**revenue** *n* gain, income, interest, money, proceeds, profits, receipts, returns, takings, yield.

**reverberate** *vb* boom, echo, pulsate, resonate, resound, ring, rumble, throb, thunder, vibrate.

**revere** *vb* admire, adore, adulate, beatify, esteem, feel reverence for, glorify, honour, idolize, pay homage to, praise, respect, reverence, value, venerate, worship. *Opp* DESPISE.

**reverence** *n* admiration, adoration, adulation, awe, deference, devotion, esteem, glorification, homage, honour, idolization, praise, respect, veneration, worship.

**reverent** *adj* adoring, awed, awe-struck, deferential, devoted, devout, pious, prayerful, religious, respectful, reverential, solemn, worshipful. *Opp* IRREVERENT.

**reverie** *n inf* brown study, day-dream, dream, fantasy, meditation.

**reverse** *adj* back, back-to-front, backward, contrary, converse, inverse, inverted, opposite, rear. • *n* **1** antithesis, contrary, converse, opposite. **2** back, rear, underside, verso, wrong side. **3** defeat, difficulty, disaster, failure, mishap, misfortune, problem, reversal, setback, *inf* upset, vicissitude. • *vb* **1** change, invert, overturn, transpose, turn round, turn upsidedown. **2** *reverse a car.* back, drive backwards, go into reverse. **3** *reverse a decision.* abandon, annul, cancel, countermand, invalidate, negate, nullify, overturn, quash, recant, repeal, rescind, retract, revoke, undo.

**review** *n* **1** examination, *inf* look back, *inf* post-mortem, reappraisal, reassessment, recapitulation, reconsideration, re-examination, report, retrospective, study, survey. **2** *book review.* appreciation, assessment, commentary, criticism, critique, evaluation, judgement, notice, *inf* write-up. • *vb* **1** appraise, assess, consider, evaluate, examine, *inf* go over, inspect, reassess, recapitulate, reconsider, re-examine, scrutinize, study, survey, take stock of, *inf* weigh up. **2** *review a book.* criticize, write a review of.

**revise** *vb* **1** adapt, alter, change, correct, edit, emend, improve, modify, overhaul, *inf* polish up, reconsider, rectify, *inf* redo, *inf* rehash, rephrase, revamp, reword, rework, rewrite, update. **2** *revise for exams.* brush up, *inf* cram, learn, study, *inf* swot.

**revival** *n* advance, progress, quickening, reanimation, reawakening, rebirth, recovery, renaissance, renewal, restoration, resurgence, resurrection, resuscitation, return, revitalization, upsurge.

**revive** *vb* **1** awaken, come back to life, *inf* come round, *inf* come to, quicken, rally, reawaken, recover, resurrect, rouse, waken. *Opp* RELAPSE. **2** bring back to life, *inf* cheer up, freshen, invigorate, refresh, renew, restore, resuscitate, revitalize, strengthen. *Opp* WEARY.

**revolt** *n* civil war, coup, coup d'état, insurrection, mutiny, putsch, rebellion, reformation, revolution, rising, *inf* take-over, uprising. • *vb* **1** disobey, dissent, mutiny, rebel, riot, rise up. **2** appal, disgust, nauseate, offend, outrage, repel, sicken, upset.

**revolting** ▷ OFFENSIVE.

**revolution** *n* **1** ▷ REVOLT. **2** circuit, cycle, gyration, orbit, rotation, spin, turn. **3** change, reorganization, reorientation, shift, transformation, *inf* turn-about, upheaval, *inf* upset, *inf* U-turn.

**revolutionary** *adj* **1** insurgent, mutinous, rebel, rebellious, seditious, subversive. **2** *revolu-tionary ideas.* avant-garde, challenging, creative, different, experimental, extremist, innovative, new, novel, progressive, radical, *inf* unheard-of, upsetting. *Opp* CONSERVATIVE. • *n* anarchist, extremist, freedom fighter, insurgent, mutineer, rebel, terrorist.

**revolve** *vb* circle, go round, gyrate, orbit, pirouette, pivot, reel, rotate, spin, swivel, turn, twirl, wheel, whirl.

**revulsion** *n* abhorrence, aversion, disgust, hatred, loathing, nausea, outrage, repugnance.

**reward** *n* award, bonus, bounty, compensation, decoration, favour, honour, medal, payment, prize, recompense, remuneration, requital, return, tribute. *Opp* PUNISHMENT. • *vb* compensate, decorate, give a reward to, honour, recompense, remunerate, repay. *Opp* PENALIZE, PUNISH. **rewarding** ▷ PROFITABLE, WORTHWHILE.

**rhapsodize** *vb* be expansive, effuse, enthuse, *inf* go into raptures.

**rhetoric** *n derog* bombast, eloquence, expressiveness, *inf* gift of the gab, grandiloquence, magniloquence, oratory, rhetorical language, *derog* speechifying.

**rhetorical** *adj* [*most synonyms derog*] artifical, bombastic, florid, *inf* flowery, fustian, grandiloquent, grandiose, highflown, insincere, oratorical, ornate, pretentious, verbose, wordy.

**rhyme** *n* doggerel, jingle, poem. ▷ VERSE.

**rhythm** *n* accent, beat, measure, metre, movement, pattern, pulse, stress, tempo, throb, time.

**rhythmic** *adj* beating, measured, metrical, predictable, pulsing, regular, repeated, steady, throbbing. *Opp* IRREGULAR.

**ribald** *adj* bawdy, coarse, disrespectful, earthy, naughty, racy, rude, scurrilous, *inf* smutty, vulgar. ▷ OBSCENE.

**ribbon** *n* band, braid, line, strip, stripe, tape, trimming. **in ribbons** ▷ RAGGED.

**rich** *adj* **1** affluent, *inf* flush, *inf* loaded, moneyed, opulent, plutocratic, prosperous, wealthy, *inf* well-heeled, well-off, well-to-do. *Opp* POOR. **2** *rich furnishings*. costly, elaborate, expensive, lavish, luxurious, precious, priceless, splendid, sumptuous, valuable. **3** *rich land*. fecund, fertile, fruitful, lush, productive. **4** *a rich harvest*. abundant, ample, bountiful, copious, plenteous, plentiful, profuse, prolific, teeming. **5** *rich colours*. deep, full, intense, strong, vibrant, vivid, warm. **6** *rich food*. cloying, creamy, fat, fattening, fatty, full-flavoured, heavy, highly-flavoured, luscious, sumptuous, sweet. **rich person** billionaire, capitalist, millionaire, plutocrat, tycoon.

**riches** *plur n* affluence, fortune, means, money, opulence, plenty, possessions, prosperity, resources, wealth.

**rickety** *adj* dilapidated, flimsy, frail, insecure, ramshackle, shaky, tottering, tumbledown, unsteady, wobbly. ▷ WEAK.

**rid** *vb* clear, deliver (from), free, purge, rescue, save. **get rid of** ▷ DESTROY, REMOVE.

**riddle** *n* **1** *inf* brain-teaser, conundrum, enigma, mystery, *inf* poser, problem, puzzle, question. **2** filter, screen, sieve. • *vb* **1** filter, screen, sieve, sift, strain. **2** *riddle with holes*. honeycomb, *inf* pepper, perforate, pierce, puncture.

**ride** *n* ▷ JOURNEY. • *vb* **1** *ride a bike*. control, handle, manage, sit on, steer. **2** *ride on a bike*. be carried, free-wheel, pedal. **3** *ride on a horse*. amble, canter, gallop, trot. ▷ TRAVEL.

**ridge** *n* arête, bank, crest, edge, embankment, escarpment. ▷ HILL.

**ridicule** *n* badinage, banter, burlesque, caricature, contumely, derision, invective, jeering, jibing, lampoon, laughter, mockery, parody, raillery, *inf* ribbing, sarcasm, satire, scorn, sneers, taunts, teasing. • *vb* be sarcastic, be satirical about, burlesque, caricature, chaff, deride, gibe at, guy, hold up to ridicule, jeer at, jibe at, joke about, lampoon, laugh at, make fun of, make jokes about, mimic, mock, parody, pillory, *inf* poke fun at, *inf* rib, satirize, scoff at, *inf* send up, sneer at, subject to ridicule, *inf* take the mickey, taunt, tease, travesty.

**ridiculous** *adj* absurd, amusing, comic, comical, *inf* crazy, *inf* daft, eccentric, farcical, foolish, grotesque, hilarious, illogical, irrational, laughable, ludicrous, mad, nonsensical, preposterous, senseless, silly, unbelievable, unreasonable, weird, *inf* zany. ▷ FUNNY, STUPID. *Opp* SENSIBLE.

**rife** *adj* abundant, common, endemic, prevalent, widespread.

**rift** *n* 1 breach, break, chink, cleft, crack, fracture, gap, gulf, opening, split. 2 *a rift between friends.* alienation, conflict, difference, disagreement, disruption, division, opposition, schism, separation.

**rig** *n* 1 ▷ RIGGING. 2 *oil rig.* platform. 3 [*inf*] *sporting rig.* apparatus, clothes, equipment, gear, kit, outfit, stuff, tackle. • *vb* **rig out** equip, fit out, kit out, outfit, provision, set up, supply.

**rigging** *n* rig, tackle. □ *halyards, ropes and pulleys, sails.*

**right** *adj* 1 decent, ethical, fair, good, honest, honourable, just, law-abiding, lawful, moral, principled, responsible, righteous, right-minded, upright, virtuous. 2 *right answers.* accurate, apposite, appropriate, apt, correct, exact, factual, faultless, fitting, genuine, perfect, precise, proper, sound, suitable, true, truthful, valid, veracious. 3 *the right way.* advantageous, beneficial, best, convenient, good, normal, preferable, preferred, recommended, sensible, usual. 4 *your right side.* right-hand, starboard [= *right facing bow of ship*]. 5 *right wing in politics.* conservative, fascist, reactionary, Tory. *Opp* LEFT, WRONG. • *n* 1 decency, equity, ethics, fairness, goodness, honesty, integrity, justice, morality, propriety, reason, truth, virtue. 2 *right to free speech.* entitlement, facility, freedom, liberty, prerogative, privilege. 3 *right to give orders.* authority, commission, franchise, licence, position, power, title. • *vb* 1 amend, correct, make amends for, put right, rectify, redress, remedy, repair, set right. 2 pick up, set upright, stand upright, straighten up.

**righteous** *adj* blameless, ethical, God-fearing, good, guiltless, *derog* holier-than-thou, honest, just, law-abiding, moral, pure, *derog* sanctimonious, upright, upstanding, virtuous. *Opp* SINFUL.

**rightful** *adj* authorized, *Lat* bona fide, correct, just, lawful, legal, legitimate, licensed, licit, proper, real, true, valid. *Opp* ILLEGAL.

**rigid** *adj* 1 adamantine, firm, hard, inelastic, inflexible, set, solid, steely, stiff, strong, unbending, wooden. ▷ OBSTINATE. 2 *rigid discipline.* harsh, intransigent, punctilious, stern, strict, uncompromising, unkind, unrelenting, unyielding. ▷ RIGOROUS. *Opp* FLEXIBLE.

**rigorous** *adj* 1 conscientious, demanding, exact, exacting, meticulous, painstaking, precise, punctilious, rigid, scrupulous, strict, stringent, structured, thorough, tough, uncompromising, undeviating, unsparing, unswerving. *Opp* LAX. **2** *rigorous climate.* extreme, hard, harsh, inclement, inhospitable, severe, unfriendly, unpleasant. ▷ COLD. *Opp* MILD.

**rim** *n* brim, brink, circumference, edge, lip, perimeter, periphery.

**rind** *n* crust, husk, outer layer, peel, skin.

**ring** *n* **1** annulus, band, bracelet, circle, circlet, collar, corona, eyelet, girdle, halo, hoop, loop, O, ringlet. **2** *boxing ring.* arena, enclosure, rink. **3** *drugs ring.* association, band, gang, mob, organization, syndicate. ▷ GROUP. **4** *the ring of a bell.* boom, buzz, chime, clang, clink, *inf* ding-a-ling, jangle, jingle, knell, peal, ping, resonance, reverberation, tinkle, tintinnabulation, tolling. **5** *give me a ring sometime. inf* bell, *inf* buzz, call, *inf* tinkle. • *vb* **1** bind, circle, embrace, encircle, enclose, encompass, gird, surround. **2** boom, buzz, chime, clang, clink, jangle, jingle, peal, ping, resonate, resound, reverberate, sound (the knell), tinkle, toll. **3** call, *inf* give a buzz, phone, ring up, telephone.

**rinse** *vb* bathe, clean, cleanse, drench, flush, sluice, swill, wash.

**riot** *n* affray, anarchy, brawl, chaos, commotion, demonstration, disorder, disturbance, fracas, fray, hubbub, imbroglio, insurrection, lawlessness, mass protest, mêlé, mutiny, pandemonium, *inf* punch-up, revolt, rioting, riotous behaviour, rising, row, *inf* rumpus, *inf* shindy, strife, tumult, turmoil, unrest, uproar, violence. • *vb* brawl, create a riot, *inf* go on the rampage, *inf* go wild, mutiny, rampage, rebel, revolt, rise up, run riot, *inf* take to the streets. ▷ FIGHT.

**riotous** *adj* anarchic, boisterous, chaotic, disorderly, lawless, mutinous, noisy, obstreperous, rampageous, rebellious, rowdy, tumultuous, uncivilized, uncontrollable, undisciplined, ungovernable, unrestrained, unruly, uproarious, violent, wild. *Opp* ORDERLY.

**rip** *vb* gash, lacerate, pull apart, rend, rupture, shred, slit, split, tear.

**ripe** *adj* mature, mellow, ready to use.

**ripen** *vb* age, become riper, come to maturity, develop, mature, mellow.

**ripple** *n* ▷ WAVE. • *vb* agitate, disturb, make waves, purl, ruffle, stir.

**rise** *n* **1** acclivity, ascent, bank, camber, climb, elevation, hill, hump, incline, ramp, ridge, slope. **2** *a rise in prices.* escalation, gain, increase, increment, jump, leap, upsurge, upswing, upturn, upward movement. • *vb* **1** arise, ascend, climb, fly up, go up, jump, leap, levitate, lift, lift off, mount, soar, spring, take off. **2** get to your feet, get up, stand up. **3** *prices rise each year.* escalate, grow, increase, spiral. **4** *cliffs rise above us.* loom, stand out, stick up, tower. **rise up** ▷ REBEL.

**risk** *n* **1** chance, likelihood, possibility. **2** danger, gamble, hazard, peril, speculation, uncertainty, venture. • *vb* **1** chance, dare, endanger, hazard, imperil, jeopardize. **2** *risk money.* gamble, speculate, venture.

**risky** *adj inf* chancy, *inf* dicey, hazardous, *inf* iffy, perilous, precarious, unsafe. ▷ DANGEROUS. *Opp* SAFE.

**ritual** *n* ceremonial, ceremony, custom, formality, liturgy, observance, practice, rite, routine, sacrament, service, set procedure, solemnity, tradition.

**rival** *n* adversary, antagonist, challenger, competitor, contender, contestant, enemy, opponent, opposition. • *vb* **1** challenge, compete with, contend with, contest, emulate, oppose, struggle with, undercut, vie with. *Opp* COOPERATE. **2** be as good as, compare with, equal, match, measure up to.

**rivalry** *n* antagonism, competition, competitiveness, conflict, contention, feuding, opposition, strife. *Opp* COOPERATION.

**river** *n* brook, rivulet, stream, watercourse, waterway. □ *channel, confluence, delta, estuary, lower reaches, mouth, source, tributary, upper reaches.*

**road** *n* roadway, route, way. □ *alley, arterial road, avenue, boulevard, bridle-path, bridle-way, bypass, byway, byroad, cart-track, causeway, clearway, crescent, cul-de-sac, drive, driveway, dual carriageway,* Amer *freeway, highway, lane, motorway, one-way street, path, pathway, ring road, service road, side-road, side-street, slip-road, street, thoroughfare, tow-path, track, trail, trunk road,* old use *turnpike.*

**roam** *vb* amble, drift, meander, prowl, ramble, range, rove, saunter, stray, stroll, *inf* traipse, travel, walk, wander.

**roar** *vb* bellow, cry out, growl, howl, shout, snarl, thunder, yell, yowl. ▷ SOUND.

**rob** *vb* burgle, *inf* con, defraud, hold up, loot, *inf* mug, *old use* mulct, pick pockets, pilfer from, pillage, plunder, ransack, rifle, steal from. ▷ STEAL.

**robber** *n* bandit, brigand, burglar, cat burglar, *inf* con-man, defrauder, embezzler, *old use* highwayman, housebreaker, looter, *inf* mugger, pickpocket, pirate, shoplifter, swindler, thief.

**robbery** *n* breaking and entering, burglary, *inf* con, confidence trick, embezzlement, fraud, hijacking, *inf* hold-up, larceny, looting, *inf* mugging, pilfering, pillage, plunder, sacking, *inf* scrumping, shoplifting, stealing, *inf* stick-up, theft, thieving.

**robe** *n* cloak, dress, frock, gown. □ *bathrobe, caftan, cassock, dressing-gown, habit, housecoat, kimono, peignoir, surplice, vestment.* • *vb* ▷ DRESS.

**robot** *n* android, automated machine, automaton, bionic man, bionic woman, computerized machine, mechanical man.

**robust** *adj* **1** athletic, brawny, fit, *inf* hale and hearty, hardy, healthy, hearty, muscular, powerful, rugged, sound, strong, sturdy, tough, vigorous. **2** durable, serviceable, strongly-made, well-made. *Opp* WEAK.

**rock** *n* boulder, crag, ore, outcrop, scree, stone. □ *igneous, metamorphic, sedimentary.* □ *basalt, chalk, clay, flint, gneiss, granite, gravel, lava, limestone, marble, obsidian, pumice, quartz, sandstone, schist, shale, slate, tufa, tuff.* • *vb* **1** lurch, move to and fro, pitch, reel, roll, shake, sway, swing, toss, totter, wobble. **2** ▷ SHOCK.

**rocky** *adj* **1** barren, inhospitable, pebbly, rough, rugged, stony. **2** ▷ UNSTEADY.

**rod** *n* bar, baton, cane, dowel, pole, rail, shaft, spoke, staff, stick, strut, wand.

**rogue** *n* blackguard, charlatan, cheat, *inf* con-man, fraud, *old* *use* knave, mischief-maker, *inf* quack, rapscallion, rascal, ruffian, scoundrel, swindler, trickster, villain, wastrel, wretch. ▷ CRIMINAL.

**role** *n* **1** character, impersonation, lines, part, portrayal. **2** *role in a business*. contribution, duty, function, job, position, post, task.

**roll** *n* **1** cylinder, drum, reel, scroll, spool, tube. **2** catalogue, directory, index, inventory, list, listing, record, register. • *vb* **1** go round, gyrate, move round, revolve, rotate, run, somersault, spin, tumble, turn, twirl, whirl. **2** coil, curl, furl, make into a roll, twist, wind, wrap. **3** *roll the lawn*. flatten, level off, level out, smooth. **4** *ship rolled in the storm*. lumber, lurch, pitch, reel, rock, stagger, sway, toss, totter, wallow, welter. **rolling** ▷ WAVY. **roll in, roll up** ▷ ARRIVE.

**romance** *n* **1** idyll, love story, novel. ▷ WRITING. **2** adventure, colour, excitement, fascination, glamour, mystery. **3** affair, amour, attachment, intrigue, liaison, love affair, relationship.

**romantic** *adj* **1** colourful, dreamlike, exotic, fabulous, fairy-tale, glamorous, idyllic, nostalgic, picturesque. **2** *romantic feelings*. affectionate, amorous, emotional, erotic, loving, passionate, *inf* sexy, *inf* soppy, tender. **3** *romantic fiction*. emotional, escapist, heartwarming, nostalgic, reassuring, sentimental, *derog* sloppy, tender, unrealistic. **4** *romantic ideals*. chimerical, *inf* head in the clouds, idealistic, illusory, impractical, improbable, quixotic, starry-eyed, unworkable, Utopian, visionary. *Opp* REALISTIC.

**room** *n* **1** *inf* elbow-room, freedom, latitude, leeway, margin, scope, space, territory. **2** *a room in a house*. apartment, cell, *old use* chamber. □ *ante-room, attic, audience chamber, bathroom, bedroom, boudoir, cell, cellar, chapel, classroom, cloakroom, conservatory, corridor, dining-room, dormitory, drawing-room, dressing-room, gallery, guest-room, hall, kitchen, kitchenette, laboratory, landing, larder, laundry, lavatory, library, living-room, loft, lounge, music-room, nursery, office, pantry, parlour, passage, play-room, porch, salon, saloon, scullery, sick-room, sitting-room, spare room, state-room, store-room, studio, study, toilet, utility room, waiting-room, ward, washroom, WC, workroom, workshop.*

**roomy** *adj* capacious, commodious, large, sizeable, spacious, voluminous. ▷ BIG. *Opp* SMALL.

**root** *n* **1** radicle, rhizome, rootlet, tap root, tuber. **2** *the root of a problem*. base, basis, bottom, cause, foundation, fount, origin, seat, source, starting-point. • *vb* **root out** ▷ REMOVE.

**rope** *n* cable, cord, line, strand, string. □ *halyard, hawser, lanyard, lariat, lasso, tether*. • *vb* bind, hitch, lash, moor, tether, tie. ▷ FASTEN.

**rot** *n* **1** corrosion, corruption, decay, decomposition, deterioration, disintegration, dry rot, mould, mouldiness, putrefaction, wet rot. **2** *What rot!* ▷ NONSENSE. • *vb* become rotten, corrode, crumble, decay, decompose, degenerate, deteriorate, disintegrate, fester, go bad, *inf* go off, perish, putrefy, rust, spoil.

**rota** *n* list, roster, schedule, timetable.

**rotary** *adj* gyrating, revolving, rotating, rotatory, spinning, turning, twirling, twisting, whirling.

**rotate** *vb* **1** go round, gyrate, have a rotary movement, move round, pirouette, pivot, reel, revolve, roll, spin, swivel, turn, turn anticlockwise, turn clockwise, twiddle, twirl, twist, wheel, whirl. **2** *rotate duties.* alternate, pass round, share out, take in turn, take turns.

**rotten** *adj* **1** bad, corroded, crumbling, decayed, decaying, decomposed, disintegrating, foul, mouldering, mouldy, *inf* off, over-ripe, perished, putrid, rusty, smelly, tainted, unfit for consumption, unsound. *Opp* SOUND. **2** ▷ IMMORAL.

**rough** *adj* **1** broken, bumpy, coarse, craggy, irregular, knobbly, jagged, lumpy, pitted, rocky, rugged, rutted, stony, uneven. **2** *rough skin.* bristly, callused, chapped, coarse, hairy, harsh, leathery, ragged, scratchy, shaggy, unshaven, wrinkled. **3** *a rough sea.* agitated, choppy, stormy, tempestuous, turbulent, violent, wild. **4** *a rough voice.* cacophonous, discordant, grating, gruff, harsh, hoarse, husky, rasping, raucous, strident, unmusical, unpleasant. *Opp* SMOOTH. **5** *rough manners, a rough fellow.* badly-behaved, bluff, blunt, brusque, churlish, ill-bred, impolite, loutish, rowdy, rude, surly, *inf* ugly, uncivil, uncivilized, undisciplined, unfriendly. **6** *rough treatment.* brutal, cruel, painful, ruffianly,thuggish, violent. **7** *rough work.* amateurish, careless, clumsy, crude, hasty, imperfect, inept, *inf* rough and ready, unfinished, unpolished, unskilful. **8** *a rough estimate.* approximate, general, hasty, imprecise, inexact, sketchy, vague. *Opp* EXACT, GENTLE, SMOOTH.

**roughly** *adv* about, approximately, around, close to, nearly.

**round** *adj* **1** [*two-dimensional*] annular, circular, curved, disc-shaped, hoop-shaped, orbicular, ring-shaped. **2** [*three-dimensional*] ball-shaped, bulbous, cylindrical, globelike, globoid, globular, orb-shaped, spherical, spheroid. **3** *a round figure.* ample, full, plump, rotund, rounded, well-padded. ▷ FAT. • *n* bout, contest, game, heat, stage. • *vb* skirt, travel round, turn. **round off** ▷ COMPLETE. **round on** ▷ ATTACK. **round the bend** ▷ MAD. **round the clock** ▷ CONTINUOUS. **round up** ▷ ASSEMBLE.

**roundabout** *adj* circuitous, circular, devious, indirect, long, meandering, oblique, rambling, tortuous, twisting, winding. *Opp* DIRECT. • *n* **1** carousel, merry-go-round, *old use* whirligig. **2** traffic island. **round-shouldered** *adj* hunch-backed, humpbacked, stooping.

**rouse** *vb* **1** arise, arouse, awaken, call, get up, wake up. **2** *rouse to a frenzy.* agitate, animate, electrify, excite, galvanize, goad, incite, inflame, provoke, spur on, stimulate, stir up, *inf* wind up, work up.

**rout** *vb* conquer, crush, overpower, overwhelm, put to flight, *inf* send packing. ▷ DEFEAT.

**route** *n* course, direction, itinerary, journey, path, road, way.

**routine** *adj* accustomed, commonplace, customary, everyday, familiar, habitual, normal, ordinary, perfunctory, planned, run-of-the-mill, scheduled, uneventful, well-rehearsed. • *n* **1** course of action, custom, *inf* drill, habit, method, pattern, plan, practice, procedure, schedule, system, way. **2** *comedy routine.* act, number, performance, programme, set piece.

**row** *n* **1** [rhyme with *crow*] chain, column, cordon, file, line, queue, rank, sequence, series, string, tier. **2** [rhyme with *cow*] ado, commotion, fuss, hubbub, hullabaloo, *inf* racket, rumpus, tumult, uproar. ▷ NOISE. **3** altercation, argument, controversy, disagreement, dispute, fight, fracas, *inf* ructions, *inf* slanging match, squabble. ▷ QUARREL. • *vb* **1** [rhyme with *crow*] *row a boat.* move, propel, scull. **2** [rhyme with *cow*] ▷ QUARREL.

**rowdy** *adj* badly-behaved, boisterous, disorderly, ill-disciplined, irrepressible, lawless, obstreperous, riotous, rough, turbulent, undisciplined, unruly, violent, wild. ▷ NOISY. *Opp* QUIET.

**royal** *adj* imperial, kingly, majestic, princely, queenly, regal, stately. • *n* [*inf*] member of royal family. □ *consort, Her/His Majesty, Her/His Royal Highness, king, monarch, prince, princess, queen, queen mother, regent, sovereign.* ▷ NOBLE.

**rub** *vb* **1** caress, knead, massage, smooth, stroke. **2** abrade, chafe, graze, scrape, wear away. **3** *rub clean.* buff, burnish, polish, scour, scrub, shine, wipe. **rub it in** ▷ EMPHASIZE. **rub out** ▷ ERASE. **rub up the wrong way** ▷ ANNOY.

**rubbish** *n* **1** debris, detritus, dregs, dross, filth, flotsam and jetsam, garbage, junk, leavings, *inf* left-overs, litter, lumber, muck, *inf* odds and ends, offal, offcuts, refuse, rejects, rubble, scrap, slops, sweepings, trash, waste. **2** ▷ NONSENSE.

**rubble** *n* broken bricks, debris, fragments, remains, ruins, wreckage.

**ruddy** *adj* fresh, flushed, glowing, healthy, red, sunburnt.

**rude** *adj* **1** abrupt, abusive, bad-mannered, bad-tempered, blasphemous, blunt, boorish, brusque, cheeky, churlish, coarse, common, condescending, contemptuous, discourteous, disparaging, disrespectful, foul, graceless, gross, ignorant, ill-bred, ill-mannered, impertinent, impolite, improper, impudent, in bad taste, inconsiderate, indecent, insolent, insulting, loutish, mocking, naughty, oafish, offensive, off-hand, patronizing, peremptory, personal (*remarks*), saucy, scurrilous, shameless, tactless, unchivalrous, uncivil, uncomplimentary, uncouth, ungracious, *old use* unmannerly, unprintable, vulgar. ▷ OBSCENE. *Opp* POLITE. **2** *rude workmanship.* awkward, basic, bumbling, clumsy, crude, inartistic, primitive, rough, rough-hewn, simple, unpolished, unskilful, unsophisticated, unsubtle. *Opp* SOPHISTICATED. **be rude to** ▷ INSULT.

**rudeness** *n* abuse, *inf* backchat, bad manners, boorishness, *inf* cheek, churlishness, condescension, contempt, discourtesy, disrespect, ill-breeding, impertinence, impudence, incivility, insolence, insults, oafishness, tactlessness, uncouthness, vulgarity.

**rudiments** *plur n* basic principles, basics, elements, essentials, first principles, foundations, fundamentals.

**rudimentary** *adj* basic, crude, elementary, embryonic, immature, initial, introductory, preliminary, primitive, provisional, undeveloped. *Opp* ADVANCED.

**ruffian** *n inf* brute, bully, desperado, gangster, hoodlum, hooligan, lout, mugger, rogue, scoundrel, thug, *inf* tough, villain, *inf* yob.

**ruffle** *vb* 1 agitate, disturb, ripple, stir. 2 *ruffle your hair*. derange, disarrange, dishevel, disorder, *inf* mess up, rumple, tangle, tousle. 3 *ruffle your composure*. annoy, confuse, disconcert, disquiet, fluster, irritate, *inf* nettle, *inf* rattle, *inf* throw, unnerve, unsettle, upset, vex, worry. *Opp* SMOOTH.

**rug** *n* blanket, coverlet, mat, matting.

**rugged** *adj* 1 bumpy, craggy, irregular, jagged, pitted, rocky, rough, stony, uneven. 2 *rugged conditions*. arduous, difficult, hard, harsh, onerous, rough, severe, tough. 3 *rugged good looks*. burly, hardy, husky, muscular, robust, rough, strong, sturdy, ungraceful, unpolished, weather-beaten.

**ruin** *n* bankruptcy, breakdown, collapse, *inf* crash, destruction, downfall, end, failure, fall, ruination, undoing, wreck. • *vb* damage, demolish, destroy, devastate, flatten, overthrow, shatter, spoil, wreck. **ruins** debris, havoc, remains, rubble, wreckage.

**ruined** *adj* crumbling, derelict, dilapidated, fallen down, in ruins, ramshackle, ruinous, tumbledown, uninhabitable, unsafe, wrecked.

**ruinous** *adj* 1 apocalyptic, calamitous, cataclysmic, catastrophic, crushing, destructive, devastating, dire, disastrous, fatal, harmful, injurious, pernicious, shattering. 2 ▷ RUINED.

**rule** *n* 1 axiom, code, decree, *plur* guidelines, law, ordinance, practice, precept, principle, regulation, ruling, statute. 2 administration, ascendancy, authority, command, control, domination, dominion, empire, government, influence, jurisdiction, management, mastery, oversight, power, regime, reign, sovereignty, supervision, supremacy, sway. 3 *as a general rule*. convention, custom, norm, routine, standard. • *vb* 1 administer, be the ruler of, command, control, direct, dominate, govern, guide, hold sway, lead, manage, predominate, reign, run, superintend. 2 adjudicate, decide, decree, deem, determine, find, judge, pronounce, resolve. **rule out** ▷ EXCLUDE.

**ruler** *n* administrator, *inf* Big Brother, law-maker, leader, manager. □ *autocrat, Caesar, demagogue, despot, dictator, doge, emir, emperor, empress, governor, kaiser, king, lord, monarch, potentate, president, prince, princess, queen, rajah, regent, satrap, sovereign, sultan, suzerain, triumvirate* [= three ruling jointly], *tyrant, tzar, viceroy*. ▷ CHIEF.

**rumour** *n* chat, *inf* chit-chat, gossip, hearsay, *inf* low-down, news, prattle, report, scandal, *inf* tittle-tattle, whisper.

**run** *n* 1 canter, dash, gallop, jog, marathon, race, sprint, trot. 2 *a run in the car*. drive, excursion, jaunt, journey, joyride, ride, *inf* spin, trip. 3 *run of bad luck*. chain, sequence, series, stretch. 4 *chicken run*. compound, coop, enclosure, pen. • *vb* 1 bolt, canter, career, dash, gallop, hare, hurry, jog, race, rush, scamper, scoot, scurry, scuttle, speed, sprint, tear, trot. 2 *buses run hourly*. go, operate, ply, provide a service, travel. 3 *car runs well*. behave, function, perform, work. **4** *water runs downhill.* cascade, dribble, flow, gush, leak, pour, spill, stream, trickle. **5** *Who runs the country?* administer, conduct, control, direct, govern, look after, maintain, manage, rule, supervise. **run across** ▷ MEET. **run after** ▷ PURSUE. **run away** ▷ ESCAPE. **run into** ▷ MEET.

**runner** *n* **1** athlete, competitor, entrant, hurdler, jogger, participant, sprinter. **2** courier, dispatch-rider, errand-boy, errand-girl, messenger. **3** *plant sends out runners.* offshoot, shoot, sprout, sucker, tendril.

**runny** *adj* fluid, free-flowing, liquid, running, thin, watery. *Opp* SOLID, VISCOUS.

**rupture** *n* **1** breach, break, burst, cleavage, fracture, puncture, rift, split. **2** *rupture between friends.* break-up, disunity, schism, separation. **3** [*medical*] hernia.
• *vb* break, burst, fracture, part, separate, split.

**rural** *adj* agrarian, agricultural, Arcadian, bucolic, countrified, pastoral, rustic, sylvan. *Opp* URBAN.

**rush** *n* **1** bustle, dash, haste, hurry, panic, pressure, race, scramble, speed, turmoil, urgency. **2** *rush of water.* cataract, flood, gush, spate, surge. **3** *rush of people.* charge, onslaught, stampede. • *vb* bolt, burst, bustle, canter, career, charge, dash, fly, gallop, *inf* get a move on, hare, hasten, hurry, jog, make haste, move fast, race, run, scamper, *inf* scoot, scramble, scurry, scuttle, shoot, speed, sprint, stampede, *inf* step on it, *inf* tear, trot, *inf* zoom.

**rust** *vb* become rusty, corrode, crumble away, oxidize, rot.

**rustic** *adj* **1** ▷ RURAL. **2** *rustic simplicity.* artless, clumsy, crude, naïve, *derog* oafish, plain, rough, simple, uncomplicated, uncultured, unpolished, unsophisticated.

**rusty** *adj* **1** corroded, oxidized, rotten, tarnished. **2** [*inf*] *My French is rusty.* dated, forgotten, out of practice, unused.

**rut** *n* **1** channel, furrow, groove, indentation, pothole, track, trough, wheel-mark. **2** *in a rut.* dead end, habit, pattern, routine, treadmill.

**ruthless** bloodthirsty, brutal, callous, cruel, dangerous, ferocious, fierce, hard, heartless, inexorable, inhuman, merciless, pitiless, relentless, sadistic, unfeeling, unrelenting, unsympathetic, vicious, violent. *Opp* MERCIFUL.

# S

**sabotage** *n* disruption, vandalism, wilful damage, wrecking.
• *vb* cripple, damage, destroy, disable, disrupt, incapacitate, put out of action, *inf* throw a spanner in the works (of), vandalize, wreck.

**sack** *n* **1** bag, pouch. **2** *inf* the boot, *inf* the chop, dismissal, firing, redundancy, *inf* your cards.
• *vb* **1** *inf* axe, discharge, dismiss, *inf* fire, give someone notice, *inf* give someone the boot, *inf* give someone the chop, *inf* give someone the sack, lay off, make redundant. **2** ▷ DESTROY, PLUNDER. **get the sack** be dismissed, be sacked, *inf* get your cards, get your marching orders, lose your job.

**sacred** *adj* blessed, blest, consecrated, dedicated, divine, godly, hallowed, holy, religious, revered, sacrosanct, sanctified, venerable, venerated. *Opp* SECULAR.

**sacrifice** *n* immolation, oblation, offering, propitiation, votive offering. • *vb* 1 immolate, kill, offer up, slaughter, yield up. 2 abandon, forfeit, forgo, give up, let go, lose, relinquish, renounce, surrender.

**sacrilege** *n* blasphemy, desecration, disrespect, heresy, impiety, irreverence, profanation.

**sacrilegious** *adj* atheistic, blasphemous, disrespectful, heretical, impious, irreligious, irreverent, profane, ungodly. ▷ WICKED. *Opp* REVERENT.

**sacrosanct** *adj* inviolable, inviolate, protected, respected, secure, untouchable. ▷ SACRED.

**sad** *adj* 1 abject, *inf* blue, broken-hearted, careworn, cheerless, crestfallen, dejected, depressed, desolate, despairing, desperate, despondent, disappointed, disconsolate, discontented, discouraged, disgruntled, disheartened, disillusioned, dismal, dispirited, dissatisfied, distressed, doleful, dolorous, *inf* down, downcast, downhearted, dreary, forlorn, friendless, funereal, gloomy, glum, grave, grief-stricken, grieving, grim, guilty, heartbroken, *inf* heavy, heavy-hearted, homesick, hopeless, in low spirits, *inf* in the doldrums, joyless, lachrymose, lonely, *inf* long-faced, *inf* low, lugubrious, melancholy, miserable, moody, moping, morose, mournful, pathetic, penitent, pessimistic, piteous, pitiable, pitiful, plaintive, poignant, regretful, rueful, saddened, serious, sober, sombre, sorrowful, sorry, tearful, troubled, unhappy, unsatisfied, upset, wistful, woebegone, woeful, wretched. 2 *sad news.* calamitous, deplorable, depressing, disastrous, discouraging, dispiriting, distressing, grievous, heartbreaking, heart-rending, lamentable, morbid, moving, painful, regrettable, *inf* tear-jerking, touching, tragic, unfortunate, unsatisfactory, unwelcome, upsetting. 3 *a sad state of disrepair.* ▷ UNSATISFACTORY. *Opp* HAPPY.

**sadden** *vb* aggrieve, *inf* break someone's heart, deject, depress, disappoint, discourage, dishearten, dismay, dispirit, distress, grieve, make sad, upset. *Opp* CHEER.

**sadistic** *adj* barbarous, beastly, brutal, inhuman, monstrous, perverted, pitiless, ruthless, vicious. ▷ CRUEL.

**sadness** *n* bleakness, care, dejection, depression, desolation, despair, despondency, disappointment, disillusionment, dissatisfaction, distress, dolour, gloom, glumness, grief, heartbreak, heaviness, homesickness, hopelessness, joylessness, loneliness, melancholy, misery, moping moroseness, mournfulness, pessimism, poignancy, regret, ruefulness, seriousness, soberness, sombreness, sorrow, tearfulness, trouble, unhappiness, wistfulness, woe. *Opp* HAPPINESS.

**safe** *adj* 1 defended, foolproof, guarded, immune, impregnable, invulnerable, protected, secured, shielded. ▷ SECURE. *Opp* VULNERABLE. 2 *inf* alive and well, *inf* all right, *inf* in one piece, intact, sound, undamaged, unharmed, unhurt, uninjured, unscathed, well, whole. 3 *safe drivers.* cautious, circumspect, dependable, reliable, trustworthy.

**4** *safe pets.* docile, friendly, harmless, innocuous, tame. **5** *safe to drink.* decontaminated, drinkable, eatable, fit for human consumption, fresh, good, non-poisonous, non-toxic, pasteurized, potable, pure, purified, uncontaminated, unpolluted, wholesome. **6** *safe vehicle.* airworthy, roadworthy, seaworthy, tried and tested. *Opp* DANGEROUS. **make safe** ▷ SECURE. **safe keeping** care, charge, custody, guardianship, keeping, protection.

**safeguard** *vb* care for, defend, keep safe, look after, protect, shelter, shield.

**safety** *n* **1** cover, immunity, invulnerability, protection, refuge, sanctuary, security, shelter. **2** *safety of air travel.* dependability, harmlessness, reliability.

**sag** *vb* be limp, bend, dip, droop, fall, flop, hang down, sink, slump. ▷ DROP.

**sail** *n* **1** canvas. □ *foresail, gaffsail, jib, lateen sail, lugsail, mainsail, mizzen, spinnaker, spritsail, topsail.* **2** cruise, sea-passage, voyage. ▷ JOURNEY. • *vb* **1** captain, navigate, paddle, pilot, punt, row, skipper, steer. **2** cruise, go sailing, put to sea, set sail, steam. ▷ TRAVEL.

**sailor** *n* mariner, seafarer, *old use* sea dog, seaman. □ *able seaman, bargee, boatman, boatswain, bosun, captain, cox, coxswain,* plur *crew, deck-hand, helmsman, mate, midshipman, navigator, pilot, rating, rower, skipper, yachtsman, yachtswoman.*

**saintly** *adj* angelic, blessed, blest, chaste, godly, holy, innocent, moral, pious, pure, religious, righteous, seraphic, sinless, virginal, virtuous. ▷ GOOD. *Opp* SATANIC.

**sake** *n* account, advantage, behalf, benefit, gain, good, interest, welfare.

**salary** *n* compensation, earnings, emolument, income, pay, payment, remuneration, stipend, wages.

**sale** *n* marketing, selling, trade, traffic, transaction, vending. □ *auction, bazaar, closing-down sale, fair, jumble sale, market, rummage sale, spring sale.*

**salesperson** *n* assistant, auctioneer, representative, salesman, saleswoman, shop-boy, shop-girl, shopkeeper.

**saliva** *n inf* dribble, *inf* spit, spittle, sputum.

**sallow** *adj* anaemic, bloodless, colourless, etiolated, pale, pallid, pasty, unhealthy, wan, yellowish.

**salt** *adj* brackish, briny, saline, salted, salty, savoury. *Opp* FRESH.

**salubrious** *adj* health-giving, healthy, hygienic, invigorating, nice, pleasant, refreshing, sanitary, wholesome. *Opp* UNHEALTHY.

**salute** *n* acknowledgement, gesture, greeting, salutation, wave. • *vb* accost, acknowledge, address, greet, hail, honour, pay respects to, recognize. ▷ GESTURE.

**salvage** *n* **1** reclamation, recovery, rescue, retrieval, salvation, saving. **2** recyclable material, waste. • *vb* conserve, preserve, reclaim, recover, recycle, redeem, rescue, retrieve, reuse, save, use again.

**salvation** *n* deliverance, escape, help, preservation, redemption, rescue, saving, way out. *Opp* DAMNATION.

**salve** *n* balm, cream, demulcent, embrocation, emolient, liniment, lotion, ointment, unguent. • *vb* alleviate, appease, assuage, comfort, ease, mitigate, mollify, soothe.

**same** *adj* **1** actual, identical, selfsame. **2** *two women wearing the same jacket.* analogous, comparable, consistent, corresponding, duplicate, equal, equivalent, indistinguishable, interchangeable, matching, parallel, similar, synonymous [= *having same meaning*], twin, unaltered, unchanged, uniform, unvaried. *Opp* DIFFERENT.

**sample** *n* bit, demonstration, example, foretaste, free sample, illustration, indication, instance, model, pattern, representative piece, selection, snippet, specimen, taste, trailer (*of film*), trial offer. • *vb* experience, inspect, take a sample of, taste, test, try.

**sanatorium** *n* clinic, convalescent home, hospital, nursing home, rest-home.

**sanctify** *vb* beatify, bless, canonize, consecrate, hallow, justify, purify.

**sanctimonious** *adj* canting, holier-than-thou, hypocritical, insincere, moralizing, pharisaical, pietistic, *sl* pi, pious, self-righteous, sententious, *inf* smarmy, smug, superior, unctuous.

**sanction** *n* agreement, approval, authorization, *inf* blessing, confirmation, consent, encouragement, endorsement, legalization, licence, permission, ratification, support, validation. • *vb* agree to, allow, approve, authorize, confirm, consent to, endorse, *inf* give your blessing to, give permission for, legalize, legitimize, licence, permit, ratify, support, validate.

**sanctity** *adj* divinity, godliness, grace, holiness, piety, sacredness, saintliness.

**sanctuary** *n* **1** asylum, haven, protection, refuge, retreat, safety, shelter. **2** *wildlife sanctuary.* conservation area, park, preserve, reservation, reserve. **3** *a holy sanctuary.* chapel, church, holy of holies, holy place, sanctum, shrine, temple.

**sands** *plur n* beach, seaside, shore, *poet* strand.

**sane** *adj inf* all there, balanced, *Lat* compos mentis, *inf* in your right mind, level-headed, lucid, normal, of sound mind, rational, reasonable, sensible, sound, stable, well-balanced. *Opp* MAD.

**sanguine** *adj* buoyant, cheerful, confident, expectant, hopeful, *inf* looking on the bright side, optimistic, positive. *Opp* PESSIMISTIC.

**sanitary** *adj* aseptic, bacteria-free, clean, disinfected, germfree, healthy, hygienic, pure, salubrious, sterile, sterilized, uncontaminated, unpolluted, wholesome. *Opp* UNHEALTHY.

**sanitation** *n* drainage, drains, lavatories, sanitary arrangements, sewage disposal, sewers.

**sap** *n* fluid, life-blood, moisture, vigour, vitality, vital juices. • *vb* bleed, drain. ▷ EXHAUST.

**sarcasm** *n* acerbity, asperity, contumely, derision, irony, malice, mockery, ridicule, satire, scorn.

**sarcastic** *adj* acerbic, acidulous, biting, caustic, contemptuous, cutting, demeaning, derisive, disparaging, hurtful, ironic, ironical, mocking, satirical, scathing, sharp, sneering, spiteful, taunting, trenchant, venomous, vitriolic, withering, wounding. ▷ HUMOROUS.

**sardonic** *adj* bitter, black, cruel, cynical, grim, heartless, malicious, mordant, wry. ▷ HUMOROUS.

**sash** *n* band, belt, cummerbund, girdle, waistband.

**satanic** *adj* demonic, devilish, diabolical, fiendish, hellish, infernal, Mephistophelian. ▷ WICKED. *Opp* SAINTLY.

**satchel** *n* bag, pouch, school-bag, shoulder-bag.

**satellite** *n* **1** moon, planet. **2** *man-made satellite.* spacecraft, sputnik.

**satire** *n* burlesque, caricature, derision, invective, irony, lampoon, mockery, parody, ridicule, satirical comedy, scorn, *inf* send-up, *inf* spoof, *inf* take-off, travesty. ▷ WRITING.

**satirical** *adj* critical, derisive, disparaging, disrespectful, ironic, irreverent, mocking, scornful. ▷ HUMOROUS, SARCASTIC.

**satirize** *vb* be satirical about, burlesque, caricature, criticize, deride, hold up to ridicule, lampoon, laugh at, make fun of, mimic, mock, parody, pillory, *inf* send up, *inf* take off, travesty. ▷ RIDICULE.

**satisfaction** *n* comfort, content, contentment, delight, enjoyment, fulfilment, gratification, happiness, joy, pleasure, pride, self-satisfaction. *Opp* DISSATISFACTION.

**satisfactory** *adj* acceptable, adequate, *inf* all right, competent, fair, *inf* good enough, *inf* not bad, passable, pleasing, satisfying, sufficient, suitable, tolerable, *inf* up to scratch. *Opp* UNSATISFACTORY.

**satisfy** *vb* appease, assuage, comfort, comply with, content, fill, fulfil, gratify, make happy, meet, pacify, placate, please, put an end to, quench, sate, satiate, serve (*a need*), settle, slake (*thirst*), solve, supply. *Opp* FRUSTRATE. **satisfied** ▷ CONTENT.

**saturate** *vb* drench, fill, impregnate, permeate, soak, souse, steep, suffuse, waterlog, wet.

**sauce** *n* **1** condiment, gravy, ketchup, relish. **2** ▷ INSOLENCE.

**saucepan** *n* cauldron, pan, pot, skillet, stockpot.

**savage** *adj* **1** barbarian, barbaric, cannibal, heathen, pagan, primitive, uncivilized, uncultivated, uneducated. *Opp* CIVILIZED. **2** *savage beasts.* feral, fierce, undomesticated, untamed, wild. **3** *savage attack.* angry, atrocious, barbarous, beastly, bestial, blistering, bloodthirsty, bloody, brutal, callous, cold-blooded, cruel, demonic, diabolical, ferocious, fierce, heartless, inhuman, merciless, murderous, pitiless, ruthless, sadistic, unfeeling, vicious, violent. *Opp* TAME. • *n* barbarian, beast, brute, cannibal, fiend, savage person. • *vb* attack, bite, claw, lacerate, maul, mutilate.

**save** *vb* **1** be sparing with, collect, conserve, economize, hoard, hold back, hold on to, invest, keep, *inf* lay aside, *inf* put by, put in a safe place, reserve, retain, scrape together, set aside, *inf* stash away, store up, take care of, use wisely. *Opp* WASTE. **2** *save from captivity.* bail out, deliver, free, liberate, ransom, redeem, release, rescue, set free. **3** *save from destruction.* recover, retrieve, salvage. **4** *save from danger.* defend, deliver, guard, keep safe, preserve, protect, safeguard, screen, shelter, shield. **5** *saved me from looking a fool.* check, deter, preclude, prevent, spare, stop. *Opp* ABANDON.

**saving** *n* **1** economizing, frugality, parsimony, prudence, *inf* scrimping and scraping, thrift. **2** cut, discount, economy, reduction.

**savings** *n* capital, funds, investments, *inf* nest-egg, reserves, resources, riches, wealth.

**saviour** *n* **1** champion, defender, deliverer, *inf* friend in need, guardian, liberator, rescuer. **2** [*theological*] Christ, Our Lord, The Messiah, The Redeemer.

**savour** *n* flavour, piquancy, relish, smell, tang, taste, zest. • *vb* appreciate, delight in, enjoy, relish, smell, taste.

**savoury** *adj* appetizing, delicious, flavoursome, piquant, salty. ▷ TASTY. *Opp* SWEET.

**saw** *n* **1** □ *chain-saw, hack-saw, jigsaw, ripsaw.* **2** [*old use*] *just an old saw.* ▷ SAYING. • *vb* ▷ CUT.

**say** *vb* affirm, allege, announce, answer, articulate, assert, asseverate, aver, *old use* bruit abroad, *inf* come out with, comment, communicate, convey, declare, disclose, divulge, ejaculate, enunciate, exclaim, express, intimate, maintain, mention, mouth, phrase, pronounce, *inf* put it about, read out, recite, rejoin, remark, repeat, reply, report, respond, retort, reveal, signify, state, suggest, tell, utter. ▷ SPEAK, TALK, TELL.

**saying** *n* adage, aphorism, apophthegm, axiom, catch-phrase, catchword, clichA, dictum, epigram, expression, formula, maxim, motto, phrase, precept, proverb, quotation, remark, *old use* saw, slogan, statement, tag, truism, watchword.

**scab** *n* clot of blood, crust, sore.

**scale** *n* **1** dandruff, flake, plate, scurf. **2** *remove scale from teeth.* caking, coating, crust, deposit, encrustation, *inf* fur, plaque, tartar. **3** *the scale on a thermometer.* calibration, gradation, graduation. **4** *the social scale.* hierarchy, ladder, order, ranking, spectrum. **5** *small/large scale.* proportion, ratio. ▷ SIZE. **6** *musical scale.* sequence, series. □ *chromatic scale, diatonic scale, major scale, minor scale.* • *vb* ascend, clamber up, climb, go up, mount. **scales** balance, weighing-machine.

**scamper** *vb* dash, frisk, frolic, gambol, hasten, hurry, play, romp, run, rush, scuttle.

**scan** *vb* **1** check, examine, explore, eye, gaze at, investigate, look at, pore over, scrutinize, search, stare at, study, survey, view, watch. **2** *scan the papers.* flip through, glance at, read quickly, skim, thumb through.

**scandal** *n* **1** discredit, disgrace, dishonour, disrepute, embarrassment, ignominy, infamy, notoriety, obloquy, outrage, reproach, sensation, shame. **2** calumny, defamation, gossip, innuendo, libel, rumour, slander, slur, *inf* smear, *inf* tittle-tattle.

**scandalize** *vb* affront, appal, disgust, horrify, offend, outrage, shock, upset.

**scandalous** *adj* **1** disgraceful, disgusting, dishonourable, disreputable, ignominious, immodest, immoral, improper, indecent, indecorous, infamous, licentious, notorious, outrageous, shameful, shocking, sinful, sordid, unmentionable, unspeakable, wicked. **2** *a scandalous lie.* calumnious, defamatory, libellous, scurrilous, slanderous, untrue.

**scansion** *n* metre, prosody, rhythm. ▷ VERSE.

**scanty** *adj* **1** inadequate, insufficient, meagre, mean, *sl* measly, *inf* mingy, minimal, scant, scarce, *inf* skimpy, sparing, sparse, stingy. ▷ SMALL. *Opp* PLENTIFUL. **2** *scanty clothes*. indecent, revealing, *inf* see-through, thin.

**scapegoat** *n* dupe, *sl* fall guy, *inf* front, whipping-boy, victim.

**scar** *n* blemish, brand, burn, cicatrice, cicatrix, cut, disfigurement, injury, mark, scab, scratch. ▷ WOUND. • *vb* blemish, brand, burn, damage, deface, disfigure, injure, leave a scar on, mark, scratch, spoil.

**scarce** *adj inf* few and far between, *inf* hard to come by, *inf* hard to find, inadequate, infrequent, in short supply, insufficient, lacking, meagre, rare, scant, scanty, sparse, *inf* thin on the ground, uncommon, unusual. *Opp* PLENTIFUL.

**scarcely** *adv* barely, hardly, only just.

**scarcity** *n* dearth, famine, inadequacy, insufficiency, lack, need, paucity, poverty, rarity, shortage, want. *Opp* PLENTY.

**scare** *n* alarm, jolt, shock, start. ▷ FRIGHT. • *vb* **1** alarm, dismay, intimidate, make someone afraid, *inf* make someone jump, menace, panic, shake, shock, startle, terrorize, threaten, unnerve. ▷ FRIGHTEN. *Opp* REASSURE.

**scarf** *n* headscarf, muffler, shawl, stole.

**scary** *adj* [*inf*] creepy, eerie, hair-raising, horrible, scaring, unnerving. ▷ FRIGHTENING.

**scathing** *adj* biting, caustic, critical, humiliating, mordant, satirical, savage, scornful, tart, withering. *Opp* COMPLIMENTARY.

**scatter** *vb* **1** break up, disband, disintegrate, dispel, disperse, divide, send in all directions, separate. **2** *scatter seeds*. broadcast, disseminate, intersperse, shed, shower, sow, spread, sprinkle, strew, throw about. *Opp* GATHER.

**scatterbrained** *adj* absent-minded, careless, crazy, disorganized, forgetful, frivolous, hare-brained, inattentive, muddled, *inf* not with it, *inf* scatty, thoughtless, unreliable, unsystematic, vague. ▷ SILLY.

**scavenge** *vb* forage, rummage, scrounge, search.

**scenario** *n* design, framework, layout, outline, plan, scheme, storyline, structure, summary.

**scene** *n* **1** area, background, context, locale, locality, location, place, position, setting, site, situation, spot, whereabouts. **2** *a beautiful scene*. picture, sight, spectacle. ▷ SCENERY. **3** *scene from a film*. act, chapter, *inf* clip, episode, part, section, sequence. **4** *a nasty scene*. altercation, argument, *inf* carry-on, commotion, disturbance, furore, fuss, quarrel, row, tantrum, *inf* to-do, *inf* upset.

**scenery** *n* **1** landscape, outlook, panorama, prospect, scene, terrain, view, vista. **2** *stage scenery*. backdrop, flats, set, setting.

**scenic** *adj* attractive, beautiful, breathtaking, grand, impressive, lovely, panoramic, picturesque, pretty, spectacular.

**scent** *n* **1** aroma, bouquet, fragrance, nose, odour, perfume, redolence, smell. **2** after-shave, eau de cologne, lavender water, perfume. **3** *an animal's scent*. spoor, track, trail. • *vb* ▷ SMELL. **scented** ▷ SMELLING.

**sceptic** *n* agnostic, cynic, doubter, *inf* doubting Thomas, disbeliever, scoffer, unbeliever. *Opp* BELIEVER.

**sceptical** *adj* agnostic, cynical, disbelieving, distrustful, doubting, dubious, incredulous, mistrustful, questioning, scoffing, suspicious, uncertain, unconvinced, unsure. *Opp* CONFIDENT.

**scepticism** *n* agnosticism, cynicism, disbelief, distrust, doubt, dubiety, incredulity, lack of confidence, mistrust, suspicion. *Opp* FAITH.

**schedule** *n* agenda, calendar, diary, itinerary, list, plan, programme, register, scheme, timetable. • *vb* appoint, arrange, assign, book, earmark, fix a time, organize, outline, plan, programme, time, timetable.

**scheme** *n* **1** approach, blueprint, design, draft, idea, method, plan, procedure, programme, project, proposal, scenario, strategy, system. **2** *a dishonest scheme.* conspiracy, *inf* dodge, intrigue, machinations, manoeuvre, plot, *inf* ploy, *inf* racket, ruse, stratagem, subterfuge, tactic. **3** *colour scheme.* arrangement, design. • *vb* collude, connive, conspire, *inf* cook something up, *inf* hatch a plot, intrigue, machinate, manoeuvre, plan, plot.

**scholar** *n* academic, *inf* egghead, expert, highbrow, intellectual, professor, pundit, savant. ▷ PUPIL.

**scholarly** *adj* **1** academic, bookish, *inf* brainy, *inf* deep, erudite, highbrow, intellectual, knowledgeable, learned, lettered, widely-read. **2** *scholarly treatise.* documented, researched, rigorous, scientific, well-argued, well-informed.

**scholarship** *n* **1** academic achievement, education, erudition, intellectual attainment, knowledge, learning, research, schooling, scientific rigour, wisdom. **2** *a scholarship to Oxford.* award, bursary, endowment, exhibition, fellowship, grant.

**school** *n* **1** educational institution. □ *academy, boarding-school, coeducational school, college, comprehensive (school), first school, grammar school, high school, infant school, junior school, kinde garten, nursery school, playgroup preparatory school, primary school, public school, secondary school, seminary.* **2** *a school of whales.* shoal. ▷ GROUP. • *vb* ▷ EDUCATE.

**science** *n* organized knowledge, systematic study. □ *acoustics, aer nautics, agricultural science, anatomy, anthropology, artifical intel gence, astronomy, astrophysics, behavioural science, biochemistry, biology, biophysics, botany, chemistry, climatology, computer science, cybernetics, dietetics, domest science, dynamics, earth science, ec logy, economics, electronics, engin eering, entomology, environmenta science, food science, genetics, geographical science, geology, geophy ics, hydraulics, information techn logy, life science, linguistics, materials science, mathematics, mechanics, medical science, metal lurgy, meteorology, microbiology, mineralogy, ornithology, pathology, pharmacology, physics, phys ology, political science, psycholog robotics, sociology, space technology, telecommunications, thermodynamics, toxicology, veterinary s ence, zoology.*

**scientific** *adj* analytical, method ical, meticulous, orderly, organized, precise, rational, regulated, rigorous, systematic.

**scientist** *n inf* boffin, researcher, scientific expert, technologist.

**scintillating** *adj* brilliant, cleve coruscating, dazzling, effervescent, flashing, glittering, liveley, sparkling, vivacious, witty. *Opp* DULL.

**scoff** *vb* **1** belittle, be sarcastic, be scornful, deride, disparage, gibe, jeer, jibe, laugh, mock, *inf* poke fun, ridicule, sneer, taunt, tease. **2** ▷ EAT.

**scold** *vb* admonish, berate, blame, *inf* carpet, castigate, censure, chide, criticize, disapprove of, find fault with, *inf* jump down some-one's throat, *inf* lecture, *inf* nag, rate, rebuke, reprehend, reprim-and, reproach, reprove, *inf* slate, *inf* tell off, *inf* tick off, upbraid.

**scoop** *n* **1** bailer, ladle, shovel, spoon. **2** *news scoop.* exclusive, inside story, *inf* latest, revelation. *vb* dig, excavate, gouge, hollow, scrape, shovel, spoon.

**scope** *n* **1** ambit, area, breadth, compass, competence, extent, field, limit, range, reach, span, sphere, terms of reference. **2** *scope for expansion.* capacity, chance, *inf* elbow-room, freedom, latitude, leeway, liberty, opportunity, out-let, room, space, spread.

**scorch** *vb* blacken, brand, burn, char, heat, roast, sear, singe.

**score** *n* **1** account, amount, count, marks, points, reckoning, result, sum, tally, total. **2** *score on furni-ture.* cut, groove, incision, line, mark, nick, scrape, scratch, slash. • *vb* **1** account for, achieve, add up, *inf* chalk up, earn, gain, *inf* knock up, make, tally, win. **2** *score a groove.* cut, engrave, gouge, incise, mark, scrape, scratch, slash. **3** *score music.* orchestrate, write out. **settle a score** ▷ RETALIATE.

**scorn** *n* contempt, contumely, deri-sion, detestation, disdain, disgust, dislike, dismissal, disparagement, disrespect, jeering, mockery, rejec-tion, ridicule, scoffing, sneering, taunt-ing. *Opp* ADMIRATION. • *vb* be scornful about, contemn, deride, despise, disapprove of, dis-dain, dislike, dismiss, disparage, hate, insult, jeer at, laugh at, look down on, make fun of, mock, reject, ridicule, *inf* scoff at, sneer at, spurn, taunt, *inf* turn up your nose at. *Opp* ADMIRE.

**scornful** *adj* condescending, con-temptuous, contumelious, deprec-ative, derisive, disdainful, dismiss-ive, disparaging, disrespectful, haughty, insulting, jeering, mock-ing, patronizing, sarcastic, satir-ical, scathing, scoffing, sneering, *inf* snide, *inf* snooty, supercilious, superior, taunting, withering. *Opp* RESPECTFUL.

**scoundrel** *n* blackguard, blighter, bounder, cad, good-for-nothing, heel, knave, miscreant, rascal, rogue, ruffian, scallywag, scamp, villain, wretch.

**scour** *vb* **1** abrade, buff up, burn-ish, clean, cleanse, polish, rub, scrape, scrub, shine, wash. **2** *scour the house.* comb, forage through, hunt through, rake through, ran-sack, rummage through, search, *inf* turn upside down.

**scourge** *n* **1** affliction, bane, curse, evil, misery, misfortune, plague, torment, woe. **2** ▷ WHIP. • *vb* beat, belt, flagellate, flog, horsewhip, lash, whip.

**scout** *n* lookout, spy. • *vb* explore, get information, hunt around, investigate, look about, recon-noitre, search, *inf* snoop, spy.

**scowl** *vb* frown, glower, grimace, *inf* look daggers, lower.

**scraggy** *adj* bony, emaciated, gaunt, lanky, lean, scrawny, skinny, starved, thin, underfed. *Opp* PLUMP.

**scramble** *n* commotion, confusion, *inf* free-for-all, haste, hurry, mêlé, race, rush, scrimmage, struggle. • *vb* 1 clamber, climb, crawl, grope, move awkwardly, scrabble. 2 *scramble for gold*. compete, contend, dash, fight, hasten, hurry, jostle, push, run, rush, scuffle, strive, struggle, tussle, vie. 3 *scramble a message*. confuse, jumble, mix up.

**scrap** *n* 1 atom, bit, crumb, fraction, fragment, grain, hint, iota, jot, mite, molecule, morsel, particle, piece, rag, scintilla, shard, shred, sliver, snippet, speck, trace. 2 *inf* junk, leavings, litter, odds and ends, offcuts, refuse, rejects, remains, remnants, residue, rubbish, salvage, waste. 3 *a friendly scrap*. argument, quarrel, scuffle, *inf* set-to, squabble, tiff, tussle, wrangle. ▷ FIGHT. • *vb* 1 abandon, cancel, discard, *inf* ditch, drop, give up, jettison, throw away, write off. 2 *scrap over trifles*. argue, bicker, flare up, quarrel, spar, squabble, tussle, wrangle. ▷ FIGHT.

**scrape** *n* 1 abrasion, graze, injury, laceration, scratch, scuff, wound. 2 *an awkward scrape*. difficulty, escapade, *inf* kettle of fish, piece of mischief, plight, prank, predicament, trouble. • *vb* 1 abrade, bark, bruise, damage, graze, injure, lacerate, scratch, scuff, skin, wound. 2 *scrape clean*. clean, file, rasp, rub, scour, scrub. **scrape together** ▷ COLLECT.

**scrappy** *adj* bitty, careless, disjointed, fragmentary, hurriedly done, imperfect, incomplete, inconclusive, sketchy, slipshod, unfinished, unpolished, unsatisfactory. *Opp* PERFECT.

**scratch** *n* abrasion, damage, dent, gash, gouge, graze, groove, indentation, injury, laceration, line, mark, score, scoring, scrape, scuff, wound. • *vb* abrade, claw at, cut, damage the surface of, dent, gash, gouge, graze, groove, incise, injure, lacerate, mark, rub, scarify, score, scrape, scuff, wound. **up to scratch** ▷ SATISFACTORY.

**scrawl** *vb* doodle, scribble, write hurriedly. ▷ WRITE.

**scream** *n* & *vb* bawl, caterwaul, cry, howl, roar, screech, shout, shriek, squeal, wail, yell, yowl.

**screen** *n* 1 blind, curtain, divider, partition. 2 camouflage, concealment, cover, disguise, protection, shelter, shield, smokescreen. 3 *sift through a screen*. filter, mesh, riddle, sieve, strainer. • *vb* 1 divide, partition off, subdivide, wall off. 2 camouflage, cloak, conceal, cover, disguise, guard, hide, mask, protect, safeguard, shade, shelter, shield, shroud, veil. 3 *screen employees for security*. *inf* check out, examine, investigate, process, sift out, vet.

**screw** *n* 1 bolt, screw-bolt. 2 rotation, spiral, turn, twist. • *vb* rotate, turn, twist.

**screw down** ▷ FASTEN. **screw up** ▷ BUNGLE, TWIST.

**scribble** *vb* ▷ SCRAWL.

**scribe** *n* amanuensis, clerk, copyist, secretary, transcriber, writer.

**script** *n* 1 calligraphy, handwriting, penmanship. 2 *script of a play*. libretto, screenplay, text, words.

**scripture** *n* bible, holy writ, sacred writings, Word of God. □ *Bhagavad-Gita, inf the Good Book, the Gospel, Holy Bible, Koran, Upanishad.*

**scrounge** *vb* beg, cadge, importune.

**scrub** *vb* 1 brush, clean, rub, scour, wash. 2 ▷ CANCEL.

**scruffy** *adj* bedraggled, dirty, dishevelled, disordered, dowdy, frowsy, messy, ragged, scrappy, shabby, slatternly, slovenly, tatty, ungroomed, unkempt, untidy, worn out. *Opp* SMART.

**scruple** *n* compunction, conscience, doubt, hesitation, misgiving, qualm, reluctance, *inf* second thought. • *vb* [*usu neg*] be reluctant, have a con-science (about), have scruples (about), hesitate, hold back (from), *inf* think twice (about).

**scrupulous** *adj* 1 careful, cautious, conscientious, diligent, exacting, fastidious, *inf* finicky, meticulous, minute, neat, painstaking, precise, punctilious, rigid, rigorous, strict, systematic, thorough. 2 *scrupulous honesty*. ethical, fair-minded, honest, honourable, just, moral, principled, proper, upright, upstanding. *Opp* UNSCRUPULOUS.

**scrutinize** *vb* analyse, check, examine, *inf* go over with a toothcomb, inspect, investigate, look closely at, *inf* probe, sift, study.

**scrutiny** *n* analysis, examination, inspection, investigation, probing, search, study.

**sculpture** *n* three-dimensional art. □ *bas-relief, bronze, bust, carving, caryatid, cast, effigy, figure, figurine, maquette, marble, moulding, plaster cast, relief, statue, statuette.* • *vb* carve, cast, chisel, fashion, form, hew, model, mould, *inf* sculpt, shape.

**scum** *n* dirt, film, foam, froth, impurities, suds.

**scurrilous** *adj* abusive, calumnious, coarse, defamatory, derogatory, disparaging, foul, indecent, insulting, libellous, low, obscene, offensive, opprobrious, scabrous, shameful, slanderous, vile, vulgar.

**sea** *adj* aquatic, marine, maritime, nautical, naval, ocean-going, oceanic, salt-water, seafaring, seagoing. • *n inf* briny, *poet* deep, lake, *old use* main, ocean.

**seal** *n* 1 sea-lion, walrus. 2 *royal seal*. badge, coat of arms, crest, emblem, escutcheon, impression, imprint, mark, monogram, sign, stamp, symbol, token. • *vb* 1 close, fasten, lock, make airtight, make watertight, plug, secure, shut, stick down, stop up. 2 *seal an agreement*. affirm, authenticate, *inf* clinch, conclude, confirm, corroborate, decide, endorse, finalize, guarantee, ratify, settle, sign, validate, verify.

**seam** *n* 1 join, stitching. 2 *seam of coal*. bed, layer, lode, stratum, thickness, vein.

**seamy** *adj* disreputable, distasteful, nasty, repulsive, shameful, sordid, squalid, unattractive, unpleasant, unsavoury, unwholesome.

**search** *n* check, enquiry, examination, hunt, inspection, investigation, look, *inf* probe, pursuit, quest, scrutiny. • *vb* 1 cast about, explore, ferret about, hunt, investigate, *inf* leave no stone unturned, look, nose about, poke about, prospect, pry, seek. 2 *search suspects*. check, examine, *inf* frisk, inspect, scrutinize. 3 *search a house*. comb, go through, ransack, rifle, rummage through, scour. **searching** ▷ INQUISITIVE, THOROUGH.

**seaside** *n* beach, coast, coastal resort, sands, sea-coast, sea-shore, shore.

**season** *n* period, phase, time. • *vb* 1 add seasoning to, flavour, *inf* pep up, salt, spice. 2 *season wood.* age, harden, mature, ripen.

**seasonable** *adj* appropriate, apt, convenient, favourable, fitting, normal, opportune, propitious, suitable, timely, well-timed.

**seasoning** *n* additives, condiments, flavouring, relish, zest. □ *dressing, herbs, mustard, pepper, relish, salt, sauce, spice, vinegar.*

**seat** *n* 1 place, sitting-place. □ *armchair, bench, carver, chair, chaise longue, couch, deck-chair, dining-chair, easy chair,* Fr *fauteuil, form, pew, pillion, pouffe, reclining chair, rocking-chair, saddle, settee, settle, sofa, squab, stall, stool, throne, window seat.* 2 *a country seat.* ▷ RESIDENCE. 3 ▷ BUTTOCKS. **seat yourself** ▷ SIT.

**secateurs** *plur n* clippers, cutters, pruning shears.

**secluded** *adj* cloistered, concealed, cut off, hidden, inaccessible, isolated, lonely, monastic, *inf* off the beaten track, private, remote, retired, screened, sequestered, sheltered, shut away, solitary, unfrequented, unvisited. *Opp* PUBLIC.

**seclusion** *n* concealment, hiding, isolation, loneliness, privacy, remoteness, retirement, separation, shelter, solitariness.

**second** *adj* added, additional, alternative, another, complementary, duplicate, extra, following, further, later, matching, next, other, repeated, subsequent, twin. • *n* 1 flash, instant, *inf* jiffy, moment, *inf* tick, *inf* twinkling, *inf* wink. 2 *second in a fight.* assistant, deputy, helper, *inf* number two, *inf* right-hand man, right-hand woman, second-in-command, *inf* stand-in, subordinate, supporter, understudy, vice-. • *vb* 1 aid, assist, back, encourage, give approval to, help, promote, side with, sponsor, support. 2 *second to another job.* move, reassign, relocate, shift, transfer.

**secondary** *adj* 1 alternative. ancillary, auxiliary, *inf* backup, extra, inessential, inferior, lesser, lower, minor, nonessential, reinforcing, reserve, second, second-rate, spare, subordinate, subsidiary, supporting, supportive, supplementary, unimportant. 2 *secondary sources.* copied, derivative, second-hand, unoriginal.

**second-hand** *adj* 1 *inf* hand-me-down, old, used, worn. *Opp* NEW. 2 *second-hand experience.* indirect, secondary, vicarious. *Opp* DIRECT.

**second-rate** *adj* commonplace, indifferent, inferior, low-grade, mediocre, middling, ordinary, poor, second-best, second-class, undistinguished, unexciting, uninspiring.

**secret** *adj* 1 clandestine, concealed, covert, disguised, hidden, *inf* hushed up, *inf* hush-hush, invisible, private, secluded, shrouded, stealthy, undercover, underground, unknown. ▷ SECRETIVE. 2 *secret papers.* classified, confidential, inaccessible, intimate, personal, restricted, sensitive, top-secret, undisclosed, unpublished. 3 *secret meanings.* arcane, cryptic, encoded, esoteric, incomprehensible, mysterious, occult, recondite. 4 *secret about his private life.* ▷ SECRETIVE. *Opp* OPEN, PUBLIC.

**secretary** *n* amanuensis, clerk, filing-clerk, personal assistant, scribe, shorthand-typist, stenographer, typist, word-processor operator.

**secrete** *vb* 1 cloak, conceal, cover up, disguise, enshroud, hide, mask, put away, put into hiding. 2 *secrete fluid.* discharge, emit, excrete, exude, give off, leak, ooze, produce, release.

**secretion** *n* discharge, emission, escape, excretion, leakage, release.

**secretive** *adj* close-lipped, enigmatic, furtive, mysterious, quiet, reserved, reticent, secret, shifty, silent, taciturn, tight-lipped, uncommunicative, unforthcoming, withdrawn. *Opp* COMMUNICATIVE.

**sect** *n* cult, denomination, faction, order, party, persuasion. ▷ GROUP.

**sectarian** *adj* bigoted, clannish, cliquish, cultic, denominational, dogmatic, exclusive, factional, fanatical, inflexible, narrow, narrow-minded, partial, partisan, prejudiced, rigid, schismatic.

**section** *n* bit, branch, chapter, compartment, component, department, division, element, fraction, fragment, group, instalment, leg (*of journey*), part, passage, piece, portion, quarter, sample, sector, segment, slice, stage, subdivision, subsection.

**sector** *n* area, district, division, part, quarter, region, zone. ▷ SECTION.

**secular** *adj* civil, earthly, lay, material, mundane, non-religious, temporal, terrestrial, worldly. *Opp* RELIGIOUS.

**secure** *adj* 1 cosy, defended, guarded, immune, impregnable, invulnerable, protected, safe, sheltered, shielded, snug, unharmed, unhurt, unscathed. 2 *secure doors.* bolted, burglarproof, closed, fast, fastened, fixed, foolproof, immovable, locked, shut, solid, tight, unyielding. 3 *secure faith.* certain, confident, firm, stable, steady, strong, sure, unquestioning. • *vb* 1 defend, guard, make safe, preserve, protect, shelter, shield. 2 anchor, attach, bolt, close, fix, lock, make fast, screw down, tie down. ▷ FASTEN. 3 *secure a loan.* acquire, be promised, come by, gain, get, obtain, procure, win.

**sedate** *adj* calm, collected, composed, controlled, conventional, cool, decorous, deliberate, dignified, equable, even-tempered, formal, grave, imperturbable, level-headed, peaceful, *derog* prim, proper, quiet, sensible, serene, serious, slow, sober, solemn, staid, strait-laced, tranquil, unruffled. *Opp* LIVELY. • *vb* calm, put to sleep, tranquillize, treat with sedatives.

**sedative** *adj* anodyne, calming, lenitive, narcotic, relaxing, soothing, soporific, tranquillizing. • *n* anodyne, barbiturate, calmative, depressant, narcotic, opiate, sleeping-pill, soporific, tranquillizer.

**sedentary** *adj* desk-bound, immobile, inactive, seated, sitting down. *Opp* ACTIVE.

**sediment** *n* deposit, dregs, grounds, lees, precipitate, remains, residue, *inf* sludge.

**sedition** *n* agitation, incitement, insurrection, mutiny, rabble-rousing, revolt, treachery, treason. ▷ REBELLION.

**seduce** *vb* 1 allure, beguile, charm, corrupt, deceive, decoy, deprave, ensnare, entice, inveigle, lead astray, lure, mislead, tempt. 2 debauch, deflower, dishonour, rape, ravish, *old use* ruin, violate.

**seduction** *n* 1 allurement, attraction, charm, temptation. 2 rape, ravishing. ▷ SEX.

**seductive** *adj* alluring, appealing, attractive, bewitching, captivating, charming, coquettish, enchanting, enticing, flirtatious, inviting, irresistible, persuasive, provocative, tantalizing, tempting, *inf* sexy. *Opp* REPULSIVE.

**see** *vb* **1** behold, catch sight of, descry, discern, discover, distinguish, espy, glimpse, identify, look at, make out, mark, note, notice, observe, perceive, recognize, regard, sight, spot, spy, view, watch, witness. **2** *see what someone means.* appreciate, apprehend, comprehend, fathom, follow, *inf* get the hang of, grasp, know, perceive, realize, take in, understand. **3** *see problems ahead.* anticipate, conceive, envisage, foresee, foretell, imagine, picture, visualize. **4** *see what can be done.* consider, decide, discover, investigate, mull over, reflect on, think about, weigh up. **5** *see a play.* attend, be a spectator at, watch. **6** *seeing him tonight.* court, go out with, *inf* have a date with, meet, socialize with, visit, woo. **7** *see you home.* accompany, conduct, escort. **8** *saw fighting in the war.* endure, experience, go through, suffer, survive, undergo. **9** *Guess who I saw today!* encounter, face, meet, run into, talk to, visit. **see to** ▷ ORGANIZE.

**seed** *n* **1** egg, embryo, germ, ovule, ovum, semen, spawn, sperm, spore. **2** *seed in fruit.* pip, pit, stone. • *vb* ▷ SOW.

**seek** *vb* aim at, apply for, ask for, aspire to, beg for, demand, desire, go after, hope for, hunt for, inquire after, look for, pursue, quest after, request, search for, solicit, strive after, try for, want, wish for.

**seem** *vb* appear, feel, give an impression of being, have an appearance of being, look, preten to be, sound.

**seep** *vb* dribble, drip, exude, flow leak, ooze, percolate, run, soak, trickle.

**seer** *n* clairvoyant, fortune-teller, oracle, prophet, prophetess, psychic, sibyl, soothsayer, vatici ator.

**seethe** *vb* be agitated, be angry, boil, bubble, erupt, foam, froth u rise, simmer, stew, surge.

**segment** *n* bit, compartment, department, division, element, fraction, fragment, part, piece, po tion, quarter, section, sector, slic subdivision, subsection, wedge.

**segregate** *vb* compartmentalize, cut off, exclude, isolate, keep apart, put apart, separate, seques ter, set apart, shut out.

**segregation** *n* **1** apartheid, discrimination. **2** isolation, quarantine, seclusion, separation.

**seize** *vb* **1** abduct, apprehend, arrest, capture, catch, clutch, *inf* collar, detain, grab, grasp, gri hold, *inf* nab, pluck, possess, snatch, take, take into custody, take prisoner. **2** *seize a country.* annex, invade. **3** *seize property.* appropriate, commandeer, confiscate, hijack, impound, steal, take away. *Opp* RELEASE. **seize up** ▷ STICK.

**seizure** *n* **1** abduction, annexation, appropriation, arrest, capture, confiscation, hijacking, inva sion, sequestration, theft, usurpation. **2** [*medical*] apoplexy, attack, convulsion, epileptic fit, f paroxysm, spasm, stroke.

**seldom** *adv* infrequently, occasio ally, rarely.

**select** *adj* best, choice, chosen, élite, excellent, exceptional, exclusive, favoured, finest, first-class, first-rate, *inf* hand-picked, preferred, prime, privileged, rare, selected, special, top-quality. *Opp* ORDINARY. • *vb* appoint, cast (*actor for role*), choose, decide on, elect, nominate, opt for, pick, prefer, settle on, single out, vote for.

**selection** *n* **1** choice, option, pick, preference. **2** *a selection of goods.* assortment, range, variety. **3** *selection from the classics.* excerpts, extracts, passages, quotations.

**selective** *adj* careful, *inf* choosy, discerning, discriminating, particular, specialized. *Opp* COMPREHENSIVE, IMPERCEPTIVE.

**self-confident** *adj* assertive, assured, collected, cool, fearless, independent, outgoing, poised, positive, self-assured, self-possessed, self-reliant, sure of yourself. ▷ BOLD. *Opp* SELF-CONSCIOUS.

**self-conscious** *adj* awkward, bashful, blushing, coy, diffident, embarrassed, ill at ease, insecure, nervous, reserved, self-effacing, sheepish, shy, uncomfortable, unnatural. ▷ TIMID. *Opp* SELF-CONFIDENT.

**self-contained** *adj* **1** complete, independent, separate. **2** aloof, cold, reserved, self-reliant, uncommunicative, undemonstrative, unemotional.

**self-control** *n* calmness, composure, coolness, patience, resolve, restraint, self-command, self-denial, self-discipline, self-possession, self-restraint, willpower.

**self-denial** *n* abstemiousness, fasting, moderation, self-abnegation, self-sacrifice, temperance, unselfishness. *Opp* SELF-INDULGENCE.

**self-employed** *adj* freelance, independent.

**self-esteem** *n* **1** ▷ SELF-RESPECT. **2** arrogance, *inf* big-headedness, conceit, egotism, overconfidence, self-admiration, self-importance, self-love, smugness, vanity.

**self-explanatory** *adj* apparent, axiomatic, blatant, clear, conspicuous, eye-catching, flagrant, glaring, inescapable, manifest, obvious, patent, plain, recognizable, self-evident, understandable, unmistakable, visible.

**self-governing** *adj* autonomous, free, independent, sovereign.

**self-important** *adj* arrogant, bombastic, conceited, grandiloquent, haughty, magisterial, ostentatious, pompous, pontifical, pretentious, self-centred, sententious, smug, *inf* snooty, *inf* stuck-up, supercilious, vainglorious.

**self-indulgence** *n* extravagance, gluttony, greed, hedonism, pleasure, profligacy, self-gratification. ▷ SELFISHNESS. *Opp* SELF-DENIAL.

**self-indulgent** *adj* dissipated, epicurean, extravagant, gluttonous, gourmandizing, greedy, hedonistic, immoderate, intemperate, pleasure-loving, profligate, sybaritic. ▷ SELFISH. *Opp* ABSTEMIOUS.

**selfish** *adj* acquisitive, avaricious, covetous, demanding, egocentric, egotistic, grasping, greedy, inconsiderate, mean, mercenary, miserly, self-absorbed, self-centred, self-indulgent, self-interested, self-seeking, self-serving, *inf* stingy, thoughtless, uncaring, ungenerous, unhelpful, unsympathetic, worldly. *Opp* UNSELFISH.

**selfishness** *n* acquisitiveness, avarice, covetousness, egotism, greed, meanness, miserliness, niggardliness, possessiveness, self-indulgence, self-interest, self-love, self-regard, *inf* stinginess, thoughtlessness.

**self-reliant** *adj* autonomous, independent, self-contained, self-sufficient, self-supporting.

**self-respect** *n Fr* amour propre, dignity, honour, integrity, morale, pride, self-confidence, self-esteem.

**self-righteous** *adj* complacent, *inf* goody-goody, *inf* holier-than-thou, mealy-mouthed, pharisaical, pietistic, pious, pompous, priggish, proud, sanctimonious, self-important, self-satisfied, sleek, smug, superior, vain.

**self-sufficient** *adj* autonomous, independent, self-reliant, self-supporting.

**self-willed** *adj* determined, dogged, forceful, headstrong, inflexible, intractable, intransigent, *inf* mulish, obstinate, *inf* pig-headed, single-minded, *inf* stiff-necked, stubborn, uncontrollable, uncooperative, wilful.

**sell** *vb* **1** auction, barter, deal in, exchange, give in part-exchange, handle, hawk, *inf* keep, *inf* knock down, offer for sale, peddle, *inf* put under the hammer, retail, sell off, stock, tout, trade, *inf* trade in (*traded in my car*), traffic in, vend. **2** *sell hard*. advertise, market, merchandise, package, promote, *inf* push.

**seller** *n* dealer, merchant, stockist, supplier, trader, vendor. □ *agent, barrow-boy, broker,* old use *colporteur, costermonger,* old use *hawker, market-trader, pedlar,* inf *rep, representative, retailer, salesman, salesperson, saleswoman, shop assistant, shopkeeper, storekeeper, street trader, tradesman, traveller, wholesaler*. ▷ SHOP.

**seminal** *adj* basic, constructive, creative, fertile, formative, imaginative, important, influential, innovative, new, original, primary, productive.

**send** *vb* **1** address, consign, convey, deliver, direct, dispatch, fax, forward, mail, post, remit, ship, transmit. **2** *send a rocket to the moon*. fire, launch, project, propel, release, shoot. **send away** ▷ DISMISS. **send down** ▷ IMPRISON. **send for** ▷ SUMMON. **send-off** *n* ▷ GOODBYE. **send out** ▷ EMIT. **send round** ▷ CIRCULATE. **send up** ▷ PARODY.

**senile** *adj* declining, doddery, *inf* in your dotage, old, *derog* past it.

**senior** *adj* chief, elder, higher, high-ranking, major, older, principal, revered, superior, well-established. *Opp* JUNIOR.

**sensation** *n* **1** awareness, feeling, perception, sense. **2** *affair caused a sensation*. commotion, excitement, furore, outrage, scandal, stir, thrill.

**sensational** *adj* **1** blood-curdling, hair-raising, lurid, melodramatic, overwritten, scandal-mongering, shocking, startling, stimulating, violent. **2** [*inf*] *a sensational result*. amazing, astonishing, astounding, breathtaking, electrifying, exciting, extraordinary, *inf* fabulous, *inf* fantastic, *inf* great, incredible, marvellous, remarkable, spectacular, spine-tingling, stirring, superb, surprising, thrilling, unbelievable, unexpected, wonderful.

**sense** *n* **1** awareness, consciousness, faculty, feeling, sensation. □ *hearing, sight, smell, taste,* *touch.* **2** brains, cleverness, gumption, intellect, intelligence, intuition, judgement, logic, *inf* nous, perception, reason, reasoning, understanding, wisdom, wit. **3** *the sense of a message.* coherence, connotations, denotation, *inf* drift, gist, import, intelligibility, interpretation, meaning, message, point, purport, significance, signification, substance. • *vb* be aware (of), detect, discern, divine, feel, guess, *inf* have a hunch, intuit, notice, perceive, *inf* pick up vibes, realize, respond to, suspect, understand. ▷ FEEL, HEAR, SEE, SMELL, TASTE. **make sense of** ▷ UNDERSTAND.

**senseless** *adj* **1** anaesthetized, asleep, comatose, insensate, insensible, knocked out, numb, *inf* out like a light, stunned, unconscious. **2** absurd, crazy, fatuous, meaningless, pointless, purposeless, silly. ▷ STUPID.

**sensible** *adj* **1** calm, commonsense, commonsensical, cool, discreet, discriminating, intelligent, judicious, level-headed, logical, prudent, rational, realistic, reasonable, reasoned, sage, sane, serious-minded, sound, straightforward, thoughtful, wise. *Opp* STUPID. **2** *sensible phenomena.* corporeal, existent, material, palpable, perceptible, physical, real, tangible, visible. **3** *sensible clothes.* comfortable, functional, *inf* no-nonsense, practical, useful. *Opp* FASHIONABLE, IMPRACTICAL. **sensible of** acquainted with, alert to, alive to, appreciative of, aware of, cognizant of, in touch with, mindful of, responsive to, *inf* wise to.

**sensitive** *adj* **1** considerate, perceptive, reactive, receptive, responsive, susceptible, sympathetic, tactful, thoughtful, understanding. **2** *a sensitive temperament.* emotional, hypersensitive, impressionable, temperamental, thin-skinned, touchy, volatile, vulnerable. **3** *sensitive skin.* delicate, fine, fragile, painful, soft, sore, tender. **4** *a sensitive topic.* confidential, controversial, delicate, *inf* tricky, secret. *Opp* INSENSITIVE. **sensitive to** affected by, attuned to, aware of, considerate of, perceptive about, receptive to, responsive to, understanding about.

**sensual** *adj* animal, bodily, carnal, fleshly, physical, pleasure-loving, self-indulgent, voluptuous, worldly. ▷ SEXY. *Opp* ASCETIC.

**sensuous** *adj* beautiful, emotional, gratifying, lush, luxurious, rich, richly embellished.

**sentence** *n* **1** exclamation, question, statement, thought, utterance. **2** decision, judgement, pronouncement, punishment, ruling. • *vb* condemn, pass judgement on, pronounce sentence on.

**sentiment** *n* **1** attitude, belief, idea, judgement, opinion, outlook, thought, view. **2** *sentiment of a poem.* emotion, feeling, sensibility.

**sentimental** *adj* **1** compassionate, emotional, nostalgic, romantic, soft-hearted, sympathetic, tearful, tender, warm-hearted, *inf* weepy. **2** [*derog*] gushing, *inf* gushy, indulgent, insincere, maudlin, mawkish, *inf* mushy, overdone, over-emotional, *inf* sloppy, *inf* soppy, *inf* sugary, tearjerking, *inf* treacly, unrealistic, *sl* yucky. *Opp* CYNICAL.

**sentimentality** *n* bathos, emotionalism, insincerity, *inf* kitsch, mawkishness, nostalgia, *inf* slush.

**sentry** *n* guard, lookout, patrol, picket, sentinel, watch, watchman.

**separable** *adj* detachable, distinguishable, fissile, removable.

**separate** *adj* apart, autonomous, cloistered, cut off, detached, different, discrete, disjoined, distinct, divided, divorced, fenced off, free-standing, independent, individual, isolated, particular, peculiar, secluded, segregated, separated, shut off, solitary, unattached, unconnected, unique, unrelated, unshared, withdrawn. • *vb* 1 break up, cut off, detach, disconnect, disengage, disentangle, disjoin, dismember, dissociate, divide, fence off, fragment, hive off, isolate, keep apart, part, pull apart, segregate, sever, split, sunder, take apart, uncouple, unfasten, unhook, unravel, unyoke. 2 *Thepaths separate here.* bifurcate, branch, diverge, fork. 3 *separated the grain from the chaff.* abstract, distinguish, filter out, remove, set apart, sift out, winnow. 4 *He separated from his partner.* become estranged, disband, divorce, part company, *inf* split up. *Opp* COMBINE, UNITE.

**separation** *n* 1 amputation, cutting off, detachment, disconnection, dismemberment, dissociation, division, fission, fragmentation, parting, rift, severance, splitting. *Opp* CONNECTION. 2 *separation of partners.* break, *inf* break-up, divorce, estrangement, rift, split. *Opp* UNION.

**septic** *adj* diseased, festering, infected, inflamed, poisoned, purulent, putrefying, putrid, suppurating.

**sequel** *n* consequence, continuation, development, *inf* follow-up, issue, outcome, result, upshot.

**sequence** *n* 1 arrangement, chain, concatenation, course, cycle, line, order, procession, programme, progression, range, row, run, sequence, series, set, string, succession, train. 2 *a sequence from a film. inf* clip, episode, excerpt, extract, scene, section.

**serene** *adj* 1 calm, idyllic, peaceful, placid, pleasing, quiet, restful, still, tranquil, unclouded, undisturbed, unperturbed, unruffled, untroubled. 2 *serene temperament* collected, composed, contented, cool, easy-going, equable, even-tempered, imperturbable, pacific, peaceable, poised, self-possessed, *inf* unflappable. *Opp* BOISTEROUS, EXCITABLE.

**series** *n* 1 arrangement, chain, concatenation, course, cycle, line, order, procession, programme, progression, range, row, run, sequence, set, string, succession, train. 2 *TV series.* mini-series, serial, *inf* soap, soap-opera.

**serious** *adj* 1 dignified, grave, grim, humourless, long-faced, pensive, poker-faced, sedate, sober, solemn, sombre, staid, stern, straight-faced, thoughtful, unsmiling. *Opp* CHEERFUL. 2 *serious discussion.* deep, earnest heavy, honest, important, intellectual, momentous, profound, significant, sincere, weighty. 3 *serious illness.* acute, appalling, awful, calamitous, critical, dangerous, dreadful, frightful, ghastly, grievous, hideous, horrible, *inf* life-and-death, nasty, severe, shocking, terrible, unfortunate, unpleasant, urgent, violent. *Opp* TRIVIAL. 4 *serious worker.* careful, committed, conscientious, diligent, hard-working.

**sermon** *n* address, discourse, homily, lecture, lesson, talk.

**serpentine** *adj* labyrinthine, meandering, roundabout, sinuous snaking, tortuous, twisting, vermicular, winding. *Opp* STRAIGHT.

**serrated** *adj* cogged, crenellated, denticulate, indented, jagged, notched, saw-like, toothed, zigzag. *Opp* STRAIGHT.

**servant** *n* assistant, attendant, *derog* dogsbody, *inf* domestic, *derog* drudge, helper, *derog* hireling, *derog* menial, *old use* servitor, *inf* skivvy, slave, *old use* vassal. □ *au pair, barmaid, barman, batman,* inf *boots, butler, chamber-maid,* inf *char, charwoman, chauffeur, chef, cleaner,* old use *coachman, commissionaire, cook,* inf *daily, errand boy, factotum,* derog *flunkey, footman, governess, groom, home help, houseboy, housemaid, housekeeper, kitchenmaid,* derog *lackey, lady-in-waiting, maid, maidservant, major-domo, manservant, nanny, page, parlour-maid,* old use *postilion,* old use *retainer,* plur *retinue, scout, scullery maid,* old use *scullion,* old use *seneschal, slave, steward, stewardess, valet, waiter, waitress.*

**serve** *vb* **1** aid, accommodate, assist, attend, *inf* be at someone's beck and call, further, help, look after, minister to, wait upon, work for. **2** *serve in the forces.* be employed, do your duty, enlist, fight, sign on. **3** *serve goods.* deal out, distribute, dole out, give out, make available, provide, sell, supply. **4** *serve at table.* carve, *inf* dish up, officiate, wait. **5** *serve a sentence.* complete, endure, go through, pass, spend, survive.

**service** *n* **1** aid, assistance, benefit, favour, help, kindness, office. **2** *service of the community.* attendance (on), employment (by), ministering (to), work (for). **3** *a bus service.* business, organization, provision, system, timetable. **4** *give the car a service.* check-over, maintenance, overhaul, repair, servicing. **5** *church service.* ceremony, liturgy, meeting, rite, ritual, worship. □ *baptism, christening, communion, compline, evensong, funeral, marriage, Mass, matins, Requiem Mass, vespers.* • *vb service a vehicle.* check, maintain, mend, overhaul, repair, tune.

**serviceable** *adj* dependable, durable, functional, hard-wearing, lasting, practical, strong, tough, usable.

**servile** *adj* abject, acquiescent, base, *inf* boot-licking, craven, cringing, deferential, fawning, flattering, grovelling, humble, ingratiating, low, menial, obsequious, slavish, submissive, subservient, sycophantic, *inf* time-serving, toadying, unctuous. *Opp* BOSSY. **be servile** ▷ GROVEL.

**serving** *n* helping, plateful, portion, ration.

**session** *n* **1** assembly, conference, discussion, hearing, meeting, sitting. **2** *a session at the baths.* period, term, time.

**set** *adj* **1** *set price.* advertised, agreed, arranged, defined, definite, fixed, prearranged, predetermined, prepared, scheduled, standard. **2** *set in your ways.* established, invariable, predictable, regular, stable, unchanging, unvarying. • *n* **1** batch, bunch, category, class, clique, collection, combination, kind, series, sort. ▷ GROUP. **2** *a TV set.* apparatus, receiver. **3** *set for a play.* scene, scenery, setting, stage. • *vb* **1** arrange, assign, deploy, deposit, dispose, lay, leave, locate, lodge, park, place, plant, *inf* plonk, put, position, rest, set down, set out, settle, situate, stand, station. **2** *set a clock.* adjust, correct, put right, rectify, regulate. **3** *set a post in concrete.* embed, fasten, fix. **4** *set like concrete.* become firm, congeal, *inf* gel, harden, *inf* jell, stiffen, take shape. **5** *set a problem.* ask, express, formulate, frame, phrase, pose, present, put forward, suggest, write. **6** *set a target.* allocate, allot, appoint, decide, designate, determine, establish, identify, name, ordain, prescribe, settle. **set about** ▷ ATTACK, BEGIN. **set free** ▷ LIBERATE. **set off** ▷ DEPART, EXPLODE. **set on** ▷ ATTACK. **set on fire** ▷ IGNITE. **set out** ▷ DEPART. **set up** ▷ ESTABLISH.

**set-back** *n inf* blow, check, complication, defeat, delay, difficulty, disappointment, hindrance, *inf* hitch, hold-up, impediment, misfortune, obstacle, problem, relapse, reverse, snag, upset.

**settee** *n* chaise longue, couch, sofa.

**setting** *n* **1** background, context, environment, environs, frame, habitat, locale, location, place, position, site, surroundings. **2** *setting for a play.* backcloth, backdrop, scene, scenery, set.

**settle** *vb* **1** arrange, conclude, deal with, decide, organize, put in order, straighten out. **2** alight, come to rest, land, light, *inf* make yourself comfortable, *inf* park yourself, pause, rest, roost, sit down. **3** *settle things in place.* assign, deploy, deposit, dispose, lay, locate, lodge, park, place, plant, position, put, rest, set, set down, situate, stand, station. **4** *the dust settled.* calm down, clear, compact, go down, sink, subside. **5** *settle what to do.* agree, choose, decide, establish, fix. **6** *settle differences.* end, negotiate, put an end to, reconcile, resolve, sort out, square. **7** *settle debts.* clear, discharge, pay, pay off. **8** *settle new territory.* become established in, colonize, immigrate, make your home in, occupy, people, set up home in, stay in.

**settlement** *n* **1** camp, colony, community, encampment, kibbutz, outpost, post, town, village. **2** agreement, arbitration, arrangement, contract, payment.

**settler** *n* colonist, frontiersman, immigrant, newcomer, pioneer, squatter.

**sever** *vb* **1** amputate, break, cut off, detach, disconnect, disjoin, part, remove, separate, split, terminate. ▷ CUT. **2** *sever a relationship.* abandon, break off, discontinue, end, put an end to, suspend, terminate.

**several** *adj* assorted, certain, different, divers, a few, a handful of, many, miscellaneous, a number of, some, sundry, a variety of, various.

**severe** *adj* **1** aloof, brutal, cold, cold-hearted, cruel, disapproving, dour, exacting, forbidding, glowering, grave, grim, hard, harsh, inexorable, merciless, obdurate, pitiless, relentless, rigorous, stern, stony, strict, unbending, uncompromising, unkind, unsmiling, unsympathetic, unyielding. **2** *severe illness.* acute, critical, dangerous, drastic, fatal, great, intense, keen, life-threatening, mortal, nasty, serious, sharp, terminal, troublesome. **3** *severe penalties.* draconian, extreme, maximum, oppressive, punitive, stringent. **4** *severe weather.* adverse, bad, inclement, violent, *inf* wicked. ▷ COLD, STORMY. **5** *a severe challenge.* arduous, demanding, difficult, onerous, punishing, taxing, tough. **6** *severe style*. austere, bare, chaste, plain, simple, spartan, stark, unadorned. *Opp* FRIENDLY, MILD, ORNATE.

**sew** *vb* baste, darn, hem, mend, repair, stitch, tack.

**sewage** *n* effluent, waste.

**sewer** *n* drain, drainage, sanitation, septic tank, soak-away.

**sewing** *n* dressmaking, embroidery, mending, needlepoint, needlework, tapestry.

**sex** *n* **1** gender, sexuality. **2** carnal knowledge, coition, coitus, congress, consummation of marriage, copulation, coupling, fornication, *inf* going to bed, incest, intercourse, intimacy, love-making, masturbation, mating, orgasm, perversion, rape, seduction, sexual intercourse, sexual relations, union. **have sex (with)** be intimate (with), consummate marriage, copulate (with), couple (with), fornicate (with), have sexual intercourse (with), make love (to), mate (with), rape, ravish, *sl* screw, seduce, unite (with).

**sexism** *n inf* chauvinism, discrimination, prejudice.

**sexual** *adj* **1** genital, procreative, progenitive, reproductive. **2** ▷ SEXY.

**sexuality** *n* gender. □ *bisexuality, hermaphroditism, heterosexuality, homosexuality.*

**sexy** *adj* **1** amorous, carnal, concupiscent, erotic, lascivious, lecherous, *derog* lewd, libidinous, *derog* lubricious, lustful, passionate, provocative, *derog* prurient, *inf* randy, seductive, sensual, sexual, *inf* sultry, venereal, voluptuous. **2** attractive, *inf* beddable, desirable, *sl* dishy, flirtatious. **3** *sexy books*. aphrodisiac, arousing, pornographic, *sl* raunchy, salacious, *inf* steamy, suggestive, titillating, *inf* torrid. ▷ OBSCENE.

**shabby** *adj* **1** bedraggled, dilapidated, dingy, dirty, dowdy, drab, faded, frayed, *inf* grubby, mangy, *inf* moth-eaten, ragged, run-down, *inf* scruffy, seedy, tattered, *inf* tatty, threadbare, unattractive, worn, worn-out. *Opp* SMART. **2** *shabby behaviour*. base, contemptible, despicable, disagreeable, discreditable, dishonest, dishonourable, disreputable, ignoble, *inf* low-down, mean, nasty, shameful, shoddy, unfair, unfriendly, ungenerous, unkind, unworthy. *Opp* HONOURABLE.

**shack** *n* cabin, hovel, hut, lean-to, shanty, shed.

**shade** *n* **1** ▷ SHADOW. **2** awning, blind, canopy, covering, curtain, parasol, screen, shelter, shield, umbrella, Venetian blind. **3** *a shade of blue*. colour, hue, intensity, tinge, tint, tone. **4** *shades of meaning*. degree, difference, nicety, nuance, variation. • *vb* **1** camouflage, conceal, cover, hide, mask, obscure, protect, screen, shield, shroud, veil. **2** *shade with pencil*. black out, block in, crosshatch, darken, fill in, make dark.

**shadow** *n* **1** darkness, dimness, dusk, gloom, obscurity, penumbra, semi-darkness, shade, umbra. **2** *The sun casts shadows*. outline, shape, silhouette. **3** *a shadow of doubt*. ▷ HINT. • *vb* dog, follow, hunt, *inf* keep tabs on, keep watch on, pursue, stalk, *inf* tag onto, tail, track, trail, watch.

**shadowy** *adj* **1** dark, dim, faint, hazy, ill-defined, indefinite, indistinct, nebulous, obscure, unclear, unrecognizable, vague. ▷ GHOSTLY. **2** ▷ SHADY.

**shady** *adj* **1** *poet* bosky, cool, dark, dim, dusky, gloomy, leafy,

shaded, shadowy, sheltered, sunless. *Opp* SUNNY. **2** *a shady character*. devious, dishonest, disreputable, dubious, *inf* fishy, questionable, shifty, suspicious, unreliable, untrustworthy.
*Opp* HONEST.

**shaft** *n* **1** arrow, column, handle, helve, pillar, pole, post, rod, shank, stanchion, stem, stick, upright. **2** duct, mine, pit, tunnel, well, working. **3** *shaft of light*. beam, gleam, laser, pencil, ray, streak.

**shaggy** *adj* bushy, dishevelled, fibrous, fleecy, hairy, hirsute, matted, rough, tousled, unkempt, unshorn, untidy, woolly.
*Opp* SMOOTH.

**shake** *vb* **1** convulse, heave, jump, quake, quiver, rattle, rock, shiver, shudder, sway, throb, totter, tremble, vibrate, waver, wobble. **2** *shake your umbrella*. agitate, brandish, flourish, gyrate, jar, jerk, *inf* jiggle, *inf* joggle, jolt, oscillate, sway, swing, twirl, twitch, vibrate, wag, *inf* waggle, wave, *inf* wiggle. **3** *The bad news shook us*. alarm, distress, disturb, frighten, perturb, *inf* rattle, shock, startle, *inf* throw, unnerve, unsettle, upset. ▷ SURPRISE.

**shaky** *adj* **1** decrepit, dilapidated, feeble, flimsy, frail, insecure, precarious, ramshackle, rickety, rocky, shaking, unreliable, unsound, unsteady, weak, wobbly. **2** *a shaky voice*. faltering, quavering, quivering, trembling, tremulous. **3** *a shaky start*. nervous, tentative, uncertain, underconfident, unimpressive, unpromising. *Opp* STEADY, STRONG.

**shallow** *adj* **1** *shallow water*. [There are no apt synonyms for this sense.] **2** *shallow argument*. empty, facile, foolish, frivolous, glib, insincere, puerile, silly, simple, *inf* skin-deep, slight, superficial, trivial, unconvincing, unscholarly, unthinking.
*Opp* DEEP.

**sham** *adj* artificial, bogus, counterfeit, ersatz, fake, false, fictitious, fraudulent, imitation, make-believe, mock, *inf* pretend, pretended, simulated, synthetic.
• *n* counterfeit, fake, fiction, fraud, hoax, imitation, make-believe, pretence, *inf* put-up job, simulation.
• *vb* counterfeit, fake, feign, imitate, make believe, pretend, simulate.

**shambles** *plur n* **1** battlefield, scene of carnage, slaughterhouse. **2** [*inf*] chaos, confusion, devastation, disorder, mess, muddle, *inf* pigsty, *inf* tip.

**shame** *n* **1** chagrin, degradation, discredit, disgrace, dishonour, distress, embarrassment, guilt, humiliation, ignominy, infamy, loss of face, mortification, obloquy, opprobrium, remorse, stain, stigma, vilification. **2** *a shame to mistreat him so*. outrage, pity, scandal, wickedness. • *vb* abash, chagrin, chasten, discomfit, disconcert, discountenance, disgrace, embarrass, humble, humiliate, make someone ashamed, mortify, *inf* put someone in his/her place, *inf* show someone up.

**shamefaced** *adj* **1** abashed, ashamed, chagrined, *inf* hang-dog, humiliated, mortified, penitent, *inf* red-faced, remorseful, repentant, sorry. **2** bashful, coy, embarrassed, modest, self-conscious, sheepish, shy, timid.
*Opp* SHAMELESS.

**shameful** *adj* **1** *a shameful crime*. base, contemptible, deplorable, disgraceful, infamous, ignoble, low, mean, outrageous, reprehensible, scandalous, unworthy. ▷ WICKED. **2** *shameful to be found out.* compromising, degrading, demeaning, discreditable, dishonourable, embarrassing, humiliating, ignominious, *sl* infra dig, inglorious, lowering, mortifying, undignified. *Opp* HONOURABLE.

**shameless** *adj* **1** barefaced, bold, brazen, cheeky, cool, defiant, flagrant, hardened, impenitent, impudent, incorrigible, insolent, unabashed, unashamed, unrepentant. **2** *shameless nudity.* frank, honest, immodest, indecorous, improper, open, rude, shocking, unblushing, unconcealed, undisguised, unselfconscious. *Opp* SHAMEFACED.

**shape** *n* **1** body, build, figure, physique, profile, silhouette. **2** *geometrical shape.* configuration, figure, form, format, model, mould, outline, pattern. □ [two-dimensional] *circle, diamond, ellipse, heptagon, hexagon, lozenge, oblong, octagon, oval, parallelogram, pentagon, polygon, quadrant, quadrilateral, rectangle, rhomboid, rhombus, ring, semicircle, square, trapezium, trapezoid, triangle.* [three-dimensional] *cone, cube, cylinder, decahedron, hemisphere, hexahedron, octahedron, polyhedron, prism, pyramid, sphere.* • *vb* adapt, adjust, carve, cast, cut, fashion, form, frame, give shape to, model, mould, *inf* sculpt, sculpture, whittle.

**shapeless** *adj* **1** amorphous, formless, indeterminate, irregular, nebulous, undefined, unformed, unstructured, vague. **2** *a shapeless figure.* deformed, distorted, *inf* dumpy, flat, misshapen, twisted, unattractive, unshapely. *Opp* SHAPELY.

**shapely** *adj* attractive, comely, *inf* curvaceous, elegant, good-looking, graceful, neat, trim, *inf* voluptuous, well-proportioned. *Opp* SHAPELESS.

**share** *n* allocation, allotment, allowance, bit, cut, division, due, fraction, helping, part, percentage, piece, portion, proportion, quota, ration, serving, *sl* whack. • *vb* **1** allocate, allot, apportion, deal out, distribute, divide, dole out, *inf* go halves or shares (with), halve, partake of, portion out, ration out, share out, split. **2** *share work.* be involved, cooperate, join, participate, take part. **shared** ▷ JOINT.

**sharp** *adj* **1** acute, arrow-shaped, cutting, fine, jagged, keen, knife-edged, needle-sharp, pointed, razor-sharp, sharpened, spiky, tapering. **2** *sharp bend, drop.* abrupt, acute, angular, hairpin, marked, precipitous, sheer, steep, sudden, surprising, unexpected, vertical. **3** *sharp focus.* clear, defined, distinct, focused, well-defined. **4** *a sharp storm.* extreme, heavy, intense, serious, severe, sudden, violent. **5** *sharp frost.* biting, bitter, keen, nippy. ▷ COLD. **6** *sharp pain.* acute, excruciating, painful, stabbing, stinging. **7** *sharp rejoinder.* acerbic, acid, acidulous, barbed, biting, caustic, critical, cutting, hurtful, incisive, malicious, mocking, mordant, sarcastic, sardonic, scathing, spiteful, tart, trenchant, unkind, venomous, vitriolic. **8** *sharp mind.* acute, agile, alert, artful, astute, bright, clever, crafty, *inf* cute, discerning, incisive, intelligent, observant, penetrating, perceptive, probing, quick-witted, searching, shrewd, *inf* smart. **9** *sharp eyes.* attentive, eagle-eyed, observant, *inf* peeled (*keep your eyes peeled*), quick, watchful, wide-open. **10** *sharp taste, smell.* acid, acrid, bitter, caustic, hot, piquant, pungent, sour, spicy, tangy, tart. **11** *sharp sound.* clear, detached, ear-splitting, high, high-pitched, penetrating, piercing, shrieking, shrill, staccato, strident. *Opp* BLUNT, DULL, SLIGHT.

**sharpen** *vb* file, grind, hone, make sharp, strop, whet. *Opp* BLUNT.

**sharpener** *n* file, grindstone, hone, pencil-sharpener, strop, whetstone.

**shatter** *vb* blast, break, break up, burst, crack, crush, dash to pieces, demolish, destroy, disintegrate, explode, pulverize, shiver, smash, *inf* smash to smithereens, splinter, split, wreck. **shattered** ▷ SURPRISED, WEARY.

**sheaf** *n* bundle, bunch, file, ream.

**shear** *vb* clip, strip, trim. ▷ CUT.

**sheath** *n* casing, covering, scabbard, sleeve.

**sheathe** *vb* cocoon, cover, encase, enclose, put away, put in a sheath, wrap.

**shed** *n* hut, hutch, lean-to, outhouse, penthouse, potting-shed, shack, shelter, storehouse. • *vb* abandon, cast off, discard, drop, let fall, moult, pour off, scatter, shower, spill, spread, throw off. **shed light** ▷ SHINE.

**sheen** *n* brightness, burnish, glaze, gleam, glint, gloss, lustre, patina, polish, radiance, reflection, shimmer, shine.

**sheep** *n* ewe, lamb, mutton, ram, wether.

**sheepish** *adj* abashed, ashamed, bashful, coy, docile, embarrassed, guilty, meek, mortified, reticent, self-conscious, self-effacing, shamefaced, shy, timid. *Opp* SHAMELESS.

**sheer** *adj* **1** absolute, arrant, complete, downright, out-and-out, plain, pure, simple, thoroughgoing, total, unadulterated, unalloyed, unmitigated, unmixed, unqualified, utter. **2** *a sheer cliff.* abrupt, perpendicular, precipitous, steep, vertical. **3** *sheer silk.* diaphanous, filmy, fine, flimsy, gauzy, gossamer, *inf* see-through, thin, translucent, transparent.

**sheet** *n* **1** bedsheet, duvet cover. **2** [*paper*] folio, leaf, page. **3** [*glass, etc*] pane, panel, plate. **4** [*ice, etc*] blanket, coating, covering, film, lamina, layer, membrane, skin, veneer. **5** [*water*] area, expanse, surface.

**shelf** *n* ledge, shelving.

**shell** *n* **1** carapace (*of tortoise*), case, casing, covering, crust, exterior, façade, hull, husk, outside, pod. **2** *fired shells at them.* cartridge, projectile. • *vb* attack with gunfire, barrage, bomb, bombard, fire at, shoot at, strafe.

**shellfish** *n* bivalve, crustacean, mollusc. □ *barnacle, clam, cockle, conch, crab, crayfish, cuttlefish, limpet, lobster, mussel, oyster, prawn, scallop, shrimp, whelk, winkle.*

**shelter** *n* **1** asylum, cover, haven, lee, protection, refuge, safety, sanctuary, security. **2** barrier, concealment, cover, fence, hut, roof, screen, shield. **3** *seek shelter for the night.* accommodation, lodging, home, housing, resting-place. ▷ HOUSE. **4** *air-raid shelter.* bunker. • *vb* **1** defend, enclose, guard, keep safe, protect, safeguard, screen, secure, shade, shield. **2** *shelter a runaway.* accommodate, give shelter to, harbour, hide, *inf* put up. **sheltered** ▷ QUIET.

**shelve** *vb* 1 defer, hold in abeyance, lay aside, postpone, put off, put on ice. 2 ▷ SLOPE.

**shield** *n* 1 barrier, bulwark, defence, guard, protection, safeguard, screen, shelter. 2 *a warrior's shield.* buckler, *heraldry* escutcheon. • *vb* cover, defend, guard, keep safe, protect, safeguard, screen, shade, shelter.

**shift** *n* 1 adjustment, alteration, change, move, switch, transfer, transposition. 2 *night shift.* crew, gang, group, period, *inf* stint, team, workforce. • *vb* adjust, alter, budge, change, displace, reposition, switch, transfer, transpose. ▷ MOVE. **shift for yourself** ▷ MANAGE.

**shiftless** *adj* idle, indolent, ineffective, inefficient, inept, irresponsible, lazy, unambitious, unenterprising. *Opp* RESOURCEFUL.

**shifty** *adj* artful, canny, crafty, cunning, deceitful, designing, devious, dishonest, evasive, *inf* foxy, furtive, scheming, secretive, *inf* shady, *inf* slippery, sly, treacherous, tricky, untrustworthy, wily. *Opp* STRAIGHTFORWARD.

**shimmer** *vb* flicker, glimmer, glisten, ripple. ▷ SHINE.

**shine** *n* brightness, burnish, coruscation, glaze, gleam, glint, gloss, glow, luminosity, lustre, patina, phosphorescence, polish, radiance, reflection, sheen, shimmer, sparkle, varnish. • *vb* 1 beam, be luminous, blaze, coruscate, dazzle, emit light, flare, flash, glare, gleam, glint, glisten, glitter, glow, phosphoresce, radiate, reflect, scintillate, shed light, shimmer, sparkle, twinkle. 2 *used to shine at maths.* be brilliant, be clever, do well, excel, *inf* make your mark, stand out. 3 *shine your shoes.* brush, buff up, burnish, clean, polish, rub up. **shining** ▷ BRIGHT, CONSPICUOUS.

**shingle** *n* 1 gravel, pebbles, stones. 2 *roofing shingle.* tile.

**shiny** *adj* bright, brilliant, burnished, gleaming, glistening, glossy, glowing, luminous, lustrous, phosphorescent, polished, reflective, rubbed, shimmering, shining, sleek, smooth. *Opp* DULL.

**ship** *n* boat. ▷ VESSEL. • *vb* carry, ls *inf* cart, convey, deliver, ferry, freight, move, send, transport.

**shirk** *vb* avoid, dodge, duck, evade, get out of, neglect, shun. **shirk work** be lazy, malinger, *inf* skive, slack.

**shiver** *n* flutter, frisson, quiver, rattle, shake, shudder, thrill, tremor, vibration. • *vb* chatter, flap, flutter, quake, quaver, quiver, rattle, shake, shudder, tremble, twitch, vibrate.

**shock** *n* 1 blow, collision, concussion, impact, jolt, thud. 2 *came as a shock. inf* bombshell, surprise, *inf* thunderbolt. 3 *state of shock.* dismay, distress, fright, trauma, upset. • *vb* 1 alarm, amaze, astonish, astound, confound, daze, dismay, distress, dumbfound, frighten, *inf* give someone a turn, jar, jolt, numb, paralyse, petrify, rock, scare, shake, stagger, startle, stun, stupefy, surprise, *inf* throw, traumatize, unnerve. 2 *Sadism shocks us.* appal, disgust, horrify, nauseate, offend, outrage, repel, revolt, scandalize, sicken.

**shoddy** *adj* 1 cheap, flimsy, *inf* gimcrack, inferior, jerry-built, meretricious, nasty, poor quality, *inf* rubbishy, second-rate, shabby, *sl* tacky, *inf* tatty, tawdry, *inf* trashy. 2 *shoddy work.* careless, messy, negligent, slipshod, *inf* sloppy, slovenly, untidy. *Opp* SUPERIOR.

**shoe** *n* plur footwear. □ *boot, bootee, brogue, clog, espadrille,* inf *flip-flop, galosh, gum-boot,* inf *lace-up, moccasin, plimsoll, pump, sabot, sandal,* inf *slip-on, slipper, trainer, wader, wellington.*

**shoemaker** *n* bootmaker, cobbler.

**shoot** *n* branch, bud, new growth, offshoot, sprout, sucker, twig. • *vb* **1** *shoot a gun.* aim, discharge, fire. **2** *shoot the enemy.* aim at, bombard, fire at, gun down, hit, hunt, kill, *inf* let fly at, open fire on, *inf* pick off, shell, snipe at, strafe, *inf* take pot-shots at. **3** *shoot from your chair.* bolt, dart, dash, fly, hurtle, leap, move quickly, race, run, rush, speed, spring, streak. **4** *plants shoot in the spring.* bud, burgeon, develop, flourish, grow, put out shoots, spring up, sprout.

**shop** *n* boutique, cash-and-carry, department store, *old use* emporium, establishment, market, outlet, retailer, seller, store, wholesaler. □ *baker, betting shop, bookshop, butcher, chandler, chemist, confectioner, couturier, creamery, dairy, delicatessen, draper, fishmonger, florist, garden-centre, greengrocer, grocer, haberdasher, herbalist, hypermarket, ironmonger, jeweller, launderette, minimarket, newsagent, off-licence, outfitter, pawnbroker, pharmacy, post office, poulterer, stationer, supermarket, tailor, take-away, tobacconist, toyshop, video shop, vintner, watchmaker.*

**shopkeeper** *n* dealer, merchant, retailer, salesgirl, salesman, saleswoman, stockist, storekeeper, supplier, trader, tradesman.

**shopper** *n* buyer, customer, patron.

**shopping** *n* **1** buying, *inf* spending-spree. **2** goods, purchases.

**shopping-centre** *n* arcade, complex, hypermarket, mall, precinct.

**shore** *n* bank, beach, coast, edge, foreshore, sands, seashore, seaside, shingle, strand. • *vb* **shore up** ▷ SUPPORT.

**short** *adj* **1** diminutive, *inf* dumpy, dwarfish, little, midget, *fem* petite, *derog* pint-sized, slight, small, squat, *inf* stubby, *inf* stumpy, stunted, tiny, *inf* wee, undergrown. **2** *a short visit.* brief, cursory, curtailed, ephemeral, fleeting, momentary, passing, quick, short-lived, temporary, transient, transitory. **3** *a short book.* abbreviated, abridged, compact, concise, cut, pocket, shortened, succinct. **4** *in short supply.* deficient, inadequate, insufficient, lacking, limited, low, meagre, scanty, scarce, sparse, wanting. **5** *a short manner.* abrupt, bad-tempered, blunt, brusque, cross, curt, gruff, grumpy, impolite, irritable, laconic, sharp, snappy, taciturn, terse, testy, uncivil, unfriendly, unkind, unsympathetic. *Opp* EXPANSIVE, LONG, PLENTIFUL, TALL. **cut short** ▷ SHORTEN.

**shortage** *n* absence, dearth, deficiency, deficit, insufficiency, lack, paucity, poverty, scarcity, shortfall, want. *Opp* PLENTY.

**shortcoming** *n* bad habit, defect, deficiency, drawback, failing, failure, fault, flaw, foible, imperfection, limitation, vice, weakness, weak point.

**shorten** *vb* abbreviate, abridge, compress, condense, curtail, cut, cut down, cut short, diminish, dock, lop, précis, prune, reduce, shrink, summarize, take up (*clothes*), telescope, trim, truncate. *Opp* LENGTHEN.

**shortly** *adv old use* anon, before long, by and by, directly, presently, soon.

**short-sighted** *adj* **1** myopic, near-sighted. **2** unadventurous, unimaginative, without vision.

**short-tempered** *adj* abrupt, acerbic, brusque, crabby, cross, crusty, curt, gruff, irascible, irritable, peevish, peremptory, shrewish, snappy, testy, touchy, waspish. *Opp* BAD-TEMPERED.

**shot** *n* **1** ball, bullet, discharge, missile, pellet, projectile, round, *inf* slug. **2** *heard a shot.* bang, blast, crack, explosion, report. **3** *a first-class shot.* marksman, markswoman, sharpshooter. **4** *give it a shot.* attempt, chance, *inf* crack, effort, endeavour, *inf* go, hit, kick, *inf* stab, stroke, try. **5** *photographic shot.* angle, photograph, picture, scene, sequence, snap, snapshot.

**shout** *vb* bawl, bellow, *inf* belt, call, cheer, clamour, cry out, exclaim, howl, rant, roar, scream, screech, shriek, talk loudly, vociferate, whoop, yell, yelp, yowl. *Opp* WHISPER.

**shove** *vb inf* barge, crowd, drive, elbow, hustle, impel, jostle, nudge, press, prod, push, shoulder, thrust.

**shovel** *vb* clear, dig, scoop, shift.

**show** *n* **1** drama, performance, play, presentation, production. ▷ ENTERTAINMENT. **2** *flower show.* competition, demonstration, display, exhibition, *inf* expo, exposition, fair, presentation. **3** *show of strength.* appearance, demonstration, façade, illusion, impression, pose, pretence, threat. **4** *just for show.* affectation, exhibitionism, flamboyance, ostentation, pretentiousness, showing off. • *vb* **1** bare, betray, demonstrate, display, divulge, exhibit, expose, make public, make visible, manifest, open up, present, produce, reveal, uncover. **2** *Let your feelings show.* appear, be seen, be visible, catch the eye, come out, emerge, make an appearance, materialize, *inf* peep through, stand out, stick out. **3** *show the way.* conduct, direct, escort, guide, indicate, lead, point out, steer, usher. **4** *show kindness.* accord, bestow, confer, grant, treat with. **5** *This photo shows us at work.* depict, give a picture of, illustrate, picture, portray, represent, symbolize. **6** *Show me how.* clarify, describe, elucidate, explain, instruct, make clear, teach, tell. **7** *Tests show I was right.* attest, bear out, confirm, corroborate, demonstrate, evince, exemplify, manifest, prove, substantiate, verify, witness. **show off** ▷ BOAST. **show up** ▷ ARRIVE, HUMILIATE.

**showdown** *n* confrontation, crisis, *inf* decider, decisive encounter, *inf* moment of truth.

**shower** *n* **1** drizzle, sprinkling. ▷ RAIN. **2** douche, shower-bath. • *vb* **1** deluge, drop, rain, spatter, splash, spray, sprinkle. **2** *shower with gifts.* heap, inundate, load, overwhelm.

**show-off** *n inf* big-head, boaster, braggart, conceited person, egotist, exhibitionist, *inf* poser, poseur, *inf* showman, swaggerer.

**showy** *adj* bright, conspicuous, elaborate, fancy, flamboyant, flashy, florid, fussy, garish, gaudy, lavish, *inf* loud, lurid, ornate, ostentatious, *inf* over the top, pretentious, striking, trumpery, vulgar. *Opp* DISCREET.

**shred** *n* atom, bit, fragment, grain, hint, iota, jot, piece, scintilla, scrap, sliver, snippet, speck, trace. • *vb* cut to shreds, destroy, grate, rip up, scrap, tear. **shreds** rags, ribbons, strips, tatters.

**shrewd** *adj* acute, artful, astute, calculating, *inf* canny, clever, crafty, cunning, discerning, discriminating, *inf* foxy, ingenious, intelligent, knowing, observant, perceptive, percipient, perspicacious, quick-witted, sage, sharp, sly, smart, wily, wise. *Opp* STUPID.

**shriek** *vb* cry, scream, screech, squawk, squeal.

**shrill** *adj* ear-splitting, high, high-pitched, jarring, penetrating, piercing, piping, screaming, screeching, screechy, sharp, shrieking, strident, treble, whistling. *Opp* GENTLE, SONOROUS.

**shrine** *n* altar, chapel, holy of holies, holy place, place of worship, reliquary, sanctum, tomb. ▷ CHURCH.

**shrink** *vb* **1** become smaller, contract, decrease, diminish, dwindle, lessen, make smaller, narrow, reduce, shorten. ▷ SHRIVEL. *Opp* EXPAND. **2** *shrink with fear.* back off, cower, cringe, flinch, hang back, quail, recoil, retire, shy away, wince, withdraw. *Opp* ADVANCE.

**shrivel** *vb* become parched, become wizened, dehydrate, desiccate, curl, droop, dry out, dry up, pucker up, wilt, wither, wrinkle. ▷ SHRINK.

**shroud** *n* blanket, cloak, cloud, cover, mantle, mask, pall, veil, winding-sheet. • *vb* camouflage, cloak, conceal, cover, disguise, enshroud, envelop, hide, mask, screen, swathe, veil, wrap up.

**shrub** *n* bush, tree. □ *berberis, blackthorn, broom, bryony, buckthorn, buddleia, camellia, daphne, forsythia, gorse, heather, hydrangea, japonica, jasmine, lavender, lilac, myrtle, privet, rhododendron, rosemary, rue, viburnum.*

**shudder** *vb* be horrified, convulse, jerk, quake, quiver, rattle, shake, shiver, squirm, tremble, vibrate.

**shuffle** *vb* **1** confuse, disorganize, intermix, intersperse, jumble, mix, mix up, rearrange, reorganize. **2** *shuffle along.* drag your feet, scrape, shamble, slide. ▷ WALK.

**shun** *vb* avoid, disdain, eschew, flee, *inf* give the cold shoulder to, keep clear of, rebuff, reject, shy away from, spurn, steer clear of, turn away from. *Opp* SEEK.

**shut** *vb* bolt, close, fasten, latch, lock, push to, replace, seal, secure, slam. **shut in** ▷ CONFINE, IMPRISON. **shut off** ▷ ISOLATE. **shut out** ▷ EXCLUDE. **shut up** ▷ CONFINE, IMPRISON, SILENCE.

**shutter** *n* blind, louvre, screen.

**shy** *adj* apprehensive, backward, bashful, cautious, chary, coy, diffident, hesitant, inhibited, introverted, modest, *inf* mousy, nervous, reserved, reticent, retiring, self-conscious, self-effacing, sheepish, timid, timorous, underconfident, wary, withdrawn. *Opp* ASSERTIVE, UNINHIBITED. • *vb* ▷ THROW.

**sibling** *n* brother, sister, twin. ▷ FAMILY.

**sick** *adj* **1** afflicted, ailing, bedridden, diseased, ill, indisposed, infirm, *inf* laid up, *inf* poorly, *inf* queer, sickly, *inf* under the weather, unhealthy, unwell. ▷ ILL. **2** airsick, bilious, carsick, likely to vomit, nauseated, nauseous, queasy, seasick, squeamish. **3** *sick* *of rudeness.* annoyed (by), bored (with), disgusted (by), distressed (by), *inf* fed up (with), glutted (with), nauseated (by), sated (with), sickened (by), tired, troubled (by), upset (by), weary. 4 [*inf*] *a sick joke.* ▷ MORBID. **be sick** ▷ VOMIT.

**sicken** *vb* 1 *inf* catch a bug, fail, fall ill, take sick, weaken. 2 appal, be sickening to, disgust, make someone sick, nauseate, offend, repel, revolt, *inf* turn someone off, *inf* turn someone's stomach. **sickening** ▷ REPULSIVE.

**sickly** *adj* 1 ailing, anaemic, delicate, drawn, feeble, frail, ill, pale, pallid, *inf* peaky, unhealthy, wan, weak. ▷ ILL. *Opp* HEALTHY. 2 *sickly sentiment.* cloying, maudlin, mawkish, *inf* mushy, nasty, nauseating, obnoxious, *inf* off-putting, syrupy, treacly, unpleasant. *Opp* REFRESHING.

**sickness** *n* biliousness, nausea, queasiness, vomiting. ▷ ILLNESS.

**side** *n* 1 *sides of a cube.* elevation, face, facet, flank, surface. 2 *side of the road.* border, boundary, brim, brink, edge, fringe, limit, margin, perimeter, rim, verge. 3 *sides in a debate.* angle, aspect, attitude, perspective, point of view, position, school of thought, slant, standpoint, view, viewpoint. 4 *sides in a quarrel.* army, camp, faction, interest, party, sect, team. • *vb* **side with** ally with, favour, form an alliance with, *inf* go along with, join up with, partner, prefer, support, team up with. ▷ HELP.

**sidestep** *vb* avoid, circumvent, dodge, *inf* duck, evade, skirt round.

**sidetrack** *vb* deflect, distract, divert.

**sideways** *adj* 1 crabwise, indirect, lateral, oblique. 2 *a sideways glance.* covert, sidelong, sly, *inf* sneaky, unobtrusive.

**siege** *n* blockade. • *vb* ▷ BESIEGE.

**sieve** *n* colander, riddle, screen, strainer. • *vb* ▷ SIFT.

**sift** *vb* 1 filter, riddle, screen, separate, sieve, strain. 2 *sift evidence.* analyse, examine, investigate, pick out, review, scrutinize, select, sort out, weed out, winnow.

**sigh** *n* breath, exhalation, murmur, suspiration. ▷ SOUND.

**sight** *n* 1 eyesight, seeing, vision, visual perception. 2 *within sight.* field of vision, gaze, range, view, visibility. 3 *a brief sight of it.* glimpse, look. 4 *an impressive sight.* display, exhibition, scene, show, show-piece, spectacle. • *vb* behold, descry, discern, distinguish, espy, glimpse, make out, notice, observe, perceive, recognize, see, spot. **catch sight of** ▷ SEE.

**sightseer** *n* globe-trotter, holiday-maker, tourist, tripper, visitor.

**sign** *n* 1 augury, forewarning, hint, indication, indicator, intimation, omen, pointer, portent, presage, warning. ▷ SIGNAL. 2 *sign that someone was here.* clue, *inf* giveaway, indication, manifestation, marker, proof, reminder, spoor (of animal), suggestion, symptom, token, trace, vestige. 3 *put up a sign.* advertisement, notice, placard, poster, publicity, signboard. 4 *identifying sign.* badge, brand, cipher, device, emblem, flag, hieroglyph, ideogram, ideograph, insignia, logo, mark, monogram, rebus, symbol, trademark. • *vb* 1 autograph, countersign, endorse, inscribe, write. 2 ▷ SIGNAL. **sign off** ▷ FINISH. **sign on** ▷ ENLIST. **sign over** ▷ TRANSFER.

**signal** *n* **1** communication, cue, gesticulation, gesture, *inf* goahead, indication, motion, sign, signal, *inf* tip-off, token, warning. □ *alarm-bell, beacon, bell, burglar-alarm, buzzer, flag, flare, gong, green light, indicator, password, red light, reveille, rocket, semaphore signal, siren, smoke-signal, tocsin,* old use *trafficator, traffic-lights, warning-light, whistle, winker.* **2** *radio signal.* broadcast, emission, output, transmission, waves. • *vb* beckon, communicate, flag, gesticulate, give or send a signal, indicate, motion, notify, sign, wave. ▷ GESTURE.

**signature** *n* autograph, endorsement, mark, name.

**signet** *n* seal, stamp.

**significance** *n* denotation, force, idea, implication, import, importance, message, point, purport, relevance, sense, signification, usefulness, value, weight. ▷ MEANING.

**significant** *adj* **1** eloquent, expressive, indicative, informative, knowing, meaningful, pregnant, revealing, suggestive, symbolic, *inf* tell-tale. **2** *significant event.* big, consequential, considerable, historic, important, influential, memorable, newsworthy, noteworthy, relevant, salient, serious, sizeable, valuable, vital, worthwhile. *Opp* INSIGNIFICANT.

**signify** *vb* **1** announce, be a sign of, betoken, communicate, connote, convey, denote, express, foretell, impart, imply, indicate, intimate, make known, reflect, reveal, signal, suggest, symbolize, tell, transmit. **2** *It doesn't signify.* be significant, count, matter, merit consideration.

**signpost** *n* finger-post, pointer, road-sign, sign.

**silence** *n* **1** calm, calmness, hush, noiselessness, peace, quiet, quietness, quietude, soundlessness, stillness, tranquillity. *Opp* NOISE. **2** *Her silence puzzled us.* dumbness, muteness, reticence, speechlessness, taciturnity, uncommunicativeness. • *vb* **1** gag, hush, keep quiet, make silent, muzzle, repress, shut up, suppress. **2** *silence engine noise.* damp, deaden, muffle, mute, quieten, smother, stifle. **Silence!** Be quiet! Be silent! *inf* Hold your tongue! Hush! Keep quiet! *inf* Pipe down! Shut up! Stop talking!

**silent** *adj* **1** hushed, inaudible, muffled, muted, noiseless, quiet, soundless. **2** dumb, laconic, *inf* mum, reserved, reticent, speechless, taciturn, tight-lipped, tongue-tied, uncommunicative, unforthcoming, voiceless. **3** *silent listeners.* attentive, rapt, restrained, still. **4** *silent agreement.* implied, implicit, mute, tacit, understood, unexpressed, unspoken, unuttered. *Opp* EXPLICIT, NOISY, TALKATIVE. **be silent** keep quiet, *inf* pipe down, say nothing, *inf* shut up.

**silhouette** *n* contour, form, outline, profile, shadow, shape.

**silky** *adj* delicate, fine, glossy, lustrous, satiny, sleek, smooth, soft, velvety.

**silly** *adj* **1** absurd, asinine, brainless, childish, crazy, daft, *inf* dopey, *inf* dotty, fatuous, feather-brained, feeble-minded, flighty, foolish, *old use* fond, frivolous, grotesque, *inf* half-baked, hare-brained, idiotic, ill-advised, illogical, immature, impractical, imprudent, inadvisable, inane, infantile, irrational, *inf* jokey, laughable, light-hearted, ludicrous, mad, meaningless, mindless, misguided, naïve, nonsensical, playful, pointless, preposterous, ridiculous, scatter-brained, *inf* scatty, senseless, shallow, simple, simple-minded, simplistic, *inf* soppy, stupid, thoughtless, unintelligent, unreasonable, unsound, unwise, wild, witless. *Opp* SERIOUS, WISE. **2** [*inf*] *knocked silly*. ▷ UNCONSCIOUS.

**silt** *n* alluvium, deposit, mud, ooze, sediment, slime, sludge.

**silvan** *adj* arboreal, leafy, tree-covered, wooded.

**similar** *adj* akin, alike, analogous, comparable, compatible, congruous, co-ordinating, corresponding, equal, equivalent, harmonious, homogeneous, identical, indistinguishable, like, matching, parallel, related, resembling, the same, toning, twin, uniform, well-matched. *Opp* DIFFERENT.

**similarity** *n* affinity, closeness, congruity, correspondence, equivalence, homogeneity, kinship, likeness, match, parallelism, relationship, resemblance, sameness, similitude, uniformity. *Opp* DIFFERENCE.

**simmer** *vb* boil, bubble, cook, seethe, stew.

**simple** *adj* **1** artless, basic, candid, childlike, elementary, frank, fundamental, guileless, homely, honest, humble, ingenuous, innocent, lowly, modest, *derog* naïve, natural, simple-minded, *derog* silly, sincere, unaffected, unassuming, unpretentious, unsophisticated. *Opp* SOPHISTICATED. **2** *simple instructions*. clear, comprehensible, direct, easy, fool-proof, intelligible, lucid, straightforward, uncomplicated, understandable. *Opp* COMPLEX. **3** *a simple dress*. austere, classical, plain, severe, stark, unadorned, unembellished. *Opp* ORNATE.

**simplify** *vb* clarify, explain, make simple, paraphrase, prune, *inf* put in words of one syllable, streamline, unravel, untangle. *Opp* COMPLICATE.

**simplistic** *adj* [*always derog*] facile, inadequate, naïve, oversimple, oversimplified, shallow, silly, superficial.

**simulate** *vb* act, counterfeit, dissimulate, enact, fake, feign, imitate, *inf* mock up, play-act, pretend, reproduce, sham.

**simultaneous** *adj* coinciding, concurrent, contemporaneous, parallel, synchronized, synchronous.

**sin** *n* blasphemy, corruption, depravity, desecration, devilry, error, evil, fault, guilt, immorality, impiety, iniquity, irreverence, misdeed, offence, peccadillo, profanation, sacrilege, sinfulness, transgression, *old use* trespass, ungodliness, unrighteousness, vice, wickedness, wrong, wrongdoing. • *vb* be guilty of sin, blaspheme, do wrong, err, fall from grace, go astray, lapse, misbehave, offend, stray, transgress.

**sincere** *adj* candid, direct, earnest, frank, genuine, guileless, heartfelt, honest, open, real, serious, simple, *inf* straight, straightforward, true, truthful, unaffected, unfeigned, upright, wholehearted. *Opp* INSINCERE.

**sincerity** *n* candour, directness, earnestness, frankness, genuineness, honesty, honour, integrity, openness, straightforwardness, trustworthiness, truthfulness, uprightness.

**sinewy** *adj* brawny, muscular, strapping, tough, wiry. ▷ STRONG.

**sinful** *adj* bad, blasphemous, corrupt, damnable, depraved, erring, evil, fallen, guilty, immoral, impious, iniquitous, irreligious, irreverent, profane, sacrilegious, ungodly, unholy, unrighteous, vile, wicked, wrong, wrongful. *Opp* RIGHTEOUS.

**sing** *vb* carol, chant, chirp, chorus, croon, descant, hum, intone, serenade, trill, vocalize, warble, whistle, yodel.

**singe** *vb* blacken, burn, char, scorch, sear.

**singer** *n* songster, vocalist. □ *alto, balladeer, baritone, bass, carol-singer, castrato, counter-tenor,* plur *choir, choirboy, choirgirl, chorister,* plur *chorus, coloratura, contralto, crooner,* It *diva, folk singer, minstrel, opera singer, pop star, precentor,* It *prima donna, soloist, soprano, tenor, treble, troubadour.*

**single** *adj* **1** exclusive, individual, isolated, lone, odd, one, only, personal, separate, singular, sole, solitary, unique, unparalleled. **2** *a single person.* celibate, *inf* free, unattached, unmarried. • *vb* **single out** ▷ CHOOSE.

**single-handed** *adj* alone, independent, solitary, unaided, unassisted, without help.

**single-minded** *adj* dedicated, determined, devoted, dogged, *derog* fanatical, *derog* obsessive, persevering, resolute, steadfast, tireless, unswerving, unwavering.

**singular** *adj* **1** ▷ SINGLE. **2** abnormal, curious, different, distinct, eccentric, exceptional, extraordinary, odd, outstanding, peculiar, rare, remarkable, strange, unusual. ▷ DISTINCTIVE. *Opp* COMMON.

**sinister** *adj* **1** dark, disquieting, disturbing, evil, forbidding, foreboding, frightening, gloomy, inauspicious, malevolent, malignant, menacing, minatory, ominous, threatening, upsetting. **2** *sinister motives.* bad, corrupt, criminal, dishonest, furtive, illegal, nefarious, questionable, *inf* shady, suspect, treacherous, unworthy, villainous.

**sink** *n* basin, stoup, washbowl. • *vb* **1** collapse, decline, descend, diminish, disappear, droop, drop, dwindle, ebb, fade, fail, fall, go down, go lower, plunge, set (*sun sets*), slip down, subside, vanish, weaken. **2** be engulfed, be submerged, founder, go down, go under. **3** *sink a ship.* scupper, scuttle. **4** *sink a borehole.* bore, dig, drill, excavate.

**sinner** *n* evil-doer, malefactor, miscreant, offender, reprobate, transgressor, wrongdoer.

**sip** *vb* drink, lap, sample, taste.

**sit** *vb* **1** be seated, perch, rest, seat (yourself), settle, squat, take a seat, *inf* take the weight off your feet. **2** *sit for a portrait.* pose. **3** *sit an exam.* be a candidate in, *inf* go in for, take, write. **4** *Parliament sat for 12 hours.* assemble, be in session, convene, gather, get together, meet.

**site** *n* area, campus, ground, location, place, plot, position, setting, situation, spot. • *vb* ▷ SITUATE.

**sitting-room** *n* drawing-room, living-room, lounge.

**situate** *vb* build, establish, found, install, locate, place, position, put, set up, site, station.

**situation** *n* **1** area, locale, locality, location, place, position, setting, site, spot. **2** *an awkward situation.* case, circumstances, condition, *inf* kettle of fish, plight, position, predicament, state of affairs. **3** *situations vacant*. employment, job, place, position, post.

**size** *n* amount, area, bigness, breadth, bulk, capacity, depth, dimensions, extent, gauge, height, immensity, largeness, length, magnitude, mass, measurement, proportions, scale, scope, volume, weight, width. ▷ MEASURE. • *vb* **size up** ▷ ASSESS.

**sizeable** *adj* considerable, decent, generous, largish, significant, worthwhile. ▷ BIG.

**skate** *vb* glide, skim, slide.

**skeleton** *n* bones, frame, framework, structure.

**sketch** *n* **1** description, design, diagram, draft, drawing, outline, picture, plan, *inf* rough, skeleton, vignette. **2** *comic sketch*. performance, playlet, scene, skit, turn. • *vb* depict, draw, indicate, outline, portray, represent. **sketch out** ▷ OUTLINE.

**sketchy** *adj* bitty, crude, cursory, hasty, hurried, imperfect, incomplete, inexact, perfunctory, rough, scrappy, undeveloped, unfinished, unpolished. *Opp* DETAILED, PERFECT.

**skid** *vb* aquaplane, glide, go out of control, slide, slip.

**skilful** *adj* able, accomplished, adept, adroit, apt, artful, capable, competent, consummate, crafty, cunning, deft, dexterous, experienced, expert, gifted, handy, ingenious, masterful, masterly, practised, professional, proficient, qualified, shrewd, smart, talented, trained, versatile, versed, workmanlike. ▷ CLEVER. *Opp* UNSKILFUL.

**skill** *n* ability, accomplishment, adroitness, aptitude, art, artistry, capability, cleverness, competence, craft, cunning, deftness, dexterity, experience, expertise, facility, flair, gift, handicraft, ingenuity, knack, mastery, professionalism, proficiency, prowess, shrewdness, talent, technique, training, versatility, workmanship.

**skilled** *adj* experienced, expert, qualified, trained, versed. ▷ SKILFUL.

**skim** *vb* **1** aquaplane, coast, fly, glide, move lightly, plane, sail, skate, ski, skid, slide, slip. **2** *skim a book*. dip into, leaf through, look through, read quickly, scan, skip, thumb through.

**skin** *n* casing, coat, coating, complexion, covering, epidermis, exterior, film, fur, hide, husk, integument, membrane, outside, peel, pelt, rind, shell, surface. • *vb* excoriate, flay, pare, peel, shell, strip.

**skin-deep** *adj* insubstantial, shallow, superficial, trivial, unimportant.

**skinny** *adj* bony, emaciated, gaunt, half-starved, lanky, scraggy, wasted. ▷ THIN.

**skip** *vb* **1** bound, caper, cavort, dance, frisk, gambol, hop, jump, leap, prance, romp, spring. **2** *skip the boring bits*. avoid, forget, ignore, leave out, miss out, neglect, omit, overlook, pass over, skim through. **3** *skip lessons*. be absent from, cut, miss, play truant from.

**skirmish** *n* brush, fight, fray, scrimmage, *inf* set-to, tussle. • *vb* ▷ FIGHT.

**skirt** *vb* avoid, border, bypass, circle, encircle, go round, pass round, *inf* steer clear of, surround.

**skit** *n* burlesque, parody, satire, sketch, spoof, *inf* take-off.

**sky** *n* air, atmosphere, *poet* blue, *poet* empyrean, *poet* firmament, *poet* heavens, space, stratosphere, *poet* welkin.

**slab** *n* block, chunk, hunk, lump, piece, slice, wedge, *inf* wodge.

**slack** *adj* **1** drooping, limp, loose, sagging, soft. *Opp* TIGHT. **2** *slack attitude.* careless, dilatory, disorganized, easy-going, flaccid, idle, inattentive, indolent, lax, lazy, listless, neglectful, negligent, permissive, relaxed, remiss, slothful, unbusinesslike, uncaring, undisciplined. *Opp* RIGOROUS. **3** *slack trade.* inactive, quiet, slow, slow-moving, sluggish. *Opp* BUSY. • *vb* be lazy, idle, malinger, neglect your duty, shirk, *inf* skive.

**slacken** *vb* **1** ease off, loosen, relax, release. **2** *slacken speed.* abate, decrease, ease, lessen, lower, moderate, reduce, slow down.

**slacker** *n inf* good-for-nothing, idler, lazy person, malingerer, *sl* skiver, sluggard.

**slake** *vb* allay, assuage, cool, ease, quench, relieve, satisfy.

**slam** *vb* **1** bang, shut. ps5 **2** [*inf*] ▷ CRITICIZE.

**slander** *n* backbiting, calumny, defamation, denigration, insult, libel, lie, misrepresentation, obloquy, scandal, slur, smear, vilification. • *vb* blacken the name of, calumniate, defame, denigrate, disparage, libel, malign, misrepresent, slur, smear, spread tales about, tell lies about, traduce, vilify.

**slanderous** *adj* abusive, calumnious, cruel, damaging, defamatory, disparaging, false, hurtful, insulting, libellous, lying, malicious, mendacious, scurrilous, untrue, vicious.

**slang** *n* argot, cant, jargon. • *vb* ▷ INSULT. **slanging match** ▷ QUARREL.

**slant** *n* **1** angle, bevel, camber, cant, diagonal, gradient, incline, list, pitch, rake, ramp, slope, tilt. **2** *slant on a problem.* approach, attitude, perspective, point of view, standpoint, view, viewpoint. **3** *slant to the news.* bias, distortion, emphasis, imbalance, one-sidedness, prejudice. • *vb* **1** be at an angle, be skewed, incline, lean, shelve, slope, tilt. **2** *slant the news.* bias, colour, distort, prejudice, twist, weight. **slanting** ▷ OBLIQUE.

**slap** *vb* smack, spank. ▷ HIT.

**slash** *vb* gash, slit. ▷ CUT.

**slaughter** *n* bloodshed, butchery, carnage, killing, massacre, murder. • *vb* annihilate, butcher, massacre, murder, slay. ▷ KILL.

**slaughterhouse** *n* abattoir, shambles.

**slave** *n old use* bondslave, drudge, serf, thrall, vassal. ▷ SERVANT. • *vb* drudge, exert yourself, grind away, labour, *inf* sweat, toil, *inf* work your fingers to the bone. ▷ WORK.

**slave-driver** *n* despot, hard task-master, tyrant.

**slaver** *vb* dribble, drool, foam at the mouth, salivate, slobber, spit.

**slavery** *n* bondage, captivity, enslavement, serfdom, servitude, subjugation, thraldom, vassalage. *Opp* FREEDOM.

**slavish** *adj* **1** abject, cringing, fawning, grovelling, humiliating, menial, obsequious, servile, submissive. **2** *slavish imitation.* close, flattering, strict, sycophantic, unimaginative, unoriginal. *Opp* INDEPENDENT.

**slay** *vb* assassinate, bump off, butcher, destroy, dispatch, execute, exterminate, *inf* finish off, martyr, massacre, murder, put down, put to death, slaughter. ▷ KILL.

**sleazy** *adj* cheap, contemptible, dirty, disreputable, low-class, mean, mucky, run-down, seedy, slovenly, sordid, squalid, unprepossessing.

**sledge** *n* bob-sleigh, sled, sleigh, toboggan.

**sleek** *adj* **1** brushed, glossy, graceful, lustrous, shining, shiny, silken, silky, smooth, soft, trim, velvety, well-groomed. *Opp* UNTIDY. **2** *a sleek look.* complacent, contented, fawning, self-satisfied, *inf* slimy, *inf* smarmy, smug, suave, thriving, unctuous, well-fed.

**sleep** *n inf* beauty sleep, catnap, coma, dormancy, doze, *inf* forty winks, hibernation, *sl* kip, *inf* nap, repose, rest, *inf* shut-eye, siesta, slumber, snooze, torpor, unconsciousness. • *vb* aestivate, be sleeping, be unconscious, catnap, *inf* doss down, doze, *inf* drop off, drowse, fall asleep, go to bed, *inf* have forty winks, hibernate, *sl* kip, *inf* nod off, rest, slumber, snooze, *inf* take a nap. **sleeping** ▷ ASLEEP.

**sleepiness** *n* drowsiness, lassitude, lethargy, somnolence, tiredness, torpor.

**sleepless** *adj* awake, conscious, disturbed, insomniac, restless, *inf* tossing and turning, wakeful, watchful, wide awake. *Opp* ASLEEP.

**sleepwalker** *n* noctambulist, somnambulist.

**sleepy** *adj* **1** comatose, *inf* dopey, drowsy, heavy, lethargic, ready to sleep, sluggish, somnolent, soporific, tired, torpid, weary. **2** *a sleepy village.* boring, dull, inactive, quiet, restful, slowmoving, unexciting. *Opp* LIVELY.

**slender** *adj* **1** fine, graceful, lean, narrow, slender, slight, svelte, sylphlike, trim. ▷ THIN. **2** *slender thread.* feeble, fragile, tenuous. **3** *slender means.* inadequate, meagre, scanty, small. *Opp* FAT, LARGE.

**slice** *n* carving, layer, piece, rasher, shaving, sliver, wedge. • *vb* carve, shave off. ▷ CUT.

**slick** *adj* **1** adroit, artful, clever, cunning, deft, dextrous, efficient, quick, skilful, smart. **2** *a slick talker.* glib, meretricious, plausible, *inf* smarmy, smooth, smug, specious, suave, superficial, *inf* tricky, unctuous, untrustworthy, urbane, wily. **3** *slick hair.* glossy, oiled, plastered down, shiny, sleek, smooth.

**slide** *n* **1** avalanche, landslide, landslip. **2** *photographic slide.* transparency. • *vb* aquaplane, coast, glide, glissade, plane, skate, ski, skid, skim, slip, slither, toboggan.

**slight** *adj* **1** imperceptible, inadequate, inconsequential, inconsiderable, insignificant, insufficient, little, minor, negligible, scanty, slim (*chance*), small, superficial, trifling, trivial, unimportant. **2** *slight build.* delicate, diminutive, flimsy, fragile, frail, petite, sickly, slender, slim, svelte, sylphlike, thin, tiny, weak. *Opp* BIG. • *n, vb* ▷ INSULT.

**slightly** *adv* hardly, moderately, only just, scarcely. *Opp* VERY.

**slim** *adj* **1** fine, graceful, lean, narrow, slender, svelte, sylphlike, trim. ▷ THIN. **2** *a slim chance.* little, negligible, remote, slight, unlikely. • *vb* become slimmer, diet, lose weight, reduce.

**slime** *n* muck, mucus, mud, ooze, sludge.

**slimy** *adj* clammy, greasy, mucous, muddy, oily, oozy, *inf* slippy, slippery, slithery, *inf* squidgy, *inf* squishy, wet.

**sling** *vb* cast, *inf* chuck, fling, heave, hurl, launch, *inf* let fly, lob, pelt, pitch, propel, shoot, shy, throw, toss.

**slink** *vb* creep, edge, move guiltily, prowl, skulk, slither, sneak, steal.

**slinky** *adj* [*inf*] *a slinky dress.* clinging, close-fitting, graceful, *inf* sexy, sinuous, sleek.

**slip** *n* **1** accident, *inf* bloomer, blunder, error, fault, *Fr* faux pas, impropriety, inaccuracy, indiscretion, lapse, miscalculation, mistake, oversight, *inf* slip of the pen, slip of the tongue, *inf* slip-up. **2** *slip of paper.* note, piece, sheet, strip. • *vb* **1** aquaplane, coast, glide, glissade, move out of control, skate, ski, skid, skim, slide, slip, slither, stumble, trip. **2** *slipped into the room.* creep, edge, move quietly, slink, sneak, steal. **give someone the slip** ▷ ESCAPE. **let slip** ▷ REVEAL. **slip away, slip the net** ▷ ESCAPE. **slip up** ▷ BLUNDER.

**slippery** *adj* **1** glassy, greasy, icy, lubricated, oily, slimy, *inf* slippy, slithery, smooth, wet. **2** *a slippery customer.* crafty, cunning, devious, evasive, *inf* hard to pin down, shifty, sly, *inf* smarmy, smooth, sneaky, specious, *inf* tricky, unreliable, untrustworthy, wily.

**slipshod** *adj* careless, disorganized, lax, messy, slapdash, *inf* sloppy, slovenly, untidy.

**slit** *n* aperture, breach, break, chink, cleft, crack, cut, fissure, gap, gash, hole, incision, opening, rift, slot, split, tear, vent. • *vb* cut, gash, slice, split, tear.

**slither** *vb* creep, glide, *inf* skitter, slide, slink, slip, snake, worm.

**sliver** *n* chip, flake, shard, shaving, snippet, splinter, strip. ▷ PIECE.

**slobber** *vb* dribble, drool, salivate, slaver.

**slogan** *n* battle-cry, catch-phrase, catchword, jingle, motto, war-cry, watchword. ▷ SAYING.

**slope** *n* angle, bank, bevel, camber, cant, gradient, hill, incline, pitch, rake, ramp, scarp, slant, tilt. □ [upwards] *acclivity, ascent, rise.* □ [downwards] *decline, declivity, descent, dip, drop, fall.* • *vb* ascend, bank, decline, descend, dip, fall, incline, lean, pitch, rise, shelve, slant, tilt, tip. **sloping** ▷ OBLIQUE.

**sloppy** *adj* **1** liquid, messy, runny, *inf* sloshy, slushy, *inf* splashing about, squelchy, watery, wet. **2** *sloppy work.* careless, dirty, disorganized, lax, messy, slapdash, slipshod, slovenly, unsystematic, untidy. **3** ▷ SENTIMENTAL.

**slot** *n* **1** aperture, breach, break, channel, chink, cleft, crack, cut, fissure, gap, gash, groove, hole, incision, opening, rift, slit, split, vent. **2** *slot on a schedule.* place, position, space, spot, time.

**sloth** *n* apathy, idleness, indolence, inertia, laziness, lethargy, sluggishness, torpor.

**slouch** *vb* droop, hunch, loaf, loll, lounge, sag, shamble, slump, stoop.

**slovenly** *adj* careless, *inf* couldn't-care-less, disorganized, lax, messy, shoddy, slapdash, slatternly, *inf* sloppy, slovenly, thoughtless, unmethodical, untidy. *Opp* CAREFUL.

**slow** *adj* **1** careful, cautious, crawling, dawdling, delayed, deliberate, dilatory, gradual, lagging, late, lazy, leisurely, lingering, loitering, measured, moderate, painstaking, plodding, protracted, slow-moving, sluggardly, sluggish, steady, tardy, torpid, unhurried, unpunctual. **2** *slow learner*. backward, dense, dim, dull, obtuse, *inf* thick. ▷ STUPID. **3** *slow worker*. phlegmatic, reluctant, unenthusiastic, unwilling. ▷ SLUGGISH. *Opp* FAST. • *vb* **slow down** brake, decelerate, *inf* ease up, go slower, hold back, reduce speed. **be slow** ▷ DAWDLE, DELAY.

**sludge** *n* mire, muck, mud, ooze, precipitate, sediment, silt, slime, slurry, slush.

**sluggish** *adj* apathetic, dull, idle, inactive, indolent, inert, lazy, lethargic, lifeless, listless, phlegmatic, slothful, torpid, unresponsive. ▷ SLOW. *Opp* LIVELY.

**sluice** *vb* flush, rinse, swill, wash.

**slumber** *n, vb* ▷ SLEEP.

**slump** *n* collapse, crash, decline, depression, dip, downturn, drop, fall, falling-off, plunge, recession, trough. *Opp* BOOM. • *vb* **1** collapse, crash, decline, dive, drop, fall off, plummet, plunge, recede, sink, slip, *inf* take a nosedive, worsen. *Opp* PROSPER. **2** *slump in a chair*. be limp, collapse, droop, flop, hunch, loll, lounge, sag, slouch, subside.

**slur** *n* affront, aspersion, calumny, imputation, innuendo, insinuation, insult, libel, slander, smear, stigma. • *vb* garble, lisp, mumble.

**slurry** *n* mud, ooze, slime.

**sly** *adj* artful, *inf* canny, *inf* catty, conniving, crafty, cunning, deceitful, designing, devious, disingenuous, *inf* foxy, furtive, guileful, insidious, knowing, scheming, secretive, *inf* shifty, shrewd, *inf* sneaky, *inf* snide, stealthy, surreptitious, treacherous, tricky, underhand, wily. *Opp* CANDID, OPEN.

**smack** *vb* pat, slap, spank. ▷ HIT.

**small** *adj* **1** *inf* baby, bantam, compact, concise, cramped, diminutive, *inf* dinky, dwarf, exiguous, fine, fractional, infinitesimal, lean, lilliputian, little, microscopic, midget, *inf* mini, miniature, minuscule, minute, narrow, petite, *inf* pint-sized, *inf* pocket-sized, *inf* poky, portable, pygmy, short, slender, slight, *inf* teeny, thin, tiny, toy, undergrown, undersized, *inf* wee, *inf* weeny. **2** *small helpings*. inadequate, insufficient, meagre, mean, *inf* measly, miserly, modest, niggardly, parsimonious, *inf* piddling, scanty, skimpy, stingy, ungenerous, unsatisfactory. **3** *a small problem*. inconsequential, insignificant, minor, negligible, nugatory, slim (*chance*), slight, trifling, trivial, unimportant. *Opp* BIG. **small arms** ▷ WEAPON.

**small-minded** *adj* bigoted, grudging, hidebound, illiberal, intolerant, narrow, narrow-minded, old-fashioned, parochial, petty, prejudiced, rigid, selfish, trivial, unimaginative. ▷ MEAN. *Opp* BROAD-MINDED.

**smart** *adj* **1** acute, adept, artful, astute, bright, clever, crafty, *inf* cute, discerning, ingenious, intelligent, perceptive, perspicacious, quick, quickwitted, shrewd, *sl* streetwise. *Opp* DULL. **2** *smart appearance*. bright, chic, clean, dapper, *inf* dashing, elegant, fashionable, fresh, modish, *inf* natty, neat, *inf* posh, *inf* snazzy, *Fr* soigné, spruce, stylish, tidy, trim, well-dressed, well-groomed, well-looked-after. *Opp* SCRUFFY. **3** *a smart pace.* brisk, *inf* cracking, fast, forceful, quick, rapid, *inf* rattling, speedy, swift. **4** *a smart blow.* painful, sharp, stinging, vigorous. • *vb* ▷ HURT.

**smash** *vb* **1** *smash to pieces.* crumple, crush, demolish, destroy, shatter, squash, wreck. ▷ BREAK. **2** *smash into a wall.* bang, bash, batter, bump, collide, crash, hammer, knock, pound, ram, slam, strike, thump, wallop. ▷ HIT.

**smear** *n* **1** blot, daub, mark, smudge, stain, streak. **2** *a smear on your name.* aspersion, calumny, defamation, imputation, innuendo, insinuation, libel, slander, slur, stigma, vilification. • *vb* **1** dab, daub, plaster, rub, smudge, spread, wipe. **2** *smear a reputation.* attack, besmirch, blacken, calumniate, defame, discredit, libel, malign, slander, stigmatize, tarnish, vilify.

**smell** *n* odour, redolence. □ [pleasant] *aroma, bouquet, fragrance, incense, nose, scent.* □ [unpleasant] *fetor, mephitis, miasma, perfume,* inf *pong, pungency, reek, stench, stink, whiff.* • *vb* **1** *inf* get a whiff of, scent, sniff. **2** *onions smell. inf* hum, *inf* pong, reek, stink, whiff.

**smelling** *adj* [*Smelling* is usually used in combination: sweet-smelling, etc.] **1** *pleasant-smelling.* aromatic, fragrant, musky, odoriferous, odorous, perfumed, redolent, scented, spicy. **2** *unpleasant-smelling.* fetid, foul, gamy, *inf* high, malodorous, mephitic, miasmic, musty, noisome, *inf* off, *sl* pongy, pungent, putrid, rank, reeking, rotten, smelly, stinking, *inf* whiffy. *Opp* ODOURLESS.

**smelly** *adj* ▷ SMELLING.

**smile** *n, vb* beam, grin, leer, laugh, simper, smirk, sneer.

**smoke** *n* **1** air pollution, exhaust, fog, fumes, gas, smog, steam, vapour. **2** cigar, cigarette, cheroot, *inf* fag, pipe, tobacco. • *vb* **1** emit smoke, fume, reek, smoulder. **2** *smoke cigars.* inhale, puff at.

**smoky** *adj* clouded, dirty, foggy, grimy, hazy, sooty. *Opp* CLEAR.

**smooth** *adj* **1** even, flat, horizontal, level, plane, regular, unbroken, unruffled. **2** *smooth sea* calm, peaceful, placid, quiet, restful. **3** *a smooth finish.* burnished, glassy, glossy, polished, satiny, shiny, silken, silky, sleek, soft, velvety. **4** *smooth progress.* comfortable, easy, effortless, fluent, steady, uncluttered, uneventful, uninterrupted, unobstructed. **5** *a smooth taste.* agreeable, bland, mellow, mild, pleasant, soft, soothing **6** *a smooth mixture.* creamy, flowing, runny. **7** *a smooth talker.* convincing, facile, glib, insincere, plausible, polite, self-assured, self-satisfied, slick, smug, sophisticated, suave, untrustworthy, urbane. *Opp* ROUGH. • *vb* buff up, burnish, even out, file, flatten, iron, level, level off, plane, polish press, roll out, sand down, sandpaper.

**smother** *vb* **1** asphyxiate, choke, cover, kill, snuff out, stifle, strangle, suffocate, throttle. **2** ▷ SUPPRESS.

**smoulder** *vb* burn, smoke. **smouldering** ▷ ANGRY.

**smudge** *vb* blot, blur, dirty, mark, smear, stain, streak.

**smug** *adj* complacent, conceited, *inf* holier-than-thou, pleased, priggish, self-important, selfrighteous, self-satisfied, sleek, superior. *Opp* HUMBLE.

**snack** *n* bite, *inf* elevenses, *inf* nibble, refreshments. ▷ MEAL.

**snack-bar** *n* buffet, café, cafeteria, fast-food restaurant, transport café.

**snag** *n* catch, complication, difficulty, drawback, hindrance, hitch, impediment, obstacle, obstruction, problem, set-back, *inf* stumbling-block. • *vb* catch, jag, rip, tear.

**snake** *n* ophidian, serpent. □ *adder, anaconda, boa constrictor, cobra, copperhead, flying-snake, grass snake, mamba, python, rattlesnake, sand snake, sea snake, sidewinder, tree snake, viper.* • *vb* crawl, creep, meander, twist and turn, wander, worm, zigzag. **snaking** ▷ TWISTY.

**snap** *adj* ▷ SUDDEN. • *vb* **1** break, crack, fracture, give way, part, split. **2** *snap your fingers.* click, crack, pop. **3** *dog snapped at me.* bite, gnash, nip, snatch. **4** *snap orders.* bark, growl, *inf* jump down someone's throat, snarl, speak angrily. **snare** *n* ambush, booby-trap, *old use* gin, noose, springe, trap. • *vb* capture, catch, decoy, ensnare, entrap, net, trap.

**snarl** *vb* **1** bare the teeth, growl. **2** *snarl up rope.* confuse, entangle, jam, knot, tangle, twist.

**snatch** *vb* **1** catch, clutch, grab, grasp, lay hold of, pluck, seize, take, wrench away, wrest away. **2** abduct, kidnap, remove, steal.

**sneak** *vb* **1** creep, move stealthily, prowl, skulk, slink, stalk, steal. **2** [*inf*] *sneak on someone. sl* grass, inform (against), report, *sl* snitch, *inf* tell tales (about).

**sneaking** *adj* furtive, half-formed, intuitive, lurking, nagging, *inf* niggling, persistent, private, secret, uncomfortable, unconfessed, undisclosed, unproved, worrying.

**sneaky** *adj* cheating, contemptible, crafty, deceitful, despicable, devious, dishonest, furtive, *inf* low-down, mean, nasty, shady, *inf* shifty, sly, treacherous, underhand, unorthodox, unscrupulous, untrustworthy. *Opp* STRAIGHTFORWARD.

**sneer** *vb* be contemptuous, be scornful, boo, curl your lip, hiss, hoot, jeer, laugh, mock, scoff, sniff. **sneer at** ▷ DENIGRATE, RIDICULE.

**sniff** *vb* **1** *inf* get a whiff of, scent, smell. **2** ▷ SNIVEL. **3** ▷ SNEER.

**snigger** *vb* chuckle, giggle, laugh, snicker, titter.

**snip** *vb* clip, dock, nick, nip. ▷ CUT.

**snipe** *vb* fire, shoot, *inf* take potshots. **snipe at** ▷ CRITICIZE.

**snippet** *n* fragment, morsel, particle, scrap, shred, snatch. ▷ PIECE.

**snivel** *vb* blubber, cry, grizzle, mule, sob, sniff, sniffle, snuffle, whimper, whine, *inf* whinge. ▷ WEEP.

**snobbish** *adj* affected, condescending, disdainful, élitist, haughty, highfalutin, *inf* hoity-toity, lofty, lordly, patronizing, pompous, *inf* posh, presumptuous, pretentious, *inf* putting on airs, self-important, smug, *inf* snooty, *inf* stuck-up, supercilious, superior, *inf* toffee-nosed. ▷ CONCEITED. *Opp* UNPRETENTIOUS.

**snoop** *vb* be inquisitive, butt in, do detective work, interfere, intrude, investigate, meddle, *inf* nose about, pry, sneak, spy, *inf* stick your nose in.

**snooper** *n* busybody, detective, investigator, meddler, sneak, spy.

**snout** *n* face, muzzle, nose, nozzle, proboscis, trunk.

**snub** *vb* be rude to, brush off, cold-shoulder, disdain, humiliate, insult, offend, *inf* put someone down, rebuff, reject, scorn, *inf* squash.

**snuff** *vb* extinguish, put out. **snuff it** ▷ DIE. **snuff out** ▷ KILL.

**snug** *adj* **1** comfortable, *inf* comfy, cosy, enclosed, friendly, intimate, protected, reassuring, relaxed, relaxing, restful, safe, secure, sheltered, soft, warm. **2** *a snug fit.* close-fitting, exact, well-tailored.

**soak** *vb* bathe, *inf* dunk, drench, immerse, marinate, penetrate, permeate, pickle, saturate, souse, steep, submerge, wet thoroughly. **soaked, soaking** ▷ WET. **soak up** ▷ ABSORB.

**soar** *vb* **1** ascend, climb, float, fly, glide, hang, hover, rise, tower. **2** *prices soared.* escalate, increase, rise, rocket, shoot up, spiral.

**sob** *vb* blubber, cry, gasp, snivel, *inf* sob your heart out, whimper. ▷ WEEP.

**sober** *adj* **1** calm, clear-headed, composed, dignified, grave, in control, level-headed, lucid, peaceful, quiet, rational, sedate, sensible, serene, serious, solemn, steady, subdued, tranquil, unexciting. *Opp* SILLY. **2** *sober habits.* abstemious, moderate, *inf* on the wagon, restrained, self-controlled, staid, teetotal, temperate. *Opp* DRUNK. **3** *sober dress.* colourless, drab, dull, plain, sombre.

**sociable** *adj* affable, approachable, *old use* clubbable, companionable, convivial, extroverted, friendly, gregarious, hospitable, neighbourly, outgoing, warm, welcoming. ▷ SOCIAL. *Opp* UNFRIENDLY, WITHDRAWN.

**social** *adj* **1** civilized, collaborative, gregarious, organized. **2** *social events.* collective, communal, community, general, group, popular, public. ▷ SOCIABLE. *Opp* SOLITARY. • *n* dance, disco, *inf* do, gathering, *inf* get-together, party, reception, reunion, soirée. ▷ PARTY.

**socialize** *vb* associate, be sociable, entertain, fraternize, get together, *inf* go out together, join in, keep company, mix, relate.

**society** *n* **1** civilization, the community, culture, the human family, mankind, nation, people, the public. **2** *the society of our friends.* camaraderie, companionship, company, fellowship, friendship, togetherness. **3** *a secret society, etc.* academy, alliance, association, brotherhood, circle, club, confraternity, fraternity, group, guild, league, organization, sisterhood, sodality, sorority, union.

**sofa** *n* chaise longue, couch, settee, sofa bed. ▷ SEAT.

**soft** *adj* **1** compressible, crumbly, cushiony, elastic, flabby, flexible, floppy, limp, malleable, mushy, plastic, pliable, pliant, pulpy, spongy, springy, squashable, squashy, squeezable, supple, tender, yielding. **2** *soft ground.* boggy, marshy, muddy, sodden, waterlogged. **3** *a soft bed.* comfortable, cosy. **4** *soft texture.* downy, feathery, fleecy, fluffy, furry, satiny, silky, sleek, smooth, velvety. **5** *soft music.* dim, faint, low, mellifluous, muted, peaceful, quiet, relaxing, restful, soothing, subdued. **6** *soft breeze.* balmy, delicate, gentle, light, mild, pleasant, warm. **7** [*inf*] *a soft option.* easy, undemanding.

8 *soft feelings*. ▷ SOFT-HEARTED. *Opp* HARD, HARSH, VIOLENT.

**soften** *vb* 1 abate, alleviate, buffer, cushion, deaden, decrease, deflect, diminish, lower, make softer, mellow, mitigate, moderate, muffle, pacify, palliate, quell, quieten, reduce the impact of, subdue, temper, tone down, turn down. 2 dissolve, fluff up, lighten, liquefy, make softer, melt. 3 *soften in attitude*. become softer, concur, ease up, give in, give way, *inf* let up, relax, succumb, weaken, yield. *Opp* HARDEN, INTENSIFY.

**soft-hearted** *adj* benign, compassionate, conciliatory, easygoing, generous, indulgent, kind-hearted, *derog* lax, lenient, merciful, permissive, sentimental, *inf* soft, sympathetic, tender, tender-hearted, tolerant, understanding. ▷ KIND. *Opp* CRUEL.

**soggy** *adj* drenched, dripping, heavy (*soil*), saturated, soaked, sodden, *inf* sopping, wet through. ▷ WET. *Opp* DRY.

**soil** *n* clay, dirt, earth, ground, humus, land, loam, marl, topsoil. • *vb* befoul, besmirch, blacken, contaminate, defile, dirty, make dirty, muddy, pollute, smear, stain, sully, tarnish.

**solace** *n* comfort, consolation, reassurance, relief. • *vb* ▷ CONSOLE.

**soldier** *n* fighter, fighting man, fighting woman, *old use* man at arms, serviceman, servicewoman, warrior. □ *cadet, cavalryman, centurion, commando, conscript, guardsman, gunner, infantryman, lancer, marine, mercenary, NCO, officer, paratrooper, private, recruit, regular, rifleman, sapper, sentry, trooper,* plur *troops, warrior*. ▷ FIGHTER, RANK. • *vb* **soldier on** ▷ PERSIST.

**sole** *adj* exclusive, individual, lone, one, only, single, singular, solitary, unique.

**solemn** *adj* 1 earnest, gloomy, glum, grave, grim, long-faced, reserved, sedate, serious, sober, sombre, staid, straight-faced, thoughtful, unsmiling. *Opp* CHEERFUL. 2 *a solemn occasion*. august, awe-inspiring, awesome, ceremonial, ceremonious, dignified, ecclesiastical, formal, grand, holy, important, imposing, impressive, liturgical, momentous, pompous, religious, ritualistic, stately. *Opp* FRIVOLOUS.

**solicit** *vb* appeal for, ask for, beg, entreat, importune, petition, seek.

**solicitous** *adj* 1 attentive, caring, concerned, considerate, sympathetic. 2 ▷ ANXIOUS.

**solid** *adj* 1 concrete, hard, impenetrable, impermeable, rigid, unmoving. 2 *a solid crowd*. compact, crowded, dense, jammed, packed. 3 *solid gold*. authentic, genuine, pure, real, unadulterated, unalloyed, unmixed. 4 *a solid hour*. continual, continuous, entire, unbroken, uninterrupted, unrelieved, whole. 5 *solid foundations*. firm, fixed, immovable, robust, sound, stable, steady, stout, strong, sturdy, substantial, unbending, unyielding, well-made. 6 *a solid shape*. cubic, rounded, spherical, thick, three-dimensional. 7 *solid evidence*. authoritative, cogent, coherent, convincing, genuine, incontrovertible, indisputable, irrefutable, provable, proven, real, sound, tangible, weighty. 8 *solid support*. complete, dependable, effective, like-minded, reliable, stalwart, strong, trustworthy, unanimous, undivided, united, unwavering, vigorous. *Opp* FLUID, FRAGMENTARY, WEAK.

**solidarity** *n* accord, agreement, coherence, cohesion, concord, harmony, like-mindedness, unanimity, unity. *Opp* DISUNITY.

**solidify** *vb* cake, clot, coagulate, congeal, crystallize, freeze, harden, jell, set, thicken. *Opp* LIQUEFY.

**soliloquy** *n* monologue, speech.

**solitary** *adj* **1** alone, anti-social, cloistered, companionless, friendless, isolated, lonely, lonesome, reclusive, unsociable, withdrawn. **2** *a solitary survivor.* individual, one, only, single, sole. **3** *a solitary place.* desolate, distant, hidden, inaccessible, isolated, out-of-the-way, private, remote, secluded, sequestered, unfrequented, unknown. *Opp* NUMEROUS, PUBLIC, SOCIAL. • *n* anchorite, hermit, *inf* loner, recluse.

**solitude** *n* aloneness, friendlessness, isolation, loneliness, privacy, remoteness, retirement, seclusion.

**solo** *adv* alone, individually, on your own, unaccompanied.

**soloist** *n* performer, player, singer. ▷ MUSIC.

**soluble** *adj* **1** explicable, manageable, solvable, tractable, understandable. **2** *soluble in water.* dispersing, dissolving, melting. *Opp* INSOLUBLE.

**solution** *n* **1** answer, clarification, conclusion, denouement, elucidation, explanation, explication, key, outcome, resolution, solving, unravelling, working out. **2** *a chemical solution.* blend, compound, emulsion, infusion, mixture, suspension.

**solve** *vb* answer, clear up, *inf* crack, decipher, elucidate, explain, explicate, figure out, find the solution to, interpret, puzzle out, resolve, unravel, work out.

**solvent** *adj* creditworthy, in credit, profitable, reliable, self-supporting, solid, sound, viable. *Opp* BANKRUPT.

**sombre** *adj* black, bleak, cheerless, dark, dim, dismal, doleful, drab, dreary, dull, funereal, gloomy, grave, grey, joyless, lowering, lugubrious, melancholy, morose, mournful, serious, sober. ▷ SAD. *Opp* CHEERFUL.

**somewhat** *adv* fairly, moderately, *inf* pretty, quite, rather, *inf* sort of.

**song** *n* air, *inf* ditty, *inf* hit, lyric, number, tune. □ *anthem, aria, ballad, blues, calypso, cantata, canticle, carol, chant, chorus, descant, folk-song, hymn, jingle,* Ger *lied* [plur *lieder*], *lullaby, madrigal, nursery rhyme, plainsong, pop song, psalm, reggae, rock, serenade, shanty, soul, spiritual, wassail.*

**sonorous** *adj* deep, full, loud, powerful, resonant, resounding, reverberant, rich, ringing. *Opp* SHRILL.

**soon** *adv old use* anon, *inf* any minute now, before long, *inf* in a minute, presently, quickly, shortly, straight away.

**sooner** *adv* **1** before, earlier. **2** *sooner have tea than coffee.* preferably, rather.

**soot** *n* dirt, grime.

**soothe** *vb* allay, appease, assuage, calm, comfort, compose, ease, mollify, pacify, quiet, relieve, salve, settle, still, tranquillize.

**soothing** *adj* **1** balmy, balsamic, comforting, demulcent, emollient, healing, lenitive, mild, palliative. **2** *soothing music.* calming, gentle, peaceful, pleasant, reassuring, relaxing, restful, serene.

**sophisticated** *adj* **1** adult, *sl* cool, cosmopolitan, cultivated, cultured, elegant, fashionable, *inf* grown-up, mature, polished, *inf* posh, *derog* pretentious, refined, stylish, urbane, worldly. *Opp* UNSOPHISTICATED. **2** *sophisticated ideas*. advanced, clever, complex, complicated, elaborate, hard to understand, ingenious, intricate, involved, subtle. *Opp* PRIMITIVE, SIMPLE.

**soporific** *adj* boring, deadening, hypnotic, sedative, sleep-inducing, sleepy, somnolent. *Opp* LIVELY.

**sorcerer** *n* conjuror, enchanter, enchantress, magician, magus, medicine man, necromancer, sorceress, *old use* warlock, witch, witch-doctor, wizard.

**sorcery** *n* black magic, charms, conjuring, diabolism, incantations, magic, *inf* mumbo jumbo, necromancy, the occult, spells, voodoo, witchcraft, wizardry.

**sordid** *adj* **1** dingy, dirty, disreputable, filthy, foul, miserable, *inf* mucky, nasty, offensive, polluted, putrid, ramshackle, seamy, seedy, *inf* sleazy, *inf* slummy, squalid, ugly, unclean, undignified, unpleasant, unsanitary, wretched. *Opp* CLEAN. **2** *sordid dealings*. avaricious, base, corrupt, covetous, degenerate, despicable, dishonourable, ignoble, ignominious, immoral, mean, mercenary, rapacious, selfish, *inf* shabby, shameful, unethical, unscrupulous. *Opp* HONOURABLE.

**sore** *adj* **1** aching, burning, chafing, delicate, hurting, inflamed, painful, raw, red, sensitive, smarting, stinging, tender. **2** aggrieved, hurt, irked, *inf* peeved, *inf* put out, resentful, upset, vexed. ▷ ANNOYED. • *n* abrasion, abscess, boil, bruise, burn, carbuncle, gall, gathering, graze, infection, inflammation, injury, laceration, pimple, rawness, redness, scrape, spot, swelling, ulcer. ▷ WOUND. **make sore** abrade, burn, bruise, chafe, chap, gall, graze, hurt, inflame, lacerate, redden, rub.

**sorrow** *n* **1** affliction, anguish, dejection, depression, desolation, despair, desperation, despondency, disappointment, discontent, disgruntlement, dissatisfaction, distress, dolour, gloom, glumness, grief, heartache, heartbreak, heaviness, homesickness, hopelessness, loneliness, melancholy, misery, misfortune, mourning, sad feelings, sadness, suffering, tearfulness, tribulation, trouble, unhappiness, wistfulness, woe, wretchedness. *Opp* HAPPINESS. **2** *sorrow for wrongdoing*. apologies, guilt, penitence, regret, remorse, repentance. • *vb* agonize, be sorrowful, be sympathetic, bewail, grieve, lament, mourn, weep. *Opp* REJOICE.

**sorrowful** *adj* broken-hearted, dejected, disconsolate, distressed, doleful, grief stricken, heartbroken, long faced, lugubrious, melancholy, miserable, mournful, regretful, rueful, saddened, sombre, tearful, unhappy, upset, woebegone, woeful, wretched. ▷ SAD, SORRY. *Opp* HAPPY.

**sorry** *adj* **1** apologetic, ashamed, conscience-stricken, contrite, guilt-ridden, penitent, regretful, remorseful, repentant, shamefaced. **2** *sorry for the homeless*. compassionate, concerned, merciful, pitying, sympathetic, understanding.

**sort** *n* **1** brand, category, class, classification, description, form, genre, group, kind, make, mark, nature, quality, set, type, variety. **2** breed, class, family, genus, race, species, strain, stock, variety. • *vb* arrange, assort, catalogue, categorize, classify, divide, file, grade, group, order, organize, put in order, rank, systematize, tidy. *Opp* MIX. **sort out 1** choose, *inf* put on one side, segregate, select, separate, set aside. **2** *sort out a problem.* attend to, clear up, cope with, deal with, find an answer to, grapple with, handle, manage, organize, put right, resolve, solve, straighten out, tackle.

**soul** *n* **1** psyche, spirit. **2** [*inf*] *poor soul!* ▷ PERSON.

**soulful** *adj* deeply felt, eloquent, emotional, expressive, fervent, heartfelt, inspiring, moving, passionate, profound, sincere, spiritual, stirring, uplifting, warm. *Opp* SOULLESS.

**soulless** *adj* cold, inhuman, insincere, mechanical, perfunctory, routine, spiritless, superficial, trite, unemotional, unfeeling, uninspiring, unsympathetic. *Opp* SOULFUL.

**sound** *adj* **1** durable, fit, healthy, hearty, *inf* in good shape, robust, secure, solid, strong, sturdy, tough, undamaged, uninjured, unscathed, vigorous, well, whole. **2** *sound food.* eatable, edible, fit for human consumption, good, wholesome. **3** *sound ideas.* balanced, coherent, commonsense, correct, convincing, judicious, logical, orthodox, prudent, rational, reasonable, reasoned, sane, sensible, well-founded, wise. **4** *a sound business.* dependable, established, profitable, recognized, reliable, reputable, safe, secure, trustworthy, viable. *Opp* BAD, WEAK. • *n* din, noise, resonance, timbre, tone. □ [Most of these words can be used as either *nouns* or *verbs.*] *bang, bark, bawl, bay, bellow, blare, blast, bleat, bleep, boo, boom, bray, buzz, cackle, caw, chime, chink, chirp, chirrup, chug, clack, clamour, clang, clank, clap, clash, clatter, click, clink, cluck, coo, crack, crackle, crash, creak, croak, croon, crow, crunch, cry, drone, echo, explosion, fizz, grate, grizzle, groan, growl, grunt, gurgle, hiccup, hiss, honk, hoot, howl, hum, jabber, jangle, jeer, jingle, lisp, low, miaow, moan, moo, murmur, neigh, patter, peal, ping, pip, plop, pop, purr, quack, rattle, report, reverberation, ring, roar, rumble, rustle, scream, screech, shout, shriek, sigh, sizzle, skirl, slam, slurp, smack, snap, snarl, sniff, snore, snort, sob, splutter, squawk, squeak, squeal, squelch, swish, throb, thud, thump, thunder, tick, ting, tinkle, toot, trumpet, twang, tweet, twitter, wail, warble, whimper, whine, whinny, whir, whistle, whiz, whoop, woof, yap, yell, yelp, yodel, yowl.* ▷ NOISE. • *vb* **1** become audible, be heard, echo, make a noise, resonate, resound, reverberate. **2** *sound a signal.* activate, cause, create, make, make audible, produce, pronounce, set off, utter. **sound out** check, examine, inquire into, investigate, measure, plumb, probe, research, survey, test, try.

**soup** *n* broth, consommé, stock.

**sour** *n* **1** acid, acidic, acidulous, bitter, citrus, lemony, pungent, sharp, tangy, tart, unripe, vinegary. **2** *sour milk.* bad, curdled, *inf* off, rancid, stale, turned. **3** *sour remarks.* acerbic, bad-tempered, bitter, caustic, cross, crusty, curmudgeonly, cynical, disaffected, disagreeable, grudging, grumpy, ill-natured, irritable, jaundiced, peevish, petulant, snappy, testy, unpleasant.

**source** *n* **1** author, begetter, cause, creator, derivation, informant, initiator, originator, root, starting-point. **2** *source of river*. head, origin, spring, start, well-head, well-spring. ▷ BEGINNING.

**souvenir** *n* heirloom, keepsake, memento, relic, reminder.

**sovereign** *adj* **1** absolute, all-powerful, dominant, highest, royal, supreme, unlimited. **2** *sovereign state*. autonomous, independent, self-governing. • *n* emperor, empress, king, monarch, prince, princess, queen. ▷ RULER.

**sow** *vb* broadcast, disseminate, plant, scatter, seed, spread.

**space** *adj* extraterrestrial, interplanetary, interstellar, orbiting. • *n* **1** emptiness, endlessness, ionosphere, infinity, stratosphere, the universe. **2** *space to move about*. *inf* elbow-room, expanse, freedom, latitude, leeway, margin, room, scope, spaciousness. **3** *an empty space*. area, blank, break, chasm, concourse, distance, duration, gap, hiatus, hole, intermission, interval, lacuna, lapse, opening, place, spell, stretch, time, vacuum, wait. • *vb space things out*. ▷ ARRANGE.

**spacious** *adj* ample, broad, capacious, commodious, extensive, large, open, roomy, sizeable, vast, wide. ▷ BIG. *Opp* SMALL.

**span** *n* breadth, compass, distance, duration, extent, interval, length, period, reach, scope, stretch, term, width. • *vb* arch over, bridge, cross, extend across, go over, pass over, reach over, straddle, stretch over, traverse.

**spank** *vb* slap, slipper, smack. ▷ HIT, PUNISH.

**spar** *vb* box, exchange blows, scrap, shadow-box. ▷ FIGHT.

**spare** *adj* **1** additional, auxiliary, extra, free, inessential, in reserve, leftover, odd, remaining, superfluous, supernumerary, supplementary, surplus, unnecessary, unneeded, unused, unwanted. *Opp* NECESSARY. **2** *a spare figure*. ▷ THIN. • *vb* **1** be merciful to, deliver, forgive, free, have mercy on, let go, let off, liberate, pardon, redeem, release, reprieve, save. **2** *spare money, time, etc*. afford, allow, donate, give, give up, manage, part with, provide, sacrifice. **sparing** ▷ ECONOMICAL, MISERLY.

**spark** *n* flash, flicker, gleam, glint, scintilla, sparkle. • *vb* **spark off** ignite, kindle. ▷ PROVOKE.

**sparkle** *vb* burn, coruscate, flash, flicker, gleam, glint, glitter, reflect, scintillate, shine, spark, twinkle, wink. ▷ LIGHTEN.

**sparkling** *adj* **1** brilliant, flashing, glinting, glittering, scintillating, shining, shiny, twinkling. ▷ BRIGHT. *Opp* DULL. **2** *sparkling drinks*. aerated, bubbling, bubbly, carbonated, effervescent, fizzy, foaming.

**sparse** *adj inf* few and far between, inadequate, light, little, meagre, scanty, scarce, scattered, sparing, spread out, thin, *inf* thin on the ground. *Opp* PLENTIFUL.

**spartan** *adj* abstemious, ascetic, austere, bare, bleak, disciplined, frugal, hard, harsh, plain, rigid, rigorous, severe, simple, stern, strict. *Opp* LUXURIOUS.

**spasm** *n* attack, contraction, convulsion, eruption, fit, jerk, outburst, paroxysm, seizure, *plur* throes, twitch.

**spasmodic** *adj inf* by fits and starts, erratic, fitful, intermittent, interrupted, irregular, jerky, occasional, *inf* on and off, periodic, sporadic. *Opp* CONTINUOUS, REGULAR.

**spate** *n* cataract, flood, flow, gush, inundation, onrush, outpouring, rush, torrent.

**spatter** *vb* bespatter, besprinkle, daub, pepper, scatter, shower, slop, speckle, splash, splatter, spray, sprinkle.

**speak** *vb* answer, argue, articulate, ask, communicate, complain, converse, declaim, declare, deliver a speech, discourse, ejaculate, enunciate, exclaim, express yourself, fulminate, harangue, hold a conversation, hold forth, object, *inf* pipe up, plead, pronounce words, read aloud, recite, say something, soliloquize, *inf* speechify, talk, tell, use your voice, utter, verbalize, vocalize, voice. ▷ SAY, TALK. **speak about** ▷ MENTION. **speak to** ▷ ADDRESS. **speak your mind** be honest, say what you think, speak honestly, speak out, state your opinion, voice your thoughts.

**speaker** *n* lecturer, mouthpiece, orator, public speaker, spokesperson.

**spear** *n* assegai, harpoon, javelin, lance, pike.

**special** *adj* **1** different, distinguished, exceptional, extraordinary, important, infrequent, momentous, notable, noteworthy, odd, *inf* out-of-the-ordinary, rare, red-letter (*day*), remarkable, significant, strange, uncommon, unconventional, unorthodox, unusual. *Opp* ORDINARY. **2** *Petrol has a special smell.* characteristic, distinctive, idiosyncratic, memorable, peculiar, singular, unique, unmistakable. **3** *my special chair.* especial, individual, particular, personal. **4** *a special tool for the job.* bespoke, proper, specific, specialized.

**specialist** *n* **1** authority, connoisseur, expert, fancier (*pigeon fancier*), master, professional, *inf* pundit, researcher. **2** [*medical*] consultant. ▷ MEDICINE.

**speciality** *n inf* claim to fame, expertise, field, forte, genius, *inf* line, specialization, special knowledge, special skill, strength, strong point, talent.

**specialize** *vb* **specialize in** be a specialist in, be best at, concentrate on, devote yourself to, have a reputation for.

**specialized** *adj* esoteric, expert, specialist, unfamiliar.

**species** *n* breed, class, genus, kind, race, sort, type, variety.

**specific** *adj* clear-cut, defined, definite, detailed, exact, explicit, express, fixed, identi-fied, individual, itemized, known, named, particular, peculiar, precise, predetermined, special, specified, unequivocal. *Opp* GENERAL.

**specify** *vb* be specific about, define, denominate, detail, enumerate, establish, identify, itemize, list, name, particularize, *inf* set out, spell out, stipulate.

**specimen** *n* exemplar, example, illustration, instance, model, pattern, representative, sample.

**specious** *adj* casuistic, deceptive, misleading, plausible, seductive.

**speck** *n* bit, crumb, dot, fleck, grain, mark, mite, *old use* mote, particle, speckle, spot, trace.

**speckled** *adj* blotchy, brindled, dappled, dotted, flecked, freckled, mottled, patchy, spattered (with), spotted, spotty, sprinkled (with), stippled.

**spectacle** *n* ceremonial, ceremony, colourfulness, display, exhibition, extravaganza, grandeur, magnificence, ostentation, pageantry, parade, pomp, show, sight, spectacular effects, splendour. **spectacles** ▷ GLASS.

**spectacular** *adj* beautiful, breathtaking, colourful, dramatic, elaborate, eye-catching, impressive, magnificent, *derog* ostentatious, sensational, showy, splendid, stunning.

**spectator** *n* plur audience, beholder, bystander, *plur* crowd, eye-witness, looker-on, observer, onlooker, passer-by, viewer, watcher, witness.

**spectre** *n* apparition, ghost, phantom, presentiment, vision, wraith. ▷ SPIRIT.

**spectrum** *n* compass, extent, gamut, orbit, range, scope, series, span, spread, sweep, variety.

**speculate** *vb* **1** conjecture, consider, hypothesize, make guesses, meditate, ponder, reflect, ruminate, surmise, theorize, weigh up, wonder. ▷ THINK. **2** *speculate in shares.* gamble, invest speculatively, *inf* play the market, take a chance, wager.

**speculative** *adj* **1** abstract, based on guesswork, conjectural, doubtful, *inf* gossipy, hypothetical, notional, suppositional, suppositious, theoretical, unfounded, uninformed, unproven, untested. *Opp* PROVEN. **2** *speculative investments. inf* chancy, *inf* dicey, *inf* dodgy, hazardous, *inf* iffy, risky, uncertain, unpredictable, unreliable, unsafe. *Opp* SAFE.

**speech** *n* **1** articulation, communication, declamation, delivery, diction, elocution, enunciation, expression, pronunciation, speaking, talking, using words, utterance. **2** dialect, idiolect, idiom, jargon, language, parlance, register, tongue. **3** *a public speech.* address, discourse, disquisition, harangue, homily, lecture, oration, paper, presentation, sermon, *inf* spiel, talk, tirade. **4** *speech in a play.* dialogue, lines, monologue, soliloquy.

**speechless** *adj* dumb, dumbfounded, dumbstruck, inarticulate, *inf* mum, mute, nonplussed, silent, thunderstruck, tongue-tied, voiceless. *Opp* TALKATIVE.

**speed** *n* **1** pace, rate, tempo, velocity. **2** alacrity, briskness, celerity, dispatch, expeditiousness, fleetness, haste, hurry, quickness, rapidity, speediness, swiftness. • *vb* **1** *inf* belt, *inf* bolt, *inf* bowl along, canter, career, dash, dart, flash, flit, fly, gallop, *inf* go like the wind, hasten, hurry, hurtle, make haste, move quickly, *inf* nip, *inf* put your foot down, race, run, rush, shoot, sprint, stampede, streak, tear, *inf* zoom. **2** *speed on the road.* break the speed limit, go too fast. **speed up** ▷ ACCELERATE.

**speedy** *adj* **1** expeditious, fast, *old use* fleet, nimble, quick, rapid, swift. **2** *a speedy exit.* hasty, hurried, immediate, precipitate, prompt, unhesitating. *Opp* SLOW.

**spell** *n* **1** bewitchment, charm, conjuration, conjuring, enchantment, incantation, magic formula, sorcery, witchcraft, witchery. **2** *the spell of the theatre.* allure, captivation, charm, enthralment, fascination, glamour, magic. **3** *a spell of rain.* interval, period, phase, season. **4** *a spell at the wheel.* session, stint, stretch, term, time, tour of duty, turn, watch. • *vb* augur, bode, foretell, indicate, mean, portend, presage, signal, signify, suggest. **spell out** ▷ CLARIFY.

**spellbound** *adj* bewitched, captivated, charmed, enchanted, enthralled, entranced, fascinated, hypnotized, mesmerized, overcome, overpowered, transported.

**spend** *vb* 1 *inf* blue, consume, *inf* cough up, disburse, expend, exhaust, *inf* fork out, fritter, *inf* get through, invest, *inf* lash out, pay out, *inf* shell out, *inf* splash out, *inf* splurge, squander. 2 *spend time*. devote, fill, occupy, pass, use up, waste.

**spendthrift** *n inf* big spender, prodigal, profligate, wasteful person, wastrel. *Opp* MISER.

**sphere** *n* 1 ball, globe, globule, orb, spheroid. 2 *sphere of influence*. area, department, discipline, domain, field, province, range, scope, speciality, subject, territory. 3 *social sphere*. caste, class, domain, milieu, position, rank, society, station, stratum, walk of life.

**spherical** *adj* ball-shaped, globe-shaped, globular, rotund, round, spheric, spheroidal.

**spice** *n* 1 flavouring, piquancy, relish, seasoning. □ *allspice, bayleaf, capsicum, cardamom, cassia, cayenne, chilli, cinnamon, cloves, coriander, curry powder, ginger, grains of paradise, juniper, mace, nutmeg, paprika, pepper, pimento, poppy seed, saffron, sesame, turmeric*. 2 *add spice to life*. colour, excitement, gusto, interest, *inf* lift, *inf* pep, sharpness, stimulation, vigour, zest.

**spicy** *adj* aromatic, fragrant, gingery, highly flavoured, hot, peppery, piquant, pungent, seasoned, spiced, tangy, zestful. *Opp* BLAND.

**spike** *n* barb, nail, pin, point, projection, prong, skewer, spine, stake, tine. • *vb* impale, perforate, pierce, skewer, spear, spit, stab, stick.

**spill** *vb* 1 overturn, slop, splash about, tip over, upset. 2 brim, flow, overflow, run, pour. 3 *lorry spilled its load*. discharge, drop, scatter, shed, tip.

**spin** *vb* 1 gyrate, pirouette, revolve, rotate, swirl, turn, twirl, twist, wheel, whirl. 2 *head was spinning*. be giddy, reel, suffer vertigo, swim. **spin out** ▷ PROLONG.

**spindle** *n* axis, axle, pin, rod, shaft.

**spine** *n* 1 backbone, spinal column, vertebrae. 2 *hedgehog's spines*. barb, bristle, needle, point, prickle, prong, quill, spike, spur, thorn.

**spineless** *adj* cowardly, craven, faint-hearted, feeble, helpless, irresolute, *inf* lily-livered, pusillanimous, *inf* soft, timid, unheroic, weedy, *inf* wimpish. ▷ WEAK. *Opp* BRAVE.

**spiral** *adj* cochlear, coiled, corkscrew, turning, whorled. • *n* coil, curl, helix, screw, whorl. • *vb* 1 turn, twist. 2 *spiralling prices*. ▷ FALL, RISE.

**spire** *n* flèche, pinnacle, steeple.

**spirit** *n* 1 *Lat* anima, breath, mind, psyche, soul. 2 *supernatural spirits*. apparition, *inf* bogy, demon, devil, genie, ghost, ghoul, gremlin, hobgoblin, imp, incubus, nymph, phantasm, phantom, poltergeist, *poet* shade, shadow, spectre, *inf* spook, sprite, sylph, vision, visitant, wraith, zombie. 3 *spirit of a poem*. aim, atmosphere, essence, feeling, heart, intention, meaning, mood, purpose, sense. 4 *fighting spirit*.

animation, bravery, cheerfulness, confidence, courage, daring, determination, dynamism, energy, enthusiasm, fire, fortitude, *inf* get-up-and-go, *inf* go, *inf* guts, heroism, liveliness, mettle, morale, motivation, optimism, pluck, resolve, valour, verve, vivacity, will-power, zest. **5** ▷ ALCOHOL.

**spirited** *adj* active, animated, assertive, brave, brisk, buoyant, courageous, daring, determined, dynamic, energetic, enterprising, enthusiastic, frisky, gallant, *inf* gutsy, intrepid, lively, mettlesome, plucky, positive, resolute, sparkling, sprightly, vigorous, vivacious. *Opp* SPIRITLESS.

**spiritless** *adj* apathetic, cowardly, defeatist, despondent, dispirited, dull, irresolute, lacklustre, languid, lethargic, lifeless, listless, melancholy, negative, passive, slow, unenterprising, unenthusiastic. *Opp* SPIRITED.

**spiritual** *adj* devotional, divine, eternal, heavenly, holy, incorporeal, inspired, other-worldly, religious, sacred, unworldly, visionary. *Opp* TEMPORAL.

**spit** *n* **1** dribble, saliva, spittle, sputum. • *vb* dribble, expectorate, salivate, splutter. **spit out** ▷ DISCHARGE. **spitting image** ▷ TWIN.

**spite** *n* animosity, animus, antagonism, *inf* bitchiness, bitterness, *inf* cattiness, gall, grudge, hate, hatred, hostility, ill-feeling, ill will, malevolence, malice, maliciousness, malignity, rancour, resentment, spleen, venom, vindictiveness. • *vb* ▷ ANNOY.

**spiteful** *adj* acid, acrimonious, *inf* bitchy, bitter, *inf* catty, cruel, cutting, hateful, hostile, hurtful, ill-natured, invidious, malevolent, malicious, nasty, poisonous, punitive, rancorous, resentful, revengeful, sharp, *inf* snide, sour, unforgiving, venomous, vicious, vindictive. *Opp* KIND.

**splash** *vb* **1** bespatter, besprinkle, shower, slop, *inf* slosh, spatter, spill, splatter, spray, sprinkle, squirt, wash. **2** *splash about in water.* bathe, dabble, paddle, wade. **3** *splash news across the front page.* blazon, display, exhibit, flaunt, *inf* plaster, publicize, show, spread. **splash out** ▷ SPEND.

**splay** *vb* make a V-shape, slant, spread.

**splendid** *adj* admirable, awe-inspiring, beautiful, brilliant, costly, dazzling, dignified, elegant, fine, first-class, glittering, glorious, gorgeous, grand, great, handsome, imposing, impressive, lavish, luxurious, magnificent, majestic, marvellous, noble, ornate, *derog* ostentatious, palatial, *inf* posh, refulgent, regal, resplendent, rich, royal, *derog* showy, spectacular, *inf* splendiferous, stately, sublime, sumptuous, *inf* super, superb, supreme, wonderful. ▷ EXCELLENT.

**splendour** *n* beauty, brilliance, ceremony, costliness, display, elegance, *inf* glitter, glory, grandeur, luxury, magnificence, majesty, nobility, ostentation, pomp, pomp and circumstance, refulgence, richness, show, spectacle, stateliness, sumptuousness.

**splice** *vb* bind, conjoin, entwine, join, knit, marry, tie together, unite.

**splinter** *n* chip, flake, fragment, shard, shaving, shiver, sliver. • *vb* chip, crack, fracture, shatter, shiver, smash, split. ▷ BREAK.

**split** *n* **1** break, chink, cleavage, cleft, crack, cranny, crevice, fissure, furrow, gash, groove, leak, opening, rent, rift, rip, rupture, slash, slit, tear. **2** breach, dichotomy, difference, dissension, divergence of opinion, division, divorce, estrangement, schism, separation. ▷ QUARREL. • *vb* **1** break up, disintegrate, divide, divorce, go separate ways, move apart, separate. **2** *split logs.* burst, chop, cleave, crack, rend, rip apart, rip open, slash, slice, slit, splinter, tear. ▷ CUT. **3** *split profits.* allocate, allot, apportion, distribute, divide, halve, share. **4** *road splits.* bifurcate, branch, diverge, fork. **split on** ▷ INFORM.

**spoil** *vb* **1** blight, blot, blotch, bungle, damage, deface, destroy, disfigure, *inf* dish, harm, injure, *inf* make a mess of, mar, *inf* mess up, ruin, stain, undermine, undo, upset, vitiate, worsen, wreck. *Opp* IMPROVE. **2** *food spoiled in the heat.* become useless, curdle, decay, decompose, go bad, *inf* go off, moulder, perish, putrefy, rot, *inf* turn. **3** *spoil children.* coddle, cosset, dote on, indulge, make a fuss of, mollycoddle, over-indulge, pamper.

**spoken** *adj* oral, unwritten, verbal, *Lat* viva voce. *Opp* WRITTEN.

**spokesperson** *n* mouthpiece, representative, spokesman, spokeswoman.

**sponge** *vb* **1** clean, cleanse, mop, rinse, sluice, swill, wash, wipe. **2** [*inf*] *sponge on friends.* be dependent (on), cadge (from), scrounge (from).

**spongy** *adj* absorbent, compressible, elastic, giving, porous, soft, springy, yielding. *Opp* SOLID.

**sponsor** *n inf* angel, backer, benefactor, donor, patron, promoter, supporter. • *vb* back, be a sponsor of, finance, fund, help, patronize, promote, subsidize, support, underwrite.

**sponsorship** *n* aegis, auspices, backing, benefaction, funding, guarantee, patronage, promotion, support.

**spontaneous** *adj* **1** *inf* ad lib, extempore, impromptu, impulsive, *inf* off-the-cuff, unplanned, unpremeditated, unprepared, unrehearsed, voluntary. **2** *a spontaneous reaction.* automatic, instinctive, instinctual, involuntary, mechanical, natural, reflex, unconscious, unconstrained, unforced, unthinking. *Opp* PREMEDITATED.

**spooky** *adj* creepy, eerie, frightening, ghostly, haunted, mysterious, scary, uncanny, unearthly, weird.

**spool** *n* bobbin, reel.

**spoon** *n* dessert-spoon, ladle, tablespoon, teaspoon.

**spoon-feed** *vb* cosset, help, indulge, mollycoddle, pamper, spoil.

**spoor** *n* footprints, scent, traces, track.

**sporadic** *adj* errratic, fitful, intermittent, irregular, occasional, periodic, scattered, separate, unpredictable.

**sport** *n* **1** activity, amusement, diversion, enjoyment, entertainment, exercise, fun, games, pastime, play, pleasure, recreation. □ *aerobics, angling, archery, athletics, badminton, base-ball, basketball, billiards, blood sports, bobsleigh, bowls, boxing, canoeing, climbing, cricket, croquet, cross-country, curling, darts, decathlon, discus, fishing, football, gliding, golf, gymnastics, hockey, hunting, hurdling, ice-hockey, javelin, jog-*

*ging, keep-fit, lacrosse, marathon, martial arts, mountaineering, netball, orienteering, pentathlon,* inf *ping-pong, polo, pool, potholing, quoits, racing, rock-climbing, roller-skating, rounders, rowing, Rugby, running, sailing, shooting, shot, show-jumping, skating, skiing, skin-diving, sky-diving, snooker, soccer, squash, street-hockey, surfing, surf-riding, swimming, table-tennis, tennis, tobogganing, trampolining, volley-ball, water-polo, water-skiing, windsurfing, winter sports, wrestling, yachting.* ▷ ATHLETICS, RACE. **2** badinage, banter, humour, jesting, joking, merriment, raillery, teasing. • *vb* **1** caper, cavort, divert yourself, frisk about, frolic, gambol, lark about, rollick, romp, skip about. **2** *sport new clothes.* display, exhibit, flaunt, show off, wear.

**sporting** *adj* considerate, fair, generous, good-humoured, honourable, sportsmanlike.

**sportive** *adj* coltish, frisky, kittenish, light-hearted, playful, waggish. ▷ SPRIGHTLY.

**sportsperson** *n* contestant, participant, player, sportsman, sportswoman.

**sporty** *adj* **1** active, athletic, energetic, fit, vigorous. **2** *sporty clothes.* casual, informal, *inf* loud, rakish, showy, *inf* snazzy.

**spot** *n* **1** blemish, blot, blotch, discoloration, dot, fleck, mark, patch, smudge, speck, speckle, stain, stigma. **2** *spot on the skin.* birthmark, boil, freckle, *plur* impetigo, mole, naevus, pimple, pock, pockmark, *plur* rash, sty, whitlow, *sl* zit. **3** *spots of rain.* bead, blob, drop. **4** *spot for a picnic.* locale, locality, location, neighbourhood, place, point, position, scene, setting, site, situation. **5** *an awkward spot.* difficulty, dilemma, embarrassment, mess, predicament, quandary, situation. **6** *spot of bother.* bit, small amount, *inf* smidgen. • *vb* **1** blot, discolour, fleck, mark, mottle, smudge, spatter, speckle, splash, spray, stain. **2** ▷ SEE.

**spotless** *adj* **1** clean, fresh, immaculate, laundered, unmarked, unspotted, white. **2** *spotless reputation.* blameless, faultless, flawless, immaculate, innocent, irreproachable, pure, unblemished, unsullied, untarnished, *inf* whiter than white.

**spotty** *adj* blotchy, dappled, flecked, freckled, mottled, pimply, pock-marked, pocky, spattered, speckled, speckly, *inf* splodgy, spotted.

**spouse** *n* better half, *old use* helpmate, husband, partner, wife.

**spout** *n* duct, fountain, gargoyle, geyser, jet, lip, nozzle, outlet, rose (*of watering-can*), spray, waterspout. • *vb* **1** discharge, emit, erupt, flow, gush, jet, pour, shoot, spew, spit, spurt, squirt, stream. **2** ▷ TALK.

**sprawl** *vb* **1** flop, lean back, lie, loll, lounge, recline, relax, slouch, slump, spread out, stretch out. **2** be scattered, branch out, spread, straggle.

**spray** *n* **1** drizzle, droplets, fountain, mist, shower, splash, sprinkling. **2** *spray of flowers.* arrangement, bouquet, branch, bunch, corsage, posy, sprig. **3** *spray for paint.* aerosol, atomizer, spray-gun, sprinkler, vaporizer. • *vb* diffuse, disperse, scatter, shower, spatter, splash, spread in droplets, sprinkle.

**spread** *n* 1 broadcasting, broadening, development, diffusion, dispensing, dispersal, dissemination, distribution, expansion, extension, growth, increase, passing on, proliferation, promotion, promulgation. 2 *spread of a bird's wings*. breadth, compass, extent, size, span, stretch, sweep. 3 ▷ MEAL. • *vb* 1 arrange, display, lay out, open out, unfold, unfurl, unroll. 2 broaden, enlarge, expand, extend, fan out, get bigger, get longer, get wider, lengthen, *inf* mushroom, proliferate, straggle, widen. 3 *spread news*. advertise, broadcast, circulate, diffuse, dispense, disperse, disseminate, distribute, divulge, give out, make known, pass on, pass round, proclaim, promote, promulgate, publicize, publish, scatter, sow, transmit. 4 *spread butter*. apply, cover a surface with, smear.

**spree** *n inf* binge, debauch, escapade, *inf* fling, frolic, *inf* orgy, outing, revel. ▷ REVELRY.

**sprightly** *adj* active, agile, animated, brisk, *inf* chipper, energetic, jaunty, lively, nimble, *inf* perky, playful, quickmoving, spirited, sportive, spry, vivacious. *Opp* LETHARGIC.

**spring** *n* 1 bounce, buoyancy, elasticity, give, liveliness, resilience. 2 *clock spring*. coil, mainspring. 3 *spring of water*. fount, fountain, geyser, source (*of river*), spa, well, well-spring. • *vb* bounce, bound, hop, jump, leap, pounce, vault. **spring from** come from, derive from, proceed from, stem from. **spring up** appear, arise, burst forth, come up, develop, emerge, germinate, grow, shoot up, sprout.

**springy** *adj* bendy, elastic, flexible, pliable, resilient, spongy, stretchy, supple. *Opp* RIGID.

**sprinkle** *vb* drip, dust, pepper, scatter, shower, spatter, splash, spray, strew.

**sprint** *vb* dash, *inf* hare, race, speed, *inf* tear. ▷ RUN.

**sprout** *n* bud, shoot. • *vb* bud, come up, develop, emerge, germinate, grow, shoot up, spring up.

**spruce** *adj* clean, dapper, elegant, groomed, *inf* natty, neat, *inf* posh, smart, tidy, trim, well-dressed, well-groomed, *inf* well-turned-out. *Opp* SCRUFFY. • *vb* **spruce up** ▷ TIDY.

**spur** *n* 1 encouragement, goad, impetus, incentive, incitement, inducement, motivation, motive, prod, prompting, stimulus, urging. 2 *motorway spur*. branch, projection. • *vb* animate, egg on, encourage, impel, incite, motivate, pressure, pressurize, prick, prod, prompt, provide an incentive, stimulate, urge.

**spurn** *vb* disown, give (someone) the cold shoulder, jilt, rebuff, reject, renounce, repel, repudiate, repulse, shun, snub, turn your back on.

**spy** *n* contact, double agent, fifth columnist, *sl* grass, infiltrator, informant, informer, *inf* mole, private detective, secret agent, snooper, stool-pigeon, undercover agent. • *vb* 1 be a spy, be engaged in spying, eavesdrop, gather intelligence, inform, *inf* snoop. 2 ▷ SEE. **spy on** keep under surveillance, *inf* tail, trail, watch.

**spying** *n* counter-espionage, detective work, eavesdropping, espionage, intelligence, snooping, surveillance.

**squabble** *vb* argue, bicker, clash, *inf* row, wrangle. ▷ QUARREL.

**squalid** *adj* **1** dingy, dirty, disgusting, filthy, foul, mean, mucky, nasty, poverty-stricken, repulsive, run-down, *inf* sleazy, slummy, sordid, ugly, uncared-for, unpleasant, unsalubrious, wretched. *Opp* CLEAN. **2** *squalid behaviour.* corrupt, degrading, dishonest, dishonourable, disreputable, immoral, scandalous, *inf* shabby, shameful, unethical, unworthy. *Opp* HONOURABLE.

**squander** *vb inf* blow, *inf* blue, dissipate, *inf* fritter, misuse, spend unwisely, *inf* splurge, use up, waste. *Opp* SAVE.

**square** *adj* **1** perpendicular, rectangular, right-angled. **2** *a square deal. inf* above-board, decent, equitable, ethical, fair, honest, honourable, proper, *inf* right and proper, *inf* straight. • *n* **1** piazza, plaza. **2** [*inf*] *an old-fashioned square.* bourgeois, conformist, conservative, conventional person, die-hard, *inf* fuddy-duddy, *inf* old fogy, *inf* stick-in-the-mud, traditionalist. • *vb square an account.* ▷ SETTLE. **squared** chequered, criss-crossed, marked in squares.

**squash** *vb* **1** compress, crumple, crush, flatten, mangle, mash, pound, press, pulp, smash, stamp on, tamp down, tread on. **2** *squash into a room.* cram, crowd, pack, push, ram, shove, squeeze, stuff, thrust, wedge. **3** *squash an uprising.* control, put down, quash, quell, repress, suppress. **4** *squash with a look.* humiliate, *inf* put down, silence, snub.

**squashy** *adj* mashed up, mushy, pulpy, shapeless, soft, spongy, squelchy, yielding. *Opp* FIRM.

**squat** *adj* burly, dumpy, plump, podgy, short, stocky, thick, thick-set. *Opp* TALL. • *vb* crouch, sit.

**squeamish** *adj inf* choosy, dainty, fastidious, finicky, over-scrupulous, particular, *inf* pernickety, prim, *inf* prissy, prudish, scrupulous.

**squeeze** *vb* **1** clasp, compress, crush, embrace, enfold, exert pressure on, flatten, grip, hug, mangle, pinch, press, squash, stamp on, tread on, wring. **2** cram, crowd, pack, push, ram, shove, squash, stuff, tamp, thrust, wedge. **squeeze out** expel, extrude, force out.

**squirm** *vb* twist, wriggle, writhe.

**squirt** *vb* ejaculate, eject, gush, jet, send out, shoot, spit, spout, splash, spray, spurt.

**stab** *n* **1** blow, cut, jab, prick, puncture, thrust, wound, wounding. **2** *stab of pain.* ▷ PAIN. • *vb* bayonet, cut, injure, jab, lance, perforate, pierce, puncture, skewer, spike, stick, thrust, transfix, wound. **have a stab at** ▷ TRY.

**stability** *n* balance, constancy, durability, equilibrium, firmness, immutability, permanence, reliability, solidity, soundness, steadiness, strength. *Opp* INSTABILITY.

**stabilize** *vb* balance, become stable, give stability to, keep upright, make stable, settle. *Opp* UPSET.

**stable** *adj* **1** balanced, firm, fixed, solid, sound, steady, strong, sturdy. **2** constant, continuing, durable, established, immutable, lasting, long-lasting, permanent, predictable, resolute, steadfast, unchanging, unwavering. **3** *a stable personality.* balanced, even-tempered, reasonable, sane, sensible. *Opp* UNSTABLE.

**stack** *n* 1 accumulation, heap, hill, hoard, mound, mountain, pile, quantity, stock, stockpile, store. 2 chimney, pillar, smokestack. 3 *stack of hay. old use* cock, haycock, haystack, rick, stook. • *vb* accumulate, amass, assemble, build up, collect, gather, heap, load, mass, pile, *inf* stash away, stockpile.

**stadium** *n* amphitheatre, arena, ground, sports-ground.

**staff** *n* 1 baton, cane, crook, crosier, flagstaff, pike, pole, rod, sceptre, shaft, stake, standard, stave, stick, token, wand. 2 *staff of a business.* assistants, crew, employees, *old use* hands, personnel, officers, team, workers, workforce. • *vb* man, provide with staff, run.

**stage** *n* 1 apron, dais, performing area, platform, podium, proscenium, rostrum. 2 *stage of a journey.* juncture, leg, phase, period, point, time. • *vb* arrange, *inf* get up, mount, organize, perform, present, produce, *inf* put on, set up, stage-manage.

**stagger** *vb* 1 falter, lurch, pitch, reel, rock, stumble, sway, teeter, totter, walk unsteadily, waver, wobble. 2 *price staggered us.* alarm, amaze, astonish, astound, confuse, dismay, dumbfound, flabbergast, shake, shock, startle, stun, stupefy, surprise, worry.

**stagnant** *adj* motionless, sluggish, stale, standing, static, still, without movement. *Opp* MOVING.

**stagnate** *vb* achieve nothing, become stale, be stagnant, degenerate, deteriorate, idle, languish, stand still, stay still, vegetate. *Opp* PROGRESS.

**stain** *n* 1 blemish, blot, blotch, discoloration, mark, smear, smutch, speck, spot. 2 *a wood stain.* colouring, dye, paint, pigment, tinge, tint, varnish. • *vb* 1 blacken, blemish, blot, contaminate, dirty, discolour, make dirty, mark, smudge, soil, tarnish. 2 *stain your reputation.* besmirch, damage, defile, disgrace, shame, spoil, sully, taint. 3 *stain wood.* colour, dye, paint, tinge, tint, varnish.

**stair** *n* riser, step, tread. **stairs** escalator, flight of stairs, staircase, stairway, steps.

**stake** *n* 1 paling, palisade, pike, pile, pillar, pole, post, rod, sceptre, shaft, stake, standard, stave, stick, upright. 2 *a gambler's stake.* bet, pledge, wager. • *vb* 1 fasten, hitch, secure, tether, tie up. 2 *stake a claim.* establish, put on record, state. 3 *stake my life on it.* bet, chance, gamble, hazard, risk, venture, wager. **stake out** define, delimit, demarcate, enclose, fence in, mark off, outline.

**stale** *adj* 1 dry, hard, limp, mouldy, musty, *inf* off, old, *inf* past its best, tasteless. 2 *stale ideas.* banal, clichéd, familiar, hackneyed, old-fashioned, out-of-date, overused, stock, threadbare, *inf* tired, trite, uninteresting, unoriginal, worn out. *Opp* FRESH.

**stalemate** *n* deadlock, impasse, standstill.

**stalk** *n* branch, shaft, shoot, stem, trunk, twig. • *vb* 1 chase, dog, follow, haunt, hound, hunt, pursue, shadow, tail, track, trail. 2 *stalk about.* prowl, rove, stride, strut. ▷ WALK.

**stall** *n* booth, compartment, kiosk, stand, table. • *vb* be obstructive, delay, hang back, haver, hesitate, pause, *inf* play for time, postpone, prevaricate, procrastinate, put off, stonewall, stop, temporize, waste time.

**stalwart** *adj* courageous, dependable, determined, faithful, indomitable, intrepid, redoubtable, reliable, resolute, robust, staunch, steadfast, sturdy, tough, trustworthy, valiant. ▷ BRAVE, STRONG. *Opp* WEAK.

**stamina** *n* endurance, energy, *inf* grit, indomitability, resilience, staunchness, staying power, *inf* stickability.

**stammer** *vb* falter, hesitate, hem and haw, splutter, stumble, stutter. ▷ TALK.

**stamp** *n* 1 brand, die, hallmark, impression, imprint, print, punch, seal. 2 *stamp of genius*. characteristic, mark, sign. 3 *stamp on a letter*. franking, postage stamp. • *vb* 1 bring down, strike, thump. 2 *stamp a mark*. brand, emboss, engrave, impress, imprint, label, mark, print, punch. **stamp on** ▷ SUPPRESS. **stamp out** ▷ ELIMINATE.

**stampede** *n* charge, dash, flight, panic, rout, rush, sprint. • *vb* 1 bolt, career, charge, dash, gallop, panic, run, rush, sprint, *inf* take to your heels, tear. 2 *stampede cattle*. frighten, panic, rout, scatter.

**stand** *n* 1 base, pedestal, rack, support, tripod, trivet. 2 booth, kiosk, stall. 3 grandstand, terraces. • *vb* 1 arise, get to your feet, get up, rise. 2 *Stand it on the floor*. arrange, deposit, erect, locate, place, position, put up, set up, situate, station, upend. 3 *Trees stand along the avenue*. be, be situated, exist. 4 *My offer stands*. be unchanged, continue, remain valid, stay. 5 *Can't stand onions*. abide, bear, endure, put up with, suffer, tolerate, *inf* wear. **stand by** ▷ SUPPORT. **stand for** ▷ SYMBOLIZE. **stand in for** ▷ DEPUTIZE. **stand out** ▷ SHOW. **stand up for** ▷ PROTECT, SUPPORT. **stand up to** ▷ RESIST.

**standard** *adj* accepted, accustomed, approved, average, basic, classic, common, conventional, customary, definitive, established, everyday, familiar, habitual, normal, official, ordinary, orthodox, popular, prevailing, prevalent, recognized, regular, routine, set, staple (*diet*), stock, traditional, typical, universal, usual. *Opp* UNUSUAL. • *n* 1 archetype, benchmark, criterion, example, exemplar, gauge, grade, guide, guideline, ideal, level of achievement, measure, measurement, model, paradigm, pattern, requirement, rule, sample, specification, touchstone, yardstick. 2 average, level, mean, norm. 3 *standard of a regiment*. banner, colours, ensign, flag, pennant. 4 *lamp standard*. column, pillar, pole, post, support, upright. **standards** ▷ MORALITY.

**standardize** *vb* average out, conform to a standard, equalize, homogenize, normalize, regiment, stereotype, systematize.

**standoffish** *adj* aloof, antisocial, cold, cool, distant, frosty, haughty, remote, reserved, reticent, retiring, secretive, self-conscious, *inf* snooty, taciturn, unapproachable, uncommunicative, unforthcoming, unfriendly, unsociable, withdrawn. *Opp* FRIENDLY.

**standpoint** *n* angle, attitude, belief, opinion, perspective, point of view, position, stance, vantage point, view, viewpoint.

**standstill** *n inf* dead end, deadlock, halt, *inf* hold-up, impasse, jam, stalemate, stop, stoppage.

**staple** *adj* basic, chief, important, main, principal. ▷ STANDARD.

**star** *n* 1 celestial body, sun. □ *asteroid, comet, evening star,* *falling star, lodestar, morning star, nova, shooting star, supernova.* **2** asterisk, pentagram. **3** *TV star.* attraction, big name, celebrity, *It* diva, *inf* draw, idol, leading lady, leading man, personage, *It* prima donna, starlet, superstar. ▷ PERFORMER.

**starchy** *adj* aloof, conventional, formal, prim, stiff. ▷ UNFRIENDLY.

**stare** *vb* gape, *inf* gawp, gaze, glare, goggle, look fixedly, peer. **stare at** contemplate, examine, eye, scrutinize, study, watch.

**stark** *adj* **1** austere, bare, bleak, depressing, desolate, dreary, gloomy, grim. **2** *stark contrast.* absolute, clear, complete, obvious, perfect, plain, sharp, sheer, thoroughgoing, total, unqualified, utter.

**start** *n* **1** beginning, birth, commencement, creation, dawn, establishment, founding, fount, inauguration, inception, initiation, institution, introduction, launch, onset, opening, origin, outset, point of departure, setting out, spring, springboard. *Opp* FINISH. **2** *an unfair start.* advantage, edge, head-start, opportunity. **3** *bank-loan gave me a start.* assistance, backing, financing, help, *inf* leg-up, *inf* send-off, sponsorship. **4** *a nasty start.* jump, shock, surprise. • *vb* **1** depart, embark, *inf* get going, *inf* get under way, *sl* hit the road, *inf* kick off, leave, move off, proceed, set off, set out. **2** *start something.* activate, beget, begin, commence, create, embark on, engender, establish, found, *inf* get cracking on, *inf* get off the ground, *inf* get the ball rolling, give birth to, inaugurate, initiate, instigate, institute, introduce, launch, open, originate, pioneer, set in motion, set up. *Opp* FINISH. **3** *start at sudden noise.* blench, draw back, flinch, jerk, jump, quail, recoil, shy, spring up, twitch, wince. **make someone start** ▷ STARTLE.

**startle** *vb* agitate, alarm, catch unawares, disturb, frighten, give you a start, jolt, make you jump, make you start, scare, shake, shock, surprise, take aback, take by surprise, upset. **startling** ▷ SURPRISING.

**starvation** *n* deprivation, famine, hunger, malnutrition, undernourishment, want.

**starve** *vb* die of starvation, go hungry, go without, perish. **starve yourself** diet, fast, go on hunger strike, refuse food. **starving** ▷ HUNGRY.

**state** *n* **1** *plur* circumstances, condition, fitness, health, mood, *inf* shape, situation. **2** agitation, excitement, *inf* flap, panic, plight, predicament, *inf* tizzy. **3** *a sovereign state.* land, nation. ▷ COUNTRY. • *vb* affirm, announce, assert, asseverate, aver, communicate, declare, express, formulate, proclaim, put into words, report, specify, submit, testify, voice. ▷ SAY, SPEAK, TALK.

**stately** *adj* august, dignified, distinguished, elegant, formal, grand, imperial, imposing, impressive, lofty, majestic, noble, pompous, regal, royal, solemn, splendid, striking. *Opp* INFORMAL. **stately home** ▷ MANSION.

**statement** *n* account, affirmation, announcement, annunciation, assertion, bulletin, comment, communication, communiqué, declaration, disclosure, explanation, message, notice, proclamation, proposition, report, testament, testimony, utterance.

**statesman** *n* diplomat, politician.

**static** *adj* constant, fixed, immobile, immovable, inert, invariable, motionless, passive, stable, stagnant, stationary, steady, still, unchanging, unmoving. *Opp* MOBILE, VARIABLE.

**station** *n* 1 calling, caste, class, degree, employment, level, location, occupation, place, position, post, rank, situation, standing, status. 2 *fire station.* base, depot, headquarters, office. 3 *radio station.* channel, company, transmitter, wavelength. 4 *railway station.* halt, platform, stopping-place, terminus, train station. • *vb* assign, garrison, locate, place, position, put, site, situate, spot, stand.

**stationary** *adj* at a standstill, at rest, halted, immobile, immovable, motionless, parked, pausing, standing, static, still, stock-still, unmoving. *Opp* MOVING.

**stationery** *n* paper, office supplies, writing materials.

**statistics** *n* data, figures, information, numbers.

**statue** *n* carving, figure, statuette. ▷ SCULPTURE.

**statuesque** *adj* dignified, elegant, imposing, impressive, poised, stately, upright.

**stature** *n* 1 build, height, size, tallness. 2 *artist of international stature.* esteem, greatness, recognition. ▷ STATUS.

**status** *n* class, degree, eminence, grade, importance, level, position, prestige, prominence, rank, reputation, significance, standing, station, stature, title.

**staunch** *adj* ▷ STEADFAST.

**stay** *n* 1 holiday, *old use* sojourn, stop, stop-over, visit. 2 *stay of execution.* ▷ DELAY. 3 ▷ SUPPORT. • *vb* 1 *old use* bide, carry on, continue, endure, *inf* hang about, hold out, keep on, last, linger, live on, loiter, persist, remain, survive, *old use* tarry, wait. 2 *stay in a hotel. old use* abide, be accommodated, be a guest, be housed, board, dwell, live, lodge, reside, settle, *old use* sojourn, stop, visit. 3 *stay judgement.* ▷ DELAY.

**steadfast** *adj* committed, constant, dedicated, dependable, determined, devoted, faithful, firm, loyal, patient, persevering, reliable, resolute, resolved, single-minded, sound, stalwart, staunch, steady, true, trustworthy, trusty, unchanging, unfaltering, unflinching, unswerving, unwavering. ▷ STRONG. *Opp* UNRELIABLE.

**steady** *adj* 1 balanced, confident, fast, firm, immovable, poised, safe, secure, settled, solid, stable, substantial. 2 *a steady flow.* ceaseless, changeless, consistent, constant, continuous, dependable, endless, even, incessant, invariable, never-ending, nonstop, perpetual, persistent, regular, reliable, repeated, rhythmic, *inf* round-the-clock, unbroken, unchanging, undeviating, unfaltering, unhurried, uniform, uninterrupted, unrelieved, unremitting, unvarying. *Opp* UNSTEADY. 3 ▷ STEADFAST. • *vb* 1 balance, brace, hold steady, keep still, make steady, secure, stabilize, support. 2 *steady your nerves.* calm, control, soothe, tranquillize.

**steal** *vb* 1 annex, appropriate, arrogate, burgle, commandeer, confiscate, embezzle, expropriate, *inf* filch, hijack, *inf* knock off, *inf* lift, loot, *inf* make off with, misappropriate, *inf* nick, peculate, pick pockets, pilfer, pillage, *inf* pinch, pirate, plagiarize, plunder, poach, purloin, *inf* rip you off, rob, seize, shop-lift, *inf* sneak, *inf* snitch, *inf* swipe, take, thieve, usurp, walk off with. **2** *steal quietly upstairs.* creep, move stealthily, slink, slip, sneak, tiptoe.

**stealing** *n* robbery, theft, thieving. □ *break-in, burglary, embezzlement, fraud, hijacking, housebreaking, larceny, looting, misappropriation, mugging, peculation, pilfering, pillage, piracy, plagiarism, plundering, poaching, purloining, scrumping, shop-lifting.*

**stealthy** *adj* clandestine, concealed, covert, disguised, furtive, imperceptible, inconspicuous, quiet, secret, secretive, *inf* shifty, sly, *inf* sneaky, surreptitious, underhand, unobtrusive. *Opp* BLATANT.

**steam** *n* condensation, haze, mist, smoke, vapour.

**steamy** *adj* **1** blurred, clouded, cloudy, fogged over, foggy, hazy, misted over, misty. **2** *a steamy atmosphere.* close, damp, humid, moist, muggy, *inf* sticky, sultry, sweaty, sweltering. **3** [*inf*] *steamy sex scenes.* ▷ SEXY.

**steep** *adj* **1** abrupt, bluff, headlong, perpendicular, precipitous, sharp, sheer, sudden, vertical. *Opp* GRADUAL. **2** *steep prices.* ▷ EXPENSIVE. • *vb* ▷ SOAK.

**steeple** *n* pinnacle, point, spire.

**steer** *vb* be at the wheel, control, direct, drive, guide, navigate, pilot. **steer clear of** ▷ AVOID.

**stem** *n* peduncle, shoot, stalk, stock, trunk, twig. • *vb* arise, come, derive, develop, emanate, flow, issue, originate, proceed, result, spring, sprout. **2** *to stem the flow.* ▷ CHECK.

**stench** *n* mephitis, *inf* pong, reek, stink. ▷ SMELL.

**step** *n* **1** footfall, footstep, pace, stride, tread. **2** doorstep, rung, stair, tread. **3** *a step forward.* advance, move, movement, progress, progression. **4** *step in a process.* action, initiative, manoeuvre, measure, phase, procedure, stage. • *vb* put your foot, stamp, stride, trample, tread. ▷ WALK. **steps** ladder, stairs, staircase, stairway, stepladder. **step down** ▷ RESIGN. **step in** ▷ ENTER, INTERVENE. **step on it** ▷ HURRY. **step up** ▷ INCREASE. **take steps** ▷ BEGIN.

**stereoscopic** *adj* solid-looking, three-dimensional, *inf* 3-D.

**stereotype** *n* formula, model, pattern, stereotyped idea.

**stereotyped** *adj* clichéd, conventional, formalized, hackneyed, predictable, standard, standardized, stock, typecast, unoriginal.

**sterile** *adj* **1** arid, barren, childless, dry, fruitless, infertile, lifeless, unfruitful, unproductive. **2** *sterile bandage.* antiseptic, aseptic, clean, disinfected, germfree, hygienic, pure, sanitary, sterilized, uncontaminated, uninfected, unpolluted. **3** *a sterile attempt.* abortive, fruitless, hopeless, pointless, unprofitable, useless. *Opp* FERTILE, FRUITFUL, SEPTIC.

**sterilize** *vb* **1** clean, cleanse, decontaminate, depurate, disinfect, fumigate, make sterile, pasteurize, purify. **2** *sterilize animals.* castrate, caponize, emasculate, geld, neuter, perform a vasectomy on, spay, vasectomize.

**stern** *adj* adamant, austere, authoritarian, critical, dour, forbidding, frowning, grim, hard, harsh, inflexible, obdurate, resolute, rigid, rigorous, severe, strict, stringent, tough, unbending, uncompromising, unrelenting, unremitting. ▷ SERIOUS. *Opp* SOFT-HEARTED. • *n stern of ship*. aft, back, rear end.

**stew** *n* casserole, goulash, hash, hot-pot, ragout. • *vb* boil, braise, casserole, simmer. ▷ COOK.

**steward, stewardess** *ns* 1 attendant, waiter. ▷ SERVANT. 2 marshal, officer, official.

**stick** *n* branch, stalk, twig. □ *bar, baton, cane, club, hockey-stick, pike, pole, rod, staff, stake, walking-stick, wand*. • *vb* 1 bore, dig, impale, jab, penetrate, pierce, pin, poke, prick, prod, punch, puncture, run through, spear, spike, spit, stab, thrust, transfix. 2 *stick with glue*. adhere, affix, agglutinate, bind, bond, cement, cling, coagulate, fuse together, glue, gum, paste, solder, weld. ▷ FASTEN. 3 *stick in your mind*. be fixed, continue, endure, keep on, last, linger, persist, remain, stay. 4 *stick in mud*. become trapped, get bogged down, seize up, jam, wedge. 5 ▷ TOLERATE. **stick at** ▷ PERSIST. **stick in** ▷ PENETRATE. **stick out** ▷ PROTRUDE. **stick together** ▷ UNITE. **stick up** ▷ PROTRUDE. **stick up for** ▷ DEFEND. **stick with** ▷ SUPPORT.

**sticky** *adj* 1 adhesive, glued, gummed, self-adhesive. 2 *sticky paint*. gluey, glutinous, *inf* gooey, gummy, tacky, viscous. 3 *sticky weather*. clammy, close, damp, dank, humid, moist, muggy, steamy, sultry, sweaty. *Opp* DRY.

**stiff** *adj* 1 compact, dense, firm, hard, heavy, inelastic, inflexible, rigid, semi-solid, solid, solidified, thick, tough, unbending, unyielding, viscous. 2 *stiff joints*. arthritic, immovable, painful, paralysed, rheumatic, taut, tight. 3 *a stiff task*. arduous, challenging, difficult, exacting, exhausting, hard, laborious, tiring, tough, uphill. 4 *stiff opposition*. determined, dogged, obstinate, powerful, resolute, stubborn, unyielding, vigorous. 5 *a stiff manner*. artificial, awkward, clumsy, cold, forced, formal, graceless, haughty, inelegant, laboured, mannered, pedantic, self-conscious, standoffish, starchy, stilted, *inf* stuffy, tense, turgid, ungainly, unnatural, wooden. 6 *stiff penalties*. cruel, drastic, excessive, harsh, hurtful, merciless, pitiless, punishing, punitive, relentless, rigorous, severe, strict. 7 *stiff wind*. brisk, fresh, strong. 8 *stiff drink*. alcoholic, potent, strong. *Opp* EASY, RELAXED, SOFT.

**stiffen** *vb* become stiff, clot, coagulate, congeal, dry out, harden, jell, set, solidify, thicken, tighten, toughen.

**stifle** *vb* 1 asphyxiate, choke, smother, strangle, suffocate, throttle. 2 *stifle laughter*. check, control, curb, dampen, deaden, keep back, muffle, restrain, suppress, withhold. 3 *stifle free speech*. crush, destroy, extinguish, kill off, quash, repress, silence, stamp out, stop.

**stigma** *n* blot, brand, disgrace, dishonour, mark, reproach, shame, slur, stain, taint.

**stigmatize** *vb* brand, condemn, defame, denounce, disparage, label, mark, pillory, slander, vilify.

**still** *adj* at rest, calm, even, flat, hushed, immobile, inert, lifeless, motionless, noiseless, pacific, peaceful, placid, quiet, restful, serene, silent, smooth, soundless, stagnant, static, stationary, tranquil, unmoving, unruffled, untroubled, windless. *Opp* ACTIVE,

NOISY. • *vb* allay, appease, assuage, calm, lull, make still, pacify, quieten, settle, silence, soothe, subdue, suppress, tranquillize. *Opp* AGITATE.

**stimulant** *n* anti-depressant, drug, *inf* pick-me-up, restorative, *inf* reviver, *inf* shot in the arm, tonic. ▷ STIMULUS.

**stimulate** *vb* activate, arouse, awaken, cause, encourage, excite, fan, fire, foment, galvanize, goad, incite, inflame, inspire, instigate, invigorate, kindle, motivate, prick, prompt, provoke, quicken, rouse, set off, spur, stir up, titillate, urge, whet. *Opp* DISCOURAGE.

**stimulating** *adj* arousing, challenging, exciting, exhilarating, inspirational, inspiring, interesting, intoxicating, invigorating, provocative, provoking, rousing, stirring, thought-provoking, titillating. *Opp* UNINTERESTING.

**stimulus** *n* challenge, encouragement, fillip, goad, incentive, inducement, inspiration, prompting, provocation, spur, stimulant. *Opp* DISCOURAGEMENT.

**sting** *n* bite, prick, stab. ▷ PAIN. • *vb* **1** bite, nip, prick, wound. **2** smart, tingle. ▷ HURT.

**stingy** *adj* **1** avaricious, cheeseparing, close, close-fisted, covetous, mean, mingy, miserly, niggardly, parsimonious, penny-pinching, tight-fisted, ungenerous. **2** *stingy helpings.* inadequate, insufficient, meagre, *inf* measly, scanty. ▷ SMALL. *Opp* GENEROUS.

**stink** *n, vb* ▷ SMELL.

**stipulate** *vb* demand, insist on, make a stipulation, require, specify.

**stipulation** *n* condition, demand prerequisite, proviso, requirement, specification.

**stir** *n* ▷ COMMOTION. • *vb* **1** agitate, beat, blend, churn, mingle, mix, move about, scramble, whisk. **2** *stir from sleep.* arise, bestir yourself, *inf* get a move on, *inf* get going, get up, move, rise, *inf* show signs of life, *inf* stir your stumps. **3** *stir emotions.* activate, affect, arouse, awaken, challenge, disturb, electrify, excite, exhilarate, fire, impress, inspire, kindle, move, resuscitate, revive, rouse, stimulate, touch, upset.

**stirring** *adj* affecting, arousing, challenging, dramatic, electrifying, emotional, emotion-charged, emotive, exciting, exhilarating, heady, impassioned, inspirational inspiring, interesting, intoxicating, invigorating, moving, provocative, provoking, rousing, spirited stimulating, stirring, thought-provoking, thrilling, titillating, touching. *Opp* UNEXCITING.

**stitch** *vb* darn, mend, repair, sew, tack.

**stock** *adj* accustomed, banal, clichéd, common, commonplace, conventional, customary, expected, hackneyed, ordinary, predictable, regular, routine, run-of-the-mill, set, standard, staple, stereotyped, *inf* tired, traditional, trite, unoriginal, usual. *Opp* UNEXPECTED. • *n* **1** cache, hoard, reserve, reservoir, stockpile, store, supply. **2** *stock of a shop.* commodities, goods, merchandise, range, wares. **3** *farm stock.* animals, beasts, cattle, flocks, herds, livestock. **4** *ancient stock.* ancestry, blood, breed, descent, dynasty, extraction, family, forebears, genealogy, line, lineage parentage, pedigree. **5** *meat stock.* broth, soup. • *vb* carry, deal in, handle, have available, *inf* keep, keep in stock, market, offer, provide, sell, supply, trade in. **out of stock** sold out, unavailable. **take stock** ▷ REVIEW.

**stockade** *n* fence, paling, palisade, wall.

**stockings** *n* nylons, panti-hose, socks, tights.

**stockist** *n* merchant, retailer, seller, shopkeeper, supplier.

**stocky** *adj* burly, compact, dumpy, heavy-set, short, solid, squat, stubby, sturdy, thickset. *Opp* THIN.

**stodgy** *adj* **1** filling, heavy, indigestible, lumpy, soggy, solid, starchy. *Opp* SUCCULENT. **2** *a stodgy lecture.* boring, dull, ponderous, *inf* stuffy, tedious, tiresome, turgid, unexciting, unimaginative, uninteresting. *Opp* LIVELY.

**stoical** *adj* calm, cool, disciplined, impassive, imperturbable, long-suffering, patient, philosophical, phlegmatic, resigned, stolid, uncomplaining, unemotional, unexcitable, *inf* unflappable. *Opp* EXCITABLE.

**stoke** *vb* fuel, keep burning, mend, put fuel on, tend.

**stole** *n* cape, shawl, wrap.

**stolid** *adj* bovine, dull, heavy, immovable, impassive, lumpish, phlegmatic, unemotional, unexciting, unimaginative, wooden. ▷ STOICAL. *Opp* LIVELY.

**stomach** *n* abdomen, belly, *inf* guts, *inf* insides, *derog* paunch, *derog* pot, *inf* tummy. • *vb* ▷ TOLERATE.

**stomach-ache** *n* colic, *inf* collywobbles, *inf* gripes, *inf* tummy-ache.

**stone** *n* **1** boulder, cobble, *plur* gravel, pebble, rock, *plur* scree. ▷ ROCK. **2** block, flagstone, sett, slab. **3** *a memorial stone.* gravestone, headstone, memorial, monolith, obelisk, tablet. **4** *precious stone.* ▷ JEWEL. **5** *stone in fruit.* pip, pit, seed.

**stony** *adj* **1** pebbly, rocky, rough, shingly. **2** *stony silence.* adamant, chilly, cold, coldhearted, expressionless, frigid, hard, *inf* hard-boiled, heartless, hostile, icy, indifferent, insensitive, merciless, pitiless, steely, stony-hearted, uncaring, unemotional, unfeeling, unforgiving, unfriendly, unresponsive, unsympathetic.

**stooge** *n* butt, dupe, *inf* fall-guy, lackey, puppet.

**stoop** *vb* **1** bend, bow, crouch, duck, hunch your shoulders, kneel, lean, squat. **2** condescend, degrade yourself, deign, humble yourself, lower yourself, sink.

**stop** *n* **1** ban, cessation, close, conclusion, end, finish, halt, shutdown, standstill, stoppage, termination. **2** *a stop for refreshments.* break, destination, pause, resting-place, stage, station, stopover, terminus. **3** *a stop at a hotel.* holiday, *old use* sojourn, stay, vacation, visit. • *vb* **1** break off, call a halt to, cease, conclude, cut off, desist from, discontinue, end, finish, halt, *inf* knock off, leave off, *inf* pack in, pause, quit, refrain from, rest from, suspend, terminate. **2** *stop the flow.* bar, block, check, curb, cut off, delay, frustrate, halt, hamper, hinder, immobilize, impede, intercept, interrupt, *inf* nip in the bud, obstruct, put a stop to, stanch, staunch, stem, suppress, thwart. **3** *stop in a hotel.* be a guest, have a holiday, *old use* sojourn, spend time, stay, visit. **4** *stop a gap*. *inf* bung up, close, fill in, plug, seal. **5** *stop a thief*. arrest, capture, catch, detain, hold, seize. **6** *the rain stopped*. be over, cease, come to an end, finish, peter out. **7** *the bus stopped*. come to rest, draw up, halt, pull up.

**stopper** *n* bung, cork, plug.

**store** *n* **1** accumulation, cache, fund, hoard, quantity, reserve, reservoir, stock, stockpile, supply. ▷ STOREHOUSE. **2** *a grocery store*. outlet, retail business, retailers, supermarket. ▷ SHOP. • *vb* accumulate, aggregate, deposit, hoard, keep, lay by, lay in, lay up, preserve, put away, reserve, save, set aside, *inf* stash away, stockpile, stock up, stow away.

**storehouse** *n* depository, repository, storage, store, store-room. □ *armoury, arsenal, barn, cellar, cold-storage, depot, granary, larder, pantry, safe, silo, stock-room, strong-room, treasury, vault, warehouse.*

**storey** *n* deck, floor, level, stage, tier.

**storm** *n* **1** disturbance, onslaught, outbreak, outburst, stormy weather, tempest, tumult, turbulence. □ *blizzard, cloudburst, cyclone, deluge, dust-storm, electrical storm, gale, hailstorm, hurricane, mistral, monsoon, rainstorm, sandstorm, simoom, sirocco, snowstorm, squall, thunderstorm, tornado, typhoon, whirlwind.* **2** *storm of protest*. ▷ CLAMOUR. • *vb* ▷ ATTACK.

**stormy** *adj* angry, blustery, choppy, fierce, furious, gusty, raging, rough, squally, tempestuous, thundery, tumultuous, turbulent, vehement, violent, wild, windy. *Opp* CALM.

**story** *n* **1** account, anecdote, chronicle, fiction, history, narration, narrative, plot, recital, record, scenario, tale, yarn. □ *allegory, children's story, crime story, detective story, epic, fable, fairy-tale, fantasy, folk-tale, legend, mystery, myth, novel, parable, romance, saga, science fiction, SF, thriller,* inf *whodunit.* **2** *story in newspaper*. article, dispatch, exclusive, feature, news item, piece, report, scoop. **3** falsehood, *inf* fib, lie, tall story, untruth.

**storyteller** *n* author, biographer, narrator, raconteur, teller.

**stout** *adj* **1** *inf* beefy, big, bulky, burly, *inf* chubby, corpulent, fleshy, heavy, *inf* hulking, overweight, plump, portly, solid, stocky, *inf* strapping, thick-set, *inf* tubby, well-built. ▷ FAT. *Opp* THIN. **2** *stout rope*. durable, reliable, robust, sound, strong, sturdy, substantial, thick, tough. **3** *stout fighter*. bold, brave, courageous, fearless, gallant, heroic, intrepid, plucky, resolute, spirited, valiant. *Opp* WEAK.

**stove** *n* boiler, cooker, fire, furnace, heater, oven, range.

**stow** *vb* load, pack, put away, *inf* stash away, store.

**straggle** *vb* be dispersed, be scattered, dangle, dawdle, drift, fall behind, lag, loiter, meander, ramble, scatter, spread out, stray, string out, trail, wander. **straggling** ▷ DISORGANIZED, LOOSE.

**straight** *adj* **1** aligned, direct, flat, linear, regular, smooth, true, unbending, undeviating, unswerving. **2** neat, orderly, organized, right, *inf* shipshape, sorted out, spruce, tidy. **3** *a straight sequence*. consecutive, continuous, non-stop, perfect, sustained, unbroken, uninterrupted, unrelieved. **4** ▷ STRAIGHTFORWARD. *Opp* CROOKED, INDIRECT, UNTIDY.

**straight away** at once, directly, immediately, instantly, now, without delay.

**straighten** *vb* disentangle, make straight, put straight, rearrange, sort out, tidy, unbend, uncurl, unravel, untangle, untwist.

**straightforward** *adj* blunt, candid, direct, easy, forthright, frank, genuine, honest, intelligible, lucid, open, plain, simple, sincere, straight, truthful, uncomplicated. *Opp* DEVIOUS.

**strain** *n* 1 anxiety, difficulty, effort, exertion, hardship, pressure, stress, tension, worry. 2 *genetic strain.* ▷ ANCESTRY. • *vb* 1 haul, heave, make taut, pull, stretch, tighten, tug. 2 *strain to succeed.* attempt, endeavour, exert yourself, labour, make an effort, strive, struggle, toil, try. 3 *strain yourself.* exercise, exhaust, overtax, *inf* push to the limit, stretch, tax, tire out, weaken, wear out, weary. 4 *strain a muscle.* damage, hurt, injure, overwork, pull, rick, sprain, tear, twist, wrench, 5 *strain liquid.* clear, drain, draw off, filter, percolate, purify, riddle, screen, separate, sieve, sift.

**strained** *adj* 1 artificial, constrained, distrustful, embarrassed, false, forced, insincere, self-conscious, stiff, tense, uncomfortable, uneasy, unnatural. 2 *strained look.* drawn, tired, weary. 3 *strained interpretation.* far-fetched, incredible, laboured, unlikely, unreasonable. *Opp* NATURAL, RELAXED.

**strainer** *n* colander, filter, riddle, sieve.

**strand** *n* fibre, filament, string, thread, wire. • *vb* 1 abandon, desert, forsake, leave stranded, lose, maroon. 2 *strand a ship.* beach, ground, run aground, wreck. **stranded** ▷ AGROUND, HELPLESS.

**strange** *adj* 1 abnormal, astonishing, atypical, bizarre, curious, eerie, exceptional, extraordinary, fantastic, *inf* funny, grotesque, irregular, odd, out-of-the-ordinary, outré, peculiar, quaint, queer, rare, remarkable, singular, surprising, surreal, uncommon, unexpected, unheard-of, unique, unnatural, untypical, unusual. 2 *strange neighbours. inf* cranky, eccentric, sinister, unconventional, weird, *inf* zany. 3 *a strange problem.* baffling, bewildering, inexplicable, insoluble, mysterious, mystifying, perplexing, puzzling, unaccountable. 4 *strange places.* alien, exotic, foreign, little-known, off the beaten track, outlandish, outof-the-way, remote, unexplored, unmapped. 5 *strange experience.* different, fresh, new, novel, unaccustomed, unfamiliar. *Opp* FAMILIAR, ORDINARY.

**strangeness** *n* abnormality, bizarreness, eccentricity, eeriness, extraordinariness, irregularity, mysteriousness, novelty, oddity, oddness, outlandishness, peculiarity, quaintness, queerness, rarity, singularity, unconventionality, unfamiliarity.

**stranger** *n* alien, foreigner, guest, newcomer, outsider, visitor.

**strangle** *vb* 1 asphyxiate, choke, garotte, smother, stifle, suffocate, throttle. 2 *strangle a cry.* ▷ SUPPRESS.

**strangulate** *vb* bind, compress, constrict, squeeze.

**strangulation** *n* asphyxiation, garotting, suffocation.

**strap** *n* band, belt, strop, tawse, thong, webbing. • *vb* ▷ FASTEN.

**stratagem** *n* artifice, device, *inf* dodge, manoeuvre, plan, ploy, ruse, scheme, subterfuge, tactic, trick.

**strategic** *adj* advantageous, critical, crucial, deliberate, key, planned, politic, tactical, vital.

**strategy** *n* approach, design, manoeuvre, method, plan, plot, policy, procedure, programme, *inf* scenario, scheme, tactics.

**stratum** *n* layer, seam, table, thickness, vein.

**straw** *n* corn, stalks, stubble.

**stray** *adj* **1** abandoned, homeless, lost, roaming, roving, wandering. **2** *stray bullets.* accidental, casual, chance, haphazard, isolated, lone, occasional, odd, random, single. • *vb* **1** get lost, get separated, go astray, meander, move about aimlessly, ramble, range, roam, rove, straggle, wander. **2** *stray from the point.* deviate, digress, diverge, drift, get off the subject, *inf* go off at a tangent, veer.

**streak** *n* **1** band, bar, dash, line, mark, score, smear, stain, stria, striation, strip, stripe, stroke, vein. **2** *a selfish streak.* component, element, strain, touch, trace. **3** *streak of good luck.* period, run, series, spate, spell, stretch, time. • *vb* **1** mark with streaks, smear, smudge, stain, striate. **2** *streak past.* dart, dash, flash, fly, gallop, hurtle, move at speed, rush, *inf* scoot, speed, sprint, tear, *inf* whip, zoom.

**streaky** *adj* barred, lined, smeary, smudged, streaked, striated, stripy, veined.

**stream** *n* **1** beck, brook, brooklet, burn, channel, freshet, *poet* rill, river, rivulet, streamlet, watercourse. **2** cascade, cataract, current, deluge, effluence, flood, flow, fountain, gush, jet, outpouring, rush, spate, spurt, surge, tide, torrent. • *vb* cascade, course, deluge, flood, flow, gush, issue, pour, run, spill, spout, spurt, squirt, surge, well.

**streamer** *n* banner, flag, pennant, pennon, ribbon.

**streamlined** *adj* **1** aerodynamic, elegant, graceful, hydrodynamic, sleek, smooth. **2** ▷ EFFICIENT.

**street** *n* avenue, roadway, terrace. ▷ ROAD.

**strength** *n* **1** brawn, capacity, condition, energy, fitness, force, health, might, muscle, power, resilience, robustness, sinew, stamina, stoutness, sturdiness, toughness, vigour. **2** *strength of purpose.* *inf* backbone, commitment, courage, determination, firmness, *inf* grit, perseverance, persistence, resolution, resolve, spirit, tenacity. *Opp* WEAKNESS.

**strengthen** *vb* **1** bolster, boost, brace, build up, buttress, encourage, fortify, harden, hearten, increase, make stronger, prop up, reinforce, stiffen, support, tone up, toughen. **2** *strengthen an argument.* back up, consolidate, corroborate, enhance, justify, substantiate. *Opp* WEAKEN.

**strenuous** *adj* **1** arduous, backbreaking, burdensome, demanding, difficult, exhausting, gruelling, hard, laborious, punishing, stiff, taxing, tough, uphill. *Opp* EASY. **2** *strenuous efforts.* active, committed, determined, dogged, dynamic, eager, energetic, herculean, indefatigable, laborious, pertinacious, resolute, spirited, strong, tenacious, tireless, unremitting, vigorous, zealous. *Opp* CASUAL.

**stress** *n* **1** anxiety, difficulty, distress, hardship, pressure, strain, tenseness, tension, trauma, worry. **2** accent, accentuation, beat, emphasis, importance, significance, underlining, urgency, weight. • *vb* **1** accent, accentuate, assert, draw attention to, emphasize, feature, highlight, insist on, lay stress on, mark, put stress on, repeat, spotlight, underline, underscore. **2** *stressed by work.* burden, distress, overstretch, pressurize, pressure, push to the limit, tax, weigh down.

**tressful** *adj* anxious, difficult, taxing, tense, tiring, traumatic, worrying. *Opp* RELAXED.

**tretch** *n* **1** period, spell, stint, term, time, tour of duty. **2** *stretch of country.* area, distance, expanse, length, span, spread, sweep, tract. • *vb* **1** broaden, crane (*your neck*), dilate, distend, draw out, elongate, enlarge, expand, extend, flatten out, inflate, lengthen, open out, pull out, spread out, swell, tauten, tighten, widen. **2** *stretch into the distance.* be unbroken, continue, disappear, extend, go, reach out, spread. **3** *stretch resources.* overextend, overtax, *inf* push to the limit, strain, tax.

**trew** *vb* disperse, distribute, scatter, spread, sprinkle.

**trict** *adj* **1** austere, authoritarian, autocratic, firm, harsh, merciless, *inf* no-nonsense, rigorous, severe, stern, stringent, tyrannical, uncompromising. *Opp* EASYGOING. **2** *strict rules.* absolute, binding, defined, *inf* hard and fast, inflexible, invariable, precise, rigid, stringent, tight, unchangeable. *Opp* FLEXIBLE. **3** *strict truthfulness.* accurate, complete, correct, exact, meticulous, perfect, precise, right, scrupulous.

**stride** *n* pace, step. • *vb* ▷ WALK.

**strident** *adj* clamorous, discordant, grating, harsh, jarring, loud, noisy, raucous, screeching, shrill, unmusical. *Opp* SOFT.

**strife** *n* animosity, arguing, bickering, competition, conflict, discord, disharmony, dissention, enmity, friction, hostility, quarrelling, rivalry, unfriendliness. ▷ FIGHT. *Opp* CO-OPERATION.

**strike** *n* **1** go-slow, industrial action, stoppage, walk-out, withdrawal of labour. **2** assault, attack, bombardment. • *vb* **1** bang against, bang into, beat, collide with, hammer, impel, knock, rap, run into, smack, smash into, *inf* thump, *inf* whack. ▷ ATTACK, HIT. **2** *strike a match.* ignite, light. **3** *tragedy struck us forcibly.* affect, afflict, *inf* come home to, impress, influence. **4** *clock struck one.* chime, ring, sound. **5** *strike for more pay. inf* come out, *inf* down tools, stop work, take industrial action, withdraw labour, work to rule. **6** *strike a flag, tent.* dismantle, lower, pull down, remove, take down.

**striking** *adj* affecting, amazing, arresting, conspicuous, distinctive, extraordinary, glaring, imposing, impressive, memorable, noticeable, obvious, out-of-the-ordinary, outstanding, prominent, showy, stunning, telling, unmistakable, unusual. *Opp* INCONSPICUOUS.

**string** *n* **1** cable, cord, fibre, line, rope, twine. **2** chain, file, line, procession, progression, queue, row, sequence, series, stream, succession, train. • *vb string together* connect, join, line up, link, thread. **stringed instruments** strings. □ *banjo, cello,* *clavichord, double-bass,* inf *fiddle, guitar, harp, harpsichord, lute, lyre, piano, sitar, spinet, ukulele, viola, violin, zither.*

**stringy** *adj* chewy, fibrous, gristly, sinewy, tough. *Opp* TENDER.

**strip** *n* band, belt, fillet, lath, line, narrow piece, ribbon, shred, slat, sliver, stripe, swathe. • *vb* **1** bare, clear, decorticate, defoliate, denude, divest, *old use* doff, excoriate, flay, lay bare, peel, remove the covering, remove the paint, remove the skin, skin, uncover. *Opp* COVER. **2** *strip to the waist.* bare yourself, disrobe, expose yourself, get undressed, uncover yourself. *Opp* DRESS. **strip down** ▷ DISMANTLE. **strip off** ▷ UNDRESS.

**stripe** *n* band, bar, chevron, line, ribbon, streak, striation, strip, stroke, swathe.

**striped** *adj* banded, barred, lined, streaky, striated, stripy.

**strive** *vb* attempt, *inf* do your best, endeavour, make an effort, strain, struggle, try. ▷ FIGHT.

**stroke** *n* **1** action, blow, effort, knock, move, swipe. ▷ HIT. **2** *a stroke of the pen.* flourish, gesture, line, mark, movement, sweep. **3** [*medical*] apoplexy, attack, embolism, fit, seizure, spasm, thrombosis. • *vb* caress, fondle, massage, pass your hand over, pat, pet, rub, soothe, touch.

**stroll** *n, vb* amble, meander, saunter, wander. ▷ WALK.

**strong** *adj* **1** durable, hard, hardwearing, heavy-duty, impregnable, indestructible, permanent, reinforced, resilient, robust, sound, stout, substantial, thick, unbreakable, well-made. **2** *strong physique.* athletic, *inf* beefy, *inf* brawny, burly, fit, *inf* hale and hearty, hardy, *inf* hefty, *inf* husky, mighty, muscular, powerful, robust, sinewy, stalwart, *inf* strapping, sturdy, tough, well-built, wiry. **3** *strong personality.* assertive, committed, determined, domineering, dynamic, energetic, forceful, independent, reliable, resolute, stalwart, steadfast, *inf* stout, strong-minded, strong-willed, tenacious, vigorous. ▷ STUBBORN. **4** *strong commitment.* active, assiduous, deep-rooted, deep-seated, *derog* doctrinaire, *derog* dogmatic, eager, earnest, enthusiastic, fervent, fierce, firm, genuine, intense, keen, loyal, passionate, positive, rabid, sedulous, staunch, true, vehement, zealous. **5** *strong government.* decisive, dependable, *derog* dictatorial, fearless, *derog* tyrannical, unswerving, unwavering. **6** *strong measures.* aggressive, draconian, drastic, extreme, harsh, high-handed, ruthless, severe, tough, unflinching, violent. **7** *a strong army.* formidable, invincible, large, numerous, powerful, unconquerable, well-armed, well-equipped, well-trained. **8** *strong colour, light.* bright, brilliant, clear, dazzling, garish, glaring, vivid. **9** *strong taste, smell.* concentrated, highly-flavoured, hot, intense, noticeable, obvious, overpowering, prominent, pronounced, pungent, sharp, spicy, unmistakable. **10** *strong evidence.* clear-cut, cogent, compelling, convincing, evident, influential, persuasive, plain, solid, telling, undisputed. **11** *strong drink.* alcoholic, concentrated, intoxicating, potent, undiluted. *Opp* WEAK.

**stronghold** *n* bastion, bulwark, castle, citadel, *old use* fastness, fort, fortification, fortress, garrison.

**structure** *n* 1 arrangement, composition, configuration, constitution, design, form, formation, *inf* make-up, order, organization, plan, shape, system. 2 complex, construction, edifice, erection, fabric, framework, pile, superstructure. ▷ BUILDING. • *vb* arrange, build, construct, design, form, frame, give structure to, organize, shape, systematize.

**struggle** *n* 1 challenge, difficulty, effort, endeavour, exertion, labour, problem. 2 ▷ FIGHT. • *vb* 1 endeavour, exert yourself, labour, make an effort, move violently, strain, strive, toil, try, work hard, wrestle, wriggle about, writhe about. 2 *struggle through mud.* flail, flounder, stumble, wallow. 3 ▷ FIGHT.

**stub** *n* butt, end, remains, remnant, stump. • *vb* ▷ HIT.

**stubble** *n* 1 stalks, straw. 2 beard, bristles, *inf* five-o'clock shadow, hair, roughness.

**stubbly** *adj* bristly, prickly, rough, unshaven.

**stubborn** *adj* defiant, determined, difficult, disobedient, dogged, dogmatic, headstrong, inflexible, intractable, intransigent, mulish, obdurate, obstinate, opinionated, persistent, pertinacious, *inf* pig-headed, recalcitrant, refractory, rigid, self-willed, tenacious, uncompromising, uncooperative, uncontrollable, unmanageable, unreasonable, unyielding, wayward, wilful. *Opp* AMENABLE.

**stuck** *adj* 1 bogged down, cemented, fast, fastened, firm, fixed, glued, immovable. 2 *stuck on a problem.* baffled, beaten, held up, *inf* stumped, *inf* stymied.

**stuck-up** *adj* arrogant, *inf* big-headed, bumptious, *inf* cocky, conceited, condescending, *inf* high-and-mighty, patronizing, proud, self-important, snobbish, *inf* snooty, supercilious, *inf* toffee-nosed. *Opp* MODEST.

**student** *n* apprentice, disciple, learner, postgraduate, pupil, scholar, schoolchild, trainee, undergraduate.

**studied** *adj* calculated, conscious, contrived, deliberate, intentional, planned, premeditated.

**studious** *adj* academic, assiduous, attentive, bookish, brainy, earnest, hard-working, intellectual, scholarly, serious-minded, thoughtful.

**study** *vb* 1 analyse, consider, contemplate, enquire into, examine, give attention to, investigate, learn about, look closely at, peruse, ponder, pore over, read carefully, research, scrutinize, survey, think about, weigh. 2 *study for exams. inf* cram, learn, *inf* mug up, read, *inf* swot, work.

**stuff** *n* 1 ingredients, matter, substance. 2 fabric, material, textile. ▷ CLOTH. 3 *all sorts of stuff.* accoutrements, articles, belongings, *inf* bits and pieces, *inf* clobber, effects, *inf* gear, impedimenta, junk, objects, *inf* paraphernalia, possessions, *inf* tackle, things. • *vb* 1 compress, cram, crowd, force, jam, pack, press, push, ram, shove, squeeze, stow, thrust, tuck. 2 *stuff a cushion.* fill, line, pad. **stuff yourself** ▷ EAT.

**stuffing** *n* 1 filling, lining, padding, quilting, wadding. 2 *stuffing in poultry.* forcemeat, seasoning.

**stuffy** *adj* 1 airless, close, fetid, fuggy, fusty, heavy, humid, muggy, musty, oppressive, stale, steamy, stifling, suffocating, sultry, unventilated, warm. *Opp* AIRY. **2** [*inf*] *a stuffy old bore.* boring, conventional, dreary, dull, formal, humourless, narrow-minded, old-fashioned, pompous, prim, staid, *inf* stodgy, strait-laced. *Opp* LIVELY.

**stumble** *vb* **1** blunder, flounder, lurch, miss your footing, reel, slip, stagger, totter, trip, tumble. ▷ WALK. **2** *stumble in speech.* become tongue-tied, falter, hesitate, pause, stammer, stutter.

**stumbling-block** *n* bar, difficulty, hindrance, hurdle, impediment, obstacle, snag.

**stump** *vb* baffle, bewilder, *inf* catch out, confound, confuse, defeat, *inf* flummox, mystify, outwit, perplex, puzzle, *inf* stymie. **stump up** ▷ PAY.

**stun** *vb* **1** daze, knock out, knock senseless, make unconscious. **2** amaze, astonish, astound, bewilder, confound, confuse, dumbfound, flabbergast, numb, shock, stagger, stupefy. **stunning** ▷ BEAUTIFUL, STUPENDOUS.

**stunt** *n inf* dare, exploit, feat, trick. • *vb stunt growth.* ▷ CHECK.

**stupendous** *adj* amazing, colossal, enormous, exceptional, extraordinary, huge, incredible, marvellous, miraculous, notable, phenomenal, prodigious, remarkable, *inf* sensational, singular, special, staggering, stunning, tremendous, unbelievable, wonderful. *Opp* ORDINARY.

**stupid** *adj* [*Most synonyms derog*] **1** addled, bird-brained, bone-headed, bovine, brainless, clueless, cretinous, dense, dim, doltish, dopey, drippy, dull, dumb, empty-headed, feather-brained, feeble-minded, foolish, gormless, half-witted, idiotic, ignorant, imbecilic, imperceptive, ineducable, lacking, lumpish, mindless, moronic, naïve, obtuse, puerile, senseless, silly, simple, simple-minded, slow, slow in the uptake, slow-witted, subnormal, thick, thick-headed, thick-skulled, thickwitted, unintelligent, unthinking, unwise, vacuous, weak in the head, witless. **2** *a stupid thing to do.* absurd, asinine, barmy, crack-brained, crass, crazy, fatuous, feeble, futile, half-baked, hare-brained, ill-advised, inane, irrational, irrelevant, irresponsible, laughable, ludicrous, lunatic, mad, nonsensical, pointless, rash, reckless, ridiculous, risible, scatter-brained, thoughtless, unjustifiable. **3** *stupid after a knock on the head.* dazed, in a stupor, semi-conscious, sluggish, stunned, stupefied. *Opp* INTELLIGENT. **stupid person** ▷ FOOL.

**stupidity** *n* absurdity, crassness, denseness, dullness, *inf* dumbness, fatuity, fatuousness, folly, foolishness, futility, idiocy, ignorance, imbecility, inanity, lack of intelligence, lunacy, madness, mindlessness, naïvety, pointlessness, recklessness, silliness, slowness, thoughtlessness. *Opp* INTELLIGENCE.

**stupor** *n* coma, daze, inertia, lassitude, lethargy, numbness, shock, state of insensibility, torpor, trance, unconsciousness.

**sturdy** *adj* **1** athletic, brawny, burly, hardy, healthy, hefty, husky, muscular, powerful, robust, stalwart, stocky, *inf* strapping, vigorous, well-built. **2** *sturdy shoes, etc.* durable, solid, sound, substantial, tough, well-made. **3** *sturdy opposition.* determined, firm, indomitable, resolute, staunch, steadfast, uncompromising, vigorous. ▷ STRONG. *Opp* WEAK.

**stutter** *vb* stammer, stumble. ▷ TALK.

**style** *n* **1** dash, elegance, flair, flamboyance, panache, polish, refinement, smartness, sophistication, stylishness, taste. **2** *not my style*. approach, character, custom, habit, idiosyncrasy, manner, method, way. **3** *style in writing*. diction, mode, phraseology, phrasing, register, sentence structure, tenor, tone, wording. **4** *style in clothes*. chic, cut, design, dress-sense, fashion, look, mode, pattern, shape, tailoring, type, vogue.

**stylish** *adj Fr* à la mode, chic, *inf* classy, contemporary, *inf* dapper, elegant, fashionable, modern, modish, *inf* natty, *inf* posh, smart, *inf* snazzy, sophisticated, *inf* trendy, up-to-date. *Opp* OLD-FASHIONED.

**subconscious** *adj* deep-rooted, hidden, inner, intuitive, latent, repressed, subliminal, suppressed, unacknowledged, unconscious. *Opp* CONSCIOUS.

**subdue** *vb* check, curb, hold back, keep under, moderate, quieten, repress, restrain, suppress, temper. ▷ SUBJUGATE.

**subdued** *adj* **1** chastened, crestfallen, depressed, downcast, grave, reflective, repressed, restrained, serious, silent, sober, solemn, thoughtful. ▷ SAD. *Opp* EXCITED. **2** *subdued music*. calm, hushed, low, mellow, muted, peaceful, placid, quiet, soft, soothing, toned down, tranquil, unobtrusive.

**subject** *adj* **1** captive, dependent, enslaved, oppressed, ruled, subjugated. **2** *subject to interference*. exposed, liable, prone, susceptible, vulnerable. *Opp* FREE. • *n* **1** citizen, dependant, national, passport-holder, taxpayer, voter. **2** *subject for discussion*. affair, business, issue, matter, point, proposition, question, theme, thesis, topic. **3** *subject of study*. area, branch of knowledge, course, discipline, field. □ *anatomy, archaeology, architecture, art, astronomy, biology, business, chemistry, computing, craft, design, divinity, domestic science, drama, economics, education, electronics, engineering, English, environmental science, ethnology, etymology, geography, geology, heraldry, history, languages, Latin, law, linguistics, literature, mathematics, mechanics, medicine, metallurgy, metaphysics, meteorology, music, natural history, oceanography, ornithology, penology, pharmacology, pharmacy, philology, philosophy, photography, physics, physiology, politics, psychology, religious studies, science, scripture, social work, sociology, sport, surveying, technology, theology, topology, zoology*. • *vb* **1** *subject a thing to scrutiny*. expose, lay open, submit. **2** ▷ SUBJUGATE.

**subjective** *adj* biased, emotional, *inf* gut (*reaction*), idiosyncratic, individual, instinctive, intuitive, personal, prejudiced, self-centred. *Opp* OBJECTIVE.

**subjugate** *vb* beat, conquer, control, crush, defeat, dominate, enslave, enthral, *inf* get the better of, master, oppress, overcome, overpower, overrun, put down, quash, quell, subdue, subject, tame, triumph over, vanquish.

**sublimate** *vb* channel, convert, divert, idealize, purify, redirect, refine.

**sublime** *adj* ecstatic, elated, elevated, exalted, great, heavenly, high, high-minded, lofty, noble, spiritual, transcendent. *Opp* BASE.

**submerge** *vb* 1 cover with water, dip, drench, drown, *inf* dunk, engulf, flood, immerse, inundate, overwhelm, soak, swamp. **2** dive, go down, go under, plummet, sink, subside.

**submission** *n* **1** acquiescence, capitulation, compliance, giving in, surrender, yielding. ▷ SUBMISSIVENESS. **2** contribution, entry, offering, presentation, tender. **3** *a legal submission.* argument, claim, contention, idea, proposal, suggestion, theory.

**submissive** *adj* accommodating, acquiescent, amenable, biddable, *derog* boot-licking, compliant, deferential, docile, humble, meek, obedient, obsequious, passive, pliant, resigned, servile, slavish, supine, sycophantic, tame, tractable, unassertive, uncomplaining, unresisting, weak, yielding. *Opp* ASSERTIVE.

**submissiveness** *n* acquiescence, assent, compliance, deference, docility, humility, meekness, obedience, obsequiousness, passivity, resignation, servility, submission, subservience, tameness.

**submit** *vb* **1** accede, bow, capitulate, concede, give in, *inf* knuckle under, succumb, surrender, yield. **2** *submit a proposal.* advance, enter, give in, hand in, offer, present, proffer, propose, propound, put forward, state, suggest. **submit to** ▷ ACCEPT, OBEY.

**subordinate** *adj* inferior, junior, lesser, lower, menial, minor, secondary, subservient, subsidiary. • *n* aide, assistant, dependant, employee, inferior, junior, menial, *inf* underling. ▷ SERVANT.

**subscribe** *vb* **subscribe to** **1** contribute to, covenant to, donate to, give to, patronize, sponsor, support. **2** *subscribe to a magazine.* be a subscriber to, buy regularly, pay a subscription to. **3** *subscribe to a theory.* advocate, agree with, approve of, *inf* back, believe in, condone, consent to, endorse, *inf* give your blessing to.

**subscriber** *n* patron, regular customer, sponsor, supporter.

**subscription** *n* fee, due, payment, regular contribution, remittance.

**subsequent** *adj* coming, consequent, ensuing, following, future, later, next, resultant, resulting, succeeding, successive. *Opp* PREVIOUS.

**subside** *vb* **1** abate, calm down, decline, decrease, die down, diminish, dwindle, ebb, fall, go down, lessen, melt away, moderate, quieten, recede, shrink, slacken, wear off. **2** *subside into a chair.* collapse, descend, lower yourself, settle, sink. *Opp* RISE.

**subsidiary** *adj* additional, ancillary, auxiliary, complementary, contributory, inferior, lesser, minor, secondary, subordinate, supporting.

**subsidize** *vb* aid, back, finance, fund, give subsidy to, maintain, promote, sponsor, support, underwrite.

**subsidy** *n* aid, backing, financial help, funding, grant, maintenance, sponsorship, subvention, support.

**substance** *n* **1** actuality, body, concreteness, corporeality, reality, solidity. **2** chemical, fabric, make-up, material, matter, stuff. **3** *substance of an argument.* core, essence, gist, import, meaning, significance, subject-matter, theme. **4** [*old use*] *a person of substance.* ▷ WEALTH.

**substandard** *adj inf* below par, disappointing, inadequate, inferior, poor, shoddy, unworthy.

**substantial** *adj* **1** durable, hefty, massive, solid, sound, stout, strong, sturdy, well-built, well-made. **2** big, consequential, considerable, generous, great, large, significant, sizeable, worthwhile. *Opp* FLIMSY, SMALL.

**substitute** *adj* **1** acting, deputy, relief, reserve, stand-by, surrogate, temporary. **2** alternative, ersatz, imitation. • *n* alternative, deputy, locum, proxy, relief, replacement, reserve, stand-in, stopgap, substitution, supply, surrogate, understudy. • *vb* **1** change, exchange, interchange, replace, *inf* swop, *inf* switch. **2** *substitute for an absentee.* act as a substitute, cover, deputize, double, stand in, supplant, take the place of, take over the role of, understudy.

**subtle** *adj* **1** delicate, elusive, faint, fine, gentle, mild, slight, unobtrusive. **2** *subtle argument.* arcane, clever, indirect, ingenious, mysterious, recondite, refined, shrewd, sophisticated, tactful, understated. ▷ CUNNING. *Opp* OBVIOUS.

**subtract** *vb* debit, deduct, remove, take away, take off. *Opp* ADD.

**suburban** *adj* residential, outer, outlying.

**suburbs** *n* fringes, outer areas, outskirts, residential areas, suburbia.

**subversive** *adj* challenging, disruptive, insurrectionary, questioning, radical, seditious, traitorous, treacherous, treasonous, undermining, unsettling. ▷ REVOLUTIONARY. *Opp* CONSERVATIVE, ORTHODOX.

**subvert** *vb* challenge, corrupt, destroy, disrupt, overthrow, overturn, pervert, ruin, undermine, upset, wreck.

**subway** *n* tunnel, underpass.

**succeed** *vb* **1** accomplish your objective, *inf* arrive, be a success, do well, flourish, *inf* get on, *inf* get to the top, *inf* make it, prosper, thrive. **2** be effective, *inf* catch on, produce results, work. **3** be successor to, come after, follow, inherit from, replace, take over from. *Opp* FAIL. **succeeding** ▷ SUBSEQUENT.

**success** *n* **1** fame, good fortune, prosperity, wealth. **2** *success of a plan.* accomplishment, achievement, attainment, completion, effectiveness, successful outcome. **3** *a great success. inf* hit, *inf* sensation, triumph, victory, *inf* winner. *Opp* FAILURE.

**successful** *adj* **1** booming, effective, effectual, flourishing, fruitful, lucrative, money-making, productive, profitable, profit-making, prosperous, rewarding, thriving, useful, well-off. **2** best-selling, celebrated, famed, famous, high-earning, leading, popular, top, unbeaten, victorious, well-known, winning. *Opp* UNSUCCESSFUL.

**succession** *n* chain, flow, line, procession, progression, run, sequence, series, string.

**successive** *adj* consecutive, continuous, in succession, succeeding, unbroken, uninterrupted.

**successor** *n* heir, inheritor, replacement.

**succinct** *adj* brief, compact, concise, condensed, epigrammatic, pithy, short, terse, to the point. *Opp* WORDY.

**succulent** *adj* fleshy, juicy, luscious, moist, mouthwatering, palatable, rich.

**succumb** *vb* accede, be overcome, capitulate, give in, give up, give way, submit, surrender, yield. *Opp* SURVIVE.

**suck** *vb* **suck up** absorb, draw up, pull up, soak up. **suck up to** ▷ FLATTER.

**sudden** *adj* **1** abrupt, brisk, hasty, hurried, impetuous, impulsive, precipitate, quick, rash, *inf* snap, swift, unconsidered, unplanned, unpremeditated. *Opp* SLOW. **2** *a sudden shock.* acute, sharp, startling, surprising, unannounced, unexpected, unforeseeable, unforeseen, unlooked-for. *Opp* PREDICTABLE.

**suds** *n* bubbles, foam, froth, lather, soapsuds.

**sue** *vb* **1** indict, institute legal proceedings against, proceed against, prosecute, summons, take legal action against. **2** *sue for peace.* ▷ ENTREAT.

**suffer** *vb* **1** bear, cope with, endure, experience, feel, go through, live through, *inf* put up with, stand, tolerate, undergo, withstand. **2** *suffer from a wound.* ache, agonize, feel pain, hurt, smart. **3** *suffer for a crime.* atone, be punished, make amends, pay.

**suffice** *vb* answer, be sufficient, *inf* do, satisfy, serve.

**sufficient** *adj* adequate, enough, satisfactory. *Opp* INSUFFICIENT.

**suffocate** *vb* asphyxiate, choke, smother, stifle, stop breathing, strangle, throttle.

**sugar** *n* □ *brown sugar, cane sugar, caster sugar, demerara, glucose, granulated sugar, icing sugar, lump sugar, molasses, sucrose, sweets, syrup, treacle.* • *vb* sweeten.

**sugary** *adj* **1** glazed, iced, sugared, sweetened. ▷ SWEET. **2** *sugary sentiments.* cloying, honeyed, sickly. ▷ SENTIMENTAL.

**suggest** *vb* **1** advise, advocate, counsel, moot, move, propose, propound, put forward, raise, recommend, urge. **2** call to mind, communicate, hint, imply, indicate, insinuate, intimate, make you think (of), mean, signal.

**suggestion** *n* **1** advice, counsel, offer, plan, prompting, proposal, recommendation, urging. **2** breath, hint, idea, indication, intimation, notion, suspicion, touch, trace.

**suggestive** *adj* **1** evocative, expressive, indicative, reminiscent, thought-provoking. **2** ▷ INDECENT.

**suicidal** *adj* **1** hopeless, *inf* kamikaze, self-destructive. **2** ▷ DESOLATE.

**suit** *n* outfit. ▷ CLOTHES. • *vb* **1** accommodate, be suitable for, conform to, fill your needs, fit in with, gratify, harmonize with, match, please, satisfy, tally with. *Opp* DISPLEASE. **2** *That colour suits you.* become, fit, look good on.

**suitable** *adj* acceptable, applicable, apposite, appropriate, apt, becoming, befitting, congenial, convenient, correct, decent, decorous, fit, fitting, handy, *old use* meet, opportune, pertinent, proper, relevant, right, satisfactory, seemly, tasteful, timely, well-chosen, well-judged, well-timed. *Opp* UNSUITABLE.

**sulk** *vb* be sullen, brood, mope, pout.

**sullen** *adj* **1** anti-social, bad-tempered, brooding, churlish, crabby, cross, disgruntled, dour, glum, grim, grudging, ill-humoured, lugubrious, moody, morose, *inf* out of sorts, petulant, pouting, resentful, silent, sour, stubborn, sulking, sulky, surly, uncommunicative, unforgiving, unfriendly, unhappy, unsociable. ▷ SAD. **2** *a sullen sky*. cheerless, dark, dismal, dull, gloomy, grey, leaden, sombre. *Opp* CHEERFUL.

**sultry** *adj* **1** close, hot, humid, *inf* muggy, oppressive, steamy, stifling, stuffy, warm. *Opp* COLD. **2** *sultry beauty*. erotic, mysterious, passionate, provocative, seductive, sensual, sexy, voluptuous.

**sum** *n* aggregate, amount, number, quantity, reckoning, result, score, tally, total, whole. • *vb* **sum up** ▷ SUMMARIZE.

**summarize** *vb* abridge, condense, digest, encapsulate, give the gist, make a summary, outline, précis, *inf* recap, recapitulate, reduce, review, shorten, simplify, sum up. *Opp* ELABORATE.

**summary** *n* abridgement, abstract, condensation, digest, epitome, gist, outline, précis, recapitulation, reduction, résumé, review, summation, summing-up, synopsis.

**summery** *adj* bright, sunny, tropical, warm. ▷ HOT. *Opp* WINTRY.

**summit** *n* **1** apex, crown, head, height, peak, pinnacle, point, top. *Opp* BASE. **2** *summit of success*. acme, apogee, climax, culmination, high point, zenith. *Opp* NADIR.

**summon** *vb* **1** command, demand, invite, order, send for, subpoena. **2** assemble, call, convene, convoke, gather together, muster, rally.

**sunbathe** *vb* bake, bask, *inf* get a tan, sun yourself, tan.

**sunburnt** *adj* blistered, bronzed, brown, peeling, tanned, weather-beaten.

**sundry** *adj* assorted, different, *old use* divers, miscellaneous, mixed, various.

**sunken** *adj* **1** submerged, underwater, wrecked. **2** *sunken cheeks*. concave, depressed, drawn, hollow, hollowed.

**sunless** *adj* cheerless, cloudy, dark, dismal, dreary, dull, gloomy, grey, overcast, sombre. *Opp* SUNNY.

**sunlight** *n* daylight, sun, sunbeams, sunshine.

**sunny** *adj* **1** bright, clear, cloudless, fair, fine, summery, sunlit, sunshiny, unclouded. *Opp* SUNLESS. **2** ▷ CHEERFUL.

**sunrise** *n* dawn, day-break.

**sunset** *n* dusk, evening, gloaming, nightfall, sundown, twilight.

**sunshade** *n* awning, canopy, parasol.

**superannuated** *adj* **1** discharged, *inf* pensioned off, *inf* put out to grass, old, retired. **2** discarded, disused, obsolete, thrown out, worn out. ▷ OLD.

**superannuation** *n* annuity, pension.

**superb** *adj* admirable, excellent, fine, first-class, first-rate, grand, impressive, marvellous, superior. ▷ SPLENDID. *Opp* INFERIOR.

**superficial** *adj* **1** cosmetic, external, exterior, on the surface, outward, shallow, skin-deep, slight, surface, unimportant. **2** careless, casual, cursory, desultory, facile, frivolous, hasty, hurried, inattentive, lightweight, *inf* nodding (*acquaintance*), oversimplified, passing, perfunctory, simple-minded, simplistic, sweeping (*generalization*), trivial, unconvincing, uncritical, undiscriminating, unquestioning, unscholarly, unsophisticated. *Opp* ANALYTICAL, DEEP.

**superfluous** *adj* excess, excessive, extra, needless, redundant, spare, superabundant, surplus, unnecessary, unneeded, unwanted. *Opp* NECESSARY.

**superhuman** *adj* **1** god-like, herculean, heroic, phenomenal, prodigious. **2** *superhuman powers*. divine, higher, metaphysical, supernatural.

**superimpose** *vb* overlay, place on top of.

**superintend** *vb* administer, be in charge of, be the supervisor of, conduct, control, direct, look after, manage, organize, oversee, preside over, run, supervise, watch over.

**superior** *adj* **1** better, *inf* classier, greater, higher, higher-born, loftier, more important, more impressive, nobler, senior, *inf* up-market. **2** *superior quality*. choice, exclusive, fine, first-class, first-rate, select, top, unrivalled. **3** *superior attitude*. arrogant, condescending, contemptuous, disdainful, élitist, haughty, *inf* high-and-mighty, lofty, paternalistic, patronizing, self-important, smug, snobbish, *inf* snooty, stuck-up, supercilious. *Opp* INFERIOR.

**superlative** *adj* best, choicest, consummate, excellent, finest, first-rate, incomparable, matchless, peerless, *inf* tip-top, *inf* top-notch, unrivalled, unsurpassed. ▷ SUPREME.

**supernatural** *adj* abnormal, ghostly, inexplicable, magical, metaphysical, miraculous, mysterious, mystic, occult, other-worldly, paranormal, preternatural, psychic, spiritual, uncanny, unearthly, unnatural, weird.

**superstition** *n* delusion, illusion, myth, *inf* old wives' tale, superstitious belief.

**superstitious** *adj* credulous, groundless, illusory, irrational, mythical, traditional, unfounded, unprovable.

**supervise** *vb* administer, be in charge of, be the supervisor of, conduct, control, direct, govern, invigilate (*an exam*), *inf* keep an eye on, lead, look after, manage, organize, oversee, preside over, run, superintend, watch over.

**supervision** *n* administration, conduct, control, direction, government, invigilation, management, organization, oversight, running, surveillance.

**supervisor** *n* administrator, chief, controller, director, executive, foreman, *inf* gaffer, head, inspector, invigilator, leader, manager, organizer, overseer, superintendent, timekeeper.

**supine** *adj* **1** face upwards, flat on your back, prostrate, recumbent. *Opp* PRONE. **2** ▷ PASSIVE.

**supplant** *vb* displace, dispossess, eject, expel, oust, replace, supersede, *inf* step into the shoes of, *inf* topple, unseat.

**supple** *adj* bending, *inf* bendy, elastic, flexible, flexile, graceful, limber, lithe, plastic, pliable, pliant, resilient, soft. *Opp* RIGID.

**supplement** *n* **1** additional payment, excess, surcharge. **2** *a newspaper supplement, etc*. addendum, addition, annexe, appendix, codicil, continuation, endpiece, extra, insert, postscript, sequel. • *vb* add to, augment, boost, complement, extend, reinforce, *inf* top up.

**supplementary** *adj* accompanying, added, additional, ancillary, auxiliary, complementary, excess, extra, new, supportive.

**supplication** *n* appeal, entreaty, petition, plea, prayer, request, solicitation.

**supplier** *n* dealer, provider, purveyor, retailer, seller, shopkeeper, vendor, wholesaler.

**supply** *n* cache, hoard, quantity, reserve, reservoir, stock, stockpile, store. **2** equipment, food, necessities, provisions, rations, shopping. **3** *a regular supply.* delivery, distribution, provision, provisioning. • *vb* cater to, contribute, deliver, distribute, donate, endow, equip, feed, furnish, give, hand over, pass on, produce, provide, purvey, sell, stock.

**support** *n* **1** aid, approval, assistance, backing, back-up, bolstering, contribution, cooperation, donation, encouragement, fortifying, friendship, help, interest, loyalty, patronage, protection, reassurance, reinforcement, sponsorship, succour. **2** brace, bracket, buttress, crutch, foundation, frame, pillar, post, prop, sling, stanchion, stay, strut, substructure, trestle, truss, underpinning. **3** *financial support.* expenses, funding, keep, maintenance, subsistence, upkeep. • *vb* **1** bear, bolster, buoy up, buttress, carry, give strength to, hold up, keep up, prop up, provide a support for, reinforce, shore up, strengthen, underlie, underpin. **2** *support someone in trouble.* aid, assist, back, be faithful to, champion, comfort, defend, encourage, favour, fight for, give support to, help, rally round, reassure, side with, speak up for, stand by, stand up for, stay with, *inf* stick up for, *inf* stick with, take someone's part. **3** *support a family.* bring up, feed, finance, fund, keep, look after, maintain, nourish, provide for, sustain. **4** *support a charity.* be a supporter of, be interested in, contribute to, espouse (*a cause*), follow, give to, patronize, pay money to, sponsor, subsidize, work for. **5** *support a point of view.* accept, adhere to, advocate, agree with, allow, approve, argue for, confirm, corroborate, defend, endorse, explain, justify, promote, ratify, substantiate, uphold, validate, verify. *Opp* SUBVERT, WEAKEN. **support yourself** lean, rest.

**supporter** *n* **1** adherent, admirer, advocate, aficionado, apologist, champion, defender, devotee, enthusiast, *inf* fan, fanatic, follower, seconder, upholder, voter. **2** ally, assistant, collaborator, helper, henchman, second.

**supportive** *adj* caring, concerned, encouraging, helpful, favourable, heartening, interested, kind, loyal, positive, reassuring, sustaining, sympathetic, understanding. *Opp* SUBVERSIVE.

**suppose** *vb* **1** accept, assume, believe, conclude, conjecture, expect, guess, infer, judge, postulate, presume, presuppose, speculate, surmise, suspect, take for granted, think. **2** daydream, fancy, fantasize, hypothesize, imagine, maintain, postulate, pretend, theorize. **supposed** ▷ HYPOTHETICAL, PUTATIVE. **supposed to** due to, expected to, having a duty to, meant to, required to.

**supposition** *n* assumption, belief, conjecture, fancy, guess, *inf* guesstimate, hypothesis, inference, notion, opinion, presumption, speculation, surmise, theory, thought.

**suppress** *vb* **1** conquer, *inf* crack down on, crush, end, finish off, halt, overcome, overthrow, put an end to, put down, quash, quell, stamp out, stop, subdue. 2 *suppress emotion.* bottle up, censor, choke back, conceal, cover up, hide, hush up, keep quiet about, keep secret, muffle, mute, obstruct, prohibit, repress, restrain, silence, smother, stamp on, stifle, strangle.

**supremacy** *n* ascendancy, dominance, domination, dominion, lead, mastery, predominance, pre-eminence, sovereignty, superiority.

**supreme** *adj* best, choicest, consummate, crowning, culminating, excellent, finest, first-rate, greatest, highest, incomparable, matchless, outstanding, paramount, peerless, predominant, pre-eminent, prime, principal, superlative, surpassing, *inf* tip-top, top, *inf* top-notch, ultimate, unbeatable, unbeaten, unparalleled, unrivalled, unsurpassable, unsurpassed.

**sure** *adj* 1 assured, certain, confident, convinced, decided, definite, persuaded, positive. 2 *sure to come.* bound, certain, compelled, obliged, required. 3 *a sure fact.* accurate, clear, convincing, guaranteed, indisputable, inescapable, inevitable, infallible, proven, reliable, true, unchallenged, undeniable, undisputed, undoubted, verifiable. 4 *a sure ally.* dependable, effective, established, faithful, firm, infallible, loyal, reliable, resolute, safe, secure, solid, steadfast, steady, trustworthy, trusty, undeviating, unerring, unfailing, unfaltering, unflinching, unswerving, unwavering. *Opp* UNCERTAIN.

**surface** *n* 1 coat, coating, covering, crust, exterior, façade, integument, interface, outside, shell, skin, veneer. 2 *cube has six surfaces.* face, facet, plane, side. 3 *a working surface.* bench, table, top, worktop. • *vb* 1 appear, arise, *inf* come to light, come up, *inf* crop up, emerge, materialize, rise, *inf* pop up. 2 coat, cover, laminate, veneer.

**surfeit** *n* excess, flood, glut, overabundance, overindulgence, oversupply, plethora, superfluity, surplus.

**surge** *n* burst, gush, increase, onrush, onset, outpouring, rush, upsurge. ▷ WAVE. • *vb* billow, eddy, flow, gush, heave, make waves, move irresistibly, push, roll, rush, stampede, stream, sweep, swirl, well up.

**surgery** *n* 1 biopsy, operation. 2 *a doctor's surgery.* clinic, consulting room, health centre, infirmary, medical centre, sick-bay.

**surly** *adj* bad-tempered, boorish, cantankerous, churlish, crabby, cross, crotchety, *inf* crusty, curmudgeonly, dyspeptic, gruff, *inf* grumpy, ill-natured, ill-tempered, irascible, miserable, morose, peevish, rough, rude, sulky, sullen, testy, touchy, uncivil, unfriendly, ungracious, unpleasant. *Opp* FRIENDLY.

**surmise** *vb* assume, believe, conjecture, expect, fancy, gather, guess, hypothesize, imagine, infer, judge, postulate, presume, presuppose, sense, speculate, suppose, suspect, take for granted, think.

**surpass** *vb* beat, better, do better than, eclipse, exceed, excel, go beyond, leave behind, *inf* leave standing, outclass, outdistance, outdo, outperform, outshine, outstrip, overshadow, top, transcend, worst.

**surplus** *n* balance, excess, extra, glut, oversupply, remainder, residue, superfluity, surfeit.

**surprise** *n* 1 alarm, amazement, astonishment, consternation, dismay, incredulity, stupefaction, wonder. 2 *a complete surprise.* blow, *inf* bolt from the blue, *inf* bombshell, *inf* eye-opener, jolt, shock. • *vb* 1 alarm, amaze, astonish, astound, disconcert, dismay, dumbfound, flabbergast, nonplus, rock, shock, stagger, startle, stun, stupefy, *inf* take aback, take by surprise, *inf* throw. 2 *surprise someone doing wrong.* capture, catch out, *inf* catch red-handed, come upon, detect, discover, take unawares.

**surprised** *adj* alarmed, amazed, astonished, astounded, disconcerted, dismayed, dumbfounded, flabbergasted, incredulous, *inf* knocked for six, nonplussed, *inf* shattered, shocked, speechless, staggered, startled, struck dumb, stunned, taken aback, taken by surprise, *inf* thrown, thunderstruck.

**surprising** *adj* alarming, amazing, astonishing, astounding, disconcerting, extraordinary, frightening, incredible, *inf* offputting, shocking, staggering, startling, stunning, sudden, unexpected, unforeseen, unlooked-for, unplanned, unpredictable, upsetting. *Opp* PREDICTABLE.

**surrender** *n* capitulation, giving in, resignation, submission. • *vb* 1 acquiesce, capitulate, *inf* cave in, collapse, concede, fall, *inf* give in, give up, give way, give yourself up, resign, submit, succumb, *inf* throw in the towel, *inf* throw up the sponge, yield. 2 *surrender your ticket.* deliver up, give up, hand over, part with, relinquish. 3 *surrender your rights.* abandon, cede, renounce, waive.

**surreptitious** *adj* clandestine, concealed, covert, crafty, disguised, furtive, hidden, private, secret, secretive, shifty, sly, *inf* sneaky, stealthy, underhand. *Opp* BLATANT.

**surround** *vb* besiege, beset, cocoon, cordon off, encircle, enclose, encompass, engulf, environ, girdle, hem in, hedge in, ring, skirt, trap, wrap.

**surrounding** *adj* adjacent, adjoining, bordering, local, nearby, neighbouring.

**surroundings** *n* ambience, area, background, context, environment, location, milieu, neighbourhood, setting, vicinity.

**surveillance** *n* check, observation, reconnaissance, scrutiny, supervision, vigilance, watch.

**survey** *n* appraisal, assessment, census, count, evaluation, examination, inquiry, inspection, investigation, review, scrutiny, study, triangulation. • *vb* 1 appraise, assess, estimate, evaluate, examine, inspect, investigate, look over, review, scrutinize, study, view, weigh up. 2 do a survey of, map out, measure, plan out, plot, reconnoitre, triangulate.

**survival** *n* continuance, continued existence, persistence.

**survive** *vb* 1 *inf* bear up, carry on, continue, endure, keep going, last, live, persist, remain. 2 *survive disaster.* come through, live through, outlast, outlive, pull through, weather, withstand. *Opp* SUCCUMB.

**susceptible** *adj* affected (by), disposed, given, inclined, liable, open, predisposed, prone, responsive, sensitive, vulnerable. *Opp* RESISTANT.

**suspect** *adj* doubtful, dubious, inadequate, questionable, *inf* shady, suspected, suspicious, unconvincing, unreliable, unsatisfactory, untrustworthy. • *vb* 1 call into question, disbelieve, distrust, doubt, have suspicions about, mistrust. 2 *suspect that she's lying*. believe, conjecture, consider, guess, imagine, infer, presume, speculate, suppose, surmise, think.

**suspend** *vb* 1 dangle, hang, swing. 2 *suspend work*. adjourn, break off, defer, delay, discontinue, freeze, hold in abeyance, hold up, interrupt, postpone, put off, *inf* put on ice, shelve. 3 *suspend from duty*. debar, dismiss, exclude, expel, lay off, lock out, send down.

**suspense** *n* anticipation, anxiety, apprehension, doubt, drama, excitement, expectancy, expectation, insecurity, irresolution, nervousness, not knowing, tension, uncertainty, waiting.

**suspicion** *n* 1 apprehension, apprehensiveness, caution, distrust, doubt, dubiety, dubiousness, *inf* funny feeling, guess, hesitation, *inf* hunch, impression, misgiving, mistrust, presentiment, qualm, scepticism, uncertainty, wariness. 2 *suspicion of a smile*. glimmer, hint, inkling, shadow, suggestion, tinge, touch, trace.

**suspicious** *adj* 1 apprehensive, *inf* chary, disbelieving, distrustful, doubtful, dubious, incredulous, in doubt, mistrustful, sceptical, uncertain, unconvinced, uneasy, wary. *Opp* TRUSTFUL. 2 *suspicious character*. disreputable, dubious, *inf* fishy, peculiar, questionable, *inf* shady, suspect, suspected, unreliable, untrustworthy. *Opp* TRUSTWORTHY.

**sustain** *vb* 1 continue, develop, elongate, extend, keep alive, keep going, keep up, maintain, prolong. 2 ▷ SUPPORT.

**sustenance** *n* eatables, edibles, food, foodstuffs, nourishment, nutriment, provender, provisions, rations, *old use* victuals.

**swag** *n* booty, loot, plunder, takings.

**swagger** *vb* parade, strut. ▷ WALK.

**swallow** *vb* consume, *inf* down, gulp down, guzzle, ingest, take down. ▷ DRINK, EAT. **swallow up** absorb, assimilate, enclose, enfold make disappear. ▷ SWAMP.

**swamp** *n* bog, fen, marsh, marshland, morass, mud, mudflats, quagmire, quicksand, saltmarsh, *old use* slough, wetlands. • *vb* deluge, drench, engulf, envelop, flood, immerse, inundate, overcome, overwhelm, sink, submerge, swallow up.

**swampy** *adj* boggy, marshy, muddy, soft, soggy, unstable, waterlogged, wet. *Opp* DRY, FIRM.

**swarm** *n* cloud, crowd, hive, horde, host, multitude. ▷ GROUP. • *vb* cluster, congregate, crowd, flock, gather, mass, move in a swarm, throng. **swarm up** ▷ CLIMB. **swarm with** ▷ TEEM.

**swarthy** *adj* brown, dark, dark-complexioned, dark-skinned, dusky, tanned.

**swashbuckling** *adj* adventurous aggressive, bold, daredevil, daring, dashing, *inf* macho, manly, swaggering. *Opp* TIMID.

**sway** *vb* 1 bend, fluctuate, lean from side to side, oscillate, rock, roll, swing, undulate, wave. 2 *sway opinions*. affect, bias, bring round, change (someone's mind), convert, convince, govern, influence, persuade, win over. **3** *sway from a chosen path*. divert, go off course, swerve, veer, waver.

**swear** *vb* **1** affirm, asseverate, attest, aver, avow, declare, give your word, insist, pledge, promise, state on oath, take an oath, testify, vouchsafe, vow. **2** blaspheme, curse, execrate, imprecate, use swearwords, utter profanities.

**swearword** *n inf* bad language, blasphemy, curse, execration, expletive, *inf* four-letter word, imprecation, oath, obscenity, profanity, swearing.

**sweat** *vb* **1** *inf* glow, perspire, swelter. **2** ▷ WORK.

**sweaty** *adj* clammy, damp, moist, perspiring, sticky, sweating.

**sweep** *vb* brush, clean, clear, dust, tidy up. **sweep along** ▷ MOVE. **sweep away** ▷ REMOVE. **sweeping** ▷ GENERAL, SUPERFICIAL.

**sweet** *adj* **1** aromatic, fragrant, honeyed, luscious, mellow, perfumed, sweetened, sweetscented, sweet-smelling. **2** [*derog*] *sickly sweet*. cloying, saccharine, sentimental, sickening, sickly, sugary, syrupy, treacly. **3** *sweet sounds*. dulcet, euphonious, harmonious, heavenly, mellifluous, melodious, musical, pleasant, silvery, soothing, tuneful. **4** *a sweet nature*. affectionate, amiable, attractive, charming, dear, endearing, engaging, friendly, genial, gentle, gracious, lovable, lovely, nice, pretty, unselfish, winning. *Opp* ACID, BITTER, NASTY, SAVOURY. • *n* **1** *inf* afters, dessert, pudding. **2** [*usu plur*] *old use* bon-bons, *Amer* candy, confectionery, *inf* sweeties, *old use* sweetmeats. □ *acid drop, barley sugar, boiled sweet, bull's-eye, butterscotch, candy, candyfloss, caramel, chewing-gum, chocolate, fondant, fruit pastille, fudge, humbug, liquorice, lollipop, marshmallow, marzipan, mint, nougat, peppermint, rock, toffee, Turkish delight.*

**sweeten** *vb* **1** make sweeter, sugar. **2** *sweeten your temper*. appease, assuage, calm, mellow, mollify, pacify, soothe.

**sweetener** *n* □ *artificial sweetener, honey, saccharine, sugar, sweetening, syrup.*

**swell** *vb* **1** balloon, become bigger, belly, billow, blow up, bulge, dilate, distend, enlarge, expand, fatten, fill out, grow, increase, inflate, mushroom, puff up, rise. **2** *swell numbers*. augment, boost, build up, extend, increase, make bigger, raise, step up. *Opp* SHRINK.

**swelling** *n* blister, boil, bulge, bump, distension, enlargement, excrescence, hump, inflammation, knob, lump, node, nodule, prominence, protrusion, protuberance, tumescence, tumour.

**sweltering** *adj* humid, muggy, oppressive, steamy, sticky, stifling, sultry, torrid, tropical. ▷ HOT.

**swerve** *vb* career, change direction, deviate, diverge, dodge about, sheer off, swing, take avoiding action, turn aside, veer, wheel.

**swift** *adj* agile, brisk, expeditious, fast, *old use* fleet, fleet-footed, hasty, hurried, nimble, *inf* nippy, prompt, quick, rapid, speedy, sudden. *Opp* SLOW.

**swill** *vb* **1** bathe, clean, rinse, sponge down, wash. **2** ▷ DRINK.

**swim** *vb* bathe, dive in, float, go swimming, *inf* take a dip.

**swimming-bath** *n* baths, leisure-pool, lido, swimming-pool.

**swim-suit** *n* bathing-costume, bathing-dress, bathing-suit, bikini, swimwear, trunks.

**swindle** *n* cheat, chicanery, *inf* con, confidence trick, deception, double-dealing, fraud, knavery, *inf* racket, *inf* rip-off, *inf* sharp practice, *inf* swizz, trickery. • *vb inf* bamboozle, cheat, *inf* con, cozen, deceive, defraud, *inf* diddle, *inf* do, double-cross, dupe, exploit, *inf* fiddle, *inf* fleece, fool, gull, hoax, hoodwink, mulct, *inf* pull a fast one on you, *inf* rook, *inf* take you for a ride, trick, *inf* welsh (*on a bet*).

**swindler** *n* charlatan, cheat, cheater, *inf* con man, counterfeiter, double-crosser, extortioner, forger, fraud, hoaxer, impostor, knave, mountebank, quack, racketeer, scoundrel, *inf* shark, trickster, *inf* twister.

**swing** *n* change, fluctuation, movement, oscillation, shift, variation. • *vb* **1** be suspended, dangle, flap, fluctuate, hang loose, move from side to side, move to and fro, oscillate, revolve, rock, roll, sway, swivel, turn, twirl, wave about. **2** *swing opinion*. affect, bias, bring round, change (someone's mind), convert, convince, govern, influence, persuade, win over. **3** *support swung to the opposition*. change, move across, shift, transfer, vary. **4** *swing from a path*. deviate, divert, go off course, swerve, veer, waver, zigzag.

**swipe** *vb* **1** lash out at, strike, swing at. ▷ HIT. **2** ▷ STEAL.

**swirl** *vb* boil, churn, circulate, curl, eddy, move in circles, seethe, spin, surge, twirl, twist, whirl.

**switch** *n* circuit-breaker, light-switch, power-point. • *vb* change, divert, exchange, redirect, replace, reverse, shift, substitute, *inf* swap, transfer, turn.

**swivel** *vb* gyrate, pirouette, pivot, revolve, rotate, spin, swing, turn, twirl, wheel.

**swoop** *vb* descend, dive, drop, fall, fly down, lunge, plunge, pounce. **swoop on** ▷ RAID.

**sword** *n* blade, broadsword, cutlass, dagger, foil, kris, rapier, sabre, scimitar.

**sycophantic** *adj* flattering, servile, *inf* smarmy, toadyish, unctuous. ▷ FLATTER.

**syllabus** *n* course, curriculum, outline, programme of study.

**symbol** *n* mark, sign, token. □ *badge, brand, character, cipher, coat of arms, crest, emblem, figure, hieroglyph, ideogram, ideograph, image, insignia, letter, logo, logotype, monogram, motif, number, numeral, pictogram, pictograph, trademark.*

**symbolic** *adj* allegorical, emblematic, figurative, meaningful, metaphorical, representative, significant, suggestive, symptomatic, token (*gesture*).

**symbolize** *vb* be a sign of, betoken, communicate, connote, denote, epitomize, imply, indicate, mean, represent, signify, stand for, suggest.

**symmetrical** *adj* balanced, even, proportional, regular. *Opp* ASYMMETRICAL.

**sympathetic** *adj* benevolent, caring, charitable, comforting, compassionate, commiserating, concerned, consoling, empathetic, friendly, humane, interested, kind-hearted, kindly, merciful, pitying, soft-hearted, solicitous, sorry, supportive, tender, tolerant, understanding, warm. *Opp* UNSYMPATHETIC.

**sympathize** *vb inf* be on the same wavelength, be sorry, be sympathetic, comfort, commiserate, condole, console, empathize, feel, grieve, have sympathy, identify (with), mourn, pity, respond, show sympathy, understand.

**sympathy** *n* affinity, commiseration, compassion, concern, condolence, consideration, empathy, feeling, fellow-feeling, kindness, mercy, pity, rapport, solicitousness, tenderness, understanding.

**symptom** *n* characteristic, evidence, feature, indication, manifestation, mark, marker, sign, warning, warning-sign.

**symptomatic** *adj* characteristic, indicative, representative, suggestive, typical.

**synthesis** *n* amalgamation, blend, coalescence, combination, composite, compound, fusion, integration, union. *Opp* ANALYSIS.

**synthetic** *adj* artificial, bogus, concocted, counterfeit, ersatz, fabricated, fake, *inf* made-up, man-made, manufactured, mock, *inf* phoney, simulated, spurious, unnatural. *Opp* GENUINE, NATURAL.

**syringe** *n* hypodermic, needle.

**system** *n* **1** network, organization, *inf* set-up, structure. **2** approach, arrangement, logic, method, methodology, *Lat* modus operandi, order, plan, practice, procedure, process, routine, rules, scheme, technique. **3** *system of government.* constitution, regime. **4** *system of knowledge.* categorization, classification, code, discipline, philosophy, science, set of principles, theory.

**systematic** *adj* according to plan, businesslike, categorized, classified, codified, constitutional, co-ordinated, logical, methodical, neat, ordered, orderly, organized, planned, rational, regimented, routine, scientific, structured, tidy, well-arranged, well-organized, well-rehearsed, well-run. *Opp* UNSYSTEMATIC.

**systematize** *vb* arrange, catalogue, categorize, classify, codify, make systematic, organize, rationalize, regiment, standardize, tabulate.

# T

**table** *n* **1** bench, board, counter, desk, gate-leg table, kitchen table, worktop. **2** *table of information.* agenda, catalogue, chart, diagram, graph, index, inventory, list, register, schedule, tabulation, timetable. • *vb* bring forward, lay on the table, offer, proffer, propose, submit.

**tablet** *n* **1** capsule, drop, lozenge, medicine, pastille, pellet, pill. **2** *tablet of soap.* bar, block, chunk, piece, slab. **3** *tablet of stone.* gravestone, headstone, memorial, plaque, plate, tombstone.

**taboo** *adj* banned, censored, disapproved of, forbidden, interdicted, prohibited, proscribed, unacceptable, unlawful, unmentionable, unnamable. ▷ RUDE. • *n* anathema, ban, curse, interdiction, prohibition, proscription, taboo subject.

**tabulate** *vb* arrange as a table, catalogue, index, list, pigeon-hole, set out in columns, systematize.

**tacit** *adj* implicit, implied, silent, undeclared, understood, unexpressed, unsaid, unspoken, unvoiced.

**taciturn** *adj* mute, quiet, reserved, reticent, silent, tight-lipped, uncommunicative, unforthcoming. *Opp* TALKATIVE.

**tack** *n* **1** drawing-pin, nail, pin, tintack. **2** *the wrong tack.* approach, bearing, course, direction, heading, line, policy, procedure, technique. • *vb* **1** nail, pin. ▷ FASTEN. **2** sew, stitch. **3** *tack in a yacht.* beat against the wind, change course, go about, zigzag. **tack on** ▷ ADD.

**tackle** *n* **1** accoutrements, apparatus, *inf* clobber, equipment, fittings, gear, implements, kit, outfit, paraphernalia, rig, rigging, tools. **2** *a football tackle.* attack, block, challenge, interception, intervention. • *vb* **1** address (yourself to), apply yourself to, attempt, attend to, combat, *inf* come to grips with, concentrate on, confront, cope with, deal with, engage in, face up to, focus on, get involved in, grapple with, handle, *inf* have a go at, manage, set about, settle down to, sort out, take on, undertake. **2** *tackle an opponent.* attack, challenge, intercept, stop, take on.

**tacky** *adj* adhesive, gluey, *inf* gooey, gummy, sticky, viscous, wet. *Opp* DRY.

**tact** *n* adroitness, consideration, delicacy, diplomacy, discernment, discretion, finesse, judgement, perceptiveness, politeness, savoir-faire, sensitivity, tactfulness, thoughtfulness, understanding. *Opp* TACTLESSNESS.

**tactful** *adj* adroit, appropriate, considerate, courteous, delicate, diplomatic, discreet, judicious, perceptive, polite, politic, sensitive, thoughtful, understanding. *Opp* TACTLESS.

**tactical** *adj* artful, calculated, clever, deliberate, planned, politic, prudent, shrewd, skilful, strategic.

**tactics** *n* approach, campaign, course of action, design, device, manoeuvre, manoeuvring, plan, ploy, policy, procedure, ruse, scheme, stratagem, strategy.

**tactless** *adj* blundering, blunt, boorish, bungling, clumsy, discourteous, gauche, heavyhanded, hurtful, impolite, impolitic, inappropriate, inconsiderate, indelicate, indiscreet, inept, insensitive, maladroit, misjudged, thoughtless, uncivil, uncouth, undiplomatic, unkind. ▷ RUDE. *Opp* TACTFUL.

**tactlessness** *n* boorishness, clumsiness, gaucherie, indelicacy, indiscretion, ineptitude, insensitivity, lack of diplomacy, misjudgement, thoughtlessness, uncouthness. ▷ RUDENESS. *Opp* TACT.

**tag** *n* **1** docket, label, marker, name tag, price tag, slip, sticker, tab, ticket. **2** *a Latin tag.* ▷ SAYING. • *vb* identify, label, mark, ticket. **tag along with** ▷ FOLLOW.

**tail** *n* appendage, back, brush (*of fox*), buttocks, end, extremity, rear, rump, scut (*of rabbit*), tail-end. • *vb* dog, follow, hunt, pursue, shadow, stalk, track, trail. **tail off** ▷ DECLINE.

**taint** *vb* **1** adulterate, contaminate, defile, dirty, infect, poison, pollute, soil. **2** *taint a reputation.* besmirch, blacken, blemish, damage, dishonour, harm, ruin, slander, smear, spoil, stain, sully, tarnish.

**take** *vb* **1** acquire, bring, carry away, *inf* cart off, catch, clasp, clutch, fetch, gain, get, grab, grasp, grip, hold, pick up, pluck, remove, secure, seize, snatch, transfer. **2** *take prisoners*. abduct, arrest, capture, catch, corner, detain, ensnare, entrap, secure. **3** *take property*. appropriate, get away with, pocket. ▷ STEAL. **4** *take 2 from 4*. deduct, eliminate, subtract, take away. **5** *take passengers*. accommodate, carry, contain, have room for, hold. **6** *take a partner*. accompany, conduct, convey, escort, ferry, guide, lead, transport. **7** *take a taxi*. engage, hire, make use of, travel by, use. **8** *take a subject*. have lessons in, learn about, read, study. **9** *can't take pain*. abide, accept, bear, brook, endure, receive, *inf* stand, *inf* stomach, suffer, tolerate, undergo, withstand. **10** *take food, drink*. consume, drink, eat, have, swallow. **11** *It takes courage to own up*. necessitate, need, require, use up. **12** *take a new name*. adopt, assume, choose, select. **take aback** ▷ SURPRISE. **take after** ▷ RESEMBLE. **take against** ▷ DISLIKE. **take back** ▷ WITHDRAW. **take in** ▷ ACCOMMODATE, DECEIVE, UNDERSTAND. **take life** ▷ KILL. **take off** ▷ IMITATE. **take off, take out** ▷ REMOVE. **take on, take up** ▷ UNDERTAKE. **take over** ▷ USURP. **take part** ▷ PARTICIPATE. **take place** ▷ HAPPEN. **take to task** ▷ REPRIMAND. **take up** ▷ BEGIN, OCCUPY.

**take-over** *n* amalgamation, combination, incorporation, merger.

**takings** *n* earnings, gains, gate, income, proceeds, profits, receipts, revenue.

**tale** *n* account, anecdote, chronicle, narration, narrative, relation, report, *sl* spiel, story, yarn. ▷ WRITING.

**talent** *n* ability, accomplishment, aptitude, brilliance, capacity, expertise, facility, faculty, flair, genius, gift, ingenuity, knack, *inf* know-how, prowess, skill, strength, versatility.

**talented** *adj* able, accomplished, artistic, brilliant, distinguished, expert, gifted, inspired, proficient, skilful, skilled, versatile. ▷ CLEVER. *Opp* UNSKILFUL.

**talisman** *n* amulet, charm, fetish, mascot.

**talk** *n* **1** baby-talk, *inf* blarney, chat, *inf* chin-wag, *inf* chit-chat, confabulation, conference, conversation, dialogue, discourse, discussion, gossip, intercourse, language, palaver, *inf* powwow, *inf* tattle, *inf* tittle-tattle, words. **2** *a public talk*. address, diatribe, exhortation, harangue, lecture, oration, *inf* peptalk, presentation, sermon, speech, tirade. • *vb* **1** address one another, articulate ideas, commune, communicate, confer, converse, deliver a speech, discourse, discuss, enunciate, exchange views, have a conversation, *inf* hold forth, lecture, negotiate, *inf* pipe up, pontificate, preach, pronounce words, say something, sermonize, speak, tell, use language, use your voice, utter, verbalize, vocalize. □ *babble, bawl, bellow, blab, blether, blurt out, breathe, burble, call out, chat, chatter, clamour, croak, cry, drawl, drone, gabble, gas, gibber, gossip, grunt, harp, howl, intone, jabber, jaw, jeer, lisp, maunder, moan, mumble, murmur, mutter, natter, patter, prattle, pray, preach, rabbit on, rant, rasp, rattle on, rave, roar, scream, screech, shout, shriek, slur, snap, snarl, speak in an undertone, splutter, spout, squeal, stammer, stutter, utter, vociferate, wail, whimper, whine, whinge, whisper, witter, yell.* **2** *talk French*.

communicate in, express yourself in, pronounce, speak. **3** *get someone to talk.* confess, give information, *inf* grass, inform, *inf* let on, *inf* spill the beans, *inf* squeal, *inf* tell tales. ▷ SAY, SPEAK. **talk about** ▷ DISCUSS. **talk to** ▷ ADDRESS.

**talkative** *adj* articulate, *inf* chatty, communicative, effusive, eloquent, expansive, garrulous, glib, gossipy, long-winded, loquacious, open, prolix, unstoppable, verbose, vocal, voluble, wordy. *Opp* TACITURN. **talkative person** chatter-box, *sl* gas-bag, gossip, *sl* wind-bag.

**tall** *adj* colossal, giant, gigantic, high, lofty, soaring, towering. ▷ BIG. *Opp* SHORT.

**tally** *n* addition, count, reckoning, record, sum, total. • *vb* **1** accord, agree, coincide, concur, correspond, match up, square. **2** *tally up the bill.* add, calculate, compute, count, reckon, total, work out.

**tame** *adj* **1** amenable, biddable, broken in, compliant, disciplined, docile, domesticated, gentle, manageable, meek, mild, obedient, safe, subdued, submissive, tamed, tractable, trained. **2** *tame animals.* approachable, bold, fearless, friendly, sociable, unafraid. **3** *a tame story.* bland, boring, dull, feeble, flat, insipid, lifeless, tedious, unadventurous, unexciting, uninspiring, uninteresting, vapid, *inf* wishy-washy. *Opp* EXCITING, WILD. • *vb* break in, conquer, curb, discipline, domesticate, house-train, humble, keep under, make tame, master, mollify, mute, quell, repress, subdue, subjugate, suppress, temper, tone down, train.

**tamper** *vb* **tamper with** alter, *inf* fiddle about with, interfere with, make adjustments to, meddle with, tinker with.

**tan** *n* sunburn, suntan. • *vb* bronze, brown, burn, colour, darken, get tanned.

**tang** *n* acidity, *inf* bite, *inf* edge, *inf* nip, piquancy, pungency, savour, sharpness, spiciness, zest.

**tangible** *adj* actual, concrete, corporeal, definite, material, palpable, perceptible, physical, positive, provable, real, solid, substantial, tactile, touchable. *Opp* INTANGIBLE.

**tangle** *n* coil, complication, confusion, jumble, jungle, knot, labyrinth, mass, maze, mesh, mess, muddle, scramble, twist, web. • *vb* **1** complicate, confuse, entangle, entwine, *inf* foul up, intertwine, interweave, muddle, ravel, scramble, *inf* snarl up, twist. **2** *tangle fish in a net.* catch, enmesh, ensnare, entrap, trap. *Opp* DISENTANGLE, FREE. **3** *tangle with criminals.* become involved with, confront, cross. **tangled** ▷ DISHEVELLED, INTRICATE.

**tangy** *adj* acid, appetizing, bitter, fresh, piquant, pungent, refreshing, sharp, spicy, strong, tart. *Opp* BLAND.

**tank** *n* **1** aquarium, basin, cistern, reservoir. **2** *army tank.* armoured vehicle.

**tanned** *adj* brown, sunburnt, suntanned, weather-beaten.

**tantalize** *vb* bait, entice, frustrate, *inf* keep on tenterhooks, lead on, plague, provoke, taunt, tease, tempt, titillate, torment.

**tap** *n* **1** *Amer* faucet, spigot, stopcock, valve. **2** knock, rap. • *vb* knock, rap, strike. ▷ HIT.

**tape** *n* **1** band, belt, binding, braid, fillet, ribbon, strip, stripe.

2 audiotape, cassette, magnetic tape, tape-recording, videotape.

**taper** *n* candle, lighter, spill. • *vb* attenuate, become narrower, narrow, thin.

**taper off** ▷ DECLINE.

**target** *n* 1 aim, ambition, end, goal, hope, intention, objective, purpose. 2 *target of attack.* butt, object, quarry, victim.

**tariff** *n* 1 charges, menu, price-list, schedule. 2 *tariff on imports.* customs, duty, excise, impost, levy, tax, toll.

**tarnish** *vb* 1 blacken, corrode, dirty, discolour, soil, spoil, stain, taint. 2 *tarnish a reputation.* blemish, blot, calumniate, defame, denigrate, disgrace, dishonour, mar, ruin, spoil, stain, sully.

**tarry** *vb* dawdle, delay, *inf* hang about, hang back, linger, loiter, pause, procrastinate, temporize, wait.

**tart** *adj* 1 acid, acidic, acidulous, astringent, biting, citrus, harsh, lemony, piquant, pungent, sharp, sour, tangy. 2 *a tart rejoinder.* ▷ SHARP. *Opp* BLAND, SWEET. • *n* 1 flan, pastry, pasty, patty, pie, quiche, tartlet, turnover. 2 ▷ PROSTITUTE.

**task** *n* activity, assignment, burden, business, charge, chore, duty, employment, enterprise, errand, imposition, job, mission, requirement, test, undertaking, work. **take to task** ▷ REPRIMAND.

**taste** *n* 1 character, flavour, relish, savour. 2 bit, bite, morsel, mouthful, nibble, piece, sample, titbit. 3 *an acquired taste.* appetite, appreciation, choice, fancy, fondness, inclination, judgement, leaning, liking, partiality, preference. 4 *a person of taste.* breeding, cultivation, culture, discernment, discretion, discrimination, education, elegance, fashion sense, finesse, good judgement, perception, perceptiveness, polish, refinement, sensitivity, style, tastefulness. • *vb* nibble, relish, sample, savour, sip, test, try. **in bad taste** ▷ TASTELESS. **in good taste** ▷ TASTEFUL.

**tasteful** *adj* aesthetic, artistic, attractive, charming, *Fr* comme il faut, correct, cultivated, decorous, dignified, discerning, discreet, discriminating, elegant, fashionable, in good taste, judicious, *inf* nice, polite, proper, refined, restrained, sensitive, smart, stylish, tactful, well-judged. *Opp* TASTELESS.

**tasteless** *adj* 1 cheap, coarse, crude, *inf* flashy, garish, gaudy, graceless, improper, inartistic, in bad taste, indecorous, indelicate, inelegant, injudicious, in poor taste, *inf* kitsch, *inf* loud, meretricious, ugly, unattractive, uncouth, uncultivated, undiscriminating, unfashionable, unimaginative, unpleasant, unrefined, unseemly, unstylish, vulgar. *Opp* TASTEFUL. 2 *tasteless food.* bland, characterless, flavourless, insipid, mild, uninteresting, watered-down, watery, weak, *inf* wishy-washy. *Opp* TASTY.

**tasty** *adj* appetizing, delectable, delicious, flavoursome, luscious, *inf* mouth-watering, *inf* nice, palatable, *inf* scrumptious, toothsome, *sl* yummy. □ *acid, bitter, creamy, fruity, hot, meaty, peppery, piquant, salty, savoury, sharp, sour, spicy, sugary, sweet, tangy, tart. Opp* TASTELESS.

**tattered** *adj* frayed, ragged, rent, ripped, shredded, tatty, threadbare, torn, worn out. *Opp* SMART.

**tatters** *plur n* bits, pieces, rags, ribbons, shreds, torn pieces.

**tatty** *adj* **1** frayed, old, patched, ragged, ripped, scruffy, shabby, tattered, torn, threadbare, untidy, worn out. **2** ▷ TAWDRY. *Opp* SMART.

**taunt** *vb* annoy, goad, insult, jeer at, reproach, tease, torment. ▷ RIDICULE.

**taut** *adj* firm, rigid, stiff, strained, stretched, tense, tight. *Opp* SLACK.

**tautological** *adj* long-winded, otiose, pleonastic, prolix, redundant, repetitious, repetitive, superfluous, tautologous, verbose, wordy. *Opp* CONCISE.

**tautology** *n* duplication, long-windedness, pleonasm, prolixity, repetition, verbiage, verbosity, wordiness.

**tavern** *n old use* alehouse, bar, hostelry, inn, *inf* local, pub, public house.

**tawdry** *adj inf* Brummagem, cheap, common, eye-catching, fancy, *inf* flashy, garish, gaudy, inferior, meretricious, poor quality, showy, tasteless, tatty, tinny, vulgar, worthless. *Opp* TASTEFUL.

**tax** *n* charge, due, duty, imposition, impost, levy, tariff, *old use* tribute. □ *airport tax, community charge, corporation tax, customs, death duty, estate duty, excise, income tax, poll tax, property tax, rates,* old use *tithe, toll, value added tax. vb* **1** assess, exact, impose a tax on, levy a tax on. **2** *tax someone's patience.* burden, exhaust, make heavy demands on, overwork, pressure, pressurize, strain, try. ▷ TIRE. **tax with** accuse of, blame for, censure for, charge with, reproach for, reprove for.

**taxi** *n* cab, *old use* hackney carriage, minicab.

**teach** *vb* advise, brainwash, coach, counsel, demonstrate to, discipline, drill, edify, educate, enlighten, familiarize with, give lessons in, ground in, impart knowledge to, implant knowledge in, inculcate habits in, indoctrinate, inform, instruct, lecture, school, train, tutor.

**teacher** *n* adviser, educator, guide. □ *coach, counsellor, demonstrator, don, governess, guru, headteacher, housemaster, housemistress, instructor, lecturer, maharishi, master, mentor, mistress, pedagogue, preacher, preceptor, professor, pundit, schoolmaster, schoolmistress, schoolteacher, trainer, tutor.*

**teaching** *n* **1** education, guidance, instruction, training. □ *brainwashing, briefing, coaching, computer-aided learning, counselling, demonstration, familiarization, grounding, indoctrination, lecture, lesson, practical, preaching, rote learning, schooling, seminar, tuition, tutorial, work experience, workshop.* **2** *religious teachings.* doctrine, dogma, gospel, precept, principle, tenet.

**team** *n* club, crew, gang, *inf* line up, side. ▷ GROUP.

**tear** *n* **1** [rhymes with *fear*] droplet, tear-drop. [*plur*] *inf* blubbering, crying, sobs, weeping. **2** [rhymes with *bear*] cut, fissure, gap, gash, hole, laceration, opening, rent, rip, slit, split. • *vb* claw, gash, lacerate, mangle, pierce, pull apart, rend, rip, rive, rupture, scratch, sever, shred, slit, snag, split. **shed tears** ▷ WEEP.

**tearful** *adj inf* blubbering, crying, emotional, in tears, lachrymose, snivelling, sobbing, weeping, *inf* weepy, wet-cheeked, whimpering. ▷ SAD.

**tease** *vb inf* aggravate, annoy, badger, bait, chaff, goad, harass, irritate, laugh at, make fun of, mock, *inf* needle, *inf* nettle, pester, plague, provoke, *inf* pull someone's leg, *inf* rib, tantalize, taunt, torment, vex, worry. ▷ RIDICULE.

**teasing** *n* badinage, banter, chaffing, joking, mockery, provocation, raillery, *inf* ribbing, ridicule, taunts.

**technical** *adj* **1** complicated, detailed, esoteric, expert, professional, specialized. **2** *technical skill.* engineering, industrial, mechanical, technological, scientific.

**technician** *n* engineer, mechanic, skilled worker, *plur* technical staff.

**technique** *n* **1** approach, dodge, knack, manner, means, method, mode, procedure, routine, system, trick, way. **2** *an artist's technique.* art, artistry, cleverness, craft, craftsmanship, expertise, facility, *inf* know-how, proficiency, skill, talent, workmanship.

**technological** *adj* advanced, automated, computerized, electronic, scientific.

**tedious** *adj* banal, boring, dreary, *inf* dry-as-dust, dull, endless, *inf* humdrum, irksome, laborious, long-drawn-out, long-winded, monotonous, prolonged, repetitious, slow, soporific, tiresome, tiring, unexciting, uninteresting, vapid, wearing, wearisome, wearying. *Opp* INTERESTING.

**tedium** *n* boredom, dreariness, dullness, ennui, long-windedness, monotony, repetitiousness, slowness, tediousness.

**teem** *vb* **1** abound (in), be alive (with), be full (of), be infested, be overrun (by), *inf* bristle, *inf* crawl, proliferate, seethe, swarm with. **2** ▷ RAIN.

**teenager** *n* adolescent, boy, girl, juvenile, minor, youngster, youth.

**teetotal** *adj* abstemious, abstinent, *sl* on the wagon, restrained, self-denying, self-disciplined, temperate.

**teetotaller** *n* abstainer, non-drinker.

**telegram** *n* cable, cablegram, fax, telex, wire.

**telepathic** *adj* clairvoyant, psychic.

**telephone** *n inf* blower, carphone, handset, phone. • *vb inf* buzz, call, dial, *inf* give someone a buzz, *inf* give someone a call, *inf* give someone a tinkle, phone, ring, ring up.

**telescope** *vb* abbreviate, collapse, compress, elide, shorten.

**telescopic** *adj* adjustable, collapsible, expanding, extending, retractable.

**televise** *vb* broadcast, relay, send out, transmit.

**television** *n inf* the box, monitor, receiver, *inf* small screen, *inf* telly, video.

**tell** *vb* **1** acquaint with, advise, announce, assure, communicate, describe, disclose, divulge, explain, impart, inform, make known, narrate, notify, portray, promise, recite, recount, rehearse, relate, reveal, utter. ▷ SPEAK, TALK. **2** *tell the difference.* calculate, comprehend, decide, discover, discriminate, distinguish, identify, notice, recognize, see. **3** *told me what to do.* command, direct, instruct, order. **tell off** ▷ REPRIMAND.

**teller** *n* **1** author, narrator, raconteur, storyteller. **2** *teller in a bank.* bank clerk, cashier.

**telling** *adj* considerable, effective, influential, potent, powerful, significant, striking, weighty.

**temper** *n* 1 attitude, character, disposition, frame of mind, humour, *inf* make-up, mood, personality, state of mind, temperament. 2 *watch your temper*. anger, churlishness, fit of anger, fury, hot-headedness, ill-humour, irascibility, irritability, *inf* paddy, passion, peevishness, petulance, rage, surliness, tantrum, unpredictability, volatility, *inf* wax, wrath. 3 *keep your temper*. calmness, composure, *sl* cool, coolness, equanimity, sang-froid, self-control, self-possession. • *vb* 1 assuage, lessen, mitigate, moderate, modify, modulate, reduce, soften, soothe, tone down. 2 *temper steel*. harden, strengthen, toughen.

**temperament** *n* attitude, character, *old use* complexion, disposition, frame of mind, *old use* humour, *inf* make-up, mood, nature, personality, spirit, state of mind, temper.

**temperamental** *adj* 1 characteristic, congenital, constitutional, inherent, innate, natural. 2 *temperamental moods*. capricious, changeable, emotional, erratic, excitable, explosive, fickle, highly-strung, impatient, inconsistent, inconstant, irascible, irritable, mercurial, moody, neurotic, passionate, sensitive, touchy, undependable, unpredictable, unreliable, *inf* up and down, variable, volatile.

**temperance** *n* abstemiousness, continence, moderation, self-discipline, self-restraint, sobriety, teetotalism.

**temperate** *adj* calm, controlled, disciplined, moderate, reasonable, restrained, self-possessed, sensible, sober, stable, steady. *Opp* EXTREME.

**tempest** *n* cyclone, gale, hurricane, tornado, tumult, typhoon, whirlwind. ▷ STORM.

**tempestuous** *adj* fierce, furious, tumultuous, turbulent, vehement, violent, wild. ▷ STORMY. *Opp* CALM.

**temple** *n* church, house of god, mosque, pagoda, place of worship, shrine, synagogue.

**tempo** *n* beat, pace, rate, rhythm, pulse, speed.

**temporal** *adj* earthly, fleshly, impermanent, material, materialistic, mortal, mundane, non-religious, passing, secular, sublunary, terrestrial, transient, transitory, worldly. *Opp* SPIRITUAL.

**temporary** *adj* 1 brief, ephemeral, evanescent, fleeting, fugitive, impermanent, interim, makeshift, momentary, passing, provisional, short, short-lived, short-term, stop-gap, transient, transitory. 2 *temporary captain*. acting. *Opp* PERMANENT.

**tempt** *vb* allure, attract, bait, bribe, cajole, captivate, coax, decoy, entice, fascinate, inveigle, lure, offer incentives, persuade, seduce, tantalize, woo. **tempting** ▷ APPETIZING, ATTRACTIVE.

**temptation** *n* allure, allurement, appeal, attraction, cajolery, coaxing, draw, enticement, fascination, inducement, lure, persuasion, pull, seduction, snare, wooing.

**tenable** *adj* arguable, believable, conceivable, credible, creditable, defendable, defensible, feasible, justifiable, legitimate, logical, plausible, rational, reasonable, sensible, sound, supportable, understandable, viable. *Opp* INDEFENSIBLE.

**tenacious** *adj* determined, dogged, firm, intransigent, obdurate, obstinate, persistent, pertinacious, resolute, single-minded, steadfast, strong, stubborn, tight, uncompromising, unfaltering, unshakeable, unswerving, unwavering, unyielding. *Opp* WEAK.

**tenant** *n* inhabitant, leaseholder, lessee, lodger, occupant, occupier, resident.

**tend** *vb* **1** attend to, care for, cherish, cultivate, guard, keep, *inf* keep an eye on, look after, manage, mind, minister to, mother, protect, supervise, take care of, watch. **2** *tend the sick.* nurse, treat. **3** *tend to fall asleep.* be biased, be disposed, be inclined, be liable, be prone, have a tendency, incline.

**tendency** *n* bias, disposition, drift, inclination, instinct, leaning, liability, partiality, penchant, predilection, predisposition, proclivity, proneness, propensity, readiness, susceptibility, trend.

**tender** *adj* **1** dainty, delicate, fleshy, fragile, frail, green, immature, soft, succulent, vulnerable, weak, young. **2** *tender meat.* chewable, eatable, edible. **3** *a tender wound.* aching, inflamed, painful, sensitive, smarting, sore. **4** *a tender love-song.* emotional, heartfelt, moving, poignant, romantic, sentimental, stirring, touching. **5** *tender care.* affectionate, amorous, caring, compassionate, concerned, considerate, fond, gentle, humane, kind, loving, merciful, pitying, soft-hearted, sympathetic, tender-hearted, warm-hearted. *Opp* TOUGH, UNSYMPATHETIC.

**tense** *adj* **1** rigid, strained, stretched, taut, tight. **2** *a tense person.* anxious, apprehensive, edgy, excited, fidgety, highly-strung, intense, jittery, jumpy, *inf* keyed-up, nervous, on edge, *inf* on tenterhooks, overwrought, restless, strained, stressed, *inf* strung up, touchy, uneasy, *sl* uptight, worried. **3** *a tense situation.* exciting, fraught, *inf* nail-biting, nerve-racking, stressful, worrying. *Opp* RELAXED.

**tension** *n* **1** pull, strain, stretching, tautness, tightness. **2** *the tension of waiting.* anxiety, apprehension, edginess, excitement, nervousness, stress, suspense, unease, worry. *Opp* RELAXATION.

**tent** *n* □ *bell tent, big-top, frame tent, marquee, ridge tent, tepee, trailer tent, wigwam.*

**tentative** *adj* cautious, diffident, doubtful, experimental, exploratory, half-hearted, hesitant, inconclusive, indecisive, indefinite, nervous, preliminary, provisional, shy, speculative, timid, uncertain, uncommitted, unsure, *inf* wishy-washy. *Opp* DECISIVE.

**tenuous** *adj* attenuated, fine, flimsy, fragile, insubstantial, slender, slight, weak. ▷ THIN. *Opp* STRONG.

**tepid** *adj* **1** lukewarm, warm. **2** *a tepid response.* ▷ APATHETIC.

**term** *n* **1** duration, period, season, span, spell, stretch, time. **2** *a school term. Amer* semester, session. **3** *technical terms.* appellation, designation, epithet, expression, name, phrase, saying, title, word. **terms 1** conditions, particulars, provisions, provisos, specifications, stipulations. **2** *a hotel's terms.* charges, fees, prices, rates, schedule, tariff.

**terminal** *adj* deadly, fatal, final, incurable, killing, lethal, mortal. • *n* **1** keyboard, VDU, workstation. **2** *passenger terminal.*

airport, terminus. 3 *electric terminal.* connection, connector, coupling.

**terminate** *vb* bring to an end, cease, come to an end, discontinue, end, finish, *inf* pack in, phase out, stop, *inf* wind up. ▷ END. *Opp* BEGIN.

**terminology** *n* argot, cant, choice of words, jargon, language, nomenclature, phraseology, special terms, technical language, vocabulary.

**terminus** *n* destination, last stop, station, terminal, termination.

**terrain** *n* country, ground, land, landscape, territory, topography.

**terrestrial** *adj* earthly, mundane, ordinary, sublunary.

**terrible** *adj* 1 acute, appalling, awful, *inf* beastly, distressing, dreadful, fearful, fearsome, formidable, frightening, frightful, ghastly, grave, gruesome, harrowing, hideous, horrendous, horrible, horrific, horrifying, insupportable, intolerable, loathsome, nasty, nauseating, outrageous, revolting, shocking, terrific, terrifying, unbearable, vile. 2 ▷ BAD.

**terrific** *adj Terrific* may mean *causing terror* (▷ TERRIBLE). It is more often used *informally* of anything which is *extreme* in its own way: *a terrific problem* ▷ EXTREME; *terrific size* ▷ BIG; *a terrific party* ▷ EXCELLENT; *a terrific storm* ▷ VIOLENT.

**terrify** *vb* appal, dismay, horrify, *inf* make your blood run cold, petrify, shock, terrorize. ▷ FRIGHTEN. **terrified** ▷ FRIGHTENED. **terrifying** ▷ FRIGHTENING.

**territory** *n* area, colony, *old use* demesne, district, domain, dominion, enclave, jurisdiction, land, neighbourhood, precinct, preserve, province, purlieu, region, sector, sphere, state, terrain, tract, zone. ▷ COUNTRY.

**terror** *n* alarm, awe, consternation, dismay, dread, fright, *inf* funk, horror, panic, shock, trepidation. ▷ FEAR.

**terrorist** *n* assassin, bomber, desperado, gunman, hijacker.

**terrorize** *vb* browbeat, bully, coerce, cow, intimidate, menace, persecute, terrify, threaten, torment, tyrannize. ▷ FRIGHTEN.

**terse** *adj* abrupt, brief, brusque, compact, concentrated, concise, crisp, curt, epigrammatic, incisive, laconic, pithy, short, *inf* short and sweet, *inf* snappy, succinct, to the point. *Opp* VERBOSE.

**test** *n* analysis, appraisal, assay, assessment, audition, *inf* check-over, *inf* check-up, evaluation, examination, inspection, interrogation, investigation, probation, quiz, screen-test, trial, *inf* try-out. • *vb* analyse, appraise, assay, assess, audition, check, evaluate, examine, experiment with, inspect, interrogate, investigate, probe, *inf* put someone through their paces, put to the test, question, quiz, screen, try out.

**testify** *vb* affirm, attest, aver, bear witness, declare, give evidence, proclaim, state on oath, swear, vouch, witness.

**testimonial** *n* character reference, commendation, recommendation, reference.

**testimony** *n* affidavit, assertion, declaration, deposition, evidence, statement, submission.

**tether** *n* chain, cord, fetter, halter, lead, leash, painter, restraint, rope. • *vb* chain up, fetter, keep on a tether, leash, restrain, rope, secure, tie up. ▷ FASTEN.

**text** *n* **1** argument, content, contents, matter, subject matter, wording. **2** *a literary text.* book, textbook, work. ▷ WRITING. **3** *a text from scripture.* line, motif, passage, quotation, sentence, theme, topic, verse.

**textile** *n* fabric, material, stuff. ▷ CLOTH.

**texture** *n* appearance, composition, consistency, feel, grain, quality, surface, tactile quality, touch, weave.

**thank** *vb* acknowledge, express thanks, say thank you, show gratitude.

**thankful** *adj* appreciative, contented, glad, grateful, happy, indebted, pleased, relieved. *Opp* UNGRATEFUL.

**thankless** *adj* bootless, futile, profitless, unappreciated, unrecognized, unrewarded, unrewarding. *Opp* PROFITABLE.

**thanks** *plur n* acknowledgement, appreciation, gratefulness, gratitude, recognition, thanksgiving. **thanks to** as a result of, because of, owing to, through.

**thaw** *vb* become liquid, defrost, de-ice, heat up, melt, soften, uncongeal, unfreeze, unthaw, warm up. *Opp* FREEZE.

**theatre** *n* **1** auditorium, hall, opera-house, playhouse. **2** acting, dramaturgy, histrionic arts, show business, thespian arts. □ *ballet, masque, melodrama, mime, musical, music-hall, opera, pantomime, play.* ▷ DRAMA, ENTERTAINMENT, PERFORMANCE.

**theatrical** *adj* **1** dramatic, histrionic, thespian. **2** [*derog*] *a theatrical exit.* affected, artificial, calculated, demonstrative, exaggerated, forced, *inf* hammy, melodramatic, ostentatious, overacted, overdone, *inf* over the top, pompous, self-important, showy, stagy, stilted, unconvincing, unnatural. *Opp* NATURAL.

**theft** *n* burglary, larceny, pilfering, robbery, thievery. ▷ STEALING.

**theme** *n* **1** argument, core, essence, gist, idea, issue, keynote, matter, point, subject, text, thesis, thread, topic. **2** *a musical theme.* air, melody, motif, subject, tune.

**theology** *n* divinity, religion, religious studies.

**theoretical** *adj* abstract, academic, conjectural, doctrinaire, hypothetical, ideal, notional, pure (*science*), putative, speculative, suppositious, unproven, untested. *Opp* PRACTICAL, PROVEN.

**theorize** *vb* conjecture, form a theory, guess, hypothesize, speculate.

**theory** *n* **1** argument, assumption, belief, conjecture, explanation, guess, hypothesis, idea, notion, speculation, supposition, surmise, thesis, view. **2** *theory of a subject.* laws, principles, rules, science. *Opp* PRACTICE.

**therapeutic** *adj* beneficial, corrective, curative, healing, healthy, helpful, medicinal, remedial, restorative, salubrious. *Opp* HARMFUL.

**therapist** *n* counsellor, healer, physiotherapist, psychoanalyst, psychotherapist.

**therapy** *n* cure, healing, remedy, tonic, treatment. □ *chemotherapy, group therapy, hydrotherapy, hypnotherapy, occupational therapy, physiotherapy, psychotherapy, radiotherapy.* ▷ MEDICINE.

**therefore** *adv* accordingly, consequently, hence, so, thus.

**thesis** *n* **1** argument, assertion, contention, hypothesis, idea, opinion, postulate, premise, premiss, proposition, theory, view. **2** *a research thesis.* disquisition, dissertation, essay, monograph, paper, tract, treatise.

**thick** *adj* **1** broad, *inf* bulky, chunky, stout, sturdy, wide. ▷ FAT. **2** *a thick layer.* deep, heavy, substantial, woolly. **3** *a thick crowd.* compact, dense, impassable, impenetrable, numerous, packed, solid. **4** *thick liquid.* clotted, coagulated, concentrated, condensed, firm, glutinous, heavy, jellied, sticky, stiff, viscid, viscous. **5** *thick growth.* abundant, bushy, luxuriant, plentiful. **6** *thick with visitors.* alive, bristling, *inf* chock-full, choked, covered, crammed, crawling, crowded, filled, full, jammed, swarming, teeming. *Opp* THIN.

**thicken** *vb* coagulate, clot, concentrate, condense, congeal, firm up, gel, jell, reduce, solidify, stiffen.

**thickness** *n* **1** breadth, density, depth, fatness, viscosity, width. **2** *a thickness of paint, rock.* coating, layer, seam, stratum.

**thief** *n* bandit, brigand, burglar, cat-burglar, cutpurse, embezzler, footpad, highwayman, housebreaker, kleptomaniac, looter, mugger, peculator, pickpocket, pilferer, pirate, plagiarist, poacher, purloiner, robber, safe-cracker, shoplifter, stealer, swindler. ▷ CRIMINAL.

**thieving** *adj* dishonest, light-fingered, rapacious.
• *n* ▷ STEALING.

**thin** *adj* **1** anorexic, attenuated, bony, cadaverous, emaciated, fine, flat-chested, gangling, gaunt, lanky, lean, narrow, pinched, rangy, scraggy, scrawny, skeletal, skinny, slender, slight, slim, small, spare, spindly, underfed, undernourished, underweight, wiry. *Opp* FAT. **2** *a thin layer.* delicate, diaphanous, filmy, fine, flimsy, gauzy, insubstantial, light, *inf* see-through, shallow, sheer (*silk*), superficial, translucent, wispy. **3** *a thin crowd.* meagre, scanty, scarce, scattered, sparse. **4** *thin liquid.* dilute, flowing, fluid, runny, sloppy, watery, weak. **5** *thin atmosphere.* rarefied. **6** *a thin excuse.* feeble, implausible, tenuous, transparent, unconvincing. *Opp* DENSE, STRONG, THICK.
• *vb* dilute, water down, weaken. **thin out 1** become less dense, decrease, diminish, disperse. **2** make less dense, prune, reduce, trim, weed out.

**thing** *n* **1** apparatus, artefact, article, body, contrivance, device, entity, gadget, implement, invention, item, object, utensil. **2** affair, circumstance, deed, event, eventuality, happening, incident, occurrence, phenomenon. **3** *a thing on your mind.* concept, detail, fact, factor, feeling, idea, point, statement, thought. **4** *a thing to be done.* act, action, chore, deed, job, responsibility, task. **5** [*inf*] *a thing about snakes.* aversion, fixation, *inf* hang-up, mania, neurosis, obsession, passion, phobia, preoccupation. **things 1** baggage, belongings, clothing, equipment, *inf* gear, luggage, possessions, *inf* stuff. **2** *How are things?* circumstances, conditions, life.

**think** *vb* **1** attend, brood, chew things over, cogitate, concentrate, consider, contemplate, day-dream, deliberate, dream, dwell (on), fantasize, give thought (to), meditate, *inf* mull over, muse, ponder, *inf* rack your brains, reason, reflect, remind yourself of, reminisce, ruminate, use your intelligence, work things out, worry. 2 *Do you think it's true?* accept, admit, assume, be convinced, believe, be under the impression, conclude, deem, estimate, feel, guess, have faith, imagine, judge, presume, reckon, suppose, surmise. **think better of** ▷ RECONSIDER. **thinking** ▷ INTELLIGENT, THOUGHTFUL. **think up** ▷ DEVISE.

**thinker** *n inf* brain, innovator, intellect, inventor, *inf* mastermind, philosopher, sage, savant, scholar.

**thirst** *n* 1 drought, dryness, thirstiness. 2 *thirst for knowledge.* appetite, craving, desire, eagerness, hunger, itch, longing, love (of), lust, passion, urge, wish, yearning, *inf* yen. • *vb* be thirsty, crave, have a thirst, hunger, long, strive (after), wish, yearn. **thirst for** ▷ WANT.

**thirsty** *adj* 1 arid, dehydrated, dry, *inf* gasping, panting, parched. 2 *thirsty for news.* avid, craving, desirous, eager, greedy, hankering, itching, longing, voracious, yearning.

**thorn** *n* barb, bristle, needle, prickle, spike, spine.

**thorny** *adj* 1 barbed, bristly, prickly, scratchy, sharp, spiky, spiny. 2 ▷ DIFFICULT.

**thorough** *adj* 1 assiduous, attentive, careful, comprehensive, conscientious, deep, detailed, diligent, efficient, exhaustive, extensive, full, *inf* in-depth, methodical, meticulous, minute, observant, orderly, organized, painstaking, particular, penetrating, probing, scrupulous, searching, systematic, thoughtful, watchful. *Opp* SUPERFICIAL. 2 *a thorough rascal.* absolute, arrant, complete, downright, out-and-out, perfect, proper, sheer, thoroughgoing, total, unmitigated, unmixed, unqualified, utter.

**thought** *n* 1 *inf* brainwork, brooding, *inf* brown study, cerebration, cogitation, concentration, consideration, contemplation, daydreaming, deliberation, intelligence, introspection, meditation, mental activity, musing, pensiveness, ratiocination, rationality, reason, reasoning, reflection, reverie, rumination, study, thinking, worrying. 2 *a clever thought.* belief, concept, conception, conclusion, conjecture, conviction, idea, notion, observation, opinion. 3 *no thought of gain.* aim, design, dream, expectation, hope, intention, objective, plan, prospect, purpose, vision. 4 *a kind thought.* attention, concern, consideration, kindness, solicitude, thoughtfulness.

**thoughtful** *adj* 1 absorbed, abstracted, anxious, attentive, brooding, contemplative, dreamy, grave, introspective, meditative, pensive, philosophical, rapt, reflective, serious, solemn, studious, thinking, wary, watchful, worried. 2 *thoughtful work.* careful, conscientious, diligent, exhaustive, intelligent, methodical, meticulous, observant, orderly, organized, painstaking, rational, scrupulous, sensible, systematic, thorough. 3 *a thoughtful kindness.* attentive, caring, compassionate, concerned, considerate, friendly, good-natured, helpful, obliging, public-spirited, solicitous, unselfish. ▷ KIND. *Opp* THOUGHTLESS.

**thoughtless** *adj* 1 absent-minded, careless, forgetful, hasty, heedless, ill-considered, impetuous, inadvertent, inattentive, injudicious, irresponsible, mindless, negligent, rash, reckless, *inf* scatter-brained, unobservant, unthinking. ▷ STUPID. **2** *a thoughtless insult.* cruel, heartless, impolite, inconsiderate, insensitive, rude, selfish, tactless, uncaring, undiplomatic, unfeeling. ▷ UNKIND. *Opp* THOUGHTFUL.

**thrash** *vb* beat, birch, cane, flay, flog, lash, scourge, whip. ▷ DEFEAT, HIT.

**thread** *n* **1** fibre, filament, hair, strand. □ *cotton, line, silk, string, thong, twine, wool, yarn.* **2** *thread of a story.* argument, continuity, course, direction, drift, line of thought, plot, story line, tenor, theme. • *vb* put on a thread, string together. **thread your way** file, pass, pick your way, wind.

**threadbare** *adj* frayed, old, ragged, scruffy, shabby, tattered, tatty, worn, worn-out.

**threat** *n* **1** commination, intimidation, menace, warning. **2** *threat of rain.* danger, forewarning, intimation, omen, portent, presage, risk, warning.

**threaten** *vb* **1** browbeat, bully, cow, intimidate, make threats against, menace, pressurize, terrorize. ▷ FRIGHTEN. *Opp* REASSURE. **2** *clouds threaten rain.* forebode, foreshadow, forewarn of, give warning of, portend, presage, warn of. **3** *the recession threatens jobs.* endanger, imperil, jeopardize, put at risk.

**threatening** *adj* forbidding, grim, impending, looming, menacing, minatory, ominous, portentous, sinister, stern, *inf* ugly, unfriendly, worrying. *Opp* SUPPORTIVE.

**three** *n* triad, trio, triplet, triumvirate.

**three-dimensional** *adj* in the round, rounded, sculptural, solid, stereoscopic.

**threshold** *n* **1** doorstep, doorway, entrance, sill. **2** *threshold of a new era.* ▷ BEGINNING.

**thrifty** *adj* careful, *derog* close-fisted, economical, frugal, *derog* mean, *derog* niggardly, parsimonious, provident, prudent, skimping, sparing. *Opp* EXTRAVAGANT.

**thrill** *n* adventure, *inf* buzz, excitement, frisson, *inf* kick, pleasure, sensation, shiver, suspense, tingle, titillation, tremor. • *vb* arouse, delight, electrify, excite, galvanize, rouse, stimulate, stir, titillate. **thrilling** ▷ EXCITING.

**thriller** *n* crime story, detective story, mystery, *inf* whodunit. ▷ WRITING.

**thrive** *vb* be vigorous, bloom, boom, burgeon, *inf* come on, develop strongly, do well, expand, flourish, grow, increase, *inf* make strides, prosper, succeed. *Opp* DIE. **thriving** ▷ PROSPEROUS, VIGOROUS.

**throat** *n* gullet, neck, oesophagus, uvula, windpipe.

**throaty** *adj* deep, gravelly, gruff, guttural, hoarse, husky, rasping, rough, thick.

**throb** *vb* beat, palpitate, pound, pulsate, pulse, vibrate.

**throe** *n* convulsion, effort, fit, labour, *plur* labour-pains, pang, paroxysm, spasm. ▷ PAIN.

**thrombosis** *n* blood-clot, embolism.

**throng** *n* assembly, crowd, crush, gathering, horde, jam, mass, mob, multitude, swarm. ▷ GROUP.

**throttle** *vb* asphyxiate, choke, smother, stifle, strangle, suffocate. ▷ KILL.

**throw** *vb* 1 bowl, *inf* bung, cast, *inf* chuck, fling, heave, hurl, launch, lob, pelt, pitch, propel, put (*the shot*), send, *inf* shy, *inf* sling, toss. 2 *throw light*. cast, project, shed. 3 *throw a rider*. dislodge, floor, shake off, throw down, throw off, unseat, upset. 4 ▷ DISCONCERT. **throw away** ▷ DISCARD. **throw out** ▷ EXPEL. **throw up** ▷ PRODUCE, VOMIT.

**throw-away** *adj* 1 cheap, disposable. 2 *throw-away remark*. casual, offhand, passing, unimportant.

**thrust** *vb* butt, drive, elbow, force, impel, jab, lunge, plunge, poke, press, prod, propel, push, ram, send, shoulder, shove, stab, stick, urge.

**thug** *n* assassin, *inf* bully-boy, delinquent, desperado, gangster, *inf* hoodlum, hooligan, killer, mugger, *inf* rough, ruffian, *inf* tough, trouble-maker, vandal, *inf* yob. ▷ CRIMINAL.

**thunder** *n* clap, crack, peal, roll, rumble. ▷ SOUND.

**thunderous** *adj* booming, deafening, reverberant, reverberating, roaring, rumbling. ▷ LOUD.

**thus** *adv* accordingly, consequently, for this reason, hence, so, therefore.

**thwart** *vb* baffle, baulk, block, check, foil, frustrate, hinder, impede, obstruct, prevent, stand in the way of, stop, stump.

**ticket** *n* 1 coupon, pass, permit, token, voucher. 2 *price ticket*. docket, label, marker, tab, tag.

**ticklish** *adj* 1 *inf* giggly, responsive to tickling, sensitive. 2 *a ticklish problem*. awkward, delicate, difficult, risky, *inf* thorny, touchy, tricky.

**tide** *n* current, drift, ebb and flow, movement, rise and fall.

**tidiness** *n* meticulousness, neatness, order, orderliness, organization, smartness, system. *Opp* DISORDER.

**tidy** *adj* 1 neat, orderly, presentable, shipshape, smart, *inf* spick and span, spruce, straight, trim, uncluttered, well-groomed, well-kept. 2 *tidy habits*. businesslike, careful, house-proud, methodical, meticulous, organized, systematic, well-organized. *Opp* UNTIDY. • *vb* arrange, clean up, groom, make tidy, neaten, put in order, rearrange, reorganize, set straight, smarten, spruce up, straighten, titivate. *Opp* MUDDLE.

**tie** *vb* 1 bind, chain, do up, hitch, interlace, join, knot, lash, moor, rope, secure, splice, tether, truss up. ▷ FASTEN. *Opp* UNTIE. 2 *tie in a race*. be equal, be level, be neck and neck, draw.

**tier** *n* course (*of bricks*), layer, level, line, order, range, rank, row, stage, storey, stratum, terrace.

**tight** *adj* 1 close, fast, firm, fixed, immovable, secure, snug. 2 *a tight lid*. airtight, close-fitting, hermetic, impermeable, impervious, leak-proof, sealed, waterproof, watertight. 3 *tight supervision*. harsh, inflexible, precise, rigorous, severe, strict, stringent. 4 *tight ropes*. rigid, stiff, stretched, taut, tense. 5 *a tight space*. compact, constricted, crammed, cramped, crowded, dense, inadequate, limited, packed, small. 6 ▷ DRUNK. 7 ▷ MISERLY. *Opp* FREE, LOOSE.

**tighten** *vb* 1 become tighter, clamp down, close, close up, constrict, harden, make tighter, squeeze, stiffen, tense. ▷ FASTEN. 2 *tighten ropes*. pull tighter, stretch, tauten. 3 *tighten screws*.

give another turn to, screw up. *Opp* LOOSEN.

**till** *vb* cultivate, dig, farm, plough, work.

**tilt** *vb* **1** angle, bank, cant, careen, heel over, incline, keel over, lean, list, slant, slope, tip. **2** *tilt with lances.* joust, thrust. ▷ FIGHT.

**timber** *n* beam, board, boarding, deal, lath, log, lumber, plank, planking, post, softwood, tree, tree trunk. ▷ WOOD.

**time** *n* **1** date, hour, instant, juncture, moment, occasion, opportunity, point. **2** duration, interval, period, phase, season, semester, session, spell, stretch, term, while. □ *aeon, century, day, decade, eternity, fortnight, hour, lifetime, minute, month, second, week, weekend, year.* **3** *time of Nero.* age, days, epoch, era, period. **4** *time in music.* beat, measure, rhythm, tempo. • *vb* **1** choose a time for, estimate, fix a time for, judge, organize, plan, schedule, timetable. **2** *time a race.* clock, measure the time of.

**timeless** *adj* ageless, deathless, eternal, everlasting, immortal, immutable, indestructible, permanent, unchanging, undying, unending.

**timely** *adj* appropriate, apt, fitting, suitable.

**timepiece** *n* □ *chronometer, clock, digital clock, digital watch, hourglass, stop-watch, sundial, timer, watch, wrist-watch.*

**timetable** *n* agenda, calendar, curriculum, diary, list, programme, roster, rota, schedule.

**timid** *adj* afraid, apprehensive, bashful, chicken-hearted, cowardly, coy, diffident, fainthearted, fearful, modest, *inf* mousy, nervous, pusillanimous, reserved, retiring, scared, sheepish, shrinking, shy, spineless, tentative, timorous, unadventurous, unheroic, wimpish. ▷ FRIGHTENED. *Opp* BOLD.

**tingle** *n* **1** itch, itching, pins and needles, prickling, stinging, throb, throbbing, tickle, tickling. **2** *a tingle of excitement.* quiver, sensation, shiver, thrill. • *vb* itch, prickle, sting, tickle.

**tinker** *vb* dabble, fiddle, fool about, interfere, meddle, *inf* mess about, *inf* play about, tamper, try to mend, work amateurishly.

**tinny** *adj* cheap, flimsy, inferior, insubstantial, poor-quality, shoddy, tawdry.

**tinsel** *n* decoration, glitter, gloss, show, sparkle, tinfoil.

**tint** *n* colour, colouring, dye, hue, shade, stain, tincture, tinge, tone, wash.

**tiny** *adj* diminutive, dwarf, imperceptible, infinitesimal, insignificant, lilliputian, microscopic, midget, *inf* mini, miniature, minuscule, minute, negligible, pygmy, *inf* teeny, unimportant, *inf* wee, *inf* weeny. ▷ SMALL. *Opp* BIG.

**tip** *n* **1** apex, cap, crown, end, extremity, ferrule, finial, head, nib, peak, pinnacle, point, sharp end, summit, top, vertex. **2** *tip for a waiter. inf* baksheesh, gift, gratuity, inducement, money, *inf* perk, present, reward, service-charge, *inf* sweetener. **3** *useful tips.* advice, clue, forecast, hint, information, pointer, prediction, suggestion, tip-off, warning. **4** *rubbish tip.* dump, rubbish-heap. • *vb* **1** careen, incline, keel, lean, list, slant, slope, tilt. **2** drop off, dump, empty, pour out, spill, unload, upset. **3** *tip a waiter.* give a tip to, remunerate, reward. **tip over** ▷ OVERTURN.

**tire** *vb* **1** become bored, become tired, flag, grow weary, weaken. **2** debilitate, drain, enervate, exhaust, fatigue, *inf* finish, *sl* knacker, make tired, overtire, sap, *inf* shatter, *inf* take it out of, tax, wear out, weary. *Opp* REFRESH. **tired** ▷ WEARY. **tired of** bored with, *inf* fed up with, impatient with, sick of. **tiring** ▷ EXHAUSTING.

**tiredness** *n* drowsiness, exhaustion, fatigue, inertia, jet-lag, lassitude, lethargy, listlessness, sleepiness, weariness.

**tireless** *adj* determined, diligent, dogged, dynamic, energetic, hard-working, indefatigable, persistent, pertinacious, resolute, sedulous, unceasing, unfaltering, unflagging, untiring, unwavering, vigorous. *Opp* LAZY.

**tiresome** *adj* **1** boring, dull, monotonous, tedious, tiring, unexciting, uninteresting, wearisome, wearying. *Opp* EXCITING. **2** *tiresome delays.* annoying, bothersome, distracting, exasperating, inconvenient, infuriating, irksome, irritating, maddening, petty, troublesome, trying, unwelcome, upsetting, vexatious, vexing.

**tiring** *adj* debilitating, demanding, difficult, exhausting, fatiguing, hard, laborious, strenuous, taxing, wearying. *Opp* REFRESHING.

**tissue** *n* **1** fabric, material, structure, stuff, substance. **2** *tissue-paper.* □ *lavatory paper, napkin, paper handkerchief, serviette, toilet paper, tracing-paper.*

**title** *n* **1** caption, heading, headline, inscription, name, rubric. **2** appellation, designation, form of address, office, position, rank, status. □ *Baron, Baroness, Count, Countess, Dame, Doctor, Dr, Duchess, Duke, Earl, Lady, Lord, Marchioness, Marquis, Master, Miss, Mr, Mrs, Ms, Professor, Rev, Reverend, Sir, Viscount, Viscountess.* ▷ RANK, ROYAL. **3** *title to an inheritance.* claim, deed, entitlement, interest, ownership, possession, prerogative, right. • *vb* call, designate, entitle, give a title to, label, name, tag.

**titled** *adj* aristocratic, noble, upper class.

**titter** *vb* chortle, chuckle, giggle, snicker, snigger. ▷ LAUGH.

**titular** *adj* formal, nominal, official, putative, *inf* so-called, theoretical, token. *Opp* ACTUAL.

**toast** *vb* **1** brown, grill. ▷ COOK. **2** *toast a guest.* drink a toast to, drink the health of, drink to, honour, pay tribute to, raise your glass to.

**tobacco** *n* □ *cigar, cigarette, pipe tobacco, plug, snuff.*

**together** *adv* all at once, at the same time, collectively, concurrently, consecutively, continuously, cooperatively, hand in hand, in chorus, in unison, jointly, shoulder to shoulder, side by side, simultaneously.

**toil** *n inf* donkey work, drudgery, effort, exertion, industry, labour, work. • *vb* drudge, exert yourself, grind away, *inf* keep at it, labour, *inf* plug away, *inf* slave away, struggle, *inf* sweat. ▷ WORK.

**toilet** *n* **1** convenience, latrine, lavatory, *sl* loo, *old use* privy, urinal, water closet, WC. **2** [*old use*] *make your toilet.* dressing, grooming, making up, washing.

**token** *adj* cosmetic, dutiful, emblematic, insincere, nominal, notional, perfunctory, representative, superficial, symbolic. *Opp* GENUINE. • *n* **1** badge, emblem, evidence, expression, indication, mark, marker, proof, reminder, sign, symbol, testimony. **2** *a token of esteem.* keepsake, memento, reminder, souvenir. **3** *a bus token.* coin, counter, coupon, disc, voucher.

**tolerable** *adj* **1** acceptable, allowable, bearable, endurable, sufferable, supportable. **2** *tolerable food.* adequate, all right, average, fair, mediocre, middling, *inf* OK, ordinary, passable, satisfactory. *Opp* INTOLERABLE.

**tolerance** *n* **1** broad-mindedness, charity, fairness, forbearance, forgiveness, lenience, open-mindedness, openness, patience, permissiveness. **2** *tolerance of others.* acceptance, sufferance, sympathy (towards), toleration, understanding. **3** *tolerance in moving parts.* allowance, clearance, deviation, fluctuation, play, variation.

**tolerant** *adj* big-hearted, broad-minded, charitable, easygoing, fair, forbearing, forgiving, generous, indulgent, *derog* lax, lenient, liberal, magnanimous, open-minded, patient, permissive, *derog* soft, sympathetic, understanding, unprejudiced. *Opp* INTOLERANT.

**tolerate** *vb* abide, accept, admit, bear, brook, concede, condone, countenance, endure, *inf* lump (*I'll have to lump it!*), make allowances for, permit, *inf* put up with, sanction, *inf* stick, *inf* stand, *inf* stomach, suffer, *inf* take, undergo, *inf* wear, weather.

**toll** *n* charge, dues, duty, fee, levy, payment, tariff, tax. • *vb* chime, peal, ring, sound, strike.

**tomb** *n* burial-chamber, burial-place, catacomb, crypt, grave, gravestone, last resting-place, mausoleum, memorial, monument, sepulchre, tombstone, vault.

**tonality** *n* key, tonal centre.

**tone** *n* **1** accent, colouring, expression, feel, inflection, intonation, manner, modulation, note, phrasing, pitch, quality, sonority, sound, timbre. **2** *tone of a poem, place.* air, atmosphere, character, effect, feeling, mood, spirit, style, temper, vein. **3** *colour tone.* colour, hue, shade, tinge, tint, tonality. **tone down** ▷ SOFTEN. **tone in** ▷ HARMONIZE. **tone up** ▷ STRENGTHEN.

**tongue** *n* dialect, idiom, language, parlance, patois, speech, talk, vernacular.

**tongue-tied** *adj* dumb, dumbfounded, inarticulate, *inf* lost for words, mute, silent, speechless.

**tonic** *n* boost, cordial, dietary supplement, fillip, *inf* pick-me-up, refresher, restorative, stimulant.

**tool** *n* apparatus, appliance, contraption, contrivance, device, gadget, hardware, implement, instrument, invention, machine, mechanism, utensil, weapon. □ [carpentry] *auger, awl, brace and bit, bradawl, chisel, clamp, cramp, drill, file, fretsaw, gimlet, glass-paper, hacksaw, hammer, jigsaw, mallet, pincers, plane, pliers, power-drill, rasp, sander, sandpaper, saw, screwdriver, spokeshave, T-square, vice, wrench.* □ [gardening] *billhook, dibber, fork, grass-rake, hoe, lawnmower, mattock, pruning knife, pruning shears, rake, roller, scythe, secateurs, shears, sickle, spade, Strimmer, trowel.* □ [various] *axe, bellows, chainsaw, chopper, clippers, crowbar, cutter, hatchet, jack, ladder, lever, penknife, pick, pickaxe, pitchfork, pocket-knife, scis-*

*sors, shovel, sledge-hammer, spanner, tape-measure, tongs, tweezers.*

**tooth** *n* □ *canine, eye-tooth, fang, incisor, molar, tusk, wisdom tooth.* **false teeth** bridge, denture, dentures, plate.

**toothed** *adj* cogged, crenellated, denticulate, indented, jagged, serrated. *Opp* SMOOTH.

**top** *adj inf* ace, best, choicest, finest, first, foremost, greatest, highest, incomparable, leading, maximum, most, peerless, preeminent, prime, principal, supreme, topmost, unequalled, winning. *n* **1** acme, apex, apogee, crest, crown, culmination, head, height, high point, peak, pinnacle, summit, tip, vertex, zenith. **2** *top of a table.* surface. **3** *top of a jar.* cap, cover, covering, lid, stopper. *Opp* BOTTOM. • *vb* **1** complete, cover, decorate, finish off, garnish, surmount. **2** beat, be higher than, better, cap, exceed, excel, outdo, outstrip, surpass, transcend.

**topic** *n* issue, matter, point, question, subject, talking-point, text, theme, thesis.

**topical** *adj* contemporary, current, recent, timely, up-to-date.

**topography** *n* features, geography, *inf* lie of the land.

**topple** *vb* **1** bring down, fell, knock down, overturn, throw down, tip over, upset. **2** collapse, fall, overbalance, totter, tumble. **3** *topple a rival.* oust, overthrow, unseat. ▷ DEFEAT.

**torch** *n* bicycle lamp, brand, electric lamp, flashlight, lamp, *old use* link.

**torment** *n* affliction, agony, anguish, distress, harassment, misery, ordeal, persecution, plague, scourge, suffering, torture, vexation, woe, worry, wretchedness. ▷ PAIN. • *vb* afflict, annoy, bait, be a torment to, bedevil, bother, bully, distress, harass, inflict pain on, intimidate, *inf* nag, persecute, pester, plague, tease, torture, vex, victimize, worry. ▷ HURT.

**torpid** *adj* apathetic, dormant, dull, inactive, indolent, inert, lackadaisical, languid, lethargic, lifeless, listless, passive, phlegmatic, slothful, slow, slow-moving, sluggish, somnolent, spiritless. *Opp* LIVELY.

**torrent** *n* cascade, cataract, deluge, downpour, effusion, flood, flow, gush, inundation, outpouring, overflow, rush, spate, stream, tide.

**torrential** *adj* copious, heavy, relentless, soaking, teeming, violent.

**tortuous** *adj* bent, circuitous, complicated, contorted, convoluted, corkscrew, crooked, curling, curvy, devious, indirect, involved, labyrinthine, mazy, meandering, roundabout, serpentine, sinuous, turning, twisted, twisting, twisty, wandering, winding, zigzag. *Opp* DIRECT, STRAIGHT.

**torture** *n* **1** cruelty, degradation, humiliation, inquisition, persecution, punishment, torment. **2** affliction, agony, anguish, distress, misery, pain, plague, scourge, suffering. • *vb* **1** be cruel to, brainwash, bully, cause pain to, degrade, dehumanize, humiliate, hurt, inflict pain on, intimidate, persecute, rack, torment, victimize. **2** *tortured by doubts.* afflict, agonize, annoy, bedevil, bother, distress, harass, *inf* nag, pester, plague, tease, vex, worry.

**toss** *vb* **1** bowl, cast, *inf* chuck, fling, flip, heave, hurl, lob, pitch, shy, sling, throw. **2** *toss about in a storm.* bob, dip, flounder, lurch, move restlessly, pitch, plunge, reel, rock, roll, shake, twist and turn, wallow, welter, writhe, yaw.

**total** *adj* 1 complete, comprehensive, entire, full, gross, overall, whole. 2 *total disaster.* absolute, downright, out-and-out, outright, perfect, sheer, thorough, thoroughgoing, unalloyed, unmitigated, unqualified, utter. • *n* aggregate, amount, answer, lot, sum, totality, whole. • *vb* 1 add up to, amount to, come to, make. 2 add up, calculate, compute, count, find the sum of, find the total of, reckon up, totalize, *inf* tot up, work out.

**totalitarian** *adj* absolute, arbitrary, authoritarian, autocratic, despotic, dictatorial, fascist, illiberal, one-party, oppressive, tyrannous, undemocratic, unrepresentative. *Opp* DEMOCRATIC.

**totter** *vb* dodder, falter, reel, rock, stagger, stumble, teeter, topple, tremble, waver, wobble. ▷ WALK.

**touch** *n* 1 feeling, texture, touching. 2 brush, caress, contact, dab, pat, stroke, tap. 3 *an expert's touch.* ability, capability, experience, expertise, facility, feel, flair, gift, knack, manner, sensitivity, skill, style, technique, understanding, way. 4 *a touch of salt.* bit, dash, drop, hint, intimation, small amount, suggestion, suspicion, taste, tinge, trace. • *vb* 1 be in contact with, brush, caress, contact, cuddle, dab, embrace, feel, finger, fondle, graze, handle, hit, kiss, lean against, manipulate, massage, nuzzle, pat, paw, pet, push, rub, stroke, tap, tickle. 2 *touch the emotions.* affect, arouse, awaken, concern, disturb, impress, influence, inspire, move, stimulate, stir, upset. 3 *touch 100 m.p.h.* attain, reach, rise to. 4 *I can't touch her skill.* be in the same league as, *inf* come up to, compare with, equal, match, parallel, rival. **touched** ▷ EMOTIONAL, MAD. **touching** ▷ EMOTIONAL. **touch off** ▷ BEGIN, IGNITE. **touch on** ▷ MENTION. **touch up** ▷ IMPROVE.

**touchy** *adj* edgy, highly strung, hypersensitive, irascible, irritable, jittery, jumpy, nervous, oversensitive, peevish, querulous, quick-tempered, sensitive, short-tempered, snappy, temperamental, tense, testy, tetchy, thin-skinned, unpredictable, waspish.

**tough** *adj* 1 durable, hard-wearing, indestructible, lasting, rugged, sound, stout, strong, substantial, unbreakable, well-built, well-made. 2 *tough physique.* *inf* beefy, brawny, burly, hardy, muscular, robust, stalwart, strong, sturdy. 3 *tough opposition.* invulnerable, merciless, obdurate, obstinate, resilient, resistant, resolute, ruthless, stiff, stubborn, tenacious, unyielding. 4 *a tough taskmaster.* cold, cool, *inf* hard-boiled, hardened, *inf* hard-nosed, inhuman, severe, stern, stony, uncaring, unsentimental, unsympathetic. 5 *tough meat.* chewy, hard, gristly, leathery, rubbery, uneatable. 6 *tough work.* arduous, demanding, difficult, exacting, exhausting, gruelling, hard, laborious, stiff, strenuous, taxing, troublesome. 7 *a tough problem.* baffling, intractable, *inf* knotty, mystifying, perplexing, puzzling, *inf* thorny. *Opp* EASY, TENDER, WEAK.

**toughen** *vb* harden, make tougher, reinforce, strengthen.

**tour** *n* circular tour, drive, excursion, expedition, jaunt, journey, outing, peregrination, ride, trip. • *vb* do the rounds of, explore, go round, make a tour of, visit. ▷ TRAVEL.

**tourist** *n* day-tripper, holiday-maker, sightseer, traveller, tripper, visitor.

**tournament** *n* championship, competition, contest, event, match, meeting, series.

**tow** *vb* drag, draw, haul, lug, pull, trail, tug.

**tower** *n* □ *belfry, campanile, castle, fort, fortress, keep, minaret, pagoda, skyscraper, spire, steeple, turret.* • *vb* ascend, dominate, loom, rear, rise, soar, stand out, stick up.

**towering** *adj* **1** colossal, gigantic, high, huge, imposing, lofty, mighty, soaring. ▷ TALL. **2** *a towering rage.* extreme, fiery, immoderate, intemperate, intense, mighty, overpowering, passionate, unrestrained, vehement, violent.

**town** *n* borough, city, community, conurbation, municipality, settlement, township, urban district, village.

**toxic** *adj* dangerous, deadly, harmful, lethal, noxious, poisonous. *Opp* HARMLESS.

**trace** *n* **1** clue, evidence, footprint, *inf* give-away, hint, indication, intimation, mark, remains, sign, spoor, token, track, trail, vestige. **2** ▷ BIT. • *vb* **1** detect, discover, find, get back, recover, retrieve, seek out, track down. ▷ TRACK. **2** *trace an outline.* copy, draw, go over, make a copy of, mark out, sketch. **kick over the traces** ▷ REBEL.

**track** *n* **1** footmark, footprint, mark, scent, spoor, trace, trail, wake (*of ship*). **2** *a farm track.* bridle-path, bridleway, cart-track, footpath, path, route, trail, way. ▷ ROAD. **3** *a racing track.* circuit, course, dirt-track, race-track. **4** *railway track.* branch, branch line, line, mineral line, permanent way, rails, railway, route, tramway. • *vb* chase, dog, follow, hound, hunt, pursue, shadow, stalk, tail, trace, trail. **make tracks** ▷ DEPART. **track down** ▷ TRACE.

**trade** *n* **1** barter, business, buying and selling, commerce, dealing, exchange, industry, market, marketing, merchandising, trading, traffic, transactions. **2** *a skilled trade.* calling, career, craft, employment, job, *inf* line, occupation, profession, pursuit, work. • *vb* buy and sell, do business, have dealings, market goods, merchandise, retail, sell, traffic (in). **trade in** ▷ EXCHANGE. **trade on** ▷ EXPLOIT.

**trader** *n* broker, buyer, dealer, merchant, retailer, roundsman, salesman, seller, shopkeeper, stockist, supplier, tradesman, trafficker (*in illegal goods*), vendor.

**tradition** *n* **1** convention, custom, habit, institution, practice, rite, ritual, routine, usage. **2** *popular tradition.* belief, folklore.

**traditional** *adj* **1** accustomed, conventional, customary, established, familiar, habitual, historic, normal, orthodox, regular, time-honoured, typical, usual. *Opp* UNCONVENTIONAL. **2** *traditional stories.* folk, handed down, old, oral, popular, unwritten. *Opp* MODERN.

**traffic** *n* conveyance, movements, shipping, transport, transportation. ▷ VEHICLE. • *vb* ▷ TRADE.

**tragedy** *n* adversity, affliction, *inf* blow, calamity, catastrophe, disaster, misfortune. *Opp* COMEDY.

**tragic** *adj* **1** appalling, awful, calamitous, catastrophic, depressing, dire, disastrous, dreadful, fatal, fearful, hapless, ill-fated, ill-omened, ill-starred, inauspicious, lamentable, terrible, tragical, unfortunate, unlucky. **2** *a tragic expression.* bereft, distressed, funereal, grief-stricken, hurt, pathetic, piteous, pitiful, sorrowful, woeful, wretched. ▷ SAD. *Opp* COMIC.

**trail** *n* **1** evidence, footmarks, footprints, marks, scent, signs, spoor, traces, wake (*of ship*). **2** path, pathway, route, track. ▷ ROAD. • *vb* **1** dangle, drag, draw, haul, pull, tow. **2** chase, follow, hunt, pursue, shadow, stalk, tail, trace, track down. **3** ▷ DAWDLE.

**train** *n* **1** carriage, coach, diesel, *inf* DMU, electric train, express, intercity, local train, railcar, steam train, stopping train. **2** *train of servants.* cortège, entourage, escort, followers, guard, line, retainers, retinue, staff, suite. **3** *train of events.* ▷ SEQUENCE. • *vb* **1** coach, discipline, drill, educate, instruct, prepare, school, teach, tutor. **2** do exercises, exercise, *inf* get fit, practise, prepare yourself, rehearse, *inf* work out. **3** ▷ AIM.

**trainee** *n* apprentice, beginner, cadet, learner, *inf* L-driver, novice, pupil, starter, student, tiro, unqualified person.

**trainer** *n* coach, instructor, teacher, tutor.

**trait** *n* attribute, characteristic, feature, idiosyncrasy, peculiarity, property, quality, quirk.

**traitor** *n* apostate, betrayer, blackleg, collaborator, defector, deserter, double-crosser, fifth columnist, informer, *inf* Judas, quisling, renegade, turncoat.

**tramp** *n* **1** hike, march, trek, trudge, walk. **2** *a homeless tramp.* beggar, *inf* destitute person, *inf* dosser, *inf* down and out, drifter, homeless person, rover, traveller, vagabond, vagrant, wanderer. • *vb inf* footslog, hike, march, plod, stride, toil, traipse, trek, trudge, *sl* yomp. ▷ WALK.

**trample** *vb* crush, flatten, squash, *inf* squish, stamp on, step on, tread on, walk over.

**trance** *n inf* brown study, daydream, daze, dream, ecstasy, hypnotic state, rapture, reverie, semiconsciousness, spell, stupor, unconsciousness.

**tranquil** *adj* **1** calm, halcyon (*days*), peaceful, placid, quiet, restful, serene, still, undisturbed, unruffled. *Opp* STORMY. **2** *a tranquil mood.* collected, composed, dispassionate, *inf* laid-back, sedate, sober, unemotional, unexcited, untroubled. *Opp* EXCITED.

**tranquillizer** *n* barbiturate, bromide, narcotic, opiate, sedative.

**transaction** *n* agreement, bargain, business, contract, deal, negotiation, proceeding.

**transcend** *vb* beat, exceed, excel, outdo, outstrip, rise above, surpass, top.

**transcribe** *vb* copy out, render, reproduce, take down, translate, transliterate, write out.

**transfer** *vb* bring, carry, change, convey, deliver, displace, ferry, hand over, make over, move, pass on, pass over, relocate, remove, second, shift, sign over, take, transplant, transport, transpose.

**transform** *vb* adapt, alter, change, convert, improve, metamorphose, modify, mutate, permute, rebuild, reconstruct, remodel, revolutionize, transfigure, translate, transmogrify, transmute, turn.

**transformation** *n* adaptation, alteration, change, conversion, improvement, metamorphosis, modification, mutation, reconstruction, revolution, transfiguration, transition, translation, transmogrification, transmutation, *inf* turn-about.

**transgression** *n* crime, error, fault, lapse, misdeed, misdemeanour, offence, sin, wickedness, wrongdoing.

**transient** *adj* brief, ephemeral, evanescent, fleeting, fugitive, impermanent, momentary, passing, *inf* quick, short, short-lived, temporary, transitory. *Opp* PERMANENT.

**transit** *n* conveyance, journey, movement, moving, passage, progress, shipment, transfer, transportation, travel.

**transition** *n* alteration, change, change-over, conversion, development, evolution, modification, movement, progress, progression, shift, transformation, transit.

**translate** *vb* change, convert, decode, elucidate, explain, express, gloss, interpret, make a translation, paraphrase, render, reword, spell out, transcribe. ▷ TRANSFORM.

**translation** *n* decoding, gloss, interpretation, paraphrase, rendering, transcription, transliteration, version.

**translator** *n* interpreter, linguist.

**transmission** *n* **1** broadcasting, communication, diffusion, dissemination, relaying, sending out. **2** *transmission of goods.* carriage, carrying, conveyance, dispatch, sending, shipment, shipping, transfer, transference, transport, transportation.

**transmit** *vb* **1** convey, dispatch, disseminate, forward, pass on, post, send, transfer, transport. **2** *transmit a message.* broadcast, cable, communicate, emit, fax, phone, radio, relay, telephone, telex, wire. *Opp* RECEIVE.

**transparent** *adj* **1** clear, crystalline, diaphanous, filmy, gauzy, limpid, pellucid, *inf* see-through, sheer, translucent. **2** *transparent honesty.* ▷ CANDID.

**transplant** *vb* displace, move, relocate, reposition, resettle, shift, transfer, uproot.

**transport** *n* carrier, conveyance, haulage, removal, shipment, shipping, transportation. □ *aircraft, barge, boat, bus, cable-car, car, chair-lift, coach, cycle, ferry, horse, lorry, Metro, minibus,* old use *omnibus, ship, space-shuttle, taxi, train, tram, van.* □ *air, canal, railway, road, sea, waterways.* ▷ VEHICLE, VESSEL. • *vb* **1** bear, carry, convey, fetch, haul, move, remove, send, shift, ship, take, transfer. **2** ▷ DEPORT.

**transpose** *vb* change, exchange, interchange, metathesize, move round, rearrange, reverse, substitute, swap, switch, transfer.

**transverse** *adj* crosswise, diagonal, oblique.

**trap** *n* ambush, booby-trap, deception, gin, mantrap, net, noose, pitfall, ploy, snare, trick. • *vb* ambush, arrest, capture, catch, catch out, corner, deceive, dupe, ensnare, entrap, inveigle, net, snare, trick.

**trappings** *plur n* accessories, accompaniments, accoutrements, adornments, appointments, decorations, equipment, finery, fittings, furnishings, *inf* gear, ornaments, paraphernalia, *inf* things, trimmings.

**trash** *n* **1** debris, garbage, junk, litter, refuse, rubbish, sweepings, waste. **2** ▷ NONSENSE.

**travel** *n* globe-trotting, moving around, peregrination, touring, tourism, travelling, wandering. □ *cruise, drive, excursion, expedition, exploration, flight, hike, holiday, journey, march, migration, mission, outing, pilgrimage, ramble, ride, safari, sail, sea-passage, tour, trek, trip, visit, voyage, walk.* • *vb inf* gad about, *inf* gallivant, journey, make a trip, move, proceed, progress, roam, *poet* rove, voyage, wander. ▷ GO. □ *aviate, circumnavigate (the world), commute, cruise, cycle, drive, emigrate, fly, free-wheel, hike, hitchhike, march, migrate, motor, navigate, paddle, pedal, pilot, punt, ramble, ride, row, sail, shuttle, steam, tour, trek, walk.*

**traveller** *n* **1** astronaut, aviator, commuter, cosmonaut, cyclist, driver, flyer, migrant, motorcyclist, motorist, passenger, pedestrian, sailor, voyager, walker. **2** *a company traveller. inf* rep, representative, salesman, saleswoman. **3** *overseas travellers.* explorer, globe-trotter, hiker, hitchhiker, holidaymaker, pilgrim, rambler, stowaway, tourist, tripper, wanderer, wayfarer. **4** *live as travellers.* gypsy, itinerant, nomad, tinker, tramp, vagabond.

**travelling** *adj* homeless, itinerant, migrant, migratory, mobile, nomadic, peripatetic, restless, roaming, roving, touring, vagrant, wandering.

**treacherous** *adj* **1** deceitful, disloyal, double-crossing, double-dealing, duplicitous, faithless, false, perfidious, sneaky, unfaithful, untrustworthy. **2** *treacherous conditions.* dangerous, deceptive, hazardous, misleading, perilous, risky, shifting, unpredictable, unreliable, unsafe, unstable. *Opp* LOYAL, RELIABLE.

**treachery** *n* betrayal, dishonesty, disloyalty, double-dealing, duplicity, faithlessness, infidelity, perfidy, untrustworthiness. ▷ TREASON. *Opp* LOYALTY.

**tread** *vb* **tread on** crush, squash underfoot, stamp on, step on, trample, walk on. ▷ WALK

**treason** *n* betrayal, high treason, mutiny, rebellion, sedition. ▷ TREACHERY.

**treasure** *n* cache, cash, fortune, gold, hoard, jewels, riches, treasure trove, valuables, wealth. • *vb* adore, appreciate, cherish, esteem, guard, keep safe, love, prize, rate highly, value, venerate, worship.

**treasury** *n* bank, exchequer, hoard, repository, storeroom, treasure-house, vault.

**treat** *n* entertainment, gift, outing, pleasure, surprise. • *vb* **1** attend to, behave towards, care for, look after, use. **2** *treat a topic.* consider, deal with, discuss, tackle. **3** *treat a patient, wound.* cure, dress, give treatment to, heal, medicate, nurse, prescribe medicine for, tend. **4** *treat food.* process. **5** *treat a friend to dinner.* entertain, give a treat, pay for, provide for, regale.

**treatise** *n* disquisition, dissertation, essay, monograph, pamphlet, paper, thesis, tract. ▷ WRITING.

**treatment** *n* **1** care, conduct, dealing (with), handling, management, manipulation, organization, reception, usage, use. **2** *treatment of illness.* cure, first aid, healing, nursing, remedy, therapy. ▷ MEDICINE, THERAPY.

**treaty** *n* agreement, alliance, armistice, compact, concordat, contract, covenant, convention, *inf* deal, entente, pact, peace, protocol, settlement, truce, understanding.

**tree** *n* bush, sapling, standard. □ *bonsai, conifer, cordon, deciduous tree, espalier, evergreen, pollard, standard.* □ *ash, banyan, baobab, bay, beech, birch, cacao, cedar, chestnut, cypress, elder, elm, eucalyptus, fir, fruit-tree, gum-tree, hawthorn, hazel, holly, horse-chestnut, larch, lime, maple, oak, olive, palm, pine, plane, poplar, redwood, rowan, sequoia, spruce, sycamore, tamarisk, tulip tree, willow, yew.*

**tremble** *vb* quail, quake, quaver, quiver, rock, shake, shiver, shudder, vibrate, waver.

**tremendous** *adj* alarming, appalling, awful, fearful, fearsome, frightening, frightful, horrifying, shocking, startling, terrible, terrific. ▷ BIG, EXCELLENT, REMARKABLE.

**tremor** *n* **1** agitation, hesitation, quavering, quiver, shaking, trembling, vibration. **2** earthquake, seismic disturbance.

**tremulous** *adj* **1** agitated, anxious, excited, frightened, jittery, jumpy, nervous, timid, uncertain. *Opp* CALM. **2** quivering, shaking, shivering, trembling, *inf* trembly, vibrating. *Opp* STEADY.

**trend** *n* **1** bent, bias, direction, drift, inclination, leaning, movement, shift, tendency. **2** *latest trend.* craze, *inf* fad, fashion, mode, *inf* rage, style, *inf* thing, vogue, way.

**trendy** *adj inf* all the rage, contemporary, fashionable, *inf* in, latest, modern, stylish, up-to-date, voguish. *Opp* OLD-FASHIONED.

**trespass** *vb* encroach, enter illegally, intrude, invade.

**trial** *n* **1** case, court martial, enquiry, examination, hearing, inquisition, judicial proceeding, lawsuit, tribunal. **2** attempt, check, *inf* dry run, experiment, rehearsal, test, testing, trial run, *inf* try-out. **3** *a sore trial.* affliction, burden, difficulty, hardship, nuisance, ordeal, *sl* pain in the neck, *inf* pest, problem, tribulation, trouble, worry.

**triangular** *adj* three-cornered, three-sided.

**tribe** *n* clan, dynasty, family, group, horde, house, nation, pedigree, people, race, stock, strain.

**tribute** *n* accolade, appreciation, commendation, compliment, eulogy, glorification, homage, honour, panegyric, praise, recognition, respect, testimony. **pay tribute to** ▷ HONOUR.

**trick** *n* **1** illusion, legerdemain, magic, sleight of hand. **2** *deceitful trick.* cheat, *inf* con, deceit, deception, fraud, hoax, imposture, joke, *inf* leg-pull, manoeuvre, ploy, practical joke, prank, pretence, ruse, scheme, stratagem, stunt, subterfuge, swindle, trap, trickery, wile. **3** *clever trick.* art, craft, device, dodge, expertise, gimmick, knack, *inf* know-how, secret, skill, technique. **4** *a trick of speech.* characteristic, habit, idiosyncrasy, mannerism, peculiarity, way. • *vb inf* bamboozle, bluff, catch out, cheat, *inf* con, cozen, deceive, defraud, *inf* diddle, dupe, fool, hoax, hoodwink, *inf* kid, mislead, outwit, *inf* pull your leg, swindle, *inf* take in.

**trickery** *n* bluffing, cheating, chicanery, deceit, deception, dishonesty, double-dealing, duplicity, fraud, *inf* funny business, guile, *inf* hocus-pocus, *inf* jiggery-pokery, knavery, *inf* skulduggery, slyness, swindling, trick.

**trickle** *vb* dribble, drip, drizzle, drop, exude, flow slowly, leak, ooze, percolate, run, seep. *Opp* GUSH.

**trifle** *vb* behave frivolously, dabble, fiddle, fool about, play about. **trifling** ▷ TRIVIAL.

**trill** *vb* sing, twitter, warble, whistle.

**trim** *adj* compact, neat, orderly, *inf* shipshape, smart, spruce, tidy, well-groomed, well-kept, well-ordered. *Opp* UNTIDY. • *vb* **1** clip, crop, cut, dock, pare down, prune, shape, shear, shorten, snip, tidy. **2** ▷ DECORATE.

**trip** *n* day out, drive, excursion, expedition, holiday, jaunt, journey, outing, ride, tour, visit, voyage. • *vb* **1** blunder, catch your foot, fall, stagger, stumble, totter, tumble. **2** *trip along.* caper, dance, frisk, gambol, run, skip. **make a trip** ▷ TRAVEL.

**trite** *adj* banal, commonplace, ordinary, pedestrian, predictable, uninspired, uninteresting.

**triumph** *n* **1** accomplishment, achievement, conquest, coup, *inf* hit, knockout, master-stroke, *inf* smash hit, success, victory, *inf* walk-over, win. **2** *return in triumph.* celebration, elation, exultation, joy, jubilation, rapture. • *vb* be victorious, carry the day, prevail, succeed, take the honours, win. **triumph over** ▷ DEFEAT.

**triumphant** *adj* **1** conquering, dominant, successful, victorious, winning. *Opp* UNSUCCESSFUL. **2** boastful, *inf* cocky, elated, exultant, gleeful, gloating, immodest, joyful, jubilant, proud, triumphal.

**trivial** *adj inf* fiddling, *inf* footling, frivolous, inconsequential, inconsiderable, inessential, insignificant, little, meaningless, minor, negligible, paltry, pettifogging, petty, *inf* piddling, *inf* piffling, silly, slight, small, superficial, trifling, trite, unimportant, worthless. *Opp* IMPORTANT.

**trophy** *n* **1** booty, loot, mementoes, rewards, souvenirs, spoils. **2** *a sporting trophy.* award, cup, laurels, medal, palm, prize.

**trouble** *n* **1** adversity, affliction, anxiety, burden, difficulty, distress, grief, hardship, illness, inconvenience, misery, misfortune, pain, problem, sadness, sorrow, suffering, trial, tribulation, unhappiness, vexation, worry. **2** *crowd trouble.* bother, commotion, conflict, discontent, discord, disorder, dissatisfaction, disturbance, fighting, fuss, misbehaviour, misconduct, naughtiness, row, strife, turmoil, unpleasantness, unrest, violence. **3** *engine trouble.* breakdown, defect, failure, fault, malfunction. **4** *took the trouble to get it right.* care, concern, effort, exertion, labour, pains, struggle, thought. • *vb* afflict, agitate, alarm, anguish, annoy, bother, cause trouble to, concern, discommode, distress, disturb, exasperate, grieve, harass, *inf* hassle, hurt, impose on, inconvenience, interfere with, irk, irritate, molest, nag, pain, perturb, pester, plague, *inf* put out, ruffle, threaten, torment, upset, vex, worry. *Opp* REASSURE. **troubled** ▷ WORRIED.

**troublemaker** *n Fr* agent provocateur, agitator, criminal, culprit, delinquent, hooligan, malcontent, mischief-maker, offender, rabble-rouser, rascal, ringleader, ruffian, scandalmonger, *inf* stirrer, vandal, wrongdoer.

**troublesome** *adj* annoying, badly-behaved, bothersome, disobedient, disorderly, distressing, inconvenient, irksome, irritating, naughty, *inf* pestiferous, pestilential, rowdy, tiresome, trying, uncooperative, unruly, upsetting, vexatious, vexing, wearisome, worrisome, worrying. *Opp* HELPFUL.

**trousers** *n inf* bags, breeches, corduroys, culottes, denims, dungarees, jeans, jodhpurs, *old use* knickerbockers, *inf* Levis, overalls, *Amer* pants, plus-fours, shorts, ski-pants, slacks, *Scot* trews, trunks.

**truancy** *n* absenteeism, desertion, malingering, shirking, *inf* skiving.

**truant** *n* absentee, deserter, dodger, idler, malingerer, runaway, shirker, *inf* skiver. **play truant** be absent, desert, malinger, *inf* skive, stay away.

**truce** *n* agreement, armistice, cease-fire, moratorium, pact, peace, suspension of hostilities, treaty.

**true** *adj* **1** accurate, actual, authentic, confirmed, correct, exact, factual, faithful, faultless, flawless, genuine, literal, proper, real, realistic, right, veracious, verified, veritable. *Opp* FALSE. **2** *a true friend*. constant, dedicated, dependable, devoted, faithful, firm, honest, honourable, loyal, reliable, responsible, sincere, staunch, steadfast, steady, trustworthy, trusty, upright. **3** *the true owner*. authorized, legal, legitimate, rightful, valid. **4** *true aim*. accurate, exact, perfect, precise, *inf* spot-on, unerring, unswerving. *Opp* INACCURATE.

**truncheon** *n* baton, club, cudgel, staff, stick.

**trunk** *n* **1** bole, shaft, stalk, stem, stock. **2** *a person's trunk*. body, frame, torso. **3** *an elephant's trunk*. nose, proboscis. **4** *a clothes' trunk*. box, case, casket, chest, coffer, crate, locker, suitcase.

**trust** *n* **1** assurance, belief, certainty, certitude, confidence, conviction, credence, faith, reliance. **2** *a position of trust*. responsibility, trusteeship. • *vb* **1** *inf* bank on, believe in, be sure of, confide in, count on, depend on, have confidence in, have faith in, *inf* pin your hopes on, rely on. **2** assume, expect, hope, imagine, presume, suppose, surmise. *Opp* DOUBT.

**trustful** *adj* confiding, credulous, gullible, innocent, trusting, unquestioning, unsuspecting, unsuspicious, unwary. *Opp* DISTRUSTFUL.

**trustworthy** *adj* constant, dependable, ethical, faithful, honest, honourable, *inf* loyal, moral, on the level, principled, reliable, responsible, *inf* safe, sensible, sincere, steadfast, steady, straightforward, true, *old use* trusty, truthful, upright. *Opp* DECEITFUL.

**truth** *n* **1** facts, reality. *Opp* LIE. **2** accuracy, authenticity, correctness, exactness, factuality, genuineness, integrity, reliability, truthfulness, validity, veracity, verity. **3** *an accepted truth*. axiom, fact, maxim, truism.

**truthful** *adj* accurate, candid, correct, credible, earnest, factual, faithful, forthright, frank, honest, proper, realistic, reliable, right, sincere, *inf* straight, straightforward, true, trustworthy, valid, veracious, unvarnished. *Opp* DISHONEST.

**try** *n* attempt, *inf* bash, *inf* crack, effort, endeavour, experiment, *inf* go, *inf* shot, *inf* stab, test, trial. • *vb* **1** aim, attempt, endeavour, essay, exert yourself, make an effort, strain, strive, struggle, venture. **2** *try something new.* appraise, *inf* check out, evaluate, examine, experiment with, *inf* have a go at, *inf* have a stab at, investigate, test, try out, undertake. **trying** ▷ ANNOYING, TIRESOME. **try someone's patience** ▷ ANNOY.

**tub** *n* barrel, bath, butt, cask, drum, keg, pot, vat.

**tube** *n* capillary, conduit, cylinder, duct, hose, main, pipe, spout, tubing.

**tuck** *vb* cram, gather, insert, push, put away, shove, stuff. *tuck in* ▷ EAT.

**tuft** *n* bunch, clump, cluster, tuffet, tussock.

**tug** *vb* drag, draw, haul, heave, jerk, lug, pluck, pull, tow, twitch, wrench, *inf* yank.

**tumble** *vb* **1** collapse, drop, fall, flop, pitch, roll, stumble, topple, trip up. **2** *tumble things into a heap.* disarrange, dump, jumble, mix up, rumple, shove, spill, throw carelessly, toss.

**tumbledown** *adj* badly maintained, broken down, crumbling, decrepit, derelict, dilapidated, ramshackle, rickety, ruined, shaky, tottering.

**tumult** *n* ado, agitation, chaos, commotion, confusion, disturbance, excitement, ferment, fracas, frenzy, hubbub, hullabaloo, rumpus, storm, tempest, upheaval, uproar, welter.

**tumultuous** *adj* agitated, boisterous, confused, excited, frenzied, hectic, passionate, stormy, tempestuous, turbulent, unrestrained, unruly, violent, wild. *Opp* CALM.

**tune** *n* air, melody, motif, song, strain, theme. • *vb* adjust, calibrate, regulate, set, temper.

**tuneful** *adj inf* catchy, euphonious, mellifluous, melodic, melodious, musical, pleasant, singable, sweet-sounding. *Opp* TUNELESS.

**tuneless** *adj* atonal, boring, cacophonous, discordant, dissonant, harsh, monotonous, unmusical. *Opp* TUNEFUL.

**tunnel** *n* burrow, gallery, hole, mine, passage, passageway, shaft, subway, underpass. • *vb* burrow, dig, excavate, mine, penetrate.

**turbulent** *adj* **1** agitated, boisterous, confused, disordered, excited, hectic, passionate, restless, seething, turbid, unrestrained, violent, volatile, wild. **2** *a turbulent crowd.* badly-behaved, disorderly, lawless, obstreperous, riotous, rowdy, undisciplined, unruly. **3** *turbulent weather.* blustery, bumpy, choppy (*sea*), rough, stormy, tempestuous, violent, wild, windy. *Opp* CALM.

**turf** *n* grass, grassland, green, lawn, *poet* sward.

**turgid** *adj* affected, bombastic, flowery, fulsome, grandiose, high-flown, overblown, pompous, pretentious, stilted, wordy. *Opp* ARTICULATE.

**turmoil** *n inf* bedlam, chaos, commotion, confusion, disorder, disturbance, ferment, *inf* hubbub, *inf* hullabaloo, pandemonium, riot row, rumpus, tumult, turbulence, unrest, upheaval, uproar, welter. *Opp* CALM.

**turn** *n* 1 circle, coil, curve, cycle, loop, pirouette, revolution, roll, rotation, spin, twirl, twist, whirl. 2 angle, bend, change of direction, corner, deviation, *inf* dogleg, hairpin bend, junction, loop, meander, reversal, shift, turning-point, *inf* U-turn, zigzag. 3 *your turn in a game.* chance, *inf* go, innings, opportunity, shot, stint. 4 *a comic turn.* ▷ PERFORMANCE. 5 *a nasty turn.* ▷ ILLNESS. • *vb* 1 circle, coil, curl, gyrate, hinge, loop, move in a circle, orbit, pivot, revolve, roll, rotate, spin, spiral, swivel, twirl, twist, whirl, wind, yaw. 2 bend, change direction, corner, deviate, divert, go round a corner, negotiate a corner, steer, swerve, veer, wheel. 3 *turn a pumpkin into a coach.* adapt, alter, change, convert, make, modify, remake, remodel, transfigure, transform. 4 *turn to and fro.* squirm, twist, wriggle, writhe. **turn aside** ▷ DEVIATE. **turn down** ▷ REJECT. **turn into** ▷ BECOME. **turn off** ▷ DEVIATE, DISCONNECT, REPEL. **turn on** ▷ ATTRACT, CONNECT. **turn out** ▷ EXPEL, HAPPEN, PRODUCE. **turn over** ▷ CONSIDER, OVERTURN. **turn tail** ▷ ESCAPE. **turn up** ▷ ARRIVE, DISCOVER.

**turning-point** *n* crisis, crossroads, new direction, revolution, watershed.

**turnover** *n* business, cash-flow, efficiency, output, production, productivity, profits, revenue, throughput, yield.

**twiddle** *vb* fiddle with, fidget with, fool with, mess with, twirl, twist.

**twig** *n* branch, offshoot, shoot, spray, sprig, sprout, stalk, stem, stick, sucker, tendril.

**twilight** *n* dusk, evening, eventide, gloaming, gloom, halflight, nightfall, sundown, sunset.

**twin** *adj* balancing, corresponding, double, duplicate, identical, indistinguishable, *inf* look-alike, matching, paired, similar, symmetrical. • *n* clone, counterpart, double, duplicate, *inf* look-alike, match, pair, *inf* spitting image.

**twirl** *vb* 1 gyrate, pirouette, revolve, rotate, spin, turn, twist, wheel, whirl, wind. 2 *twirl an umbrella.* brandish, twiddle, wave.

**twist** *n* 1 bend, coil, curl, kink, knot, loop, tangle, turn, zigzag. 2 *a twist to a story.* revelation, surprise ending. • *vb* 1 bend, coil, corkscrew, curl, curve, loop, revolve, rotate, screw, spin, spiral, turn, weave, wind, wreathe, wriggle, writhe, zigzag. 2 *twist ropes.* entangle, entwine, intertwine, interweave, tangle. 3 *twist a lid off.* jerk, wrench, wrest. 4 *twist out of shape.* buckle, contort, crinkle, crumple, distort, screw up, warp, wrinkle. 5 *twist meaning.* alter, change, falsify, misquote, misrepresent. **twisted** ▷ CONFUSED, PERVERTED, TWISTY.

**twisty** *adj* bending, bendy, circuitous, coiled, contorted, crooked, curving, curvy, *inf* in and out, indirect, looped, meandering, misshapen, rambling, roundabout, serpentine, sinuous, snaking, tortuous, twisted, twisting, *inf* twisting and turning, winding, zigzag. *Opp* STRAIGHT.

**twitch** *n* blink, convulsion, flutter, jerk, jump, spasm, tic, tremor. • *vb* fidget, flutter, jerk, jump, start, tremble.

**two** *n* couple, duet, duo, match, pair, twosome.

**type** *n* 1 category, class, classification, description, designation, form, genre, group, kind, mark, set, sort, species, variety. 2 *He was the very type of evil.* embodiment, epitome, example, model, pattern, personification, standard. **3** *printed in large type.* characters, font, fount, lettering, letters, print, printing, typeface.

**typical** *adj* **1** characteristic, distinctive, particular, representative, special. **2** *a typical day.* average, conventional, normal, ordinary, orthodox, predictable, standard, stock, unsurprising, usual. *Opp* UNUSUAL.

**tyrannical** *adj* absolute, authoritarian, autocratic, *inf* bossy, cruel, despotic, dictatorial, domineering, harsh, high-handed, illiberal, imperious, oppressive, overbearing, ruthless, severe, totalitarian, tyrannous, undemocratic, unjust. *Opp* DEMOCRATIC, LIBERAL.

**tyrant** *n* autocrat, despot, dictator, *inf* hard taskmaster, oppressor, slave-driver. ▷ RULER.

# U

**ugly** *adj* **1** deformed, disfigured, disgusting, dreadful, frightful, ghastly, grim, grisly, grotesque, gruesome, hideous, horrible, *inf* horrid, ill-favoured, loathsome, misshapen, monstrous, nasty, objectionable, odious, offensive, repellent, repulsive, revolting, shocking, sickening, terrible. **2** *ugly furniture.* displeasing, inartistic, inelegant, plain, tasteless, unattractive, unpleasant, unprepossessing, unsightly. **3** *an ugly mood.* angry, cross, dangerous, forbidding, hostile, menacing, ominous, sinister, surly, threatening, unfriendly. *Opp* BEAUTIFUL.

**ulterior** *adj* concealed, covert, hidden, personal, private, secondary, secret, undeclared, underlying, undisclosed, unexpressed. *Opp* OVERT.

**ultimate** *adj* **1** closing, concluding, eventual, extreme, final, furthest, last, terminal, terminating. **2** *ultimate truth.* basic, fundamental, primary, root, underlying.

**umpire** *n* adjudicator, arbiter, arbitrator, judge, linesman, moderator, official, *inf* ref, referee.

**unable** *adj* impotent, incompetent, powerless, unfit, unprepared, unqualified. *Opp* ABLE.

**unacceptable** *adj* bad, forbidden, illegal, improper, inadequate, inadmissible, inappropriate, inexcusable, insupportable, intolerable, invalid, taboo, unsatisfactory, unsuitable, wrong. *Opp* ACCEPTABLE.

**unaccompanied** *adj* alone, lone, single-handed, sole, solo, unaided, unescorted.

**unaccountable** *adj* ▷ INEXPLICABLE.

**unaccustomed** *adj* ▷ STRANGE.

**unadventurous** *adj* **1** cautious, cowardly, spiritless, timid, unimaginative. **2** *an unadventurous life.* cloistered, limited, protected, sheltered, unexciting. *Opp* ADVENTUROUS.

**unalterable** *adj* ▷ IMMUTABLE.

**unambiguous** *adj* ▷ DEFINITE.

**unanimous** *adj* ▷ UNITED.

**unasked** *adj* ▷ UNINVITED.

**unassuming** *adj* ▷ MODEST.

**unattached** *adj* autonomous, *inf* available, free, independent, separate, single, uncommitted, unmarried, *inf* unspoken for.

**unattractive** *adj* characterless, colourless, displeasing, dull, inartistic, inelegant, nasty, objectionable, *inf* off-putting, plain, repellent, repulsive, tasteless, uninviting, unpleasant, unprepossessing, unsightly. ▷ UGLY. *Opp* ATTRACTIVE.

**unauthorized** *adj* illegal, illegitimate, illicit, irregular, unapproved, unlawful, unofficial. *Opp* OFFICIAL.

**unavoidable** *adj* certain, compulsory, destined, fated, fixed, ineluctable, inescapable, inevitable, inexorable, mandatory, necessary, obligatory, predetermined, required, sure, unalterable.

**unaware** *adj* ▷ IGNORANT.

**unbalanced** *adj* **1** asymmetrical, irregular, lopsided, off-centre, shaky, uneven, unstable, wobbly. **2** biased, bigoted, one-sided, partial, partisan, prejudiced, unfair, unjust. **3** *unbalanced mind.* ▷ MAD.

**unbearable** *adj* insufferable, insupportable, intolerable, overpowering, overwhelming, unacceptable, unendurable. *Opp* TOLERABLE.

**unbeatable** *adj* ▷ INVINCIBLE.

**unbecoming** *adj* dishonourable, improper, inappropriate, indecorous, indelicate, offensive, tasteless, unattractive, unbefitting, undignified, ungentlemanly, unladylike, unseemly, unsuitable. *Opp* DECOROUS.

**unbelievable** *adj* ▷ INCREDIBLE.

**unbelieving** *adj* ▷ INCREDULOUS.

**unbend** *vb* **1** straighten, uncurl, untwist. **2** loosen up, relax, rest, unwind.

**unbending** *adj* ▷ INFLEXIBLE.

**unbiased** *adj* balanced, disinterested, enlightened, even-handed, fair, impartial, independent, just, neutral, non-partisan, objective, open-minded, reasonable, *inf* straight, unbigoted, undogmatic, unprejudiced. *Opp* BIASED.

**unbreakable** *adj* ▷ INDESTRUCTIBLE.

**unbroken** *adj* ▷ CONTINUOUS, WHOLE.

**uncalled-for** *adj* ▷ UNNECESSARY.

**uncared-for** *adj* ▷ DERELICT.

**uncaring** *adj* ▷ CALLOUS.

**unceasing** *adj* ▷ CONTINUOUS.

**uncertain** *adj* **1** ambiguous, arguable, *inf* chancy, confusing, conjectural, cryptic, enigmatic, equivocal, hazardous, hazy, *inf* iffy, imprecise, incalculable, inconclusive, indefinite, indeterminate, problematical, puzzling, questionable, risky, speculative, *inf* touch and go, unclear, unconvincing, undecided, undetermined, unforeseeable, unknown, unresolved, woolly. **2** *uncertain what to believe.* agnostic, ambivalent, doubtful, dubious, *inf* hazy, insecure, *inf* in two minds, self-questioning, unconvinced, undecided, unsure, vague, wavering. **3** *an uncertain climate.* changeable, erratic, fitful, inconstant, irregular, precarious, unpredictable, unreliable, unsettled, variable. *Opp* CERTAIN.

**unchanging** *adj* ▷ CONSTANT.

**uncharitable** *adj* ▷ UNKIND.

**uncivilized** *adj* anarchic, antisocial, backward, barbarian, barbaric, barbarous, brutish, crude, disorganized, illiterate, Philistine, primitive, savage, uncultured, uneducated, unenlightened, unsophisticated, wild. *Opp* CIVILIZED.

**unclean** *adj* ▷ DIRTY.

**unclear** *adj* ▷ UNCERTAIN.

**unclothed** *adj* ▷ NAKED.

**uncomfortable** *adj* **1** bleak, cold, comfortless, cramped, hard, inconvenient, lumpy, painful.

2 *uncomfortable clothes.* formal, restrictive, stiff, tight, tight-fitting. 3 *an uncomfortable silence.* awkward, distressing, embarrassing, nervous, restless, troubled, uneasy, worried. *Opp* COMFORTABLE.

**uncommon** *adj* ▷ UNUSUAL.

**uncommunicative** *adj* ▷ TACITURN.

**uncomplimentary** *adj* censorious, critical, deprecatory, depreciatory, derogatory, disapproving, disparaging, pejorative, scathing, slighting, unfavourable, unflattering. ▷ RUDE. *Opp* COMPLIMENTARY.

**uncompromising** *adj* ▷ INFLEXIBLE.

**unconcealed** *adj* ▷ OBVIOUS.

**unconditional** *adj* absolute, categorical, complete, full, outright, total, unequivocal, unlimited, unqualified, unreserved, unrestricted, wholehearted, *inf* with no strings attached. *Opp* CONDITIONAL.

**uncongenial** *adj* alien, antipathetic, disagreeable, incompatible, unattractive, unfriendly, unpleasant, unsympathetic. *Opp* CONGENIAL.

**unconquerable** *adj* ▷ INVINCIBLE.

**unconscious** *adj* 1 anaesthetized, *inf* blacked-out, comatose, concussed, *inf* dead to the world, insensible, *inf* knocked out, *inf* knocked silly, oblivious, *inf* out for the count, senseless, sleeping. 2 blind, deaf, ignorant, oblivious, unaware. 3 *unconscious humour.* accidental, inadvertent, unintended, unintentional, unwitting. 4 *an unconscious reaction.* automatic, *sl* gut, impulsive, instinctive, involuntary, reflex, spontaneous, unthinking. 5 *an unconscious desire.* repressed, subconscious, subliminal, suppressed. *Opp* CONSCIOUS.

**unconsciousness** *n inf* blackout, coma, faint, oblivion, sleep.

**uncontrollable** *adj* ▷ UNDISCIPLINED.

**unconventional** *adj* abnormal, atypical, *inf* cranky, eccentric, exotic, futuristic, idiosyncratic, independent, inventive, nonconforming, non-standard, odd, off-beat, original, peculiar, progressive, revolutionary, strange, surprising, unaccustomed, unorthodox, *inf* way-out, wayward, weird, zany. *Opp* CONVENTIONAL.

**unconvincing** *adj* implausible, improbable, incredible, invalid, spurious, unbelievable, unlikely. *Opp* PERSUASIVE.

**uncooperative** *adj* lazy, obstructive, recalcitrant, selfish, unhelpful, unwilling. *Opp* COOPERATIVE.

**uncover** *vb* bare, come across, detect, dig up, disclose, discover, disrobe, exhume, expose, locate, reveal, show, strip, take the wraps off, undress, unearth, unmask, unveil, unwrap. *Opp* COVER.

**undamaged** *adj* ▷ PERFECT.

**undefended** *adj* defenceless, exposed, helpless, insecure, unarmed, unfortified, unguarded, unprotected, vulnerable, weaponless. *Opp* SECURE.

**undemanding** *adj* ▷ EASY.

**undemonstrative** *adj* ▷ ALOOF.

**underclothes** *plur n* lingerie, *inf* smalls, underclothing, undergarments, underthings, underwear, *inf* undies. □ *bra, braces, brassière, briefs, camiknickers, corset, drawers, garter, girdle, knickers, panties, pantihose, pants, petticoat, slip, suspenders, tights,* *trunks, underpants, underskirt, vest.*

**undercurrent** *n* atmosphere, feeling, hint, sense, suggestion, trace, undertone.

**underestimate** *vb* belittle, depreciate, dismiss, disparage, minimize, miscalculate, misjudge, underrate, undervalue.
*Opp* EXAGGERATE.

**undergo** *vb* bear, be subjected to, endure, experience, go through, live through, put up with, *inf* stand, submit yourself to, suffer, withstand.

**underground** *adj* **1** buried, hidden, subterranean, sunken. **2** clandestine, revolutionary, secret, subversive, unofficial, unrecognized.

**undergrowth** *n* brush, bushes, ground cover, plants, vegetation.

**undermine** *vb* burrow under, destroy, dig under, erode, excavate, mine under, ruin, sabotage, sap, subvert, tunnel under, undercut, weaken, wear away.

**underprivileged** *adj* deprived, destitute, disadvantaged, downtrodden, impoverished, needy, oppressed. ▷ POOR.
*Opp* PRIVILEGED.

**undersea** *adj* subaquatic, submarine, underwater.

**understand** *vb* **1** appreciate, apprehend, be conversant with, *inf* catch on, comprehend, *inf* cotton on to, decipher, decode, fathom, figure out, follow, gather, *inf* get, *inf* get to the bottom of, grasp, interpret, know, learn, make out, make sense of, master, perceive, realize, recognize, see, take in, *inf* twig. **2** *understand animals.* be in sympathy with, empathize with, sympathize with.

**understanding** *n* **1** ability, acumen, brains, cleverness, discernment, insight, intellect, intelligence, judgement, penetration, perceptiveness, percipience, sense, wisdom. **2** *understanding of a problem.* appreciation, apprehension, awareness, cognition, comprehension, grasp, knowledge. **3** *understanding between people.* accord, agreement, compassion, consensus, consent, consideration, empathy, fellow feeling, harmony, kindness, mutuality, sympathy, tolerance. **4** *a formal understanding.* arrangement, bargain, compact, contract, deal, entente, pact, settlement, treaty.

**understate** *vb* belittle, *inf* make light of, minimize, *inf* play down, *inf* soft-pedal. *Opp* EXAGGERATE.

**undertake** *vb* **1** agree, attempt, consent, covenant, guarantee, pledge, promise, try. **2** *undertake a task.* accept responsibility for, address, approach, attend to, begin, commence, commit yourself to, cope with, deal with, embark on, grapple with, handle, manage, tackle, take on, take up.

**undertaking** *n* **1** affair, business, enterprise, project, task, venture. **2** agreement, assurance, contract, guarantee, pledge, promise, vow.

**undervalue** *vb* belittle, depreciate, dismiss, disparage, minimize, miscalculate, misjudge, underestimate, underrate.

**underwater** *adj* subaquatic, submarine, undersea.

**undeserved** *adj* unearned, unfair, unjustified, unmerited, unwarranted.

**undesirable** *adj* ▷ OBJECTIONABLE.

**undisciplined** *adj* anarchic, chaotic, disobedient, disorderly, disorganized, intractable, rebellious, uncontrollable, uncontrolled, ungovernable, unmanageable, unruly, unsystematic, untrained, wild, wilful. *Opp* OBEDIENT.

**undiscriminating** *adj* ▷ IMPERCEPTIVE.

**undisguised** *adj* ▷ OBVIOUS.

**undistinguished** *adj* ▷ ORDINARY.

**undo** *vb* **1** detach, disconnect, disengage, loose, loosen, open, part, separate, unbind, unbuckle, unbutton, unchain, unclasp, unclip, uncouple, unfasten, unfetter, unhook, unleash, unlock, unpick, unpin, unscrew, unseal, unshackle, unstick, untether, untie, unwrap, unzip. **2** *undo someone's good work.* annul, cancel out, destroy, mar, nullify, quash, reverse, ruin, spoil, undermine, vitiate, wipe out, wreck.

**undoubted** *adj* ▷ INDISPUTABLE.

**undoubtedly** *adv* certainly, definitely, doubtless, indubitably, of course, surely, undeniably, unquestionably.

**undress** *vb* disrobe, divest yourself, *inf* peel off, shed your clothes, strip off, take off your clothes, uncover yourself.

**undressed** *adj* ▷ NAKED.

**undue** *adj* ▷ EXCESSIVE.

**undying** *adj* ▷ ETERNAL.

**uneasy** *adj* anxious, apprehensive, awkward, concerned, distressed, distressing, disturbed, edgy, fearful, insecure, jittery, nervous, restive, restless, tense, troubled, uncomfortable, unsettled, upsetting, worried.

**uneducated** *adj* ▷ IGNORANT.

**unemotional** *adj* apathetic, clinical, cold, cool, dispassionate, frigid, hard-hearted, heartless, impassive, indifferent, objective, unfeeling, unmoved, unresponsive. *Opp* EMOTIONAL.

**unemployed** *adj* jobless, laid off, on the dole, out of work, redundant, *inf* resting, unwaged. ▷ IDLE.

**unendurable** *adj* ▷ UNBEARABLE

**unenthusiastic** *adj* ▷ APATHETIC UNINTERESTED.

**unequal** *adj* **1** different, differing disparate, dissimilar, uneven, varying. **2** *unequal treatment.* biased, prejudiced, unjust. **3** *an unequal contest.* ill-matched, one-sided, unbalanced, uneven, unfair *Opp* EQUAL, FAIR.

**unequalled** *adj* incomparable, inimitable, matchless, peerless, supreme, surpassing, un-matched, unparalleled, unrivalled, unsurpassed.

**unethical** *adj* ▷ IMMORAL.

**uneven** *adj* **1** bent, broken, bumpy, crooked, irregular, jagged jerky, pitted, rough, rutted, undulating, wavy. **2** *an uneven rhythm.* erratic, fitful, fluctuating, inconsistent, spasmodic, unpredictable, variable, varying. **3** *an uneven load.* asymmetrical, lopsided, unsteady. **4** *uneven contest.* ill-matched, one-sided, unbalanced, unequal, unfair. *Opp* EVEN.

**uneventful** *adj* ▷ UNEXCITING.

**unexciting** *adj* boring, dreary, dry, dull, humdrum, monotonous, predictable, quiet, repetitive, routine, soporific, straightforward, tedious, trite, uneventful, uninspiring, uninteresting, vapid, wearisome. ▷ ORDINARY. *Opp* EXCITING.

**unexpected** *adj* accidental, chance, fortuitous, sudden, surprising, unforeseen, unhoped-for, unlooked-for, unplanned, unpredictable, unusual. *Opp* PREDICTABLE.

**unfair** *adj* ▷ UNJUST.

**unfaithful** *adj* deceitful, disloyal, double-dealing, duplicitous, faithless, false, fickle, inconstant, perfidious, traitorous, treacherous, treasonable, unreliable, untrue, untrustworthy. *Opp* FAITHFUL.

**unfaithfulness** *n* **1** duplicity, perfidy, treachery, treason. **2** adultery, infidelity.

**unfamiliar** *adj* ▷ STRANGE.

**unfashionable** *adj* dated, obsolete, old-fashioned, *inf* out, outmoded, passé, superseded, unstylish. *Opp* FASHIONABLE.

**unfasten** *vb* ▷ UNDO.

**unfavourable** *adj* **1** adverse, attacking, contrary, critical, disapproving, discouraging, hostile, ill-disposed, inauspicious, negative, opposing, uncomplimentary, unfriendly, unhelpful, unkind, unpromising, unpropitious, unsympathetic. **2** *an unfavourable reputation.* bad, undesirable, unenviable, unsatisfactory. *Opp* FAVOURABLE.

**unfeeling** *adj* ▷ CALLOUS.

**unfinished** *adj* imperfect, incomplete, rough, sketchy, uncompleted, unpolished. *Opp* PERFECT.

**unfit** *adj* **1** ill-equipped, inadequate, incapable, incompetent, unsatisfactory, useless. **2** *unfit for family viewing.* improper, inappropriate, unbecoming, unsuitable, unsuited. **3** *an unfit athlete.* feeble, flabby, out of condition, unhealthy. ▷ ILL. *Opp* FIT.

**unflagging** *adj* ▷ TIRELESS.

**unflinching** *adj* ▷ RESOLUTE.

**unforeseen** *adj* ▷ UNEXPECTED.

**unforgettable** *adj* ▷ MEMORABLE.

**unforgivable** *adj* inexcusable, mortal (*sin*), reprehensible, shameful, unjustifiable, unpardonable, unwarrantable. *Opp* FORGIVABLE.

**unfortunate** *adj* ▷ UNLUCKY.

**unfriendly** *adj* aggressive, aloof, antagonistic, antisocial, cold, cool, detached, disagreeable, distant, forbidding, frigid, haughty, hostile, ill-disposed, ill-natured, impersonal, indifferent, inhospitable, menacing, nasty, obnoxious, offensive, remote, reserved, rude, sour, standoffish, *inf* starchy, stern, supercilious, threatening, unapproachable, uncivil, uncongenial, unenthusiastic, unforthcoming, unkind, unneighbourly, unresponsive, unsociable, unsympathetic, unwelcoming. *Opp* FRIENDLY.

**ungainly** *adj* ▷ AWKWARD.

**ungodly** *adj* ▷ IRRELIGIOUS.

**ungovernable** *adj* ▷ UNDISCIPLINED.

**ungrateful** *adj* displeased, ill-mannered, rude, selfish, unappreciative, unthankful. *Opp* GRATEFUL.

**unhappy** *adj* **1** dejected, depressed, dispirited, down, downcast, gloomy, miserable, mournful, sorrowful. ▷ SAD. **2** *unhappy about losing.* bad-tempered, disaffected, discontented, disgruntled, disillusioned, displeased, dissatisfied, *inf* fed up, *inf* grumpy, morose, sulky, sullen, unsatisfied. **3** *an unhappy choice.* ▷ UNSATISFACTORY.

**unhealthy** *adj* **1** ailing, debilitated, delicate, diseased, feeble, frail, infected, infirm, *inf* poorly, sick, sickly, suffering, unwell, valetudinary, weak. ▷ ILL. **2** *unhealthy conditions.* deleterious, detrimental, dirty, harmful, insalubrious, insanitary, noxious, polluted, unhygienic, unwholesome. *Opp* HEALTHY.

**unheard-of** *adj* ▷ UNUSUAL.

**unhelpful** *adj* disobliging, inconsiderate, negative, slow, uncivil, uncooperative, unwilling. *Opp* HELPFUL.

**unhygienic** *adj* ▷ UNHEALTHY.

**unidentifiable** *adj* anonymous, camouflaged, disguised, hidden, undetectable, unidentified, unknown, unrecognizable. *Opp* IDENTIFIABLE.

**unidentified** *adj* anonymous, incognito, mysterious, nameless, unfamiliar, unknown, unmarked, unnamed, unrecognized, unspecified. *Opp* SPECIFIC.

**uniform** *adj* consistent, even, homogeneous, identical, indistinguishable, predictable, regular, same, similar, single, standard, unbroken, unvaried, unvarying. *Opp* DIFFERENT. • *n* costume, livery, outfit.

**unify** *vb* amalgamate, bring together, coalesce, combine, consolidate, fuse, harmonize, integrate, join, merge, unite, weld together. *Opp* SEPARATE.

**unimaginative** *adj* banal, boring, clichéd, derivative, dull, hackneyed, inartistic, insensitive, obvious, ordinary, prosaic, stale, trite, ugly, uninspired, uninteresting, unoriginal. *Opp* IMAGINATIVE.

**unimportant** *adj* ephemeral, forgettable, immaterial, inconsequential, inconsiderable, inessential, insignificant, irrelevant, lightweight, minor, negligible, peripheral, petty, secondary, slight, trifling, trivial, valueless, worthless. ▷ SMALL. *Opp* IMPORTANT.

**uninhabitable** *adj* condemned, in bad repair, unliveable, unusable. *Opp* HABITABLE.

**uninhabited** *adj* abandoned, deserted, desolate, empty, tenantless, uncolonized, unoccupied, unpeopled, unpopulated, untenanted, vacant.

**uninhibited** *adj* abandoned, candid, casual, easygoing, frank, informal, natural, open, outgoing, outspoken, relaxed, spontaneous, unbridled, unconstrained, unrepressed, unreserved, unrestrained, unselfconscious, wild. *Opp* REPRESSED.

**unintelligent** *adj* ▷ STUPID.

**unintelligible** *adj* ▷ INCOMPREHENSIBLE.

**unintentional** *adj* accidental, fortuitous, inadvertent, involuntary, unconscious, unintended, unplanned, unwitting. *Opp* INTENTIONAL.

**uninterested** *adj* apathetic, bored, incurious, indifferent, lethargic, passive, phlegmatic, unconcerned, unenthusiastic, uninvolved, unresponsive. *Opp* INTERESTED.

**uninteresting** *adj* boring, dreary, dry, dull, flat, monotonous, obvious, predictable, tedious, unexciting, uninspiring, vapid, wearisome. ▷ ORDINARY. *Opp* INTERESTING.

**uninterrupted** *adj* ▷ CONTINUOUS.

**uninvited** *adj* **1** unasked, unbidden, unwelcome. **2** *an uninvited comment.* gratuitous, unsolicited, voluntary.

**uninviting** *adj* ▷ UNATTRACTIVE.

**union** *n* **1** alliance, amalgamation, association, coalition, confederation, conjunction, federation, integration, joining together, merger, unanimity, unification, unity. **2** amalgam, blend, combination, combining, compound, fusion, grafting, marrying, mixture, synthesis, welding. 3 marriage, matrimony, partnership, wedlock.

**unique** *adj* distinctive, incomparable, lone, *inf* one-off, peculiar, peerless, *inf* second to none, single, singular, unequalled, unparalleled, unrepeatable, unrivalled.

**unit** *n* component, constituent, element, entity, item, module, part, piece, portion, section, segment, whole.

**unite** *vb* ally, amalgamate, associate, blend, bring together, coalesce, collaborate, combine, commingle, confederate, connect, consolidate, conspire, cooperate, couple, federate, fuse, go into partnership, harmonize, incorporate, integrate, interlock, join, join forces, link, link up, marry, merge, mingle, mix, stick together, tie up, unify, weld together. ▷ MARRY. *Opp* SEPARATE.

**united** *adj* agreed, allied, coherent, collective, common, concerted, coordinated, corporate, harmonious, integrated, joint, like-minded, mutual, *inf* of one mind, shared, *inf* solid, unanimous, undivided. *Opp* DISUNITED. **be united** ▷ AGREE.

**unity** *n* accord, agreement, coherence, concord, consensus, harmony, integrity, like-mindedness, oneness, rapport, solidarity, unanimity, wholeness. *Opp* DISUNITY.

**universal** *adj* all-embracing, all-round, boundless, common, comprehensive, cosmic, general, global, international, omnipresent, pandemic, prevailing, prevalent, total, ubiquitous, unbounded, unlimited, widespread, worldwide.

**universe** *n* cosmos, creation, the heavens, *old use* macrocosm.

**unjust** *adj* biased, bigoted, indefensible, inequitable, one-sided, partial, partisan, prejudiced, undeserved, unfair, unjustified, unlawful, unmerited, unreasonable, unwarranted, wrong, wrongful. *Opp* JUST.

**unjustifiable** *adj* excessive, immoderate, indefensible, inexcusable, unacceptable, unconscionable, unforgivable, unjust, unreasonable, unwarranted. *Opp* JUSTIFIABLE.

**unkind** *adj* abrasive, *inf* beastly, callous, caustic, cold-blooded, discourteous, disobliging, hard, hard-hearted, harsh, heartless, hurtful, ill-natured, impolite, inconsiderate, inhuman, inhumane, insensitive, malevolent, malicious, mean, merciless, nasty, pitiless, relentless, rigid, rough, ruthless, sadistic, savage, selfish, severe, sharp, spiteful, stern, tactless, thoughtless, uncaring, uncharitable, unchristian, unfeeling, unfriendly, unpleasant, unsympathetic, unthoughtful, vicious. ▷ ANGRY, CRITICAL, CRUEL. *Opp* KIND.

**unknown** *adj* 1 anonymous, disguised, incognito, mysterious, nameless, strange, unidentified, unnamed, unrecognized, unspecified. 2 *an unknown country*. alien, foreign, uncharted, undiscovered, unexplored, unfamiliar, unmapped. 3 *an unknown actor*. humble, insignificant, little-known, lowly, obscure, undistinguished, unheard-of, unimportant. *Opp* FAMOUS.

**unlawful** *adj* ▷ ILLEGAL.

**unlikely** *adj* 1 dubious, far-fetched, implausible, improbable, incredible, suspect, suspicious, *inf* tall (*story*), unbelievable, unconvincing, unthinkable. 2 *an* *unlikely possibility*. distant, doubtful, faint, *inf* outside, remote, slight. *Opp* LIKELY.

**unlimited** *adj* ▷ BOUNDLESS.

**unload** *vb* disburden, discharge, drop off, *inf* dump, empty, offload, take off, unpack. *Opp* LOAD.

**unloved** *adj* abandoned, discarded, forsaken, hated, loveless, lovelorn, neglected, rejected, spurned, uncared-for, unvalued, unwanted. *Opp* LOVED.

**unlucky** *adj* **1** accidental, calamitous, chance, disastrous, dreadful, tragic, unfortunate, untimely, unwelcome. **2** *an unlucky person*. *inf* accident-prone, hapless, luckless, unhappy, unsuccessful, wretched. **3** *an unlucky number*. cursed, ill-fated, ill-omened, ill-starred, inauspicious, jinxed, ominous, unfavourable. *Opp* LUCKY.

**unmanageable** *adj* ▷ UNDISCIPLINED.

**unmarried** *adj inf* available, celibate, *inf* free, single, unwed. **unmarried person** bachelor, celibate, spinster.

**unmentionable** *adj* ▷ TABOO.

**unmistakable** *adj* ▷ DEFINITE, OBVIOUS.

**unnamed** *adj* ▷ UNIDENTIFIED.

**unnatural** *adj* **1** abnormal, bizarre, eccentric, eerie, extraordinary, fantastic, freak, freakish, inexplicable, magic, magical, odd, outlandish, preternatural, queer, strange, supernatural, unaccountable, uncanny, unusual, weird. **2** *unnatural feelings*. callous, cold-blooded, cruel, hard-hearted, heartless, inhuman, inhumane, monstrous, perverse, perverted, sadistic, savage, stony-hearted, unfeeling, unkind. **3** *unnatural behaviour*. actorish, affected, bogus, contrived, fake, feigned, forced, insincere, laboured, mannered, *inf* out of character, overdone, *inf* phoney, pretended, *inf* pseudo, *inf* put on, self-conscious, stagey, stiff, stilted, theatrical, uncharacteristic, unspontaneous. **4** *unnatural materials*. artificial, fabricated, imitation, man-made, manufactured, simulated, synthetic. *Opp* NATURAL.

**unnecessary** *adj* dispensable, excessive, expendable, extra, inessential, needless, nonessential, redundant, supererogatory, superfluous, surplus, uncalled-for, unjustified, unneeded, unwanted, useless. *Opp* NECESSARY.

**unobtrusive** *adj* ▷ INCONSPICUOUS.

**unofficial** *adj* friendly, informal, *inf* off the record, private, secret, unauthorized, unconfirmed, undocumented, unlicensed. *Opp* OFFICIAL.

**unorthodox** *adj* ▷ UNCONVENTIONAL.

**unpaid** *adj* **1** due, outstanding, owing, payable, unsettled. **2** *unpaid work*. honorary, unremunerative, unsalaried, voluntary.

**unpalatable** *adj* disgusting, distasteful, inedible, nasty, nauseating, *inf* off, rancid, sickening, sour, tasteless, unacceptable, unappetizing, uneatable, unpleasant. *Opp* PALATABLE.

**unparalleled** *adj* ▷ UNEQUALLED.

**unpardonable** *adj* ▷ UNFORGIVABLE.

**unplanned** *adj* ▷ SPONTANEOUS.

**unpleasant** *adj* abhorrent, abominable, antisocial, appalling, atrocious, awful, bad-tempered, beastly, bitter, coarse, crude, despicable, detestable, diabolical, dirty, disagreeable, disgusting, displeasing, distasteful, dreadful, evil, execrable, fearful, fearsome, filthy, foul, frightful, ghastly, grim, grisly, gruesome, harsh, hateful, *inf* hellish, hideous, horrible, horrid, horrifying, improper, indecent, inhuman, irksome, loathsome, *inf* lousy, malevolent, malicious, mucky, nasty, nauseating, objectionable, obnoxious, odious, offensive, *inf* off-putting, repellent, repugnant, repulsive, revolting, rude, shocking, sickening, sickly, sordid, sour, spiteful, squalid, terrible, ugly, unattractive, uncouth, undesirable, unfriendly, unkind, unpalatable, unsavoury, unwelcome, upsetting, vexing, vicious, vile, vulgar. ▷ BAD. *Opp* PLEASANT.

**unpopular** *adj* despised, disliked, friendless, hated, ignored, *inf* in bad odour, minority (*interests*), out of favour, rejected, shunned, unfashionable, unloved, unwanted. *Opp* POPULAR.

**unpredictable** *adj* changeable, surprising, uncertain, unexpected, unforeseeable, variable. *Opp* PREDICTABLE.

**unprejudiced** *adj* ▷ UNBIASED.

**unpremeditated** *adj* ▷ SPONTANEOUS.

**unprepared** *adj* *inf* caught napping, caught out, ill-equipped, surprised, taken off-guard, unready. ▷ READY.

**unpretentious** *adj* humble, modest, plain, simple, straightforward, unaffected, unassuming, unostentatious, unsophisticated. *Opp* PRETENTIOUS.

**unproductive** *adj* **1** ineffective, fruitless, futile, pointless, unprofitable, unrewarding, useless, valueless, worthless. **2** *an unproductive garden.* arid, barren, infertile, sterile, unfruitful. *Opp* PRODUCTIVE.

**unprofessional** *adj* amateurish, casual, incompetent, inefficient, inexpert, lax, negligent, shoddy, *inf* sloppy, unethical, unfitting, unprincipled, unseemly, unskilful, unskilled, unworthy. *Opp* PROFESSIONAL.

**unprofitable** *adj* futile, loss-making, pointless, uncommercial, uneconomic, ungainful, unproductive, unremunerative, unrewarding, worthless. *Opp* PROFITABLE.

**unprovable** *adj* doubtful, inconclusive, questionable, undemonstrable, unsubstantiated, unverifiable. *Opp* PROVABLE, PROVEN.

**unpunctual** *adj* behindhand, belated, delayed, detained, last-minute, late, overdue, tardy, unreliable. *Opp* PUNCTUAL.

**unravel** *vb* disentangle, free, solve, sort out, straighten out, undo, untangle.

**unreal** *adj* chimerical, false, fanciful, illusory, imaginary, imagined, make-believe, nonexistent, phantasmal, *inf* pretend, *inf* pseudo, sham. ▷ HYPOTHETICAL. *Opp* REAL.

**unrealistic** *adj* **1** inaccurate, non-representational, unconvincing, unlifelike, unnatural, unrecognizable. **2** *unrealistic ideas.* delusory, fanciful, idealistic, impossible, impracticable, impractical, over-ambitious, quixotic, romantic, silly, visionary, unreasonable, unworkable. **3** *unrealistic prices.* ▷ EXCESSIVE. *Opp* REALISTIC.

**unreasonable** *adj* ▷ IRRATIONAL.

**unrecognizable** *adj* ▷ UNIDENTIFIABLE.

**unrelated** *adj* **1** different, independent, unconnected, unlike. **2** ▷ IRRELEVANT. *Opp* RELATED.

**unreliable** *adj* **1** deceptive, false, flimsy, implausible, inaccurate, misleading, suspect, unconvincing. **2** *unreliable friends.* changeable, disreputable, fallible, fickle, inconsistent, irresponsible, treacherous, undependable, unpredictable, unsound, unstable, untrustworthy. *Opp* RELIABLE.

**unrepentant** *adj* brazen, confirmed, conscienceless, hardened, impenitent, incorrigible, incurable, inveterate, irredeemable, shameless, unapologetic, unashamed, unblushing, unreformable, unregenerate. *Opp* REPENTANT.

**unripe** *adj* green, immature, sour, unready. *Opp* RIPE.

**unrivalled** *adj* ▷ UNEQUALLED.

**unruly** *adj* ▷ UNDISCIPLINED.

**unsafe** *adj* ▷ DANGEROUS.

**unsatisfactory** *adj* defective, deficient, disappointing, displeasing, dissatisfying, faulty, frustrating, imperfect, inadequate, incompetent, inefficient, inferior, insufficient, lacking, not good enough, poor, *inf* sad (*state of affairs*), unacceptable, unhappy, unsatisfying, *inf* wretched. *Opp* SATISFACTORY.

**unscrupulous** *adj* amoral, conscienceless, corrupt, *inf* crooked, cunning, dishonest, dishonourable, immoral, improper, self-interested, shameless, *inf* slippery, sly, unconscionable, unethical, untrustworthy. *Opp* SCRUPULOUS.

**unseemly** *adj* ▷ UNBECOMING.

**unseen** *adj* ▷ INVISIBLE.

**unselfish** *adj* altruistic, caring, charitable, considerate, disinterested, generous, humanitarian, kind, liberal, magnanimous, open-handed, philanthropic, public-spirited, self-effacing, selfless, self-sacrificing, thoughtful, ungrudging, unstinting. *Opp* SELFISH.

**unsightly** *adj* ▷ UGLY.

**unskilful** *adj* amateurish, bungled, clumsy, crude, incompetent, inept, inexpert, maladroit, *inf* rough and ready, shoddy, unprofessional. *Opp* SKILFUL.

**unskilled** *adj* inexperienced, unqualified, untrained. *Opp* SKILLED.

**unsociable** *adj* ▷ UNFRIENDLY.

**unsophisticated** *adj* artless, childlike, guileless, ingenuous, innocent, lowbrow, naïve, plain, provincial, simple, simple-minded, straightforward, unaffected, uncomplicated, unostentatious, unpretentious, unrefined, unworldly. *Opp* SOPHISTICATED.

**unsound** *adj* ▷ WEAK.

**unspeakable** *adj* dreadful, indescribable, inexpressible, nameless, unutterable.

**unspecified** *adj* ▷ UNIDENTIFIED.

**unstable** *adj* capricious, changeable, fickle, inconsistent, inconstant, mercurial, shifting, unpredictable, unsteady, variable, volatile. *Opp* STABLE.

**unsteady** *adj* **1** flimsy, frail, insecure, precarious, rickety, *inf* rocky, shaky, tottering, unbalanced, unsafe, unstable, wobbly. **2** changeable, erratic, inconstant, intermittent, irregular, variable. **3** *an unsteady light.* flickering, fluctuating, quavering, quivering, trembling, tremulous, wavering. *Opp* STEADY.

**unsuccessful** *adj* **1** abortive, failed, fruitless, futile, ill-fated, ineffective, ineffectual, loss-making, sterile, unavailing, unlucky, unproductive, unprofitable, unsatisfactory, useless, vain, worthless. **2** *unsuccessful contest-*

*ants*. beaten, defeated, foiled, hapless, losing, luckless, vanquished. *Opp* SUCCESSFUL.

**unsuitable** *adj* ill-chosen, ill-judged, ill-timed, inapposite, inappropriate, incongruous, inept, irrelevant, mistaken, unbefitting, unfitting, unhappy, unsatisfactory, unseasonable, unseemly, untimely. *Opp* SUITABLE.

**unsure** *adj* ▷ UNCERTAIN.

**unsurpassed** *adj* ▷ UNEQUALLED.

**unsuspecting** *adj* ▷ CREDULOUS.

**unsympathetic** *adj* apathetic, cool, cold, dispassionate, hard-hearted, heartless, impassive, indifferent, insensitive, neutral, pitiless, reserved, ruthless, stony, stony-hearted, unaffected, uncaring, uncharitable, unconcerned, unfeeling, uninterested, unkind, unmoved, unpitying, unresponsive. *Opp* SYMPATHETIC.

**unsystematic** *adj* anarchic, chaotic, confused, disorderly, disorganized, haphazard, illogical, jumbled, muddled, *inf* shambolic, *inf* sloppy, unmethodical, unplanned, unstructured, untidy. *Opp* SYSTEMATIC.

**unthinkable** *adj* ▷ INCONCEIVABLE.

**unthinking** *adj* ▷ THOUGHTLESS.

**untidy** *adj* **1** careless, chaotic, cluttered, confused, disorderly, disorganized, haphazard, *inf* higgledy-piggledy, in disarray, jumbled, littered, *inf* messy, muddled, *inf* shambolic, slapdash, *inf* sloppy, slovenly, *inf* topsy-turvy, unsystematic, upside-down. **2** *untidy hair*. bedraggled, blowzy, dishevelled, disordered, rumpled, scruffy, shabby, tangled, tousled, uncared-for, uncombed, ungroomed, unkempt. *Opp* TIDY.

**untie** *vb* cast off (*boat*), disentangle, free, loosen, release, unbind, undo, unfasten, unknot, untether.

**untried** *adj* experimental, innovatory, new, novel, unproved, untested. *Opp* ESTABLISHED.

**untroubled** *adj* carefree, peaceful, straightforward, undisturbed, uninterrupted, unruffled.

**untrue** *adj* ▷ FALSE.

**untrustworthy** *adj* ▷ DISHONEST.

**untruthful** *adj* ▷ LYING.

**unused** *adj* blank, clean, fresh, intact, mint (*condition*), new, pristine, unopened, untouched, unworn. *Opp* USED.

**unusual** *adj* abnormal, atypical, curious, *inf* different, exceptional, extraordinary, *inf* freakish, *inf* funny, irregular, odd, out of the ordinary, peculiar, queer, rare, remarkable, singular, strange, surprising, uncommon, unconventional, unexpected, unfamiliar, *inf* unheard-of, *inf* unique, unnatural, unorthodox, untypical, unwonted. *Opp* USUAL.

**unutterable** *adj* ▷ INDESCRIBABLE.

**unwanted** *adj* ▷ UNNECESSARY.

**unwarranted** *adj* ▷ UNJUSTIFIABLE.

**unwary** *adj* ▷ CARELESS.

**unwavering** *adj* ▷ RESOLUTE.

**unwelcome** *adj* disagreeable, unacceptable, undesirable, uninvited, unpopular, unwanted. *Opp* WELCOME.

**unwell** *adj* ▷ ILL.

**unwholesome** *adj* ▷ UNHEALTHY.

**unwieldy** *adj* awkward, bulky, clumsy, cumbersome, inconvenient, ungainly, unmanageable. *Opp* HANDY, PORTABLE.

**unwilling** *adj* averse, backward, disinclined, grudging, half-hearted, hesitant, ill-disposed, indisposed, lazy, loath, opposed, reluctant, resistant, slow, uncooperative, unenthusiastic, unhelpful. *Opp* WILLING.

**unwise** *adj inf* daft, foolhardy, foolish, ill-advised, ill-judged, illogical, imperceptive, impolitic, imprudent, inadvisable, indiscreet, inexperienced, injudicious, irrational, irresponsible, mistaken, obtuse, perverse, rash, reckless, senseless, short-sighted, silly, stupid, thoughtless, unintelligent, unreasonable. *Opp* WISE.

**unworldly** *adj* ▷ SPIRITUAL.

**unworthy** *adj* contemptible, despicable, discreditable, dishonourable, disreputable, ignoble, inappropriate, mediocre, second-rate, shameful, substandard, undeserving, unsuitable. *Opp* WORTHY.

**unwritten** *adj* oral, spoken, verbal, *inf* word-of-mouth. *Opp* WRITTEN.

**unyielding** *adj* ▷ INFLEXIBLE.

**upbringing** *n* breeding, bringing-up, care, education, instruction, nurture, raising, rearing, teaching, training.

**update** *vb* amend, bring up to date, correct, modernize, review, revise.

**upgrade** *vb* enhance, expand, improve, make better.

**upheaval** *n* chaos, commotion, confusion, disorder, disruption, disturbance, revolution, *inf* to-do, turmoil.

**uphill** *adj* arduous, difficult, exhausting, gruelling, hard, laborious, stiff, strenuous, taxing, tough.

**uphold** *vb* back, champion, defend, endorse, maintain, preserve, protect, stand by, support, sustain.

**upkeep** *n* care, conservation, keep, maintenance, operation, preservation, running, support.

**uplifting** *adj* civilizing, edifying, educational, enlightening, ennobling, enriching, humanizing, improving, spiritual. *Opp* SHAMEFUL.

**upper** *adj* elevated, higher, raised, superior, upstairs.

**uppermost** *adj* dominant, highest, loftiest, supreme, top, topmost.

**upright** *adj* **1** erect, on end, perpendicular, vertical. **2** *an upright judge.* conscientious, fair, good, high-minded, honest, honourable, incorruptible, just, moral, principled, righteous, *inf* straight, true, trustworthy, upstanding, virtuous. • *n* column, pole, post, vertical.

**uproar** *n inf* bedlam, brawling, chaos, clamour, commotion, confusion, din, disorder, disturbance, furore, *inf* hubbub, *inf* hullabaloo, *inf* a madhouse, noise, outburst, outcry, pandemonium, *inf* racket, riot, row, *inf* ructions, *inf* rumpus, tumult, turbulence, turmoil.

**uproot** *vb* deracinate, destroy, eliminate, eradicate, extirpate, get rid of, *inf* grub up, pull up, remove, root out, tear up, weed out.

**upset** *vb* **1** capsize, destabilize, overturn, spill, tip over, topple. **2** *upset a plan.* affect, alter, change, confuse, defeat, disorganize, disrupt, hinder, interfere with, interrupt, jeopardize, overthrow, spoil. **3** *upset feelings.* agitate, alarm, annoy, disconcert, dismay, distress, disturb, excite, fluster, frighten, grieve, irritate, offend, perturb, *inf* rub up the wrong way, ruffle, scare, unnerve, worry.

**upside-down** *adj* inverted, *inf* topsy-turvy, upturned, wrong way up.

**upstart** *n Fr* nouveau riche, social climber, *inf* yuppie.

**up-to-date** *adj* **1** advanced, current, latest, modern, new, present-day, recent. **2** contemporary, fashionable, *inf* in, modish, stylish, *inf* trendy. *Opp* OLD-FASHIONED.

**upward** *adj* ascending, going up, rising, uphill. *Opp* DOWNWARD.

**urban** *adj* built-up, densely populated, metropolitan, suburban. *Opp* RURAL.

**urge** *n* compulsion, craving, desire, drive, eagerness, hunger, impetus, impulse, inclination, instinct, *inf* itch, longing, pressure, thirst, wish, yearning, *inf* yen. • *vb* accelerate, advise, advocate, appeal to, beg, beseech, *inf* chivvy, compel, counsel, drive, *inf* egg on, encourage, entreat, exhort, force, goad, impel, implore, importune, incite, induce, invite, move on, nag, persuade, plead with, press, prod, prompt, propel, push, recommend, solicit, spur, stimulate. *Opp* DISCOURAGE.

**urgent** *adj* **1** acute, compelling, compulsive, dire, essential, exigent, high-priority, immediate, imperative, important, inescapable, instant, necessary, pressing, top-priority, unavoidable. **2** *an urgent cry for help.* eager, earnest, forceful, importunate, insistent, persistent, persuasive, solicitous.

**usable** *adj* acceptable, current, fit to use, functional, functioning, operating, operational, serviceable, valid, working.

**use** *n* advantage, application, benefit, employment, function, necessity, need, *inf* point, profit, purpose, usefulness, utility, value, worth. • *vb* **1** apply, administer, deal with, employ, exercise, exploit, handle, make use of, manage, operate, put to use, utilize, wield, work. **2** consume, drink, eat, exhaust, expend, spend, use up, waste, wear out.

**used** *adj* cast-off, *inf* hand-me-down, second-hand, soiled. *Opp* UNUSED.

**useful** *adj* **1** advantageous, beneficial, constructive, good, helpful, invaluable, positive, profitable, salutary, valuable, worthwhile. **2** *a useful tool.* convenient, effective, efficient, handy, powerful, practical, productive, utilitarian. **3** *a useful player.* capable, competent, effectual, proficient, skilful, successful, talented. *Opp* USELESS.

**useless** *adj* **1** fruitless, futile, hopeless, pointless, unavailing, unprofitable, unsuccessful, vain, worthless. **2** *inf* broken down, *inf* clapped out, dead, dud, impractical, ineffective, inefficient, unusable. **3** *a useless player.* incapable, incompetent, ineffectual, lazy, unhelpful, unskilful, unsuccessful, untalented. *Opp* USEFUL.

**usual** *adj* accepted, accustomed, average, common, conventional, customary, everyday, expected, familiar, general, habitual, natural, normal, official, ordinary, orthodox, predictable, prevalent, recognized, regular, routine, standard, stock, traditional, typical, unexceptional, unsurprising, well-known, widespread, wonted. *Opp* UNUSUAL.

**usurp** *vb* appropriate, assume, commandeer, seize, steal, take, take over.

**utensil** *n* appliance, device, gadget, implement, instrument, machine, tool.

**utter** *vb* articulate, *inf* come out with, express, pronounce, voice. ▷ SPEAK, TALK.

# V

**vacancy** *n* job, opening, place, position, post, situation.

**vacant** *adj* **1** available, bare, blank, clear, empty, free, hollow, open, unfilled, unused, usable, void. **2** abandoned, deserted, uninhabited, unoccupied, untenanted. **3** *a vacant look.* absent-minded, abstracted, blank, deadpan, dreamy, expressionless, far-away, fatuous, inattentive, vacuous. *Opp* BUSY.

**vacate** *vb* abandon, depart from, desert, evacuate, get out of, give up, leave, quit, withdraw from.

**vacuous** *adj* apathetic, blank, empty-headed, expressionless, inane, mindless, uncomprehending, unintelligent, vacant. ▷ STUPID. *Opp* ALERT.

**vacuum** *n* emptiness, space, void.

**vagary** *n* caprice, fancy, fluctuation, quirk, uncertainty, unpredictability, *inf* ups and downs, whim.

**vagrant** *n* beggar, destitute person, *inf* down-and-out, homeless person, itinerant, tramp, traveller, vagabond, wanderer, wayfarer.

**vague** *adj* **1** ambiguous, ambivalent, broad, confused, diffuse, equivocal, evasive, general, generalized, imprecise, indefinable, indefinite, inexact, loose, nebulous, uncertain, unclear, undefined, unspecific, unsure, *inf* woolly. **2** amorphous, blurred, dim, hazy, ill-defined, indistinct, misty, shadowy, unrecognizable. **3** absent-minded, careless, disorganized, forgetful, inattentive, scatter-brained, thoughtless. *Opp* DEFINITE.

**vain** *adj* **1** arrogant, *inf* bigheaded, boastful, *inf* cocky, conceited, egotistical, haughty, narcissistic, proud, self-important, self-satisfied, *inf* stuck-up, vainglorious. *Opp* MODEST. **2** *a vain attempt.* abortive, fruitless, futile, ineffective, pointless, senseless, unavailing, unproductive, unrewarding, unsuccessful, useless, worthless. *Opp* SUCCESSFUL.

**valiant** *adj* bold, brave, courageous, doughty, gallant, heroic, plucky, stalwart, stout-hearted, valorous. *Opp* COWARDLY.

**valid** *adj* acceptable, allowed, approved, authentic, authorized, *Lat* bona fide, convincing, current, genuine, lawful, legal, legitimate, official, permissible, permitted, proper, ratified, rightful, sound, suitable, usable. *Opp* INVALID.

**validate** *vb* authenticate, authorize, certify, endorse, legalize, legitimize, make valid, ratify.

**valley** *n* canyon, chasm, coomb, dale, defile, dell, dingle, glen, gorge, gulch, gully, hollow, pass, ravine, vale.

**valour** *n* bravery, courage, pluck.

**valuable** *adj* **1** costly, dear, expensive, generous, irreplaceable, precious, priceless, prized, treasured, valued. **2** *valuable advice.* advantageous, beneficial, constructive, esteemed, helpful, invaluable, positive, profitable, useful, worthwhile. *Opp* WORTHLESS.

**value** *n* **1** cost, price, worth. **2** advantage, benefit, importance, merit, significance, use, usefulness. • *vb* **1** assess, estimate the value of, evaluate, price, *inf* put a figure on. **2** appreciate, care for, cherish, esteem, *inf* have a high regard for, *inf* hold dear, love, prize, respect, treasure.

**vandal** *n* barbarian, delinquent, hooligan, looter, marauder, Philistine, raider, ruffian, savage, thug, trouble-maker.

**vanish** *vb* clear, clear off, disappear, disperse, dissolve, dwindle, evaporate, fade, go away, melt away, pass. *Opp* APPEAR.

**vanity** *n* arrogance, *inf* bigheadedness, conceit, egotism, narcissism, pride, self-esteem.

**vaporize** *vb* dry up, evaporate, turn to vapour.

**vapour** *n* exhalation, fog, fumes, gas, haze, miasma, mist, smoke, steam.

**variable** *adj* capricious, changeable, erratic, fickle, fitful, fluctuating, fluid, inconsistent, inconstant, mercurial, mutable, protean, shifting, temperamental, uncertain, unpredictable, unreliable, unstable, unsteady, *inf* up-and-down, vacillating, varying, volatile, wavering. *Opp* INVARIABLE.

**variation** *n* alteration, change, conversion, deviation, difference, discrepancy, diversification, elaboration, modification, permutation, variant.

**variety** *n* **1** alteration, change, difference, diversity, unpredictability, variation. **2** array, assortment, blend, collection, combination, jumble, medley, miscellany, mixture, multiplicity. **3** brand, breed, category, class, form, kind, make, sort, species, strain, type.

**various** *adj* assorted, contrasting, different, differing, dissimilar, diverse, heterogeneous, miscellaneous, mixed, *inf* motley, multifarious, several, sundry, varied, varying. *Opp* SIMILAR.

**vary** *vb* **1** change, deviate, differ, fluctuate, go up and down, vacillate. **2** *vary your speed.* adapt, adjust, alter, convert, modify, reset, switch, transform, upset. *Opp* STABILIZE. **varied, varying** ▷ VARIOUS.

**vast** *adj* boundless, broad, colossal, enormous, extensive, gigantic, great, huge, immeasurable, immense, infinite, interminable, large, limitless, mammoth, massive, measureless, monumental, never-ending, titanic, tremendous, unbounded, unlimited, voluminous, wide. ▷ BIG. *Opp* SMALL.

**vault** *n* basement, cavern, cellar, crypt, repository, strongroom, undercroft. • *vb* bound over, clear, hurdle, jump, leap, leapfrog, spring over.

**veer** *vb* change direction, dodge, swerve, tack, turn, wheel.

**vegetable** *adj* growing, organic. • *n* □ *artichoke, asparagus, aubergine, bean, beet, beetroot, broad bean, broccoli, Brussels sprout, butter bean, cabbage, carrot, cauliflower, celeriac, celery, chicory, courgette, cress, cucumber, garlic, kale, kohlrabi, leek, lettuce, marrow, mushroom, onion, parsnip, pea, pepper, potato, pumpkin, radish, runner bean, salsify, shallot, spinach, sugar beet, swede, sweetcorn, tomato, turnip, watercress, zucchini.*

**vegetate** *vb* be inactive, do nothing, *inf* go to seed, idle, lose interest, stagnate.

**vegetation** *n* foliage, greenery, growing things, growth, plants, undergrowth, weeds.

**vehement** *adj* animated, ardent, eager, enthusiastic, excited, fervent, fierce, forceful, heated, impassioned, intense, passionate, powerful, strong, urgent, vigorous, violent. *Opp* APATHETIC.

**vehicle** *n* conveyance.
□ *ambulance,* inf *buggy, bulldozer, bus, cab, camper, caravan, carriage, cart,* old use *charabanc, chariot, coach, dump truck, dustcart, estate car, fire-engine, float, gig, go-kart, hearse, horse-box, jeep, juggernaut, lorry, minibus, minicab,* inf *motor, motor car,* old use *omnibus, pantechnicon, patrol-car, pick-up, removal van, rickshaw, scooter, sedan-chair, side-car, sledge, snow-plough, stage-coach, steam-roller, tank, tanker, taxi, traction-engine, tractor, trailer, tram, transporter, trap, trolley-bus, truck, tumbrel, van, wagon,* sl *wheels.* ▷ CAR, CYCLE, TRAIN, VESSEL.

**veil** *vb* camouflage, cloak, conceal, cover, disguise, hide, mask, shroud.

**vein** *n* **1** artery, blood vessel, capillary. **2** *mineral vein.* bed, course, deposit, line, lode, seam, stratum. **3** ▷ MOOD.

**veneer** *n* coating, covering, finish, gloss, layer, surface. • *vb* ▷ COVER.

**venerable** *adj* aged, ancient, august, dignified, esteemed, estimable, honourable, honoured, old, respectable, respected, revered, reverenced, sedate, venerated, worshipped, worthy of respect.

**venerate** *vb* adore, esteem, hero-worship, honour, idolize, look up to, pay homage to, respect, revere, reverence, worship.

**vengeance** *n* reprisal, retaliation, retribution, revenge, *inf* tit for tat.

**vengeful** *adj* avenging, bitter, rancorous, revengeful, spiteful, unforgiving, vindictive. *Opp* FORGIVING.

**venom** *n* poison, toxin.

**venomous** *adj* deadly, lethal, poisonous, toxic.

**vent** *n* aperture, cut, duct, gap, hole, opening, orifice, outlet, passage, slit, slot, split. • *vb* articulate, express, give vent to, let go, make known, release, utter, ventilate, voice. ▷ SPEAK.

**ventilate** *vb* aerate, air, freshen, oxygenate.

**venture** *n* enterprise, experiment, gamble, risk, speculation, undertaking. • *vb* **1** bet, chance, dare, gamble, put forward, risk, speculate, stake, wager. **2** *venture out.* dare to go, risk going.

**venturesome** *adj* adventurous, bold, courageous, daring, doughty, fearless, intrepid.

**venue** *n* meeting-place, location, rendezvous.

**verbal** *adj* **1** lexical, linguistic. **2** *a verbal message.* oral, said, spoken, unwritten, vocal, word-of-mouth. *Opp* WRITTEN.

**verbatim** *adj* exact, faithful, literal, precise, word for word.

**verbose** *adj* diffuse, garrulous, long-winded, loquacious, prolix, rambling, repetitious, talkative, voluble. ▷ WORDY. *Opp* CONCISE.

**verbosity** *n inf* beating about the bush, circumlocution, diffuseness, garrulity, long-windedness, loquacity, periphrasis, prolixity, repetition, verbiage, wordiness.

**verdict** *n* adjudication, assessment, conclusion, decision, finding, judgement, opinion, sentence.

**verge** *n* bank, boundary, brim, brink, edge, hard shoulder, kerb, lip, margin, roadside, shoulder, side, threshold, wayside.

**verifiable** *adj* demonstrable, provable.

**verify** *vb* affirm, ascertain, attest to, authenticate, bear witness to, check out, confirm, corroborate, demonstrate the truth of, establish, prove, show the truth of, substantiate, support, uphold, validate, vouch for.

**verisimilitude** *n* authenticity, realism.

**vermin** *plur n* parasites, pests.

**vernacular** *adj* common, everyday, indigenous, local, native, ordinary, popular, vulgar.

**versatile** *adj* adaptable, all-purpose, all-round, flexible, gifted, multi-purpose, resourceful, skilful, talented.

**verse** *n* lines, metre, rhyme, stanza. □ *blank verse, Chaucerian stanza, clerihew, couplet, free verse, haiku, hexameter, limerick, ottava rima, pentameter, quatrain, rhyme royal, sestina, sonnet, Spenserian stanza, terza rima, triolet, triplet, vers libre, villanelle.* ▷ POEM.

**versed** *adj* accomplished, competent, experienced, expert, knowledgeable, practised, proficient, skilled, taught, trained.

**version** *n* **1** account, description, portrayal, reading, rendition, report, story. **2** adaptation, interpretation, paraphrase, rendering, translation. **3** design, form, kind, mark, model, style, type, variant.

**vertical** *adj* erect, perpendicular, precipitous, sheer, upright. *Opp* HORIZONTAL.

**vertigo** *n* dizziness, giddiness, light-headedness.

**very** *adv* acutely, enormously, especially, exceedingly, extremely, greatly, highly, *inf* jolly, most, noticeably, outstandingly, particularly, really, remarkably, *inf* terribly, truly, uncommonly, unusually.

**vessel** *n* **1** ▷ CONTAINER. **2** bark, boat, craft, ship. □ *aircraftcarrier, barge, bathysphere, battleship, brigantine, cabin cruiser, canoe, catamaran, clipper, coaster, collier, coracle, corvette, cruise-liner, cruiser, cutter, destroyer, dhow, dinghy, dredger, dugout, ferry, freighter, frigate, galleon, galley, gondola, gunboat, houseboat, hovercraft, hydrofoil, hydroplane, icebreaker, junk, kayak, ketch, landing-craft, launch, lifeboat, lighter, lightship, liner, longboat, lugger, man-of-war, merchant ship, minesweeper, motor boat, narrow-boat, oil-tanker, packet-ship, paddle-steamer, pontoon, power-boat, pram, privateer, punt, quinquereme, raft, rowing-boat, sailing-boat, sampan, schooner, skiff, sloop, smack, speed-boat, steamer, steamship, submarine, tanker, tender, torpedo boat, tramp steamer, trawler, trireme, troopship, tug, warship, whaler, wind-jammer, yacht, yawl.*

**vet** *vb inf* check out, examine, investigate, review, scrutinize.

**veteran** *adj* experienced, mature, old, practised. • *n* experienced soldier, ex-serviceman, ex-servicewoman, old hand, old soldier, survivor.

**veto** *n* ban, block, embargo, prohibition, proscription, refusal, rejection, *inf* thumbs down. • *vb* ban, bar, blackball, block, disallow, dismiss, forbid, prohibit, proscribe, quash, refuse, reject, rule out, say no to, turn down, vote against. *Opp* APPROVE.

**vex** *vb inf* aggravate, annoy, bother, displease, exasperate, harass, irritate, provoke, *inf* put out, trouble, upset, worry. ▷ ANGER.

**viable** *adj* achievable, feasible, operable, possible, practicable, practical, realistic, reasonable, supportable, sustainable, usable, workable. *Opp* IMPRACTICAL.

**vibrant** *adj* alert, alive, dynamic, electric, energetic, lively, living, pulsating, quivering, resonant, thrilling, throbbing, trembling, vibrating, vivacious. *Opp* LIFELESS.

**vibrate** *vb* fluctuate, judder, oscillate, pulsate, quake, quiver, rattle, resonate, reverberate, shake, shiver, shudder, throb, tremble, wobble.

**vibration** *n* juddering, oscillation, pulsation, quivering, rattling, resonance, reverberation, shaking, shivering, shuddering, throbbing, trembling, tremor.

**vicarious** *adj* delegated, deputed, indirect, second-hand, surrogate.

**vice** *n* **1** badness, corruption, degeneracy, degradation, depravity, evil, evil-doing, immorality, iniquity, lechery, profligacy, promiscuity, sin, venality, villainy, wickedness, wrongdoing. **2** bad habit, blemish, defect, failing, fault, flaw, foible, imperfection, shortcoming, weakness.

**vicinity** *n* area, district, environs, locale, locality, neighbourhood, outskirts, precincts, proximity, purlieus, region, sector, territory, zone.

**vicious** *adj* **1** atrocious, barbaric, barbarous, beastly, bloodthirsty, brutal, callous, cruel, diabolical, fiendish, heinous, hurtful, inhuman, merciless, monstrous, murderous, pitiless, ruthless, sadistic, savage, unfeeling, vile, violent. **2** *a vicious character*. bad, *inf* bitchy, *inf* catty, depraved, evil, heartless, immoral, malicious, mean, perverted, rancorous, sinful, spiteful, venomous, villainous, vindictive, vitriolic, wicked. **3** *vicious animals*. aggressive, bad-tempered, dangerous, ferocious, fierce, snappy, untamed, wild. **4** *a vicious wind*. cutting, nasty, severe, sharp, unpleasant. *Opp* GENTLE.

**vicissitude** *n* alteration, change, flux, instability, mutability, mutation, shift, uncertainty, unpredictability, variability.

**victim** *n* **1** casualty, fatality, injured person, patient, sufferer, wounded person. **2** *sacrificial victim*. martyr, offering, prey, sacrifice.

**victimize** *vb* bully, cheat, discriminate against, exploit, intimidate, oppress, persecute, *inf* pick on, prey on, take advantage of, terrorize, torment, treat unfairly, *inf* use. ▷ CHEAT.

**victor** *n* champion, conqueror, prizewinner, winner. *Opp* LOSER.

**victorious** *adj* champion, conquering, first, leading, prevailing, successful, top, top-scoring, triumphant, unbeaten, undefeated, winning. *Opp* UNSUCCESSFUL.

**victory** *n* achievement, conquest, knockout, mastery, success, superiority, supremacy, triumph, *inf* walk-over, win. *Opp* DEFEAT.

**vie** *vb* compete, contend, strive, struggle.

**view** *n* **1** aspect, landscape, outlook, panorama, perspective, picture, prospect, scene, scenery, seascape, spectacle, townscape, vista.

2 angle, look, perspective, sight, vision. 3 *political views.* attitude, belief, conviction, idea, judgement, notion, opinion, perception, position, thought. • *vb* 1 behold, consider, contemplate, examine, eye, gaze at, inspect, observe, perceive, regard, scan, stare at, survey, witness. 2 *view TV.* look at, see, watch.

**viewer** *n plur* audience, observer, onlooker, spectator, watcher, witness.

**viewpoint** *n* angle, perspective, point of view, position, slant, standpoint.

**vigilant** *adj* alert, attentive, awake, careful, circumspect, eagle-eyed, observant, on the watch, on your guard, *inf* on your toes, sharp, wakeful, wary, watchful, wideawake. *Opp* NEGLIGENT.

**vigorous** *adj* active, alive, animated, brisk, dynamic, energetic, fit, flourishing, forceful, full-blooded, *inf* full of beans, growing, hale and hearty, healthy, lively, lusty, potent, prosperous, red-blooded, robust, spirited, strenuous, strong, thriving, virile, vital, vivacious, zestful. *Opp* FEEBLE.

**vigour** *n* animation, dynamism, energy, fitness, force, forcefulness, gusto, health, life, liveliness, might, potency, power, robustness, spirit, stamina, strength, verve, *inf* vim, virility, vitality, vivacity, zeal, zest.

**vile** *adj* bad, base, contemptible, degenerate, depraved, despicable, disgusting, evil, execrable, filthy, foul, hateful, horrible, immoral, loathsome, low, nasty, nauseating, obnoxious, odious, offensive, perverted, repellent, repugnant, repulsive, revolting, sickening, sinful, ugly, vicious, wicked.

**vilify** *vb* abuse, calumniate, defame, denigrate, deprecate, disparage, revile, *inf* run down, slander, *inf* smear, speak evil of, traduce, vituperate.

**villain** *n* blackguard, criminal, evil-doer, malefactor, mischief-maker, miscreant, reprobate, rogue, scoundrel, sinner, wretch. ▷ CRIMINAL.

**villainous** *adj* bad, corrupt, criminal, dishonest, evil, sinful, treacherous, vile. ▷ WICKED.

**vindictive** *adj* avenging, malicious, nasty, punitive, rancorous, revengeful, spiteful, unforgiving, vengeful, vicious. *Opp* FORGIVING.

**vintage** *adj* choice, classic, fine, good, high-quality, mature, mellowed, old, seasoned, venerable.

**violate** *vb* 1 breach, break, contravene, defy, disobey, disregard, flout, ignore, infringe, overstep, sin against, transgress. 2 *violate someone's privacy.* abuse, desecrate, disturb, invade, profane. 3 [*of men*] *violate a woman.* assault, attack, debauch, dishonour, force yourself on, rape, ravish.

**violation** *n* breach, contravention, defiance, flouting, infringement, invasion, offence (against), transgression.

**violent** *adj* 1 acute, damaging, dangerous, destructive, devastating, explosive, ferocious, fierce, forceful, furious, hard, harmful, intense, powerful, rough, ruinous, savage, severe, strong, swingeing, tempestuous, turbulent, uncontrollable, vehement, wild. 2 *violent behaviour.* barbaric, berserk, bloodthirsty, brutal, cruel, desperate, frenzied, headstrong, homicidal, murderous, riotous, rowdy, ruthless, uncontrolled, unruly, vehement, vicious, wild. *Opp* GENTLE.

**VIP** *n* celebrity, dignitary, important person.

**virile** *adj derog* macho, manly, masculine, potent, vigorous.

**virtue** *n* **1** decency, fairness, goodness, high-mindedness, honesty, honour, integrity, justice, morality, nobility, principle, rectitude, respectability, righteousness, right-mindedness, sincerity, uprightness, worthiness. **2** advantage, asset, good point, merit, quality, *inf* redeeming feature, strength. **3** *sexual virtue.* abstinence, chastity, honour, innocence, purity, virginity. *Opp* VICE.

**virtuoso** *n* expert, genius, maestro, prodigy, showman, *inf* wizard. ▷ MUSICIAN.

**virtuous** *adj* blameless, chaste, decent, ethical, exemplary, fair, God-fearing, good, *derog* goody-goody, high-minded, high-principled, honest, honourable, innocent, irreproachable, just, law-abiding, moral, noble, principled, praiseworthy, pure, respectable, right, righteous, right-mindedness, sincere, *derog* smug, spotless, trustworthy, uncorrupted, unimpeachable, unsullied, upright, virginal, worthy. *Opp* WICKED.

**virulent** *adj* **1** dangerous, deadly, lethal, life-threatening, noxious, pernicious, poisonous, toxic, venomous. **2** *virulent abuse.* acrimonious, bitter, hostile, malicious, malign, malignant, mordant, nasty, spiteful, splenetic, vicious, vitriolic.

**viscous** *adj* gluey, sticky, syrupy, thick, viscid. *Opp* RUNNY.

**visible** *adj* apparent, clear, conspicuous, detectable, discernible, distinct, evident, manifest, noticeable, observable, obvious, open, perceivable, perceptible, plain, recognizable, unconcealed, undisguised, unmistakable. *Opp* INVISIBLE.

**vision** *n* **1** eyesight, perception, sight. **2** apparition, chimera, daydream, delusion, fantasy, ghost, hallucination, illusion, mirage, phantasm, phantom, spectre, spirit, wraith. **3** *a man of vision.* far-sightedness, foresight, imagination, insight, spirituality, understanding.

**visionary** *adj* dreamy, fanciful, far-sighted, futuristic, idealistic, imaginative, impractical, mystical, prophetic, quixotic, romantic, speculative, transcendental, unrealistic, Utopian. • *n* dreamer, idealist, mystic, poet, prophet, romantic, seer.

**visit** *n* **1** call, *old use* sojourn, stay, stop, visitation. **2** day out, excursion, outing, trip. • *vb* call on, come to see, *inf* descend on, *inf* drop in on, go to see, *inf* look up, make a visit to, pay a call on, *inf* pop in on, stay with. **visit regularly** ▷ HAUNT.

**visitor** *n* **1** caller, *plur* company, guest. **2** holiday-maker, sightseer, tourist, traveller, tripper. **3** *a visitor from abroad.* alien, foreigner, migrant, visitant.

**visor** *n* protector, shield, sunshield.

**vista** *n* landscape, outlook, panorama, prospect, scene, scenery, seascape, view.

**visualize** *vb* conceive, dream up, envisage, imagine, picture.

**vital** *adj* **1** alive, animate, animated, dynamic, energetic, exuberant, life-giving, live, lively, living, sparkling, spirited, sprightly, vigorous, vivacious, zestful. *Opp* LIFELESS. **2** *vital information.* compulsory, current, crucial, essential, fundamental, imperative, important, indispensable, mandatory, necessary, needed, relevant, requisite. *Opp* INESSENTIAL.

**vitality** *n* animation, dynamism, energy, exuberance, *inf* go, life, liveliness, *inf* sparkle, spirit, sprightliness, stamina, strength, vigour, *inf* vim, vivacity, zest.

**vitriolic** *adj* abusive, acid, biting, bitter, caustic, cruel, destructive, hostile, hurtful, malicious, savage, scathing, vicious, vindictive, virulent.

**vituperate** *vb* abuse, berate, calumniate, censure, defame, denigrate, deprecate, disparage, reproach, revile, *inf* run down, slander, upbraid, vilify.

**vivacious** *adj* animated, bubbly, cheerful, ebullient, energetic, high-spirited, light-hearted, lively, merry, positive, spirited, sprightly. *Opp* LETHARGIC.

**vivid** *adj* **1** bright, brilliant, colourful, dazzling, fresh, *derog* gaudy, gay, gleaming, glowing, intense, rich, shining, showy, strong, vibrant. **2** *a vivid description.* clear, detailed, graphic, imaginative, lifelike, lively, memorable, powerful, realistic, striking. *Opp* LIFELESS.

**vocabulary** *n* **1** diction, lexis, words. **2** dictionary, glossary, lexicon, phrase book, word-list.

**vocal** *adj* **1** oral, said, spoken, sung, voiced. **2** *vocal in discussion.* communicative, forthcoming, loquacious, outspoken, talkative, vociferous. *Opp* TACITURN.

**vocation** *n* calling, career, employment, job, life's work, occupation, profession, trade.

**vogue** *n* craze, fad, fashion, *inf* latest thing, mode, rage, style, taste, trend. **in vogue** ▷ FASHIONABLE.

**voice** *n* accent, articulation, expression, idiolect, inflexion, intonation, singing, sound, speaking, speech, tone, utterance. • *vb* ▷ SPEAK.

**void** *adj* **1** blank, empty, unoccupied, vacant. **2** *a void contract.* annulled, cancelled, inoperative, invalid, not binding, unenforceable, useless. • *n* blank, emptiness, nothingness, space, vacancy, vacuum.

**volatile** *adj* **1** explosive, sensitive, unstable. **2** *volatile moods.* changeable, erratic, fickle, flighty, inconstant, lively, mercurial, temperamental, unpredictable, *inf* up and down, variable. *Opp* STABLE.

**volley** *n* barrage, bombardment, burst, cannonade, fusillade, salvo, shower.

**voluble** *adj* chatty, fluent, garrulous, glib, loquacious, talkative. ▷ WORDY.

**volume** **1** *old use* tome. ▷ BOOK. **2** *volume of a container.* aggregate, amount, bulk, capacity, dimensions, mass, measure, quantity, size.

**voluminous** *adj* ample, billowing, bulky, capacious, cavernous, enormous, extensive, gigantic, great, huge, immense, large, mammoth, massive, roomy, spacious, vast. ▷ BIG. *Opp* SMALL.

**voluntary** *adj* **1** elective, free, gratuitous, optional, spontaneous, unpaid, willing. *Opp* COMPULSORY. **2** *a voluntary act.* conscious, deliberate, intended, intentional, planned, premeditated, wilful. *Opp* INVOLUNTARY.

**volunteer** *vb* **1** be willing, offer, propose, put yourself forward. **2** ▷ ENLIST.

**voluptuous** *adj* 1 hedonistic, luxurious, pleasure-loving, self-indulgent, sensual, sybaritic, 2 *voluptuous figure.* attractive, buxom, *inf* curvaceous, desirable, erotic, sensual, *inf* sexy, shapely, *inf* well-endowed.

**vomit** *vb* be sick, *inf* bring up, disgorge, *inf* heave up, *inf* puke, regurgitate, retch, *inf* spew up, *inf* throw up.

**voracious** *adj* avid, eager, fervid, gluttonous, greedy, hungry, insatiable, keen, ravenous, thirsty.

**vortex** *n* eddy, spiral, whirlpool, whirlwind.

**vote** *n* ballot, election, plebiscite, poll, referendum, show of hands. • *vb* ballot, cast your vote. **vote for** choose, elect, nominate, opt for, pick, return, select, settle on.

**vouch** *vb* **vouch for** answer for, back, certify, endorse, guarantee, speak for, sponsor, support.

**voucher** *n* coupon, ticket, token.

**vow** *n* assurance, guarantee, oath, pledge, promise, undertaking, word of honour. • *vb* declare, give an assurance, give your word, guarantee, pledge, promise, swear, take an oath.

**voyage** *n* cruise, journey, passage. • *vb* circumnavigate, cruise, sail. ▷ TRAVEL.

**vulgar** *adj* 1 churlish, coarse, common, crude, foul, gross, ill-bred, impolite, improper, indecent, indecorous, low, offensive, rude, uncouth, ungentlemanly, unladylike. ▷ OBSCENE. *Opp* POLITE. 2 *vulgar colour scheme.* crude, gaudy, inartistic, in bad taste, inelegant, insensitive, lowbrow, plebeian, tasteless, tawdry, unrefined, unsophisticated. *Opp* TASTEFUL.

**vulnerable** *adj* 1 at risk, defenceless, exposed, helpless, unguarded, unprotected, weak, wide open. 2 easily hurt, sensitive, thin-skinned, touchy. *Opp* RESILIENT.

# W

**wad** *n* bundle, lump, mass, pack, pad, plug, roll.

**wadding** *n* filling, lining, packing, padding, stuffing.

**wade** *vb* ford, paddle, splash. ▷ WALK.

**waffle** *n* evasiveness, padding, prevarication, prolixity, verbiage, wordiness. • *vb inf* beat about the bush, *inf* blather on, hedge, prattle, prevaricate.

**waft** *vb* 1 be borne, drift, float, travel. 2 bear, carry, convey, puff, transmit, transport.

**wag** *vb* bob, flap, move to and fro, nod, oscillate, rock, shake, sway, undulate, *inf* waggle, wave, *inf* wiggle.

**wage** *n* compensation, earnings, emolument, honorarium, income, pay, pay packet, recompense, remuneration, reward, salary, stipend. • *vb* carry on, conduct, engage in, fight, prosecute, pursue, undertake.

**wager** *vb* bet, gamble.

**wail** *vb* caterwaul, complain, cry, howl, lament, moan, shriek, waul, weep, *inf* yowl.

**waist** *n* middle, waistline.

**waistband** *n* belt, cummerbund, girdle.

**wait** *n* delay, halt, hesitation, hiatus, *inf* hold-up, intermission, interval, pause, postponement, rest, stay, stop, stoppage. • *vb* **1** *old use* bide, delay, halt, *inf* hang about, *inf* hang on, hesitate, hold back, keep still, linger, mark time, pause, remain, rest, *inf* sit tight, stand by, stay, stop, *old use* tarry. **2** *wait at table*. serve.

**waive** *vb* abandon, cede, disclaim, dispense with, forgo, give up, relinquish, remit, renounce, resign, sign away, surrender.

**wake** *n* **1** funeral, vigil, watch. **2** *wake of a ship*. path, track, trail, turbulence, wash. • *vb* **1** arouse, awaken, bring to life, call, disturb, galvanize, rouse, stimulate, stir, waken. **2** become conscious, bestir yourself, *inf* come to life, get up, rise, *inf* stir, wake up. **wake up to** ▷ REALIZE. **waking** ▷ CONSCIOUS.

**wakeful** *adj* alert, awake, insomniac, *inf* on the qui vive, restless, sleepless.

**walk** *n* **1** bearing, carriage, gait, stride. **2** constitutional, hike, *old use* promenade, ramble, saunter, stroll, traipse, tramp, trek, trudge, *inf* turn. **3** *a paved walk*. aisle, alley, path, pathway, pavement. • *vb* **1** be a pedestrian, travel on foot. □ *amble, crawl, creep, dodder,* inf *foot-slog, hike, hobble, limp, lope, lurch, march, mince,* sl *mooch, pace, pad, paddle, parade,* old use *perambulate, plod, promenade, prowl, ramble, saunter, scuttle, shamble, shuffle, slink, stagger, stalk, steal, step,* inf *stomp, stride, stroll, strut, stumble, swagger, tiptoe,* inf *toddle, totter, tramp, trample, traipse, tramp, trek, troop, trot, trudge, waddle, wade.* **2** *don't walk on the flowers*. stamp, step, trample, tread. **walk away with** ▷ WIN. **walk off with** ▷ STEAL. **walk out** ▷ QUIT. **walk out on** ▷ DESERT.

**walker** *n* hiker, pedestrian, rambler.

**wall** *n* □ *barricade, barrier, bulkhead, bulwark, dam, dike, divider, embankment, fence, fortification, hedge, obstacle, paling, palisade, parapet, partition, rampart, screen, sea-wall, stockade.* **wall in** ▷ ENCLOSE.

**wallet** *n* notecase, pocketbook, pouch, purse.

**wallow** *vb* **1** flounder, lie, pitch about, roll about, stagger about, tumble, wade, welter. **2** *wallow in luxury*. glory, indulge yourself, luxuriate, revel, take delight.

**wan** *adj* anaemic, ashen, bloodless, colourless, exhausted, faint, feeble, livid, pale, pallid, pasty, sickly, tired, waxen, worn.

**wand** *n* baton, rod, staff, stick.

**wander** *vb* **1** drift, go aimlessly, meander, prowl, ramble, range, roam, rove, saunter, stray, stroll, travel about, walk, wind. **2** *wander off course*. curve, deviate, digress, drift, err, go off at a tangent, stray, swerve, turn, twist, veer, zigzag. **wandering** ▷ INATTENTIVE, NOMADIC.

**wane** *vb* decline, decrease, dim, diminish, dwindle, ebb, fade, fail, *inf* fall off, grow less, lessen, peter out, shrink, subside, taper off, weaken. *Opp* STRENGTHEN.

**want** *n* **1** demand, desire, need, requirement, wish. **2** *a want of ready cash*. absence, lack, need. **3** *war against want*. dearth, famine, hunger, insufficiency, penury, poverty, privation, scarcity, shortage. • *vb* **1** aspire to, covet, crave, demand, desire, fancy, hanker after, *inf* have a yen for, hunger for, *inf* itch for, like, long for, miss, pine for, please, prefer, *inf* set your heart on, thirst after, thirst for, wish for, yearn for. **2** *want manners*. be short of, lack, need, require.

**war** *n* campaign, conflict, crusade, fighting, hostilities, military action, strife, warfare. □ *ambush, assault, attack, battle, blitz, blockade, bombardment, counter-attack, engagement, guerrilla warfare, invasion, manoeuvres, operations, resistance, siege, skirmish.* **wage war** ▷ FIGHT.

**ward** *n* charge, dependant, minor. • *vb* **ward off** avert, beat off, block, chase away, check, deflect, fend off, forestall, parry, push away, repel, repulse, stave off, thwart, turn aside.

**warder** *n* gaoler, guard, jailer, keeper, prison officer.

**warehouse** *n* depository, depot, store, storehouse.

**wares** *plur n* commodities, goods, manufactures, merchandise, produce, stock, supplies.

**warlike** *adj* aggressive, bellicose, belligerent, hawkish, hostile, militant, militaristic, pugnacious, warmongering, warring.

**warm** *adj* **1** close, hot, lukewarm, subtropical, sultry, summery, temperate, tepid, warmish. **2** *warm clothes*. cosy, thermal, thick, winter, woolly. **3** *a warm welcome*. affable, affectionate, ardent, cordial, emotional, enthusiastic, excited, fervent, friendly, genial, impassioned, kind, loving, passionate, sympathetic, warm-hearted. *Opp* COLD, UNFRIENDLY. • *vb* heat, make warmer, melt, raise the temperature of, thaw, thaw out. *Opp* COOL.

**warn** *vb* admonish, advise, alert, caution, counsel, forewarn, give a warning, give notice, inform, notify, raise the alarm, remind, *inf* tip off.

**warning** *n* **1** advance notice, augury, forewarning, hint, indication, notice, notification, omen, portent, premonition, presage, prophecy, reminder, sign, signal, threat, *inf* tip-off, *inf* word to the wise. □ *alarm, alarm-bell, beacon, bell, fire-alarm, flashing light, fog-horn, gong, hooter, red light, siren, traffic-lights, whistle.* **2** *let off with a warning*. admonition, advice, caveat, caution, reprimand.

**warp** *vb* become deformed, bend, buckle, contort, curl, curve, deform, distort, kink, twist.

**warrant** *n* authority, authorization, certification, document, entitlement, guarantee, licence, permit, pledge, sanction, search-warrant, warranty, voucher. • *vb* ▷ JUSTIFY.

**wary** *adj* alert, apprehensive, attentive, *inf* cagey, careful, chary, cautious, circumspect, distrustful, heedful, observant, on the lookout, on your guard, suspicious, vigilant, watchful. *Opp* RECKLESS.

**wash** *n old use* ablutions, bath, rinse, shampoo, shower. • *vb* **1** clean, cleanse, flush, launder, mop, rinse, scrub, shampoo, sluice, soap down, sponge down, swab down, swill, wipe. **2** bath, bathe, *old use* make your toilet, perform your ablutions, shower. **3** *The sea washes against the cliff*. break, dash, flow, lap, pound, roll, splash. **wash your hands of** ▷ ABANDON.

**washing** *n* cleaning, dirty clothes, laundry, *inf* the wash.

**washout** *n* débacle, disappointment, disaster, failure, *inf* flop.

**waste** *adj* **1** discarded, extra, superfluous, unprofitable, unusable, unused, unwanted, worthless. **2** *waste land.* bare, barren, derelict, empty, overgrown, rundown, uncared-for, uncultivated, undeveloped, unproductive, wild. • *n* **1** debris, dregs, effluent, excess, garbage, junk, leavings, *inf* left-overs, litter, offcuts, refuse, remnants, rubbish, scrap, scraps, trash, unusable material, unwanted material, wastage. **2** extravagance, indulgence, overprovision, prodigality, profligacy, self-indulgence. • *vb* be wasteful with, dissipate, fritter, misspend, misuse, over provide, *sl* splurge, squander, use up, use wastefully. *Opp* CONSERVE. **waste away** become emaciated, become thin, become weaker, mope, pine, weaken.

**wasteful** *adj* excessive, expensive, extravagant, improvident, imprudent, lavish, needless, prodigal, profligate, reckless, spendthrift, thriftless, uneconomical, unthrifty. *Opp* ECONOMICAL. **wasteful person** ▷ SPENDTHRIFT.

**watch** *n* chronometer, clock, digital watch, stop-watch, timepiece, timer, wrist-watch. • *vb* **1** attend, concentrate, contemplate, eye, gaze, heed, keep an eye open for, keep your eyes on, look at, mark, note, observe, pay attention, regard, see, spy on, stare, take notice, view. **2** *watch sheep.* care for, chaperon, defend, guard, keep an eye on, keep watch on, look after, mind, protect, safeguard, shield, superintend, supervise, take charge of, tend. **keep watch** ▷ GUARD. **on the watch** ▷ WATCHFUL. **watch your step** ▷ BEWARE.

**watcher** *n plur* audience, *inf* looker-on, observer, onlooker, spectator, viewer, witness.

**watchful** *adj* attentive, eagleeyed, heedful, observant, *inf* on the lookout, *inf* on the qui vive, on the watch, quick, sharp-eyed, vigilant. ▷ ALERT. *Opp* INATTENTIVE.

**watchman** *n* caretaker, custodian, guard, lookout, nightwatchman, security guard, sentinel, sentry, watch.

**water** *n* **1** Adam's ale, bath water, brine, distilled water, drinking water, mineral water, rainwater, sea water, spa water, spring water, tap water. **2** lake, lido, ocean, pond, pool, river, sea. ▷ STREAM. • *vb* damp, dampen, douse, drench, flood, hose, inundate, irrigate, moisten, saturate, soak, souse, spray, sprinkle, wet. **water down** ▷ DILUTE.

**waterfall** *n* cascade, cataract, chute, rapids, torrent, white water.

**waterlogged** *adj* full of water, saturated, soaked.

**waterproof** *adj* damp-proof, impermeable, impervious, water-repellent, water-resistant, watertight, weatherproof. *Opp* LEAKY. • *n* cape, groundsheet, *inf* mac, mackintosh, sou'wester.

**watertight** *adj* hermetic, sealed, sound. ▷ WATERPROOF.

**watery** *adj* **1** aqueous, bland, characterless, dilute, diluted, fluid, liquid, *inf* runny, *inf* sloppy, tasteless, thin, watered-down, weak, *inf* wishy-washy. **2** *watery eyes.* damp, moist, tear-filled, tearful, *inf* weepy. ▷ WET.

**wave** *n* **1** billow, breaker, crest, heave, ridge, ripple, roller, surf, swell, tidal wave, undulation, wavelet, *inf* white horse. **2** flourish, gesticulation, gesture, shake, sign, signal. **3** *a wave of enthusiasm.* current, flood, ground swell, outbreak, surge, tide, upsurge. **4** *a new wave*. advance, fashion, tendency, trend. **5** *radio waves*. pulse, vibration. • *vb* **1** billow, brandish, flail about, flap, flourish, fluctuate, flutter, move to and fro, ripple, shake, sway, swing, twirl, undulate, waft, wag, waggle, wiggle, zigzag. **2** gesticulate, gesture, indicate, sign, signal. **wave aside** ▷ DISMISS.

**wavelength** *n* channel, station, waveband.

**waver** *vb inf* be in two minds, be unsteady, change, falter, flicker, hesitate, quake, quaver, quiver, shake, shiver, shudder, sway, teeter, tergiversate, totter, tremble, vacillate, wobble.

**wavy** *adj* curling, curly, curving, heaving, rippling, rolling, sinuous, undulating, up and down, winding, zigzag. *Opp* FLAT, STRAIGHT.

**way** *n* **1** advance, direction, headway, journey, movement, progress, route. ▷ ROAD. **2** distance, length, measurement. **3** *a way to do something*. approach, avenue, course, fashion, knack, manner, means, method, mode, *Lat* modus operandi, path, procedure, process, system, technique. **4** *foreign ways*. custom, fashion, habit, *Lat* modus vivendi, practice, routine, style, tradition. **5** *funny ways*. characteristic, eccentricity, idiosyncrasy, oddity, peculiarity. **6** *in some ways*. aspect, circumstance, detail, feature, particular, respect.

**waylay** *vb* accost, ambush, await, buttonhole, detain, intercept, lie in wait for, pounce on, surprise. ▷ ATTACK.

**wayward** *adj* disobedient, headstrong, obstinate, self-willed, stubborn, uncontrollable, uncooperative, wilful. ▷ NAUGHTY. *Opp* COOPERATIVE.

**weak** *adj* **1** breakable, brittle, decrepit, delicate, feeble, flawed, flimsy, fragile, frail, frangible, inadequate, insubstantial, rickety, shaky, slight, substandard, tender, thin, unsafe, unsound, unsteady, unsubstantial. **2** *weak in health*. anaemic, debilitated, delicate, enervated, exhausted, feeble, flabby, frail, helpless, ill, infirm, listless, *inf* low, *inf* poorly, puny, sickly, slight, thin, tired out, wasted, weakly, *derog* weedy. **3** *a weak character*. cowardly, fearful, impotent, indecisive, ineffective, ineffectual, irresolute, poor, powerless, pusillanimous, spineless, timid, timorous, unassertive, weak-minded, wimpish. **4** *a weak position*. defenceless, exposed, unguarded, unprotected, vulnerable. **5** *weak excuses*. hollow, lame, *inf* pathetic, shallow, unbelievable, unconvincing, unsatisfactory. **6** *weak light*. dim, distant, fading, faint, indistinct, pale, poor, unclear, vague. **7** *weak tea*. dilute, diluted, tasteless, thin, watery. *Opp* STRONG.

**weaken** *vb* **1** debilitate, destroy, dilute, diminish, emasculate, enervate, enfeeble, erode, exhaust, impair, lessen, lower, make weaker, reduce, ruin, sap, soften, thin down, undermine, *inf* water down. **2** abate, become weaker, decline, decrease, dwindle, ebb, fade, flag, give in, give way, sag, wane, yield. *Opp* STRENGTHEN.

**weakling** *n* coward, *inf* milksop, *inf* pushover, *inf* runt, *inf* softie, weak person, *inf* weed, *inf* wimp.

**weakness** *n* **1** *inf* Achilles' heel, blemish, defect, error, failing, fault, flaw, flimsiness, foible, fragility, frailty, imperfection, inadequacy, mistake, shortcoming, softness, *inf* weak spot. **2** debility, decrepitude, delicacy, feebleness, impotence, incapacity, infirmity, lassitude, vulnerability. ▷ ILLNESS. **3** *a weakness for wine.* affection, fancy, fondness, inclination, liking, partiality, penchant, predilection, *inf* soft spot, taste. *Opp* STRENGTH.

**wealth** *n* **1** affluence, assets, capital, fortune, *old use* lucre, means, opulence, possessions, property, prosperity, riches, *old use* substance. ▷ MONEY. *Opp* POVERTY. **2** *a wealth of information.* abundance, bounty, copiousness, cornucopia, mine, plenty, profusion, store, treasury. *Opp* SCARCITY.

**wealthy** *adj* affluent, *inf* flush, *inf* loaded, moneyed, opulent, *joc* plutocratic, privileged, prosperous, rich, *inf* well-heeled, well-off, well-to-do. *Opp* POOR. **wealthy person** billionaire, capitalist, millionaire, plutocrat, tycoon.

**weapon** *n* bomb, gun, missile. □ *airgun, arrow, atom bomb, ballistic missile, battering-ram, battle-axe, bayonet, bazooka, blowpipe, blunderbuss, boomerang, bow and arrow, bren-gun, cannon, carbine, catapult, claymore, cosh, crossbow, CS gas, cudgel, cutlass, dagger, depth-charge, dirk, flame-thrower, foils, grenade,* old use *halberd, harpoon, H-bomb, howitzer, incendiary bomb, javelin, knuckleduster, lance, land-mine, laser beam, longbow, machete, machine-gun, mine, mortar, musket, mustard gas, napalm bomb, pike, pistol, pole-axe, rapier, revolver, rifle, rocket, sabre, scimitar, shotgun,* inf *six-shooter, sling, spear, sten-gun, stiletto, sub-machine-gun, sword, tank, tear-gas, time-bomb, tomahawk, tommy-gun, torpedo, truncheon, warhead, water-cannon.* **weapons** armaments, armoury, arms, arsenal, magazine, munitions, ordnance, weaponry. □ *artillery, automatic weapons, biological weapons, chemical weapons, firearms, missiles, nuclear weapons, small arms, strategic weapons, tactical weapons.*

**wear** *vb* **1** be dressed in, clothe yourself in, don, dress in, have on, present yourself in, put on, wrap up in. **2** *wear a smile.* adopt, assume, display, exhibit, show. **3** *wears the carpet.* damage, fray, injure, mark, scuff, wear away, weaken. **4** *wear well.* endure, last, *inf* stand the test of time, survive. **wear away** ▷ ERODE. **wear off** ▷ SUBSIDE. **wear out** ▷ WEARY.

**wearisome** *adj* boring, dreary, exhausting, monotonous, repetitive, tedious, tiring, wearying. ▷ TROUBLESOME. *Opp* STIMULATING.

**weary** *adj* bone-weary, *inf* dead beat, *inf* dog-tired, *inf* done in, drawn, drained, drowsy, enervated, exhausted, *inf* fagged, fatigued, fed up, flagging, footsore, impatient, jaded, *inf* jet-lagged, *sl* knackered, listless, prostrate, *inf* shattered, *inf* sick (of), sleepy, spent, tired out, travel-weary, wearied, *inf* whacked, worn out. *Opp* FRESH, LIVELY. • *vb* **1** debilitate, drain, enervate, exhaust, fatigue, *inf* finish, make tired, *sl* knacker, overtire, sap, *inf* shatter, *inf* take it out of, tax, tire, wear out. *Opp* REFRESH. **2** become bored, become tired, flag, grow weary, weaken.

**weather** *n* climate, the elements, meteorological conditions. □ *blizzard, breeze, cloud, cyclone, deluge, dew, downpour, drizzle, drought, fog, frost, gale, hail, haze,* *heatwave, hoar-frost, hurricane, ice, lightning, mist, rain, rainbow, shower, sleet, slush, snow, snowstorm, squall, storm, sunshine, tempest, thaw, thunder, tornado, typhoon, whirlwind, wind.* • *vb* ▷ SURVIVE. **under the weather** ▷ ILL.

**weave** *vb* **1** braid, criss-cross, entwine, interlace, intertwine, interweave, knit, plait, sew. **2** *weave a story*. compose, create, make, plot, put together. **3** *weave through a crowd*. dodge, make your way, tack, *inf* twist and turn, wind, zigzag.

**web** *n* criss-cross, lattice, mesh, net, network.

**wedding** *n* marriage, nuptials, union.

**wedge** *vb* cram, force, jam, pack, squeeze, stick.

**weep** *vb* bawl, *inf* blub, blubber, cry, *inf* grizzle, lament, mewl, moan, shed tears, snivel, sob, wail, whimper, whine.

**weigh** *vb* **1** measure the weight of. **2** *weigh evidence*. assess, consider, contemplate, evaluate, judge, ponder, reflect on, think about, weigh up. **3** *evidence weighed with the jury*. be important, carry weight, *inf* cut ice, count, have weight, matter. **weigh down** ▷ BURDEN. **weigh up** ▷ EVALUATE.

**weighing-machine** *n* balance, scales, spring-balance, weighbridge.

**weight** *n* **1** avoirdupois, burden, density, heaviness, load, mass, pressure, strain, tonnage. **2** *My voice has some weight*. authority, credibility, emphasis, force, gravity, importance, power, seriousness, significance, substance, value, worth. • *vb* ballast, bias, hold down, keep down, load, make heavy, weigh down.

**weird** *adj* **1** creepy, eerie, ghostly, mysterious, preternatural, scary, *inf* spooky, supernatural, unaccountable, uncanny, unearthly, unnatural. **2** *weird behaviour*. abnormal, bizarre, *inf* cranky, curious, eccentric, *inf* funny, grotesque, odd, outlandish, peculiar, queer, quirky, strange, unconventional, unusual, *inf* way-out, *inf* zany. *Opp* CONVENTIONAL, NATURAL.

**welcome** *adj* acceptable, accepted, agreeable, appreciated, desirable, gratifying, much-needed, *inf* nice, pleasant, pleasing, pleasurable. *Opp* UNWELCOME. • *n* greeting, hospitality, reception, salutation. • *vb* **1** give a welcome to, greet, hail, receive. **2** *They welcome criticism*. accept, appreciate, approve of, delight in, like, want.

**weld** *vb* bond, cement, fuse, join, solder, unite. ▷ FASTEN.

**welfare** *n* advantage, benefit, felicity, good, happiness, health, interest, prosperity, well-being.

**well** *adj* **1** fit, hale, healthy, hearty, *inf* in fine fettle, lively, robust, sound, strong, thriving, vigorous. **2** *All is well*. all right, fine, *inf* OK, satisfactory. • *n* fountain, shaft, source, spring, waterhole, well-spring. □ *artesian well, borehole, gusher, oasis, oil well, wishing-well.*

**well-behaved** *adj* cooperative, disciplined, docile, dutiful, good, hard-working, law-abiding, manageable, *inf* nice, polite, quiet, well-trained. ▷ OBEDIENT. *Opp* NAUGHTY.

**well-bred** *adj* courteous, courtly, cultivated, decorous, genteel, polite, proper, refined, sophisticated, urbane, well-brought-up, well-mannered. *Opp* RUDE.

**well-built** *adj* athletic, big, brawny, burly, hefty, muscular, powerful, stocky, *inf* strapping, strong, sturdy, upstanding. *Opp* SMALL.

**well-known** *adj* celebrated, eminent, familiar, famous, illustrious, noted, *derog* notorious, prominent, renowned. *Opp* UNKNOWN.

**well-meaning** *adj* good-natured, obliging, sincere, well-intentioned, well-meant. ▷ KIND. *Opp* UNKIND.

**well-off** *adj* affluent, comfortable, moneyed, prosperous, rich, *inf* well-heeled, well-to-do. *Opp* POOR.

**well-spoken** *adj* articulate, educated, polite, *inf* posh, refined, *inf* upper crust.

**wet** *adj* **1** awash, bedraggled, clammy, damp, dank, dewy, drenched, dripping, moist, muddy, saturated, sloppy, soaked, soaking, sodden, soggy, sopping, soused, spongy, submerged, waterlogged, watery, wringing. **2** *wet weather*. drizzly, humid, misty, pouring, rainy, showery, teeming. **3** *wet paint*. runny, sticky, tacky. • *n* dampness, dew, drizzle, humidity, liquid, moisture, rain. • *vb* dampen, douse, drench, irrigate, moisten, saturate, soak, spray, sprinkle, steep, water. *Opp* DRY.

**wheel** *n* circle, disc, hoop, ring. □ *bogie, castor, cog-wheel, spinning-wheel, steering-wheel*. • *vb* change direction, circle, gyrate, move in circles, pivot, spin, swerve, swing round, swivel, turn, veer, whirl.

**wheeze** *vb* breathe noisily, cough, gasp, pant, puff.

**whereabouts** *n* location, neighbourhood, place, position, site, situation, vicinity.

**whiff** *n* breath, hint, puff, smell.

**whim** *n* caprice, desire, fancy, impulse, quirk, urge.

**whine** *vb* complain, cry, *inf* grizzle, groan, moan, snivel, wail, weep, whimper, *inf* whinge.

**whip** *n* birch, cane, cat, cat-o'-nine-tails, crop, horsewhip, lash, riding-crop, scourge, switch. • *vb* **1** beat, birch, cane, flagellate, flog, horsewhip, lash, scourge, *inf* tan, thrash. ▷ HIT. **2** beat, stir vigorously, whisk.

**whirl** *vb* circle, gyrate, pirouette, reel, revolve, rotate, spin, swivel, turn, twirl, twist, wheel.

**whirlpool** *n* eddy, maelstrom, swirl, vortex, whirl.

**whirlwind** *n* cyclone, hurricane, tornado, typhoon, vortex, waterspout.

**whisk** *n* beater, mixer. • *vb* beat, mix, stir, whip.

**whiskers** *plur n* bristles, hairs, moustache.

**whisper** *n* **1** murmur, undertone. **2** *a whisper of scandal*. gossip, hearsay, rumour. • *vb* breathe, hiss, murmur, mutter. ▷ TALK.

**whistle** *n* hooter, pipe, pipes, siren. • *vb* blow, pipe.

**white** *adj* chalky, clean, cream, ivory, milky, off-white, silver, snow-white, snowy, spotless, whitish. ▷ PALE.

**whiten** *vb* blanch, bleach, etiolate, fade, lighten, pale.

**whole** *adj* coherent, complete, entire, full, healthy, in one piece, intact, integral, integrated, perfect, sound, total, unabbreviated, unabridged, unbroken, uncut, undamaged, undivided, unedited, unexpurgated, unharmed, unhurt, uninjured, unscathed. *Opp* FRAGMENTARY, INCOMPLETE.

**wholesale** *adj* comprehensive, extensive, general, global, indiscriminate, mass, total, universal, widespread. *Opp* LIMITED.

**wholesome** *adj* beneficial, good, healthful, health-giving, healthy, hygienic, nourishing, nutritious, salubrious, sanitary. *Opp* UNHEALTHY.

**wicked** *adj* abominable, *inf* awful, bad, base, beastly, corrupt, criminal, depraved, diabolical, dissolute, egregious, evil, foul, guilty, heinous, ill-tempered, immoral, impious, incorrigible, indefensible, iniquitous, insupportable, intolerable, irresponsible, lawless, lost (*soul*), machiavellian, malevolent, malicious, mischievous, murderous, naughty, nefarious, offensive, perverted, rascally, scandalous, shameful, sinful, sinister, spiteful, *inf* terrible, ungodly, unprincipled, unregenerate, unrighteous, unscrupulous, vicious, vile, villainous, violent, wrong. ▷ IRRELIGIOUS, OBSCENE. *Opp* MORAL. **wicked person** ▷ VILLAIN.

**wickedness** *n* baseness, depravity, enormity, guilt, heinousness, immorality, infamy, iniquity, irresponsibility, *old use* knavery, malice, misconduct, naughtiness, sin, sinfulness, spite, turpitude, ungodliness, unrighteousness, vileness, villainy, wrong, wrongdoing. ▷ EVIL.

**wide** *adj* **1** ample, broad, expansive, extensive, large, panoramic, roomy, spacious, vast, yawning. **2** *wide sympathies*. all-embracing, broad-minded, catholic, comprehensive, eclectic, encyclopedic, inclusive, wide-ranging. **3** *arms open wide*. extended, open, outspread, outstretched. **4** *a wide shot*. off-course, off-target. *Opp* NARROW.

**widen** *vb* augment, broaden, dilate, distend, enlarge, expand, extend, flare, increase, make wider, open out, spread, stretch.

**widespread** *adj* common, endemic, extensive, far-reaching, general, global, pervasive, prevalent, rife, universal, wholesale. *Opp* RARE.

**width** *n* beam (*of ship*), breadth, broadness, calibre (*of gun*), compass, diameter, distance across, extent, girth, range, scope, span, thickness.

**wield** *vb* **1** brandish, flourish, handle, hold, manage, ply, wave. **2** *wield power*. employ, exercise, exert, have, possess, use.

**wild** *adj* **1** *wild animals*. free, undomesticated, untamed. **2** *a wild moor*. deserted, desolate, *inf* God-forsaken, natural, overgrown, remote, rough, rugged, uncultivated, unenclosed, unfarmed, uninhabited, waste. **3** *wild behaviour*. aggressive, barbaric, barbarous, berserk, boisterous, disorderly, ferocious, fierce, frantic, hysterical, lawless, mad, noisy, obstreperous, on the rampage, out of control, rabid, rash, reckless, riotous, rowdy, savage, uncivilized, uncontrollable, uncontrolled, undisciplined, ungovernable, unmanageable, unrestrained, unruly, uproarious, violent. **4** *wild weather*. blustery, stormy, tempestuous, turbulent, violent, windy. **5** *wild enthusiasm*. eager, excited, extravagant, uninhibited, unrestrained. **6** *wild notions*. crazy, fantastic, impetuous, irrational, silly, unreasonable. **7** *a wild guess*. inaccurate, random, unthinking. *Opp* CALM, CULTIVATED, TAME.

**wilderness** *n* desert, jungle, waste, wasteland, wilds.

**wile** *n* artifice, gambit, *inf* game, machination, manoeuvre, plot, ploy, ruse, stratagem, subterfuge, trick.

**wilful** *adj* **1** calculated, conscious, deliberate, intended, intentional, premeditated, purposeful, voluntary. *Opp* ACCIDENTAL. **2** *a wilful character*. *inf* bloody-minded, determined, dogged, headstrong, immovable, intransigent, obdurate, obstinate, perverse, *inf* pig-headed, refractory, self-willed, stubborn, uncompromising, unyielding, wayward. *Opp* AMENABLE.

**will** *n* aim, commitment, desire, determination, disposition, inclination, intent, intention, longing, purpose, resolution, resolve, volition, will-power, wish. • *vb* **1** command, encourage, force, influence, inspire, persuade, require, wish. **2** *will a fortune*. bequeath, hand down, leave, pass on, settle on.

**willing** *adj* acquiescent, agreeable, amenable, assenting, complaisant, compliant, consenting, content, cooperative, disposed, docile, *inf* game, happy, helpful, inclined, pleased, prepared, obliging, ready, well-disposed. ▷ EAGER. *Opp* UNWILLING.

**wilt** *vb* become limp, droop, fade, fail, flag, flop, languish, sag, shrivel, weaken, wither. *Opp* THRIVE.

**wily** *adj* artful, astute, canny, clever, crafty, cunning, deceptive, designing, devious, disingenuous, furtive, guileful, ingenious, knowing, scheming, shifty, shrewd, skilful, sly, tricky, underhand. ▷ DISHONEST. *Opp* STRAIGHTFORWARD.

**win** *vb* **1** be the winner, be victorious, carry the day, come first, conquer, overcome, prevail, succeed, triumph. **2** *win a prize*. achieve, acquire, *inf* carry off, collect, *inf* come away with, deserve, earn, gain, get, obtain, *inf* pick up, receive, secure, *inf* walk away with. *Opp* LOSE.

**wind** *n* **1** air-current, blast, breath, breeze, current of air, cyclone, draught, gale, gust, hurricane, monsoon, puff, squall, storm, tempest, tornado, whirlwind, *poet* zephyr. **2** *wind in the stomach*. flatulence, gas, heartburn. • *vb* bend, coil, curl, curve, furl, loop, meander, ramble, reel, roll, slew, snake, spiral, turn, twine, twist, *inf* twist and turn, veer, wreathe, zigzag. **winding** ▷ TORTUOUS. **wind up** ▷ FINISH.

**window** *n* □ *casement, dormer, double-glazed window, embrasure, fanlight, French window, light, oriel, pane, sash window, skylight, shop window, stainedglass window, windscreen.*

**windswept** *adj* bare, bleak, desolate, exposed, unprotected. ▷ WINDY.

**windy** *adj* blowy, blustery, boisterous, breezy, draughty, fresh, gusty, squally, stormy, tempestuous. ▷ WINDSWEPT. *Opp* CALM.

**wink** *vb* **1** bat (*eyelid*), blink, flutter. **2** *lights winked*. flash, flicker, sparkle, twinkle.

**winner** *n* *inf* champ, champion, conquering hero, conqueror, first, medallist, prizewinner, titleholder, victor. *Opp* LOSER.

**winning** *adj* **1** champion, conquering, first, leading, prevailing, successful, top, top-scoring, triumphant, unbeaten, undefeated, victorious. *Opp* UNSUCCESSFUL. **2** *a winning smile*. ▷ ATTRACTIVE.

**wintry** *adj* arctic, icy, snowy. ▷ COLD. *Opp* SUMMERY.

**wipe** *vb* brush, clean, cleanse, dry, dust, mop, polish, rub, scour, sponge, swab, wash. **wipe out** ▷ DESTROY.

**wire** *n* **1** cable, coaxial cable, flex, lead, wiring. **2** cablegram, telegram. ▷ COMMUNICATION.

**wiry** *adj* lean, muscular, sinewy, strong, thin, tough.

**wisdom** *n* astuteness, common sense, discernment, discrimination, good sense, insight, judgement, judiciousness, penetration, perceptiveness, perspicacity, prudence, rationality, reason, sagacity, sapience, sense, understanding. ▷ INTELLIGENCE.

**wise** *adj* **1** astute, discerning, enlightened, erudite, fair, just, knowledgeable, penetrating, perceptive, perspicacious, philosophical, sagacious, sage, sensible, shrewd, sound, thoughtful, understanding, well-informed. ▷ INTELLIGENT. **2** *a wise decision.* advisable, appropriate, considered, diplomatic, expedient, informed, judicious, politic, proper, prudent, rational, reasonable, right. *Opp* UNWISE. **wise person** philosopher, pundit, sage.

**wish** *n* aim, ambition, appetite, aspiration, craving, desire, fancy, hankering, hope, inclination, *inf* itch, keenness, longing, objective, request, urge, want, yearning, *inf* yen. • *vb* ask, choose, crave, desire, hope, want, yearn. **wish for** ▷ WANT.

**wisp** *n* shred, strand, streak.

**wispy** *adj* flimsy, fragile, gossamer, insubstantial, light, streaky, thin. *Opp* SUBSTANTIAL.

**wistful** *adj* disconsolate, forlorn, melancholy, mournful, nostalgic, regretful, yearning. ▷ SAD.

**wit** *n* **1** banter, cleverness, comedy, facetiousness, humour, ingenuity, jokes, puns, quickness, quips, repartee, witticisms, wordplay. ▷ INTELLIGENCE. **2** comedian, comic, humorist, jester, joker, wag.

**witch** *n* enchantress, gorgon, hag, sibyl, sorceress, *plur* weird sisters.

**witchcraft** *n* black magic, charms, enchantment, *inf* hocus-pocus, incantation, magic, *inf* mumbo-jumbo, necromancy, the occult, occultism, sorcery, spells, voodoo, witchery, wizardry.

**withdraw** *vb* **1** abjure, call back, cancel, *inf* go back on, recall, rescind, retract, take away, take back. **2** *withdraw from the fight.* back away, back down, back out, *inf* chicken out, *inf* cry off, draw back, drop out, fall back, move back, pull back, pull out, *inf* quit, recoil, retire, retreat, run away, *inf* scratch, secede, shrink back. ▷ LEAVE. *Opp* ADVANCE, ENTER. **3** *withdraw teeth.* extract, pull out, remove, take out.

**withdrawn** *adj* bashful, diffident, distant, introverted, private, quiet, reclusive, remote, reserved, retiring, shy, silent, 5solitary, taciturn, timid, uncommunicative. *Opp* SOCIABLE.

**wither** *vb* become dry, become limp, dehydrate, desiccate, droop, dry out, dry up, fail, flag, flop, sag, shrink, shrivel, waste away, wilt. *Opp* THRIVE.

**withhold** *vb* check, conceal, control, hide, hold back, keep back, keep secret, repress, reserve, retain, suppress. *Opp* GIVE.

**withstand** *vb* bear, brave, confront, cope with, defy, endure, fight, grapple with, hold out against, last out against, oppose, *inf* put up with, resist, stand up to, *inf* stick, survive, take, tolerate, weather (*storm*). *Opp* SURRENDER.

**witness** *n* bystander, eyewitness, looker-on, observer, onlooker, spectator, viewer, watcher. • *vb* attend, behold, be present at, look on, note, notice, observe, see, view, watch. **bear witness** ▷ TESTIFY.

**witty** *adj* amusing, clever, comic, droll, facetious, funny, humorous, ingenious, intelligent, jocular, quick-witted, sarcastic, sharp-witted, waggish.

**wizard** *n* enchanter, magician, magus, sorcerer, *old use* warlock, witch-doctor.

**wobble** *vb* be unsteady, heave, move unsteadily, oscillate, quake, quiver, rock, shake, sway, teeter, totter, tremble, vacillate, vibrate, waver.

**wobbly** *adj* insecure, loose, rickety, rocky, shaky, teetering, tottering, unbalanced, unsafe, unstable, unsteady. *Opp* STEADY.

**woe** *n* affliction, anguish, dejection, despair, distress, grief, heartache, melancholy, misery, misfortune, sadness, suffering, trouble, unhappiness, wretchedness. ▷ SORROW. *Opp* HAPPINESS.

**woebegone** *adj* crestfallen, dejected, downhearted, forlorn, gloomy, melancholy, miserable, *inf* sorry for yourself, woeful, wretched. ▷ SAD. *Opp* CHEERFUL.

**woman** *n sl* bird, bride, *old use* dame, *old use* damsel, daughter, dowager, female, girl, girlfriend, *derog* hag, *derog* harridan, housewife, hoyden, *derog* hussy, lady, lass, madam, Madame, maid, *old use* maiden, matriarch, matron, mistress, mother, *derog* termagant, *derog* virago, virgin, widow, wife.

**wonder** *n* **1** admiration, amazement, astonishment, awe, bewilderment, curiosity, fascination, respect, reverence, stupefaction, surprise, wonderment. **2** *a wonder of science.* marvel, miracle, phenomenon, prodigy, *inf* sensation. • *vb* ask yourself, be curious, be inquisitive, conjecture, marvel, ponder, question yourself, speculate. ▷ THINK. **wonder at** ▷ ADMIRE.

**wonderful** *adj* amazing, astonishing, astounding, extraordinary, impressive, incredible, marvellous, miraculous, phenomenal, remarkable, surprising, unexpected, *old use* wondrous. *Opp* ORDINARY.

**woo** *vb* **1** *sl* chat up, court, make love to. **2** *woo custom.* attract, bring in, coax, cultivate, persuade, pursue, seek, try to get.

**wood** *n* **1** afforestation, coppice, copse, forest, grove, jungle, orchard, plantation, spinney, thicket, trees, woodland, woods. **2** blockboard, chipboard, deal, planks, plywood, timber. □ *balsa, beech, cedar, chestnut, ebony, elm, mahogany, oak, pine, rosewood, sandalwood, sapele, teak, walnut.*

**wooded** *adj* afforested, *poet* bosky, forested, silvan, timbered, tree-covered, woody.

**wooden** *adj* **1** ligneous, timber, wood. **2** *wooden acting.* dead, emotionless, expressionless, hard, inflexible, lifeless, rigid, stiff, stilted, unbending, unemotional, unnatural. *Opp* LIVELY.

**woodwind** *n* □ *bassoon, clarinet, cor anglais, flute, oboe, piccolo, recorder.*

**woodwork** *n* carpentry, joinery.

**woody** *adj* fibrous, hard, ligneous, tough. ▷ WOODEN.

**woolly** *adj* 1 wool, woollen. 2 *woolly toy.* cuddly, downy, fleecy, furry, fuzzy, hairy, shaggy, soft. 3 *woolly ideas.* ambiguous, blurry, confused, hazy, ill-defined, indefinite, indistinct, uncertain, unclear, unfocused, vague.

**word** *n* 1 expression, name, term. 2 ▷ NEWS. 3 ▷ PROMISE. • *vb* articulate, express, phrase. **word for word** ▷ VERBATIM.

**wording** *n* choice of words, diction, expression, language, phraseology, phrasing, style, terminology.

**wordy** *adj* chatty, diffuse, digressive, discursive, garrulous, long-winded, loquacious, pleonastic, prolix, rambling, repetitious, talkative, unstoppable, verbose, voluble, *inf* windy. *Opp* CONCISE.

**work** *n* 1 *inf* donkey-work, drudgery, effort, exertion, *inf* fag, *inf* graft, *inf* grind, industry, labour, *inf* plod, slavery, *inf* slog, *inf* spadework, strain, struggle, *inf* sweat, toil, *old use* travail. 2 *work to be done.* assignment, chore, commission, duty, errand, homework, housework, job, mission, project, responsibility, task, undertaking. 3 *regular work.* business, calling, career, employment, job, livelihood, living, métier, occupation, post, profession, situation, trade. • *vb* 1 *inf* beaver away, be busy, drudge, exert yourself, *inf* fag, *inf* grind away, *inf* keep your nose to the grindstone, labour, make efforts, navvy, *inf* peg away, *inf* plug away, *inf* potter about, *inf* slave, *inf* slog away, strain, strive, struggle, sweat, toil, travail. 2 act, be effective, function, go, operate, perform, run, succeed, thrive. 3 *work slaves hard.* drive, exploit, utilize. **working** ▷ EMPLOYED, OPERATIONAL. **work out** ▷ CALCULATE. **work up** ▷ DEVELOP, EXCITE.

**worker** *n* artisan, breadwinner, coolie, craftsman, employee, *old use* hand, labourer, member of staff, navvy, operative, operator, peasant, practitioner, servant, slave, tradesman, wage-earner, working man, working woman, workman.

**workforce** *n* employees, staff, workers.

**workmanship** *n* art, artistry, competence, craft, craftsmanship, expertise, handicraft, handiwork, skill, technique.

**workshop** *n* factory, mill, smithy, studio, workroom.

**world** *n* 1 earth, globe, planet. 2 area, circle, domain, field, milieu, sphere.

**worldly** *adj* avaricious, covetous, earthly, fleshly, greedy, human, material, materialistic, mundane, physical, profane, secular, selfish, temporal. *Opp* SPIRITUAL.

**worm** *vb* crawl, creep, slither, squirm, wriggle. writhe.

**worn** *adj* 1 frayed, moth-eaten, old, ragged, *inf* scruffy, shabby, tattered, *inf* tatty, thin, threadbare, worn-out. 2 ▷ WEARY.

**worried** *adj* afraid, agitated, agonized, alarmed, anxious, apprehensive, bothered, concerned, distraught, distressed, disturbed, edgy, fearful, *inf* fraught, fretful, guilt-ridden, insecure, nervous, nervy, neurotic, obsessed (by), on edge, overwrought, perplexed, perturbed, solicitous, tense, troubled, uncertain, uneasy, unhappy, upset, vexed.

**worry** *n* **1** agitation, anxiety, apprehension, disquiet, distress, fear, neurosis, perplexity, perturbation, tension, unease, uneasiness. **2** affliction, annoyance, bother, burden, care, concern, misgiving, problem, *plur* trials and tribulations, trouble, vexation. • *vb* **1** agitate, annoy, *inf* badger, bother, depress, disquiet, distress, disturb, exercise, *inf* hassle, irritate, molest, nag, perplex, perturb, pester, plague, tease, threaten, torment, trouble, upset, vex. *Opp* REASSURE. **2** *worry about money*. agonize, be anxious, be worried, brood, exercise yourself, feel uneasy, fret.

**worsen** *vb* **1** aggravate, exacerbate, heighten, increase, intensify, make worse. **2** *his health worsened*. decline, degenerate, deteriorate, fail, get worse, *inf* go downhill, weaken. *Opp* IMPROVE.

**worship** *n* adoration, adulation, deification, devotion, glorification, homage, idolatry, love, praise, reverence, veneration. • *vb* admire, adore, adulate, be devoted to, deify, dote on, exalt, extol, glorify, hero-worship, idolize, kneel before, lionize, laud, look up to, love, magnify, pay homage to, praise, pray to, *inf* put on a pedestal, revere, reverence, venerate.

**worth** *n* benefit, cost, good, importance, merit, quality, significance, use, usefulness, utility, value. **be worth** be priced at, cost, have a value of.

**worthless** *adj old use* bootless, dispensable, disposable, frivolous, futile, *inf* good-for-nothing, hollow, insignificant, meaningless, meretricious, paltry, pointless, poor, *inf* rubbishy, *inf* trashy, trifling, trivial, trumpery, unimportant, unproductive, unprofitable, unusable, useless, vain, valueless. *Opp* WORTHWHILE.

**worthwhile** *adj* advantageous, beneficial, considerable, enriching, fruitful, fulfilling, gainful, gratifying, helpful, important, invaluable, meaningful, noticeable, productive, profitable, remunerative, rewarding, satisfying, significant, sizeable, substantial, useful, valuable. ▷ WORTHY. *Opp* WORTHLESS.

**worthy** *adj* admirable, commendable, creditable, decent, deserving, estimable, good, honest, honourable, laudable, meritorious, praiseworthy, reputable, respectable, worth supporting, worthwhile. *Opp* UNWORTHY.

**wound** *n* damage, hurt, injury, scar, trauma. □ *amputation, bite, bruise, burn, contusion, cut, fracture, gash, graze, laceration, lesion, mutilation, puncture, scab, scald, scar, scratch, sore, sprain, stab, sting, strain, weal, welt.* • *vb* cause pain to, damage, harm, hurt, injure, traumatize. □ *amputate, bite, blow up, bruise, burn, claw, cut, fracture, gash, gore, graze, hit, impale, knife, lacerate, maim, make sore, mangle, maul, mutilate, scratch, shoot, sprain, strain, stab, sting, torture.* *Opp* HEAL, MEND.

**wrap** *n* cape, cloak, mantle, poncho, shawl, stole. • *vb* bind, bundle up, cloak, cocoon, conceal, cover, do up, encase, enclose, enfold, enshroud, envelop, hide, insulate, lag, muffle, pack, package, shroud, surround, swaddle, swathe, wind.

**wreathe** *vb* adorn, decorate, encircle, festoon, intertwine, interweave, twist, weave.

**wreck** *n* **1** hulk, shipwreck. ▷ WRECKAGE. **2** *the wreck of all my hopes.* demolition, destruction, devastation, loss, obliteration, overthrow, ruin, termination, undoing. • *vb* **1** annihilate, break up, crumple, crush, dash to pieces, demolish, destroy, devastate, ruin, shatter, smash, spoil, *inf* write off. **2** *wreck a ship.* capsize, founder, ground, scuttle, sink, shipwreck.

**wreckage** *n* bits, debris, *inf* flotsam and jetsam, fragments, pieces, remains, rubble, ruins.

**wrench** *vb* force, jerk, lever, prize, pull, rip, strain, tear, tug, twist, wrest, wring, *inf* yank.

**wrestle** *vb* grapple, strive, struggle, tussle. ▷ FIGHT.

**wretch** *n* **1** beggar, down-and-out, miserable person, pauper, unfortunate. **2** ▷ VILLAIN.

**wretched** *adj* **1** dejected, depressed, dispirited, downhearted, hapless, melancholy, miserable, pathetic, pitiable, pitiful, unfortunate. ▷ SAD. **2** ▷ UNSATISFACTORY.

**wriggle** *vb* crawl, snake, squirm, twist, waggle, wiggle, wobble, worm, writhe, zigzag.

**wring** *vb* **1** clasp, compress, crush, grip, press, shake, squeeze, twist, wrench, wrest. **2** coerce, exact, extort, extract, force.

**wrinkle** *n* corrugation, crease, crinkle, *inf* crow's feet, dimple, fold, furrow, gather, line, pleat, pucker, ridge, ripple. • *vb* corrugate, crease, crinkle, crumple, fold, furrow, gather, make wrinkles, pleat, pucker up, ridge, ripple, ruck up, rumple, screw up.

**wrinkled** *adj* corrugated, creased, crinkly, crumpled, furrowed, lined, pleated, ridged, ripply, rumpled, screwed up, shrivelled, undulating, wavy, wizened, wrinkly. *Opp* SMOOTH.

**write** *vb* be a writer, compile, compose, copy, correspond, doodle, draft, draw up, engrave, indite, inscribe, jot, note, pen, print, put in writing, record, scrawl, scribble, set down, take down, transcribe, type. **write off** ▷ CANCEL, DESTROY.

**writer** *n* **1** amanuensis, clerk, copyist, *derog* pen-pusher, scribe, secretary, typist. **2** author, bard, composer, *derog* hack, littérateur, wordsmith. □ *biographer, columnist, contributor, copy-writer, correspondent, diarist, dramatist, essayist, freelancer, ghost-writer, journalist, leader-writer, librettist, novelist, playwright, poet, reporter, scriptwriter.*

**writhe** *vb* coil, contort, jerk, squirm, struggle, thrash about, thresh about, twist, wriggle.

**writing** *n* **1** calligraphy, characters, copperplate, cuneiform, handwriting, hieroglyphics, inscription, italics, letters, longhand, notation, penmanship, printing, runes, scrawl, screed, scribble, script, shorthand. **2** authorship, composition, journalism. **3** hard copy, literature, manuscript, opus, printout, text, typescript, work. □ *article, autobiography, belles-lettres, biography, comedy, copywriting, correspondence, crime story, criticism, detective story, diary, dissertation, documentary, drama, editorial, epic, epistle, essay, fable, fairy-tale, fantasy, fiction, folk-tale, history, legal document, legend, letter, libretto, lyric, monograph, mystery, myth, newspaper column, non-fiction, novel, parable, parody, philosophy, play, poem, propaganda, prose, reportage, romance, saga, satire, science* *fiction, scientific writing, scriptwriting, SF, sketch, story, tale, thesis, thriller, tragedy, tragi-comedy, travel writing, treatise, trilogy, TV script, verse,* inf *whodunit, yarn.*

**written** *adj* documentary, *inf* in black and white, inscribed, in writing, set down, transcribed, typewritten. *Opp* SPOKEN.

**wrong** *adj* **1** base, blameworthy, corrupt, criminal, crooked, deceitful, dishonest, dishonourable, evil, felonious, illegal, illegitimate, illicit, immoral, iniquitous, irresponsible, mendacious, misleading, naughty, reprehensible, sinful, specious, unethical, unjustifiable, unlawful, unprincipled, unscrupulous, vicious, villainous, wicked, wrong-headed. **2** *wrong answers.* erroneous, fallacious, false, imprecise, improper, inaccurate, incorrect, inexact, misinformed, mistaken, unfounded, untrue. **3** *a wrong decision.* curious, ill-advised, ill-considered, ill-judged, impolitic, imprudent, injudicious, misguided, misjudged, unacceptable, unfair, unjust, unsound, unwise, wrongful. **4** *go the wrong way.* abnormal, contrary, inappropriate, incongruous, inconvenient, misleading, opposite, unconventional, undesirable, unhelpful, unsuitable, worst. **5** *Something's wrong.* amiss, broken down, defective, faulty, out of order, unusable. ▷ BAD. *Opp* RIGHT. • *vb* abuse, be unfair to, cheat, damage, do an injustice to, harm, hurt, injure, malign, maltreat, misrepresent, mistreat, traduce, treat unfairly. **do wrong** ▷ MISBEHAVE.

**wrongdoer** *n* convict, criminal, crook, culprit, delinquent, evildoer, law-breaker, malefactor, mischief-maker, miscreant, offender, sinner, transgressor.

**wrongdoing** *n* crime, delinquency, disobedience, evil, immorality, indiscipline, iniquity, malpractice, misbehaviour, mischief, naughtiness, offence, sin, sinfulness, wickedness.

**wry** *adj* **1** askew, aslant, awry, bent, contorted, crooked, deformed, distorted, lopsided, twisted, uneven. **2** *a wry sense of humour.* droll, dry, ironic, mocking, sardonic.

# Y

**yard** *n* court, courtyard, enclosure, garden, *inf* quad, quadrangle.

**yarn** *n* **1** fibre, strand, thread. **2** account, anecdote, fiction, narrative, story, tale.

**yawning** *adj* gaping, open, wide.

**yearly** *adj* annual, perennial, regular.

**yearn** *vb* ache, feel desire, hanker, have a craving, hunger, itch, long, pine. ▷ WANT.

**yellow** *adj* □ *chrome yellow, cream, gold, golden, orange, tawny.*

**yield** *n* **1** crop, harvest, output, produce, product. **2** earnings, gain, income, interest, proceeds, profit, return, revenue. • *vb* **1** acquiesce, agree, assent, bow, capitulate, *inf* cave in, cede, comply, concede, defer, give in, give up, give way, *inf* knuckle under, submit, succumb, surrender, *inf* throw in the towel, *inf* throw up the sponge. **2** *yield interest.* bear, earn, generate, pay out, produce, provide, return, supply. **yielding** ▷ FLEXIBLE, SPONGY, SUBMISSIVE.

**young** *adj* **1** baby, early, growing, immature, new-born, undeveloped, unfledged, youngish. **2** *young people*. adolescent, juvenile, pubescent, teenage, underage, youthful. **3** *young for your age*. babyish, boyish, callow, childish, girlish, *inf* green, immature, inexperienced, infantile, juvenile, naïve, puerile. • *n* brood, family, issue, litter, offspring, progeny. **young creatures** bullock, calf, chick, colt, cub, cygnet, duckling, fawn, fledgling, foal, gosling, heifer, kid, kitten, lamb, leveret, nestling, pullet, puppy, yearling. **young people** adolescent, baby, boy, *derog* brat, child, girl, infant, juvenile, *inf* kid, lad, lass, *inf* nipper, teenager, toddler, *derog* urchin, youngster, youth.

**youth** *n* **1** adolescence, babyhood, boyhood, childhood, girlhood, growing up, immaturity, infancy, minority, pubescence, *inf* salad days, *inf* teens. **2** adolescent, boy, juvenile, *inf* kid, *inf* lad, minor, stripling, teenager, youngster.

**youthful** *adj* fresh, lively, sprightly, vigorous, well-preserved, young-looking. ▷ YOUNG.

# Z

**zany** *adj* absurd, clownish, crazy, eccentric, idiotic, *in* loony, *inf* mad, madcap, playful, ridiculous, silly, *sl* wacky.

**zeal** *n derog* bigotry, earnestness, enthusiasm, fanaticism, fervour, partisanship.

**zealot** *n* bigot, extremist, fanatic, partisan, radical.

**zealous** *adj* conscientious, diligent, eager, earnest, enthusiastic, fanatical, fervent, keen, militant, obsessive, partisan, passionate. *Opp* APATHETIC.

**zenith** *n* acme, apex, apogee, climax, height, highest point, meridian, peak, pinnacle, summit, top. *Opp* NADIR.

**zero** *n cricket* duck, *tennis* love, naught, nil, nothing, nought, *sl* zilch. **zero in on** ▷ AIM.

**zest** *n* appetite, eagerness, energy, enjoyment, enthusiasm, exuberance, hunger, interest, liveliness, pleasure, thirst, zeal.

**zigzag** *adj* bendy, crooked, *inf* in and out, indirect, meandering, serpentine, twisting, winding. • *vb* bend, curve, meander, snake, tack, twist, wind.

**zone** *n* area, belt, district, domain, locality, neighbourhood, province, quarter, region, section, sector, sphere, territory, tract, vicinity.

**zoo** *n* menagerie, safari park, zoological gardens.

**zoom** *vb* career, dart, dash, hurry, hurtle, race, rush, shoot, speed, *inf* whiz, *inf* zip. ▷ MOVE.